Evans

DEVELOPMENTS IN SEDIMENTOLOGY 27

# INTERNATIONAL CLAY CONFERENCE 1978

FURTHER TITLES IN THIS SERIES
VOLUMES 1, 2, 3, 5, 8 and 9 are out of print

4   F.G. TICKELL
THE TECHNIQUES OF SEDIMENTARY MINERALOGY
6   L. VAN DER PLAS
THE IDENTIFICATION OF DETRITAL FELDSPARS
7   S. DZULYNSKI and E.K. WALTON
SEDIMENTARY FEATURES OF FLYSCH AND GREYWACKES
10   P.McL.D.DUFF, A. HALLAM and E.K. WALTON
CYCLIC SEDIMENTATION
11   C.C. REEVES Jr.
INTRODUCTION TO PALEOLIMNOLOGY
12   R.G.C. BATHURST
CARBONATE SEDIMENTS AND THEIR DIAGENESIS
13   A.A. MANTEN
SILURIAN REEFS OF GOTLAND
14   K.W. GLENNIE
DESERT SEDIMENTARY ENVIRONMENTS
15   C.E. WEAVER and L.D. POLLARD
THE CHEMISTRY OF CLAY MINERALS
16   H.H. RIEKE III and G.V. CHILINGARIAN
COMPACTION OF ARGILLACEOUS SEDIMENTS
17   M.D. PICARD and L.R. HIGH Jr.
SEDIMENTARY STRUCTURES OF EPHEMERAL STREAMS
18   G.V. CHILINGARIAN and K.H. WOLF
COMPACTION OF COARSE-GRAINED SEDIMENTS
19   W. SCHWARZACHER
SEDIMENTATION MODELS AND QUANTITATIVE STRATIGRAPHY
20   M.R. WALTER, Editor
STROMATOLITES
21   B. VELDE
CLAYS AND CLAY MINERALS IN NATURAL AND SYNTHETIC SYSTEMS
22   C.E. WEAVER and K.C. BECK
MIOCENE OF THE SOUTHEASTERN UNITED STATES
23   B.C. HEEZEN, Editor
INFLUENCE OF ABYSSAL CIRCULATION ON SEDIMENTARY
ACCUMULATIONS IN SPACE AND TIME
24   R.E. GRIM and N. GÜVEN
BENTONITES
25A   G. LARSEN and G.V. CHILINGARIAN, Editors
DIAGENESIS IN SEDIMENTS AND SEDIMENTARY ROCKS
26   T. SUDO and S. SHIMODA, Editors
CLAYS AND CLAY MINERALS OF JAPAN

DEVELOPMENTS IN SEDIMENTOLOGY 27

# INTERNATIONAL CLAY CONFERENCE 1978

*Proceedings of the VI International Clay Conference 1978 held in Oxford, 10—14 July 1978, organized by the Clay Minerals Group, Mineralogical Society, London, under the auspices of Association Internationale pour l'Etude des Argiles*

Edited by

## M.M. MORTLAND

*Michigan State University, East Lansing, Mich., U.S.A.*

and

## V.C. FARMER

*The Macaulay Institute for Soil Research, Aberdeen, Great Britain*

ELSEVIER SCIENTIFIC PUBLISHING COMPANY
AMSTERDAM — OXFORD — NEW YORK 1979

ELSEVIER SCIENTIFIC PUBLISHING COMPANY
335 Jan van Galenstraat
P.O. Box 211, 1000 AE Amsterdam, The Netherlands

*Distributors for the United States and Canada:*

ELSEVIER/NORTH-HOLLAND INC.
52, Vanderbilt Avenue
New York, N.Y. 10017

Library of Congress Cataloging in Publication Data

International Clay Conference, Oxford, 1978.
  International Clay Conference, 1978.

  (Developments in sedimentology ; 27)
  Bibliography: p.
  Includes indexes.
  1. Clay minerals--Congresses.  I. Mortland, Max
Merle, 1923-    II. Farmer, Victor Colin.  III. Min-
eralogical Society of Great Britain and Ireland.  Clay
Minerals Group.  IV. Series.
QE389.625.I57      1978    549'.6       78-20909
ISBN 0-444-41773-7

ISBN 0 - 444 - 41773 - 7 ( Vol. 27)
ISBN 0 - 444 - 41238 - 7 (Series)

© Elsevier Scientific Publishing Company, 1979
All rights reserved. No part of this publication may be reproduced, stored in a retrieval system or transmitted in any form or by any means, electronic, mechanical, photocopying, recording or otherwise, without the prior written permission of the publisher, Elsevier Scientific Publishing Company, P.O. Box 330, 1000 AH Amsterdam, The Netherlands

Printed in The Netherlands

## CONFERENCE ORGANISING COMMITTEE

| | |
|---|---|
| Chairman | MR. D. MITCHELL |
| General Secretary | MR. J.A. BAIN |
| Treasurer | DR. I.R. BASHAM |
| Convenor (Scientific Programme) | DR. R.C. MACKENZIE |
| " (Social Programme) | DR. B.S. NEUMANN |
| " (Field Excursions) | MR. R.K. HARRISON |
| Associate Editor | DR. V.C. FARMER |
| Committee member | DR. A.C.D. NEWMAN |
| " " | DR. E.J.W. WHITTAKER |
| Local Secretary (Oxford) | MRS A. HICKMAN |

Members of Sub-Committees:

DR. A.C. BISHOP
DR. C.M. BRISTOW
DR. P. BULLOCK
DR. E.C. FRESHNEY
MR. R.W. GALLOIS
DR. F.P. GLASSER
MR. D.E. HIGHLEY
MR. P.S. KEELING
MR. B.D. MITCHELL
DR. D.J. MORGAN
MRS D. NELSON
DR. C.M. RICE
DR. M.J. WILSON
MR. B.R. YOUNG

VI

## PREFACE

The 1978 meeting of the Association Internationale pour L'Etude des Argiles was held at Oxford, England, July 10-14, 1978. This Sixth International Clay Conference was organized by the Clay Minerals Group of the Mineralogical Society, London, with Mr. Denis Mitchell serving as Chairman, and Mr. J.A. Bain as General Secretary of the Organizing Committee.

This Proceedings volume contains selected papers from the large number presented at the Conference together with Introductory lectures by the Technical Session Chairmen as well as the Plenary lecture by Dr. R.C. Mackenzie of The Macaulay Institute for Soil Research of Aberdeen, Scotland, and the Presidential remarks of Dr. S.W. Bailey of the University of Wisconsin at Madison. Other papers presented are listed by title only, since a Book of Summaries covering all papers is available at £3.50 (post free) from Sixth International Clay Conference, c/o The Institute of Geological Sciences, 64-78 Gray's Inn Road, LONDON WC1X 8NG, U.K. Many of these papers have been or will be submitted to regular journals.

In contrast to previous volumes, a photo-offset process is utilized in production of these Proceedings. This required that each author act essentially as his own editor in preparing the draft of his manuscript for publication. The results of this procedure was a degree of limitation for the Editors in editing the manuscripts as well as a variety of type-faces.

All research papers were reviewed before acceptance by scientists familiar with the subject matter involved. The Editors would like to express their deepest thanks to the following scientists who participated in the review process: S.W. Bailey, J.B. Dixon, M.E. Harward, W.D. Johns, E.C. Jonas, W.D. Keller, M.B. McBride, T.J. Pinnavaia, C.B. Roth, D.C. Bain, C.M. Bristow, J. Chaussdion, C.S. Cundy, J.A. Gard, F.P. Glasser, B.A. Goodman, P.L. Hall, A. Herbillon, C.V. Jeans, G. Lagally, D.J. Morgan, F.R. Noble, R.H. Ottewill, I. Parsons, E. Paterson, B. Rand, J.H. Rayner, C.M. Rice, R.H.S. Robertson, J.B. Rowse, J.D. Russell, U. Schwertmann, J. Thorez, B. Velde, R. Wey, J.W. White, F.W. Wilburn, M.J. Wilson.

In conclusion we would like to thank the authors for their fine spirit of cooperation in preparing their manuscripts in the required form and in making alterations and corrections where necessary.

M.M. Mortland<br>
V.C. Farmer<br>
<u>Editors</u>

VIII

PRESIDENT'S OPENING ADDRESS

My Lord Mayor, Mr. Vice Chancellor, Mr. Chairman, Mr. President of the Mineralogical Society, Ladies and Gentlemen:

It is customary for the AIPEA President to preface his opening remarks by a few words in the language of the host country. This is a good custom, and I propose to go even further and to give my entire address in English. I apologize to our hosts for my obvious accent.

AIPEA is very pleased to be meeting in England this year as a guest of the Clay Minerals Group of the Mineralogical Society, London. The Clay Minerals Group has been very active in promoting meetings of clay scientists in Great Britain and in disseminating the results of their clay research. An interesting point I have noticed is their frequent sponsorship of joint meetings, often on special theme topics, with other societies in related disciplines both in Great Britain and in Europe. Their highly successful journal Clay Minerals has now become the official West European clay journal, sponsored by seven national clay societies with papers published in English, French, German, or Spanish. The authoritative monographs published by the Clay Minerals Group and the Mineralogical Society, London, have been well received by clay mineralogists throughout the world. These include "The X-ray Identification and Crystal Structures of Clay Minerals", now in its third edition, "the Infrared Spectra of Minerals", "The Electron-Optical Investigation of Clays", "The Clay Mineralogy of British Sediments", "The Differential Thermal Investigation of Clays", plus others in preparation. I must record here my own indebtedness to this publication program, as I received my introduction to practical clay mineralogy by reading the first edition of "The X-ray Identification and Crystal Structures of Clay Minerals".

In working with the Local Organizing Committee for this conference, I have been very impressed with the high degree of organization and efficiency achieved by the Chairman, Mr. Denis Mitchell, who also is Chairman of the host Clay Minerals Group, and by the General Secretary, Mr. James A. Bain. They have gathered together a very efficient and enthusiastic group of co-workers. I predict as a result of the total efforts of all these people that we will reap the benefits of a well-run conference, scientifically, socially, and in the field. The scientific program in which we are about to participate has been arranged by Dr. Robert C. Mackenzie, who also will present the plenary lecture immediately following this opening address. The Proceedings Volume will be published in approximately seven to nine months time under the editorship of Dr. Max M. Mortland and Dr. Victor C. Farmer. AIPEA expresses its gratitude to all of these people and to their assistants for their efforts in arranging this conference. We appreciate very much also the provision of the meeting and housing facilities by the University of Oxford through the kind services of Professor E.A. Vincent and Dr. E.J.W. Whittaker.

In examining the scientific program I am reminded again of the uniqueness of clay mineral meetings, whether on the national or international level. It would be difficult to find anywhere in the world such a heterogeneous group of scientists from so many disciplines, all gathered together under the same roof to discuss that common theme of interest that binds us all together - namely, clays. We include among our ranks geologists and mineralogists, soil scientists and agronomists, chemists and geochemists, biochemists, crystallographers, physicists, mathematicians ceramists, engineers of several types, anthropologists, animal and poultry scientists, nutritionists, and others. Despite our diverse backgrounds and special objectives, we share a common interest in the structures, compositions, properties, origins, occurrences, and applications of clay minerals. And it is the interaction between these multitudes of backgrounds as applied to a multitude of different aspects of clays that makes these conferences so fascinating and instructive. Many of us look forward to these clay conferences in preference to meetings within our own disciplines primarily because of this diversity of backgrounds and interests and of the fruitful cross-pollination of ideas that often results.

On behalf of AIPEA I invite all participants and guests to enjoy themselves in the formal and informal sessions of the next few days. I look forward, along with all of you, to a most successful conference.

S.W. BAILEY
AIPEA President, 1975-1978

CONTENTS

| | |
|---|---|
| PREFACE | *vii* |
| PRESIDENT'S OPENING ADDRESS | *ix* |

*Plenary Lecture*

Clay Mineralogy—whence and whither?
    R.C. MACKENZIE — 1

*Section 1: Crystal Chemistry and Structure*

Structural iron oxidation during mica expansion.
    A.D. SCOTT and A.F. YOUSSEF — 17

Distribution of octahedral ions in phlogopites and biotites.
    J.A. RAUSELL-COLOM, J. SANZ, M. FERNANDEZ and J.M. SERRATOSA — 27

Effect of texture on vermiculite structure: Lithium minerals.
    C. DE LA CALLE, R. GLAESER and H. PEZERAT — 37

Qualitative and quantitative study of a structural reorganization in montmorillonite after potassium fixation.
    A. PLANÇON, G. BESSON, J.P. GAULTIER, J. MAMY and C. TCHOUBAR — 45

The ferric analogue of pyrophyllite and related phases.
    F.V. CHUKHROV, B.B. ZVYAGIN, V.A. DRITS, A.I. GORSHKOV, L.P. ERMILOVA, E.A. GOILO and E.S. RUDNITSKAYA — 55

Mössbauer spectra of chlorites and their decomposition products.
    B.A. GOODMAN and D.C. BAIN — 65

Effect of structural $Fe^{2+}$ on visible absorption spectra of nontronite suspensions.
    W.L. ANDERSON and J.W. STUCKI — 75

An interstratified chlorite-vermiculite in weathered red shale near Toyoma, Japan.
    T. NISHIYAMA, K. OINUMA and M. SATO — 85

Other papers. — 95

*Section 2: Colloidal Properties and Surface Chemistry*

Surface properties of fibrous clay minerals (palygorskite and sepiolite).
    J.M. SERRATOSA — 99

The structure and dynamics of clay-water systems studied by neutron scattering.
    D.J. CEBULA, R.K. THOMAS and J.W. WHITE — 111

Neutron scattering studies of the dynamics of interlamellar water in montmorillonite and vermiculite.
    P.L. HALL, D.K. ROSS, J.J. TUCK and M.H.B. HAYES — 121

Selective coagulation and mixed-layer formation from sodium smectite solutions.
    E. FREY and G. LAGALY — 131

Pore size distribution in water-saturated calcium montmorillonite using low-temperature heat-flow scanning calorimetry.
L.G. HOMSHAW and J. CHAUSSIDON ............ 141

The application of X-ray photoelectron spectroscopy (XPS or ESCA) to the study of mineral surface chemistry.
M.H. KOPPELMAN and J.G. DILLARD ............ 153

Evolution of exchange properties and crystallographic characteristics of biionic K-Ca montmorillonite submitted to alternate wetting and drying.
J.P. GAULTIER and J. MAMY ............ 167

Formation of metallic silver as related to iron oxidation in K-depleted micas.
M. SAYIN, B. BEYME and H. GRAF VON REICHENBACH ............ 177

Protonation of bases in clay suspensions.
J.R. FELDKAMP and J.L. WHITE ............ 187

Infrared study of sepiolite and palygorskite surfaces.
C.J. SERNA and G.E. VANSCOYOC ............ 197

Ion exchange of the poly-amine complexes of some transition metal ions in montmorillonite.
P. PEIGNEUR, A. MAES and A. CREMERS ............ 207

X.P.S. study of the interaction of some porphyrins and metalloporphyrins with montmorillonite.
P. CANESSON, M.I. CRUZ and H. VAN DAMME ............ 217

Adsorption of Chlordimeform by montmorillonite.
J.L. PEREZ RODRIGUEZ and M.C. HERMOSIN GAVINO ............ 227

Other papers. ............ 235

*Section 3: Geology and Sedimentology*

Studies of clay minerals in sediments - a review.
T. SUDO ............ 241

Correlation between coal and clay diagenesis in the Carboniferous of the Upper Silesian Coal Basin.
J. ŚRODOŃ ............ 251

Mineralogical and geochemical transformation of clays during burial-diagenesis (catagenesis): Relation to oil generation.
A.E. FOSCOLOS and T.G. POWELL ............ 261

Clay minerals as indicators of the Cenozoic evolution of the North Atlantic Ocean.
C. LATOUCHE ............ 271

Changes in mineralogical composition of Tertiary sediments from North Sea wells.
W. KARLSSON, J. VOLLSET, K. BJØRLYKKE and P. JØRGENSEN ............ 281

The origin of clay minerals in Cenomanian littoral deposits around the Armorican massif.
J. LOUAIL, J. ESTEOULE and J. ESTEOULE-CHOUX ............ 291

A montmorillonite, kaolinite association in the Lower Cretaceous of south-east England.
D.J. MORGAN, D.E. HIGHLEY and D.J. BLAND ............ 301

Mineral distributions in sediments associated with the Alton Marine Band near Penistone, South Yorkshire.
D.A. ASHBY and M.J. PEARSON ............ 311

Petrology of K-bentonite beds in the carbonate series of the Visean and Tournaisian Stages of Belgium.
    J. THOREZ and H. PIRLET    323

Other papers.    333

*Section 4: Genesis and Synthesis*

Genesis and synthesis of clays and clay minerals: recent developments and future prospects.
    B. SIFFERT    337

Clay mineral composition and potassium status of some typical Hungarian soils.
    E.M. VARJU and P. STEFANOVITS    349

Alteration of basaltic rocks by hydrothermal activity at 100–300°C.
    H. KRISTMANNSDOTTIR    359

Clays and clay minerals of hydrothermal origin in Hawaii.
    P.F. FAN    369

Reaction series for dioctahedral smectite: The synthesis of mixed-layer pyrophyllite/smectite.
    D. EBERL    375

Stabilité des minéraux phylliteux 2/1 en conditions acides. Rôle de la composition octaédrique.
    M. ROBERT and G. VENEAU    385

Synthetic illite in the chemical system $K_2O-Al_2O_3-SiO_2-H_2O$ at 300°C and 2 kb.
    B. VELDE and A.H. WEIR    395

Biotite weathering in granites from western France.
    A. MEUNIER and B. VELDE    405

Micromorphology of halloysite produced by weathering of plagioclase in volcanic ash.
    K. TAZAKI    415

Other papers.    423

*Section 5: Applied Clay Mineralogy*

Recent developments in applied clay mineralogy.
    P. DE SOUZA-SANTOS    427

Regional appraisal of clay resources – A challenge to the clay mineralogist.
    J.A. BAIN and D.E. HIGHLEY    437

The quantitative determination of quartz in clay mixtures by infrared spectroscopy.
    R.H. ANDREWS, J.A. GIBSON and I.M. SHAW    447

Sorption properties of consolidated and compressed clays.
    E.T. STĘPKOWSKA    457

Electrophoretic phenomena as applied to the investigation of interaction between clays and anionic polyelectrolytes.
    D. RIOCHE and B. SIFFERT    465

The clay deposits of Mexico.
    L. DE PABLO GALAN    475

Other papers.    487

*Section 6: Non-crystalline and Accessory Minerals*

Non-crystalline and accessory minerals.
    U. SCHWERTMANN    491

Reversibility of lattice collapse in synthetic buserite.
M.I. TEJEDOR-TEJEDOR and E. PATERSON — 501

Intercalation compounds of $KHSi_2O_5$ and $H_2Si_2O_5$ with alkylammonium ions and alkylamines.
A. KALT, B. PERATI and R. WEY — 509

Crystallization of nordstrandite in citrate systems and in the presence of montmorillonite.
A. VIOLANTE and M.L. JACKSON — 517

Nature of hydrolytic precipitation products of aluminum as influenced by low molecular weight complexing organic acids.
N.F. NG KEE KWONG and P.M. HUANG — 527

Structural formulas of allophanes.
K. WADA — 537

Synthetic imogolite, a tubular hydroxyaluminium silicate.
V.C. FARMER and A.R. FRASER — 547

Application of Mössbauer spectroscopy to the study of iron oxides in some red and yellow/brown soil samples from New Zealand.
C.W. CHILDS, B.A. GOODMAN and G.J. CHURCHMAN — 555

Natural amorphous materials, their origin and identification procedures.
J. RIMSAITE — 567

Other papers. — 577

## Section 7: Kaolin Investigations

Methods of kaolin investigation.
W.D. KELLER — 581

Australian kaolins.
A.J. GASKIN, P.J. DARRAGH and F.C. LOUGHNAN — 591

South African kaolins.
L.J. MURRAY and R.O. HECKROODT — 601

The crystallinity index of kaolinite in relation to other properties of the kaolin mass of Karlovy Vary.
J. KONTA — 609

Rate of transformation of halloysite to metahalloysite under hydrothermal conditions.
H. MINATO and M. AOKI — 619

Reactions of salts with kaolinite at elevated temperatures. Part 2.
L. HELLER-KALLAI and M. FRENKEL — 629

Critical assessment of the joint use of various physico-chemical techniques in the study of the thermal transformation of kaolin.
B. DELMON, A.J. HERBILLON, A.J. LEONARD and M. BULENS — 639

Other papers. — 649

AUTHOR INDEX — 651

SUBJECT INDEX — 657

*Plenary Lecture*

CLAY MINERALOGY - WHENCE AND WHITHER?

R. C. Mackenzie
The Macaulay Institute for Soil Research, Craigiebuckler, Aberdeen, Scotland, UK


I am very conscious and appreciative of the honour done me by the Organizing Committee of the Conference in inviting me to give this Plenary Lecture and trust the subject matter may, if nothing else, stimulate some thought. At this point I should like, on my part, to congratulate AIPEA and its predecessor CIPEA on thirty fruitful years - it is most fitting that this meeting should be in Britain, for it was in London in 1948 that CIPEA was founded (MacEwan, 1949).

Assessment of the present position in any science involves consideration of that science as a continuous and continuing entity whose past and present hold the seeds of future development. Thus, one must agree with Paul Kruger who wrote in 1904*, "whosoever wishes to build a future must not neglect the past" - but one must also have regard to the caveat of George Brindley in 1976 that "future trends cannot be forecast by linear extrapolation of the past". Since, therefore, any prognostications regarding the future are likely to be, at best, about fifty per cent successful, the greater part of this lecture will be devoted to a brief review of selected historical aspects that may help to put the present and future into perspective.

Study of clay mineralogy, and of the personalities connected with it, suggests that a sense of vocation may have played a greater part in development of this science than in that of many others. This perhaps arises partially from the fact that the clay mineralogist must have a detailed, and frequently expert, knowledge of a range of pure and applied scientific disciplines, a sense of history, a philosophical outlook and, above all, an enthusiasm for establishing what may be regarded as the truth. But there may be more deep-seated reasons. Thus, it is interesting to note that there are similarities between the concepts of *clay* and *truth*. In the early nineteenth century the English writer Charles Lamb, under the pseudonym of Elia, wrote an essay entitled "Of Truth", which begins with the striking observation, "'What is truth?' said jesting Pilate and would not wait for an answer": he then goes on to consider the many-faceted concept known as *truth*. A similar comment and essay could well be written about *clay*, which is

---
*In his last letter to his people in South Africa: the original and an English translation are in the Kruger Museum, Pretoria, South Africa.

just as indefinable and just as open to subjective interpretation as *truth*. Why then should one study something that cannot be defined - be it *truth* by the philosopher or *clay* by the scientist - if one does not feel called upon to do so? From such sources stems, I suspect, the sense of vocation inherent in all eminent clay mineralogists.

WHAT IS CLAY?

Despite the indefinable nature of clay, the scientist by his training is always haunted by the question, echoing that of Pilate, "What is clay"? Certainly, the soil scientist or the sedimentary petrologist can define clay (or, more correctly, the clay fraction) in terms of particle size; the soil scientist has his clay soils and the petrologist his clay rocks - but these concepts are not necessarily strictly equivalent. Yet one can examine a deposit cursorily in the field and say, quite dogmatically and correctly, "That is clay", without being able to say precisely why: it may be by a combination of subjective factors that are virtually indefinable in scientific terms - such as by having one's boot become almost inextricable on a very wet day, by slipping on a moist surface, by visual inspection or by rubbing with one's fingers. I well remember several years ago considering in detail some published definitions of clay made over the centuries and even having the temerity to propose a modification. I shall not give the reference, however, since more mature consideration suggests that any such attempt is almost certainly doomed to failure on some count or other and one wonders whether our Nomenclature Committee could in fact much better the definition given by Georgius Agricola in 1546 (Bandy and Bandy, 1955): "..... mineral bodies that can be worked in the hands when they are moistened and from which mud can be made when they are saturated with water".

It must certainly have been by some such criteria that clay was recognized by the ancients who used sun-dried clay or mud bricks, frequently with a binding of straw, from time immemorial. In Middle Eastern climates some of these, now over 4000 years old, still appear as fresh as the day they were made - for example, the mud and straw bricks of the pyramid of Sesostris II (*ca* 2370 B.C.) at El-Lahun in Egypt. It is interesting to note that the Roman architect Vitruvius (or, more correctly, Marcus Vitruvius Pollo), writing about 27 B.C., concentrates mainly on sun-dried brick, which he recommends should be made in the spring or autumn so that it dries uniformly throughout - clearly thermal gradients were troublesome even then - and which should be at least two years old before being used. He also notes that "the citizens of Utica use no brick for building unless the magistrate has approved them as being dry and made five years before" (Granger, 1962). It is nice to know that bureaucratic delay is not a modern disease! Sun-dried bricks, often with a mud facing and decoration, are excellent building materials still extensively used in the Middle East, particularly in small rural communities, although in prosperous cities they are rapidly being replaced by the ubiquitous and anonymous concrete. Even in areas subjected to moderate rainfall they can be of considerable permanence, especially when protected by courses of projecting tiles or slates. Yet do we know sufficient of clay mineralogy to understand fully the forces between clay particles, with or without organic matter, that give sun-dried clay bricks and facings such permanence?

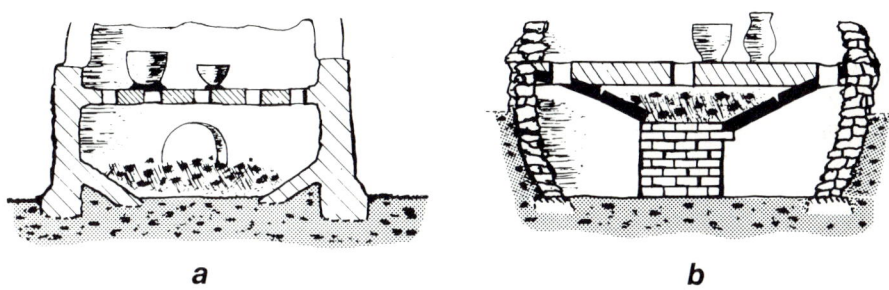

Fig. 1. Sections through ancient pottery kilns: (a) at Khafajé, 5th millenium B.C. (Aitcheson, 1962); (b) at Mohenjo-daro, 3rd millenium B.C. (Ray, 1956).

Open hearths of the eighth or seventh millenium B.C. probably yielded insufficient heat for firing clay into pottery, but closed pottery kilns capable of giving temperatures over 1000°C, dated to the fifth millenium B.C. (Fig. 1a), have been found in Iran (Aitcheson, 1962), Palestine (Kenyon, 1957, 1960) and China (Anon., 1973). Indeed, by the later part of the third millenium B.C. the Indus Valley Civilization had achieved a remarkably uniform colour in their pottery (Ray, 1956) by the use of kilns capable of giving a very uniform temperature in their firing chamber through a construction that indicates a fairly advanced technology (Fig. 1b). The permanence of inscribed pottery probably led to the use of clay tablets for recording purposes: enormous libraries of these tablets, such as that at Nineveh dating to about 700 B.C. and the Tel-el-Amarna letters dating to about 1400 B.C., have survived the centuries and have been invaluable in archaeological research. Surprisingly, fired brick seems to have been little used in the time of Vitruvius, except as a capping, and such as was made was apparently very variable in quality, for he comments that "brickwork that is not made of good clay or is too little baked shows its faults when weathered by ice and hoar-frost" (Granger, 1962). Although he does not define "good clay", some means of assessment must surely have existed.

## THE DAWN OF CLAY MINERALOGY

Early written references to clay refer essentially to the material in the form of common clay or clay used for pottery. The ancient Hebrew words Chomer (חֹמֶר) and Tit (טִיט) seem, for example, to have this inference and to be virtually interchangeable in the Bible. Likewise, the greek πηλὸs and the latin *lutum* have the inference of common clay or mud, but both languages distinguished other types - such as αργιλος or *argilla*, a white clay. It seems indeed to have been the Greeks who first differentiated earths - which embraced all fine-grained deposits and hence included clays - to a degree sufficient to be regarded as the pioneers of clay mineralogy.

The descriptions of many earths given by Theophrastus (372-287 B.C.) in his treatise "On Stones", written about 315 B.C., are such that they can be fairly definitely identified in terms of modern mineral names. Thus, Robertson (1958) has shown that Samian earth can be identified with kaolinite, Kimolian earth with montmorillonite and Melian earth possibly with a siliceous kaolinite. The description of the stone of Scaptehyle, that resembled rotten wood but was not affected on burning with oil, fits that of mountain wood or mountain cork so closely that it was probably palygorskite (Robertson, 1963) and the stone of Siphnos that was "soft enough to be turned on the lathe and carved but when dipped in oil and fired" became "extremely dark and hard" was most likely steatite (Eichholz, 1965). Theophrastus also notes that single deposits differ in quality.

Although the *Historia Naturalis* of Pliny the Elder (A.D. 23-79), written about A.D. 50, is based largely on earlier manuscripts, including that of Theophrastus, there is more detail on earths and, again, the uses to which they were put give clues to their identity. Thus, Umbrian earth has been collated (Robertson, 1949) with a kaolin mineral, *saxum* with bentonite (or sodium montmorillonite) and *Sarda* with fuller's earth (or calcium montmorillonite).

By this stage, therefore, many clay and related minerals were recognized, although it is most unlikely that their occurrence in common clay was appreciated. Two other aspects worthy of note are (a) the manner of naming earths, usually after their locality or origin, and (b) the use by Theophrastus of fire, or heat, in distinguishing minerals.

## THE NEXT 2000 YEARS

One of the best means of following the progress of clay studies, and particularly clay mineralogy, over a period of time is a critical examination of mineralogical texts published during that period. These indicate not only the rate of increase in the number of known clays, or earths, but also the uses to which they were put and, from classification systems, the deductions made as to synonymy and diversity.

The early classification scheme of Aristotle (384-322 B.C.) for minerals in general - into ορυκτά (minerals: *lit.* [things] dug) and μεταλλευτά (metallic ores) (Bandy and Bandy, 1955) - reflects the relatively small number of minerals then recognized. This was later expanded by Theophrastus to: metals, stones, earths - a system used by Pliny and indeed for the next thousand years or so until it was again expanded slightly by Avicenna (980-1036) to: stones and earths, sulphurous minerals, metals, salts. Although earths were therefore recognized as important from earliest times, little classification within the major categories was possible in the absence of chemical information and even as late as the thirteenth century Albertus Magnus (*ca* 1200-1280) simply listed earths alphabetically (see Albertus Magnus, 1541).

A more advanced period, however, began in the mid-sixteenth century with Agricola (1494-1555) (see Bandy and Bandy, 1955) who not only devised a more comprehensive classification system for "mineral bodies" but also gives very considerable insight into the clays known and used at that time. It is worth, therefore, pausing to consider the state of the art (one can hardly yet call it a science) at that time. The classification scheme

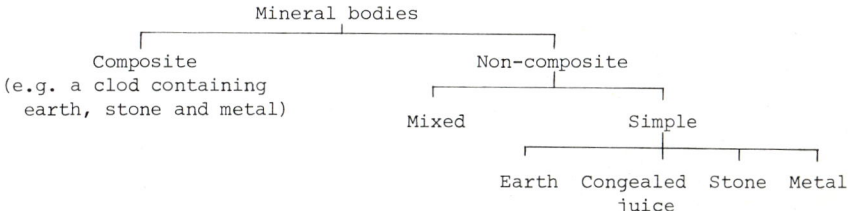

clearly shows that "mineral bodies" then, and of course earlier, were not by any means what we would term minerals but included complex mixtures. Agricola regards earths, which included soils as well as clays, as mixed minerals, but he also distinguishes simple earths from composite earths. Earths are defined in terms of feel, porosity, hardness, smoothness, colour, taste, odour and, interestingly to modern clay mineralogists, the tabular or non-tabular form of occurrence. Having some glimmerings of synonymy and diversity, he decries the common method of naming clays after their place of origin or some special characteristic, but he concludes it is the only useful method at that time. After describing the desirable qualities of earths required by farmers, artisans and physicians he discusses those available commercially; clearly the number had increased very considerably from the time of Theophrastus and Pliny and many European and Mediterranean sources were available.

Some 50 years after Agricola, Andreas Caesalpinus (1596) of Italy gave the following interesting account of the possible origins of earths as then understood: "Their origins are twofold. One is from a cooled dry exhalation that descends from the air in powder or when dissolved in rain water is converted into loam on drying: of this sort is fertile soil because it can contribute to the atmosphere a naturally rarefied juice for nurturing plants. For this reason farmers frequently turn the earth over so that it can be rarefied by the sun. The other mode of formation is from mixtures formed by humble decay and from rocks decomposed by sun or fire and then dissolved in rain water. For so are the stony mountains altered by heat and dissolved into earth by showers of rain. The diversity of rocks and other mixtures gives rise to a large variety of earths". The role of weathering is here clearly recognized: the reference to exhalations and material deposited from rain is less clear but may be related to debris from dust storms, volcanic dust and "red rain" - all of which could have been familiar to an Italian.

Although the importance of chemistry in mineralogy was appreciated by Emanuel König of Basel when he compiled his *Regnum Minerale* in 1687, the first chemical classification of minerals seems to have been that of A. F. Cronstedt (1722-1765) of Sweden, who divided earths into calcareous, siliceous, garnet-like *(Terrae granateae)* and argillaceous, making further subdivision on the basis of colour and use (Cronstedt, 1760). Chemical concepts were also supported by the great Swedish scientists T. O. Bergman (1735-1784) and J. J. Berzelius (1779-1848) but were somewhat at variance with the outlook developed in continental Europe on the basis of the views of A. G. Werner (1750-1817) at Freiburg, or rather those of his successor F. Mohs (1773-1839), who recommended purely physical criteria.

Yet mineralogists of the calibre of Richard Kirwan (1735-1812) realized that both chemical and physical criteria were essential for characterization. In the two-volume second edition of his *Elements of Mineralogy* in 1794-96, Kirwan devotes a large part of one volume to earths. He divides these primarily into a magnesian group (*Muriatic Genus*) and an aluminian group (*Argillaceous Genus*), thus being one of the earliest to distinguish, albeit unknowingly, trioctahedral and dioctahedral minerals. Further subdivision was made largely on the basis of physical characteristics such as fracture, friability, hardness, etc., although the effect of heat or the blowpipe (so effectively developed by Cronstedt and Bergman) is not neglected. In this book too we see the introduction of names still in use today - such as chlorite, talc, and kaolin - while other minerals are readily recognizable from older names - e.g. sepiolite (kiffekil [*sic*] or meerschaum), palygorskite (*suber montanum*) and montmorillonite (fuller's earth or smectis).

By the end of the eighteenth century, therefore, one can see the dawn of modern thought and it is clear that clay minerals could be distinguished sufficiently for various cases of synonymy to be recognized.

THE NOMENCLATURAL REVOLUTION

During the 19th century much knowledge accumulated on relatively pure clay minerals. Moreover, a subtle change took place in nomenclature in that the old names of earths, based on locality of origin or specific properties, became outmoded and were replaced by mineral names, usually ending in the suffix -ite. Henceforth, therefore, it would be more correct to talk of *clay minerals* than of earths and clays.

The clay minerals allophane, halloysite and nontronite were described, analysed and named by Stromeyer in 1816, Berthier in 1826 and Berthier in 1827, respectively, whilst montmorillonite was recognized as a mineral by Mauduyt in 1847. The identification of these was on a completely different basis from that of most of the other clay minerals then known, in that characterization was based essentially on laboratory studies whereas earlier identifications had been by a combination of, admittedly fairly subjective but also very effective, physical or field tests with, latterly, rudimentary chemical data. Unfortunately, the rapid change from practical to laboratory characterization was not without its drawbacks and about the middle of the century <u>one</u> misattribution of names led to considerable confusion between halloysite and montmorillonite that had repercussions for some time to come (Mackenzie, 1963). Had the criteria of the practical mineralogists of previous centuries been used, it is most unlikely this episode would ever have occurred. Indeed, this should serve as a lesson to us all - never to let our enthusiasm for new methods or techniques blind us to the fact that the old can still yield information that can occasionally prevent costly mistakes. Despite this marked whirlpool on the smooth stream of progress, this period nevertheless saw several noteworthy contributions to clay mineralogy - such as the massive study of kaolins by Brongniart in 1840-41 (Brongniart, 1840; Brongniart and Malaguti, 1841).

The difficulty of distinguishing, or establishing the purity of clay minerals in the laboratory was keenly felt by many scientists towards the end of the century. Although many instances of synonymy were recognized, mineral status was still accorded to many materials now known to be mixtures - as is clear, despite its advanced nature, from the classification of clay and related minerals by Dana in 1894. The revolution in nomenclature that had occurred during the 1800s is well illustrated by comparing this with the classification of Kirwan in 1794 (Mackenzie, 1963): within a century all the traditional names had disappeared.

DEVELOPMENT OF INVESTIGATIONAL METHODS

Since it is impossible to do full justice to this enormous subject I shall refer here only to a few highlights regarding the main methods currently in use.

In their study of kaolins, Brongniart and Malaguti (1841) concluded that some were contaminated with "silice à l'état gélatineux" which they removed by extraction with a solution of potassium hydroxide in aqueous alcohol. Their example was soon followed by others. Thus, Salvétat in 1847 (Damour and Salvétat, 1847) used sodium hydroxide to purify some montmorillonite samples but by 1851 (Salvétat, 1851) he had changed over to sodium carbonate - presumably because the hydroxide was too severe. Selective chemical techniques are therefore older than many of us realise and it is noteworthy that sodium hydroxide, potassium hydroxide and sodium carbonate are still widely used (Briner and Jackson, 1970) in the study of what might well be termed "gelatinous" components.

Later in the 19th century, Le Chatelier attempted to distinguish clay minerals on the basis of their heating curves. In his 1887 paper, which is usually, but erroneously (Mackenzie, 1978b), regarded as the first on thermal analysis, he describes an ingenious automatic recording system which he used to show that allophane, halloysite, kaolinite, pyrophyllite and montmorillonite all behaved differently when heated. Despite his strong commendation of this technique, it and its later more sensitive offspring, differential thermal analysis (DTA) (Roberts-Austen, 1899), found their major use in metallurgy for the next 25 years. Indeed, not until 1913 were clays studied by DTA (Wallach, 1913). Subsequent DTA studies on clays were very sporadic and only after the work of Norton in 1939 did the method arouse widespread interest. Today thermoanalytical methods are regarded as useful complementary techniques in most clay mineral studies.

Infrared absorption spectroscopy was, in the opening years of the twentieth century, at a very early stage of development, only single-beam instruments being known and manual plotting being standard. Despite these drawbacks, Coblentz in 1905-1906, working like a troglodyte in a basement at night to avoid interference from heat and vibration, succeeded in recording absorption spectra of solids and minerals. The first major work on clay minerals seems to have been that of Adler and his colleagues in 1950, where spectra for a set of clay minerals obtained in four laboratories using commercial double-beam instruments are compared. Now that band assignments have been established, infrared absorption spectroscopy is invaluable in assessing short range relationships between an atom and its nearest neighbours in both crystalline and highly disordered components of clays (Farmer, 1974).

Although Röntgen discovered X-rays in 1895, it was 1912 before work in M. von Laue's laboratory established that crystalline materials diffracted X-rays (Friedrich et al., 1912) and 1913 before the Bragg equation, which simplified structural studies, was published (Bragg and Bragg, 1913). X-ray diffraction established the crystalline nature of clay minerals in the 1920s and the basic structures of the main groups were elucidated over the next ten years or so (Grim, 1968). Subsequently methods of pretreatment were devised to enable precise identification of groups and species presenting difficulty (Brown, 1961) and today X-ray diffraction is the one indispensable technique that must be in every clay mineralogist's locker.

Two independent studies in 1927 (Davidson and Germer, 1927; Thomson and Reid, 1927) essentially confirmed the wave nature of electrons, as proposed by de Broglie in 1924, and laid the basis of electron diffraction. The first electron microscope was constructed in 1928-1932 by Knoll and Ruska (1932) and an instrument capable of both microscopy and diffraction in 1936 by Boersch. Continuous technological advances have improved transmission electron microscopes until now their resolution is near theoretical. Clays have long been of interest to electron microscopists and, although particle shape or thickness are not diagnostic criteria on their own, electron microscopy coupled with selected-area diffraction (Gard, 1971), and more recently electron-probe microanalysis (Jepson and Rowse, 1975), is a particularly valuable tool. The scanning electron microscope, which when coupled with electron-probe microanalysis gives essentially a three-dimensional picture of specimen surfaces and reveals their chemical composition, has recently assumed increasing importance (McHardy and Birnie, 1975).

Within the past twenty years other, perhaps less familiar, techniques have also been contributing markedly to clay mineral investigations - and are frequently referred to in the papers being presented at this Conference. For example, the potentialities of nuclear magnetic resonance (NMR) in clay studies were recognized, independently, by Ducros and Fripiat as early as 1960, Mössbauer spectroscopy was first applied to phyllosilicates in 1962 by Pollak et al., electron paramagnetic resonance (EPR) has been used fairly steadily since 1969 (Furuhata and Kuwata, 1969) and some possibilities of electron spectroscopy (ESCA), in the form of X-ray photoelectron spectroscopy (XPS), were examined by Freund and Hamich in 1971. Despite the fact that these and other investigational methods must be regarded as complementary to the major techniques, such as X-ray diffraction, each has its own quota of information to contribute.

THE PRESENT AND THE FUTURE

In view of the masterful review of Brindley (1976), which covers most of the fundamental detail and needs little if any updating, I should like here to examine only a few broad aspects of a more practical and applied nature.

First of all, although X-ray diffraction must always remain the one indispensable investigational method for crystalline clay minerals, the notable advances that have occurred in other techniques should not be minimized. In clay mineralogy it is still particularly true that even a meaningful semi-quantitative mineralogical analysis can only be obtained by integration of the results of all possible techniques (Mackenzie and Mitchell, 1966). For example, X-ray powder diffraction gives information on the crystalline minerals present and the three-dimensional arrangement of atoms in space, but it must not be forgotten that this information is statistical for a large number of crystallites and that for individual crystals electron diffraction is essential; infrared absorption spectroscopy also gives structural information, but this relates to the relationships between an atom and its neighbours and applies to both crystalline and non-crystalline (or highly disordered) materials. Thermal analysis can yield quantitative information, perhaps more readily than most other techniques, distinguish atypical dioctahedral phyllosilicates and reveal very minor amounts of certain accessory minerals (Mackenzie and Mitchell, 1972); it has, however, considerable limitations for identification. Transmission electron microscopy reveals crystal shape, aggregation and cementation; the high resolution lattice imaging now available enables even the micro-structure of individual layers and stacking irregularities (Yoshida, 1976), twinning (Akizuki and Zussman, 1978), etc., to be directly observed. Mössbauer spectroscopy has recently been giving very valuable information on the nature of the iron present, NMR on sorbed water and EPR on transition-element environments. Selective chemical techniques are invaluable in assessing highly disordered components and total chemical analysis can, at least on occasion, be used to check the possible accuracy of a mineralogical analysis otherwise derived (see Mackenzie and Mitchell, 1966). Finally, scanning electron microscopy and associated electron probe microanalysis can yield much physical and chemical information on particle surfaces.

And yet, having obtained, at very considerable cost, such a large amount of information, can we confidently state that we completely understand any particular clay and its behaviour? In general, I believe the answer must be no - not because the acquisition of such knowledge would leave us without a job, but because there are imponderables still to be assessed. For example, take interstratified minerals: we know that layers of three kinds can interstratify and that such interstratifications can now be identified (Cradwick and Wilson, 1978), but do minerals exist with four or more kinds of layers, and if so, how can they be identified and what are their properties? We know, too, that clay particles have extremely reactive surfaces - but how far are these surfaces understood? While the electron microprobe enables the chemistry of a surface to be investigated, it must be remembered that the probe penetrates to a distance of something like 1 µm, or several hundred atomic layers, so that the statistical picture obtained may not reflect the composition of the few important atomic layers at the particle surface. Here, however, we can look forward to further development of ESCA which should be able to give the information desired. Moreover, when clays are associated with organic matter, as in soils, how is the surface affected? Many soil clays are known to have a highly disordered or non-crystalline inorganic phase associated with their surfaces - whether discrete or arising from increasing disorder towards the surface we cannot yet tell. This material, however, can have an effect on clay properties disproportionate to the amount present and in some instances render knowledge of the underlying crystalline material of relatively little importance. Moreover, current information suggests that the active hydroxyl groups in the surface material can react with organic materials to give organomineral compounds with quite different properties from those of the underlying inorganic mineral and possibly also of the organic material itself. It is not surprising, therefore, that surface properties and reactions of clays are attracting so much attention.

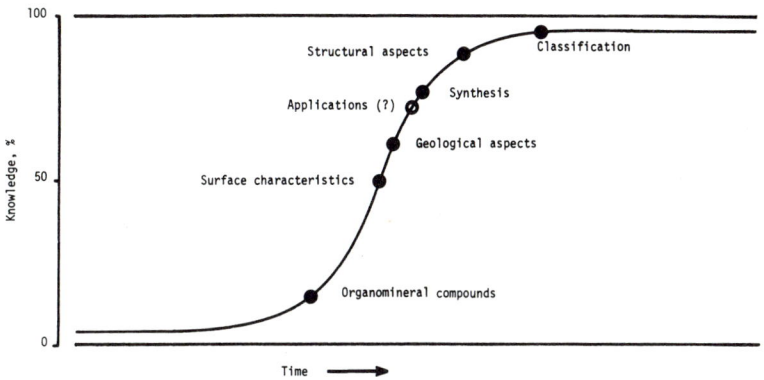

Fig. 2. Possible present state of knowledge on various aspects of clay mineralogy.

In view of these comments, it is instructive to consider the stages of development currently reached by individual aspects of clay mineralogy. It seems to me that the progress of any *one* aspect can be plotted on a knowledge/ time curve of the cumulative normal distribution type* shown in Fig. 2. Inserted on this curve is a purely subjective assessment of the stages in the various fields noted. Synthesis is difficult to place, as I doubt whether we can yet produce completely "tailor-made" minerals; yet we are possibly 80% of the way. "Applications" should probably not appear, since it is likely to follow the usual stepwise curve, receiving fresh impetus periodically from advances in other fields: I feel, however, it may be useful to consider its approximate relative position. Since there are undoubtedly as yet unrecognized aspects along the lower limb of the curve, this diagram must be regarded merely as a convenient visualization that possibly gives some indication of future trends.

The difficulties and pitfalls of prognostication are great, yet, having regard to Fig. 2, I venture to suggest that the emphasis on the characteristics of surfaces, be they crystalline, highly disordered or organomineral, is likely to remain with us for some time to come. There is, too, likely to be increasing effort to understand the formation, nature and properties of organomineral compounds, both synthetic and natural. Study of these may well require a new

---

* The equation governing this type of curve is:

$$K = \int_{-\infty}^{t} (1/\sigma\sqrt{2\pi}) \exp\{-(t-\mu)^2/2\sigma^2\} dt$$

where K is knowledge, t time, σ the standard deviation and μ the mean.

technology and already one envisages the use of such complex methods as pyrolysis-gas chromatography-mass spectrometry. Thermoanalytical and calorimetric methods - such as differential scanning calorimetry and flow microcalorimetry - have some part to play and there may even be scope for thermosonimetry, thermoacoustimetry, and thermoelectrometry (Mackenzie, 1978a). NMR, EPR, ESCA, neutron diffraction and more esoteric methods are also likely to be called in or devised and, in the final stage, the ubiquitous computer may be the only means available of assessing the overall value of all the information collected.

This may be rather a forbidding picture to some, but it should not be if we really want to understand fully the mineralogy and behaviour of clays. Undoubtedly, empirical tests could give some of the same information more readily, but the knowledge of *why* would be missing and scientific advance is possible only when reasons are known.

ACKNOWLEDGEMENTS

The author thanks several colleagues at the Macaulay Institute as well as the Rev. A.G.H. Grant and Prof. L. Heller-Kallai for assistance with matters outside his own particular field and Dr A.C. Bishop, Keeper of Minerals, for providing access to early mineralogical texts housed in the British Museum (Natural History).

REFERENCES

Adler, H. H., Kerr, P. F., Bray, E. E., Stevens, N. P., Hunt, J. M., Keller, W. D. and Pickett, E. E., 1950. Infrared spectra of reference clay minerals. API Project 49, Clay Minerals Standards, Prelim. Rep. No. 8, Columbia University New York, 146 pp.
Akizuki, M. and Zussman, J., 1978. The unit cell of talc. Mineralog. Mag., 42: 107-110.
Albertus Magnus, 1541. De Mineralibus et Rebus Metallicis. Balthassar Beck, Strasbourg.
Anon., 1973. The Genius of China: [Catalogue of] an exhibition of archaeological finds in the People's Republic of China held at the Royal Academy, London. Times Newspapers, London.
Bandy, M. C. and Bandy, J. A. (Editors), 1955. De Natura Fossilium. Spec. Pap. geol. Soc. Am., No. 63.
Berthier, P., 1826. Analyse de l'halloysite. Annls Chim. Phys., 32: 332-334.
Berthier, P., 1827. Nontronite, nouveau minéral découvert dans le département de la Dordogne. Annls Chim. Phys., 36: 22-27.
Boersch, H., 1936. Primäre und sekundäre Bild im Elektronenmikroskop, I, II. Annln Phys., 26: 631-644; 27: 75-80.
Bragg, W. H. and Bragg, W. L., 1913. The reflection of X-rays by crystals. Proc. R. Soc., A88: 428-438.
Brindley, G. W., 1976. Current and future trends in clay mineralogy - a review. Clay Minerals, 11: 257-268.
Briner, G. P. and Jackson, M. L., 1970. Mineralogical analysis of clays in soils developed from basalts in Australia. Israel J. Chem., 8: 487-500.
de Broglie, L., 1924. Tentative theory of light quanta. Phil. Mag., 47: 446-458.
Brongniart, A., 1840. Premier mémoire sur les kaolins ou argiles à porcelaine. Arch. Mus. Hist. nat., 1: 243-301.
Brongniart, A. and Malaguti, J., 1841. Second mémoire sur les kaolins ou argiles à porcelaine. Arch. Mus. Hist. nat., 2: 217-255.

Brown, G. (Editor), 1961. The X-ray identification and Crystal Structures of Clay Minerals. Mineralogical Society, London.
Caesalpinus, A., 1596. De Metallicis. Aloysius Zennetti, Rome.
Le Chatelier, H., 1887. De l'action de la chaleur sur les argiles. Bull. Soc. fr. Minér. Cristallogr., 10: 204-211.
Coblentz, W. W., 1905. Infrared absorption spectra. II Liquids and solids. Phys. Rev., 20: 337-363.
Coblentz, W. W., 1906. Infrared absorption and reflection spectra of water and minerals. Phys. Rev., 23: 125-154.
Cradwick, P. D. and Wilson, M. J., 1978. Calculated X-ray diffraction curves for the interpretation of a three-component interstratified system. Clay Minerals, 13: 53-65.
[Cronstedt, A. F.], 1760. German translation by S. Wiedemann: Versuch einer neuen Mineralogie. Rothensche Buchhandlung, Kopenhagen.
Damour, A. and Salvétat, L. A., 1847. Notice et analyses sur un hydrosilicate d'alumine trouvé à Montmorillon (Vienne). Annls Chim. Phys., 21: 376-383.
Dana, E. S., 1894. The System of Mineralogy of James Dwight Dana: Descriptive Mineralogy (6th edn.). Kegan Paul, Trench, Trübner, London.
Davisson, C. and Germer, L. H., 1927. Diffraction of electrons by a crystal of nickel. Phys. Rev., 30: 705-740.
Ducros, P., 1960. Les possibilités de la résonance magnétique nucléaire (RMN) dans les études structurales, en particulier dans le cas de la mobilité d'ions ou de molécules en phase solide. Bull. Grpe. fr. Argiles, 12: 19-23.
Eichholz, D. E. (Editor), 1965. Theophrastus De Lapidibus. Clarendon Press, Oxford.
Farmer, V. C. (Editor), 1974. The Infrared Spectra of Minerals. Mineralogical Society, London.
Freund, F. and Hamich, M., 1971. Photoelektronen- und Röntgenfluoreszenzspektroskopie an Magnesiummineralen. Fortshr. Miner., 48: 243-258.
Friedrich, W., Knipping, P. and Laue, M., 1912. Interferenzerscheinungen bei Röntgenstrahlen. Sber. bayer. Akad. Wiss., Math-phys. Klasse, pp. 303-322.
Fripiat, J. J., 1960. In discussion on Ducros (1960), q.v.
Furuhata, A. and Kuwata, K., 1969. ESR spectra of manganese (II) and copper (II) adsorbed on clay minerals and silica-alumina mixtures. Nendo Kagaku, 9: 19-27.
Gard, J. A. (Editor), 1971. The Electron-Optical Examinations of Clays. Mineralogical Society, London.
Granger, F. (Editor), 1962. Vitruvius On Architecture. Heinemann, London.
Grim, R. E., 1968. Clay Mineralogy (2nd edn.). McGraw Hill, New York.
Jepson, W. B. and Rowse, J. B., 1975. The composition of kaolinite - an electron microscope microprobe study. Clays Clay Miner., 23: 310-317.
Kenyon, K. M., 1957. Digging up Jericho. Praeger, New York.
Kenyon, K. M., 1960. Archaeology in the Holy Land. Praeger, New York.
Kirwan, R., 1794. Elements of Mineralogy (2nd edn., 2 vols). Elmsly, London.
König, E., 1687. Regnum Minerale. E. and J. König, Basel.
Knoll, M. and Ruska, H., 1932. Das Elektronenmikroskop. Z. Phys., 78: 318-339.
McHardy, W. J. and Birnie, A. C., 1975. Scanning electron microscope studies of a surface water gley. J. Soil Sci., 26: 426-431.
MacEwan, D. M. C., 1949. [Editorial in] Clay Miner. Bull., 1: 69-70.
Mackenzie, R. C., 1963. De Natura Lutorum. Clays Clay Miner., 11: 11-28.
Mackenzie, R. C., 1978a. Nomenclature in Thermal Analysis: IV. J. therm. Analysis, 13: 387-392.
Mackenzie, R. C., 1978b. De Calore: Prelude to thermal analysis. Thermochim. Acta, in press.
Mackenzie, R. C., and Mitchell, B. D., 1966. Clay Mineralogy. Earth Sci. Rev., 2: 47-91.
Mackenzie, R. C. and Mitchell, B. D., 1972. Soils. In: R. C. Mackenzie, Editor. Differential Thermal Analysis, Vol. 2. Academic Press, London and New York, pp. 267-297.
Mauduyt, -, 1847. Un mot sur un morceau de quartz d'une variété particulière, ainsi que sur une substance minérale trouvée dans le département de la Vienne. Bull. Soc. géol. Fr., 4: 168-170.

Norton, F. H., 1939. Identification of clay minerals by differential thermal analysis. J. Am. Ceram. Soc., 22: 54-63.
Pollak, H,, Coster, M. and Amelinckx, S., 1962. Mössbauer effect in biotite. Phys. Stat. Solid., 2: 1653-1659.
Ray, P., 1956. History of Chemistry in Ancient and Medieval India. Indian Chemical Society, Calcutta.
[Roberts-Austen, W. C.], 1899. Alloys. Nature, Lond., 59: 566-567.
Robertson, R. H. S., 1949. The fuller's earths of the Elder Pliny. Classical Rev., 63: 51-52.
Robertson, R. H. S., 1958. The earths of Theophrastus. Classical Rev., 8: 222-223.
Robertson, R. H. S., 1963. 'Perlite' and palygorskite in Theophrastus. Classical Rev., 13: 132
Salvétat, L. A., 1851. Analyse de quelques hydrosilicates d'alumine. Annls Chim. Phys., 31: 102-116.
[Stromeyer, F.], 1816. In: Götting. gelehr. Anz., pp. 1251-1252.
Thomson, G. P. and Reid, A., 1927. Diffraction of cathode rays by a thin film. Nature, Lond., 119: 890.
Wallach, R., 1913. Analyse thermique des argiles. C.r. hebd. Séanc. Acad. Sci., Paris, 157: 48-50.

# SECTION 1

# Crystal Chemistry and Structure

STRUCTURAL IRON OXIDATION DURING MICA EXPANSION

A. D. SCOTT AND A. F. YOUSSEF
Department of Agronomy, Iowa State University, Ames, Iowa

ABSTRACT

Determinations were made of the $Fe^{++}$ oxidation that occurred while lepidomelane samples were progressively K depleted and the expanded products were exposed to different redox conditions for specified periods. $Fe^{++}$ oxidation occurred in NaTPB-treated mica but not as rapidly as the structural $Fe^{++}$ was exposed by an expansion of the mineral -- 19% being oxidized during the week required for full expansion and 52% in the next 4 years. Changes in the mica expansion-oxidation relationship occurred when $H_2O_2$, $Na_2S_2O_4$ or $O_2$ were added, the particle size or reaction temperature were varied and successive NaCl extractions were used. Fully expanded mica samples with only 7% oxidation were prepared.

INTRODUCTION

Interlayer K replacement by hydrated cations and structural $Fe^{++}$ oxidation are closely linked aspects of mica weathering. The relationships between these processes, however, have not been fully resolved. In particular, little is known about the progress of $Fe^{++}$ oxidation during the expansion of mica.

Both Robert and Pedro (1966) and Ross and Rich (1974) related the degrees of $Fe^{++}$ oxidation in salt-treated biotite to specific stages of K depletion but not to the treatment period involved. Others (Bowen et al., 1969; Farmer et al., 1971; Mackintosh et al., 1972; Newman and Brown, 1966) were more concerned with the $Fe^{++}$ status of mica samples that were treated for maximum K removal. As shown by Reed (1963, pp. 56-59), however, the treatment period may be relevant to the mica expansion-oxidation relationship. Thus, further study was made of the changes in $Fe^{++}$ that can occur while micas are being K depleted and the expanded micas are then exposed to different conditions for various periods.

EXPERIMENTAL

Sheets of lepidomelane from Faraday Township, Ontario, Canada, were dry-ground in a water-cooled Waring Blendor and suspended in water to separate 2-5μm and 10-20μm size-fractions by sedimentation. Portions of these mica samples were treated with solutions of NaCl or NaCl-NaTPB for specified periods to replace increments of interlayer K by Na and thereby expand the mica to different degrees. Various modifications were made in the extracting solution to expose the mineral to diverse redox conditions. Some of the samples that were fully expanded by the treatments were subsequently left in the extracting solutions for various periods. The amounts of K replaced by Na were used as a measure of the expansion of the mineral and compared with the $Fe^{++}$ oxidation that occurred during and after the expansion.

The basic NaCl-NaTPB treatments were carried out by placing 0.5g mica in 10 ml $1\underline{N}$ NaCl-$0.2\underline{N}$ NaTPB-$0.01\underline{M}$ EDTA solutions for different periods in stoppered 50 ml Erlenmeyer flasks at 25°C (Smith and Scott, 1966). The reaction temperature was increased to 65°C to enhance the rate of K extraction from some samples (Ismail and Scott, 1972). Other samples were treated with 10 ml increments of the NaCl-NaTP solution that were combined with 5 ml 30% $H_2O_2$ or a mixture of 0.1g $Na_2S_2O_4$-0.25g Na citrate-0.09g $NaHCO_3$ to expose the mineral to stronger oxidizing and reducing conditions, respectively. To ensure the presence of dissolved oxygen during the treatment of some mica samples, the usual mixture of mica and NaCl-NaTPB solution in stoppered flasks was mixed by a stream of $O_2$ bubbles. The amounts of K extracte by each treatment were determined to obtain expansion-oxidation relationships that were applicable to the specific conditions involved.

To K-deplete mica samples with NaCl solutions, 0.5g mica samples were treated with successive 1000 ml increments of $1\underline{N}$ NaCl at 90°C. The NaCl solution was replaced every 30 minutes for 3 hours and every hour for another 3 hours. The number of successive treatments was varied to obtain mineral samples with different degrees of K depletion. Some of the samples that received successive treatments for 6 hours were subsequently left in $1\underline{N}$ NaCl at 90°C for 24 hours. The amount of K extracted by each NaCl increment was determined by flame emission and using standards prepared from NaCl with the same lot number.

The NaTPB- and NaCl-treated mica samples were prepared for $Fe^{++}$ determinations by terminating the K extraction with 10 ml $2\underline{N}$ KCl, adding acetone to dissolve the KTPB (final solution being 60% acetone by volume), filtering, washing the mineral with alcohol until $Cl^-$ free and then with acetone, and finally air drying the minera This procedure produced no changes in the $Fe^{++}$ of the mineral. The amount of $Fe^{++}$ in the various mineral samples was determined by the method described by Peters (1968). The applicability of this method was verified with mineral standards and interferences from the NaTPB treatments were avoided.

Total K and Fe determinations were made by digesting samples of the mica in HF and analyzing the solutions by atomic absorption. The 2-5μm and 10-20μm lepidomelane samples contained 191 and 196 me K per 100g, respectively, whereas both samples contained 275 mmoles Fe per 100g. The $Fe^{++}$ contents of these size-fractions were 248 and 257 mmoles per 100g.

## RESULTS AND DISCUSSION

### Particle size effects

The oxidation of $Fe^{++}$ that occurred in 2-5μm and 10-20μm lepidomelane particles while they were being K-depleted by NaCl-NaTPB solutions and subsequently stored in the same solutions for periods after they were fully expanded is described in Fig. 1. Curves showing the amounts of K extracted (and, therefore, the extent of expansion) by the different treatment periods are also presented in Fig. 1. It is evident that oxidation began shortly after the structural $Fe^{++}$ was exposed by an expansion of the mica, but the subsequent rate of oxidation was rather slow. Instead of the $Fe^{++}$ being oxidized as fast as it was exposed, as might be expected, only 19% of the $Fe^{++}$ in 10-20μm mica particles was oxidized by the time the mineral was fully expanded. After 4 years in an expanded state in the NaCl-NaTPB solutions, the 10-20μm mica still had only 52% of its $Fe^{++}$ oxidized. Mica expansion was required for oxidation but had no obvious impact on the rate of oxidation. Indeed, the curves indicate that some rate limiting factor other than $Fe^{++}$ exposure was operative throughout the treatment period.

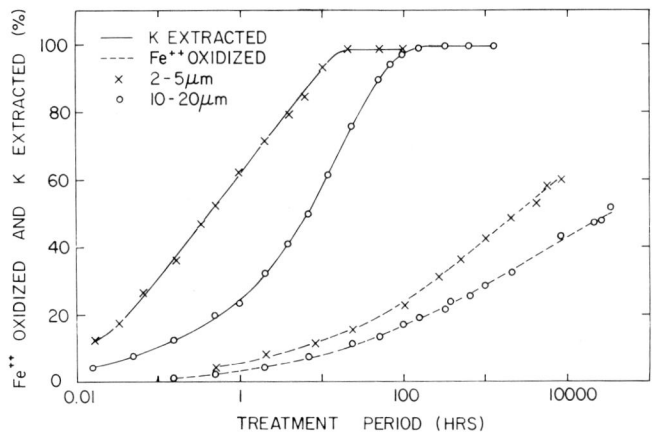

Fig. 1. K extracted and $Fe^{++}$ oxidized by treating 2-5μm and 10-20μm samples of lepidomelane with NaCl-NaTPB solutions at 25°C for different periods.

Coarse mica particles in NaCl-NaTPB solutions release K and expand in accordance with an edge weathering model (Scott, 1968). Consequently, $Fe^{++}$ in the expanded rim would be exposed to oxidation first and the progress of oxidation could be expected to depend on the development of the edge altered zone and the transfer of reactants in the expanded edges. In this event, the size of mica particles should have a major effect on the rates of both the K release and $Fe^{++}$ oxidation. Decreasing the lepidomelane particle size from 10-20μm to 2-5μm did increase the rates of the two processes (Fig. 1), but the low levels of oxidation make it difficult to tell from the relative changes whether the processes are linked by a common edge situation. It is evident, however, that oxidation in the small particles was no more limited by a lack of $Fe^{++}$ exposure than it was in the 10-20μm particles. After the mica was expanded, oxidation continued at a higher rate in the smaller particles and led to 60% of the $Fe^{++}$ being oxidized in 1 year instead of the 42% observed with 10-20μm samples. In this system, at least, the rate determining factor for oxidation was particle size dependent. Since expansion was not responsible, the lateral surface area and interlayer distances must be involved.

In a comparison of the $Fe^{++}$ oxidation that occurred in NaTPB treated size-fractions of biotite, Reed (1963, pp. 56-59) related the occurrence of less oxidation in smaller particles to the shorter degradation periods required. As shown in Fig. 1, however, a reduction in particle size can enhance the rates of both K release and oxidation. As a result, in the 1-day and 1-week periods that were required for full expansion of the 2-5μm and 10-20μm samples, respectively, 16 and 19% of the $Fe^{++}$ was oxidized. This difference is too small to make the size of particles a useful approach for minimizing oxidation.

Whereas Reed (1963) oxidized as little as 2% of the $Fe^{++}$ in biotite samples, Newman and Brown (1966) oxidized 5 to 45% (calculated) of the $Fe^{++}$ in biotite and lepidomelane samples and Bowen et al. (1969) oxidized about 30% of the $Fe^{++}$ in biotite samples when they removed most of the K in the micas with NaTPB treatments. Raman and Jackson (1966) partially degraded biotite (ca. 50%) with NaTPB treatments and found 3% of the $Fe^{++}$ was oxidized. Many of these differences can be attributed to the effects of particle size and contact periods shown in Fig. 1. With this information, expanded mica samples with various levels of oxidation can be easily prepared, but an expansion of mica without oxidation seems unlikely.

Modified redox conditions

When mica samples are treated with NaCl-NaTPB solutions, the conditions relevant to the exchange of interlayer K are well defined in terms of the replacing cation (Na), the precipitation of KTPB, etc. On the other hand, the redox conditions imposed by these treatments are ill-defined and uncertainties exist as to if or how the $Fe^{++}$ in expanded portions of the mica can respond to changes in the redox status of the solution. For information on these points, experiments were conducted with 10-20μm lepidomelane samples in NaCl-NaTPB solutions that were mixed with $H_2O_2$ or $Na_2S_2O_4$ to ensure a strong oxidizing or reducing environment. The pH of the NaTPB-mica mixtures was between 6.8 and 7.2 throughout most of the treatment periods whether or not $H_2O_2$ or $Na_2S_2O_4$ were included in the mixtures. The amounts of K extracted and $Fe^{++}$ oxidized by exposing the mica to these solutions for specific periods are given in Fig. 2. A curve describing the $Fe^{++}$ oxidation observed with 10-20μm samples in just the NaCl-NaTPB solution has been included for comparative purposes.

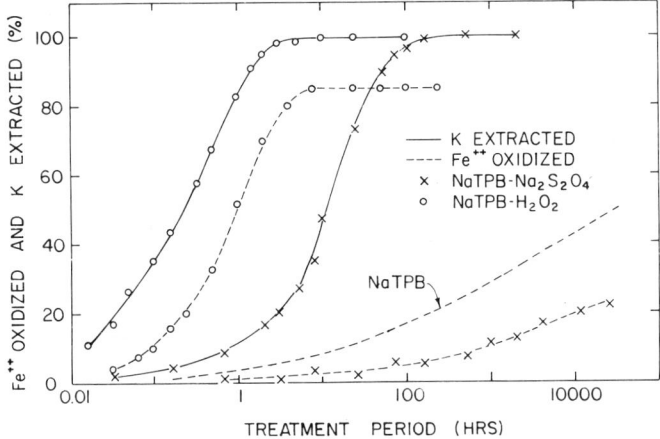

Fig. 2. K extracted and $Fe^{++}$ oxidized by treating 10-20μm lepidomelane samples at 25°C with NaCl-NaTPB solutions that contain $H_2O_2$ or $Na_2S_2O_4$.

Most of the NaTPB-exposed $Fe^{++}$ in the lepidomelane samples proved to be quite responsive to changes in the redox status of the solution. Most obvious in Fig. 2 is the response involving the rates of oxidation. In a period of just 8 hours, 86% oxidation and complete K removal was achieved with the $H_2O_2$ mixtures, whereas only 19% oxidation occurred in the week needed for full expansion of mica samples in the NaCl-NaTPB solutions. Even with $H_2O_2$ in the solution, however, part of the $Fe^{++}$ exposed by a release of K was not oxidized (52% oxidation versus 83% K extraction in 1 hour). The incomplete oxidation of exposed $Fe^{++}$ in the partially expanded samples (up to 80% expansion) was strictly a rate determined situation. If these samples were left in the NaTPB-$H_2O_2$ solution for longer periods, the rest of the exposed $Fe^{++}$ was soon oxidized. Thus, the effects of $H_2O_2$ are indicative of an increase in the rate of oxidation that would be anticipated from a higher concentration of oxidant, but they involve a higher rate of mica expansion in the NaTPB-$H_2O_2$ mixture as well.

The NaTPB-$H_2O_2$ treatments failed to oxidize 14% of the $Fe^{++}$ in this mica even though it was exposed by an exchange of Na for interlayer K and the treatment periods were extended to eliminate the effects of slow reactions. In case the total oxidation were restricted by a loss of $H_2O_2$ during the treatment periods, the 8-hour and 2-day treatments were repeated with a procedure in which successive additions of 30% $H_2O_2$ were made throughout the treatment period. No change in $Fe^{++}$ oxidation was observed. The pH of some of the NaTPB-$H_2O_2$-mica mixtures was reduced from 7.2 to 4.5 or 3.5 after 24 hours by adding HCl and the treatments were continued for another hour. This decrease in pH did not enhance the oxidation of $Fe^{++}$ in the expanded mica either. Thus, even though Robert and Pedro (1969) and Ross and Rich (1974) oxidized more of the $Fe^{++}$ in biotite samples, the maximum of 86%

oxidation in this mica would suggest that there are limits to the degree of $Fe^{++}$ oxidation that are possible with $H_2O_2$ treatments. Since the maximum level of oxidation for the NaTPB treatments without $H_2O_2$ was not established and other oxidants were not applied, it is not known how this nonoxidizable part of the structural $Fe^{++}$ relates to the redox status of the solution. The limited oxidation observed with $H_2O_2$, however, would suggest that structural alterations in the oxidized mineral--such as the Fe explusion proposed by Besson et al. (1975) for $H_2O_2$ treated vermiculite--can impose a restraint on the oxidation process. This restraint may account not only for the maximum degree of oxidation but for the $Fe^{++}$ not being oxidized as rapidly as it was exposed to a strong $H_2O_2$ solution.

$Na_2S_2O_4$ treatments have been commonly used to reduce structural $Fe^{+++}$ in layer silicates (Ross and Rich, 1974; Stucki and Roth, 1977; Veith and Jackson, 1974). Thus, $Na_2S_2O_4$ was mixed with NaCl-NaTPB solutions that were used to K-deplete 10-20μm samples of lepidomelane. As shown in Fig. 2, the $Na_2S_2O_4$ additions did not reduce any of the indigenous 18 mmoles $Fe^{+++}$ per 100g in the contracted or expanded mica samples. The oxidation of $Fe^{++}$ normally found with NaTPB-treated mica was decreased by the $Na_2S_2O_4$ but even this oxidation was not eliminated entirely. Since the oxidation did not exceed 6% while the mica was being expanded, the NaTPB-$Na_2S_2O_4$ treatment merits further consideration as a means of preparing expanded micas with minimum redox alterations. The subsequent oxidation of another 18% $Fe^{++}$ in the expanded mica raises questions about the oxidation process and the applicability of $Na_2S_2O_4$ as a long-term reducing agent that will need to be resolved

Different procedures

The occurrence of appreciable amounts of $Fe^{++}$ oxidation in expanded mica samples has been attributed to $O_2$ in the NaTPB solutions (Newman and Brown, 1966) and NaCl solutions (Farmer, et al., 1971) that were used to K-deplete the mineral. Thus, the possibility arose as to whether the oxidation reported in Fig. 1 might have been limited by the supply of $O_2$ in the 10 ml NaCl-NaTPB solution and the air enclosed in the stoppered 50 ml Erlenmeyers. To test this possibility, several 10-20μm samples of lepidomelane were mixed with the usual NaCl-NaTPB solution by a stream of $O_2$ bubbles for different periods. The effects of this $O_2$ treatment are described in Fig. 3. The mixing action of the gas enhanced the initial release of K and thereby caused a small increase in $Fe^{++}$ oxidation. Otherwise, the $O_2$ addition had little effect on the NaTPB-treated mica over a 24-week period. Since this procedure should have eliminated restrictions in oxidation due to the concentration or diffusion of $O_2$, but did not, $O_2$ was either too weak an oxidant or was not the rate limiting factor. According to Besson et al. (1975) $O_2$ is a weak oxidant, but Stucki and Roth (1977) have reported that even a brief exposure of reduced nontronite to air can cause a very rapid oxidation of $Fe^{++}$. This difference in response to $O_2$ and the comparative effects of $O_2$ and $H_2O_2$ on

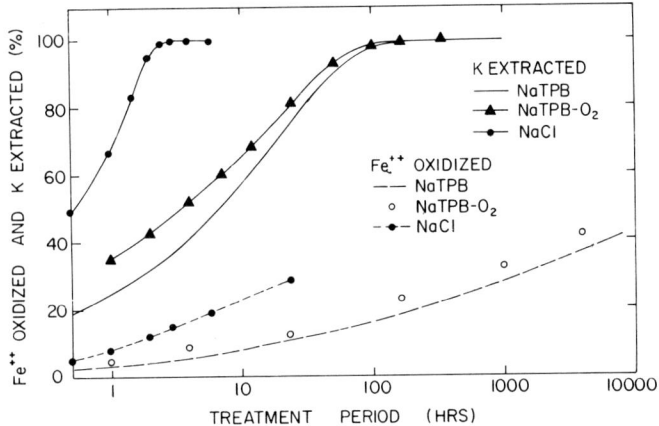

Fig. 3. K extracted and $Fe^{++}$ oxidized by treating 10-20µm lepidomelane samples with successive increments of NaCl solution at 90°C or with oxygen-aerated NaCl-NaTPB solutions at 25°C.

expanded mica reflect the relative susceptibilities of indigenous and pre-reduced $Fe^{++}$ to oxidation. Experiments with different $H_2O_2$ concentrations similar to those of Wey et al. (1966) should help delineate the degrees to which indigenous $Fe^{++}$ resists oxidation and the role of associated mineral alterations.

Since the interlayer K in trioctahedral micas is relatively easy to replace, various methods of expanding the minerals can be used to study the associated changes in $Fe^{++}$. In this investigation, successive NaCl extractions were used to replace the K by Na as in the NaTPB treatments but omit the TPB salts. By using 0.5g portions of the 10-20µm mica in 1000 ml $\underline{1N}$ NaCl at 90°C and changing the solution every 30 minutes, the interlayer K was removed much faster than it was by the NaTPB treatments (Fig. 3). This increase in rate of $Fe^{++}$ exposure prompted an increase in the rate of oxidation as it did in the particle size experiments. In a 3-hour period the mineral was fully expanded and 15% of the $Fe^{++}$ was oxidized--a level of oxidation that was comparable to the 19% in freshly expanded samples in NaTPB solutions. The oxidation of $Fe^{++}$ continued when the expanded mineral was left in the NaCl solution for 6 or 24 hours and at a slightly higher rate than that observed with NaTPB-treated samples. The limited NaCl data, however, identified too little difference in the effects of the two solutions on expanded mica to warrant a prolonged use of the cumbersome treatment with successive large volumes of hot NaCl or comparisons with other solution volumes, temperatures, changes, etc.

Comparisons of the 5 to 20% oxidation values for NaCl-treated mica in Fig. 3 with the oxidation levels reported by others for various salt-treated micas need to be made in terms of both the degree and period of expansion involved. Robert and Pedro (1969) and Ross and Rich (1974) removed increments of K from biotite samples with NaCl and $CaCl_2$ solutions, respectively, and observed more oxidation at each expansion stage than found here. Using dodecylammonium chloride at 70°C with biotite and lepidomelane samples, Mackintosh et al. (1971) also oxidized more of their $Fe^{++}$. In these instances, the differences probably stem from the use of longer treatment periods. On the other hand, Farmer et al. (1971) refluxed two biotite samples in successive increments of $BaCl_2$ and then NaCl for periods of 2 to 3 weeks to reduce the K to <0.5% and oxidized only 5 and 20% (based on their $Fe^{++}$ values) of the $Fe^{++}$. The occurrence of only 5% oxidation in a fully expanded mica after several weeks contact with successive increments of solution raises questions as to what other conditions could have limited oxidation.

Since the $Fe^{++}$ in mica was never oxidized as rapidly as it was exposed by a release of K, an effort was made to exploit the time lag in oxidation to prepare expanded mineral samples with a minimum amount of oxidation. In the particle size and NaCl extraction experiments this approach failed because the increases induced in the rate of K release were largely matched by an increase in the rate of oxidation. However, quite different results were obtained with different reaction temperatures which have a major effect on the K-release rate. K release and $Fe^{++}$ oxidation results that were obtained with 10-20μm samples in NaCl-NaTPB solutions at 25°C and 65°C are compared in Fig. 4. Raising the temperature increased the rate of K release but not the rate of oxidation. As a result, the mineral samples were fully expanded in an 8-hour period at 65° (instead of a week) while only 7% of the $Fe^{++}$ was oxidized. Why the oxidation rate did not respond to the quicker exposure of $Fe^{++}$, as it did in other experiments, is not known. Nevertheless, as a routine method of preparing K-depleted mica samples with very little oxidation, this procedure would appear to be very useful.

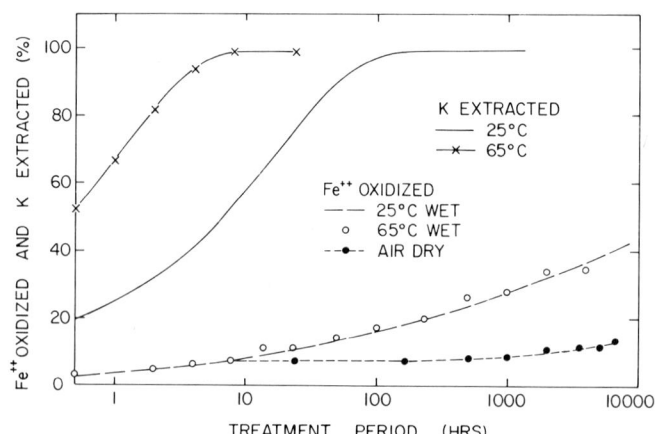

Fig. 4. K extracted and $Fe^{++}$ oxidized by treating 10-20μm lepidomelane samples with NaCl-NaTPB solutions at 25°C or 65°C and by air drying samples of 65°C-degraded mica.

The $Fe^{++}$ determinations for this investigation were carried out as soon as the K-depleted mica samples were separated and dried. Since a use of these samples at a later date was anticipated (for redeterminations of $Fe^{++}$, K exchange experiments, etc.) information on the stability of the $Fe^{++}$ in the dried samples was needed. Therefore, several mica samples were fully expanded by NaTPB treatments at 65°C for 8 hours, mixed with KCl, freed of KTPB and salts by the usual acetone-alcohol-acetone treatment, air dried and combined into one sample of several grams. Subsamples were taken after specified periods of dry storage and analyzed for $Fe^{++}$. The results are described by the air-dry curve in Fig. 4. It is obvious that the $Fe^{++}$ in the degraded mica was not altered by dry-storage periods of several weeks. However, enough change in oxidation did occur with prolonged storage (a change from 7 to 13% oxidation in 40 weeks) to preclude their long-time use without periodic analyses.

CONCLUSIONS

As soon as the expansion of lepidomelane was initiated by an exchange of Na for interlayer K, the structural $Fe^{++}$ in the mica exhibited a greater susceptibility for oxidation. The rate and degree of oxidation, however, was much more limited than was expected for the exposed $Fe^{++}$. These limitations are depicted by the curves in Fig. 5, which compare the expansion-oxidation relationships that were observed with various NaTPB-treated mica samples (the NaCl, particle size and $O_2$ experiments yielded curves comparable to the NaTPB-25°C curve) and the maximum amounts of oxidation achieved in unpublished experiments with the same mica and prolonged treatments with $H_2O_2$-NaTPB mixtures that produced specific degrees of mica expansion. The slow rate of $Fe^{++}$ oxidation (relative to its

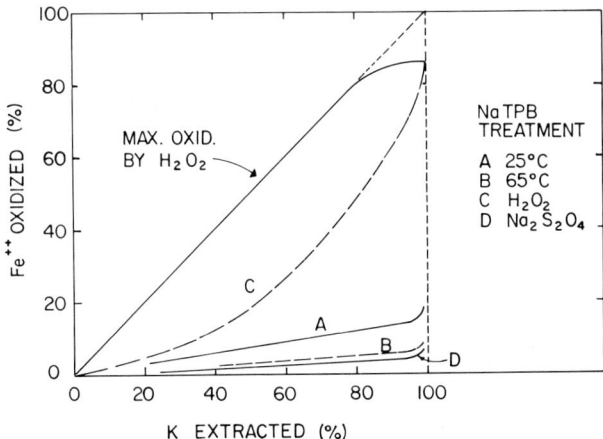

Fig. 5. Comparison of the relative degrees of $Fe^{++}$ oxidation and K extraction that occurred while 10-20μm lepidomelane samples were expanded by NaCl-NaTPB treatments at different temperatures and in the presence of $H_2O_2$ or $Na_2S_2O_4$.

exposure by mica expansion and to $Fe^{++}$ oxidation in reduced smectites) and the maximum degree of 86% oxidation may be due to associated structural alterations in the mica. The absence of a temperature effect on the oxidation rate should help us learn more about the rate controlling process.

ACKNOWLEDGEMENTS

Journal Paper No. J-9166 of the Iowa Agriculture and Home Economics Experiment Station, Ames, Iowa. Project No. 1722.

REFERENCES

Besson, G., Estrade, H., Gatineau, L., Tchoubar, C. and Mering, J., 1975. A kinetic survey of the cation exchange and of the oxidation of a vermiculite. Clays and Clay Miner. 23:318-322.

Bowen, L. H., Weed, S. B. and Stevens, J. G., 1969. Mössbauer study of micas and their potassium-depleted products. Am. Mineral. 54:72-84.

Farmer, V. C., Russell, J. D., McHardy, W. J., Newman, A. C. D., Ahlrichs, J. L. and Rimsaite, J. Y. H., 1971. Evidence for loss of protons and octahedral iron from oxidized biotites and vermiculites. Mineral. Mag. 38:121-137.

Ismail, F. T. and Scott, A. D., 1972. Temperature effects on interlayer potassium exchange in micaceous minerals. Soil Sci. Soc. A. Proc. 36:506-510.

Mackintosh, E. E., Lewis, D. G. and Greenland, D. J., 1972. Dodecylammonium-mica complexes - II. Characterization of the reaction products. Clays and Clay Miner. 20:125-134.

Newman, A. C. D. and Brown, G., 1966. Chemical change during the alteration of micas. Clay Miner. 6:297-310.

Peters, Von Arnd., 1968. Ein neues Verfahren zur Bestimmung von Eisen (11) oxid in Mineralen und Gesteinen. Neues Jahrb. Mineral. Monatshe. 3/4:119-125.

Raman, K. V. and Jackson, M. L., 1966. Layer charge relations in clay minerals of micaceous soils and sediments. Clays and Clay Miner. 14:53-68.

Reed, M. G., 1963. Kinetics of potassium release from biotite in solutions containing sodium tetraphenylboron. Unpublished Ph.D. Thesis, Library, Iowa State University, Ames, Iowa.

Robert, M. and Pedro, G., 1969. Etude des relations entre les phenomenes d'oxydatic et l'aptitude a l'ouverture dans les micas trioctaedriques. Proc. Int. Clay Conf., Tokyo 1:455-473.

Ross, G. J. and Rich, C. I., 1974. Effect of oxidation and reduction on potassium exchange of biotite. Clays and Clay Miner. 22:355-360.

Scott, A. D., 1968. Effect of particle size on interlayer potassium exchange in micas. Trans. 9th Int. Congr. Soil Sci. 2:649-660.

Smith, S. J. and Scott, A. D., 1966. Extractable potassium in Grundite illite: I. Method of extraction. Soil Sci. 102:115-122.

Stucki, J. W. and Roth, C. B., 1977. Oxidation-reduction mechanism for structural iron in nontronite. Soil Sci. Soc. Am. J. 41:808-814.

Veith, J. A. and Jackson, M. L., 1974. Iron oxidation and reduction effects on structural hydroxyl and layer charge in aqueous suspensions of micaceous vermiculites. Clays and Clay Miner. 22:345-353.

Wey, R., LeDred, R. and Schoenfelder, J., 1966. Transformation d'un mica partiellement vermiculitise en vermiculite par oxydation du fer (II). Bull. Groupe Fr. Argiles 12:107-114.

# DISTRIBUTION OF OCTAHEDRAL IONS IN PHLOGOPITES AND BIOTITES

J.A. RAUSELL-COLOM, J. SANZ, M. FERNANDEZ and J.M. SERRATOSA
Seccion de Fisico-Quimica, Instituto de Edafologia y Biologia Vegetal,
Serrano 115-dpdo., MADRID-6, SPAIN.

## ABSTRACT

IR spectra of analysed phlogopites and biotites of varying octahedral compositions and fluorine contents were recorded in the OH-stretching region (4000-3000 $cm^{-1}$) and the complex band in this region was resolved into a number of N, I and V components, each corresponding to OH groups coordinated to specific groupings of octahedral cations and vacancies. The distribution of cations and $F^-$ in the octahedral sheet of the different micas has been studied by relating the intensities of these IR components to the molar fractions of these elements as derived from the chemical analyses. Our results show that:

1) In micas having isomorphous substitutions of OH by F, the octahedral $Mg^{2+}$ ions are preferentially arranged in groupings of MgMgMg around the $F^-$ ions, and

2) In Al rich biotites having vacant octahedral sites, the $Fe^{2+}$ ions are found in groupings of the type $V(AlFe^{2+}v)$ in preference to groupings of the type $I(Fe^{2+}Fe^{2+}R^{3+})$ around the OH groups.

---

## INTRODUCTION

The IR spectra of phlogopites and biotites show, in the OH stretching region, a broad complex band from 3750 $cm^{-1}$ to 3450 $cm^{-1}$, which reflects the diversity of environments of the hydroxyl groups. Studies of the fine structure and dichroic properties of the $\gamma_{OH}$ spectral region from specimens of known composition have shown that the band could be resolved into a number of components at fixed frequencies, each of which could be assigned to OH groups coordinated to a particular combination of three divalent cations -N(A,B,C,D) components-, or of two divalent cations and one polyvalent cation -I(A,B,C) components- or of two cations and one vacancy -V components- (Serratosa and Bradley,1958; Vedder,1964; Wilkins,1967; Farmer et al,1971; Chaussidon,1973; Rousseaux et al,1973a). Estimations by Rousseaux et al. (1973a) of the relative absorption coefficients of the N and I components make possible the use of quantitative IR determinations to detect short range ordering of the octahedral cations around the OH groups.

In the present study IR evidence will be presented for certain ordering patterns consistent with evidence furnished by other spectroscopic techniques.

EXPERIMENTAL AND RESULTS

Flakes cleaved from large crystals of trioctahedral micas were mounted at 40° to the incident beam of a Perkin Elmer 225 IR spectrophotometer set to record quantitatively the $\nu_{OH}$ absorption bands in transmittance. Spectra were then replotted in absorbance, and resolved by numerical computation into N, I and V components. The frequencies of the N(A,B,C,D) and I(A,B,C) components were taken from Wilkins (1967) with half band widths of 22 cm$^{-1}$ for the N bands and of 38 cm$^{-1}$ for the I bands, the decomposition method allowing for variations of $\pm 15\%$ in $\Delta\nu_{1/2}$. For the frequencies of the V components the criteria of Farmer et al (1971) and of Rousseaux et al (1973a) have been followed on the whole, our least squares decomposition allowing for slight variations in $\nu$ and in $\Delta\nu_{1/2}$ because of uncertainties still existing for some frequencies and half band widths. Agreement between experimental and calculated profiles was better than 2%. Formulae for the specimens used (1 to 13) were derived from analyses reported by Rausell-Colom et al (1965) and by Rousseaux et al (1973b), (see Table 1).

Spectral data from other 20 micas (P1 to P20) obtained by Rousseaux (1972) have been included in the present study. Their formulae have been calculated with the same basis as for specimens 1 to 13 ( water free, $0^{2-}=11$) from the analyses reported by Rousseaux et al (1973b). Table 2 gives the integrated intensities of the component

TABLE 1

Structural formulae of micas

| M.S. | tetrahedral | | | octahedral | | | | | | | | interlayer | | | | | |
|---|---|---|---|---|---|---|---|---|---|---|---|---|---|---|---|---|---|
| | $Si^{4+}$ | $Al^{3+}$ | $Fe^{3+}$ | $Al^{3+}$ | $Fe^{3+}$ | $Ti^{4+}$ | $Mg^{2+}$ | $Fe^{2+}$ | $Mn^{2+}$ | $Li^+$ | $\Sigma oct$ | $Ca^{2+}$ | $Na^+$ | $K^+$ | $F^-$ | $Cl^-$ | $OH^-$ |
| 1 | 2.65 | 1.35 | - | 0.48 | 0.08 | 0.17 | 0.68 | 1.33 | 0.02 | 0.02 | 2.78 | - | 0.04 | 0.91 | 0.06 | 0.01 | 1.93 |
| 2 | 3.01 | 0.99 | - | - | 0.05 | 0.03 | 2.36 | 0.51 | - | 0.02 | 2.97 | 0.01 | 0.02 | 0.90 | 0.62 | 0.08 | 1.30 |
| 3 | 2.79 | 1.21 | - | 0.23 | 0.07 | 0.05 | 2.25 | 0.28 | - | 0.01 | 2.89 | 0.02 | 0.10 | 0.91 | 0.38 | 0.01 | 1.61 |
| 4 | 2.96 | 1.00 | 0.04 | - | 0.07 | 0.13 | 1.52 | 1.05 | 0.06 | 0.06 | 2.89 | 0.02 | 0.09 | 0.90 | 0.83 | 0.02 | 1.15 |
| 5 | 3.06 | 0.88 | 0.06 | - | 0.02 | 0.09 | 2.26 | 0.48 | 0.02 | 0.06 | 2.93 | 0.02 | 0.07 | 0.87 | 1.01 | 0.01 | 0.98 |
| 6 | 2.86 | 1.11 | 0.03 | - | 0.01 | 0.05 | 2.80 | 0.16 | - | 0.01 | 3.03 | 0.01 | 0.05 | 0.92 | 0.87 | 0.01 | 1.12 |
| 7 | 2.81 | 1.19 | - | 0.09 | 0.06 | 0.16 | 2.16 | 0.40 | 0.01 | - | 2.88 | - | 0.02 | 0.92 | 1.12 | 0.01 | 0.87 |
| 8 | 2.79 | 1.21 | - | 0.05 | 0.21 | 0.14 | 1.00 | 1.35 | 0.04 | 0.05 | 2.84 | 0.05 | 0.05 | 0.91 | 0.37 | - | 1.63 |
| 9 | 2.69 | 1.31 | - | 0.22 | 0.09 | 0.17 | 1.07 | 1.25 | 0.03 | 0.02 | 2.85 | 0.02 | 0.03 | 0.93 | 0.14 | - | 1.86 |
| 10 | 2.72 | 1.28 | - | 0.40 | 0.21 | 0.14 | 0.56 | 1.24 | 0.08 | 0.10 | 2.73 | 0.01 | 0.05 | 0.97 | 0.68 | - | 1.32 |
| 11 | 2.99 | 0.97 | 0.04 | - | - | 0.17 | 1.84 | 0.73 | 0.01 | 0.09 | 2.84 | 0.07 | 0.07 | 0.90 | 0.51 | - | 1.49 |
| 12 | 2.99 | 0.97 | 0.04 | - | 0.05 | 0.13 | 1.57 | 1.02 | 0.05 | 0.05 | 2.87 | 0.01 | 0.09 | 0.92 | 0.82 | - | 1.18 |
| 13 | 2.80 | 1.20 | - | 0.04 | 0.18 | 0.13 | 1.01 | 1.42 | 0.04 | 0.04 | 2.85 | 0.02 | 0.06 | 0.96 | 0.38 | - | 1.62 |
| P 1 | 2.79 | 1.21 | - | 0.17 | 0.05 | 0.05 | 2.57 | 0.13 | - | 0.01 | 2.98 | 0.02 | 0.06 | 0.86 | 0.18 | - | 1.82 |
| P 2 | 2.75 | 1.25 | - | 0.19 | 0.01 | 0.05 | 2.56 | 0.16 | - | 0.01 | 2.98 | 0.05 | 0.06 | 0.85 | 0.15 | - | 1.85 |
| P 3 | 2.83 | 1.17 | - | 0.01 | 0.03 | 0.07 | 2.86 | 0.02 | - | 0.01 | 3.00 | 0.02 | 0.10 | 0.86 | 0.47 | - | 1.53 |
| P 4 | 2.85 | 1.15 | - | 0.09 | 0.04 | 0.08 | 2.64 | 0.10 | - | 0.01 | 2.96 | 0.01 | 0.07 | 0.86 | 0.45 | - | 1.55 |
| P 5 | 2.75 | 1.25 | - | - | 0.04 | 0.08 | 2.68 | 0.05 | 0.01 | - | 2.86 | 0.18 | 0.04 | 0.90 | 0.04 | - | 1.96 |
| P 6 | 2.88 | 1.12 | - | 0.05 | 0.01 | 0.05 | 2.82 | 0.05 | - | - | 2.98 | 0.04 | 0.05 | 0.88 | 0.73 | - | 1.27 |
| P 7 | 2.88 | 1.12 | - | 0.05 | 0.02 | 0.05 | 2.70 | 0.15 | - | 0.01 | 2.98 | 0.01 | 0.04 | 0.93 | 0.67 | - | 1.33 |
| P 8 | 2.82 | 1.18 | - | 0.08 | - | 0.07 | 2.82 | 0.02 | - | - | 2.99 | 0.03 | 0.06 | 0.87 | 0.63 | - | 1.37 |
| P 9 | 2.82 | 1.18 | - | 0.14 | 0.01 | 0.05 | 2.75 | 0.03 | - | - | 2.98 | 0.02 | 0.07 | 0.88 | 0.58 | - | 1.42 |
| P 10 | 2.89 | 1.11 | - | 0.12 | 0.02 | 0.03 | 2.75 | 0.01 | - | 0.01 | 2.94 | 0.08 | 0.06 | 0.80 | 0.50 | - | 1.50 |
| P 11 | 2.88 | 1.12 | - | 0.04 | 0.01 | 0.05 | 2.85 | 0.03 | - | 0.01 | 2.99 | 0.04 | 0.06 | 0.86 | 0.70 | - | 1.30 |
| P 12 | 2.82 | 1.18 | - | 0.19 | 0.02 | 0.05 | 2.70 | 0.02 | - | 0.01 | 2.99 | 0.01 | 0.04 | 0.89 | 0.48 | - | 1.52 |
| P 13 | 2.84 | 1.16 | - | 0.08 | - | 0.02 | 2.70 | 0.15 | - | 0.01 | 2.96 | 0.07 | 0.03 | 0.94 | 0.80 | - | 1.20 |
| P 14 | 2.95 | 1.05 | - | 0.15 | - | 0.03 | 2.72 | 0.03 | - | 0.01 | 2.94 | 0.03 | 0.06 | 0.83 | 0.60 | - | 1.40 |
| P 15 | 2.85 | 1.15 | - | 0.09 | 0.01 | 0.07 | 2.72 | 0.04 | - | 0.03 | 2.96 | 0.02 | 0.02 | 0.95 | 0.73 | - | 1.27 |
| P 16 | 2.81 | 1.19 | - | 0.02 | 0.04 | 0.03 | 2.89 | 0.07 | - | - | 3.05 | 0.01 | 0.03 | 0.92 | 0.96 | - | 1.04 |
| P 17 | 2.84 | 1.16 | - | 0.04 | 0.02 | 0.05 | 2.87 | 0.03 | - | - | 3.01 | 0.02 | 0.08 | 0.88 | 0.74 | - | 1.26 |
| P 18 | 2.96 | 1.04 | - | 0.01 | 0.03 | 0.02 | 2.72 | 0.17 | 0.01 | 0.03 | 2.98 | 0.03 | 0.04 | 0.94 | 1.15 | - | 0.85 |
| P 19 | 2.88 | 1.12 | - | 0.07 | 0.01 | 0.04 | 2.86 | 0.03 | - | - | 3.01 | 0.01 | 0.04 | 0.89 | 0.70 | - | 1.30 |
| P 20 | 2.96 | 1.04 | - | 0.02 | 0.04 | 0.02 | 2.73 | 0.16 | - | 0.02 | 2.99 | 0.01 | 0.05 | 0.93 | 1.27 | - | 0.73 |

Micas 1 to 8 from Rausell-Colom et al (1965).
Micas 9 to 13 and P-1 to P-20 from analyses by Rousseaux et al (1973b).

bands for the spectra of all the micas used. Data for micas P1 to P20 are from Rousseaux (1972). Some selected spectra and their components are illustrated in Fig. 1.

TABLE 2

Integrated intensities of N and I components of IR spectra of micas. Sum of intensities normalized to 100.

| component | N | | | | I | | | component | N | | | | I | | |
|---|---|---|---|---|---|---|---|---|---|---|---|---|---|---|---|
| M.S. | $N_A$ | $N_B$ | $N_C$ | $N_D$ | $I_A$ | $I_B$ | $I_C$ | M.S. | $N_A$ | $N_B$ | $N_C$ | $N_D$ | $I_A$ | $I_B$ | $I_C$ |
| 1 | - | 7.6 | 14.2 | 2.8 | 25.1 | 47.7 | 2.7 | P4 | 58 | 4 | - | - | 30 | 8 | - |
| 2 | 33.0 | 28.4 | 7.9 | - | 21.1 | 9.6 | - | P5 | 74 | 2 | - | - | 24 | - | - |
| 3 | 18.3 | 20.8 | 9.2 | - | 23.5 | 21.0 | 7.1 | P6 | 75 | - | - | - | 21 | 4 | - |
| 4 | 9.3 | 25.0 | 26.2 | 9.7 | 13.0 | 10.9 | 5.9 | P7 | 72 | 2 | 2 | - | 24 | - | - |
| 5 | 20.1 | 26.2 | 15.8 | 1.1 | 17.2 | 14.9 | 4.9 | P8 | 67 | - | - | - | 30 | 3 | - |
| 6 | 27.0 | 23.5 | 9.7 | - | 15.9 | 20.2 | 3.7 | P9 | 60 | - | - | - | 40 | - | - |
| 7 | 8.2 | 21.8 | 8.5 | - | 14.3 | 33.1 | 14.1 | P10 | 66 | 3 | - | - | 30 | 1 | - |
| 8 | 1.4 | 12.0 | 19.7 | 10.6 | 14.4 | 26.0 | 16.0 | P11 | 75 | - | - | - | 25 | - | - |
| 9 | 3.6 | 16.7 | 21.4 | 9.0 | 14.0 | 30.0 | 5.4 | P12 | 58 | - | - | - | 42 | - | - |
| 10 | - | 2.7 | 9.7 | 7.3 | - | 64.3 | 16.1 | P13 | 47 | 9 | - | - | 35 | 9 | - |
| 11 | 17.4 | 30.8 | 21.3 | 4.4 | 9.3 | 11.9 | 4.8 | P14 | 67 | - | - | - | 33 | - | - |
| 12 | 6.6 | 25.6 | 24.9 | 13.1 | 8.0 | 14.6 | 7.2 | P15 | 66 | 6 | - | - | 28 | - | - |
| 13 | 2.0 | 13.4 | 19.6 | 8.8 | 9.0 | 24.9 | 22.3 | P16 | 67 | 3 | - | - | 27 | 3 | - |
|  |  |  |  |  |  |  |  | P17 | 70 | - | - | - | 30 | - | - |
| P1 | 60 | 5 | - | - | 30 | 5 | - | P18 | 75 | 8 | 2 | - | 15 | - | - |
| P2 | 59 | 5 | - | - | 36 | - | - | P19 | 66 | - | - | - | 34 | - | - |
| P3 | 78 | - | - | - | 18 | 4 | - | P20 | 68 | 12 | 1 | - | 19 | - | - |

Micas P-1 to P-20, data from Rousseaux (1972).

DISCUSSION

The integrated intensity of a component band should be proportional to the concentration of OH coordinated to a specific grouping of octahedral cations. Hydroxyls, thus, serve as a probe for testing the nature of the cations to which they are coordinated. If one assumes a random distibution of cations and anions in the octahedral sheet, then the ratio of the number of N and I sites for any specimen may be derived from its formula by expressions (Wedder 1964) such as

$$P(N_A)/P(I_A) = x/3y \; ; \quad P(N_A)/P(N_B) = x/3z \; ; \quad P(N_B)/P(N_C) = x/z \quad \text{etc.}$$

where x is the number of $Mg^{2+}$ ions, y the sum of the numbers of $Al^{3+}$, $Fe^{3+}$ and $Ti^{4+}$ ions, and z the number of $Fe^{2+}$ ions per three octahedral positions. Accordingly, the ratio of the band intensities becomes

$$N_A/I_A = (k_N/k_I)(x/3y) \; ; \quad N_A/N_B = x/3z \; ; \quad N_B/N_C = x/z \quad \text{etc.} \tag{1}$$

where $k_N$ and $k_I$ are the absorption coefficients of hydroxyls in N and I sites. Both $k_N$ and $k_I$ are supposed to be constant for the different N and I bands, their ratio being estimated as 1.8 by Rousseaux et al (1973a).

It follows that spectra from micas with similar x, y, and z contents should resemble each other quite closely. The spectra of Fig. 1, however, would suggest that there are marked departures from such behaviour. For example, micas 3 and 7 have similar compositions, but the intensities of the $N_A$, $I_A$ and $V_1$ components, relative to those of the $N_B$, $I_B$ and $V_2$ components, are considerably lower in the spectrum of mica 7, as if the octahedral Mg content of this mica were appreciably lower than indicated by its formula. A similar difference is seen between the spectra of micas 1 and 10. Contrarily, micas 1 and 9 have quite different contents of $Al^{3+}$ and $Mg^{2+}$, but give comparable IR spectra. To a greater or lesser extent, inconsistencies of this sort can be traced in most of the spectra, suggesting that the assumptions implicit in equation (1) are not strictly fulfilled, i.e., that the actual cation distribution is other than rand

Cation distributions may be studied by plotting the ratio of intensities of suitable bands against the ratio of the corresponding contents of octahedral cations. Si nificant departures from linearity in the correlation between the two parameters should afford clues for prevailing ordering patterns.

The method contains limitations which are worth discussing. Ratios $N_A/I_A$ implying distributions of $Mg^{2+}$ and $R^{3+,4+}$ are suitable for the phlogopites. $N_A$ and $I_A$ bands a preferred because are prominent, as they correspond to environments of high concentr tion, and also because errors in the estimation of their intensity arising from over lapping of other contiguous bands are small. For specimens with very low $R^{3+,4+}$ conte $I_A$ becomes too weak, and the ratio $N_A/I_A$ is unreliable. With increasing $Fe^{2+}$ content the conditions above are reversed, so that for most biotites ratios $N_A/I_A$ are also u reliable. Ratios between $N_A$, $N_B$ and $N_C$ bands implying distributions of $Mg^{2+}$ and $Fe^{2+}$

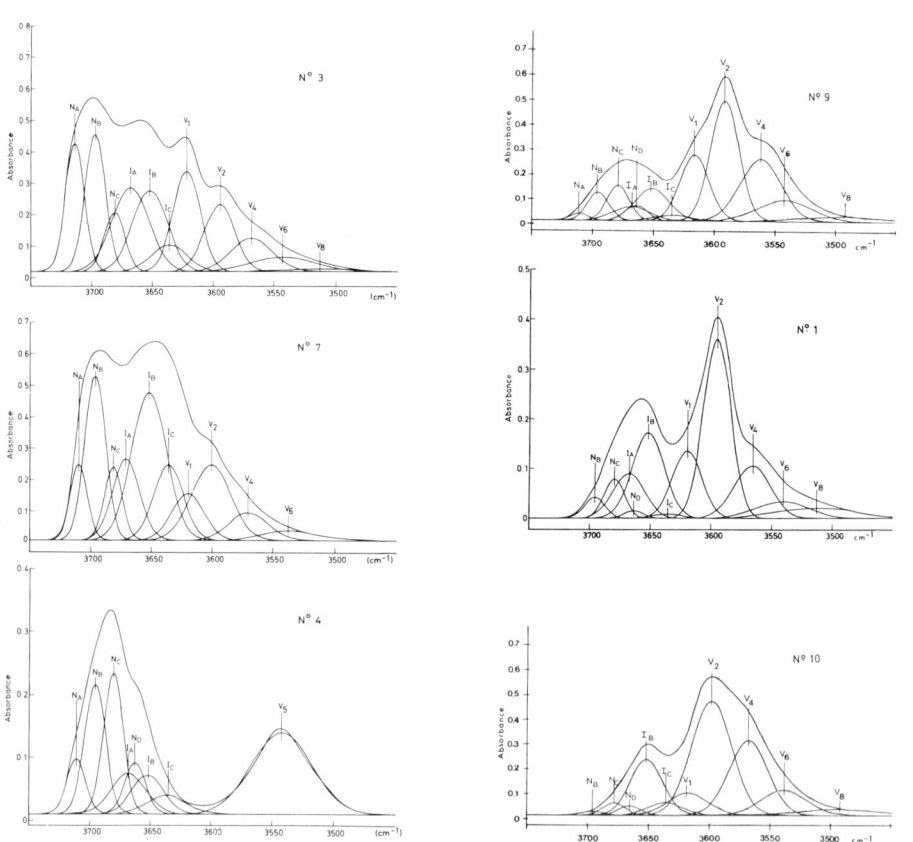

Fig. 1. Selected IR spectra and their resolution into N, I and V components. $\nu_{OH}$ spe tral region. 40° inclination

ions would complete the distribution pattern of all the cations of the octahedral structure. For the biotites, $N_B$ and $N_C$ components are favoured and ratios $N_B/N_C$ are generally adequate.

Fig. 2 shows the disposition of the experimental points in the respective plots. In both cases the departure from a straight line relationship is apparent from the scatter of the points. The full line in Fig. 2b represents a theoretical 1:1 relationship, implicit in equation 1, between the IR and compositional data. In Fig. 2a the slope of the line would correspond to a $k_I/k_N$ ratio of 1.8 as suggested by Rousseaux et al (1973a). In either case the experimental points are all below the respective lines, as if the $Mg^{2+}$ contents of all the micas were considerably lower than indicated by their respective formulae. It may be concluded that the intensity distribution on the IR spectra is not consistent with a random distribution of octahedral ions.

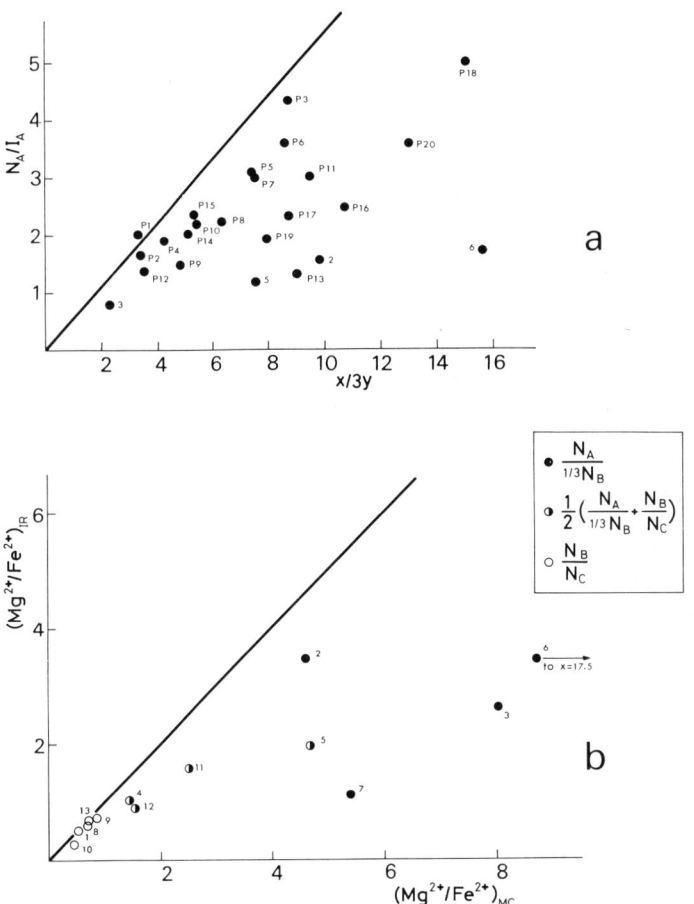

Fig. 2. Plot of ratios of intensities of IR components versus molar ratios of corresponding cations.

## Short range ordering induced by fluorine substitution

Studies by NMR of the H and F signals from phlogopites have furnished evidence on various local cation-anion associations within the octahedral sheet: $Fe^{2+}$ ions are excluded from the environment of the $F^-$ and are randomly distributed around the OH the two possible structural sites $M_1$ and $M_2$ (Sanz,1976; Sanz and Stone,1977). If OH and $F^-$ ions were not randomly distributed over the various cationic environments in

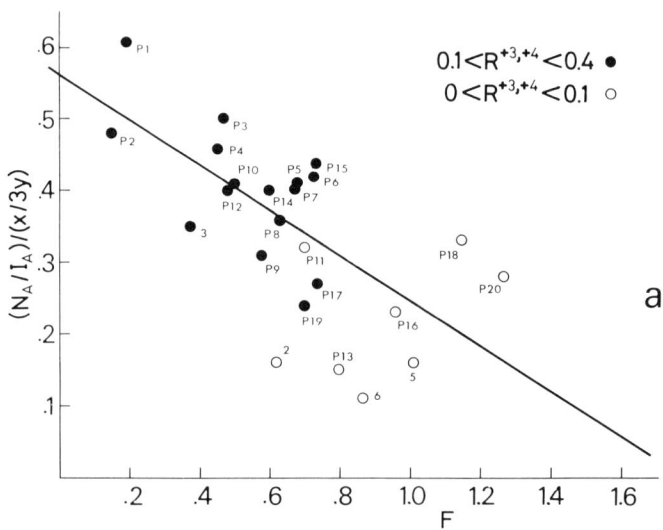

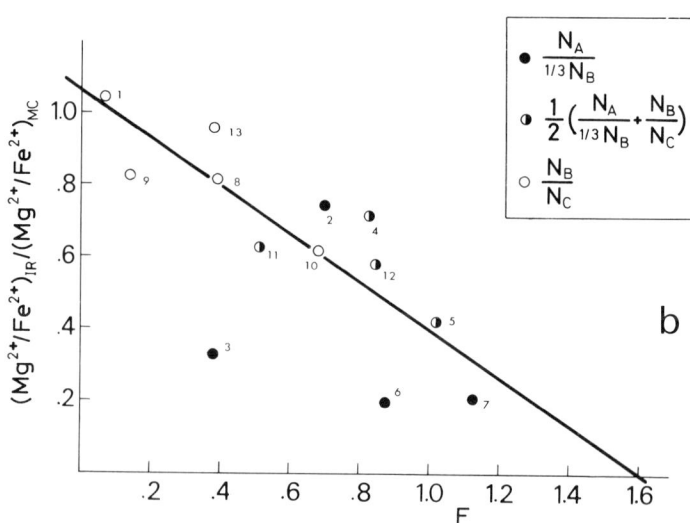

Fig. 3. Dependence of $(N_A/I_A)/(x/3y)$ anf of $(Mg^{2+}/Fe^{2+})_{IR}/(Mg^{2+}/Fe^{2+})_{MC}$ parameters on F content.

the octahedral structure, then the intensity distribution on the IR spectra should be sensitive to this fact. The plots of Fig. 3 clearly show a tendency for fluorine to segregate towards Mg-rich environments. Indeed, a random distribution should make the parameter $(Mg^{2+}/Fe^{2+})_{IR}/(Mg^{2+}/Fe^{2+})_{MC}$ independent of F content, but Fig. 3b shows this parameter to decrease as F contents increase. A plot of $(N_A/I_A)/(x/3y)$ against F content yields a similar trend (Fig. 3a). For the micas where $Cl^-$ content is known the corresponding points have been plotted in Fig. 3 at $F^-$ contents equal to $(F^- + Cl^-)$. Least squares fittings yield regression equations $y = -0.69x + 1.08$ with $R = 0.86$ for Fig. 3b, and $y = -0.31x + 0.56$ with $R = 0.66$ for Fig. 3a. One point in Fig. 3b (mica 3) falls too distant from the regression line, and has been excluded from the regression. No reason is apparent for its departure from the main trend. In Fig. 3a the points P-13, P-18, P-20, 2 and 6 depart significantly from the regression line due possibly, to errors inherent to the low content of octahedral $R^{3+,4+}$ of these micas. A weight factor of 1/2 has been attributed to these points when calculating the regression parameters and regression coefficients.

Sanz (1976) reported that F-containing phlogopites give a NMR signal indicative of a F-F dipolar interaction which is stronger than could be expected from a random distribution of F and OH over the octahedral structure. In addition to the exclusion of $Fe^{2+}$ from the fluorine environment, such observations suggested segregation of F-Mg-rich domains within the octahedral sheet. The question immediately suggests itself of how many $Mg^{2+}$ ions are, on the average, trapped by the $F^-$ ions.

A simple, but extreme, model of distribution may be considered in which all the $F^-$ ions are trapped into $3Mg^{2+}$ sites (as in synthetic fluor-phlogopites, with a F/Mg ratio of 2/3), leaving the remaining $Mg^{2+}$ ions and the other octahedral ions distributed at random around the hydroxyls in OH-rich domains. This model implies a modification of the intensity distribution in the IR spectrum in the sense that, if F is the number of fluorine ions per two octahedral sites, then

$$N_A/I_A = (k_N/k_I)(x-1.5F)/3y \quad ; \quad N_B/N_C = (x-1.5F)/z$$

which may be rewritten as

$$(N_A/I_A)/(x/3y) = (k_N/k_I)(1-1.5\tfrac{F}{x}) \quad ; \quad (N_B/N_C)/(x/z) = (3N_A/N_B)/(x/z) = 1-1.5\tfrac{F}{x}$$

A plot of $(N_A/I_A)/(x/3y)$ versus $F/x$ should yield a straight line intercepting the ordinate at $k_N/k_I$ and the abscissa at $1/1.5 = 0.66$. Similarly, a plot of $(3N_A/N_B)/(x/z)$ or of $(N_B/N_C)/(x/z)$ versus $F/x$ should also yield a straight line with intercepts at 1 and at 0.66. This may be considered as a test for the validity of the model proposed.

Fig. 4a shows that the points fit the line predicted. It yields a value of 0.57 for the ratio $k_N/k_I$, in agreement with the estimate of Rousseaux et al (1973a). Open circles represent cases where the $R^{3+,4+}$ content is rather low, so that errors inherent in the determination of y or $I_A$ may contribute to the dispersion. As before (Fig. 3a), a weight factor of 1/2 has been assigned to these points when calculating the regres-

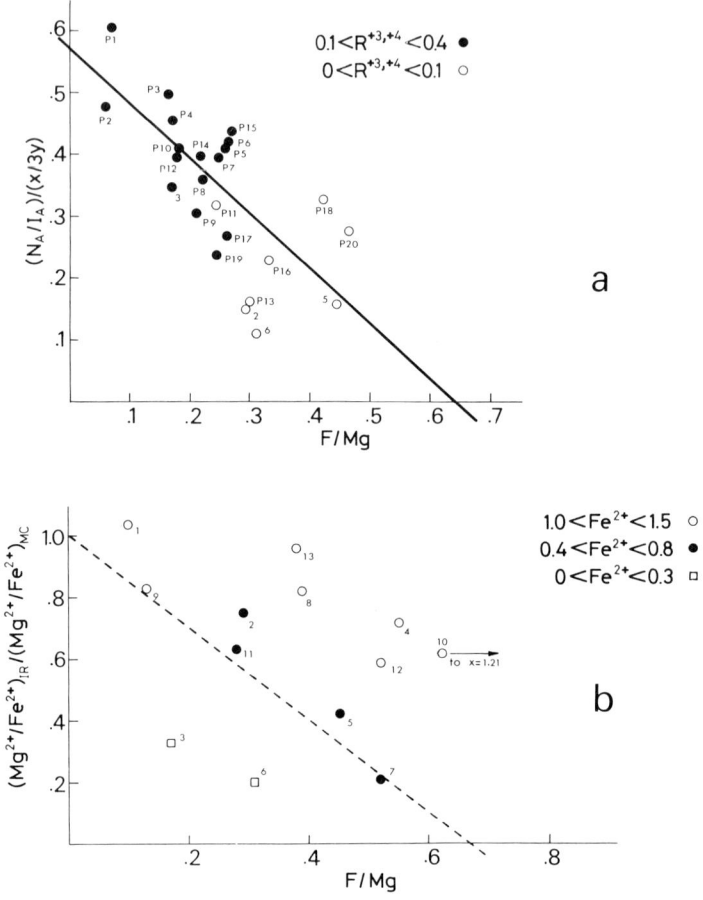

Fig. 4. Correlation between $(N_A/I_A)/(x/3y)$ or $(Mg^{2+}/Fe^{2+})_{IR}/(Mg^{2+}/Fe^{2+})_{MC}$ parameters and molar ratio F/Mg.

sion parameters. The regression coefficient is $R = 0.69$.

As for the second representation, i.e., plotting $(Mg/Fe)_{IR}/(Mg/Fe)_{MC}$ versus F/x (Fig. 4b) the result is not so encouraging at first glance: the points on the graph a just too scattered. Nevertheless, when the $Fe^{2+}$ content of the different micas is taken into account, the plot would suggest the presence of two different situations. Micas whose $Fe^{2+}$ content is lower than 0.8 ions per three octahedral positions fall close to the theoretical prediction (broken line, Fig. 4b). A second group of micas with $Fe^{2+}$ contents higher than one per three octahedral positions fall above the theo retical prediction. Two points corresponding to the micas with the lowest $Fe^{2+}$ conter (micas 3 and 6, with $Fe^{2+}$ contents of 0.28 and 0.16 respectively) fall well below the theoretical line

If any significance can be attributed to the above trends, it would lead to the conclusion that, in micas with octahedral compositions where Mg predominates, fluorine ions have a tendency for specific segregation towards $3Mg^{2+}$ sites. As the $Fe^{2+}$ increases, this tendency becomes less specific if the F content is also high, so that environments such as $MgMgR^{3+}$ and $MgMgFe^{2+}$ become acceptable to the fluorine ions. Therefore the conclusion of Sanz (1976) that $Fe^{2+}$ ions are excluded from the environment of F in phlogopites cannot be generalised to Fe-rich biotites.

Short range order around the OH

The discussion above has dealt with the extent to which the intensity distribution in the IR spectra would be consistent with fluorine segregation in Mg-rich environments. It seems pertinent to discuss, too, the validity of the second proposition in the model, i.e., the existence of a random distribution around OH rich domains.

In Fig. 4a, where the agreement with the theoretical prediction is more obvious, the points on the graph are still too scattered (R = 0.68) and, at least for certain phlogopites, the cause for the deviation cannot be traced to errors implicit in determinations of the IR and compositional variables (mainly $I_A$ intensities and $Fe^{3+}$ contents). One is thus forced to think that fluorine segregation may not be the only cause for short range ordering within the octahedral structure.

Short range order around the OH could result from requirements of local charge balance, and it appears that some spectra have features consistent with this possibility. One group of micas (2, 4, 5, 11 and 12, Table 1) all have composition where $Al_T \sim 1$ and $Al_{oct} \sim 0$. Since the negative charge in the tetrahedral layer is entirely balanced by the positive charge of interlayer K cations, the best possible arrangement for the octahedral cations should be one where local associations of the type $R^{2+}R^{2+}R^{2+}$ $R^{3+}R^{3+}v$, or $Ti^{4+}R^{2+}v$, all with a total charge of +6, were favoured. Moreover, all these micas have about the same content of $Ti^{+4}$ as of vacancies, and also that $Ti^{4+}$ is the predominant polyvalent octahedral cation. Spectrum 4 in Fig. 2 is characteristic of this group of micas, and one may see that it is resolved into strong N and V components. The band at 3540 cm$^{-1}$ has been assigned to the environment $Ti^{4+}R^{2+}v$ (Rousseaux et al 1973a).

Further generalization may be drawn from the spectra of Al-rich biotites (Fig. 1). These micas (1, 9 and 10, Table 1) have $Fe^{2+} \sim 1.3$, $R^{3+} > 0.5$, and are high in vacant octahedral positions. Consequently the component $I_C$ of the spectrum should be prominent, but the low intensity of this band clearly shows that the association $Fe^{2+}Fe^{2+}R^{3+}$ around the OH is most unfavourable. It seems as if $Fe^{2+}$ ions would rather concentrate in associations $AlFe^{2+}v$, the component $V_2$ being the most prominent band in the spectra and in association $MgFe^{2+}Al$, since the $I_B$ band is also intense. In fact, the high content of F in mica 10 should favour the concentration of $Fe^{2+}$ in associations $Fe^{2+}Fe^{2+}Fe^{2+}$ ($N_D$) and $Fe^{2+}Fe^{2+}R^{3+}$ ($I_C$) around the OH groups, but since these components

are weak (Fig. 1), it can be said that $Fe^{2+}$ ions are also concentrated in environme
$V_2$ and $I_B$ in preference to the environments $N_D$ and $I_C$. Again, these observations su
port the existence of ordering patterns around the OH, additional to the order imposed by the F ions.

REFERENCES

Chaussidon, J., 1973. Le spectre infrarouge des biotites: vibrations d'élongation basse frequence des OH du reseau. Proc. 1972 Intern. Clay Conf. Madrid. 99-106.

Farmer, V.C., Russell, J.D., Mc Hardy, W.J., Newman, A.C.D., Ahlrichs, J.L. and Rimsaite, J.Y.H., 1971. Evidence of loss of protons and octahedral iron from oxid zed biotites and vermiculites. Miner. Mag. 38, 121-137.

Rausell-Colom, J.A., Sweatman, T.R., Wells, C.B. and Norrish K., 1965. Studies in t artificial weathering of mica. In: E.G. Hallsworth and D.V. Crawford (Editors), Experimental Pedology, Butterworths, London, 40-72.

Rousseaux, J.M., 1972. Aspects cristallochimiques et cristallographiques de l'aptit a la vermiculitisation des micas trioctaedriques. Ph. D. Thesis. Université Catholique de Louvain.

Rousseaux, J.M., Gomez Laverde, C., Nathan, Y. and Rouxhet, P.G., 1973a. Correlatio between the hydroxyl stretching bands and the chemical composition of trioctahed micas. Proc. 1972 Intern. Clay Conf. Madrid. 89-98.

Rousseaux, J.M., Rouxhet, P.G., Vielvoye, L.A. and Herbillon, A., 1973b. The vermic litization of troctahedral micas. I. The K-level and its correlation with chemic composition. Clay Miner., 10, 1-16.

Sanz, J., 1976. Ordre-desordre dans le couche octaédrique des micas trioctaédriques Etude par résonance magnétique nucleaire, infrarouge et Mossbauer. Ph. D. Thesis Université Catholique de Louvain.

Sanz, J. and Stone, W.E.E., 1977. NMR study of micas. I. Distribution of $Fe^{2+}$ ions the octahedral sites. J. Chem. Phys., 67, 3739-3743.

Serratosa, J.M. and Bradley, W.F., 1958. Determination of the orientation of OH bon axes in layer silicates by infrared absorption. J. Phys. Chem., 62, 1164-1167.

Vedder, W., 1964. Correlation between infrared spectrum and chemical composition of mica. Am. Mineral., 49, 736-768.

Wilkins, R.W.T., 1967. The hydroxyl stretching region of the biotite mica spectrum. Miner. Mag., 36, 325-333.

EFFECT OF TEXTURE ON VERMICULITE STRUCTURE : LITHIUM MINERALS

C. DE LA CALLE[*], R. GLAESER and H. PEZERAT
Laboratoire de Chimie des Solides, ER 133 CNRS
Université P. et M. Curie, 75230 Paris Cedex 05
* Instituto de Quimica Inorganica "Elhuyar". CSIC. Madrid

ABSTRACT

The effect of texture on vermiculite structure has been studied with Li-vermiculite of various origine. The vermiculite swelling was measured, with varying relative humidities, for monocrystals of 1 mm diameter and for powders of 10 $\mu$m diameter. Important differences in the observed swelling were noted; some were due to the layer charge density and others to the texture of the mineral. If the relative humidity was about 40% the powders and some monocrystals exhibited a one layer water hydrate, while other monocrystals whatever their charge density showed a two layer hydrate. Scanning electron microscopy showed that monocrystals exhibiting the same swelling properties as powders have a particularly porous texture. It is concluded that in certain cases the vermiculite crystal structure could be a function of the ratio of external surface area to volume. Thus, if the ratio is large, the swelling exhibited is small.

INTRODUCTION

Dans des travaux précédents (de la Calle et al. 1975 a, 1975 b, 1978 (sous presse)), nous avons mis en évidence l'influence du polytype du mica de départ, de la nature des cations interfoliaires, de l'humidité relative et de la charge, sur les modes d'empilement des feuillets dans les vermiculites hydratées.

Le présent travail montre la nécessité d'introduire un paramètre supplémentaire de texture parmi les facteurs conditionnant le gonflement et le mode d'empilement des feuillets.

Dans cette étude, nous appellerons "poudre" des microcristaux dont le diamètre est de l'ordre de 10 à 20 microns et "monocristaux" des plaquettes dont le diamètre est de l'ordre du mm.

Nous avons remarqué (de la Calle 1977) qu'il existait un décalage de frontières des domaines d'existence des modes d'empilement en fonction de l'humidité relative suivant que l'on utilise une poudre ou un monocristal.

La taille ou plus généralement la texture des phyllosilicates, semble donc jouer un rôle dans les propriétés de gonflement et il nous a paru intéressant de rechercher le cas le plus caractéristique pour mieux cerner les paramètres en cause. Nous avons choisi le cas des vermiculites lithiques.

DESCRIPTION ET PREPARATION DES ECHANTILLONS

L'étude a été faite sur des échantillons en monocristaux et en poudre provenant de différentes origines : Llano, Santa Olalla, Benahavis et Prayssac. Nous avons utilisé par ailleurs des vermiculites préparées au laboratoire, à partir des phlogopites A et B. Nous avons fait sur ces phlogopites deux traitements parallèles, dans l'un il y a eu un passage intermédiaire par l'état magnésien, dans l'autre l'échange potassium-lithium a été réalisé directement. Nous appellerons :
- V Li (1) les vermiculites provenant de l'échange direct K par Li, avec utilisation indistincte des deux phlogopites A et B qui donnent les mêmes résultats.
- V Li (2) les vermiculites lithiques préparées à partir des mêmes phlogopites traitées préalablement avec $MgCl_2$.

Les formules structurales et les densités de charge de ces différents échantillons sont rasemblés dans le tableau I. Les modes opératoires concernant les transformations phlogopite-vermiculite ont déjà été donnés dans des précédentes publications (de la Calle et al. 1975 a).

ETUDE DU GONFLEMENT DES MINERAUX LITHIQUES

L'étude a porté sur les six échantillons de vermiculite lithiques suivants : V Li (1), V Li (2), Prayssac, Llano, Santa Olalla et Benahavis. Les échantillons étaient examinés dans des capillaires après un long séjour en présence de l'humidité choisie et scellement rapide. Pour chaque humidité, l'équilibre est atteint par désorption.

Echantillons en poudre

Dans la figure 1 a, on a représenté la variation de $d_{001}$ en fonction de l'humidité relative pour les échantillons en poudre. Sur cette figure, on observe que les vermiculites de basse charge ont tendance à présenter plus facilement l'état d'hydratation à "deux couches" que celles de haute charge. Elles présentent également un espacement $d_{001}$ plus grand que les échantillons à haute charge tant pour les états à "une couche" que pour les états à "deux couches".

Dans l'eau, les échantillons de haute charge sont hydratés à "deux couches" tandis que ceux de basse charge (Prayssac et Benahavis) se dispersent jusqu'à la perte de tout parallélisme des feuillets. La différence de comportement est manifestement en relation directe avec la différence de charge.

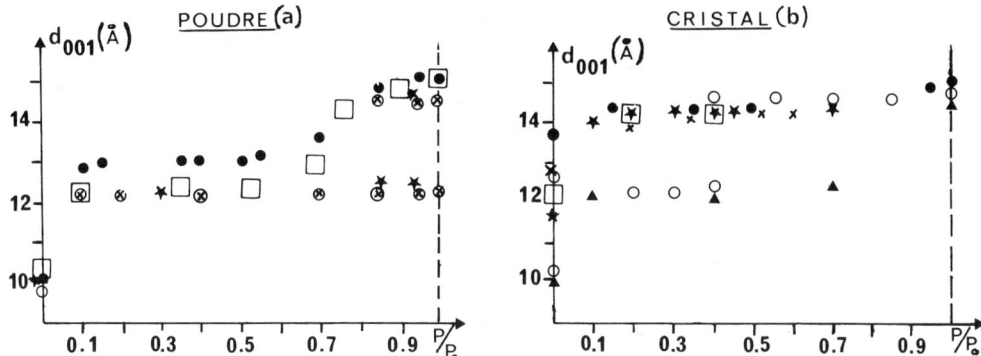

Fig. 1. Valeurs des espacements $d_{001}$ en fonction de l'humidité relative pour les échantillons de vermiculites lithiques : (a) : poudre ; (b) : cristal
× V Li (1),　○ V Li (2),　● Prayssac Li,　★ Santa Olalla Li,　□ Benahavis Li et　▲ Llano Li.

Echantillons en monocristal

Le comportement des vermiculites Li quand elles se trouvent à l'état de monocristal d'une taille moyenne supérieure au millimètre est très particulier.

La figure 1 b montre la variation de $d_{001}$ en fonction de l'humidité relative pour des monocristaux des différentes vermiculites étudiées. Les échantillons V Li (1), Santa Olalla Li, Prayssac Li et Benahavis Li ont tendance à présenter un hydrate à "deux couches" même aux très basses humidités relatives. Par contre, les échantillons V Li (2) et Llano Li ont un comportement différent.

En dessous d'une humidité relative de 70%, la Llano Li présente un hydrate à "une couche" et la V Li (2) présente ce type d'hydrate pour les humidités relatives inférieures à 40%. Pour expliquer les différences de gonflement entre les monocristaux de diverses origines, on ne peut faire appel aux différences de charge. En effet, la charge des échantillons V Li (1), V Li (2), Santa Olalla Li et Llano Li est pratiquement équivalente (cf. Tableau I) et relativement élevée. Or, ces échantillons ne se comportent pas de la même façon. Quant aux échantillons de Prayssac et de Benahavis, bien que de basse charge, ils présentent un gonflement analogue aux échantillons V Li (1) et Santa Olalla entre 5 et 100% d'humidité relative.

TABLEAU 1

Origine, formule structurale et densité de charge des échantillons

| Origine | Formule structurale | Densité de charge |
|---|---|---|
| Phlogopite A (Madagascar) | $Li_{0,88}$ $(Mg_{2,59}\ Al_{0,15}\ Fe^{3+}_{0,10}\ Fe^{2+}_{0,17}\ Ti_{0,04})$ $(Si_{2,68}\ Al_{1,32})\ O_{10}\ (OH)_2$ | 0,88 |
| Phlogopite B (Madagascar) | $Li_{0,90}$ $(Mg_{2,19}\ Al_{0,48}\ Fe^{2+}_{0,08}\ Fe^{3+}_{0,06}\ Ti_{0,04})$ $(Si_{2,77}\ Al_{1,17}\ Fe^{3+}_{0,06})\ O_{10}\ (OH)_{1,85}\ F_{0,15}$ (don de J. Mamy) | 0,90 |
| Vermiculite Llano (USA) | $Li_{0,90}$ $(Mg_{2,94}\ Ti_{0,02}\ Fe^{3+}_{0,01}\ Al_{0,1})$ $(Si_{2,78}\ Al_{1,22})\ O_{10}\ (OH)_2$ (Norrish, 1972) | 0,90 |
| Vermiculite Santa Olalla (Espagne) | $Li_{0,82}$ $(Mg_{2,59}\ Al_{0,06}\ Fe^{3+}_{0,24}\ Fe^{2+}_{0,03}\ Ti_{0,08})$ $(Si_{2,72}\ Al_{1,28})\ O_{10}\ (OH)_2$ | 0,82 |
| Vermiculite Prayssac (France) | $Li_{0,54}$ $(Si_{2,82}\ Al_{1,17})$ $(Mg_{2,36}\ Al_{0,20}\ Fe^{3+}_{0,43}\ Fe^{2+}_{0,004})\ O_{10}\ (OH)_2$ (Louis André, 1972) | 0,54 |
| Vermiculite Benahavis (Espagne) | $Li_{0,53}$ $(Mg_{2,46}\ Ti_{0,11}\ Fe^{3+}_{0,43})$ $(Si_{2,81}\ Al_{1,10}\ Fe^{3+}_{0,09})\ O_{10}\ (OH)_2$ (don de Lopez Gonzalez) | 0,53 |

Dans un premier temps, nous avions pensé qu'il s'agissait d'un effet de cinétique de gonflement. Le concept imagé de "paresse" à abandonner un état d'hydratation donné a été souvent évoqué pour justifier certaines anomalies de comportement. Si nous regroupions l'ensemble des résultats sur poudre et sur monocristaux, on pouvait poser en hypothèse que quatre échantillons en monocristaux: Santa Olalla, Prayssac, Benahavis et V Li (1) sont - dans les basses humidités - bloqués dans un état métastable sur l'ouverture à "deux couches".

L'hypothèse, à priori, nous semblait cependant peu vraisemblable car nous laissions les échantillons pendant trois semaines environ dans l'humidité choisie, avant de les examiner. Nous avons vérifié qu'il ne s'agissait effectivement pas d'un problème de cinétique en amenant pendant une longue durée un échantillon de Santa Olalla à 0% d'humidité relative avant de l'équilibrer à 20%.

L'espacement révèle alors qu'à 0% le minéral est interstratifié "zéro couche" - "une couche", et on constate qu'à 20%, il repasse à l'état "deux couches". L'état "deux couches" de certains monocristaux, de 5 à 50% d'humidité relative n'est donc pas un état métastable.

Ne pouvant évoquer non plus de différences dans le niveau et dans la localisation des charges pour expliquer les différences de gonflement constatées, nous sommes obligés d'envisager des différences de texture.

ETUDE DE LA TEXTURE

Nous poserons en hypothèse que l'équilibre entre l'hydrate à "deux couches" et celui à "une couche" n'est pas seulement - pour un minéral lithique donné - fonction de l'humidité relative, mais qu'il dépend également du rapport entre surface externe et volume des cristallites élémentaires. Nous appellerons cristallites élémentaires ceux qui n'auront pas de porosité interne susceptible de créer une interface entre les couches interfoliaires et le milieu gazeux ambiant. En d'autres termes, cela revient à considérer qu'il existe une valeur du rapport volume-surface externe, en dessous de laquelle l'hydrate à "deux couches" n'est plus stable aux faibles humidités relatives.

Pour que ce raisonnement soit cohérent avec nos données expérimentales, il faut que nos échantillons en monocristaux diffèrent dans leur texture, certains devant présenter une porosité interne plus importante que d'autres.

Nous avons donc étudié les monocristaux en microscopie à balayage afin de juger de la qualité des surfaces. Nous avons examiné en microscopie à balayage sept échantillons, soit : Llano naturelle, Llano Li, Santa Olalla naturelle, Santa Olalla Li, Benahavis naturelle, Benahavis Li et Prayssac naturelle.

Par rapport aux échantillons naturels, les échantillons lithiques présentent des bords moins nets où l'aspect stratifié est souvent absent, masqué probablement par des feuillets froissés et rabattus. Tous les échantillons lithiques dont nous disposions ayant été préalablement taillés au rasoir, la comparaison avec les éclats des minéraux naturels n'est pas très convaincante, car la taille abîme considérablement les bords des plaquettes cristallines.

Dans la figure 2, nous donnons les photographies des sections de deux minéraux naturels. Celle de la vermiculite de Prayssac montre des stratifications continues, sans trace nette de porosité. Par contre, les clichés sur la section d'une plaquette de Llano montre très peu de lignes nettes et parallèles mais un grand nombre de cavités.

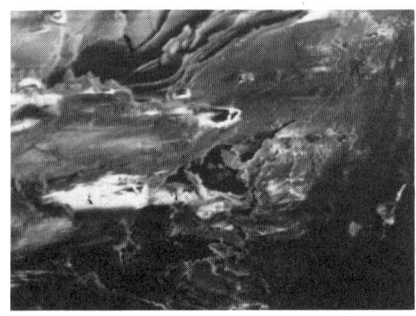 

(a) 1 cm = 20 μm        (b) 1 cm = 40 μm

Fig. 2. Photographie au microscope à balayage des bords de cristaux de vermiculite (a) Llano (b) Prayssac.

De la même façon, la vermiculite de Llano fraîchement clivée révèle une surface assez différente de celle des autres échantillons. Alors que les échantillons de Prayssac et Santa Olalla donnent de grandes surfaces bien planes, la vermiculite de Llano présente une surface très accidentée fissurée et poreuse (cf. fig. 3).

De cet aspect, on peut déduire une forte présomption que les monocristaux de Llano sont dotés d'une porosité interne relativement importante. Comme les diamètres des plaquettes de vermiculite, dans ce que nous avons appelé des monocristaux et des poudres, sont dans un rapport de l'ordre de 50, on peut admettre que les "cristallites élémentaires" du monocristal de Llano ont des tailles comparables à celles de microcristaux des poudres.

Nos essais pour déterminer en microbalance la surface spécifique des monocristaux de vermiculite de Llano et de Santa Olalla, par adsorption de krypton à la température de l'azote liquide ont été sans succès. Il nous a été impossible de mettre en évidence une variation de masse correspondant à une quelconque adsorption, compte-tenu des limites de sensibilité de l'appareillage utilisé (Thermobalance Setaram M18).

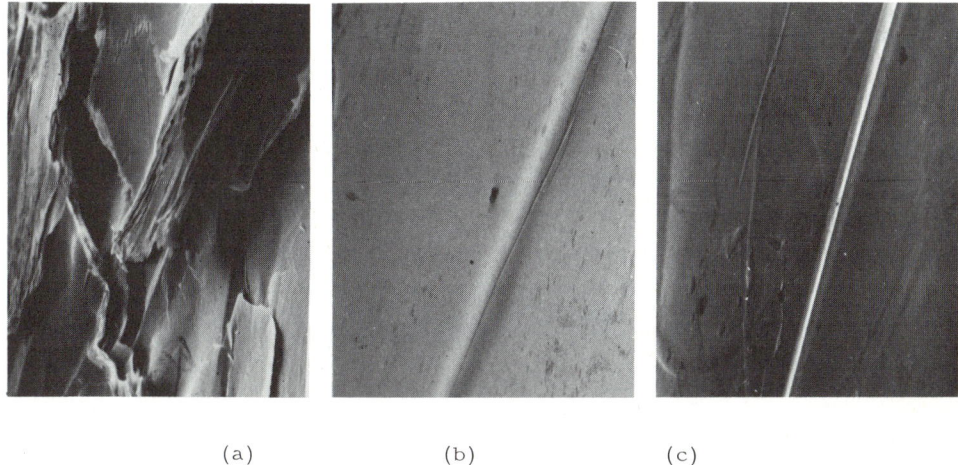

(a) (b) (c)

Fig. 3. Photographies au microscope à balayage des surfaces des cristaux
fraîchement clivés de vermiculite de : (a) Llano (b) Prayssac (c) Santa Olalla)
(1 cm = 40 μm).

DISCUSSION

Le comportement de la vermciulite de Llano Li en monocristal apparaît très
semblable à celui des échantillons en poudre, non seulement de 5 à 70% d'humidité
relative mais aussi à 0% puisqu'il y a fermeture aisée à 10 Å dans les deux cas.

L'aspect plissé et la porosité des surfaces et des sections conduisent à
considérer qu'il existe dans cet échantillon une porosité interne qui augmente
considérablement l'interface entre les couches interfoliaires et l'air ambiant,
le rapport interface sur volume pouvant atteindre le même ordre de grandeur
que celui existant dans les poudres.

L'échantillon V Li (2) ayant pour sa part été traité à 80°C par une solution
de $MgCl_2$, 2 N, relativement acide, il est certain que le cristal a été assez
fortement attaqué à tous les joints de grain, et un simple examen en microscopie
optique suffit pour révéler une surface extrêmement fissurée et abîmée.

On peut donc considérer qu'il existe une certaine similitude entre les poudres
et les monocristaux à texture poreuse et qu'en conséquence il est possible
d'établir une corrélation entre les propriétés de gonflement et la taille des
cristallites élémentaires.

Pour justifier l'hypothèse d'une influence directe de la taille de ces
cristallites sur l'hydratation des couches interfoliaires, il faut rappeler
que ces dernières sont en état de perpétuel échange avec les molécules d'eau

de l'air ambiant.

Prenant une plaquette circulaire de rayon r, on peut supposer qu'il existe à la périphérie des couches interfoliaires une zone en couronne de profondeur $\Delta r$ où le degré d'hydratation sera moindre qu'au centre du cristal, en raison des échanges avec le milieu extérieur, plus pauvre en eau. Dans cet anneau, à la périphérie des couches interfoliaires, les feuillets vont avoir tendance à se rapprocher pour donner un stade d'hydratation inférieur à celui du coeur de cristal.

Pour qu'un état d'hydratation soit stable sur l'ensemble du cristal, il est nécessaire que $\Delta r$ reste très petit par rapport à r. Sinon le cristal tendra à se stabiliser sur le stade inférieur d'hydratation.

Il est probable que pour un cation et une humidité donnée, $\Delta r$ sera quasiment une constante. La variation de r pourra donc conduire à des changements dans les états de gonflement.

Il est vraisemblable que l'évaluation des "diamètres de cohérence" des cristallites, parallèlement aux feuillets dans des échantillons en cristaux macroscopiques, doit prendre en compte la totalité des défauts qui rompe la continuité du cristal, y compris les frontières des domaines de croissance propres à chaque dislocation hélicoïdale

REMERCIEMENTS

Nous exprimons tous nos remerciements à D. Delafosse et O. Cornu pour les analyses gravimétriques et à Monsieur Froment et Madame Vignaud du laboratoire de Physique des Liquides et Electrochimie pour les photographies en microscopie électronique à balayage.

BIBLIOGRAPHIE

L. André, 1972, Contribution à l'étude des mécanismes d'échange de cations dans les vermiculites trioctaédriques. Thèse, Université Paul Sabatier, Toulouse.
C. de la Calle, H. Suquet et H. Pezerat, Glissement de feuillets accompagnant certains échanges cationiques dans les monocristaux de vermiculites, (1975a) Bull. G. fr. Argiles, 27, 31-49.
C. de la Calle, H. Suquet, J. Dubernat, H. Pezerat, Gaultier et J. Mamy, Crystal structure of two-layer vermiculites and Na, Ca-vermiculites. Proceed International Clay Conference, Mexico, (1975b), 201-209.
C. de la Calle, H. Suquet, J. Dubernat et H. Pezerat, Mode d'empilement des feuillets dans les vermiculites hydratées à "deux couches", (1978), sous presse.
K. Norrish, Factors in the weathering of mica to vermiculite, Proceed International Clay Conference, Madrid, (1972), 417-432.
C. de la Calle, 1977, Thèse, Université P. et M. Curie, Paris.

QUALITATIVE AND QUANTITATIVE STUDY OF A STRUCTURAL REORGANIZATION
IN MONTMORILLONITE AFTER POTASSIUM FIXATION

A. PLANÇON*, G. BESSON*, J.P. GAULTIER**, J. MAMY** and C. TCHOUBAR*.
* Laboratoire de Cristallographie, Université d'Orléans ; C.R.S.O.C.I, C.N.R.S.
  45045 Orléans Cedex, France.
** Station de Sciences du Sol, Centre National de Recherches Agronomiques,
  I.N.R.A., 78000 Versailles, France

ABSTRACT

The structural reorganization of a K-montmorillonite by wetting and drying cycles previously observed by Mamy and Gaultier is refined. After one hundred cycles, the mineral is almost completely stable, but a tridimensional order is not reached. In stackings, only one layer in four is translated by $-\vec{a}/3$ with respect to the adjacent one, without any rotational stacking fault (this translation corresponds to a superimposition of the tetrahedral sheet centers of two adjacent layers). One layer in four is translated by $-\vec{a}/3$ and rotated by $n\ 2\pi/6$ (n.integer) and one in two is randomly translated. The shift of the centers of tetrahedral sheets by $\pm \vec{b}/3$ does not seem to occur.

INTRODUCTION

In a previous paper, Mamy and Gaultier (1976) have shown that it is possible to induce, in a layer silicate as disordered as montmorillonite, a reorganization involving the degree of order in the stacking of the elementary layers. The process can be summarized as follows : a turbostratic montmorillonite (whose X-ray powder pattern consists of unmodulated *(hk)* bands) is transformed into a K-montmorillonite and then submitted alternatively to wetting and drying. The selected area electron diffraction and the X-ray powder patterns show that a reorganization occurs : for example, modulations appear on the *(hk)* band profiles in X-ray patterns. But certain questions have remained in the interpretation of results, concerning the nature of the reorganization : for example, the existence of $\pm \vec{b}/3$ shifts between the hexagonal holes of adjacent layers. The aim of the present paper is now to give some information concerning *i)* the nature of the reorganization and *ii)* the nature and proportion of defects remaining in montmorillonite after wetting and drying treatment.

MATERIALS and METHODS

The treatment effected by Mamy and Gaultier can be described as follows : a Wyoming montmorillonite is first saturated by potassium (this mineral is named 0 W.D. K-montmorillonite) and then alternatively dried in an oven (∼ 80°C) and shaken in water for several hours (this is named a W.D. cycle). The process is repeated more than one hundred times, but after 100 W.D. cycles, the mineral (named 100 W.D. K-montmorillonite) is almost completely stable.

It is verified that the transformation is non-destructive for the layer, and that the modification of the X-ray pattern of 100 W.D. K-montmorillonite is only due to a rearrangement of the layers.

The present paper describes the crystallographic study which has been performed on the three samples described above (namely the starting Na montmorillonite, the 0 W.D. K-montmorillonite and the 100 W.D. K-montmorillonite) in order to describe the rearrangement of the layers.

The X-ray intensities have been recorded in a way previously described by Plançon et Tchoubar (1977 b). Some recordings are shown on figure 1.

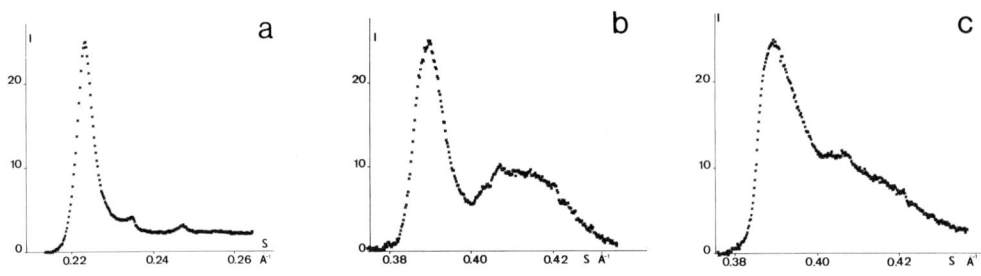

Figure 1. a)(02,11) band of the 100 W.D. K-montmorillonite. b)(20,13) band of the 100W.D. K-montmorillonite. c)(20,13) band of the 0 W.D. K-montmorillonite.

The peaks located at $s = 0.235 \text{ Å}^{-1}$, $s = 0.248 \text{ Å}^{-1}$ and $s = 0.407 \text{ Å}^{-1}$ are not modulations of the bands but reflexions from quartz and cristobalite. One keeps in mind these impurities when comparing the theoretical and experimental profiles presented in the next paragraphs. The (02,11) band of the 0 W.D. K-montmorillonite is unmodulated as is the (02,11) band of the 100 W.D. montmorillonite. The (02,11) and (20,13) bands of the starting Na montmorillonite are both unmodulated.*

---

* Remark : Much information not reproduced in this paper will be given in a future paper by Gaultier, Besson, Plançon, Tchoubar and Mamy.

STUDY OF $00l$ REFLECTIONS

Theory.

The profiles of these reflections are not sensitive to the translational or rotational defects parallel to the layer plane but depend on atomic coordinates, basal distances and the distribution of stacking thicknesses. Theoretically, (Maire and Mering, 1960 ; Kodama *et al*, 1971) this last parameter can be directly deduced from the $G_{00}(s)$ modulation function obtained from the ratio of the experimental intensities to the square of the corresponding structure factor, $F_{00}(s)$.

In the case of smectites, it is difficult to apply this method, because, among other things, the structure factor falls to zero in the region of the three first $00l$ reflections. So, we have used an indirect method (Rousseaux and Tchoubar, 1977 ; Pons *et al*., 1978) which consist in the determination of the α(M) probability distribution of stacking thicknesses by trial adjustment between the observed and calculated profiles. Then,

$$I_{00}(s) \simeq \frac{1}{4\pi s^2} |F_{00}(s)|^2 \sum_M \alpha(M) G_{00}(M,s)$$

where s is the modulus of the scattering vector (s = 2 sin θ/λ) and M is the number of layers in a stack.

Comparison of theoretical and experimental profiles

The determination of the distribution of stacking thicknesses (fig. 2) is performed by using the $001$ reflection of each sample. The atomic coordinates are those of Pezerat and Mering (1967) where the Na cation is positioned at a height of 4.15 Å. For the K-montmorillonites where the basal distance is almost the same as in muscovites, the K cation is put at the centre of each layer, following Radoslovitch (1960).

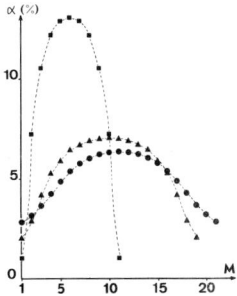

Figure 2. Distribution of stacking thicknesses : a) Na montmorillonite ($\bar{M}$ = 11), b) 0 W.D.K-montmorillonite ($\bar{M}$ = 10), c) 100 W.D. K-montmorillonite ($\bar{M}$ = 6).

The fits obtained for the 001 reflections appear on figure 3 :

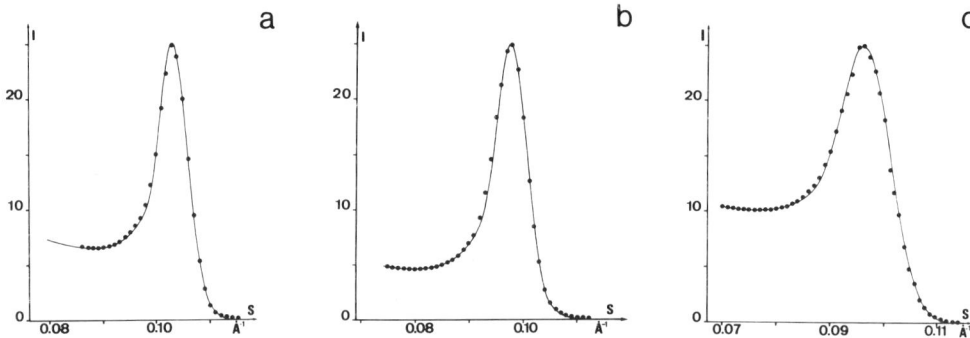

Figure 3. Experimental (dots) and calculated profiles (full line) of the 001 reflections : a) starting Na montmorillonite, b) 0 W.D. K-montmorillonite, c) 100 W.D. K-montmorillonite.

The comparison of the three montmorillonites shows $i)$ a decrease of the $\bar{M}$ value by a factor two for the 100 W.D. K-montmorillonite with respect to the Na montmorillonite. This is consistent with a BET surface which is twice the BET surface of the starting material (Mamy, 1968), $ii)$ a modification of the basal distance from 10.1 Å in the 0 W.D. K-montmorillonite to 10.0 Å in the 100 W.D. K-montmorillonite, $iii)$ a narrowing of the distribution and an almost entire disappearance in the 100 W.D. K-montmorillonite of the layers diffracting independently. In the 0 W.D. sample, these correspond either to true independent layers or to deformations of the layers (observable by electron microscopy) which have disappeared in the 100 W.D. K-montmorillonite.

On the other hand one notices that the important modification of the structure factor in the domain of the 001 reflection, associated with the distribution of stacking thicknesses leads to a shift of the top of the reflection with respect to the theoretical value $1/d_{001}$ where $d_{001}$ is the basal distance (Tettenhorst and Roberson, 1973 ; Trunz, 1976), and to an asymmetry of intensity profiles. Moreover the $\bar{M}$ value are sensibly different from the values obtained by use of the Scherrer formula ($\bar{M}$ = 15 for the Na montmorillonite and, respectively, 12 and 7 for the 0 and 100 W.D. K-montmorillonite).

STUDY of (hk) BANDS

In addition to the structure factor, the profiles of these reflections depend fundamentally on the nature and proportion of the defects caused by translation or rotation of the layers. In clays, the complex structure of the layer excludes the possibility of deducing the defects directly from the experimental intensities. So, these defects are specified by comparing the experimental and theoretical intensities calculated by using stacking models containing different kinds and proportions of possible defects (Plançon and Tchoubar, 1977 a,b).

## Description of the proposed stacking faults

The defects which can be imagined in montmorillonites must be in agreement with the observations by selected area electron diffraction (Mamy and Gaultier, 1976) : for the Na montmorillonite and the 0 W.D. K-montmorillonite, the SAD does not consist in a set of diffraction spots distributed at the points of an hexagonal lattice (SAD of a single montmorillonite layer), but the spots are distributed along a series of concentric circles. This indicates that there exist in these samples, *random rotations of layers* (turbostratic stacking). By contrast, following one hundred wetting and drying cycles, the SAD patterns show a single set of diffraction spots, forming an hexagonal array, independent of the thickness of the particles. Such an observation can only be explained by a mutual reorientation of the elementary layers. The diffracted intensities being roughly uniform in each set of six spots, at least some crystallographic axis of layers must be relatively rotated by multiples of $2\pi/6$. This is quite consistent with the pseudo-hexagonal symetry of tetrahedral sheets.

So, one could consider that, in stackings of 100 W.D. K-montmorillonite, there exist six kinds of rotated layers corresponding to *six possible relative orientations* of two adjacent layers.

It is also relevant that some mutual translations of the layers can occur without being detected by SAD. The models studied here must then take into account possible *relative translations* of layers, either $\pm \vec{b}/3$ *translations or random ones*.

## Mathematical treatment.

The method of calculation has been previously described by Plançon and Tchoubar (1977 a).

The theoretical intensity, for stackings containing g kinds of layers (or g orientations of layers) is calculated by :

$$I(s) \simeq \frac{M}{4\pi s} \text{Re}\left[\sum_{i=1}^{g} \sum_{j=1}^{g} N(\varphi) \, [F_{hk}(Z)]_i \, [F_{hk}^*(Z)]_j \, \alpha_{ij}(Z) \, T(X) \, d\varphi\right]$$

where Re corresponds to the real part of the expression between brackets

- $N(\varphi)$ is a function depending only on the degree of orientation of particles in the powder with respect to the plane of the specimen holder. For each sample this function has been determined by using the method of de Courville *et al.*(1978).

- $[F_{hk}(Z)]_i$ is the structure factor, at the Z position, for the hk rod of a layer of kind i. The x,y atomic coordinates are taken from those of muscovites (Radoslovitch, 1960) taking account of a $\beta$ rotation of tetrahedrons in the tetrahedral sheets. Such a rotation has been previously observed in beidellites (Besson and Tchoubar, 1972).

- $\alpha_{ij}(Z)$ characterizes the interference between the i and j layers. It is equivalent to the modulation function G presented in the study of $00l$ reflections and depends on the nature and proportion of defects, described below :

Let us consider a defined layer of kind i. In absence of defects, its upper first neighbouring layer is translated by $\vec{t}_i = -\vec{a}_i/3 + d_{001}\vec{n}$, where $\vec{a}_i$ is the a axis of the i layer and $\vec{n}$ is a unit vector normal to the layer plane. This corresponds to a perfect superimposition of the hexagonal holes of adjacent tetrahedral sheets of two neighbouring layers.

But there exist certain defects. Let's call P the probability that this upper first neighbouring layer is randomly rotated or translated. Because these two defects suppress the interference between the beams diffracted by each layer (Maire and Mering, 1960), the probability that this upper layer interferes with the lower one is only (1-P). On the other hand, this upper layer can be additionally translated by $\pm$ m $\vec{b}/3$ (Mamy and Gaultier, 1976) and/or rotated by n $2\pi/6$ (m and n integers).

Calling $\eta_b$ the probability that this neighbouring layer is affected by a $\pm \vec{b}/3$ translational defect and $\eta_r$ that it is affected by a + $2\pi/6$, + $2\pi/3$, $\pi$, - $2\pi/3$ or - $2\pi/6$ rotation; assuming that the rotational and translational defects are independent ; assuming that + $\vec{b}/3$ and - $\vec{b}/3$ defects are equally probable and that each of the five rotational defects are equally probable. Then, the $m_{ij}$ term (i $\neq$ j) of the Q matrix (Plançon and Tchoubar, 1977 a) which describes the interference between first neighbouring layers is equal to $(1-P)(1-\eta_b + 2\eta_b \cos 2\pi k/3)$ $(\eta_r \exp 2\pi i \vec{s}.\vec{t}i)/5$, while the $m_{ii}$ term is equal to $(1-P)(1-\eta_b + 2\eta_b \cos 2\pi k/3)$ $(1-\eta_r) \exp 2\pi i \vec{s}.\vec{t}i$

- T(X) is a function which depends on the shape and size of coherent diffraction domains in the layers. We suppose that these are circular and their radii R are adjusted at the same time as the proportion of defects.

### 100 W.D. K-montmorillonite

As explain in a previous paragraph this sample does not contain any random rotations. Moreover, the *(20,13)* band is not sensitive to $\pm \vec{b}/3$ translational defects (Plançon and Tchoubar, 1976). So, its profile can be affected only by two kinds of defects : *i)* n $2\pi/6$ rotations and *ii)* random translations, if these exist. Preliminary calculations were performed assuming no random translations. The figure 4 shows the calculated profile for the *(20,13)* band in the case of maximum disorder of n $2\pi/6$ rotations ($\eta_r$ = 0.833). This calculated profile remains obviously too much modulated with respect to the experimental one.

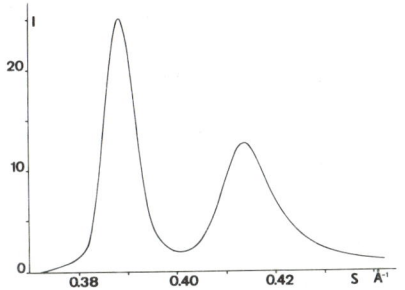

Figure 4. Calculated profile of the (20,13) band of the 100 W.D. K-montmorillonite by using a model containing only 2π/6 rotational defects.

So the presence of only n 2π/6 rotations cannot explain the experimental (20,13) band profile (a modification of the radius of the coherent domains does not improve the fit).

Thus, it is necessary to introduce the existence of defects by random translations. This leads to satisfactory results. The best fit obtained in this case is shown on the figure 5 b (with P = 0.5, $\eta_r$ = 0.5, R = 200 Å). The rotation of tetrahedrons is 5°. Another rotation of tetrahedrons (e.g. β = 0°, with the same values of P, $\eta_r$, R) leads to an important modification of the structure factor and deforms appreciably the (20,13) band profile.

The calculation of the theoretical (02,11) band profile (fig. 5a) has been accomplished with the values of P and $\eta_r$ determined from the fit of the (20,13) band, with adjustment of the $\eta_b$ value which leads to the best fit between the theoretical and experimental profiles, namely R = 250 Å and $\eta_b$ = 0 (that is to say an absence of ± $\vec{b}$/3 defects).

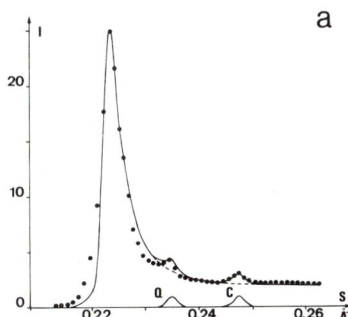
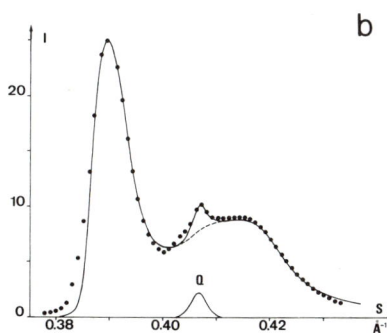

Figure 5. Experimental (dots) and calculated (full line) profiles of (02,11) and (20,13) bands of 100 W.D. K-montmorillonite (P = 0.5, $\eta_r$ = 0.5, $\eta_b$ = 0).

The reflections from quartz (Q) and cristobalite (C) impurities are added as shown on the bottom of the figures. The disagreement between the experimental and calculated intensities, at the beginning of the band can be explained by the existence of a distribution of dimensions of coherent diffraction domains (Plançon, Tchoubar, 1976)

## 0 W.D. K-montmorillonite

The wetting and drying cycles bring about the disappearance of the random rotations. So, it is foreseen that the P proportion of translational and rotational random defects of the 0 W.D. sample is greater than the P proportion in the 100 W.D. sample. The best fit between experimental and calculated profiles is effectively obtained for the *(20,13)* band with P = 0.8, $\eta_r$ = 0.5, R = 200 Å and a β rotation of tetrahedrons of 5° (fig. 6b).

For the *(02,11)* band, the best fit corresponds, as in the 100 W.D. K-montmorillonite, to an *absence* of ± $\vec{b}$/3 defects, namely P = 0.8, $\eta_r$ = 0.5, $\eta_b$ = 0; β = 5° and R = 250 Å (fig. 6a).

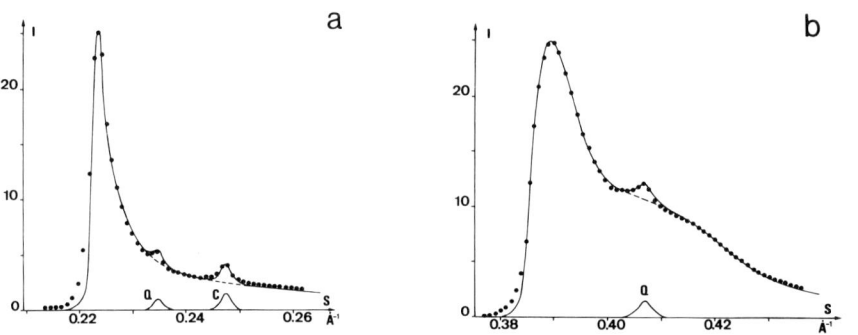

Figure 6. Experimental and calculated profiles of *(02,11)* and *(20,13)* bands of OW.D. K-montmorillonite (P = 0.8, $\eta_r$ = 0.5, $\eta_b$ = 0).

## Starting Na montmorillonite

The best fit is obtained for the starting Na montmorillonite with P ≠ 1, β = 10°, R = 200° for the *(02,11)* band (fig. 7a) and R = 200 Å for the *(20,13)* band (fig. 7b).

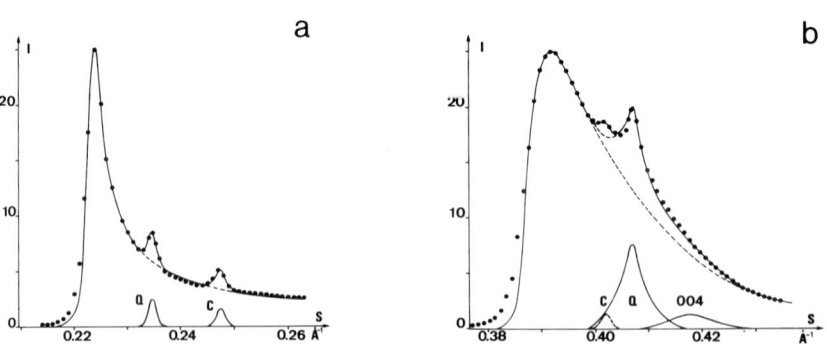

Figure 7. Experimental and calculated profiles of *(02,11)* and *(20,13)* bands of starting Na montmorillonite (P = 1). For this sample, the *004* reflection is intense and must be taken into account.

One could notice that such a value of P indicates that practically all the layers diffract independently. On the other hand, the rotation of the tetrahedrons differs from that of the K-montmorillonite, following the modification of cell parameters by cation exchange (Ravina and Low, 1977).

CONCLUSION

The present work provides information concerning the nature of the disorder in montmorillonite.

First of all, the X-ray experimental profiles of *(02,11)* band and *(20,13)* band of montmorillonites are well described by a stacking model containing variable proportions of random defects (translations and/or rotations) and n $2\pi/6$ rotations. The best fits between the experimental and calculated profiles leads to the conclusions below :

- in the stackings of starting Wyoming Na-montmorillonite, nearly all the layers are either rotated by rotations not multiples of $2\pi/6$, or translated by random translations: consequently nearly all the layers are diffracting independently (except for $00l$ reflections).

- after the potassium exchange and before any wetting and drying cycle, a partial reorganization of layers occurs. A perfect superimposition of the centers of the hexagonal holes of adjacent tetrahedral sheets of two neighbouring layers (with or without n $2\pi/6$ rotations) arises for one layer in five. Correlatively a slight decrease of the mean number of layer in the stackings is related to the potassium exchange. Also the tetrahedral rotation changes from 10° to 5°.

- afterwards, as the number of W.D. cycles proceeds, the reorganization of the layers increases, with progressive disappearance of the random rotations of layers. However, after 100 W.D. cycles, full tridimensional order is not reached, because one layer in two still exhibits a random translation with respect to its neighbour (P = 0.5) while one layer in four exhibits a superimposition of the hexagonal holes with a $2\pi/6$ rotational fault ($\eta_r$ = 0.5). So, only one layer in four is stacked without any translational or rotational fault. It seems also that a superimposition of layers where the centers of hexagonal holes are shifted by $\pm \vec{b}/3$ does not occur. On the other hand, the W.D. treatment has caused a cleavage of particles and reduced the mean number of layers in a diffracting unit to half that of the starting sample.

Finally, it should be observed that the greatest care must be taken in the interpretation of the X-ray patterns. For example, the 100 W.D. K-montmorillonite exhibits a *(02,11)* band without any modulation, while the *(20,13)* band is an obviously modulated one. The calculations show that these two effects can be explained at the same time using the <u>same</u> parameters, and that these do not necessarily involve the presence of $\pm \vec{b}/3$ translational defects.

REFERENCES

Besson, G. and Tchoubar, C., 1972. Détermination du groupe de symétrie du feuillet élémentaire de la beidellite. C.R. Acad. Sc. Paris, 275: 633-636.

De Courville, J., Tchoubar, C. and Tchoubar, D., 1978. To be published.

Gaultier, J.P., Besson, G., Plançon, A., Tchoubar, C. and Mamy, J., 1978. To be published.

Kodama, H., Gatineau, L. and Mering, J., 1971. An analysis of X-ray diffraction line profiles of micro-crystalline muscovite. Clays and Clay minerals, 19: 405-413.

Maire, J. and Mering, J., 1960. Chemistry and Physics of Carbon. In P. L. Walker (Editor). Marcel Dekker, Amsterdam, Vol. 6, pp 125-189.

Mamy, J., 1968. Ph. D. Thesis.

Mamy, J. and Gaultier, J.P., 1976. Les phénomènes de diffraction des rayonnements X et électroniques par les réseaux atomiques : application à l'étude de l'ordre cristallin dans les minéraux argileux - II. Annales Agronomiques, 27: 1-16.

Pezerat, H. and Mering, J., 1967. Recherches sur la position des cations échangeable et de l'eau dans les montmorillonites. C.R. Acad. Sc. Paris, 265: 529-532.

Plançon, A. and Tchoubar, C., 1976. Etude des fautes d'empilement dans les kaolinite partiellement désordonnées. II. J. Appl. Crystallogr., 9: 279-285.

Plançon, A. and Tchoubar, C., 1977.a. Determination of structural defects in phyllosilicates by X-ray powder diffraction - I, Clays and Clay Minerals, 25: 430-43

Plançon, A. and Tchoubar, C., 1977 b. Determination of structural defects in phyllosilicates by X-ray powder diffraction - II, Clays and Clay Minerals, 25: 436-4

Pons, C.H., Ben Brahim, J. and Tchoubar, C., 1978. To be published.

Radoslovitch, E.W., 1960. The structure of Muscovite. Acta Cryst., 13: 919-932.

Ravina, I and Low, P.F., 1977. Change of b-dimension with swelling of montmorillonite. Clays and Clay Minerals, 25: 201-204.

Rousseaux, F. and Tchoubar, D., 1977. Structural evolution of a glassy carbon as a result of thermal treatment between 1000 and 2700°C - II. Carbon, 15: 63-68.

Tettenhorst, R. and Roberson, H.E., 1973. X-ray diffraction aspects of montmorillonites. Amer. Miner., 58: 73-80.

Trunz, V., 1976. The influence of crystallite size on the apparent basal spacings of kaolinite. Clays and Clay Minerals, 24: 84-87.

# THE FERRIC ANALOGUE OF PYROPHYLLITE AND RELATED PHASES

F.V. CHUKHROV, B.B. ZVYAGIN, V.A. DRITS, A.I. GORSHKOV, L.P. ERMILOVA, E.A. GOILO, E.S. RUDNITSKAYA
Institute of Ore Geology, Petrography, Mineralogy and Geochemistry, USSR Ac. Sc., Moscow (USSR)

## ABSTRACT

The ferric analogue of pyrophyllite (ideal formula $Fe_2Si_4O_{10}(OH)_2$) and related phases that form evenly expanded or mixed-layer structures with water have been found and identified by means of electron and X-ray diffraction.

## INTRODUCTION

The phases studied have been found in deposits from Strassenschacht near Eibenstock (GDR) and on Mount Tologay (central Kazakhstan). The samples are cryptocrystalline, yellowish or greenish, and consist of disoriented flakes up to 0.03 mm in dimension. The substance is biaxial, negative, 2V small, $\alpha$ 1.686-1.688, $\beta$ 1.676-1.678, $\gamma$ 1.65-1.66; pleochroism noticeable; specific gravity between 3.01 and 2.97. The microcrystals are generally elongated.

The main components are $SiO_2$, $Fe_2O_3$ and $H_2O$ with minor $Al_2O_3$ and CaO. It is remarkable that the molecular ratio $SiO_2:Fe_2O_3$ is nearly 2:1. The DTA- and TG-characteristics display two endotherms at 130° and 500° showing the existence of two kinds of water, and a small exotherm at 835°.

## OBLIQUE-TEXTURE ELECTRON DIFFRACTION PATTERNS (OTEDP)

The structural peculiarity of the samples was first established by OTEDP's which revealed the presence of only one crystalline phase with a monoclinic unit cell: a 5.26, b 9.10, c 19.1Å, $\beta$ 95.5°. The intensities of reflexions in the second ellipse indicate that 2:1 layers are superimposed as in pyrophyllite and talc so that Si-hexagons of adjacent layers do not overlap in the normal projec-

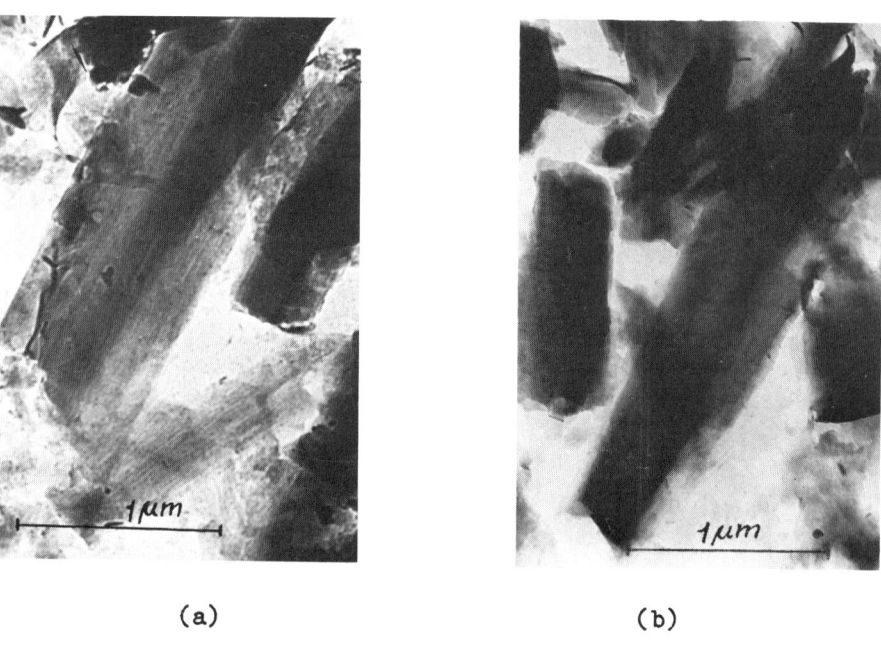

Fig. 1. Electron micrographs: a) Strassenschacht sample; b) Tologay sample.

tion on the ab plane as in micas, but are displaced to give minimal Si-Si-repulsion. In particular, the pair of coinciding reflexions 202, $\bar{1}33$ (2.46Å) is much stronger than $\bar{2}04$, 133 (2.38Å), while for micas this relation is reversed (cf. Fig's 2a, 2b and 2c). The reflexions in the first ellipse are very distinctive: three reflexions, 020 (4.55Å), $\bar{1}12$ (4.22Å), 114 (3.16Å) are clearly stronger than the others (cf. Figs 2a and 2b). This is a unique feature of a two layer monoclinic polytype 2M - $\sigma_3 \sigma_3 \tau_1 \sigma_3 \sigma_3 \tau_5 \ldots$ (Zvyagin et al., 1968) with ideal symmetry C2/c. The sequence of layers expressed by this notation was observed for pyrophyllites while talcs usually have a semirandom structure (ordered only in the projection on the ac plane) or sometimes a one-layer triclinic structure 1TC- $\sigma_2 \sigma_2 \tau_4 \ldots$ (Smolin et al., 1975). The reflexions 261, 401 in the sixth ellipse and 171, 351, 421 in the seventh have comparable intensities, this being an indication that the phyllosilicate is dioctahedral (for trioctahedral phyllosilicates the

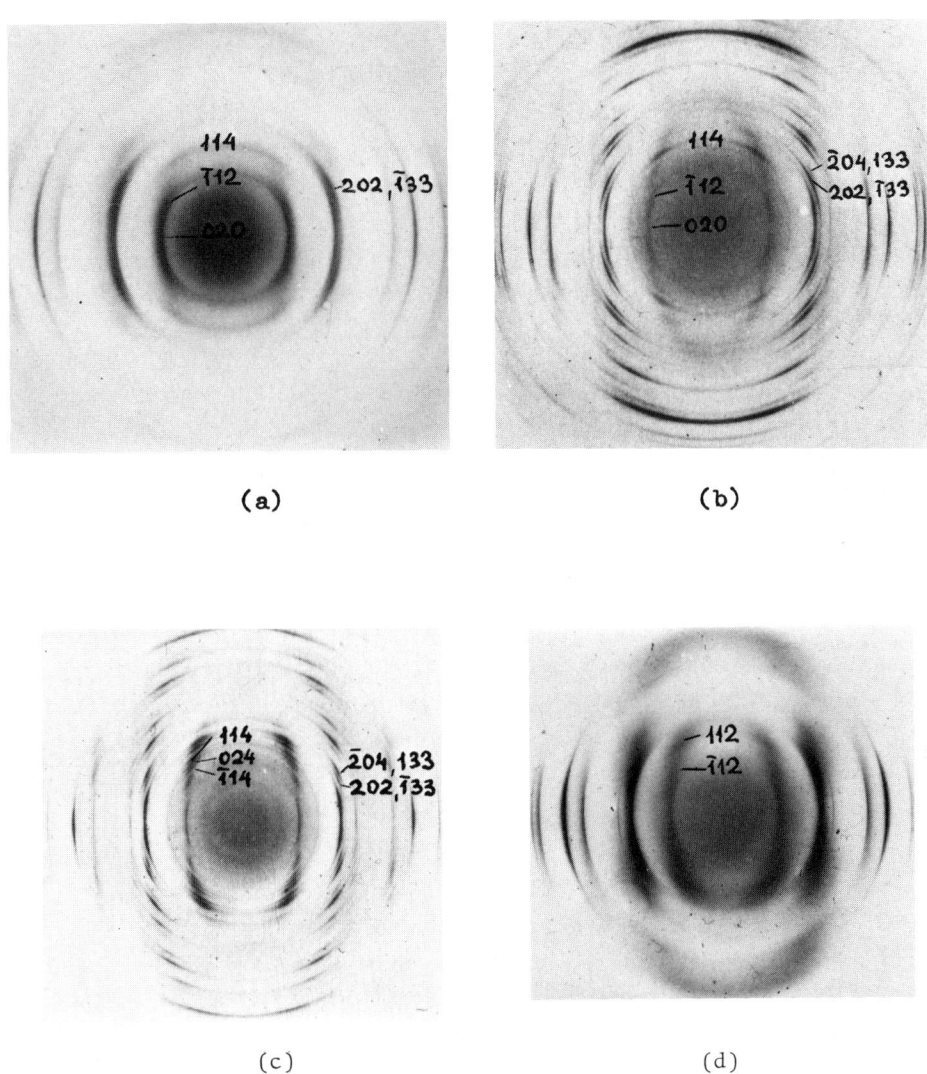

Fig. 2. Oblique-texture electron diffraction patterns of the studied samples (a), pyrophyllite, (b), muscovite (c), nontronite (d).

former are in general stronger than the latter).

All these structural data together with the evidence of the chemical composition point directly to a ferric analogue of pyrophyllite having $Fe^{3+}$ instead of Al in octahedra and consequently an ideal formula $Fe_2Si_4O_{10}(OH)_2$.

Starting with the value b 9.13Å of talc and using regression coefficients for Mg and $Fe^{3+}$ (Drits, 1969) one may evaluate what b-dimension is to be expected for the ideal composition according to the relation b = 9.13−0.062(Mg)+0.077($Fe^{3+}$), where ($Fe^{3+}$) is the actual content of $Fe^{3+}$ and (Mg) is the deficit in Mg in comparison with the formula of talc. For the new phase (Mg) = 3 and ($Fe^{3+}$) = 2, so b = 9.098Å. The calculated sp. gr. according to the indicated unit cell containing Z = 4 ideal formula units is 3.05. Such a close agreement between the calculated and experimental values supports the structure proposed for the substance as it exists under the vacuum conditions of electron diffraction.

X-RAY DATA

The X-ray study has however revealed some differences and some additional features for both samples. The powder diagram of the Strassenschacht sample has a few lines which may be interpreted with the help of the OTEDP's. One may recognize there the main reflexions mentioned above (202, $\bar{1}33$; 020 (060); $\bar{1}12$) which are related to the three-dimensional pyrophyllite-like structure. There are also basal reflexions 001 (9.6Å), 003 (3.17Å), the last probably coinciding with 114. Without the evidence given by OTEDP's it is easy to interpret such a powder diagram as resulting from a semirandom talc structure (there are slight differences in the basal spacings). The absence of OTEDP's is perhaps a reason why pyrophyllite-like structures with $Fe^{3+}$ in octahedra have not been recognized earlier in the course of current X-ray work.

TABLE 1

The X-ray powder diagram of the Strassenschacht sample

| hkl | d (Å) | I | hkl | d (Å) | I |
| --- | --- | --- | --- | --- | --- |
| 002 | 9.6 | 8 | 202, $\bar{1}33$ | 2.47 | 4 |
| 020 | 4.54 | 10 | 0.0.10 | 1.89 | 1 |
| $\bar{1}12$ | 4.25 | 2 | 150, 241, $31\bar{1}$ | 1.725 | 3 |
| 114, 006 | 3.17 | 7 | 208, $\bar{1}39$ | 1.665 | 3 |
| $\bar{2}02$, 131 | 2.62 | 4 | 060 | 1.518 | 8 |

The X-ray pattern of the oriented specimen has two strong basal reflexions at spacings 9.72 and 3.18Å and a weak peak at 12Å. After heating at 600° the small peak disappeared, the two other peaks remaining practically unchanged. Saturation with glycerol results in a small displacement of the first reflexion towards greater $\vartheta$ (9.57Å) and in broadening of the reflexion at 3.2-3.3Å. After saturation with ethylene glycol a nonintegral series of basal reflexions appears at 18.1, 9.29, 3.31-3.18Å.

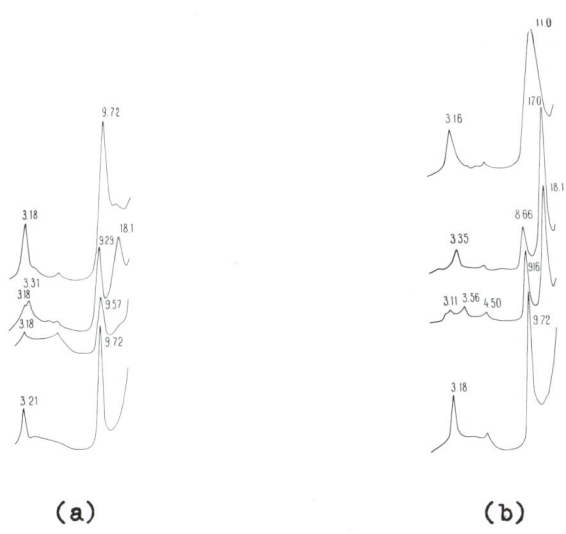

(a)          (b)

Fig. 3. X-ray diffraction patterns of oriented specimens: a) Strassenschacht sample, b) Tologay sample. Order from top to bottom: natural state; saturated with ethylene glycol; saturated with glycerol; heated at 600°.

On the basis of these data one may conclude that the main part of the Strassenschacht sample does not expand with water. Its structure is essentially the same as indicated by electron diffraction. The small quantity of interlayer water is present in a phase which is a one-layer complex ($d_{001}$ 12Å) not exceeding 5% of the substance. The interaction of the interlayers with polar liquids is greater. With glycerol and ethylene glycol about 15-20% and 40-50% respectively of interlayers expand giving a random interstratification of interlayers with and without polar liquids. However, powder hkl reflexions related to the monoclinic 2M structure are not affected by such treatments, showing that some part of the substance is a non-expanding phase.

According to the X-ray data the main part of the Tologay sample is hydrated. The first basal reflexion is at 12Å for the coarser fraction and at 11Å for the finer fraction. With ethylene glycol 80-90% of interlayers expand giving maxima at 17.0, 8.66, 5.56, 3.35Å. Under glycerol, fewer interlayers expand and the basal spacings are 18.1, 9.16, 4.50, 3.56Å. In this case also reflexions at 4.20, 3.18, 2.46Å are seen after all treatments, indicating the presence of a non swelling phase which however, does not exceed 10-20%, otherwise an additional basal reflexion at 9.6Å would be noticed in the X-ray patterns of the sample saturated with ethylene glycol.

SELECTED AREA ELECTRON DIFFRACTION DATA

All particles lying normal to the electron beam gave hexagonal spot diagrams with strong 0$\bar{2}$0 reflexions in accordance with the monoclinic lattice. Comparison with micrographs show that elongation is in the a-direction. Bent edges (Gorshkov, 1970) gave basal reflexions which are characteristic for single particles. It was established that all particles of the Strassenschacht sample give a rational 001 series of a nonexpanded lattice while the majority of particles of the Tologay sample give an irrational 001 series due to a mixed-layer structure. After saturation with ethylene glycol three kinds of basal series have been obtained: a) an integral series from the nonexpanded structure (the first reflexion at 9.6Å) - mainly for the Strassenschacht sample; b) an integral series from a one-layer complex with the first reflexion near 13 Å -

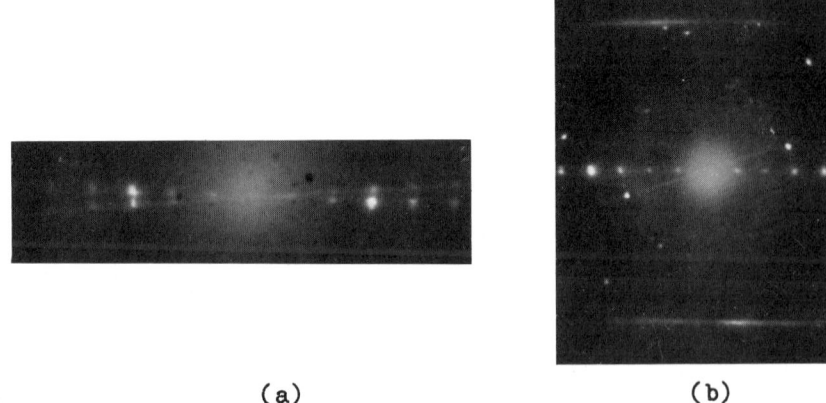

(a)                                    (b)

Fig. 4. SAD-patterns: a) rational 001-series with 001 at 9.6Å; b) rational 001-series with 001 at 13Å. Continuous streaks 201 in b) show that the particles are elongated in the a-direction.

mainly for the Tologay sample; c) a nonintegral series of mixed-layer structures with different ratios of expanded and non-expanded interlayers. It is interesting to note that basal series of different kinds have been met both for different particles and for one and the same particle.

DISCUSSION OF PHASES

The behaviour of the samples following the treatments described indicate the existence of inhomogeneities with different deviations from the ideal structure associated with vacancies and some substitutions of Al for Si accompanied by location of Ca in the interlayers. These inhomogeneities can be localized in different microcrystals or coexist in one and the same particle.

Considering the interaction of the samples with any of the liquids, three phases may be distinguished, one remaining unchanged, the two others forming, respectively, evenly expanded or mixed-layer structures. As the activity of the liquids is different increasing in the order: water - glycerol - ethylene glycol, phases defined according to their interaction with different liquids do not coincide. If combinations of interaction properties with different liquids are considered, the number of phases increases with the number of reagents applied and the distinction between the phases becomes more slight. In the case of two liquids there may be six phases, for three liquids, ten phases. We consider, however, that only the three phases that exist in nature, and that differ in their interaction with water should be accepted as original phases. Any changes of these phases under the action of glycerol and ethylene glycol may be regarded as effects of artificial transformations and the resulting mixed-layer or completely expanded structures as derivative products, like those which arise in the interaction of kaolinite with potassium acetate.

Phase 1 does not expand in water and has the above-mentioned pyrophyllite-like 2M structure. It is a close $Fe^{3+}$ -analogue of pyrophyllite. A name "ferripyrophyllite" is reserved for it until the final decision of the CNM IMA.

Phase 2 has an evenly expanded structure which exhibits a regular alternation of 2:1 - layers with single layers of water molecules and attains the 2M structure of phase 1 after dehydration: this is the principal difference between phase 2 and nontronite, which has two layers of water molecules between the 2:1-layers and usually attains on dehydration a semirandom structure sometimes displaying features of a mica-like 1M polytype (cf. Figs 2a and 2d). The stackings of layers of the mica and pyrophyllite types are crystallochemically and geometrically nonequivalent and the corresponding polytypes belong to independent families. The formation of the different layer stackings of micas and pyrophyllites requires some special although not always detectable features of structure and composition. If $Fe^{3+}$-smectites that have in the dehydrated state mica- and pyrophyllite-like structures really exist one may be sure that they are essentially different. Therefore a name "hydroferripyrophyllite" is reserved for phase 2. Such a special name is necessary to underline its peculiar properties in view of which it should not be confused with nontronite.

Phase 3 is a random interstratification of phases 1 and 2 with the ratio of water interlayers varying between 0 abd 1. It also attains the pyrophyllite-like 2M structure after removal of the interlayer water. According to the existing rules, its name should be a derivative of the names of phases 1 and 2, i.e. "ferripyrophyllite-hydroferripyrophyllite".

COMPOSITION AND OTHER PROPERTIES OF THE PHASES

The sample from Strassenschacht consists mainly of phase 1, but also contains phase 3 with a low ratio of water interlayers and a small admixture of phase 2. The sample from Tologay consists mainly of phase 2, but also contains phase 3 with a high ratio of water interlayers and a small admixture of phase 1. It may be accepted therefore that the properties of phase 1 (ferripyrophyllite) and phase 2 (hydroferripyrophyllite) are approximated respectively by the properties of the two samples. The properties of the mixed-layer phase are intermediate depending on the ratio of interlayers.

These conclusions, deduced from all the diffraction data allowed a better understanding of the results given by other methods.

Recalculating the bulk chemical analyses the following averaged structural formulas have been obtained.

Strassenschacht sample:

$Ca_{0.05} Fe^{3+}_{1.96} Mg_{0.11} (Si_{3.80} Al_{0.13} Fe^{3+}_{0.07} O_{10})(OH)_2 \cdot H_2O$

Tologay sample:

$Ca_{0.18} (Na,K)_{0.03} Fe^{3+}_{1.97} Mg_{0.02} (Si_{3.74} Al_{0.23} O_{10})(OH)_2 \cdot 1.5 H_2O$

The deviation of these formulas from the ideal are not great and may be attributed to inhomogeneities and to the presence of different phases. The differences of the formulas in the quantities of Al, Ca, $H_2O$ are in agreement with the distribution of phases in the samples. Accordingly, the ideal formula $Fe_2Si_4O_{10}(OH)_2$ may be accepted for phase 1 (ferripyrophyllite) in analogy with $Al_2Si_4O_{10}(OH)_2$ for the usual pyrophyllite. Phase 2 (hydroferripyrophyllite) may be described by an approximate formula $Ca_{0.2}Fe_{1.95}Si_{3.75}Al_{0.25}O_{10}(OH)_2 \cdot (1.5-2.0)H_2O$. The degree of replacement of Si by Al in this phase is less than in nontronites; although not great this difference is evidently essential for the swelling properties and structural ordering after removal of interlayer water.

In view of the structural data the thermal behaviour of the samples becomes clear. The endothermic effects are due to the loss of interlayer $H_2O$ and structural (OH). The shapes of the DTA- and TG-curves indicate there is no a sharp energetic boundary between the two kinds of water, presumably as a result of structural imperfections. The lower temperature of the dehydroxylation, compared with those of pyrophyllite and talc (i.e. 500° instead of 800° and 1000° resp. Mackenzie, 1957) is a consequence of the composition and lesser stability of the structure.

The IR spectra of both specimens are consistent with the proposed structures, and clearly distinguish these phases from nontronite.

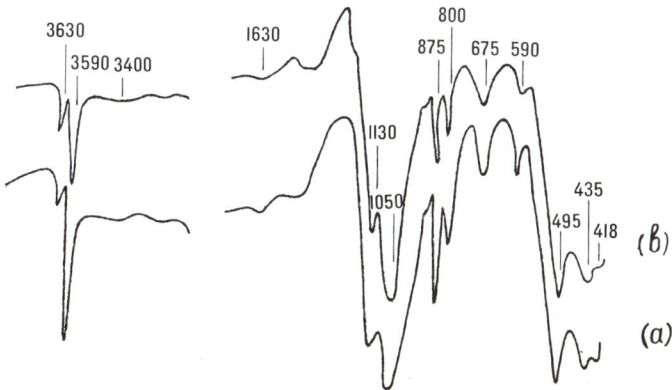

Fig. 5. IR-spectra: a) Strassenschacht sample; b) Tologay sample.

Most obviously, their OH stretching ($3590$ cm$^{-1}$) and bending ($842$ cm$^{-1}$) vibrations lie at higher frequencies than those of nontronite ($3570$ and $818$ cm$^{-1}$). The same relationship exists between these vibrations in pyrophyllite ($3675$ and $950$ cm$^{-1}$) and those in beidellite ($3661$ and $940$ cm$^{-1}$) or montmorillonite ($3630$ and $915$ cm$^{-1}$.

The formation of phase 1 (ferripyrophyllite) is possible in media with a high content of $Fe^{3+}$. If nontronite is a mineral of the hypergenesis zone, the $Fe^{3+}$-analogue of pyrophyllite is a precipitate from low-temperature hydrothermal solutions. The swelling phases are likely to be formed as a result of later transformations.

## ACKNOWLEDGEMENTS

The authors are indebted to Dr. V.C. Farmer for his crytical comments and recommendations especially concerning the IR-data.

## REFERENCES

Drits V.A. 1969. Proceedings of the Intern. Clay Conf., Tokyo, September 5-10. Israel Universities Press, Jerusalem, pp.51-59.
Farmer V.C. 1974. Infrared Spectra of Minerals, London.
Gorshkov A.I. 1970. Izvestiya AN SSSR, ser. geol., N 3.(in Russian).
Mackenzie R. (Edit.). 1957. The Differential Thermal Investigations of Clays, London.
Smolin P.P., Zvyagin B.B., Drits V.A., Sidorenko O.V., Alexandrova V.A. 1975. Crystallochimiya mineralov i geologitcheskie problemy, Nauka, Moscow, pp. 91-99. (in Russian).
Zvyagin B.B., Mishtshenko C.S. and Soboleva S.V. 1968. Crystallographiya, 13, N 4. (in Russian).

## ADDENDUM

The mineral name ferripyrophyllite has been approved by the IMA Commission on New Minerals and Mineral Names in September 1978.

MÖSSBAUER SPECTRA OF CHLORITES AND THEIR DECOMPOSITION PRODUCTS

B.A. GOODMAN and D.C. BAIN

The Macaulay Institute for Soil Research, Craigiebuckler, Aberdeen AB9 2QJ, (Scotland)

ABSTRACT

Nine trioctahedral chlorite specimens, having a wide range of compositions, have been investigated by $^{57}$Fe Mössbauer spectroscopy. In all samples the major oxidation state is ferrous with a principal component that has parameters similar to those from $Fe^{2+}$ at the site with cis OH groups in biotite. With all spectra an additional ferrous component was required for a statistically-acceptable fit to the data, but this does not necessarily correspond to a different crystallographic site. Ferric iron was also present in all samples and some spectra contained a component with isomer shift characteristic of 4-coordination in addition to the 6-coordinate component more generally observed.

Five specimens were heated to various temperatures and also subjected to the chemical treatment of Ross and Kodama (1974) with the object of forming vermiculite. Spectra were recorded at all stages and the samples exhibited a behaviour that apparently depended on the iron content of the original sample. The low-iron chlorites could be completely oxidized and partially vermiculitized but there was no evidence for the production of vermiculite with the iron-rich species.

---

INTRODUCTION

Chlorites have a 2:1 layer structure with an interlayer hydroxide sheet and general formula $(M^{2+}_{6-x} M^{3+}_{x})(Si_{4-x} M^{3+}_{x})O_{10}(OH)_{8}$. Because of isomorphous substitution within the hydroxide sheet and in the tetrahedral and octahedral positions in the 2:1 layer, a large number of species exists and a wide variation in chemical composition is encountered. Various attempts have been made by X-ray diffraction (XRD) techniques to estimate the chemical composition of chlorites and to determine tetrahedral and octahedral cation populations and, although detailed information can be obtained from a comprehensive single crystal study (e.g. Bailey and Riley, 1977), methods applicable to powder specimens (Bailey, 1972) give only an indication of the difference between the cation distributions in the two octahedral sheets and cannot provide precise information on the positions of the octahedral cations.

Although limited to the study of iron, Mössbauer spectroscopy is able to provide detailed information on its oxidation states and, through the magnitudes of the isomer shift, $\delta$, and the quadrupole splitting, $\Delta$, the types of coordination in which the iron is involved. The technique has been successfully applied to the study of several silicate minerals (Bancroft, 1973 Chapter 7) but only a few studies of chlorites have been published (Taylor et al., 1968; Hayashi et al., 1972); although the spectra obtained were not of a sufficiently high standard to permit detailed computer fitting most of the iron was reported to be in the ferrous form. Ferric iron was, however, also present and Taylor et al. found one specimen in which the ferric iron could be assigned partly to octahedral and partly to tetrahedral sites. Hayashi et al. also examined the effects of heat treatment on the Mössbauer spectra and found the amount of ferric iron to increase with increasing temperature.

The present work was undertaken with the object of gaining further information on the distribution of iron in the chlorite minerals by studying (a) specimens having a wide range of iron contents, (b) the effects of heat treatment and (c) the effects of treatment of specimens with HCl after pre-heating to various temperatures. This latter treatment has been reported by Ross and Kodama (1974) to lead to the loss of the hydroxide sheet and the formation of vermiculite if the chlorite had been heated beyond the temperature of its major DTA endotherm.

MATERIALS AND METHODS

The specimens used in this study are listed in Table 1. Ferrous iron was determined using a modification of the method of Wilson (1960) and ferric iron calculated by difference from the total iron subsequently determined on the same solution after reduction with ascorbic acid. The other oxides were determined by X-ray fluorescence spectrometry using the fusion method of Norrish and Hutton (1969) and a Philips PW 1540 spectrometer. The analysis of sample 8 is taken from Sudo (1943). X-ray diffraction patterns were recorded on a Philips 2 kW diffractometer and a 114.83 mm diameter Debye-Scherrer camera using iron-filtered CoK$\alpha$ radiation. The specimens are all trioctahedral chlorites with II b polytype structure except for specimen 8 which has a I b (90°) structure. Using the nomenclature of Bayliss (1975), samples 1-6 are varieties of clinochlore and 7-9 varieties of chamosite, samples 8 and 9 containing significant amounts of manganese. Only minor impurities have been found e.g. calcite in 1, gypsum in 5, quartz in 9 and anatase and calcite in 7 (the calcite was removed by mild acid treatment).

Vermiculitization of several samples was attempted by the method of Ross and Kodama (1974), whereby the sample is heated to dehydroxylate the hydroxide sheet, digested in 0.2 N HCl and then saturated with $Na^+$. The specimens were ground under alcohol or acetone to <10 μm and then heated at a rate of 10K per minute in a nitrogen atmosphere in a Du Pont 900 Differential Thermal Analyzer

TABLE 1

Chemical composition of chlorite specimens

| Sample | $Fe_2O_3$ | FeO | $SiO_2$ | $Al_2O_3$ | MnO | CaO | MgO | $K_2O$ | $TiO_2$ | L.O.I. at 1000C | Total |
|---|---|---|---|---|---|---|---|---|---|---|---|
| 1 | 1.05 | 2.17 | 31.0 | 18.2 | 0.10 | 1.77 | 31.0 | <0.01 | 0.05 | 14.7 | 100.04 |
| 2 | 1.75 | 1.63 | | | | | | | | | |
| 3 | 1.37 | 2.59 | | | | | | | | | |
| 4 | 2.21 | 2.28 | | | | | | | | | |
| 5 | 1.06 | 3.32 | 30.6 | 21.2 | - | 0.35 | 30.6 | <0.01 | 0.27 | 12.7 | 100.10 |
| 6 | 2.45 | 3.95 | 31.9 | 12.9 | 0.07 | 0.04 | 34.0 | 0.03 | <0.01 | 16.4 | 101.74 |
| 7 | 7.60 | 18.98 | | | | | | | | | |
| 8 | 7.45 | 31.46 | 24.50 | 16.32 | 3.33 | - | 4.59 | - | - | 11.36 | 99.01 |
| 9 | 1.61 | 34.8 | 28.6 | 18.3 | 1.25 | 0.04 | 4.49 | 0.11 | 0.02 | 10.1 | 99.32 |

Details of origin of samples:
1. Clinochlore, Ala, Piedmont. Spec. No. Ludlam 7903, I.G.S., London.
2. Clinochlore, West Chester, Pennsylvania. Spec. No. M.I. 23110. I.G.S., London.
3. Penninite, Gallenstock, Zermatt. Spec. No. 1885. 30. 6, Royal Scottish Museum.
4. Penninite, Rimpfischwange, Zermatt. Spec. No. 1924. 4. 16, Royal Scottish Museum.
5. Magnesian chlorite, Wanibuchi Mine, Shimane Prefecture, Japan.
6. Clinochlore, unknown locality. Spec. No. 799, Royal Scottish Museum.
7. Thuringite, Kragero, Telemark, Norway. Spec. No. 83514, British Museum.
8. Chamosite, Arakawa Mine, Skita Prefecture, Japan.
9. Daphnite, Wall, Redruth, Cornwall. Spec. No. 1937, 1568, British Museum.

to temperatures just below, at the peak of, and just above the endotherm produced by dehydroxylation of the hydroxide sheet. Vermiculite was indicated by an XRD peak at about 12.4 Å (room humidity 35-55%), moving to 14.9 Å when moistened.

Mössbauer spectra at room temperature were obtained with a conventional constant acceleration spectrometer (Harwell Scientific Services, Didcot, Berks), employing an Ortec Model 6200 1024-channel analyzer operating in the multiscalar mode. A $^{57}Co$ in Pd source of nominal strength 25mCi was used together with an argon-methane proportional counter as γ-ray detector. Velocity calibration was carried out with high purity metallic iron foil using the data of Preston et al. (1962). The spectra, obtained from 512 channels of the analyzer, were fitted by means of a least squares computer programme to a sum of doublets having Lorentzian line shapes. The peaks of each doublet were constrained to have equal areas and widths and a parabolic baseline was assumed for all spectra. $\chi^2$ was used as a goodness-of-fit parameter so that for a fit to be acceptable $\chi^2$ must lie between 410 and 555 for 480 degrees of freedom, the number of degrees of freedom being equivalent to the number of channels minus the number of variables being fitted. In order to minimize thickness effects the amounts of sample were adjusted to contain 3 mg Fe per $cm^2$. For some iron-rich specimens it was necessary to mix the sample with a matrix of low atomic weight atoms (sucrose was used for ease of removal at the end of the experiment) in order to ensure an even thickness of the specimens.

RESULTS

Untreated specimens

TABLE 2

Computed Mössbauer parameters for untreated chlorites

| Sample Number | Fe(II) | | | | | | | | Fe(III) | | | | | | | | χ |
|---|---|---|---|---|---|---|---|---|---|---|---|---|---|---|---|---|---|
| | Δ | δ | Γ | % | Δ | δ | Γ | % | Δ | δ | Γ | % | Δ | δ | Γ | % | |
| 1 | 2.69 | 1.13 | 0.28 | 40 | 2.46 | 1.13 | 0.41 | 21 | 0.57 | 0.39 | 0.24 | 15 | 0.82 | 0.34 | 0.76 | 24 | 413 |
| 2 | 2.68 | 1.13 | 0.28 | 39 | 2.30 | 1.15 | 0.30 | 11 | 0.78 | 0.28 | 0.78 | 50 | - | - | - | - | 526 |
| 3 | 2.72 | 1.13 | 0.29 | 36 | 2.38 | 1.15 | 0.38 | 17 | 0.53 | 0.27 | 0.71 | 42 | 1.40 | 0.40 | 0.80 | 5 | 563 |
| 4 | 2.72 | 1.13 | 0.26 | 36 | 2.41 | 1.14 | 0.34 | 20 | 0.70 | 0.25 | 0.79 | 44 | - | - | - | - | 596 |
| 5 | 2.70 | 1.13 | 0.25 | 41 | 2.51 | 1.13 | 0.36 | 51 | 0.58 | 0.32 | 0.27 | 8 | - | - | - | - | 591 |
| 6 | 2.71 | 1.13 | 0.28 | 43 | 2.37 | 1.14 | 0.34 | 19 | 0.66 | 0.24 | 0.75 | 39 | - | - | - | - | 515 |
| 7 | 2.67 | 1.14 | 0.35 | 44 | 2.39 | 1.15 | 0.55 | 33 | - | - | - | - | 1.10 | 0.36 | 0.85 | 23 | 410 |
| 8 | 2.71 | 1.13 | 0.28 | 57 | 2.42 | 1.12 | 0.39 | 28 | 0.61 | 0.37 | 0.41 | 9 | 1.07 | 0.47 | 0.61 | 7 | 594 |
| 9 | 2.69 | 1.13 | 0.28 | 63 | 2.35 | 1.12 | 0.42 | 17 | 0.72 | 0.26 | 0.61 | 11 | 1.06 | 0.37 | 0.52 | 9 | 617 |

All values in mm s$^{-1}$. Isomer shifts relative to iron metal.

Representative spectra of the chlorites are illustrated in Fig. 1 and the computed parameters in Table 2. These results show that for all specimens at least 50% of the total iron is in the ferrous state. The parameters for the major components are similar for all specimens and resemble those reported for octahedral coordination with OH groups in a cis arrangement in biotite (Goodman and Wilson, 1973; Annersten, 1974; Bancroft and Brown, 1975). The minor ferrous components, however, have somewhat larger values of Δ than do those in biotites. It is tempting to assign the major ferrous component to the 2:1 layer but, in the absence of any knowledge of the possible values from the hydroxide sheet, this would be unwise. Nevertheless, if ferrous iron is present in the hydroxide sheet, its Mössbauer parameters must be similar to those for ferrous in the octahedral sites in the 2:1 layer. In contrast, the parameters for the ferric

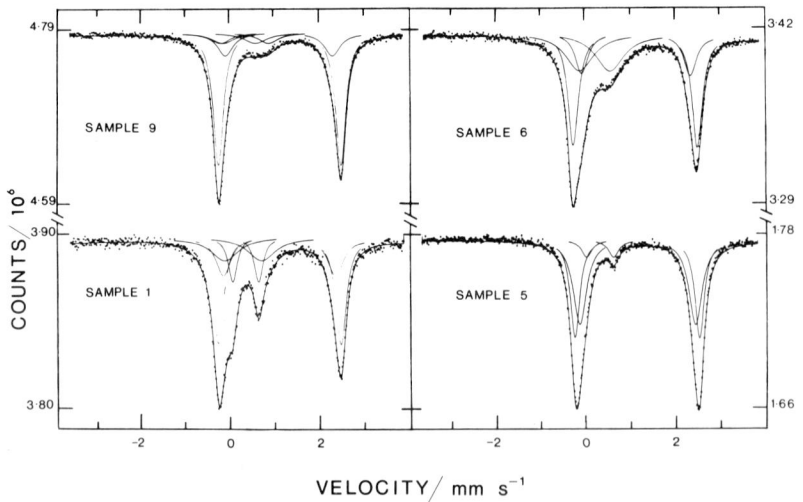

Fig. 1. Mössbauer spectra of representative chlorite specimens.

components vary appreciably through the series of samples, several of which have a component with δ appreciably lower than the values of 0.34-0.39 mm s$^{-1}$ (relative to Fe metal), which have been reported for room temperature spectra of micas containing appreciable quantitites of ferric iron in octahedral sites (Goodman, 1976). Since there is a direct relationship between coordination number and δ, those components with low δ must arise, at least in part, from the presence of tetrahedrally-coordinated ferric iron. This result is surprising, since all of the samples (for which complete analyses were obtained, Table 1) have more than enough aluminium to fill the sites in the tetrahedral layer not occupied by silicon. Thus even in samples with low ferric and high aluminium contents, iron can occupy tetrahedral sites, although it should be pointed out that there is appreciably more aluminium than ferric iron in the tetrahedral layers of all these specimens. The components with δ in the range characteristic of octahedral coordination have Δ values similar to those reported for biotites. The minor ferric component in the spectrum of specimen 8 also has a larger value of δ than any ferric component in the other spectra but, because of its low intensity, its computed parameters must be treated with some reservation.

Heated specimens

The X-ray diffraction patterns of all samples after dehydroxylation of the hydroxide sheet exhibit the decrease in the c-dimension and increase in the intensities of the odd-order reflections relative to the even-order reflections characteristic of chlorites. The b-dimensions of the iron-rich samples 7 and 9 decrease in a manner commensurate with some oxidation of the ferrous iron (Shirozu, 1962), whereas in the magnesian samples 1, 5 and 6 they increase (Brindley, 1961). The effects of heat treatment on the Mössbauer spectra will be considered separately for the low and high-iron specimens.

Low-iron chlorites. The 3 specimens selected for heat treatment were chosen because of the differences in their original spectra. Specimen 5 contained little ferric iron, whereas 1 and 6 had approximately 40% of their iron in this form. The ferric iron in specimen 1 had isomer shifts consistent with 6-coordination, but in specimen 6, δ was much lower and, as stated above, corresponds to predominantly tetrahedral coordination. The results from the heat-treated samples are given in Table 3 and representative spectra from two are illustrated in Fig. 2.

In each case heating in $N_2$ to just below the DTA endotherm resulted in some oxidation of the iron, although this had little effect on the Mössbauer parameters of the remaining ferrous iron or on the 060 spacings. There was also little change in the ferric parameters for specimen 1, but for the other two specimens, the ferric iron produced as a result of heat treatment had much larger values of Δ than any of the original ferric components. Heating to the peak of the DTA endotherm led in each case to complete oxidation, a large percentage of the ferric iron having a much larger value of Δ than in the original chlorites. The

TABLE 3

Computed Mössbauer parameters for treated chlorites

| Sample Number | Treatment | Fe(II) | | | | | | Fe(III) | | | | | | $\chi^2$ |
|---|---|---|---|---|---|---|---|---|---|---|---|---|---|---|
| | | $\Delta$ | $\delta$ | $\Gamma$ | % | | | $\Delta$ | $\delta$ | $\Gamma$ | % | $\Delta$ | $\delta$ | $\Gamma$ | % | |
| 9 | 415C in $N_2$ | 2.69 | 1.14 | 0.36 | 81 | – | – | – | 0.79 | 0.35 | 0.62 | 19 | – | – | – | – | 437 |
| | 585C in $N_2$ | 2.46 | 1.06 | 0.50 | 29 | 1.96 | 0.98 | 0.67 | 46 | 0.94 | 0.53 | 0.81 | 25 | – | – | – | – | 522 |
| | 415C in $N_2$, then HCl | 2.59 | 1.16 | 0.38 | 34 | – | – | – | – | – | – | – | – | 1.21 | 0.33 | 0.70 | 66 | 555 |
| | 585C in $N_2$, then HCl | 2.46 | 1.07 | 0.46 | 26 | 1.97 | 0.99 | 0.66 | 48 | 0.92 | 0.49 | 0.85 | 26 | – | – | – | – | 541 |
| | 610C in $O_2$ | – | – | – | – | – | – | – | – | 0.84 | 0.34 | 0.40 | 31 | 1.39 | 0.34 | 0.67 | 69 | 922 |
| | 610C in $O_2$, then HCl | – | – | – | – | – | – | – | – | 0.78 | 0.34 | 0.39 | 22 | 1.31 | 0.35 | 0.76 | 78 | 504 |
| 7 | 610C in $N_2$ | 2.57 | 1.08 | 0.41 | 21 | 2.07 | 1.00 | 0.64 | 39 | 1.08 | 0.40 | 0.72 | 39 | – | – | – | – | 541 |
| | 610C in $N_2$, then HCl | 2.52 | 1.11 | 0.46 | 18 | 2.07 | 1.01 | 0.77 | 41 | 1.10 | 0.37 | 0.80 | 42 | – | – | – | – | 429 |
| | 610C in $O_2$ | – | – | – | – | – | – | – | – | 1.24 | 0.35 | 0.73 | 100 | – | – | – | – | 649 |
| | 610C in $O_2$, then HCl | – | – | – | – | – | – | – | – | 0.94 | 0.36 | 0.54 | 42 | 1.53 | 0.34 | 0.75 | 58 | 354 |
| 6 | 510C in $N_2$ | 2.64 | 1.13 | 0.37 | 36 | – | – | – | – | 0.57 | 0.24 | 0.63 | 19 | 1.37 | 0.40 | 0.85 | 45 | 474 |
| | 650C in $N_2$ | – | – | – | – | – | – | – | – | 0.92 | 0.32 | 0.55 | 32 | 1.51 | 0.29 | 0.81 | 68 | 540 |
| | 510C in $N_2$, then HCl | 2.69 | 1.13 | 0.27 | 30 | 2.33 | 1.14 | 0.29 | 8 | 0.71 | 0.15 | 0.69 | 26 | 1.36 | 0.42 | 0.76 | 36 | 933 |
| | 650C in $N_2$, then HCl | – | – | – | – | – | – | – | – | 1.08 | 0.33 | 0.82 | 100 | – | – | – | – | 389 |
| 5 | 502C in $N_2$ | 2.68 | 1.13 | 0.26 | 51 | 2.43 | 1.14 | 0.39 | 33 | 1.36 | 0.35 | 0.86 | 17 | – | – | – | – | 492 |
| | 624C in $N_2$ | – | – | – | – | – | – | – | – | 1.11 | 0.34 | 0.54 | 43 | 1.69 | 0.33 | 0.77 | 57 | 804 |
| | 725C in $N_2$ | – | – | – | – | – | – | – | – | 1.08 | 0.36 | 0.50 | 38 | 1.57 | 0.36 | 0.73 | 62 | 486 |
| | 502C in $N_2$, then HCl | 2.65 | 1.13 | 0.27 | 63 | 2.28 | 1.08 | 0.30 | 12 | 1.05 | 0.32 | 0.89 | 25 | – | – | – | – | 706 |
| | 624C in $N_2$, then HCl | – | – | – | – | – | – | – | – | 1.19 | 0.35 | 0.71 | 100 | – | – | – | – | 519 |
| | 725C in $N_2$, then HCl | – | – | – | – | – | – | – | – | 1.19 | 0.36 | 0.68 | 100 | – | – | – | – | 442 |
| 1 | 510C in $N_2$ | 2.67 | 1.11 | 0.33 | 51 | – | – | – | – | 0.70 | 0.37 | 0.50 | 49 | – | – | – | – | 612 |
| | 655C in $N_2$ | – | – | – | – | – | – | – | – | 0.60 | 0.38 | 0.27 | 20 | 1.27 | 0.30 | 0.97 | 80 | 401 |
| | 655C in $N_2$, then HCl | – | – | – | – | – | – | – | – | 0.59 | 0.39 | 0.24 | 24 | 0.99 | 0.35 | 0.83 | 76 | 400 |

All values in mm s$^{-1}$. Isomer shifts relative to iron metal.

isomer shifts were slightly lower than the typical values for 6-coordination, thus indicating that there might be a small contribution to the spectra from components with a lower coordination number.

Heating to just below the principal endotherm followed by the acid-treatment of Ross and Kodama (1974) produced little change in either the XRD patterns or Mössbauer spectra of the chlorites (Table 3, Fig. 2). However the same treatment applied to the chlorites which had been heated beyond this endotherm resulted in some conversion to vermiculite, as indicated by an XRD peak at 14.9 Å. The Mössbauer results indicate that at least part of the ferric iron with the larger values of Δ was removed in each case as a result of the acid treatment. The Quadrupole splittings of the remaining ferric iron are similar to those observed by Goodman and Wilson (1973) for interstratified chlorite-vermiculite derived from biotite by natural weathering. Moreover, since the isomer shifts are consistent with 6-coordination the HCl treatment must have removed or caused the rearrangement of that iron having lower coordination number in the original specimen 6.

High-iron chlorites. As with the low-iron chlorites there was little change in the iron-rich specimens as a result of heating in $N_2$ to just below the major DTA endotherm. (Table 3). On further heating, however, there was only a slight increase in the amount of ferric iron (Fig. 3a), in contrast to the low-

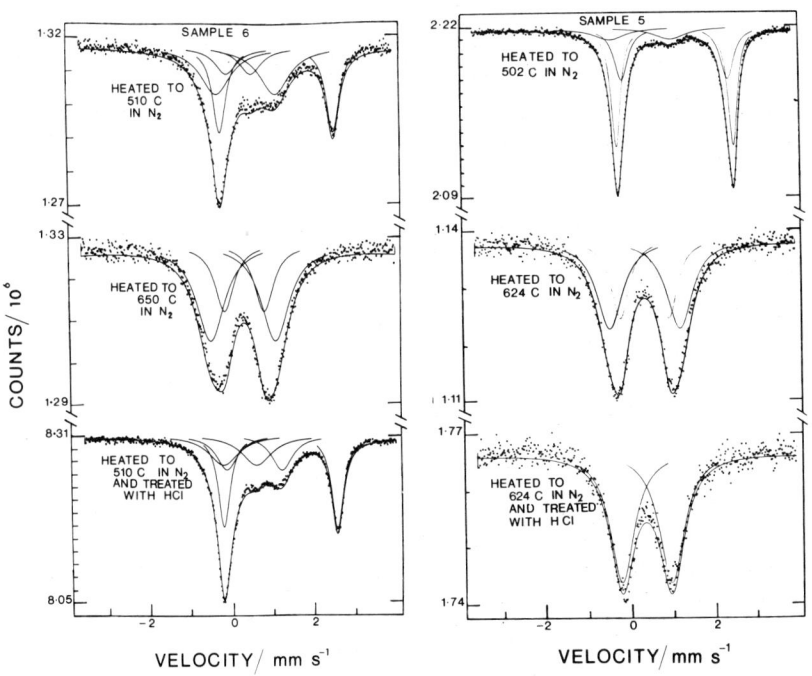

Fig. 2. Mössbauer spectra of low iron chlorites subjected to various treatments.

iron specimens where oxidation was complete. The parameters for both ferrous and ferric components are quite different from those found in the original specimens. The envelopes of ferrous peaks are much broader and in each case contain at least 2 doublets, with $\delta$ decreasing with decreasing $\Delta$, which is the behaviour expected if decreases in coordination number are accompanied by increases in asymmetry at the ferrous-containing sites. The broad ferric peaks indicate that this ion also occurs in a range of environments and the isomer shifts are greatly increased from their previous values. An explanation for this is not immediately obvious, however, since dehydroxylation of the lattice would be expected to reduce the coordination number at cation sites and, therefore, decrease the values of $\delta$. Sample 9 also became poorly diffracting to X-rays and an accurate measure of the 060 spacing could not be obtained. Treatment of these heated samples with HCl did not induce vermiculitization and produced no observable changes in the Mössbauer spectra either in the amounts of, or parameters for, any of the components. However, the structure has been disrupted to some extent because although no discrete peak for vermiculite appears on the XRD pattern, a broad 13.7 Å peak with a shoulder at 13.0 Å is present at 43% R.H.

When these chlorites were heated in oxygen to the peak of, or above, the DTA endotherm, complete oxidation occurred (e.g. Fig. 3b, Table 3), all of the iron being found in sites of 6-coordination with various levels of distortion, and treatment with HCl produced no significant changes in either the Mössbauer spectra or XRD patterns.

<u>Chemical Treatment</u>. Analyses of the HCl extracts from specimens 5, 6 and 9 heated to beyond their endotherm show that 27, 41 and 29%, respectively, of the total iron was removed. For samples 5 and 6 which have low iron contents, this represents only a small proportion of the total cations but it may be of interest that sample 6 was one that had an appreciable fraction of its iron in tetrahedral sites before treatment. With sample 9 a considerable proportion of the total cations are removed and this evidently leads to complete disruption of the structure, since the sample becomes poorly diffracting and the Mössbauer parameters for the

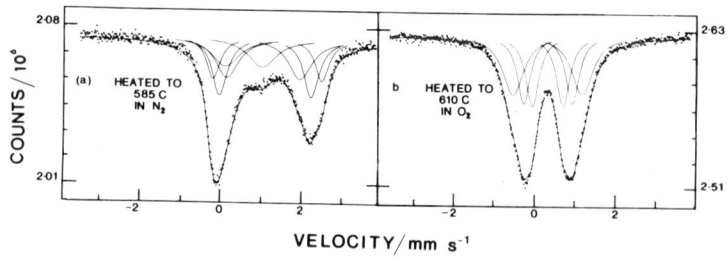

Fig. 3. Mössbauer spectra of high-iron chlorite (Sample 9) heated in $N_2$ and $O_2$ atmosphere.

remaining iron are very different from those for samples treated to just below the endotherm.

DISCUSSION

It must be considered whether these reuults on heat-treated chlorites can provide any further information on the site populations of iron in the original chlorites. The iron-rich samples undergo only partial oxidation on heating in $N_2$ to the temperature of dehydroxylation of the hydroxide sheet, whereas the low-iron specimens are completely oxidized. Since the former specimens must have contained ferrous iron in both the hydroxide sheet and the 2:1 layer (Table 1), it seems reasonable to conclude that the iron not oxidized on heating, is in the 2:1 layer and that the ferrous iron in the low-iron chlorites must have all been in the hydroxide sheet. However, this conclusion is not supported by the analytical results, since specimen 5, which was almost completely converted to vermiculite as a result of the Ross and Kodama treatment, lost only 27% of its iron on the extraction after heating beyond its major endotherm. If allowance is made for the fact that some attack on the 2:1 layer by HCl may also have occurred, then appreciably less than 50% of the ferrous iron can have been in the hydroxide sheet. Thus, some oxidation of ferrous iron in the 2:1 layer must occur on heating in $N_2$, presumably via an internal dehydrogenation reaction, $2Fe^{2+} + 2OH^- \rightarrow 2Fe^{3+} + 2O^{2-} + H_2$, although this does not occur extensively in the iron-rich specimens.

CONCLUSIONS

(i) Iron is predominantly in the ferrous form and has similar Mössbauer parameters for all of the samples investigated.
(ii) Parameters for the ferric components are more varied, some specimens having an appreciable proportion in sites with 4-coordination.
(iii) No components can be specifically assigned to the hydroxide sheets although any ferrous iron occurring there must have parameters similar to that in the 2:1 layer.
(iv) Heat treatment of high-iron specimens in a $N_2$ atmosphere results in little oxidation of the iron, although in $O_2$ complete oxidation occurs.
(v) There is no tendency for the high-iron specimens to form vermiculite as a result of heat treatment followed by acid digestion, although the structure of the highest iron sample is disrupted.
(vi) Heat treatment of the low-iron specimens in a $N_2$ atmosphere leads to complete oxidation of the iron at the dehydroxylation temperature, part of this ferric iron being removed by acid digestion. XRD shows that there is partial conversion to vermiculite as a result of this treatment.

ACKNOWLEDGEMENTS

The authors are grateful to the following for supplying specimens: Dr. H.G. Macpherson of the Royal Scottish Museum, Mr. B.R. Young of the Institute of Geological Sciences, Professor S.W. Bailey and Professor H. Hayashi.

REFERENCES

Annersten, H., 1974. Mössbauer studies of natural biotites. Am. Mineral., 59: 143-151.
Bailey, S.W., 1972. Determination of chlorite compositions by X-ray spacings and intensities. Clays Clay Miner. 20: 381-388.
Bailey, S.W. and Riley, J.F., 1977. An unusual chlorite from Western Australia. Mineral Mag., 41: 541-544.
Bancroft, G.M., 1973. Mössbauer Spectroscopy - An Introduction for Inorganic Chemists and Geochemists. McGraw Hill, London, 252pp.
Bancroft, G.M. and Brown, J.R., 1975. A Mössbauer study of coexisting hornblendes and biotites: Quantitative $Fe^{3+}/Fe^{2+}$ ratios. Am. Mineral., 60: 265-272.
Bayliss, R., 1975. Nomenclature of the trioctahedral chlorites. Can. Mineral., 13: 178-190.
Brindley, G.W., 1961. In: G. Brown (Editor), The X-ray Identification and Crystal Structure of Clay Minerals. Mineralogical Society, London, pp51-131.
Goodman, B.A., 1976. The Mössbauer spectrum of a ferrian muscovite and its implications in the assignment of sites in dioctahedral micas. Mineral. Mag., 40: 513-517.
Goodman, B.A. and Wilson, M.J., 1973. A study of the weathering of biotite using the Mössbauer effect. Mineral Mag., 39: 448-454.
Hayashi, H., Sano, H., and Shirozu, H., 1972. Mössbauer spectra of chlorites in natural and heated states. Kobutsugaku Zasshi, 10: 507-516.
Norrish, K. and Hutton, J.T., 1969. An accurate X-ray spectrographic method for the analysis of a wide range of geological samples. Geochim. Cosmochim. Acta, 33: 431-453.
Preston, R.S., Hanna, S.S., and Heberle, J., 1962. Mössbauer effect in metallic iron. Phys. Rev., 128: 2207-2218.
Ross, G.J. and Kodama, H., 1974. Experimental transformation of a chlorite into vermiculite. Clays Clay Miner., 22: 205-211.
Shirozu, H., 1962. Thermal reaction of iron chlorites. Clay Sci., 1: 108-113.
Sudo, T., 1943. On some low temperature hydrous silicates found in Japan. Bull. Chem. Soc. Jpn., 18: 281-329.
Taylor, G.L., Ruotsala, A.P., and Keeling, R.O., 1968. Analysis of iron in layer silicates by Mössbauer spectroscopy. Clays Clay Miner., 16: 381-391.
Wilson, A.D., 1960. The micro-determination of ferrous iron in silicate minerals by a volumetric and a calorimetric method. Analyst, 85: 823-827.

EFFECT OF STRUCTURAL $Fe^{2+}$ ON VISIBLE ABSORPTION SPECTRA OF NONTRONITE SUSPENSIONS

W.L. ANDERSON and J.W. STUCKI[*]
Department of Agronomy, Univ. of Illinois, Urbana, IL, 61801 USA

ABSTRACT

Color changes in nontronite suspensions due to chemical reduction of structural $Fe^{3+}$ were measured by visible spectroscopy. The most prominent change in the visible spectrum as $Fe^{2+}$ increases is the appearance of a broad absorbance band with a maximum initially at 760 nm. Continued reduction of the clay enhances the intensity of this band, and the peak maximum shifts to 725 nm when $Fe^{2+}$ concentrations exceed 60 mmoles/100 g clay. Multiple linear regression analysis of absorbance at 760 nm versus structural $Fe^{2+}$ content revealed two different linear regions of the curve. The inflexion point separating these linear segments of the curve was located at a $Fe^{2+}$ concentration of 53 mmoles/100 g clay. Correlation coefficients for the upper and lower segments were 0.97 and 0.98, respectively.

A possible explanation for the multiple linear segments of the visible absorbance versus $Fe^{2+}$ plot is that the coordination environment surrounding octahedral iron becomes altered significantly when the $Fe^{2+}$ concentration in the clay mineral reaches the vicinity of the inflexion point. This explanation is consistent with previously published information dealing with redox mechanisms for structural iron in dioctahedral smectites.

INTRODUCTION

The color of nontronite is naturally yellow but changes to green, blue-green, or black when treated with chemical reducing agents, which alter the oxidation state and possibly the site symmetry of structural iron. The particular color obtained when the clay is reduced depends on a number of factors including choice of reducing agent (Rozenson and Heller-Kallai, 1976), the amount of structural $Fe^{2+}$ produced, and perhaps other experimental conditions such as pH and presence of certain chelating agents (Stucki and Roth, 1977).

---

[*]Graduate Research Assistant and Asst. Prof. of Soil Chemistry, respectively.

The color associated with gley formation in submerged soils had been measured by visible reflectance spectroscopy and related to $Fe^{2+}$ activity in the soil (Yamanaka and Motomura, 1959). Karickhoff and Bailey (1973) demonstrated that the uv-visible spectrum of several clay minerals can be obtained either in suspension or dry state using transmission methods, and they identified bands that arise from charge transfer and intraconfigurational transitions for structural iron in the unaltered clays. However, no attempt was made to observe spectral differences during changes in oxidation state. The objective of this work was to determine spectroscopically if color changes in nontronite could be used to a) measure quantitatively in situ the structural $Fe^{2+}$ content of aqueous suspensions of nontronite, and b) monitor possible changes in the electronic and coordination configuration of the iron as it undergoes reduction.

MATERIALS AND METHODS

Studies were conducted using nontronite (Garfield, WA, API H-33a) which had been fractionated to <2μm, Na-saturated, dialyzed, oven-dried and stored in a desiccator over drierite. Suspensions of known concentration were prepared by sonifying (ca. 75 W for 2 min.) a weighed portion of the clay in either C-B buffer solution (prepared as 0.03 $\underline{M}$ $Na_3C_6H_5O_7$ and 0.12 $\underline{N}$ $HCO_3$) or doubly distilled $H_2O$. Reduced suspensions were obtained by adding weighed quantities of solid sodium dithionite ($Na_2S_2O_4$) to C-B suspensions and heating to 75 C for 15 minutes. The tubes were then stoppered with serum stoppers through which access was by syringe needles. The extent of reduction was controlled qualitatively by the amount of dithionite added. Unheated samples containing larger quantities of dithionite required ca. 24 hours to reach the same color as samples heated for 15 minutes. Dithionite treatment turned $H_2O$ suspensions black with <5 minutes of heating. The black color formed as a separate phase and could be retarded by agitation of the sample.

Absorption spectra were recorded on an Aminco DW-2 dual grating spectrophotometer in the split beam mode, using either air or an oxidized suspension in the reference beam. To avoid loss of signal due to high scattering by the turbid suspensions, all samples were placed immediately adjacent to the end-on photomultiplier tube. Ten-mm path-length cells were used except shorter path lengths were used in cases where the absorbance would otherwise exceed 1.75 - the point near which the instrument no longer yielded reliable absorbance linearity. A Beckman Model DK2-A spectrophotometer was used for reflectance measurements with $BaSO_4$ as the reference.

Ferrous and total iron were determined by a modification of the 1,10-phenanthroline method described by Roth et al. (1969). The major modifications were a) adjustment for the presence of excess reducing agent (Stucki et al., 1977);

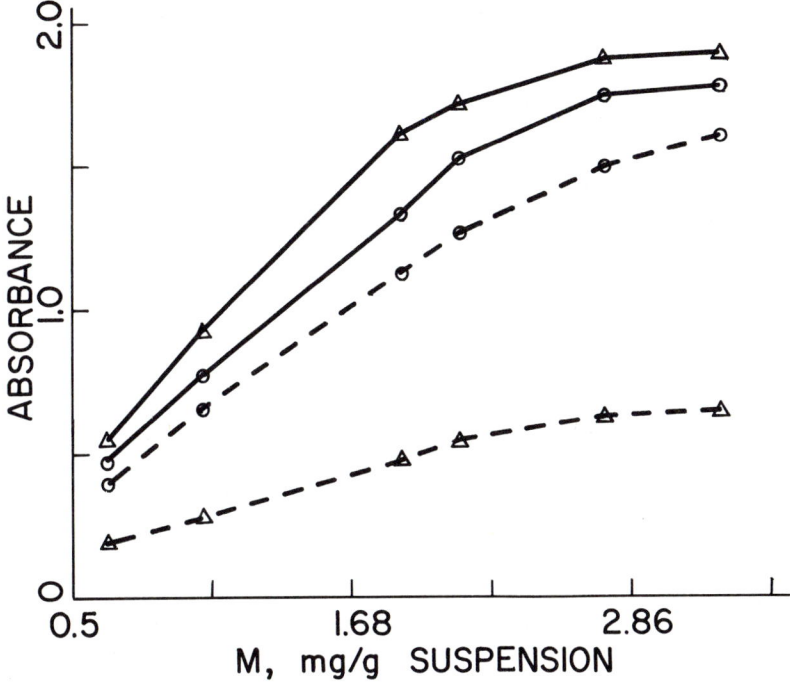

Fig. 1. Effect of clay concentration, M, on absorbance for oxidized (- -) and reduced (——) nontronite at 450 (O) and 735 (△) nm.

b) exclusion of light during analysis to inhibit rate of $Fe^{3+}$ reduction (manuscript in preparation; 3) 30 ml rather than 6 ml of 3.6 $\underline{M}$ $H_2SO_4$ were added to the digest in order to bring the pH of the buffered suspension down to 1.6; and 4) the amount of hydroxylamine hydrochloride was doubled in the flask for total iron.

RESULTS AND DISCUSSIONS

The effect of clay concentration on absorption by oxidized and reduced samples is illustrated in Figure 1. The relationship is linear for suspension concentrations below about 2.6 mg/g for both oxidized and reduced suspensions at different wavelengths, but cannot be described by a simple Beer-Lambert relationship since the slope changes with $Fe^{2+}$ content. This apparent complexity of the system most likely is due to the turbid nature of the suspension, which gives rise to much scattering and various uncertainties as to path lengths and means for calculating the absorptivity.

A dramatic difference exists in the reflectance spectra for oxidized (Figure 2b) and reduced (Figure 2a) nontronite suspensions. Most notable is the shift in wavelength of major reflectance from the yellow region for the oxidized sample

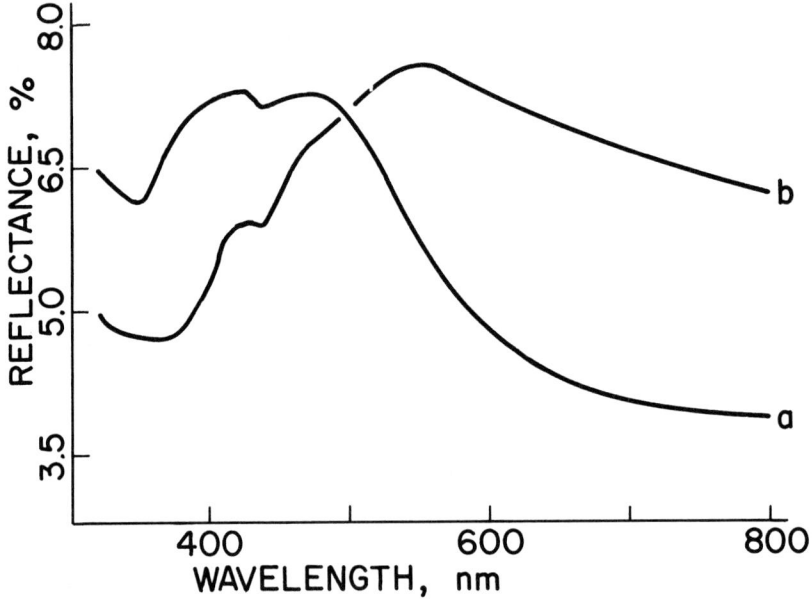

Fig. 2. Visible reflectance spectra for reduced (a) and oxidized (b) nontronite suspensions (2.2 mg/g).

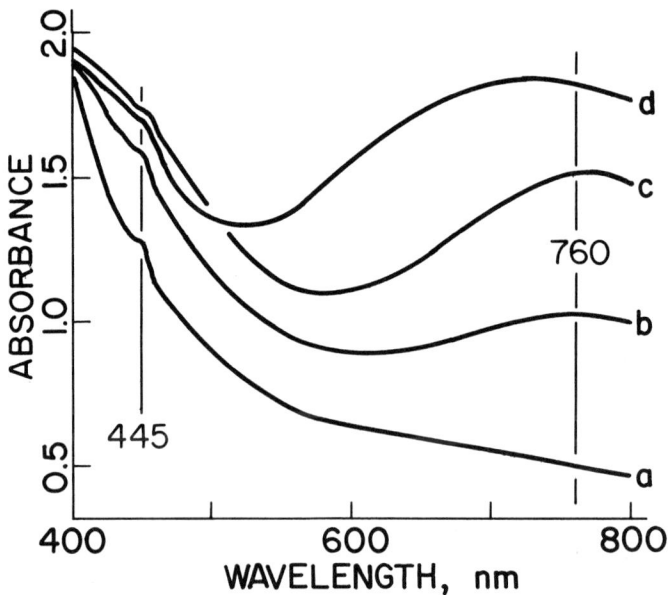

Fig. 3. Visible absorption spectra for oxidized (a) and increasingly reduced (b-d) nontronite suspensions (2.2 mg/g).

toward the blue, i.e. toward shorter wavelength with reduction. The shape of the spectrum depends on the extent to which the suspension has been reduced. Attempts to quantitatively relate reflectance to $Fe^{2+}$ content were unsuccessful because the clay concentrations required for 1,10-phenanthroline analysis yielded reflectance values near the maximum detection limit of the instrument thus introducing effects due to signal damping. These problems were avoided in transmission mode by varying the path length of the cells.

The visible absorption spectrum for nontronite suspension recorded in the transmission mode (Figure 3a) compares favorably with previously reported spectra (Karickhoff and Bailey, 1973) having two bands near 380 nm (not shown due to the relatively high clay concentration of the suspensions, but present in more dilute suspensions and in Figure 2), a prominent shoulder at 445 nm, and a weak band at 515 nm. Figure 3 reveals at least four changes in the spectrum following treatment with gradually increased amounts of sodium dithionite: 1) an apparent decrease in intensity of the band at 445 nm; 2) formation of a strong, broad absorbance in the region above 600 nm with a maximum near 760 nm; 3) at higher levels of reduction the background intensity decreases and the band at 760 nm shifts to about 725 nm; and 4) an overall increase in the background absorption as revealed by wavelengths below 520 nm.

The band at 445 nm, assigned to a $^4A^4E(G) \leftarrow {}^6A(E)$ transition in structural $Fe^{3+}$ (Karickhoff and Bailey, 1973), should be indicative of the amount of $Fe^{3+}$ actually present. The intensity above background at 445 nm was estimated for each curve in Figure 3 by extending the background line under the peak and computing a peak height ratio using the oxidized sample as the reference. Qualitative agreement is obtained between peak-height ratios and structural $Fe^{3+}$ determined by the 1,10-phenanthroline method (Table 1), but a quantitative relationship could not be derived. Perhaps more sophisticated methods for resolving the peak components would refine the differences and quantitize the relationship.

The band initially centered at 760 nm is assigned to a $Fe^{2+} \rightarrow Fe^{3+}$ charge transfer transition (Faye, 1968; Faye and Nickel, 1970) and accounts for the color change from yellow to green and blue-green. Since nontronite is rich in $Fe^{3+}$ and the $Fe^{3+}/Fe^{2+}$ ratio at maximum reduction is about 1.30, the probability that each $Fe^{2+}$ ion will be adjacent to a $Fe^{3+}$ is relatively high. The relationship between absorption intensity at 760 nm and the structural $Fe^{2+}$ content (Figure 4) is best described by two linear equations (1) and (2),

$$(Fe^{2+}) = 49 \ (A/L) - 21 \quad ; [r^2 = .987, \ 0 \leq Fe^{2+} \leq 53] \tag{1}$$

$$(Fe^{2+}) = 227 \ (A/L) - 286; \ [r^2 = .974, \ 53 \leq Fe^{2+} \leq 200] \tag{2}$$

where $(Fe^{2+})$ is the structural $Fe^{2+}$ content in mmoles/100 g clay; A, absorbance; and L, cell path length. The lines intersect at a $Fe^{2+}$ value of 53 mmoles/100 g. The fit of these equations is very good as indicated by the $r^2$ values, and attempts to curve-fit to polynomial and exponential equations for the entire range yielded less satisfactory results. Equations (1) and (2) serve as calibration curves in the specific absorbance range for which each was derived, and can be used with confidence to compute $Fe^{2+}$ content quantitatively from the absorbance index, A/L, of any suspension of nontronite at the given concentration. Of course different calibration curves must be obtained if the spectrophotometer or the clay concentration changes.

The existence of two linear segments in the A/L versus $Fe^{2+}$ relationship indicates that $Fe^{2+}$ contributes to the color in more than one way, or at least changes the nature of its contribution. In addition, one observes a shift in position of the peak maximum toward shorter wavelength as $Fe^{2+}$ increases in concentration (Figure 5), which accounts for the change in color from green to blue-green. Figure 5 could contain a trend similar to that in Figure 4 but the data are inconclusive, except that a definite downward trend exists.

Differences in intensities and peak position of the band at 760 nm are readily explained by the oxidation-reduction mechanism proposed by Stucki and Roth (1977) for structural iron in nontronite. The reduction process involves first an electron transfer from reducing agent at the interlayer surface to $Fe^{3+}$ in the structure accompanied by an increase in layer charge. Then as $Fe^{2+}$ concentration

TABLE 1

Comparison of colors and values for structural $Fe^{2+}$ and $Fe^{3+}$ obtained from 1,10-phenanthroline (method A) and peak-height ratios at 445 nm (method B) for oxidized and reduced nontronite suspensions (2.2 mg/g).

| Dithionite Added, mg | $Fe^{2+}$, mmoles/100 g | | $Fe^{3+}$, mmoles/100 g | | Color |
|---|---|---|---|---|---|
| | Method A | Method B | Method A | Method B | |
| 0 | 3.4 | - | 431.6 | - | yellow |
| 1.6 | 24.9 | 60.0 | 410.1 | 375.0 | light green |
| 2.3 | 26.4 | 62.1 | 408.6 | 372.9 | light green |
| 10.1 | 97.1 | 186.0 | 337.9 | 249.0 | dark green |
| 27.5 | 186.7 | 193.0 | 248.3 | 242.0 | blue-green |
| 20.0* | 189.4 | 229.0 | 245.6 | 206.0 | deep blue-green |

*Aged 4 days prior to analysis

exceeds 50 or 60 mmoles/100 g the layer charge remains constant due to simultaneous elimination of structural hydroxyl groups, which neutralizes the increased negative

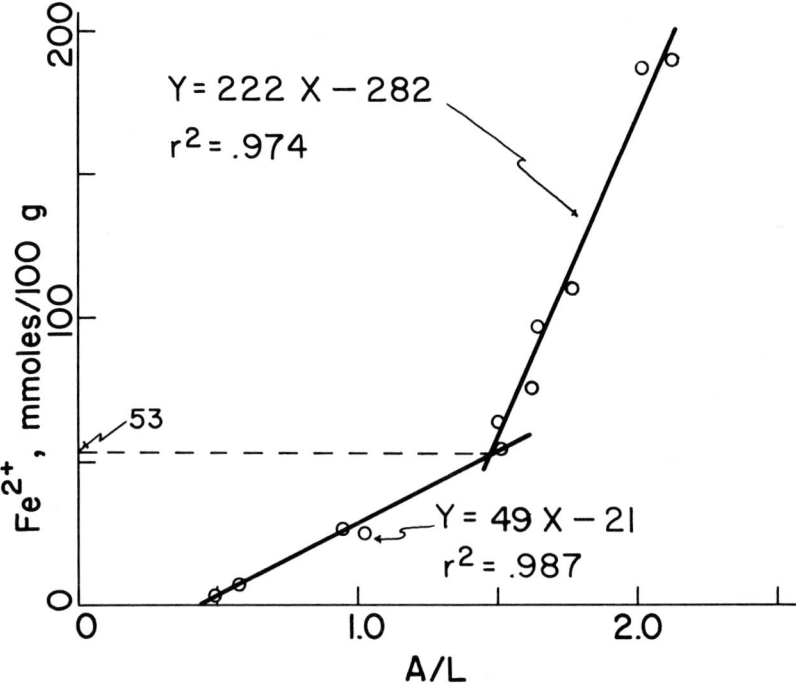

Fig. 4. Effect of structural $Fe^{2+}$ content on the absorbance index, A/L, for nontronite suspensions (2.2 mg/g).

charge on the iron due to reduction. Iron in the dehydroxylated sites is left five-coordinate. If this mechanism and the assignment of the band at 760 to $Fe^{2+} \rightarrow Fe^{3+}$ transitions are valid, then one can conclude that in the range of A/L where equation (1) applies, the band intensity and position arise from six-coordinate sites of $Fe^{2+}$ and $Fe^{3+}$; in the region of equation (2), five-coordinate sites dominate. A shift in charge transfer transition energy as a result of a change in coordination number is not surprising, and indeed would be expected due to perturbations in crystal field energy. Likewise, a change in the relationship between $Fe^{2+}$ and band intensity would be predicted if the absorptivity of iron in a five coordinate ligand field differs from that in a six-coordinate configuration. The possibility of charge transfer transitions between $Fe^{2+}$ and other metal ions in the mineral could be expected also (not necessarily at these wavelengths), although relative concentrations of these ions are so low in nontronite that a significant contribution to any intensity by such species is doubtful.

The effect of flocculation on the spectrum of oxidized nontronite revealed little or no intensity contribution in the 500–800 nm region; and a small, but measurable

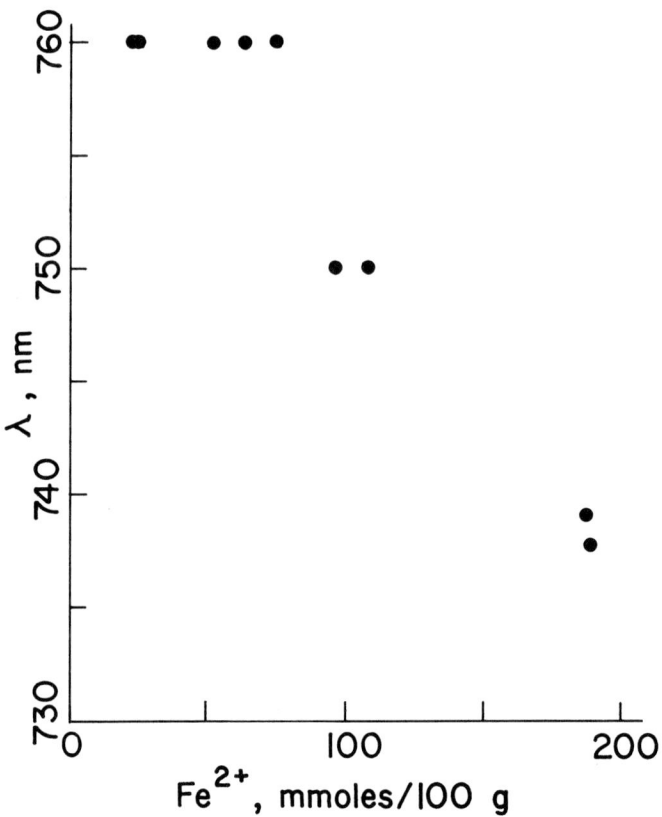

Fig. 5. Wavelength of peak maximum of the band for $Fe^{2+} \rightarrow Fe^{3+}$ charge transfer in oxidized and reduced nontronite suspensions (2.2 mg/g).

change below 500 nm. The effect due to flocculation is therefore much too small to account for the differences in background intensity observed at shorter wavelengths in Figure 3. It is believed that the source of this increased backgroun is simply overlap of the peak at 760 nm.

## SUMMARY AND CONCLUSIONS

Nontronite suspensions yield detailed spectra by visible reflectance and absorbance measurements. Major absorbance peaks are present at 445 and 760 nm, and are assigned to $Fe^{3+}$ intraconfigurational and $Fe^{2+} \rightarrow Fe^{3+}$ charge transfer transitions, respectively. As oxidized nontronite suspensions are reduced, the structural $Fe^{2+}$ content is increased, the color changes to green and blue-green, and four major differences in the spectra are observed: 1) an apparent decrease in intensity of the band at 445 nm; 2) formation of a strong, broad absorbance in the region above 600 nm with a maximum near 760 nm; 3) at higher levels of reduction the background intensity decreases and the band at 760 nm shifts to about 725 nm; and 4) an overall increase in the background absorption as revealed by wavelengths below 520 nm.

The relationship between structural $Fe^{2+}$ content and the absorbance index, A/L, is linear according to equation (1) for values below 53 mmoles $Fe^{2+}$/100 g clay; but above this value the linear interaction is best described by equation (2). Concomitant shifts in the peak position to shorter wavelengths are also observed. These spectral differences are explained by the oxidation-reduction mechanism proposed by Stucki and Roth (1977) which requires a change in layer charge in the region of equation (1) and a decrease in coordination number of iron from six to five in the region of equation (2).

REFERENCES

1 Faye, G. H., 1968. The optical absorption spectra of iron in six-coordinate sites in chlorite, biotite, phlogopite and vivianite. Some aspects of pleochroism in the sheet silicates. Can. Miner., 9:403-425.
2 Faye, G. H. and E. H. Nickel, 1970. The effect of charge transfer processes on the colour and pleochroism of amphiboles. Can. Miner., 10:616-635.
3 Karickhoff, S. W. and G. W. Bailey, 1973. Optical absorption spectra of clay minerals. Clays Clay Miner., 21:59-70.
4 Roth, C. B., M. L. Jackson, E. G. Lotse, and J. K. Syers, 1968. Ferrous-ferric ratio and C.E.C. changes on deferration of weathered micaceous vermiculites. Isr. J. Chem., 6:261-273.
5 Rozenson, I. and L. Heller-Kallai, 1976. Reduction and oxidation of $Fe^{3+}$ in dioctahedral smectite - 1: reduction with hydrazine and dithionite. Clays Clay Miner., 24:271-282.
6 Stucki, J. W. and C. B. Roth, 1977. Oxidation-reduction mechanism for structural iron in nontronite. Soil Sci. Soc. Amer. J., 41:808-814.
7 Stucki, J. W., C. B. Roth, and W. L. Anderson, 1977. Manipulation of air-sensitive clay suspensions for determination of $Fe^{3+}$, $Fe^{2+}$, and cation exchange capacity. Agronomy Abstracts, 1977:145.
8 Yamanaka, K. and S. Motomura, 1959. Studies on the gley formation of soils I. on the mechanism of the formation of active ferrous iron in soils. Soil and Plant Food, 5:134-140.

AN INTERSTRATIFIED CHLORITE-VERMICULITE IN WEATHERED RED SHALE NEAR TOYOMA, JAPAN

Tsutomu NISHIYAMA*, Kaoru OINUMA* and Mitsuo SATO**
*Natural Science Laboratory, Toyo University, Hakusan, Bunkyo-ku, Tokyo, Japan
**Department of Applied Chemistry, Faculty of Technology, Gumma University, Kiryu, Gumma, Japan

ABSTRACT

An interstratified chlorite-vermiculite altered from chlorite by weathering is found in the weathered red shale near Toyoma in Miyagi Pref., Japan. This mineral shows a very strong reflection at 14.7Å with an almost regular series of basal reflections, but does not show a reflection with a long spacing. The characteristic behaviour of the basal reflections after various treatments shows clearly the presence of chlorite and vermiculite layers in this mineral. The ratio of chlorite layer and vermiculite layer is 1:1 in the mineral. The structure of this mineral deviates from a 1:1 regular interstratification of chlorite and vermiculite layers, but is not a completely random one. Dehydration of the vermiculite layer occurs mainly at about 56-83°C and 159°C, and dehydration of the hydroxyl sheet in the chlorite layer occurs at about 555°C. The chemical formula obtained for the mineral is $11.2SiO_2$ $4.2Al_2O_3$ $8.7MgO$ $3.0Fe_2O_3$ $10.6H_2O$. Cation exchange capacity is estimated to be 60meq/100g and the calculated interlayer charge of vermiculite layer is 1.01 on the basis of a $O_{20}(OH)_4$ unit.

INTRODUCTION

Various occurrences of interstratified minerals of chlorite-vermiculite have been reported by many workers. One of the occurrences of these minerals is an alteration from chlorite by weathering. Johnson (1964) reported the occurrence of regularly interstratified chlorite-vermiculite. Some types of the series from chlorite to an interstratification of chlorite and vermiculite layers with the ratio of 1:1 were recognized in the weathered red shale near Toyoma, Miyagi Pref., Japan by Nishiyama and Oinuma (1973) and Nishiyama, et al (1973a). These interstratified minerals were altered from chlorite by weathering. This paper reports one kind of interstratified chlorite-vermiculite system different from a 1:1 regularly interstratified type and presents its detailed character and thermal behaviour.

This research was suported in part by grants-in-aid for Scientific Research

(C-254284) and Co-operative Research (A-234055) from the Ministry of Education.

## SAMPLE AND EXPERIMENTAL METHODS

### Occurrence of minerals

The Triassic Hineushi Formation is distributed in area over a distance of 4km between Yanaizu and Hazawa, near Toyoma, Miyagi Pref., Japan. The lower Hineushi Formation was deposited locally under a half-saline environment and is composed of conglomerates, red shales and black shales (Ueda, 1963). The red shales contain quartz, feldspar, hematite, illite and chlorite (Nishiyama, et al., 1973b). Weathered parts of the red shales are soft and include an interstratified chlorite-vermiculite and kaolinite besides the minerals mentioned above. The weathered red shales are rich in quartz and poor in feldspar in comparison with fresh ones. There is a general tendency for chlorite to decrease in the weathered portions. In the weathered red shale, some varieties within the series from chlorite to an interstratification of chlorite and vermiculite layers with the ratio of 1:1 were found, but discrete vermiculite was not observed. From these observations, it can be said that the interstratified chlorite-vermiculite in this area was formed from chlorite under weathering processes.

### Sample

The weathered red shale (sample No. 2-b) was collected from an outcrop of the lower Hineushi Formation at Yanaizu, near Toyoma, Miyagi Pref., Japan. The $<2\,\mu m$ fraction of the sample was investigated by means of X-ray powder diffraction analysis, DTG, TG, DTG, chemical analysis and electron microscopy. This fraction is free from chlorite and consists of the interstratified chlorite-vermiculite, illite and small amounts of kaolinite and hematite.

## RESULTS AND DISCUSSIONS

### X-ray powder diffraction

X-ray powder diffraction patterns were obtained from specimens prepared on glass slides. The untreated sample shows the reflections of the interstratified mineral described below, as well as these of illite, kaolinite and hematite (Fig. 1)

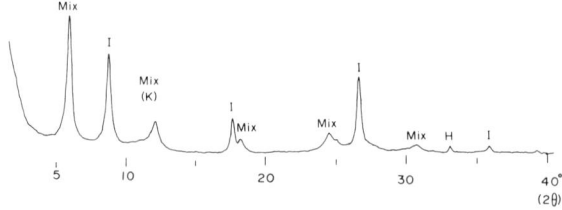

Fig. 1. X-ray powder diffraction pattern of the $<2\,\mu m$ fraction of the untreated sample. I: illite, K: kaolinite, H: hematite, Mix: an interstratified chlorite-vermiculite. X-ray is CuK$\alpha$.

After the sample was heated at 95°C in 6N hydrochloric acid for 10 min., the reflections of the interstratified mineral and hematite disappeared, but the reflections due to illite and kaolinite remained. Therefore the X-ray diffraction patterns of the interstratified clay mineral alone after various treatments could be obtained from those of the sample by subtracting the X-ray reflections of illite, kaolinite and hematite (Fig. 2).

The untreated mineral shows a very strong reflection at 14.7Å and an almost regular series of 00ℓ reflections. The basal spacing of 14.7Å is greater than those of chlorite and Mg-vermiculite. After ethylene glycol treatment, the positions of the maxima of these reflections shift to the higher angle side, the reflections of the 1st (at 14.4Å) and 2nd (at 7.24Å) orders have tails at lower angle side, while the reflection of the 4th (at 3.575Å) order has a tail at higher angle side. This indicates that this mineral contains some layers which have a nonexpandable nature and others which are expandable with ethylene glycol.

After the sample was boiled in 1N $MgCl_2$ for 30 min., the mineral shows an integral sequence of the sharp 00ℓ reflections with a basal spacing of 14.4Å. When the Mg-saturated sample was treated with glycerol, its X-ray pattern did not show a remarkable change except the slight decrease of basal spacing. This shows that the expandable layer with ethylene glycol does not expand with Mg-glycerol treatment and

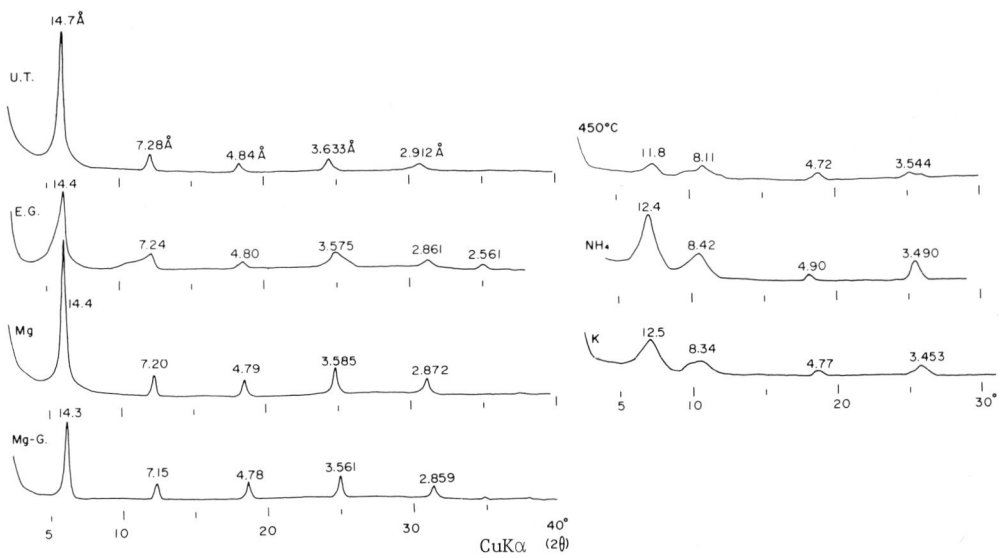

Fig. 2. X-ray powder diffraction patterns of the interstratified mineral before and after various treatments. U.T.: untreated, E.G.: treated with ethylene glycol, Mg: treated with 1N $MgCl_2$, Mg-G.: treated with glycerol after treated with 1N $MgCl_2$, 450°C: heated at 450°C for 1hr., $NH_4$: treated with 1N $NH_4Cl$, K: treated with 1N KCl.

has an interlayer character more like vermiculite than montmorillonite (Walker, 196

After heating at 450°C for 1 hr. and saturated with $NH_4^+$ (boiled in 1N $NH_4Cl$ for 30 min.) and $K^+$ (boiled in 1N KCl for 30 min.), reflections occur at about 12Å and 8Å. These reflections can be explained by the presence of a layer compone having a 14Å basal spacing and a layer component having about a 10Å basal spacing with dehydration and $NH_4$ and K saturation. This mineral coincides with neither normal vermiculite nor swelling chlorite described by Honeyborne (1951) and Stephen and MacEwan (1951). From the above observations, this mineral may be regarded as an interstratification of chlorite and vermiculite layers.

To interpret the characteristic behaviour of the basal reflections of this interstratified mineral after each treatment, the visual inspection method (MacEwan, et al., 1961) was tried as shown in Fig. 3. The basal spacing of the chlorite in the fresh red shale in the studied area was used for the chlorite layer in this figure. The diffraction pattern of the untreated mineral could be interpreted by visual inspection as that of an interstratified mineral of fully hydrated vermiculite layer (V') having a 14.8Å basal spacing (Walker, 1961) and chlorite layer (Ch). After Mg-treatment, it is expected that the fully hydrated vermiculite layer would be transformed into a normal Mg-vermiculite layer having 14.36Å basal spacing (V). In accord with this, the X-ray pattern of the Mg-saturated mineral shows a regular series of very sharp reflections having a 14.4Å basal spacing.

X-ray diffraction patterns of the mineral after ethylene glycol and Mg-glycerol treatment can be explained by the presence of the expandable vermiculite layer (V'') having a 17Å basal spacing with glycol and the vermiculite layer (V) having a 14.36Å one with Mg-glycerol, in addition to the presence of the chlorite layer (C If the basal spacing of the vermiculite layer occurs at 9.5, 10.34 and 10.34Å after heating at 450°C and saturation with $NH_4$ and K, respectively, the diffraction

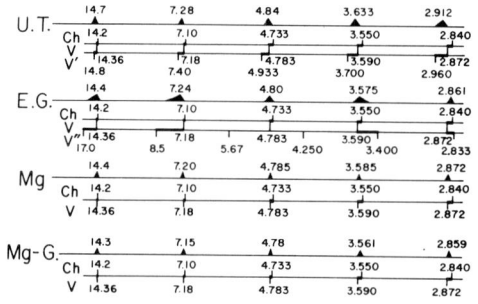
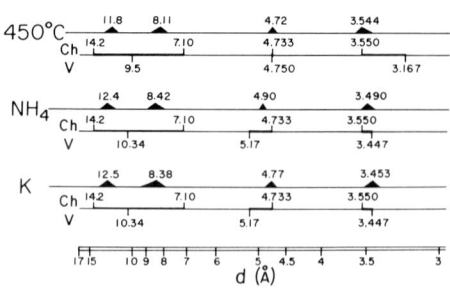

Fig. 3. Diagrammatic illustration of visual inspection method of interpreting peaks from some interstratified structures. Signs are the same as those in Fig.2. Ch: chlorite, V: vermiculite, V': fully hydrated vermiculite, V'': vermiculite having a 17Å basal spacing with ethylene glycol.

patterns after such treatments are consistent with the expected patterns by the visual inspection method.

We calculated the theoretical d spacings of the basal reflections of an interstratification of chlorite (c) and K-vermiculite (v) layers with various frequencies of occurrence of such layers (Wc and Wv) using Kakinoki and Komura's equation (Sato, 1965). From the results of these calculations, the figures showing the relation between the spacings of basal reflections and Wc( or Wv) were made for a structure of Reichweite g= 0 (completely random) and a structure with g= 1 in

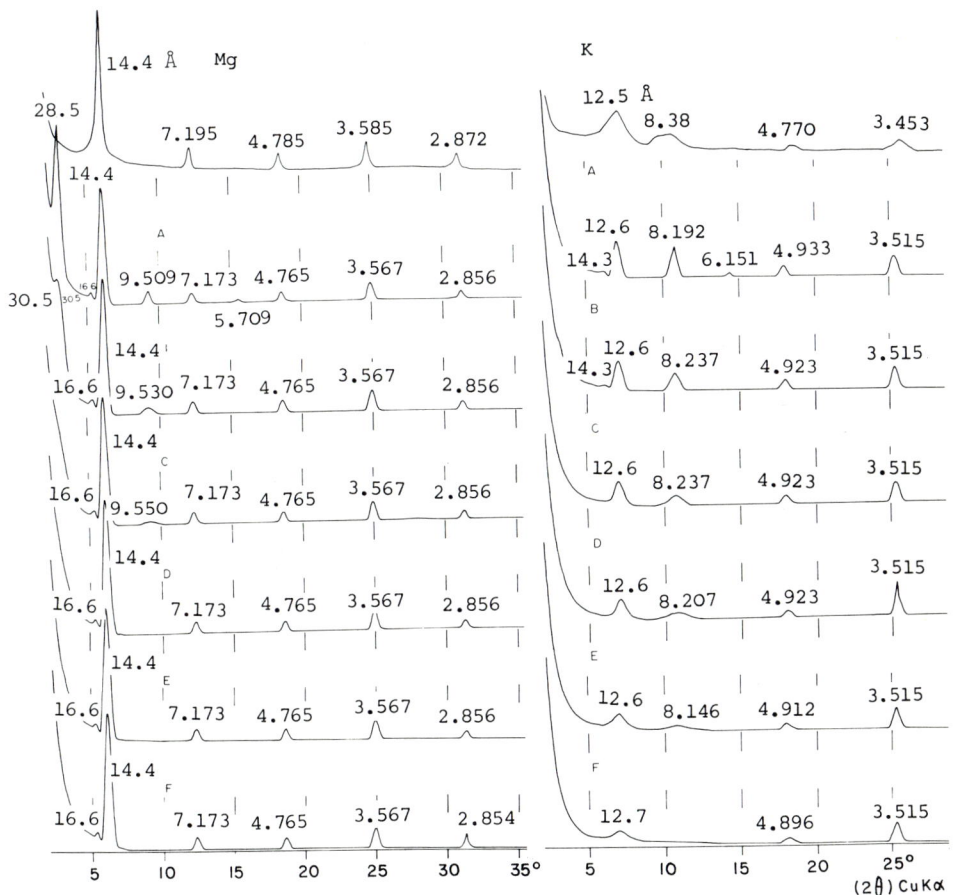

Fig. 4. X-ray diffraction patterns of the treated samples and the calculated diffraction patterns of the various interstratifications of chlorite and vermiculite layers (g= 1, Wc= 0.5, Wv= 0.5, layer number= 10 and standard deviation of layer number= 2.0). Left: interstratifications of chlorite (14.2Å) and Mg-vermiculite (14.34Å) layers, right: interstratifications of chlorite (14.2Å) and K-vermiculite (10.4Å) layers, Mg: Mg-saturated specimen, K: K-saturated specimen, A: Pcc= 0.01, Pvv= 0.01, B: Pcc= 0.1, Pvv= 0.1, C: Pcc= 0.2, Pvv= 0.2, D: Pcc= 0.3, Pvv= 0.3, E: Pcc= 0.4, Pvv= 0.4, F: Pcc= 0.5 Pvv= 0.5.

which the probability of finding chlorite layer succeeding chlorite is Pcc= 0 for Wc $<$ Wv or Pvv= 0 for Wc $>$ Wv. By referring to a figure omitted in this paper, the reflections of the mineral were in approximate agreement with those of the structure with g= 1, Wc= 0.55, Wv= 0.45, Pcc= 0.19 and Pvv= 0 except for the absence of a reflection of 24Å. We then calculated many diffraction patterns of various structures deviating from the one described above, e.g., with a smaller layer number and some random structures. From a comparison of the X-ray diffraction pattern of the mineral with calculated patterns, it may be concluded that the most favorable structure of the mineral is g= 1, Wc= 0.5, Wv= 0.5, Pcc= 0.2-0.3, Pvv= 0.2-0.3, layer number= 10 and standard deviation of layer number= 2.0 (Fig. 4). This structure is slightly different from the 1:1 regular interstratification of chlorite and vermiculite layers (g= 1, Wc= 0.5, Wv= 0.5, Pcc= 0 and Pvv= 0).

X-ray diffraction effects were also examined after heating samples at certain temperatures for 1 hr. and quenching them in air. Above 350°C, the almost regular series of the basal reflections with 14Å basal spacing disappears and the reflection at 12 and 8Å occur. This remarkable change can be explained by the contraction of basal spacing of vermiculite layer, one component of this mineral, from 14.4Å to about 10Å as observed in normal Mg-vermiculite (Hayashi and Oinuma, 1963). The reflection at 12Å increases in its intensity in the range from 550°C to 650°C and almost disappears above 750°C, while the reflection at 8Å disappears above 600°C. These changes in the reflection intensity on heating are similar to those that occur in the reflections at 14Å and 7Å of the chlorite in the fresh red shale.

Thermal behaviour

DTA and TG curves of the sample were recorded simultaneously at a heating rate of 10°C/min. with about 30mg samples (Fig. 5). Endothermic reactions are observed at 52, 69, 94, 181, 507, 567 and 851°C and they are accompanied by weight loss due to dehydration or dehydroxylation. The DTA peak at 507°C can be assigned to kaolinite contained in the sample, because the X-ray reflections due to kaolinite disappear above this temperature. Just as illite contained in a fresh red shale sample gave no distinct peak in DTA, so also illite in this sample seems to have no distinct peak. A DTA peak at 567°C disappeared in the specimen treated with hydrochloric acid; this treatment causing decomposition of the interstratified mineral. This indicates that the peak at 567°C is due to dehydroxylation of the chlorite layer in the interstratified mineral. It may be concluded that the peaks below 200°C and at 567° are mainly attributable to vermiculite and chlorite, respectively, and a peak at 815°C to dehydroxylation of silicate sheets of the mineral.

In order to investigate the detailed character of dehydration of the mineral continuous heating X-ray diffraction analysis was attempted and compared with TG and DTG. The results are shown in Fig. 6 and Table 1. The heating rate was 5°C/min. in these analyses. A reflection at 7.4Å which disappears above 425°C may be attribu

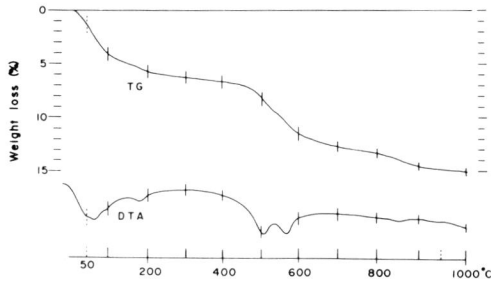

Fig. 5. DTA and TG curves of the sample ( $<2\,\mu$m). Heating rate: 10°C/min.

able to kaolinite and a DTG peak at 495°C also to kaolinite.

Four stages are apparent in the effect of heating on the X-ray reflections as shown in Table 1. Judging from various well-known stages of dehydration of vermiculite (14.8, 11.6 and 9.02Å phases; Walker, 1961), it may be said that the two DTG peaks at 56°C(-83°C) and 159°C correspond to the first and second changes in the reflections respectively and are attributable to dehydration of vermiculite layers with a double sheet of water molecules to a single sheet of water molecules and from a single sheet to ones with a small amount of water molecules in interlayer, respectively. The appearance of the fourth stage (11.4Å reflection) at about 525°C corresponds

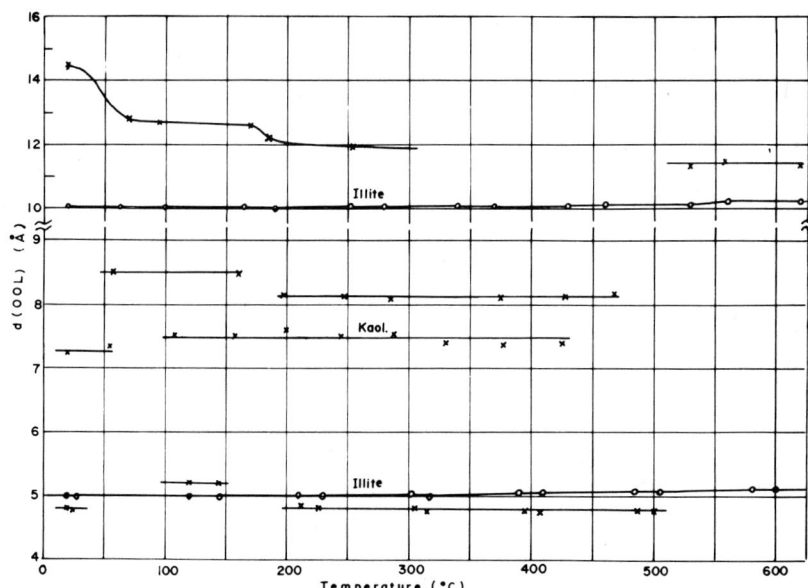

Fig. 6. Continuous heating X-ray diffraction analysis. Heating rate: 5°C/min., Illite: reflections of illite, Kaol.: reflection of kaolinite.

TABLE 1

Reflections of the interstratified chlorite-vermiculite obtained by continuous heating X-ray diffraction and the results of TG and DTG analyses.

| X-ray reflections | 14.4Å 7.2 4.8 | → 60°C → | 12.7Å 8.5 5.2 | → 150 ~ 175°C → | 11.9Å 8.1 4.75 | → 525°C → | 11.4Å |
|---|---|---|---|---|---|---|---|
| | 20°C | 74°C | | 135°C | 245°C | 525°C | 711°C |
| Weight loss | | 3.8% | | 2.0% | 1.2% | 5.7% | 5.0% |
| Peak temp. of DTG | | 56°C | | 83°C | 159°C | 495°C(K) | 555°C |

Heating rate is 5°C/min.  K: kaolinite.

to the beginning of dehydroxylation of the brucite sheet in the chlorite layer showing a DTG peak at 555°C.

Chemical analysis

Table 2 shows the chemical compositions, exchangeable cations and cation exchange capacities (CEC) of the specimen before and after Mg-treatment. The CEC was measured by Schollenberger and Simon's method and the exchangeable cations were measured in the ammonium acetate solutions after leaching the samples. From these data, we could calculate the chemical formula and the interlayer charge of vermiculite layer in this mineral under the following assumptions; (1) The ratio of chlorite layers to vermiculite layers is 0.5:0.5. (2) Mg occupies 2.8 of a total of 6 sites for the cations in the octahedral position in the interstratified mineral (from the data of chlorite in fresh red shales in this region). (3) The MgO component all belongs to the interstratified mineral. (4) The exchangeable cations all belong to the interlayer cations in vermiculite layer.

The chemical formula obtained for this interstratified mineral is 11.2$SiO_2$ 4.2$Al_2O_3$ 8.7$MgO$ 3.0$Fe_2O_3$ 10.6$H_2O$. The interlayer charges(x) of the vermiculite layer were calculated to be 1.01 for the untreated specimen and 1.61 for the Mg-saturated specimen on the basis of a $O_{20}(OH)_4$ unit. The interlayer charge of the untreated vermiculite layer corresponds to an intermediate value between 1.5-1.0 of vermiculite and 1.0-0.5 of montmorillonite (Mackenzie, 1966). The value of the Mg-saturated vermiculite layer corresponds to nearly the highest value for vermiculite. This result supports the earlier explanation of the characteristic behaviour of the X-ray basal reflections with organic reagents such as ethylene glycol and glycerol.

The results of the quantitative estimation of minerals by X-ray diffraction and chemical analysis show that the specimen consists of 27% interstratified mineral, 53% illite, 7% hematite and 12% kaolinite. Assuming illite to have a CEC of 25meq/100g (Grim, 1968), CEC of this interstratified mineral was estimated as follows; 60meq/100g for the untreated specimen and 56meq/100g for the Mg-specimen. These values are lower than a value of 85meq/100g for a 1:1 regular interstratified

TABLE 2

Cation exchange capacity (A), exchange cations (B) and chemical analysis (C).

(A)

|  | CEC (meq/100g) |
|---|---|
| Specimen | 30.0 |
| Mg-saturated sp. | 29.0 |

(B)

|  | $Fe_2O_3$ | MgO | CaO | $K_2O$ | $Na_2O$ |
|---|---|---|---|---|---|
| Specimen | 0.00% | 0.18% | 0.00% | 0.039% | 0.046% |
| Mg-saturated sp. | 0.00 | 0.36 | 0.00 | 0.012 | 0.025 |

(C)

|  | SiO | $Al_2O_3$ | $Fe_2O_3$ | FeO | MgO | CaO | $K_2O$ | $Na_2O$ | $H_2O(+)$ | $H_2O(-)$ | Total |
|---|---|---|---|---|---|---|---|---|---|---|---|
| Specimen | 40.04 | 23.02 | 10.99 | 0.67 | 4.20 | tr. | 5.09 | 0.73 | 8.30 | 6.70 | 99.74% |
| Mg-saturated sp. |  | 24.54 | 12.33 |  | 4.86 | tr. | 5.55 |  |  |  |  |

chlorite-vermiculite reported by Johnson (1964).

Electron micrograph

The particles of the interstratified mineral were observed under an electron microscope and compared with the chlorite particles in the fresh red shale. The particles of the mineral show an irregular fraction without sharp edges. The very thin portions of the particles show an inhomogeneous contrast image which may be due to the partial alteration of chlorite layers to vermiculite under weathering processes and/or dehydration of vermiculite layers under vacuum. On the other hand, the particles of the chlorite in the fresh red shale show an irregular material with sharp edges and the very thin areas show a smooth contrast image.

CONCLUSIONS

This mineral is one of various types of interstratifications of chlorite and vermiculite layers found in the weathered red shale in this area, and is formed from chlorite in the red shales under weathering processes.

This interstratified chlorite-vermiculite when untreated, shows a very strong reflection at 14.7Å with an almost regular series of basal reflections. The basal spacing is a little larger than those of chlorite and/or Mg-vermiculite. However this mineral does not show a reflection with a long spacing. The characteristic behaviour of the basal reflections after various treatments is as follows:
After ethylene glycol treatment, the basal spacing shifts to 14.4Å, however this basal reflection and the one at 7.24Å have a tail on the lower angle side and the one at 3.575Å has a tail on the higher angle side. The basal spacing is 14.4-14.3Å with both $MgCl_2$ and $MgCl_2$-glycerol treatments. The basal reflections appear at 12.5, 8.38, 4.77 and 3.453Å with KCl. After heating at 450°C, the mineral yields the reflections at 11.8, 8.11, 4.72 and 3.544Å.

The most favorable structure of this mineral is one with Reichweite g= 1, Wc= 0.5, Wv= 0.5, Pcc= 0.2-0.3 and Pvv= 0.2-0.3. This structure deviates from a 1:1 regular interstratification of chlorite and vermiculite layers, but is not completely random.

The structural change from double sheets of interlayer water molecules of the vermiculite layer to a single sheet occurs at 56-83°C and some water molecules in the single sheet are released at about 159°C. Dehydration of the hydroxyl sheet in the chlorite layer occurs at about 555°C. The chemical formula obtained for the mineral is $11.2SiO_2$ $4.2Al_2O_3$ $8.7MgO$ $3.0Fe_2O_3$ $10.6H_2O$. The cation exchange capacity is estimated to be 60meq/100g and the calculated interlayer charge of vermiculite is 1.01 on the basis of a $O_{20}(OH)_4$ unit.

REFERENCES

Grim, R.E., 1968. Clay mineralogy (second edition). McGraw-Hill, New York, 189pp.
Hayashi, H. and Oinuma, K., 1963. X-ray and infrared studues on the behaviour of clay minerals on heating. Clay Sci., 1: 134-154.
Honeyborne, D.B., 1951. Clay minerals in the Keuper Marl. Clay Minerals Bull., 1: 150-155.
Johnson, L.J., 1964. Occurrence of regularly interstratified chlorite-vermiculite as a weathering product of chlorite in a soil. Amer. Mineral., 41: 497-504.
MacEwan, D.M.C., Ruiz Amil, A. and Brown, G., 1961. Interstratified clay minerals. In: G. Brown (Editor), The X-ray identification and crystal structures of clay minerals. Mineral. Soc., London, pp. 393-445.
Mackenzie, R.C., 1966. Introduction to section I, Part 1. Proc. Internat. Clay Conf., 2, Israel, 2-4.
Nishiyama, T. and Oinuma, K., 1973. An interstratified mineral of chlorite and vermiculite in the red shale of the Triassic system, Japan. J. Clay Sci. Soc. Japan, 13: 8-14 (in Japanese, with English abstract).
Nishiyama, T., Oinuma, K. and Sato, M., 1973a. An interstratified chlorite-vermiculite in weathered red shale. J. Mineral. Soc. Japan, 11 (special No.1): 141-147 (in Japanese).
Nishiyama, T., Oinuma, K. and Ueda, F., 1973b. Mineralogical studies on the red sh of the Triassic system in the Toyoma district, Miyagi Prefecture, Northeast Japa J. Toyo Univ. General Education (Nat. Sci.) 16: 21-31 (in Japanese, with English abstract).
Sato, M., 1965. Structure of interstratified (Mixed-layer) minerals. Nature 208: 70-71.
Stephen, I. and MacEwan, D.M.C., 1951. Swelling chlorites. Geotechnique (London), 2: 82-83.
Ueda, F., 1963. The geological structure of the Permian and the Triassic systems in Toyoma and Maiya Districts, Southern Kitakami Massif, Northeast Japan. J. Toy Univ. General Education (Nat. Sci.) 4: 1-78 (in Japanese, with English abstract)
Walker, G.F., 1961. Vermiculite minerals. In: G. Brown (Editor), The X-ray identif cation and crystal structures of clay minerals. Mineral. Soc., London, pp. 297-3

OTHER PAPERS PRESENTED IN SECTION 1

ELECTROSTATIC CALCULATIONS ON A STRUCTURAL MODEL OF $2M_1$-MICA

    C.A.J. Appelo (Instituut Voor Aardwetenschappen, Vrije Universiteit, De Boelelaan 1085, Amsterdam, Netherlands.)

HYDROXYL ORIENTATIONS IN 2:1 MINERALS

    R.F. GIESE, Jr. (Department of Geological Sciences, State University of New York at Buffalo, 4240 Ridge Lea Road, Amherst NY, 14226 USA.)

VARIATION OF THE $a_1^1$, $Si-O_{apical}$ (stretch) BOND ENERGY AS A FUNCTION OF COMPOSITION FOR SOME MICA AND SMECTITE MINERALS

    B. Velde (Lab. de Petrographie, Univ. P. et M. Curie, 4 Place Jussieu, 75230 - Paris, France.)

ORGANIZATION OF INTERLAMELLAR CATIONS IN VERMICULITE

    J.F. Alcover and L. Gatineau (Centre de Recherche sur les Solides à Organisation Cristalline Imparfaite - 1.B, Rue de la Férollerie - 45045 - Orleans - France.)

FACTORS AFFECTING HYDRATED VERMICULITE STRUCTURE

    H. Pezerat, H. Suquet and C. de la Calle (Laboratoire de Chimie des Solides, ER 133 CNRS, Université Pierre et Marie Curie, 75230 Paris Cedex 05, France.)

LARGE-ANGLE X-RAY SCATTERING STUDIES OF A 20% Na-MONTMORILLONITE WATER SUSPENSION

    C.H. Pons[1], C. Tchoubar[1] and D. Tchoubar[2] ([1] Laboratoire de Cristallographie, Université d'Orleans, C.R.S.O.C.I. CNRS, Orleans, France. [2] Centre de Recherche sur les Solides à Organisation Cristalline Imparfaite, CNRS, Orleans, France.)

CHANGES IN b-DIMENSION OF SAPONITES WITH INCREASING LAYER CHARGE

    H. Suquet, E. Copin and H. Pezerat (Laboratoire de Chimie des Solides, ER 133 CNRS, Université P. et M. Curie, 75230 Paris Cedex 05, France.)

DEHYDROXYLATION OF DIFFERENTLY COORDINATED HYDROXYLS IN SMECTITES

    L. Heller-Kallai and I. Rozenson (Department of Geology, The Hebrew University, Jerusalem, Israel.)

A STUDY OF STACKING IN DIOCTAHEDRAL SMECTITES: INFLUENCE OF THE LOCALISATION OF ISOMORPHIC SUBSTITUTIONS

    G. Besson and C. Tchoubar (Laboratoire de Cristallographie, Universite d'Orleans, C.R.S.O.C.I., C.N.R.S., 45045 Orleans, France.)

EXPLANATION OF X-RAY LINE PROFILES OF ESKISEHIR SEPIOLITE USING THE PROFILE MATCHING TECHNIQUE

    A. Yucel, M. Rautureau, D. Tchoubar, C. Tchoubar (Centre de Recherche sur les Solides à Organisation Cristalline Imparfaite, 45045 Orleans Cedex, France.)

MEASUREMENT OF THE INTENSITY DIFFRACTED BY CLAYS ON AN ABSOLUTE SCALE PRINCIPLES AND ADVANTAGES OF THE METHOD

C.H. Pons[1], J. Ben Brahim[2], A. Yucel[3], D. Tchoubar[4] and C. Tchoubar[1] ([1] Laboratoire de Cristallographie, Université d'Orléans: C.R.S.O.C.I., CNRS, Orleans, France. [2] Université de Tunis, ENS, Tunisie: C.R.S.O.C.I., CNRS, Orleans, France. [3] Université d'Ankara, Faculté des Sciences, Turkey. [4] C.R.S.O.C.I., CNRS, Orleans, France.)

THE CHARACTERIZATION OF CELADONITES AND GLAUCONITES BY INFRARED SPECTROSCOPY

V.C. Farmer (Department of Spectrochemistry, The Macaulay Institute for Soil Research, Craigiebuckler, Aberdeen, Scotland.)

DEHYDROXYLATION OF DIOCTAHEDRAL SMECTITES: THE ROLE OF THE INTERLAYER CATIONS

I. Horváth (Institute of Inorganic Chemistry, Slovak Academy of Sciences, 80934 Bratislava, Czechoslovakia.)

THE IR SPECTRA OF ALTERED SODIUM SMECTITE

H.H. Rieke[1], R.B. Muter[1] and G.V. Chilingar[2] ([1] College of Mineral and Energy Resources, West Virginia University, Morgantown, West Virginia 26506, USA. [2] Petroleum Engineering Department, University of Southern California, Los Angeles, California 90007, USA.)

INTERSTRATIFIED MICA- OR CHLORITE-MONTMORILLONITE AMINE COMPLEXES

A. Ruiz Amil and F. Aragón de la Cruz (Instituto de Química Inorgánica "Elhuyar" del C.S.I.C. and Facultad de Ciencias Químicas, Ciudad Universitaria, Madrid-3, Spain.

A NEW CLAY MINERAL, SURITE - A Pb-RICH LAYER SILICATE

R. Otsuka[1], S. Tsutsumi[2], J.A. Dristas[3] and T. Sudo[4] ([1] Department of Mineral Industry, Waseda University, Tokyo, Japan. [2] Institute of Earth Science, Waseda University, Tokyo, Japan. [3] Departamento de Geologia, Universidad Nacional del Sur, Bahia Blanca, Argentina. [4] 3-20-7 Miyasaka, Setagaya-ku, Tokyo, Japan.)

THERMAL TRANSFORMATION OF CHRYSOTILE STUDIED BY HIGH-RESOLUTION ELECTRON MICROSCOPY

Helena de Souza Santos and Keiji Yada (Physics Institute, University of Sao Paulo, São Paulo, Brazil and Research Institute for Scientific Measurements, Tohoku University, Sendai, Japan.)

# SECTION 2

# Colloidal Properties and Surface Chemistry

*Chairman's Introduction*

SURFACE PROPERTIES OF FIBROUS CLAY MINERALS (PALYGORSKITE AND SEPIOLITE)

J.M. SERRATOSA
Instituto de Edafología y Biología Vegetal, C.S.I.C. Madrid (Spain)

ABSTRACT

The surface characteristics of fibrous clay minerals (palygorskite and sepiolite), and the nature of sorption active centers are discussed in relation to chemical composition, crystal structure and morphology. The general features of the evolution of surface area as a function of outgassing temperature are described. The selectivity of external surfaces and the accesibility of intracrystalline channels are discussed in terms of the size, shape and polarity of the molecules adsorbed. Finally, the reactivity of Si-OH groups at external mineral surfaces towards organic reactants, and the properties of the resulting organo-mineral derivatives are described.

INTRODUCTION

I thank the Organizing Committee of the VI International Clay Conference for the invitation to present the introductory lecture of the section on Colloidal Properties and Surface Chemistry of Clay Minerals.

The subject has been expanded greatly in recent years as regards the number of minerals involved, the range and sophistication of the techniques employed, the complexities of the problems studied and the number of laboratories participating in the work. Rather than attempting a general review in such a broad area, I feel I could concentrate in aspects of research of surface properties of fibrous clay minerals (palygorskite and sepiolite). Some aspects have been treated in the past, and others have experienced a significant development in the last three years. In my opinion the present state of knowledge deserves being brought to the consideration of the colleagues attending this Conference.

CHEMICAL AND STRUCTURAL CHARACTERISTICS

Palygorskite and sepiolite have fibrous structures consisting of talc-like ribbons parallel to the fibre axis, assembled in such way that the tetrahedral sheet is continuous throughout but inverts apical directions in adjacent ribbons, each ribbon alternating with channels along the fibre axis(Fig. 1-I). Octahedral cations present

at the edges of the ribbons, complete their coordination with water molecules (coor
dination water).Additional water molecules are hydrogen bonded to the water of coor
dination both at the external surfaces and within the channels (zeolitic water)
(Bradley, 1940; Brunauer and Preisinger, 1956).

The width of the ribbons and, consequently, the width of the channels is differe
in the two minerals. In palygorskite the width corresponds to five octahedral catio
sites per half formula unit, normally occupied by 2 $Mg^{2+}$ and 2 $Al^{3+}$ ions in a dioct
hedral pattern which seem to be in the sequence Mg-Al-Vacancy-Al-Mg (Drits and Alek
sandrova, 1966; Serna et al., 1977).Sepiolite has eight octahedral cation sites per
half formula unit, normally occupied by eight $Mg^{2+}$ ions in a trioctahedral pattern.
The tetrahedral sheet is predominantly occupied by $Si^{4+}$ ions, although chemical ana
lyses published by several authors show that tetrahedral Al is present between 0.01
0.69 per eight tetrahedral sites in palygorskite, and between 0.04-0.48 per twelve
tetrahedral sites in sepiolite (Weaver and Pollard, 1973; Mifsud et al., 1978). The
charge defficiency, however, is internally compensated to a large extent, so that
the very low c.e.c. values reported for these minerals are mainly due to charge def
ciencies confined to the external surfaces (Hénin and Caillere, 1975).

All zeolitic water and approximately half of the coordination water are lost at
around 100-300°C for sepiolite and at lower temperatures for palygorskite, at $10^{-4}$
Torr. The crystal structure (Fig. 1-II) undergoes drastical changes: opposite ribbo
approach each other by alternating rotations of contiguous structural units around
the lines of the linking oxygen atoms. This structural change is commonly referred
to as "crystal folding" (Preisinger, 1963; Nagata et al., 1974; Serna et al., 1975;
Prost, 1975; Van Scoyoc et al., 1978). As a consequence, the channel volume is lost
and, in addition, important changes in surface topography take place. At this dehyd
tion stage, folded palygorskite or sepiolite revert easily to the unfolded structure
even at normal ambient conditions.

In both minerals the rest of the coordinated water is lost between 300-500°C at
$10^{-4}$ Torr. without any other significant structural changes taking place. In palygo
kite the last stage of dehydration occurs simultaneously with dehydroxylation, while
in sepiolite dehydroxylation occurs at a higher temperature (above 700°C). Rehydrat

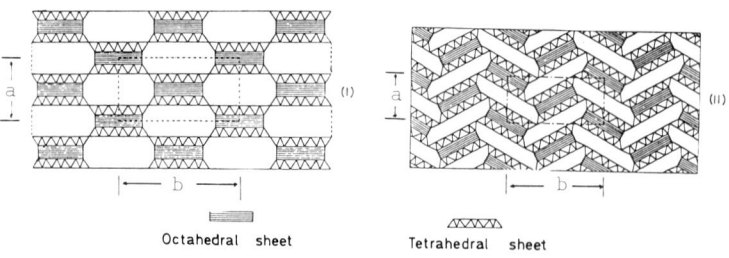

Fig. 1. Channel structure of fibrous clay minerals (I) unfolded; (II) folded.

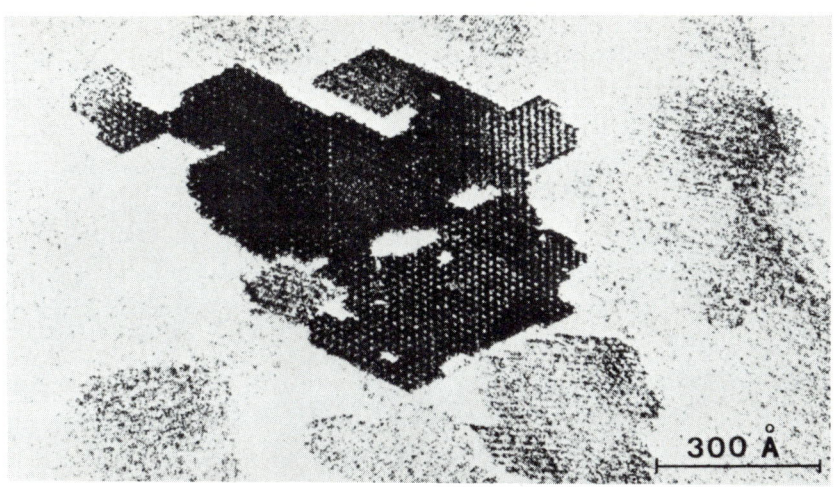

Fig. 2. High resolution electron micrograph of sepiolite. Section normal to the fibre axis (Rautureau, 1974).

of palygorskite to the unfolded structure is still possible, even when it involves the reincorporation of both coordinated water and structural hydroxyls. Rehydration of sepiolite is also possible if dehydration has not proceeded beyond the dehydroxylation temperature (Serna et al., 1975; Van Scoyoc, 1976).

## MORPHOLOGY

With the exception of sepiolite from Ampandandrava, (Madagascar), which grows fibres several centimeters long, natural sepiolites and palygorskites consist of submicroscopic fibres whose dimensions are quite variable depending on their origin: lenghts range between 0.2-2 μm., widths between 100-300 Å and thicknesses between 50-100 Å (Martin-Vivaldi and Robertson, 1971). Such morphology confers a high external surface to the crystallites. The extent to which thick fibres of sepiolite from Ampandandrava expose external surfaces is illustrated by high resolution electron micrographs (Fig. 2) of sections cut perpendicular to the fibre axis (Tchoubar et al., 1973).

## SORPTION ACTIVE CENTERS

Three kinds of active sorption centers may be distinguished in sepiolite and palygorskite:

<u>Oxygen ions</u> on the tetrahedral sheet of the ribbons. On account of the very few isomorphous substitutions present in the tetrahedral sheet of these minerals, those

oxygen ions behave as weak electron donors and the interactions with sorbed species should also be weak.

Water molecules coordinated to magnesium ions at the edges of structural ribbons (two $H_2O$ per each $Mg^{2+}$). Infrared evidences show that these coordinated water molecules are equivalent but that they are asymmetric (symmetry $C_s$); one of the protons on each molecule forms a hydrogen bond with neighbour structural oxygens (Prost, 197 Serna et al., 1977). The coordinated water do not show acidic properties such as the commonly found in the water coordinated to exchangeable cations of layer silicates. This is best demonstrated by the absence of $NH_4^+$ absorption bands in the IR spectra of sepiolite and palygorskite after $NH_3$ adsorption (Van Scoyoc et al., 1978; Serna and Van Scoyoc, 1978).

SiOH groups. The structure and morphology of palygorskite and sepiolite determine the presence of a large number of terminal silica tetrahedra on the ribbons at the external surfaces. Broken Si-O-Si bonds compensate their residual charge by accepting a proton or a hydroxyl and become Si-OH groups. These groups occur at intervals of approximately 5 Å along the fibre axis, and their abundance can be related to the dimension of the fibres and also to crystal imperfections. Large compact crystals have fewer exposed edges and less Si-OH groups. Short thin fibres with rough surface topography have a greater number of exposed Si-OH.

Relative abundances of Si-OH groups in sepiolite samples may be determined using IR and thermogravimetric methods. Ratios of intensities of IR absorption bands produced by Si-OH groups (3720 $cm^{-1}$) and by structural $Mg_3(OH)$ hydroxyls (3680 $cm^{-1}$) were obtained by Ahlrichs et al.(1975), for Ampandandrava, Vallecas and Salinelles sepiolites, which differ essentially in fibre size and crystallinity. The Ampandandrava sample, with macroscopic fibres, gave an absorbance ratio of 0,2, while Vallecas and Salinelles samples, both crystalline powders, gave ratios of 0.5 and 0.8 respectively. The weight loss at 1100°C, attributed to elimination of water from Si-OH groups, amounts to 0.6 and 1.8 per cent for Vallecas and Salinelles samples, respectively, in complete agreement with the results obtained by infrared spectrosco

In palygorskite, Si-OH groups appear to be less abundant that in sepiolite. Only recently Serna et al. (1977) have been able to detect a weak IR absorption band at 3705 $cm^{-1}$ in the IR spectra of palygorskite. The lower content of Si-OH groups is consistent with the smaller specific surface area of this mineral.

Si-OH groups interact with molecules adsorbed on external surfaces. This is demostrated by the shift to lower frequency of the $\nu_{OH}$ band at 3720 $cm^{-1}$. Even non polar compounds such as fluorolube and nujol will perturb this vibration (Ahlrichs et al. 1975).

The abundance of Si-OH groups becomes relevant for the capability of these groups to form true covalent bonds with certain organic reagents, as will be discussed below

SURFACE AREA AND POROSITY

The total surface of sepiolite and palygorskite computed from structural models is, approximately, of 800-900 $m^2/g$., this figure including the area of the channels (Serna and Van Scoyoc, 1978). The surface available, however, depends strongly on the nature (size, shape and polarity) of the molecules used as sorbate. The sorption capacity for different gases and vapours, as well as its dependence with the outgassing temperature, has been extensively studied by Barrer and Mackenzie (1954); Barrer et al. (1954); Dandy (1968, 1971); Fernández-Alvárez (1970); Serna and Fernández-Alvárez (1975); Jiménez-López et al. (1978).

Nitrogen surface area

The results of nitrogen sorption determinations on sepiolite and palygorskite will be discussed in some detail in order to compare the surface properties of the two minerals as well as the properties of different samples of each mineral.

The sorption isotherms for nitrogen are of type II of Brunauer classification. They show a slight, but definite, hysteresis at the high pressure side of the isotherm. Nitrogen sorption always reaches equilibrium rapidly (Barrer and Mackenzie, 1954) suggesting that nitrogen molecules do not penetrate significantly into the intracrystalline channels. Particle size and surface topography characteristics mainly govern the extent of nitrogen sorption, and, thus, values of the surface area measured by this method reflect the above characteristics for the different samples. The amount of nitrogen sorbed depends strongly on pretreatment, what is best explained when the water content of the sample is taken into account.

Fig. 3 shows the evolution of nitrogen surface area calculated by the BET method, for various sepiolites and palygorskites as a function of outgassing temperature (Barrer and Mackenzie, 1954; Dandy and Nadiye-Tabbiruka, 1975; Fernández-Alvárez, 1978). The surface increases first as adsorbed and zeolitic water are eliminated, until a maximun is reached. Next, a sudden drop occurs at the temperature at which half the coordination water is eliminated and crystal folding occurs. Beyond this point the surface area remains practically constant. Prior to folding, the sepiolite samples have surface areas (230-380 $m^2/g$.) considerably higher than those of the palygorskites (140-190 $m^2/g$.). Also, the drop of surface area in the palygorskites occurs at temperature which may be as much as 100°C lower than in sepiolite: this is consistent with the fact that, in the former, the coordination water is more easily lost and, consequently, crystal folding occurs earlier than in the latter.

It is also apparent from figure 3 that there are considerable differences between, various samples of the same mineral, with regards to their maximum and their minimum surface areas and also to their transition temperatures. The higher the maximum the lower the transition temperature, and the wider the drop in surface area before and after folding has taken place. The cause for this behaviour should undoubtedly be traced to differences in the particle size distribution and crystal imperfections

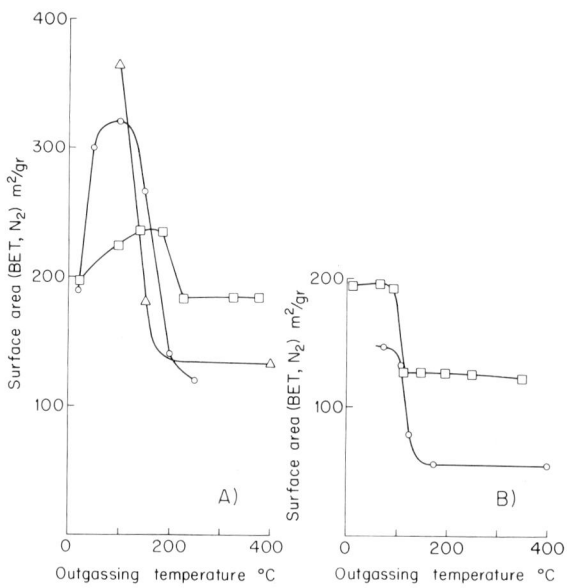

Fig. 3. Effect of outgassing temperature on surface area (BET-$N_2$) of sepiolites (A) and palygorskites (B).
Data from: □ Barrer and Mackenzie, (1954); △ Dandy and Nadiye-Tabbiruka (1975); ○ Fernández-Alvárez (1978).

of the various samples. In support to this assumption, the nitrogen surface area of Ampandandrava sepiolite (unground) at 100°C outgassing has been found to be only 79 $m^2$/g., a value which is considerably lower than the values normally found for the microcrystalline sepiolites.

Porosity

The analysis of nitrogen sorption data by the MP-method (Mikhails et al., 1968) enables the determination of that part of the specific surface which corresponds to micropores, i.e. pores having widths of 16-20 Å or less. Complementarily, the method of Pierce (1953) gives the part of the specific surface area corresponding to pores of radius greater than 15 Å (meso and macropores).

Table 1 gives data on total and differential surface areas by both methods at several outgassing temperatures, for sepiolite from Vallecas (Spain) and palygorskite from Serradilla (Spain) (Fernández-Alvárez, 1978). It is apparent from this data that prior to folding as much as 60-70 % of the surface area corresponds to micropores, and also that most of the micropores disappear at the temperature when folding takes place. Similar results were obtained by Dandy and Nadiye-Tabbiruka (1975) with sepiolite from Amboseli (Tanzania). Assuming that intracrystalline channels are not accesible to nitrogen molecules (Barrer and Mackenzie, 1954; Serna and Van Scoyoc, 1978), the above results indicate that crystal folding reduces profoundly the rugosity of the external surfaces.

TABLE 1

Comparison of the pore surfaces obtained by the method of Pierce and by the MP-method with $S_{BET}$ (Fernández-Alvárez, 1978).

| Outgassing temperatures 0°C | $S_{BET}$ $m^2/g$ | Method of Pierce | | MP-method | | |
|---|---|---|---|---|---|---|
| | | $S_{cum.}$ r>15 Å $m^2/g$ | $1 - \frac{S_{cum.}}{S_{BET}}$ | $S_{total}$ $m^2/g$ | $\Delta S$ t<4.5 Å (*) $m^2/g$ | $\frac{\Delta S}{S_{total}}$ |
| Palygorskite | | | | | | |
| 75 | 146 | 56 | 0.62 | 163 | 117 | 0.72 |
| 110 | 132 | 55 | 0.58 | 139 | 93 | 0.67 |
| 125 | 79 | 54 | 0.32 | 84 | 30 | 0.36 |
| 175 | 56 | 56 | 0.00 | 56 | 0 | 0.00 |
| Sepiolite | | | | | | |
| 100 | 321 | 112 | 0.65 | 348 | 208 | 0.59 |
| 150 | 266 | 115 | 0.57 | 286 | 162 | 0.56 |
| 200 | 140 | 111 | 0.21 | 139 | 15 | 0.11 |
| 250 | 120 | 107 | 0.11 | 120 | 0 | 0.00 |

(*) t is the statistical thickness of the adsorbed film which for parallel plates is half the pore width.

SORPTION PROPERTIES

Sorption of water and polar molecules

IR spectra of hydrated sepiolites and palygorskites clearly prove that zeolitic water, as well as water adsorbed at the external surface, are hydrogen bonded to the water coordinated to the edge $Mg^{2+}$ ions, and also that they are in interaction with the octahedral OH groups, whose OH vibration they perturb (Prost, 1975).

Both zeolitic and coordination water molecules may be replaced by $D_2O$ or $NH_3$ (Prost, 1975; Serna and Van Scoyoc, 1978), thus demonstrating that the external and intracrystalline channels are accessible to small molecules of high polarity.

Short chain primary alcohols (methanol and ethanol) penetrate considerably inside the channels. On sorption they replace zeolitic water, and, initially, they are hydrogen bonded to the coordination water molecules, but subsequently they replace part of the coordination water by succesive sorptions and desorptions. Alcohols of longer chain length, and acetone may also replace zeolitic and coordination water, but sorption is confined to open channels at the external surfaces (Serna and Van Scoyoc, 1978).

Sorption of non-polar molecules

Sorption of non-polar organic compounds on sepiolite and palygorskite seems to be restricted to external surfaces and is strongly influenced by the size and shape of the sorbate molecules. According to Barrer et al. (1954), free energy, heat and entropy changes for the n-paraffins sorbed on palygorskite are greater than for the adsorption of branched-chain paraffins. Also, the former are adsorbed in larger amounts than the latter. The fact that the entropy decrease on transferring n-paraffins from the liquid phase to the adsorbed film is initially very large, strongly suggests that the molecules fit closely surface corrugations of nearly analogous dimensions which, undoubtedly, are the open channels at the external surfaces.

Sepiolite shows a similar behaviour (Barrer et al., 1954; Serna and Fernandez-Alvárez, 1975) but its selectivity is different than that of palygorskite; thus sepiolite sorbs n- and iso-pentane equally strongly but neo-pentane less strongly. In sepiolite the channel width is larger than in palygorskite, and it seems that the molecules of the former sorbates are well accomodated into the channels at the external surface, but not so the molecules of neo-pentane.

In both minerals, selectivity is lost at outgassing temperatures above 100°C when the structure is folded and the open channels at the external surfaces collapse.

## ORGANO-MINERAL DERIVATIVES OF SEPIOLITE AND PALYGORSKITE

Under the name of organo-mineral derivatives those compounds are included in which true covalents bonds are formed between the mineral substrate and the molecules of the organic reactant. The reactivity of surface Si-OH groups towards organic reagents has been extensively studied on silica surfaces. Organic derivatives of chrysotile, montmorillonite and vermiculite have been prepared by Fripiat and Mendelovici (1968) and by Zapata et al. (1972). These minerals expose few Si-OH groups, and more groups have to be created prior to reaction if organic derivatives with substantial amounts of incorporated organic material have to be obtained.

Sepiolite contains large amounts of external Si-OH groups and reacts directly with organic reagents either in vapour phase or in solution in inert organic solvents In either case the mineral framework is essentially preserved.

Grafting reactions that have been experimented with sepiolite may be classed in two types:

a) Grafting through Si-O-Si bonds, when the mineral reacts with organochlorosilanes (Ruiz-Hitzky and Fripiat, 1976).

$$\equiv Si-OH + Cl-Si\begin{matrix}R_1\\R_2\\R_3\end{matrix} \longrightarrow \equiv Si-O-Si\begin{matrix}R_1\\R_2\\R_3\end{matrix} + HCl$$

The linkage of the organic molecules to the mineral substrate through Si-O-Si bonds is resistant to hydrolysis.

b) Grafting through Si-O-C bonds, by addition of alkyl-or phenyl-isocianates (Fernández-Hernández 1977).

$$\equiv Si-OH + O=C=N-R \longrightarrow \equiv Si-O-CO-NH-R$$

or by reaction with epoxides (Casal-Piga and Ruiz-Hitzky 1977).

$$\equiv Si\text{-}OH + R\text{-}CH\text{-}CH_2 \underset{\diagdown O \diagup}{\longrightarrow} \equiv Si\text{-}O\text{-}CH\text{-}R$$
$$\phantom{\equiv Si\text{-}OH + R\text{-}CH\text{-}CH_2 \longrightarrow \equiv Si\text{-}O\text{-}}CH_2OH$$

Grafting trough bonds Si-O-C leads to products which are easily hydrolysed, since those bonds are rather unstable in the presence of water.

Organomineral derivatives of palygorskite have been prepared (Mendelovici and Carroz-Portillo, 1976). The method of cohydrolysis (Lentz, 1964) was used to increase the number of active Si-OH groups in the surface of mineral.

Organomineral derivatives of sepiolite and palygorskite are materials of interest because they have the surface and reactivity properties corresponding to the grafted organic molecules while they preserve the mechanical properties of the mineral framework. When the molecules attached contain unsaturated groups it becomes possible to copolymerize the organomineral compound with several monomers. The addition of vinyl derivatives of sepiolite to natural or synthetic rubber and subsequent polymerization, produces elastomers with special physical properties (Ruiz-Hitzky, 1974). Finally, double bonds of vinyl or allyl derivatives can fix chemically compounds of transition metal ions such as osmium tetroxide (Barrios-Neira et al., 1978); this possibility opens ways for the preparation of new catalysts.

CONCLUDING REMARKS

The use of infrared spectroscopy has contributed substantially to the knowledge of the structure and surface properties of fibrous clay minerals. The mechanisms of interaction of sorbed molecules with the silicate surface have been well established in many cases. The accesibility of intracrystalline channels seems to be restricted to small polar molecules but this point has not been completely clarified. We believe that more quantitative work including the determination of thermodynamic parameters are necessary.

The possibility of preparation of organomineral derivatives opens ways to new research related to technical applications of these minerals.

Finally, more attention should be given to chemical composition and crystallinity of the samples in studying the sorption properties of these minerals. As pointed by Mifsud et al. (1978), it is insufficient to draw broad conclusions from the study of a few samples of a particular mineral.

REFERENCES

Ahlrichs,J.L., Serna,C. and Serratosa,J.M., 1975. Structural hydroxyls in sepiolites. Clays and Clay Minerals, 23, 119-124.
Barrer,R.M. and Mackenzie, N., 1954. Sorption by attapulgite. Part I. Availability of intracrystalline channels. J. Phys. Chem., 58, 560-568.
Barrer,R.M., Mackenzie, N. and Macleod, D.M., 1954. Sorption by attapulgite.Part II. Selectivity shown by attapulgite, sepiolite and montmorillonite for n-paraffins. J.Phys.Chem.,58, 568-572.
Barrios-Neira,J., Poncelet,G. and Fripiat,J.J., 1978. Nitrogen retention by an osmium complex supported on sepiolite. (To be published).

Bradley,W.F., 1940. Structure of attapulgite. Am. Mineral., 25, 405-410.
Brauner, K. and Preisinger, A., 1956. Struktur und Entschung des Sepioliths. Tschermaks Min. Petr. Mitt., 6, 120-140.
Casal-Piga, B. and Ruiz-Hitzky, E., 1977. Reaction of epoxides on mineral surfaces. Organic derivatives of sepiolite. Proc. Third European Clay Conf., Oslo, 35-37.
Dandy, A.J., 1968. Sorption of vapors by sepiolite. J. Phys. Chem., 72, 334-339.
Dandy, A.J., 1971. Zeolitic water content and adsorption capacity for ammonia of microporous sepiolite. J. Chem. Soc. (A), 2383-2387.
Dandy, A.J. and Nadiye-Tabbiruka, M.S., 1975. The effect of heating in vacuo on the microporosity of sepiolite. Clays and Clay Minerals, 23, 428-430.
Drits, V.A. and Aleksandrova, V.A., 1966. The crystallochemical nature of palygorski. Zap. Vses. Miner. Obshch., 95, 551-560.
Fernández-Alvárez, T., 1970. Superficie específica y estructura de poro de la sepiolta calentada a diversas temperaturas. Proc. Reunión hispano-belga de minerales de la arcilla., Madrid, 202-209.
Fernández-Alvárez, T., 1978. Capacidad adsorbente de la palygorskita y sepiolita en función de la deshidratación. I. Adsorción de nitrógeno. Clay Minerals, (in press).
Fernández-Hernández, M.N., 1977. Aplicación de la espectroscopía infrarroja al estudio de la interacción de isocianatos con sepiolita. Tesis Licen., Universidad Complutense, Madrid.
Fripiat, J.J. and Mendelovici, E., 1968. Dérivés organiques des silicates.I. Le dérivé méthylé du chrysotile. Bull. Soc. Chim., 483-492.
Henin, S. and Caillere, S., 1975. Fibrous Minerals. In: J.E. Gieseking (Editor), Soil Components, Springer Verlag, Berlin. Vol. 2, pp. 335-349.
Jiménez-López, A., López-González, J.D., Ramírez-Saénz, A., Rodriguez-Reinoso, F., Valenzuela-Calahorro, C. and Zurita-Herrera, L., 1978. Evolution of surface area in a sepiolite as a function of acid and heat treatment. (To be published).
Lentz, C.W., 1964. Silicate minerals as source of trimethyl-silyl silicates and silicate structure analysis of sodium silicate solutions. Inorg. Chem., 3, 574-579.
Martin-Vivaldi, J.L. and Robertson, R.H.S., 1971. Palygorskite and sepiolite (the hormites). In: J.A. Gard (Editor), Electron-optical Investigation of Clays. Mineralogical Society, London, pp. 255-276.
Mendelovici, E. and Carroz-Portillo, D., 1976. Organic derivatives of attapulgite.I. Infrared spectroscopy and x-ray diffraction studies. Clays and Clay Minerals, 24, 177-182.
Mifsud, A., Rautureau, M. and Fornés,V., 1978. Etude de l'eau dans la palygorskite a l'aide des analyses thermiques. Clay Minerals, (in press).
Mikhail, R.Sh., Brunauer, S. and Bodor,E.E., 1968. Investigations of a complete pore structure analysis. I. Analysis of micropores. J. Colloid Interface Sci., 26, 45-53'
Nagata, M., Shimoda, S. and Sudo, T., 1974. On dehydration of bound water of sepiolite. Clays and Clay Minerals, 22, 285-293.
Pierce, C., 1953. Computation of pore sizes from physical adsorption data. J. Phys. Chem., 57, 149-152.
Preisinger, A., 1963. Sepiolite and related compounds: its stability and application Clays and Clay Minerals, 10, 365-371.
Prost, R., 1975. Etude de l'hydratation des argiles: interactions eau-mineral and mecanisme de la rétention de l'eau. These, Université de Paris VI.
Rautureau, M. 1974. Analyse structurale de la sepiolite par microdiffraction electronique. These, Université d'Orleans.
Ruiz-Hitzky, E., 1974. Contribution a l'etude des reactions de greffage de groupemen organiques sur les surfaces minerales. Greffage de la sepiolite. Theses, Universit de Louvain.
Ruiz-Hitzky, E. and Fripiat, J.J., 1976. Organomineral derivatives obtained by reacting organochlorosilanes with the surface of silicates in organic solvents. Clays and Clay Minerals, 24, 25-30.
Serna, C. and Fernández-Alvárez, T., 1975. Adsorción de hidrocarburos en sepiolita. II. Propiedades de superficie. An. Quim., 71, 371-376.
Serna, C., Ahlrichs, J.L. and Serratosa, J.M., 1975. Folding in sepiolite crystals. Clays and Clay Minerals, 23, 452-457.

Serna, C., Van Scoyoc, G.E. and Ahlrichs, J.L., 1977. Hydroxyl groups and water in palygorskite. Am. Mineral., 62, 784-792.
Serna, C. and Van Scoyoc, G.E., 1978. Infrared study of sepiolite and palygorskite surfaces. (These Proceedings).
Tchoubar, C., Rautureau, M., Clinard, Ch.and Ragot, J.P., 1973. Technique d'inclusion appliquée a l'étude des silicates lamellaires et fibreux. J. Microscopie. 18, 147-154.
Van Scoyoc, G.E., 1976. Surface and structural properties of palygorskite. Ph. D. Thesis, Purdue University.
Van Scoyoc, G.E., Serna, C. and Ahlrichs, J.L., 1978. Structural changes in palygorskite upon dehydration. Am. Mineral., (in press).
Weaver, C.E. and Pollard, L.D., 1975. The Chemistry of Clay Minerals. Elsevier, Amsterdam, 213 pp.
Zapata, L., Castelein, J., Mercier, J.P. and Fripiat, J.J., 1972. Dérivés organiques des silicates. II. Les dérivés vinyliques et allyliques du chrysotile et de la vermiculite. Bull. Soc. Chim., 54-63.

THE STRUCTURE AND DYNAMICS OF CLAY-WATER SYSTEMS STUDIED BY NEUTRON SCATTERING

D.J. CEBULA and R.K. THOMAS
Physical Chemistry Laboratory, South Parks Road, Oxford, and
J.W. WHITE
Institut Laue-Langevin, Grenoble, France.

ABSTRACT

Three properties of Montmorillonite-water systems have been studied by neutron scattering, the long range order of clay platelets in the sol, the structure of compacted samples at low water concentrations and the dynamical properties of water in the compacted samples.

Neutron small-angle scattering is a sensitive technique for studying both long range order in colloidal dispersions and the macroscopic structure of the particles themselves. Its important features are its wide range of momentum transfer and the possibility of isotopic H-D labelling or $H_2O$-$D_2O$ contrast variation because of the different cross sections of H and D. We have made preliminary experiments to study both long and short range order in montmorillonite sols as a function of ionic strength and with different counter ions.

For the study of the structure of compacted samples neutron diffraction has the advantage that large samples can be used. This ensures homogeneity, removing any problems associated with external surface effects in the heterogeneous clay-water systems. We find, for example, that at low water concentration the larger platelets are partially oriented, their basal spacing varying with a Gaussian distribution about a mean value. The small platelets in the system are randomly oriented.

For the study of the dynamical properties, neutron quasielastic scattering has the great advantage that the signal is dominated by scattering from protons in the system. The frequency spectrum of water in the system is therefore quite easily measured with little interference from the clay substrate. In systems containing 1, 2 and 3 water layers between platelets we find that protons are moving rapidly. We describe the motion in terms of a model of three dimensional translational diffusion with simultaneous rotational motion.

## INTRODUCTION

There are several features of neutron scattering that make it a powerful technique for studying the structure and dynamics of heterogeneous systems such as clay-water mixtures. The wavelength of thermal neutrons is suitable for probing the structure of a clay both at the molecular level (neutron diffraction) and at the particulate level (neutron small-angle scattering). When compared with X-ray diffraction neutron diffraction has the advantage for heterogeneous systems that the attenuation of a neutron beam in a sample is much less than for X-rays. Thus, for a montmorillonite containing two molecular layers of water, the 'absorption' length for neutrons is about 30mm but only about 0.1 mm for X-rays. This makes it possible to use large samples for neutron diffraction ensuring that external surface effects are negligible (Cebula et al., 1978a). For studying the coarser structural features of clay-water systems, particularly in the sol or gel states, neutron scattering has two advantages. Firstly, the wavelength of the neutrons may easily be varied to avoid Bragg diffraction inside the clay particles. Secondly, the hydrogen-deuterium ratio in the solvent may be varied to alter the 'contrast' between a clay particle and its surroundings. This effect is likely to be of particular value for the study of absorption at the clay-water interface (Cebula et al., 1978b).

As well as elastic scattering experiments, measurement of the exchanges of energy between neutrons and sample makes it possible to analyse the oscillatory and diffusive motions of individual molecules in a heterogeneous system. The incoherent neutron scattering spectrum is related to the motion of individual molecules and is dominated by the protons in the system. For a sample of montmorillonite containing only one molecular layer of water the incoherent scattering is mostly from the water in the system. At present, the available energy of neutron beams and the resolution of neutron spectrometers cover the time scale $10^{-8} - 10^{-14}$ s, extending from relatively slow diffusion ($D \sim 10^{-8}$ $m^2 s^{-1}$) to quite high vibration frequencies ($\sim 2000$ $cm^{-1}$).

In this paper we describe different neutron scattering experiments on montmorillonite-water mixtures varying from the sol ($\sim 1\%$ by weight of clay) almost to the dry state (1 molecular layer of water between the platelets).

## EXPERIMENTAL

The raw clay material used was bentonite from Clay Spur, Wyoming (A.P.I.No.26). Standard techniques (Callaghan and Ottewill, 1974) were used to remove residual organic compounds and traces of undesirable heavy metal cations. Experiments were done using the Li, Na, K and Cs homoionic forms of montmorillonite.

Compacted samples with a water layer thickness of about 20 Å were prepared

by compression (Callaghan and Ottewill, 1974) and dried to a water content of
0.1 - 0.3 g $H_2O$/g clay (1 - 3 molecular layers of water between the platelets).
For the diffraction and incoherent scattering experiments typical samples were
3cm in diameter and 0.5 - 1mm thick.

The exact amount of water between the platelets was controlled throughout
each experiment by circulating air at a known relative humidity directly over
the clay. The clay was mounted in an aluminium can in which the temperature
and humidity could be accurately controlled.

Neutron diffraction experiments were done on the diffractometers 7H2R at
A.E.R.E. Harwell (Haywood and Worcester, 1973), D16 at the Institut Laue-
Langevin, Grenoble and the small angle camera, D11, also at the I.L.L. The
small-angle scattering experiments were done on D11 and the incoherent
scattering experiments on the multichopper time-of-flight spectrometer, IN5, at
the I.L.L. (Institut Laue-Langevin, 1977).

STRUCTURE OF COMPACTED MONTMORILLONITE-WATER SYSTEMS

A neutron diffraction pattern obtained from a compacted montmorillonite
containing two layers of water is shown in Figure 1. Only one reflection from
the basal planes is observed and that is broad and unsymmetrical, characteristic
of a highly disordered structure.

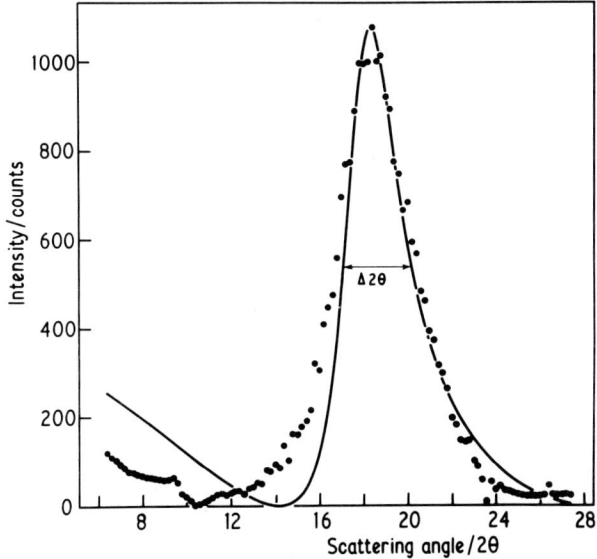

Fig. 1. Neutron diffraction pattern of Lithium Montmorillonite (T(K) = 293,
relative humidity (%) = 76, incident wavelength (Å) = 4.63, basal spacing
(Å) = 15.1).

The types of fault that might occur in such a sample are illustrated schematically in Figure 2. The diffraction peaks will be broadened particularly by regions of gross folding (C), the small size of ordered domains (D) and regions of disordered stacking (E). Assuming a model of small crystallites (D) and applying the Scherrer formula (Guinier, 1963) to the (001) reflection shows that the average ordered domain consists of seven platelets. This is an unrealistic model, however, and does not account for the asymmetry of the diffraction peak shown in Figure 1.

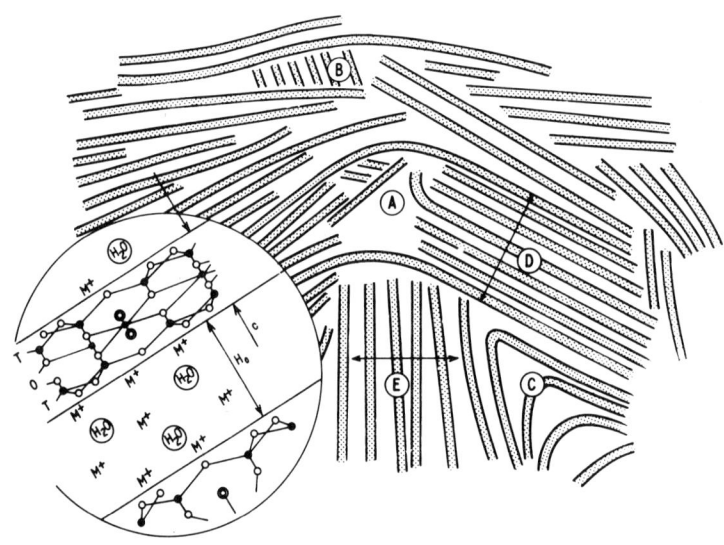

Fig. 2. Schematic diagram of a compacted montmorillonite-water sample. Clay platelets are shaded and the defects A, B, C, D and E are referred to in the text. (T and O refer to tetra- and octahedral layers; O = O, ● = Si, ■ = Aℓ, ◉ = OH, $M^+$ = counter ion; c(Å) = 9.6, $H_o$ = water space.)

The polydispersity of the platelet diameters must lead to regions with faults of types (C) and (E) as well as voids (A) and edge-to-face stacking (B). A model which includes a wider range of faults in a single parameter is one of platelets stacked irregularly along the axis of a one-dimensional lattice (Cebula et al., 1978a). A platelet lies at a distance $d(1 \pm f(\sigma))$ from the previous one, where d is the mean basal spacing and $f(\sigma)$ has a Gaussian distribution of values. $\sigma$ characterizes the halfwidth of the distribution. The advantage of using a Gaussian distribution is that an analytical expression can be derived for the diffracted intensity. The diffraction pattern is

$$I(s) \propto |F_o(s)|^2 \cdot Z(s) \cdot A(s) \tag{1}$$

where $|s| = (2\sin\theta)/\lambda$, $F_o(s)$ is the structure factor for the one-dimensional unit cell, $Z(s)$ is the Fourier transform of the disordered lattice, and $A(s)$ takes account of various instrumental corrections. A fit of this model to the observed pattern is shown in Figure 1 with a Gaussian distribution of half-width about 10% of the basal spacing. For a compacted montmorillonite with two molecular layers of water this corresponds to a spread of lattice parameters from 13.5 to 16.5 Å. This is sufficiently narrow that one, two and three layer samples may just be distinguished as discontinuities in the diffraction pattern as a function of humidity (Cebula et al., 1978a).

The model is already quite successful in predicting the main features of the diffraction pattern. Other features such as regions of gross folding could also be included and would improve further the agreement between calculated and observed patterns (Güven, 1975). A similar model has recently been applied to the (hk0) reflections (Plançon and Tchoubar, 1977).

We have also used neutron diffraction to determine the orientation of the platelets with respect to the plane of the sample (Cebula et al., 1978a). A rocking curve ($\omega$ scan) shows that a large fraction of platelets are partially oriented in the plane. The rocking curve is explained quantitatively by a mixture model in which ~ 10% of the platelets are randomly oriented and the remainder are partially oriented in the plane of the sample, with an angular spread of ~ 40° about the mean orientation. The presence of a randomly oriented fraction is confirmed by the observation of a complete Debye-Scherrer ring in the transmission diffraction pattern. We attribute the randomly oriented fraction to the small platelets in the system (Figure 2).

DYNAMICS OF WATER IN COMPACTED LITHIUM MONTMORILLONITE

Incoherent quasielastic neutron scattering experiments were done on samples containing one, two and three molecular layers of water. The motion of water under these conditions is much slower than in the bulk liquid. A high energy resolution is therefore necessary which was not available for earlier neutron scattering experiments (Hunter et al., 1971, Olejnik et al., 1970, Olejnik and White, 1972).

In the quasielastic scattering experiment, an almost monoenergetic beam of neutrons is scattered from the clay sample, the energies of the scattered neutrons and their angular distribution being analysed simultaneously. Diffusive motions of the scattering nuclei lead to energy changes in the scattered neutrons which broaden the incident beam. This 'quasielastic' broadening varies with momentum transfer, $Q$ ($Q=4\pi\sin\theta/\lambda$, where $2\theta$ is the scattering angle and $\lambda$ the incident wavelength), in a way depending on the spatial properties of the diffusive motion. Because of the large incoherent scattering cross section of the proton it is relatively easy to subtract any background scattering from the clay and the resulting spectrum is essentially that of the protons in the water layer. Since the time and length scales for the scattering interaction are short, the spectrum is directly related to the diffusive motions of the protons at the molecular level.

The quasielastic spectrum of three layers of water in a compacted lithium montmorillonite is shown in Figure 3 after subtraction of the background and the standard corrections for instrumental factors. The dashes represent the observed scattering, the continuous lines are calculated curves which are discussed in more detail below. The most important feature of the scattering is that the energy fwhm (full-width at half maximum) increases with Q. This shows unambiguously that the protons must be undergoing translational diffusion with a diffusion coefficient greater than about $10^{-10}$ $m^2$ $s^{-1}$. The broadening and therefore the rate of diffusion also increases with the thickness of the water layer.

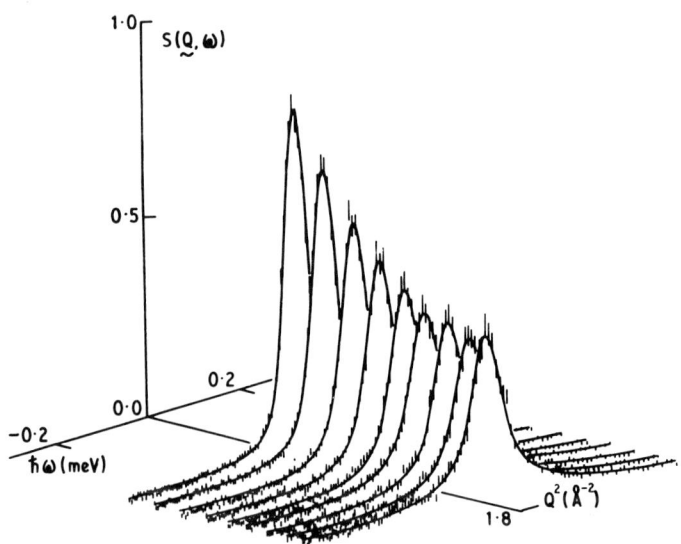

Fig. 3. Neutron quasielastic scattering from Lithium Montmorillonite containing three molecular layers of water. (T(K) = 293, relative humidity (%) = 98, spacing (Å) = 18.9).

The quantitative analysis of the quasielastic scattering in heterogeneous systems is more complicated (Gamlen et al., 1978). The protons may diffuse directly by proton transfer or by rotational and translational diffusion of water molecules. For simple translational diffusion obeying Fick's law the energy fwhm is

$$\Delta E = 2D_t \hbar Q^2 \qquad (2)$$

where $D_t$ is the diffusion coefficient. The shape of the scattering law ($S(Q,\omega)$) is Lorentzian. The translational motion of water between clay platelets will be restricted to two dimensions. For this the scattering law remains Lorentzian but Q is replaced by $Q\sin\phi$ where $\phi$ is the angle between Q and the normal to the plane of the platelets. Though Q has a definite orientation with respect to the plane of the clay sample, the structural work described above shows that there is a spread of orientation of Q with respect to the platelets. The resulting scattering law is therefore an average over a number of Lorentzian curves with different heights and widths (Dianoux et al., 1975). A further factor to be included in any comparison of a model with the experimental data is the finite resolution of the spectrometer (about 40 μeV fwhm). The resolution function is a triangle and the expression for the convolution of a Lorentzian with a triangle is analytical. The average scattering law was computed with this expression using the orientation distribution and powder fraction determined by neutron diffraction. The data were fitted separately at each scattering angle. The model of translational diffusion in two dimensions was found to fit the data only moderately well though considerably better than the simpler model of Fickian diffusion in three dimensions.

Water molecules are also expected to undergo rapid rotational diffusion. Rotational diffusion of a spherical molecule which is not undergoing translational diffusion gives a scattering law consisting approximately of an elastic peak and a Lorentzian. The fwhm of the Lorentzian is

$$\Delta E = 4\hbar D_r \qquad (3)$$

where $D_r$ is the rotational diffusion coefficient, and the ratio, p, of the intensity of the elastic peak to the total intensity is

$$p = \sin^2(Qr)/(Qr)^2 \qquad (4)$$

where r is the radius of gyration of $H_2O$ (Springer, 1972). Rotational diffusion alone does not account for the observed scattering law because the observed data does not contain a contribution from an elastic peak. However, a model allowing simultaneous rotational diffusion (using the approximate model above) and translational diffusion in three dimensions fits the data extremely well. The solid curves shown in Figure 3 were calculated using this model with values of the diffusion coefficients given in Table 1.

TABLE 1

Diffusion coefficients of water in Lithium Montmorillonite

| Number of water layers | Relative humidity % | Basal spacing/Å | $D_t$ /($m^2 s^{-1} \times 10^{-10}$) | $D_r$ /($s^{-1} \times 10^{11}$) |
|---|---|---|---|---|
| 1 | 32 | 12.5 | 3.0 | 5 |
| 2 | 58 | 15.0 | 4.0 | 5 |
| 3 | 98 | 18.9 | 5.2 | 5 |

The neutron results show that the protons move more slowly than in normal water ($D_t \sim 2.0 \; 10^{-9} m^2 s^{-1}$) but rotational diffusion is comparable. If distinct water types exist in the system (bound, free etc.), then: a) exchange between types is at a rate comparable to the diffusion, and/or b) motions of $Li^+$, say, are such as to average proton motions as seen by the scattered neutrons. It is likely that many types of water exist in the system not only as a result of it's heterogeneity but also owing to it's fault-like structure earlier described.

AGGREGATION IN CLAY SOLS

Because montmorillonite platelets are typically about 10Å thick and up to 2000 Å in diameter, it is possible by choosing the range of momentum transfer of a neutron small angle scattering experiment to study selectively adsorption at the surface of the platelets and aggregation phenomena.

Figure 4 shows the small-angle neutron scattering for a 1% weight for weight solution of lithium montmorillonite in water corrected for background and detector efficiency.

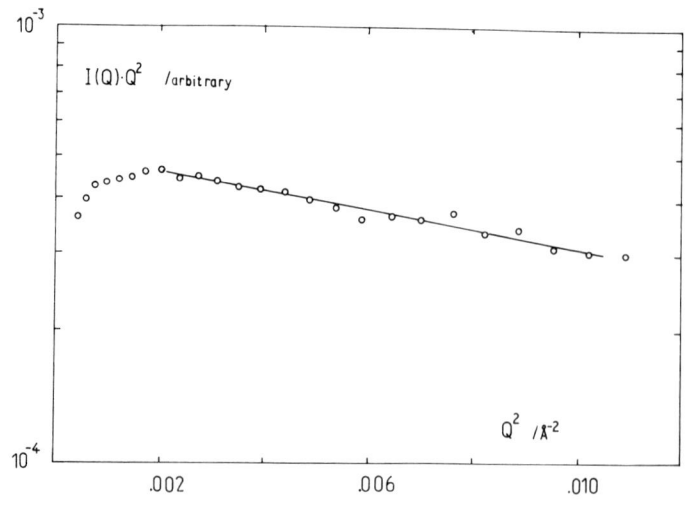

Fig.4. S.A.S. from a 1% weight for weight sol of Lithium Montmorillonite in water corrected for background and detector efficiency.

For disc like colloidal particles the small-angle scattered intensity is given by (Guinier and Fournet, 1955)

$$I(Q) \propto \exp(-Q^2 H^2/12)/Q^2 \qquad (5)$$

where R is the radius of the disc and H its thickness. The formula is valid for QH < 1 and for QR >> 1. The polydispersity in platelet diameters is not important in the range of Q used for these experiments. The linear portion of the curve in Figure 4 is well represented by equation (5), where H has a value of 10.3 Å. This is the value expected for the completely dispersed sol.

In contrast, as the counter ion is changed through sodium and potassium to caesium there is a marked change in the small-angle scattering pattern and, in the case of a caesium sol with the same concentration of montmorillonite, the small angle scattering is quite different (Figure 5).

When H becomes large the equation given above is no longer valid (QH >> 1). Instead, the scattering is often given simply by Porod's law (Guinier and Fournet, 1955) which is

$$I(Q) \propto Q^{-4} \qquad (6)$$

As shown in Figure 5 the scattering from the caesium sol obeys Porod's law very well showing that H≃40Å and that extensive aggregation has occurred with particles containing, on average, about four platelets together.

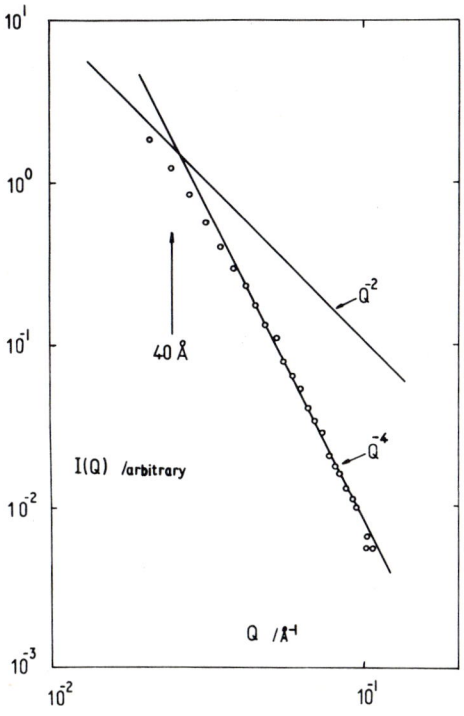

Fig. 5. Small angle scattering from a 1% weight for weight solution of caesium montmorillonite in water corrected for background and detector efficiency. The slope of the log-log plot is -4 indicating that aggregation of the particles has occurred.

Potassium montmorillonite sols show similar scattering patterns to those of caesium. The scattering from sodium is close to that of lithium at high Q but shows some low Q effects. Insofar as it has been possible to detect clustering of platelets of caesium montmorillonite, it is evident that the study of adsorption on the platelets is readily accessible through low-angle neutron scattering; the platelet surface area is of the order of 800 $m^2 g^{-1}$ and the contrast matching point for the montmorillonite mineral lies between 60 and 70% $D_2O$, depending on whether a full proton exchange or no proton exchange with the protons in the mineral occurs.

ACKNOWLEDGEMENTS

We thank Professor R.H. Ottewill and Dr S.R. Middleton (Bristol University) for the samples and helpful discussions. We also thank Dr G. Zaccai (I.L.L., Grenoble) for his help and advice. The work was supported by the Science Research Council grant B-RG-80212.

REFERENCES

Callaghan, I.C. and Ottewill, R.H. (1974) Interparticle forces in Montmorillonite gels; Disc. Faraday Soc., 57,110.
Cebula, D.J., Thomas, R.K., Middleton, S.R., Ottewill, R.H. and White, J.W., (1978a) Neutron diffraction from clay-water systems : Clays and Clay Min.in press.
Cebula, D.J., Thomas, R.K., Tabony, J., Harris, N.M. and White, J.W. (1978b) Neutron scattering from colloids : Disc. Faraday Soc., 65.
Dianoux, A.J., Volino, F. and Hervet, H. (1975) Incoherent scattering law for N.Q.E.S. : Mol. Phys., 30 (4), 1181.
Gamlen, P.H., Thomas, R.K., Trewern, T., Bomchil, G., Harris, N.M., Leslie, M., Tabony, J. and White, J.W. (1978) Structure and dynamics of $NH_3$ adsorbed on Surfaces: J.C.S. Faraday I., in press.
Guinier, A (1963) X-ray diffraction : Freeman, San Francisco.
Guinier, A. and Fournet, G. (1955), S.A.S. of X-rays: Wiley, New York.
Güven, N. (1975) Evaluation of bending effects on diffraction intensities: Clays and Clay Min., 23, 196.
Hunter, R.J., Stirling, G.C. and White, J.W. (1971) Water dynamics in clays by Q.E.N.S.: Nature Phys. Sci., 230 (17), 192.
Haywood, B.C.G. and Worcester, D.L. (1973) A simple neutron diffractometer for S.A.S.: J. Phys. E. 6, 568.
I.L.L. (1977) Facilities at the high flux beam reactor, Grenoble.
Olejnik, S., Stirling, G.C. and White, J.W. (1970) Neutron scattering studies of hydrated layer silicates. Spec. Disc. Faraday Soc., 1.
Olejnik, S. and White, J.W. (1972) Thin layers of water in Vermiculites and Montmorillonites - modification of water diffusion: Nature Phys.Sci.,236(62),15.
Plançon, A. and Tchoubar, C. (1977) Determination of structural defects in phyllosilicates by X-ray diffraction I and II : Clays and Clay Min., 25,430 & 436
Springer, T. (1972) Q.E.N.S. for the study of diffusive motions in solids and liquids : Springer Verlag, Berlin.

# NEUTRON SCATTERING STUDIES OF THE DYNAMICS OF INTERLAMELLAR WATER IN MONTMORILLONITE AND VERMICULITE

PETER L. HALL, D.K. ROSS, J.J. TUCK and M.H.B. HAYES
Department of Physics and Chemistry, University of Birmingham, Birmingham
B15 2TT, United Kingdom.

## ABSTRACT

The technique of quasi-elastic neutron scattering (QENS) spectroscopy is discussed. This technique can give detailed information on rapid ($10^{-11}$-$10^{-14}$ s) rotational and translational motions of hydrogenous molecules. Results of QENS studies of $Ca^{2+}$ and $Mg^{2+}$ - exchanged vermiculite and montmorillonite containing up to two and three layers of interlamellar water molecules are presented. The data clearly distinguish between coordinated (hydration-shell) and non-coordinated water molecules. The former do not diffuse on the observable time-scale. The latter exhibit rotational motions with a correlation time in the range $10^{-11}$ to $10^{-12}$ seconds, and somewhat slower translational diffusion. High resolution QENS measurements indicate a diffusion coefficient of $3.4 \times 10^{-6}$ $cm^2 s^{-1}$ for the non-coordinated water molecules in two-layer hydrate of Ca and Mg montmorillonite.

## INTRODUCTION

The scattering of slow neutrons by condensed matter can yield detailed information about the structure and dynamics of the scattering medium. Springer (1972) and Willis (1973) among others have provided comprehensive introductions to the field; here we briefly review the topic of incoherent quasi-elastic neutron scattering (QENS) which is of main relevance to the present study. In this technique, one observes small Doppler energy and momentum changes, $\hbar\omega$ and $\hbar Q$ respectively, which occur when neutrons are scattered by individual atoms undergoing rapid diffusive motions. These changes result in a broadening of the elastic scattering peak. The dependence of the intensity and width of the quasi-elastic broadening on Q contains detailed dynamical information such as diffusion coefficients and residence times. In particular, because of the relatively large incoherent scattering cross-section of hydrogen (Willis, 1973), QENS is particularly sensitive to the diffusion of hydrogenous liquids or hydrogenous adsorbed phases (Springer, 1972), among which may be included interlamellar water or organic molecules in clays and other layer structures.

In order to interpret the QENS spectra, it is necessary to postulate a physical model for the molecular motions, from which one can obtain the self correlation function $G_s(\underline{r},t)$. Van Hove (1954) showed that this function can then be Fourier transformed to obtain the incoherent scattering function $S_{inc}(\underline{Q},\omega)$, which describes the energy and momentum distribution of scattered neutrons. The simplest case is that of molecules undergoing continuous diffusion obeying Fick's Law with a diffusion coefficient D (units $cm^2 s^{-1}$). This model (Vineyard, 1958) predicts a scattering function of the form

$$S_{inc}(\underline{Q},\omega) = \frac{1}{\pi} \frac{DQ^2}{(DQ^2)^2 + \omega^2} \tag{1}$$

i.e. the quasi-elastic peak will be of Lorentzian shape on an energy scale, its full width at half-maximum, $\Delta E$, given by $\Delta E = 2\hbar DQ^2$ where, for elastic scattering $Q = (4\pi/\lambda_0) \sin(\theta/2)$. Here $\lambda_0$ is the incident neutron wavelength, and $\theta$ is the scattering angle. Thus a plot of $\Delta E$ versus $Q^2$ (a 'quasi-elastic broadening curve') would in this case be linear with gradient $2\hbar D$.

For many materials, the experimental broadening curves are approximately linear at low Q, but show significant deviations from linearity at high Q, the values of $\Delta E$ perhaps reaching a plateau value or even decreasing with increasing values of Q. Behaviour of this kind can be explained on the basis of jump diffusion models, such as those of Singwi and Sjolander (1960) or Chudley and Elliott (1960). Broadening curves of more complex shapes, including inflexions, also may occur. These indicate either (a) diffusion with a well-defined jump length, or (b) a complex quasi-elastic peak shape. The latter might be due either to the superimposition of two distinct types of motion, or to the occurrence of rotations about fixed axes on the observable time scale ($10^{-11} - 10^{-14}$ sec.) (Springer, 1972).

In general, QENS is sensitive to both translational and rotational motions and therefore great care is necessary in interpreting experimental data. In particular, when significant rotational contributions to the quasi-elastic scattering occur, it is no longer possible to relate the gradient of the broadening curve to the true, macroscopic diffusion coefficient, D. Values derived from equation (1) can be regarded only as "apparent" diffusion coefficients, $D_{eff}$, which will not agree with values measured by macroscopic methods if rotational components are present. The separation into rotational and centre-of-mass diffusion requires both detailed analysis of quasi-elastic peak shapes and also good spectrometer energy resolution (since the quasi-elastic scattering is invariably resolution-broadened). High resolution instruments (such as the IN10 backscattering spectrometer) and sophisticated data-fitting methods (Howells, 1975, Hall et al, 1978) have been only relatively recently developed.

The pioneering neutron scattering work on interlamellar water dynamics in $Li^+$

and $Na^+$ exchanged montmorillonite and vermiculite (Hunter et al, 1971; Olejnik and White, 1972) were mainly concerned with large water layer thicknesses and did not fulfil either of the above criteria. Their data were interpreted in terms of translational Fickian diffusion only; moreover the energy resolutions then available were insufficient to permit a separation between translational and rotational processes to be made. These studies may thus be regarded as giving a qualitative picture of interlamellar water diffusion in clays. Here we describe more recent QENS measurements on clay-water systems at both medium and high energy resolution. These studies permit, for the first time, a separation between rotational and translational motions in clay-water systems (Hall et al, 1977).

## EXPERIMENTAL
### Preparation of homo-ionic clays

The clay minerals used were montmorillonite (API No. 26, Wyoming, U.S.A.) and vermiculite (Libby, Montana, U.S.A.) and were obtained from Ward's Natural Science Establishment, Rochester, N.Y., U.S.A. Sodium exchanged montmorillonite was prepared as described by Posner and Quirk (1964), and the 0.2 - 2.0µm e.s.d. vermiculite (obtained by grinding flakes in a Glen Creston ball mill) was steeped for 10 days (with frequent replacement of electrolyte) in 1M sodium chloride solution. After removal of excess electrolyte, self supporting films were laid down by suction of the liquid through a cellulose nitrate membrane. 1 M $CaCl_2$ or $MgCl_2$ solutions as appropriate were added to the liquid over the sedimented gels, and excess electrolyte was sucked through for an extended period. After washing with water until chloride free, the sedimented clays were dried in air at $110°C$. Samples were then equilibrated at constant water vapour pressures $P/P_o$ to give up to three interlamellar water layers for montmorillonite, and two for vermiculite. The equilibrated clay samples were then cut into rectangular slabs (ca. 40mm x 40mm) of thickness 1 - 2mm (calculated to give a transmission factor of 0.90 - 0.95 for slow neutrons), and sealed in aluminium cassettes for the neutron beam experiments.

### Neutron Diffraction Studies

Neutron diffraction measurements, using both a four-circle diffractometer (D16, at the Institut Laue Langevin (ILL) Grenoble, France, incident wavelength $\lambda_o$ = 4.63Å) and a two-circle diffractometer (Guide Tube, at AERE, Harwell, U.K., incident wavelength 4.7Å) were made both to check the homogeneity of expansion and degree of platelet orientation of the samples prepared for the quasi-elastic scattering experiments. Space does not permit a full description of the results of these measurements, which will be given elsewhere. To summarise very briefly:
(i) Neutron Bragg diffraction patterns indicated homogeneous expansions, and c-axis

spacings which agreed well with those measured by X-ray diffraction. However apart from the 001 reflection, higher (00 ) reflections were rarely observed. The (00 ) peaks from neutron diffraction were in general broader than the comparable X-ray peaks.

(ii) Neutron rocking curves (rotation of sample with fixed detector position) were measured. Fig. 1 shows the (001) Bragg peak and corresponding rocking curve for $Ca^{2+}$ montmorillonite (two water layer, $d_{001} = 15.5$Å). After correction for beam attenuation effects, the corrected rocking curve (continuous curve in Fig. 1(b)) indicates a platelet orientation distribution function having a full width at half maximum intensity (FWHM) of $37°$. This value was typical of all of the samples used in this work, and is in good agreement with parallel neutron diffraction studies of monovalent cation-exchanged montmorillonite-water systems prepared by a compression technique (Cebula et al, 1978).

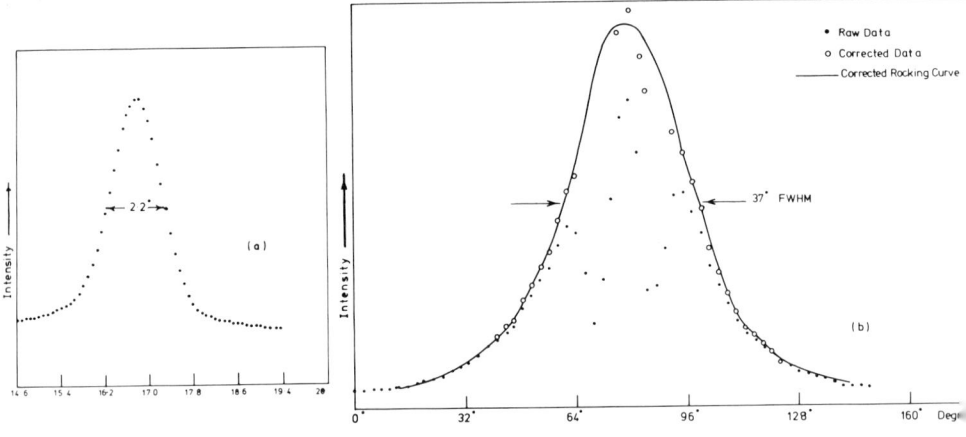

Fig. 1(a) (001) reflection of $Ca^{2+}$ montmorillonite two-layer hydrate, obtained by neutron diffraction. ($d_{001} = 15.5$Å). (b) Corresponding rocking curve (ω scan). (Data taken on D16 diffractometer).

Quasi-elastic scattering measurements

Two neutron scattering instruments at the ILL, referred to below as IN5 and IN10, were used. IN5, a multichopper time-of-flight spectrometer, in which the incident wavelength, $\lambda_o$, may be varied between 4 and 12Å, achieves energy resoluti as low as 20μeV. IN10, a high-resolution spectrometer, is based on the back scattering principle and has the highest energy resolution of the instruments available at this time (ca. 1.0μeV at $\lambda_o = 6.3$Å).

Some of the results of the present QENS measurements on clay-water systems have been reported previously (Hall et al, 1977). The time-of-flight neutron scattering spectra of $Ca^{2+}$ - and $Mg^{2+}$ - montmorillonite and - vermiculite clays, which had a variety of water layer thicknesses, were measured on IN5 at $\lambda_o = 8.5$Å and $\lambda_o = 6.0$Å. Spectra were recorded for scattering angles, $\theta$, ranging from $10°$ to $135°$. The principal experimental parameters are summarised in Table 1.

TABLE 1  Wavelengths, energy resolutions and maximum Q-values of neutron scattering experiments.

| Instrument | Incident wavelength, $\lambda_o$ (Å) | Resolution FWHM ($\mu$eV) | Maximum Q (Å$^{-1}$) |
|---|---|---|---|
| IN5 | 6.0 | 110 | 1.93 |
| IN5 | 8.5 | 38.5 | 1.36 |
| IN10 | 6.3 | 1.5 | 1.41 |

Data were analysed by a method developed in Birmingham by Hall et al (1978) and incorporated into a fitting programme QUELDA written in Fortran. The basis of the method is to directly simulate the observed time-of-flight spectrum by a model which contains a relatively small number of variable parameters. The values of the latter may be refined by a Newton-Raphson method. The main parameters obtained are (i) $\Delta E$, the FWHM of the quasi-elastic component, and (ii) X, the ratio of quasi-elastic to total (elastic plus quasi-elastic scattering).

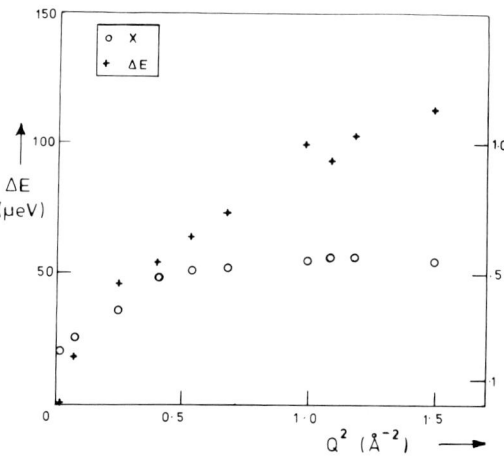

Fig. 2  Variation of $\Delta E$ and X with $Q^2$ for $Ca^{2+}$ montmorillonite two layer hydrate. Data from IN5, at $\lambda_o = 8.5$Å.

The first point that emerges from the results is that $\Delta E$ does not depend on the direction of Q relative to the sample plane normal, as would be expected for two dimensional Fickian diffusion parallel to the sample plane. The data are therefore plotted without regard to sample orientation. A typical example, for the two-layer hydrate of $Ca^{2+}$ montmorillonite recorded on IN5 using $\lambda_o$ = 8.5Å, is shown in Fig. 2. The most remarkable feature of the data is that X increases smoothly with Q up to an asymptotic value (X = .55). $\Delta E$ also increases with Q. All samples showed similar results apart from significant inter-sample variations in the asymptotic value of X. In a second series of experiments at $\lambda_o$ = 6.0Å, the range of Q was increased. In the overlapping region identical values of $\Delta E$ and X were obtained but a point of inflection now appears in the broadening curves at $Q^2 \sim 2.0$Å$^{-2}$. Table 2 summarises the apparent diffusion coefficients, $D_{eff}$, and the asymptotic values of X.

TABLE 2   Apparent diffusion coefficients and asymptotic quasi-elastic fractions for clay-water samples.

| Sample | No. of water layers | Component | $D_{eff}$ (cm$^2$s$^{-1}$ x10$^{-6}$) | Quasi-elastic Fraction, X |
|---|---|---|---|---|
| $Ca^{2+}$ montmorillonite | 2 | Broad | 13.7 | 0.55 |
| - | - | Narrow | 3.4 | - |
| $Ca^{2+}$ montmorillonite | 3 | Broad | 16.0 | 0.65 |
| $Mg^{2+}$ montmorillonite | 2 | Broad | 15.0 | 0.60 |
| - | - | Narrow | 3.4 | - |
| $Mg^{2+}$ montmorillonite | 3 | Broad | 16.0 | 0.68 |
| $Ca^{2+}$ vermiculite | 2 | Broad | 10.3 | 0.40 |

Two samples ($Ca^{2+}$ and $Mg^{2+}$ two layer hydrates) were examined on IN10. The data were analysed using the programme SAND (Howells, 1975) which permits, as for the IN5 data, a separation of the observed spectra at elastic and quasi-elastic components. Up to a value of $Q^2 = 0.6$Å$^{-2}$, the broadenings for both samples increased linearly, and lay on the same straight line which had a gradient of $3.4 \pm 0.2 \times 10^{-6}$ cm$^2$s$^{-1}$. No definite anisotropy of $\Delta E$ with respect to Q direction has yet been observed. However the broadenings measured on IN10 are approximately 5 times smaller than the comparable data from IN5. This result has been confirmed in a recent, more extensive IN10 experiment the analysis of which is at present in progress. At higher Q-values, the broadenings apparently become too large to be observed with the narrow energy window (±15μeV) used on the IN10 instrument.

DISCUSSION OF RESULTS

In discussing the interpretation of the experimental data, our physical model will first be described. The data will then be compared with the predictions of this model. It is hoped that this framework will facilitate a clear exposition of our analysis with the space available. More detailed discussions may be found elsewhere (Hall et al, 1977).

The first conclusion is that the interlamellar water molecules may be divided into two categories on the basis of their neutron scattering characteristics:
(i) a relatively tightly-bound, immobile fraction, which gives elastic scattering only;
(ii) a relatively loosely-bound, mobile fraction, exhibiting quasi-elastic broadening.

On the basis of considerations discussed below, we attribute the two categories to (i) water molecules directly coordinated to exchange cations in inner hydration shells, and (ii) molecules not so coordinated. In comparison with this distinction, any difference between surface layers and non-surface layers is not significant.

The second important conclusion is that the fastest type of diffusive motion - which gives rise to the broadest quasi-elastic component - is due to rapid reorientations of the non-coordinated water molecules. Further, the available evidence favours the conclusion that the next fastest type of motion is the macroscopic (translational) diffusion of the non-coordinated water molecules. The coordinated molecules are therefore presumed not to diffuse significantly on the neutron timescale, and will only contribute to the elastic and inelastic parts of the spectra.

## The rotational motions of the non-coordinated water

As described above, the IN5 time-of-flight data can be separated into 'quasi-elastic' and 'elastic' components. A preliminary indication that the former was due to some kind of rotational motion was the absence of any anisotropy in the broadening in preferentially orientated samples. A two-dimensional diffusion model (Stockmeyer et al, 1976), averaged over the particle orientation functions deduced from our neutron diffraction data, predicts that significant orientational dependence would be observed. Even a more realistic calculation of Fick's Law diffusion in three dimensions, in which the effect of the silicate layers is introduced by the assumption of parallel planar impervious boundaries (Hall and Ross, 1978) predicts significant anisotropy, particularly at low Q. The alternative conclusion, that a localized motion is being observed, is strongly supported by the fact that the quasi-elastic components of the IN5 spectra have zero or low intensity at low Q rising to a constant value at $Q \sim 1 \text{Å}^{-1}$. The corresponding decrease in the elastic intensity is exactly the behaviour expected for incoherent scattering from atoms diffusing within a restricted volume and is referred to as the "Elastic Incoherent Structure Factor" (EISF) (Dianoux et al, 1975). Models of this type include reorientations on the surface of a sphere (Sears, 1966) or about a single axis (Barnes, 1973). The general form of $S(\underline{Q},\omega)$ predicted by these models may be written

$$S^{rot}(\underline{Q},\omega) = A_o(Q)\delta(\omega) + \frac{1}{\pi} \sum_n A_n(Q) L_n(\omega) \qquad (2)$$

where $A_o(Q)$ is the EISF and the $L_n(\omega)$ are Lorentzians. The actual forms of $A_n(Q)$

are model dependent, but it may be noted that the position of the first minimum of $A_o(Q)$ is of the order of magnitude of the inverse of the radius of gyration of the motion. At present, though it is not possible to be very precise about the nature of the rotational motion, the Q-dependence of the observed EISF is certainly much what one would expect from the size of the water molecule. Moreover the increase in $\Delta E$ with Q implies a relatively complex isotropic or uniaxial rotational motion, rather than fixed-length jumps between adjacent layers or $180°$ rotational flips around a twofold axis (Springer, 1972).

The next feature of our model is assignment of the 'tightly' and 'loosely' bound fractions to coordinated and non-coordinated water molecules. This arises from careful consideration of the relative intensities of the elastic and quasi-elastic scattering at high Q. In all of the rotational models mentioned above, $A_o(Q)$ decreases steadily to a value close to zero with increasing Q. Now $S^{rot}_{\text{tot}}(Q,\omega)$ is normalized to unity, and therefore taking $A_o(Q) \to 0$ at high Q, the quasi-elastic/elastic ratio will be given by $R = \Sigma_{H_2O(rot)}/\Sigma_{lattice} + \Sigma_{H_2O(non\,rot)}$ if a fraction of the water molecules do not rotate on the observed time scale, so contributing a cross section $\Sigma_{H_2O}$ (non-rot) to the elastic scattering. Knowing the appropriate cross sections (Willis, 1973) and the mass of water adsorbed per gram of clay, the ratio $T = \Sigma_{H_2O(total)}/\Sigma_{lattice}$ can be calculated. The values of R found experimentally were always significantly below the calculated values of T. It was thus necessary to assume in all cases the existence of a non-rotating fraction of water molecules contributing additional elastic scattering. Taking accepted values for the c.e.c. of the montmorillonite and vermiculite (Posner and Quirk, 1964; Pick, 1972) and assuming a coordination number of 6 for the divalent exchange cations when two or more water layers are present (Shirozu and Bailey, 1966) it is found that a good correlation between experimental and calculated values of R is obtained on the assumption that the coordinated water contributes to the additional elastic scattering ($\Sigma_{H_2O(non-rot)}$) and that the non-coordinated water contribution to quasi-elastic scattering ($\Sigma_{H_2O(rot)}$) (Hall et al, 1977).

From the data in Table 2 it is seen that the rapidity of the rotational motions tend to increase with increasing water layer thickness. It is of interest to note that the rotational mobility of the non-shell water in $Ca^{2+}$ vermiculite ($d_{001} = 14.8$Å) is of the same order of magnitude as, though somewhat less than, the analogous component in $Ca^{2+}$ montmorillonite ($d_{001} = 15.5$Å). From the $\Delta E$ ($Q \to \infty$) val correlation times $\tau \sim \Delta E/2h$ (Springer, 1972) of $10^{-11} - 10^{-12}$s are obtained.

## Translational motions of the non-coordinated water

The higher energy resolution IN10 data can again be divided into a quasi-elastic plus elastic component, and a flat background. Because of the narrow energy window of this instrument, the broadest component in the peak will only contribute to the flat background. On the other hand, the quasi-elastic component now visible would have been so narrow as to have been included in the elastic component of the

IN5 profile. As stated above, we interpret this quasi-elastic component as macroscopic diffusion of the non-coordinated water molecules on a somewhat slower time scale than their rotational motions discussed above. On this basis, it is predicted that (i) the elastic component of the IN10 data (due to lattice and coordinated water) will have an intensity independent of Q; (ii) the quasi-elastic component will decrease in intensity with Q due to the decrease of $A_o(Q)$; (iii) the width, $\Delta E$ of the narrow quasi-elastic component should increase linearly with $Q^2$; and (iv) the quasi-elastic component should be somewhat anisotropic at low Q, as predicted by Hall and Ross (1978). A recent IN10 experiment on $Ca^{2+}$ montmorillonite at a wide range of relative humidities has been completed. Though comprehensive analysis of the data is still in progress, it is clear that both IN10 experiments confirm predictions (i), (ii), and (iii) while no conclusion can yet be reached regarding (iv). Accepting our hypothesis, therefore, we derive a macroscopic diffusion coefficient for the non-hydration shell water in both the two-layer $Ca^{2+}$ and $Mg^{2+}$ montmorillonites of $D = 3.4 \pm 0.2 \times 10^{-6} cm^2 s^{-1}$.

This value is considerably larger than the values for the diffusion coefficient of the hydrated exchange cations in montmorillonite and vermiculite measured by radioactive tracer techniques (Lai and Mortland, 1968). Similarly the rotational diffusion of the cation hydration shells in $Na^+$ Llano vermiculite measured by Hougardy et al (1976) by NMR techniques, is on too slow a timescale to be measured in our experiments. Both the above techniques therefore support our hypothesis that the narrow quasi-elastic component is due to macroscopic diffusion of non-coordinated water molecules. The dominance of the rotational contributions, however, may well explain the previous difficulties found in reconciling neutron scattering and NMR data from clay water systems (Hecht and Geissler, 1973).

The observation of two distinct water populations by neutron scattering is also supported by the fact that the postulated rate of exchange between shell and non-shell water molecules $\sim 10^5 sec^{-1}$ (Hougardy et al, 1976) is too slow to influence the neutron scattering data. In addition, while it is known that water in clays is much more highly dissociated than in bulk water, so that the electrical conductivity of clay-water systems is thought to be protonic in origin (Fripiat et al, 1965; Touillaux et al, 1968), nevertheless, the maximum degree of dissociation (<1%) is far too small to explain our neutron scattering results. The diffusion observed by neutron scattering is therefore molecular in character. The results are also consistent with infrared measurements (Farmer and Russell, 1971).

REFERENCES

Barnes, J.D., 1973. Inelastic neutron scattering study of the "rotator" phase transition in n-nonadecane. J. Chem. Phys., 58, 5193-5201.
Cebula, D., Thomas R.K., Middleton, S., Ottewill, R.H. and White, J.W., 1978. Neutron diffraction from clay-water systems. Clays and Clay Minerals (in press).

Chudley, C.T., and Elliott, R.J., 1960. Neutron scattering from a liquid on a jump diffusion model. Proc. Phys. Soc., 77, 353-361.

Dianoux, A.J., Volino, F. and Hervet, H., 1975. Incoherent scattering law for neutron quasi-elastic scattering in liquid crystals. Mol. Phys. 30, 1181-1194.

Farmer, V.C. and Russell, J.D., 1971. Interlayer complexes in layer silicates. The structure of water in lamellar ionic solutions. Trans. Faraday Soc. 67, 2737-2749

Fripiat, J.J., Jelli, A., Poncelet, G. and Andre, J., 1965. Thermodynamic properties of adsorbed water molecules and electrical conduction in montmorillonites and silicas. J. Phys. Chem. 69, 2185-2197.

Hall, P.L., Ross, D.K., Tuck, J.J. and Hayes, M.H.B., 1977. Dynamics of interlamella water in divalent cation exchanged expanding lattice clays. Proc. IAEA Symp. Neutron Inelastic Scattering (Vienna, October 1977), 1, 617-635.

Hall, P.L. and Ross, D.K., 1978. Incoherent neutron scattering function for molecula diffusion in lamellar systems. Mol. Phys. (in press).

Hall, P.L., Ross, D.K. and Anderson, I.S., 1978. Direct model fitting of neutron scattering data from time-of-flight spectrometers. Nucl. Instr. Meth., (in press).

Hecht, A.M. and Geissler, E., 1973. Nuclear spin relaxation in a single and double layer system of adsorbed water. J. Coll. Interface Sci., 44, 1-12.

Hougardy, J., Stone, W.E.E. and Fripiat, J.J., 1976. NMR Study of adsorbed water I. Molecular orientation and protonic motions in 2-layer hydrate of Na vermiculite. J. Chem. Phys., 64, 3840-3851.

Howells, W.S., 1975. Institut Laue Langevin (Grenoble, France). Report Nos. 75H130T and 76H122T.

Hunter, R.J., Stirling, G.C. and White, J.W., 1971. Water dynamics in clays by neutron spectroscopy. Nature Phys. Sci., 230, 92-4.

Lai, T.M. and Mortland, M.M., 1968. Cationic diffusion in clay minerals II Orientation effects. Clays and Clay Minerals, 16, 129-136.

Olejnik, S. and White, J.W., 1972. Thin layers of water in vermiculites and montmorillonites modification of water diffusion. Nauure Phys. Sci., 236, 15-16.

Pick, M.E., 1972. Ph.D. Thesis, University of Birmingham.

Posner, A.M. and Quirk, J.P., 1964. The adsorption of water from concentrated electrolyte solutions by montmorillonite and illite. Proc. Roy. Soc., 275A, 35-56.

Sears, V.F., 1966. Theory of cold neutron scattering by homonuclear diatomic liquids I. Free rotation. Can. J. Phys. 44, 1279-97. Also: II Hindered rotation, ibid, p. 1299-1311.

Shirozu, H. and Bailey, S.W., 1966. Crystal structure of a two-layer Mg-vermiculite Am. Mineral., 51, 1124-1143.

Singwi, K.S. and Sjölander, A., 1960. Diffusive motions in water and cold neutron scattering. Phys. Rev., 119, 863-861.

Springer, T., 1972. Quasi-elastic Neutron Scattering for Investigation of Diffuse Motion in Solids and Liquids, Springer-Verlag, Berlin.

Stockmeyer, R., Stortnik, H.S. and Conrad, H.M., 1976. Diffusive motions of molecules on catalytic surfaces. Proc. Conf. Neutron Scattering, Gatlinburg, Tenn., U.S.A., 1, 303-309.

Touillaux, R., Salvador, R., van der Meersche, C. and Fripiat, J.J. 1968. Study of water layers adsorbed on Na- and Ca-montmorillonite by pulsed NMR technique. Israel J. Chem., 6, 337-348.

Van Hove, L., 1954. Correlations in space and time and Born approximation scattering in systems of interacting particles. Phys. Rev., 95, 249-262.

Vineyard, G.H., 1958. Scattering of slow neutrons by a liquid. Phys. Rev., 110, 999-1010.

Willis, B.T.M., 1973. Chemical Applications of Thermal Neutron Scattering. Oxford University Press.

SELECTIVE COAGULATION AND MIXED LAYER FORMATION FROM SODIUM SMECTITE
SOLUTIONS

E. FREY and G. LAGALY
Institut für anorganische Chemie der Universität Kiel, 23 Kiel (G.F.R.)


ABSTRACT
Coagulation of colloidal solutions containing high- and low-charged smectite layers leads to different layer sequences in the coagulated crystals. Coarser fractions (0,1 - 2$\mu$m) are selectively coagulated whereas in finer fractions ( $<$ 0,1$\mu$m) mixed-layer formation predominates.

---

INTRODUCTION

An unsolved problem in colloid chemistry is the selective coagulation of one component from colloidal solutions of two or more kinds of particles. The few studies reported refer to systems with widely differing kinds of particles (Pugh and Kitchener, 1972). Separation of chemically identical species by coagulation was first realized by Iler in 1975. In a mixture of silica particles of different sizes, larger particles can be preferentially coagulated and separated from smaller ones. In general, however, it is impossible to detect selective coagulation in systems with chemically identical particles. Only one system offers particularly this possibility. Colloidal solutions of sodium smectites represent the most suitable model for the study of selectivity phenomena during coagulation by electrolytes. Their advantage of being available in a wide variety of samples with different surface charge densities but identical framework structures cannot be surpassed by any other system.

In diluted sodium salt solutions ( $<$ 0,2 m) sodium smectites expand beyond the layer separation of the quasicrystalline structure (about 2 nm; Lagaly et al., 1972; Norrish, 1973). The layers separate completely from one another and lose their parallel orientation. The process is reversible; increasing electrolyte concentration decreases the distances between the particles and forces the layers in parallel orientation and finally into the quasicrystalline state: the colloidal solution coagulates at distinct salt concentrations (critical coagulation concentrations "c.c.c.") The c.c.c. values are not subject of this discussion, but will be reported in a further paper (Frey and Lagaly,1978). Data for laponite and montmorillonite were determined by classical methods by Matijević and coworkers (Perkins et al., 1974; Schwartzen-Allen and Matijević, 1975).

Coagulation of the colloidal solution of a homogeneous sodium smectite leads to coagulates with the same layer sequence as the starting material, but probably with different particle size distribution. If a solution of two differently charged smectites is coagulated, the coagulated particles can be of different structures. For example particles formed by coagulation from a solution of a high-charged smectite A and a low-charged smectite B (in ratio about 1 : 1) can be composed of crystallites with predominantly layers A and crystallites with predominantly layers B (selective coagulation) or of crystals which contain layers A and B in regular or random distribution (mixed-layer formation, Fig. 1).

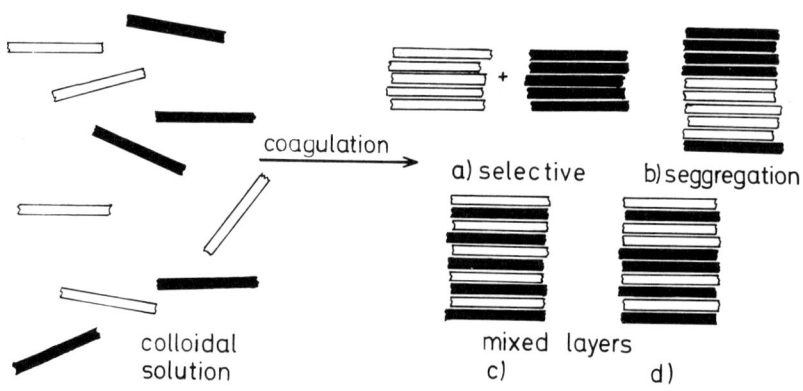

Fig. 1. Sequences of low- and high-charged layers in the coagulates of mixed colloidal solutions of low- and high-charged smectites (interlayer cations are not shown).

The kind of coagulation should be controlled by the interparticle forces especially the electrostatic terms between low- and high-charged surfaces. The present theories on colloid stability allow no definite prediction of the coagulation mechanism and the kind of the coagulates formed.

ANALYSIS OF THE LAYER SEQUENCE

Analysis of the structure of the coagulates provides a special problem. The structures shown in Figure 1 had to be distinguished even if the layers differ only slightly by the charge density. Exchange of the interlayer cations by long chain alkylammonium ions and X-ray investigation of the alkylammonium derivatives offers the unique method for layer sequence determination. Without going into details of the method (Lagaly and Weiss, 1976) Figure 2 illustrates the relations between layer sequence and variation of the basal spacings with the chain length of the alkylammonium ions. Segregation of larger numbers of identical layers in the crystallites (Fig. 1b) cannot be distinguished from selective coagulation (Fig. 1a).

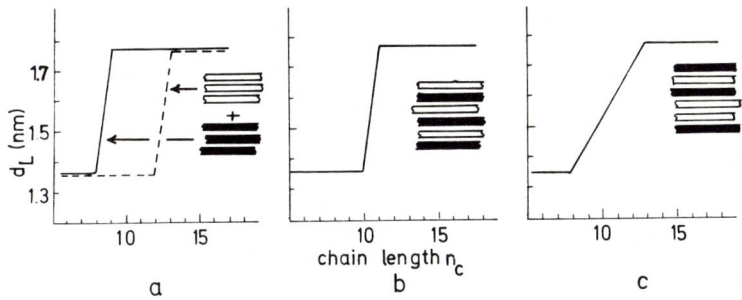

Fig. 2. Variation of the basal spacings $d_L$ of the alkylammonium ion exchanged forms of the coagulates with the alkyl chain length.

The realization of the discribed procedure with natural smectites involves the difficulty that the clays are not strongly homogeneous (Lagaly and Weiss, 1976). The starting materials A and B do not have identical layers A or B but a distribution of the layer charges $\xi$ around the mean values $\bar{\xi}_A$ and $\bar{\xi}_B$ (Fig. 4a and b). The knowledge of the charge distribution, however, allows the same conclusions to be drawn as in the ideal case.

EXPERIMENTAL

Smectites

The smectites used (a lower-charged beidellite "B 18/4", Unterrupsroth/Rhön, Germany, montmorillonite "M 1" Upton, Wyoming, AIP N° 25, Wards 48W 1250) were selected due to good quality of X-ray diagrams and overlapping charge distribution curves (Fig. 4a, b) which were considered to favor mixed-layer formation. Both sodium exchanged smectites in water show only central scattering even at high clay concentrations ( $\lesssim$ 50 g/l).

Preparation of sodium smectites

The natural samples of beidellite and montmorillonite were treated with dithionite solutions to remove the iron oxides and $H_2O_2$ to destroy organic material after Mehra and Jackson (1960) and Stul and Uytterhoeven (priv. comm.). For exchange of the calcium ions by sodium ions the samples were treated six times with sodium chloride solution (10 ml 1 m NaCl solution per g smectite). The fraction < 2 $\mu m$ separated by sedimentation was fractionated in particles < 0,1 $\mu m$ and particles 0,1 - 2 $\mu m$ (Tributh, 1970; Tanner and Jackson, 1947).

Preparation of colloidal solutions and coagulation experiments

Colloidal solutions were prepared by dispersion of 0,25 g sodium smectite in 1 l water or in 0,01 m $Na_4P_2O_7$ solution. The colloidal solutions were stable at least several days. Equal volumes of sodium beidellite and sodium montmorillonite solutions were mixed and coagulated under different conditions: fast and very slow addition of coagulating salt solutions, coagulation at different pH-values (4 $\lesssim$ pH $\lesssim$ 11), coagulation by $Ca^{2+}$ ions and alkylammonium ions, with $Li^+$ ions instead of $Na^+$ ions, by addition of ethanol or by slow evaporation of the water etc..

The coagulates were reacted with alkylammonium chloride solutions to determine the distribution of the high- and low-charged layers in the coagulates.

RESULTS

We spent much effort on coagulation experiments under a wide scale of different conditions. As a general result, the structure of t coagulates could be shown to depend principally on the particle size and not as much on the coagulation conditions.

Selective coagulation of larger particles (0,1 - 2$\mu m$)

The mixed solutions of the larger fraction of sodium beidellite and montmorillonite were selectively coagulated under all conditions mentioned above: the deposits consisted of a mixture of crystallites of higher-charged layers and crystallites of lower-charged layers.

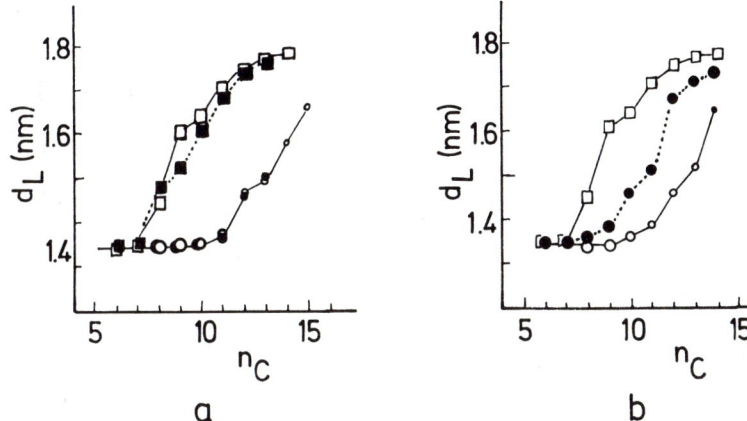

Fig. 3. Basal spacings $d_L$ of the alkylammonium derivatives
a: fraction 0.1 - 2 µm (selective coagulation)
  □ ○ high- and low-charged components of a "mechanical" mixture
  ■ ● high- and low-charged components of the coagulate
b: fraction <0.1 µm (mixed-layer formation)
  □ ○ as in Fig. a), ● coagulated sample
(changing sizes of symbols indicate changes of the intensity of the (001)-reflections with chain length.)

On first glance the basal spacings of the alkylammonium derivatives of the coagulates differ only slightly from those of a "mechanical" mixture of beidellite and montmorillonite (Fig. 3a). One might assume that the coagulate consisted of beidellite and montmorillonite crystals completely unchanged in comparison with the starting materials. The charge distribution curves, however, reveal differences and prove that the coagulated samples are similar to but not exactly identical with the starting materials (Fig. 4). The low-charged layers seggregate preferentially with each other and also the high-charged ones.

Different c.c.c. of sodium beidellite and sodium montmorillonite (0,25 - 0,29 and 0,36 - 0,44 moles NaCl/l, 0,01 m $Na_4P_2O_7$, pH ≈ 9) lead to the assumption that the separation in high- and low-charged crystallites is caused by successive coagulation with increasing salt concentration; lower concentration coagulates the lower-charged layers, higher concentration the higher-charged layers. The higher charged beidellite coagulates at lower salt concentration than montmorillonite due to fixation of gegen ions in the Stern layer (Frey and Lagaly, 1978). Fast increase of the salt concentration should favor the mixture of the layers. A complete mixture could not be observed even for very fast coagulation (colloidal solution into

2 m NaCl solution); some mixing of low- and high-charged layers, however, is indicated by the pronounced broadening of the X-ray reflections remarkably exceeding the line broadening of the coagulates from pure sodium beidellite or sodium montmorillonite solutions (Fig. 5).

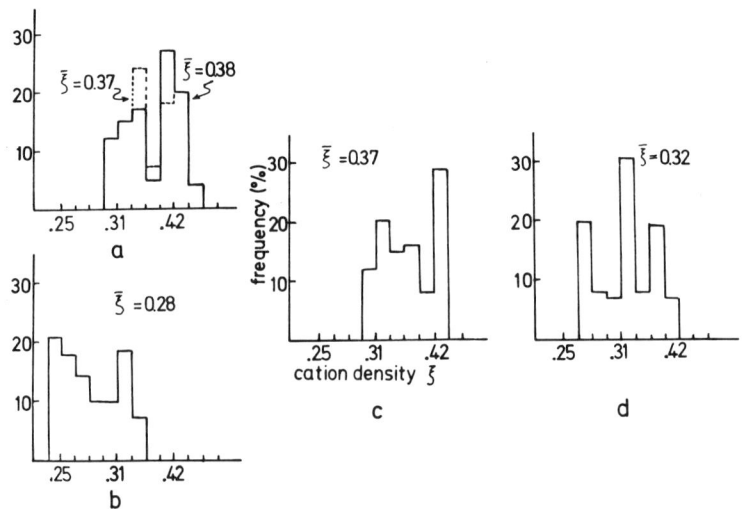

Fig. 4. Charge distribution curves:
a, b: starting beidellite (fraction 0,1 - 2 $\mu m$, dotted line: < 0,1 $\mu m$) and montmorillonite (both fractions) respectively;
c: fraction 0,1 - 2 $\mu m$ selectively coagulated, high charged component;
d: fraction < 0,1 $\mu m$, mixed-layer crystals ($\xi, \bar{\xi}$ in eq/(Si,Al)$_4$O$_{10}$).

Fig. 5. Debye-Scherrer diagrams of the decylammonium derivatives after slow (a) and fast (b) coagulation of a montmorillonite-beidellite solution (fraction 0,1 - 2 $\mu m$)

The results illustrate that beidellite-like crystals (not in any case completely identical with the starting beidellite crystals) can be selectively coagulated from colloidal beidellite-montmorillonite solutions. A typical experiment is shown in Table 1. 24 hours after addition of 0,28 moles NaCl/l to the colloidal solution (0,25 smectite g/l, 0,01 m $Na_4P_2O_7$, pH = 9) the coagulate is separated from the remaining solution which is then coagulated by addition of sodium chloride. The decylammonium derivatives indicate a nearly complete selective coagulation of the high-charged particles.

TABLE 1
Example for selective coagulation (colloidal solution: 0,125 g/l sodium beidellite + 0,125 g/l sodium montmorillonite, 0,01 m $Na_4P_2O_7$, pH = 9)

| Sample | $d_{001}$ *) | I *) | main composition |
|---|---|---|---|
| coagulate after addition of 0,28 m NaCl/l | 1,36 nm | weak | particles with beidellitic layers |
|  | 1,64 nm | strong |  |
| remaining colloidal solution after coagulation | 1,36 nm | strong | particles with montmorillonitic layers |
|  | 1,65 nm | weak |  |
| rapid coagulation with 0,5 m NaCl/l | 1,35 nm | strong | mixture of low- and high-charged particles |
|  | 1,70 nm | strong |  |

*) d- value and intensity of the (001)-reflections of the decylammonium derivatives

Mixed-layer formation of small particles (fraction < 0,1 μm)

In contrast to the behavior of larger particles coagulation of mixed solutions of sodium montmorillonite and beidellite particles < 0,1 μm leads to pronounced mixed-layer formation. Figures 3b and 4d show the spacings of the mixed-layer coagulate after alkylammonium ion exchange compared with the "mechanical" mixture of the same particles size and the resulting charge distribution curves. The mixing of the layers is further demonstrated by the powder diffraction diagrams in Figure 6. Despite some line broadening by the reduced particle size the peaks of beidellite and montmorillonite of the "mechanical" mixture can be clearly distinguished and are much different from the mixed-layer peaks of the coagulates.

The broadening and the low number of observable basal reflections allow no analysis of the statistics of the layer sequence, for example by determination of the probability coefficients. Swelling experiments give evidence of some segregation of low- and high-charged layers in the crystals. Short chain alkylammonium derivatives swell in alkanol only if the silicate contains a high number of layers with layer charges above distinct values (Lagaly and Weiss, 1971). Hexylammonium beidellite swells in butanol to 2,24 nm whereas hexylammonium montmorillonite does not.

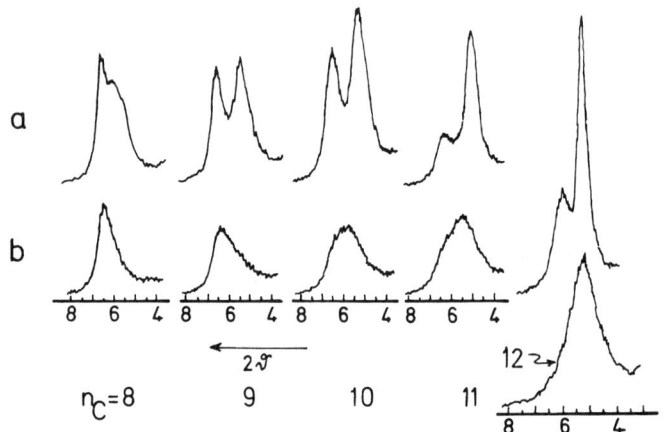

Fig. 6. Mixed-layer formation of the fraction $< 0,1 \mu m$: (001)-reflections of the alkylammonium ion exchanged "mechanical" mixture (a) and coagulate (b).

If the mixed-layer crystallites contain aggregates with large proportions of high- charged layers and aggregates with predominantly low-charged layers, reaction of the hexylammonium derivatives for example with butanol should cause a decomposition of the crystals in packets swollen with butanol and in packets with unchanged spacings of 1,35 nm.

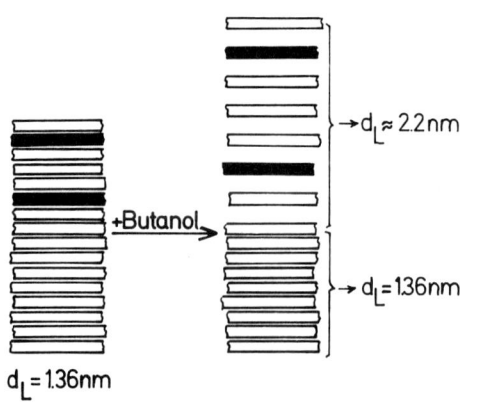

Fig. 9. Behavior of mixed-layer crystals with some seggregation of higher-charged layers and short chain alkylammonium interlayer cations against alkanols

In fact the hexylammonium mixed-layer coagulate gives under butanol two broadened (001)- reflections which center at d = 1,36 nm and ≈ 2,2 nm. This may be considered as an evidence that the low- and high-charged layers are not distributed in complete random sequence but tend to be more or less seggregated.

CONCLUSION

Smectites as models for colloidal systems were often brought into discredit. It is well known that the agreement between colloidal stability and DLVO-theory is relatively poor. But it should be kept in mind that the differences do not exceed much the differences observed in other colloidal systems. The smectites, however, represent an excellent system for the study of colloidal solutions containing differently charged particles. It could be shown for the first time that the coagulation can proceed as selective coagulation as well as mixed-layer formation. The behavior can be explained by the basical principles of the DLVO-theory (Frey and Lagaly, 1978).

ACKNOWLEDGEMENT

We are indebted to "Deutsche Forschungsgemeinschaft" for financial support.

REFERENCES

1 Frey, E. and Lagaly, G., 1978. Stability of colloidal solutions of smectites, in preparation for J. Colloid Interf. Sci.
2 Iler, R.K., 1975. Coagulation of colloidal silica by calcium ions, mechanism and effect of particle size. J. Colloid Interf. Sci., 53: 476-488.
3 Lagaly, G. und Weiss, A., 1971. n-Alkylammonium-Schichtsilicate unter prim. n-Alkanolen. Kolloid Z. Z. Polymere, 248: 968-978.
4 Lagaly, G., Stange, H. und Weiss, A., 1972. Über quasikristalline Strukturen bei der Flockung von Montmorilloniten und die Ausbildung diffuser Ionendoppelschichten in Nitrobenzol. Kolloid Z. Z. Polymere, 250: 675-682.
5 Lagaly, G. and Weiss, A., 1976. The layer charge of smectitic layer silicates. Proc. Intern. Clay Conf. Mexico, 1975, Appl. Publ. Ltd., Wilmette, Ill., USA, 1976, p. 157-172.
6 Mehra, O.P. and Jackson, M.L., 1960. Iron oxide removal from soils and clays by a dithionite-citrate system buffered with sodium bicarbonate. Clays Clay Min. 7. Natl. Conf., p. 317-327.
7 Norrish, K., 1973. Forcesbetween clay particles. Proc. Intern. Clay Conference, Madrid 1972, Div. Ciencias C.S.I.C., Madrid 1973, p. 375-384.
8 Parkins, R., Brace, R. and Matijević, E., 1974. Colloid and surface properties of clay suspensions. I. Laponite LP. J. Colloid Interf. Sci., 48: 417-426.
9 Pugh, R.J. and Kitchener, J.A., 1972. Experimental confirmation of selective coagulation in mixed colloidal suspensions. J. Colloid Interf. Sci., 38: 656-657.

10 Schwartzen-Allen, S. Lee and Matijević, E., 1975. Colloid and surface properties of clay suspensions. II. Electrophoresis and cation adsorption of montmorillonite. J. Colloid Interf. Sci., 50: 143-153.
11 Tanner, C.B. and Jackson, M.L., 1947. Nomographs of sedimentation times for soil particles under gravity or centrifugal acceleration. Soil Sci. Soc. Amer. Proc., 12: 60-65.
12 Tributh, H., 1970. Die Bedeutung der erweiterten Tonfraktionierung für die genaue Kennzeichnung des Mineralbestandes und seiner Eigenschaften. Z. Pflanzenern. Bodenkunde, 126: 117-134.

PORE SIZE DISTRIBUTION IN WATER-SATURATED CALCIUM MONTMORILLONITE USING LOW-TEMPERATURE HEAT-FLOW SCANNING CALORIMETRY.

L.G. HOMSHAW and J. CHAUSSIDON
Institut National de la Recherche Agronomique, Station de Science du Sol
Route de St Cyr - 78000 VERSAILLES - FRANCE

ABSTRACT

Low temperature differential scanning calorimetry enables a study of the distribution of water in porous solids to be made. This method has been applied to montmorillonite samples prepared in different ways. It is shown that whereas "dry" methods such as nitrogen adsorption reveal an important variation of the porosity and the accessible surface of the clay, calorimetry performed on wet samples shows a remarkable constancy of the properties of the material, thus suggesting a strong effect of water in the arrangement of the clay structural units.

INTRODUCTION

Water which is held in small pores freezes and melts at sub-zero temperatures. The depression of freezing and melting points is approximately inversely proportional to the pore Kelvin radius. For cylindrical pores the Kelvin radius is equivalent to the radius of the cylinder, whereas for parallel-sided fissures it corresponds to the wall to wall distance (Gregg 1967 p. 139).

The latent heat of water changing phase also apparently diminishes with pore size, and experimental quantitative data relating pore size to variation of the triple point and latent heat have been published by Brun et al. (1973).

The relationship between freezing point and pore size in this publication is given empirically by the following,

$$r \approx \frac{1000}{\Delta T_f} \tag{1}$$

$r$ = Kelvin radius, Å
$\Delta T_f$ = freezing point lowering, °C.

The empirical relationship which we used in this work for relating the pore Kelvin radius to the melting temperature is :

$$r = \frac{-521}{-\Delta T_m + 1,2\left[\exp{-\Delta T_m} - 1\right]}$$

r = Kelvin radius Å (2)

$\Delta T_m$ = melting point depression

This equation was obtained by combining the data of several publications (e.g. Blachère et al. 1972, Brun et al. 1973) and some of our own experimental results. From a theoretical viewpoint the endotherms are most suitable for pore size analysis if the pore form is known. Thus, equation 2 is valid for studying pores having a geometry similar to the calibration samples used to obtain the fusion curves given by Brun et al. above. However, extensive use has been made of the endotherm results, as these permit the sample water distribution to be studied after rapid freezing, thus minimizing the possibility of artefacts due to ice-crystal growth, when slow cooling rates are used (Minold et al. 1976). Having established that the cooling rate had small effect on the water location over the pore size range presented for the Ca-clay and after suitable precautions had been taken to avoid supercooling, low temperature exotherms were obtained and analyzed in accordance with equation 1. The exotherm and endotherm results thus give knowledge both of pore size and pore form.

For this application it is important to distinguish between heat exchange and phase change phenomena, and to ensure that moisture migration under a temperature gradient is eliminated or minimized. As these samples are saturated, we assume that there is negligible moisture movement in the vapour phase. The rigorous mathematical analysis of the temperature distribution within such samples is still complicated, and the results depend upon the geometry and heat flow parameters of a particular sample (Homshaw 1973). However, by suitable container design and the use of relatively small samples, in conjunction with a calorimeter having a short response time, and a slow heating rate, the temperature distribution within the sample may be considered to be uniform and proportional to the differential signal.

METHOD

An Arion microcalorimeter (type MCB) is used. This apparatus possesses an exceptionally sensitive measurement unit (50 - 60μV /mW dissipated). The fluxmeter is based on the Barberi design with an electric simulator. The sample and reference containers are chosen to be matched from a thermal point of view and measurements are made under vacuum to avoid error due to convectional heat losses. The reference material used is an empty sample container.

The sample is held in a special, vacuum-tight, highly conducting and thin walled metal container having a volume of about 0,14 $cm^3$. A chromel-alumel thermocouple, which is electrically insulated with a thin film of epoxy resin, is placed at the

bottom of a cavity in the lid and has thermal continuity with the very thin layer of
metal in intimate contact with the sample, by way of a vacuum grease junction. In
conjunction with a slow heating rate, the average temperature of the sample may be
known with a precision of about 0,05° C. A perfect thermal contact is assumed to
exist between the container and the fluxmeter-well, thanks to the thin layer of
vacuum grease which joins them.

The thermal geometry of the container permits rapid uniform cooling of the sample,
and this is achieved by quickly immersing the sample, in its container, in liquid
freon 12 at its fusion temperature (- 155° C). Cooling rates of about 40° C sec$^{-1}$ are
registered. The container is quickly transferred to its place in the fluxmeter-well
(which is held at a suitable low temperature) and tested for good thermal contact
before the system is evacuated for several hours to remove condensation. The
assemblage is then heated at a linear programmed rate of 0,1° C mn$^{-1}$ to obtain the
endotherm of fusion of a rapidly frozen sample. Exotherms corresponding to freezing
are also obtained at cooling rates of 0,1° C or 0,2° C mn$^{-1}$.

The thermogram trace between temperatures Tp and Tn is considered as being an
envelope corresponding to the summation in time of an infinite number of discrete
energy signals modulated in their intensity by the number of pores in a given class
and by the apparent variation of latent heat with pore size. Each signal occurs in
a temperature interval dT, and requires a finite time (determined principally by the
constant $\beta$, see below) to dissipate itself, mainly by conduction, through the series
of 180 thermocouples of the fluxmeter assemblage.

After completion of a reaction, be it exothermic or endothermic, the thermogram
relaxation curve assumes an exponential form about five seconds after the temperature
measuring thermocouple indicates the end of the reaction. The return function is
independent of the heating or cooling rate (0,1 to 2,0°C mn$^{-1}$), the sample size and the
sample thermal parameters, showing that the heat exchange between the sample and the
container is quasi-immediate.

Fig. 1 shows part of a thermogram lying between
temperatures $T_p$ and $T_n$ at times tp and tn. The
height of the thermogram trace from the base line
is hn at a time tn. We assume that were the
reaction to finish at time tn, the contribution
to the curve area generated by the return
function would be :

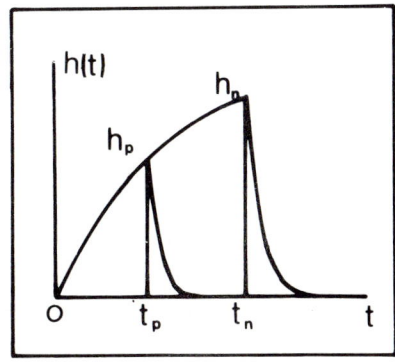

Fig. 1. See text.

$$\int_{t_n}^{t_\infty} h_n \exp\left[-\beta(t - t_n)\right] dt = \frac{h_n}{\beta} \qquad (3)$$

($\beta$ = 1,7.10$^{-2}$ sec$^{-1}$ = proportional apparent heat transfer coefficient). This is confirmed in practice.

The portion of thermogram surface due to reactions occuring between the temperatures $T_p$ and $T_n$ is $\Delta S_{p \to n}$ and

$$\Delta S_{p \to n} = \int_{t_p}^{t_n} h \, dt + \frac{h_n}{\beta} - \frac{h_p}{\beta} \qquad (4)$$

MATERIAL

The material used was Ca-saturated Greek Montmorillonite ($<2$ μm fraction) which was carefully and repeatedly washed to remove free electrolyte which is concentrated into the small pores during the freezing process (Homshaw 1978a). A series of samples of different water contents was prepared by exerting gas pressures ranging from 4 to 10 bars, for various time intervals under isothermal conditions, onto the very wet centrifuge extract. Another series of samples was prepared by hydration, either by simply leaving the dry powder in contact with saturated water vapour, or by flash-freezing a 5500 % water content suspension in rapidly stirred liquid freon (- 155° C) after it had been injected under pressure through a capillary ; the small particles of frozen suspension were then freeze-dried at low temperature before being hydrated by simply placing a drop of distilled water onto the sample and leaving it in its closed container for 1 month at 20° C..

The specimens used to obtain the $N_2$ adsorption-desorption data were aliquots of the same material which was prepared for calorimetric investigation. Immediately after the sample was removed from the vessel in which it was conditioned, the specimen to be used for $N_2$ adsorption was frozen in liquid Freon 12 at its melting point and dehydrated under vacuum at low temperature. The sample water content was evaluated by the weight loss of the sample used in the calorimetric assays after heating for 15 hours at 160° C..

RESULTS

Fig. 2 shows the exotherm (**Fig. 2a**) and the corresponding endotherm (**Fig. 2b**) of sample containing 51 % water. Much information is absent from the thermogram corresponding to the freezing cycle due to the presence of supercooling-a phenomenon which is almost unavoidable unless special techniques are used (Homshaw 1978b). The results here presented were obtained from the endothermic phase change, either after initial cooling at about 50° C sec$^{-1}$ or at slower cooling rates. In the temperature interval chosen, the evolution of the thermogram surface showed a small variation between the quickly frozen samples and those frozen at a slow rate (0,2° C mn$^{-1}$), a slight shift of the water distribution towards larger pores being discernable from direct comparison of the respective thermograms. This shift is always in the same sense but, for this material, does not significantly affect the porosity spectrum.

The two small low temperature peaks shown during the freezing cycle have already been described (Anderson 1971), however, thanks to the increased sophistication in measuring equipment, it is possible to report that their form is distinctive (Fig. 2c). The sharp change in baseline slope after the first peak is most probably related to a change in the thermal diffusivity of the sample.

Further reduction of the water content of the sample results in the disappearance of a large part of the peak corresponding to the water contained in the larger pores

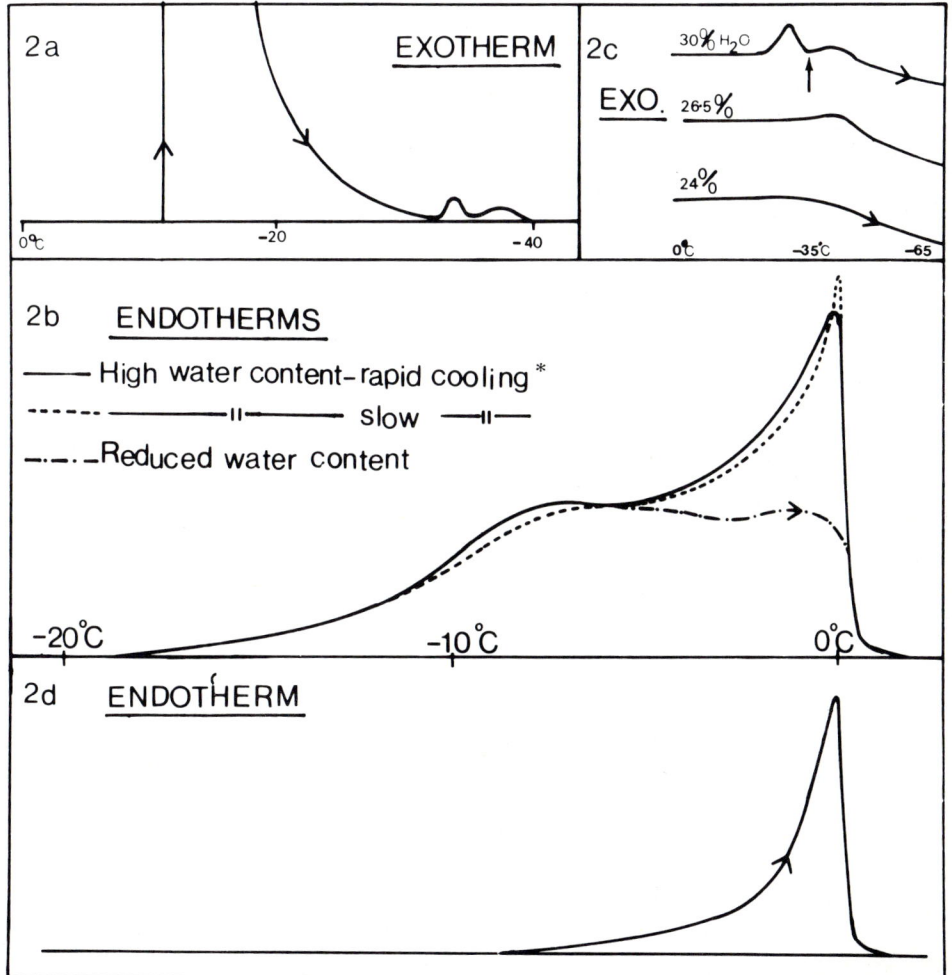

Fig. 2. Montmorillonite thermograms. 2a : exotherm, Ca-Clay - 2b : endotherms, Ca-Clay - 2c : low temperature peaks, Ca-Clay - 2d : endotherm, K-Clay.
* An equation for the endotherm trace shown in 2b is given in the appendix.

(fig. 2b) whilst, however, having negligible effect on the thermogram characteristics corresponding to the water held in pores of small size ; this is what might be expected, when considering the mechanism of water retention relative to pore size and form. Fig. 2d shows a comparable endotherm of K-Montmorillonite from which it is at once obvious that the water distribution in this sample is very different from that shown in the calcium one.

A simple mapping of the thermogram surface, taking into account the apparent variation of the latent heat of water with pore structure, as a function of temperature, thus gives a direct image of the distribution of water throughout the pore volume considered.

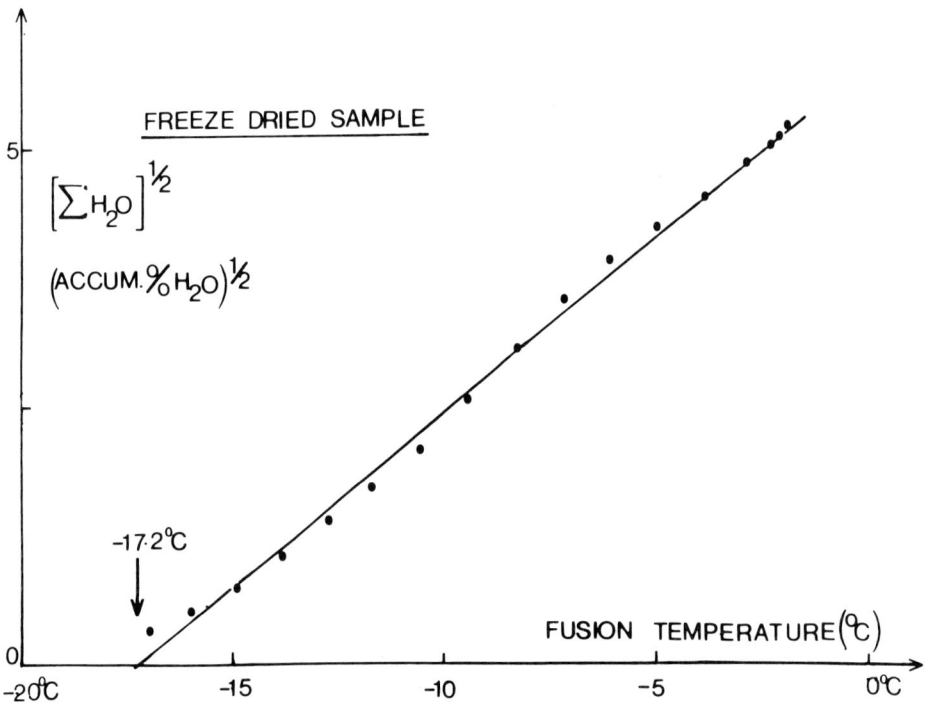

Fig. 3. Amount of water melting versus temperature of fusion.

Using the data presented above, a variety of representations may be given for the interpretation of the thermograms obtained ; Fig. 3, for example, represents a relationship between the accumulated water content and the temperature of fusion. The straight line relationship gives an intercept at $-17,2°$ C (recorded start of the endotherm $-17,1°$ C) and leads directly to considerations about pore geometry. Similarly, using relationship (2) the thermograms may be analysed to give the pore volume distribution in the region considered. This latter interpretation is presented in Fig. 4 which represents the accumulated wet pore volume as a function of pore size for a freeze-dried sample prepared in the manner delineated above.

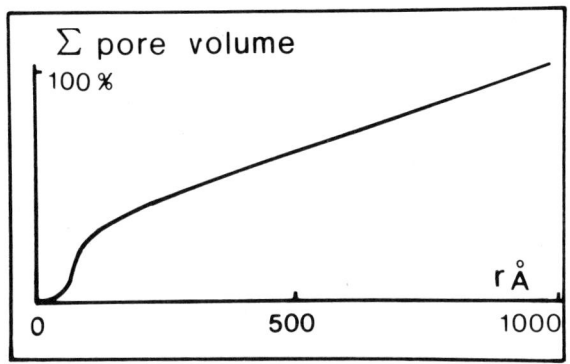

Fig. 4. Freeze-dried Ca-clay - cumulative pore size distribution.

This interesting sample had a water content of 85 % and is represented, in part, again in **Figure 5**. The sample is extraordinary in that the totality of the endotherm surface is completed before $-0,3°$ C (heating rate $0,1°$ C $mn^{-1}$). Although absolute values of fusion temperatures in the domain of pore sizes between 300 Å and 2500 Å may not be certain, the finishing temperature of the fusion reaction is sufficiently accurately known to say that, the water in the sample resides in pores of less than about 1000 Å. Due to the reciprocal nature of the

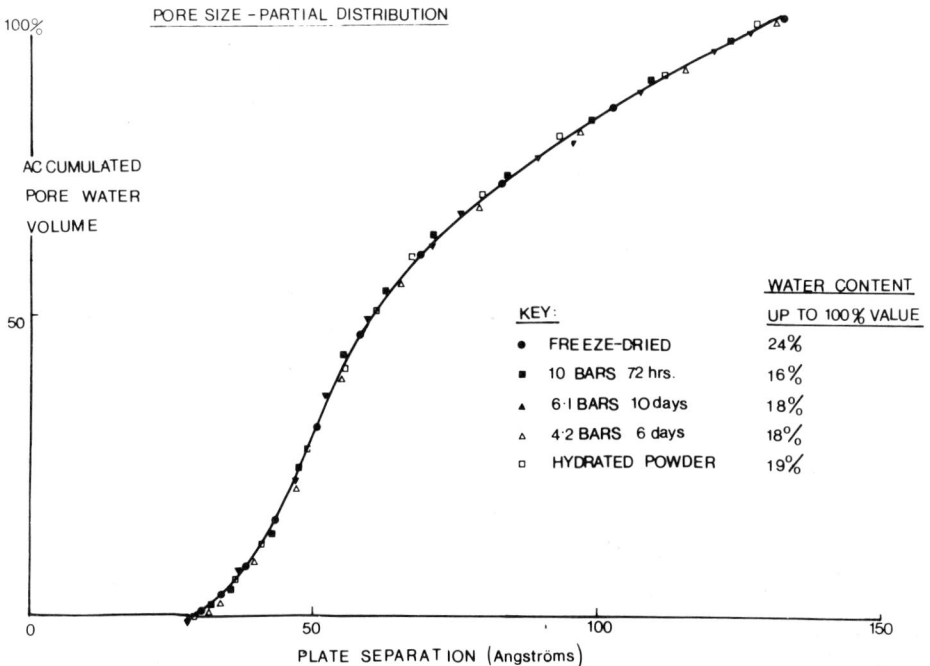

Fig. 5. Pore size partial distributions. Wet Ca-clays. (Endotherm results).

variation of fusion temperature with pore size, detailed exploration of large pore sizes with this method would be best pursued using a similar high resolution system with heating rates of $0,02°$ C mn$^{-1}$ or less.

To compare the nitrogen adsorption-desorption data with the calorimetric results the upper limit of thermogram analysis has been limited to a pore size corresponding to a fusion temperature of $- 2,4$ $(2)°$ C.

This temperature is the maximum fusion temperature for which the thermograms were utilized for accurate comparison with the $N_2$ - data. Elements of the pore size distribution are proportional to portions of the thermogram area between two discrete temperatures. Thus the points shown on the curve in **Figure 5** are the central values for each pore class size. The central value of a class of pore radii corresponds to a interval of fusion temperature of about $1°$ C. Thus more points are represented at the small pore side of the distribution due to the reciprocal relationship between pore radius and fusion temperature. Similarly the 100 % value at about 135 Å corresponds with the element of thermogram surface lying between $- 3,2°$ C and $- 2,4°$ C (average $- 2,8°$ C). The method of thermogram integration permits elements of the curve area corresponding to very small increments in fusion temperature to be evaluated ; however, $1°$ C increments together with a heating rate of $0,1°$ C mn$^{-1}$ were adequate here.

Figure 5 shows the results obtained for the samples used in this project. The amount of water (additional to the unfrozen water) measured by calorimetry required to fill all pores up to the class size equivalent to an average plate to plate distance of 135 Å, is taken as 100 % on the ordinate scale. The detectable water melting varies for different preparations, especially the freeze dried **specimen,** and is listed as percentage of anhydrous clay under "water contents up to 100 % value".

The constancy of the relative pore distribution compared with the variable water contents suggests that, according to the preparation, the samples differ in the number of layers associated in the aggregated structure.

It is also immediately apparent that, regardless of the manner of preparation of the samples, the relative pore size distribution throughout the region may be considered as invariant, leading to the hypothesis that water plays an important role in determining the porosity spectrum of this particular swelling "milieu" in the saturated state. All the measurements show an intercept at "28 Å" for the endotherm results analyzed in accordance with equation 2, suitable adjustments having been made for apparent variation of latent heat as a function of pore size. The exotherm result using equation 1 gave a similar distribution curve which was displaced linearly along the abscissa towards the larger pores by about 20 Å, thus giving a "zero pore size" intercept of about 50 Å. This latter curve better represents the actual pore size distribution of this sample in the wet state. The difference in these distributions implies that the form of the pores in this material is not exactly the same as in the calibration samples used to obtain equations 1 and 2, which is not surprising.

Comparison of the freezing and melting temperatures gives information about the pore geometry (Brun et al. 1973). Similar analyses on different porous media having various surface charge densities, give quite different "zero pore size" intercepts.

As the water content is further diminished and approaches 29 %, only the two low-temperature peaks of very small thermicity are observed. Further dehydration results in the disappearance of one peak, (see Fig. 2c) and both low temperature peaks become of negligible importance at a water content of 26 %. It is interesting to note the correspondance of this result with the near-infrared spectroscopy work of Fornès et al. (1975). Low temperature X-ray diffraction measurements on Ca-Montmorillonite at these water contents show a basal spacing compatible with the presence of two interlamellar water layers (Anderson et al. 1965).

It is reasonable to assume that the unfrozen water does not all rest in the interlamellar space and some of it is on the clay external surface. Let SE be the external surface area of the clay ($m^2 g^{-1}$). If the total surface area is 760 $m^2 g^{-1}$, one obtains (taking into account Anderson's results and assuming a packing of monolayer water of 66 $m^2$ mmole$^{-1}$) :

internal water = $(760 - SE)\dfrac{18}{66}$     mg g$^{-1}$ of dry clay. (5)

If $W_u$ is the amount of unfrozen water (mg g$^{-1}$ of dry clay) and n is the number of external unfrozen water layers, SE may be computed :

$$W_u - (760 - SE)\dfrac{18}{66} = \dfrac{18}{66} SE.n \qquad (6)$$

e.g. with n = 3

SE = 152 $m^2 g^{-1}$ for $W_u$ = 290 mg g$^{-1}$ (29 %)
SE = 97 $m^2 g^{-1}$ for $W_u$ = 260 mg g$^{-1}$ (26 %)

Equation 6, shows the marked dependency of SE on an assumed value of n and is presented to give an order of magnitude of the external surface area in the wet state. These values suggest that, in between the hydration state where the high temperature peak disappears (29 %) and the state where the low temperature peaks become too faint to be detectable (26 %), the clay platelets aggregate ; nevertheless, they retain an external surface area which is noticeably greater than those which are observed from measurements on the dry samples.

The measurements of external surface using nitrogen adsorption on the dry or freeze-dried sample at - 196° C, however, give values ranging between 30 $m^2 g^{-1}$ and 45 $m^2 g^{-1}$ according to the history of the sample. This well marked difference between the results obtained using the methods outlined above is illustrated in Fig. 6 which shows the respective nitrogen adsorption-desorption isotherms corresponding to the calorimetric data presented in Fig. 5. This leads to a conclusion that porous media which swell when in contact with water exhibit a structure which is very different from that which one may postulate exists when the sample is dry.

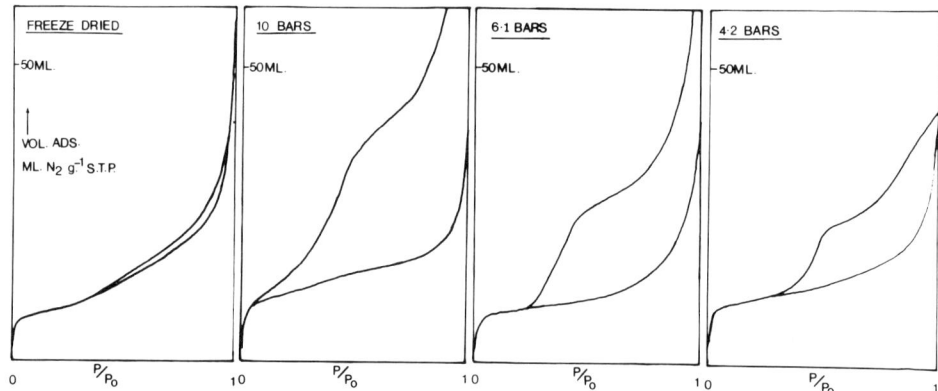

Fig. 6. $N_2$ adsorption-desorption isotherms for samples of Fig. 5.

CONCLUSION

The difference between these results may be explained by assuming that dehydration of this material results in a "condensation" of the clay structural units giving rise to two effects already observed in the electron-microscopy work of Tessier (1978) namely :

(1) Reduction of "available" surface area.
(2) Change in the lower limit of pore size distribution.

The calorimetric results on the wet samples do not indicate much change in the behaviour of the samples in the range chosen, which strongly suggests that water is a powerful agent in determining structure in such materials.

The date here presented underline the difficulty of adequately describing the behaviour of wet swelling clays using only information obtained from analysing the dry material.

ACKNOWLEDGEMENTS

We acknowledge the technical collaboration of Mrs E. Huard and thank Messrs J. Cop and P. Lemain for machining and modifying the standard calorimetric equipment.

REFERENCES

Anderson D.M., Hoekstra P. 1965. Migration of interlamellar water during freezing and thawing of Wyoming bentonite. Soil Sci. Soc. Am. Proc. 29, 5, 498-504.
Anderson D.M., Tice A.R. 1971. Low temperature phases of interfacial water in clay water systems. Soil Sci. Soc. Am. Proc. 35, 1, 47-54.
Blachère J.R., Young J.E. 1972. The freezing point of water in porous glass. J. Am. Ceram. Soc. 55, n° 6, 306-308.
Brun M., Lallemand A., Lorette G., Quinson J.F., Richard M., Eyraud L., Eyraud C. 1973. Changement d'état liquide-solide dans les milieux poreux. J. Chim. Phys. 70, n° 6, 973-996.
Fornès V., Chaussidon J. 1975. Near infrared spectroscopic studies of clay water systems. Proc. Int. Clay Cfe. Mexico 383-391.
Gregg S.J. and Sing K.S.W. 1967. Adsorption, surface area and porosity. Academic Press, London 371 pp.
Homshaw L.G. 1973. The "in situ" determination of some soil heat flow parameters... M. Sc. Thesis. Univ. Reading 190 pp.
Homshaw L.G. 1978a. Observations on the comportment of some frozen salt solutions in small pores using low temperature differential heat flux scanning calorimetry. Proceedings of the French and Italian calorimetry and thermal analysis conference (A.F.C.A.T.) Turin 1978.
Homshaw L.G. 1978b. Doctorate thesis. Paris. to be published.
Minold R., Luttge B., Kaiser W. 1976. Freeze fracturing, a new method for the investigation of dispersions by electron microscopy. Adv. in Colloid and Interf. Sci. 5, 281-335.
Tessier D., Berrier J., Robert M. 1978. Microscopic characterization of the organization of hydrated smectite. Communication n° 253, this Congress.

APPENDIX

An equation for the endotherm trace characteristic of this clay in the wet state (with $\beta = 1,7 \times 10^{-2}$ sec$^{-1}$, and using a strict linear heating rate) between its commencement and $-2,2°$ C is given by :

$$h_{mc} = K \left\{ A\Psi + B\exp\left[-C(\Psi-\mu)^2\right] + D\exp\left[-E(\Psi-\phi)^2\right] - F \right\} , \quad 7$$

$h_{mc}$ = the calculated height of the endotherm trace above its baseline (using 7),
$K$ = is a constant depending upon sample dry weight and the differential temperature measuring scale,
$\Psi = 1/\Delta Tm$,
and A, B, $\mu$, C, D, E, $\phi$, and F are numerical constants.

The regression line between $h_{mc}$ and the actual trace height, $h_m$ at a given value of $\Delta Tm$ (over 15 almost equally spaced intervals of fusion temperature from $-17°$ C to $-2,2°$ C) is :

$h_m = 1,003\ h_{mc} - 0,012$ ; correlation coefficient, r, = 0,99978.

Integration of equation 7 from $-16,5°$ C to $-2,2°$ C gives the curve surface area correct to 0,04 %. Thus the above equation permits analysis of the curve trace in temperature intervals dT. Values for the above parameters together with further analysis of this equation relative to the trace equations of other exotherms and endotherms of water in porous media are powerful tools for analyses of pore volume and form distribution, and are given in (Homshaw 1978b).

THE APPLICATION OF X-RAY PHOTOELECTRON SPECTROSCOPY (XPS OR ESCA)
TO THE STUDY OF MINERAL SURFACE CHEMISTRY

M.H. KOPPELMAN
Georgia Kaolin Company, Elizabeth, New Jersey  U.S.A.
J.G. DILLARD
Dept. of Chemistry, VPI & SU, Blacksburg, Virginia  U.S.A.

ABSTRACT

It has only been in relatively recent years that the instrumentation necessary to accurately determine the binding energy of core electrons has been available to scientists. The measurement of core electron binding energies, main photopeak configuration and satellite peak structure provide the investigator with insight into the bonding nature of the element in question. Photoejected electron escape depths of approximately 20Å make x-ray photoelectron spectroscopy an extremely valuable tool to the study of the chemistry of solid mineral surfaces.

The fundamentals, concepts, and experimental apparatus of XPS will be presented. Particular emphasis will center on a discussion of the applications of XPS to determine mineral surface composition (as compared to bulk phase analysis), iron oxidation state in clays, mineral surface dissolution mechanisms, the bonding nature of metal ions adsorbed on clays, and the quantitative measurement of adsorbed metal species.

INTRODUCTION

Knowledge of mineral surface chemistry is often obtained by inference from bulk chemical or solution reactions and spectroscopic techniques. Techniques such as x-ray powder diffraction, electron spin resonance spectroscopy, infrared, ultra-violet and visible spectroscopy (transmission), and Mössbauer spectroscopy, while highly sensitive and informative techniques, all reveal properties and information on the entire or bulk mineral phase. The mineral surface represents only a relatively small portion of a mineral's bulk, hence, only a fraction of the information obtained through these techniques may be due to surface contributions.

The use of x-ray photoelectron spectroscopy (XPS) in the study of mineral surfaces has afforded a method of direct examination of the chemistry at mineral interfaces. Applications may be classified into those with analytical intentions and those affording an insight into the chemical state (bonding) of the elements present.

It is the purpose of this paper to describe in greater detail some of the applications of XPS to examine mineral surfaces.

## Principles of x-ray photoelectron spectroscopy

One of the fundamental concepts of chemistry is that the electrons of atoms and molecules exist in orbitals of well defined energies. The technique of x-ray photoelectron spectroscopy enables one to measure the binding energies or ionization potentials of electrons in atomic or molecular orbitals. In effect, a set of "orbital ionization potentials" can be compiled.

It has been known for more than fifty years that electrons ejected from atoms by a primary excitation process carry vital information about atomic structure and chemical bonding. (Siegbahn, 1931; Siegbahn, 1967). Primary excitation sources may be x-ray quanta, UV quanta, electrons or ions. XPS specifically utilizes x-ray radiation (either Al $K_\alpha$ or Mg $K_\alpha$) to produce photoejected electrons from inner core levels. Detailed reviews of the theory and instrumentation of XPS may be found in the works of Siegbahn (1967), Jenkin (1977), and Szalkowski (1977).

The kinetic energy of the photoelectrons expelled during the XPS process is a well-defined quantity; any uncertainty or lack of definition of its energy results from the natural width of the level from which the electron has been ejected and the inherent width of the incident radiation. Measurement of the ejected electron's kinetic energy ($E_{Kin}$) enables the calculation of the binding energy ($E_B$) of that electron;

$$E_{Kin} = E_o - E_B - \phi_w \qquad (1)$$

where all binding energies are given a value of zero at the Fermi level, $\phi_w$ is the work function of the spectrometer, and $E_o$ is the energy of the incident radiation.

It is important to recognize that XPS is essentially a surface technique. It is obvious that the x-ray beam penetrates far into the bulk of the sample, but the effective sampling depth is determined by the escape depth of the photoejected electrons. This escape depth will vary from sample to sample and is dependent upon

the energy of the incident radiation, and upon the density and
crystal structure of the sample material. For most samples the
effective sampling depth is generally between 5 and 50Å (Lindau,
1974; Klasson, 1974).

In working with solid samples, differences in work functions
between the sample and the spectrometer must be taken into account.
Similarly, sample charging (buildup of positive charges on the
sample surface) of semi-conductive and non-conductive samples must
be evaluated. In the use of equation (1) for calculating the binding
energy of electrons ejected from semi-conductors or insulators, the
term $\phi_w$ may include not only the work function of the spectrometer,
but also the extent to which the sample surface charges. Discussions
of the various methods of calibration of XPS spectra can be found
in the works of Johansson (1973), Ebel (1974) and Koppelman (1976).

## Analytical applications

*Comparison of bulk and surface chemical compositions.* There are
several features that make XPS particularly attractive for analytical
applications; (1) detection of all elements in the periodic table
except hydrogen, (2) binding energy data gives insight into the
chemical nature of a species, and (3) application to surface
analysis.

The bulk sensitivity of XPS is limited to concentrations of
approximately 0.1% based on bulk percentage (Siegbahn, 1967).
If, however, the element under investigation is found primarily
in the surface (<20Å region), it may be detected by XPS in amounts
as small as 0.2% of a monolayer (Hercules, 1974; Brundle, 1973).

Koppelman's (1976) investigation focused on the comparison of
the bulk chemical composition of a Fithian illite sample with the
chemical composition of the surface region of that sample using XPS.
It was shown that for elements (Mg, Al, Si and Fe) which are constituents of the tetrahedral or octahedral layers, the Si/metal
ratios obtained by XPS were within 6% of the bulk chemical composition.
The Si/K ratio obtained by XPS, however, differed greatly (approx.
50%) from that determined by bulk chemical analysis. It was suggested that this deviation arose because no correction for elemental
depth in the sample was made. Since potassium is in the interlayer
in illite, it is at a depth considerably deeper than any metal in an
exposed octahedral or tetrahedral layer. Therefore, the relative
intensity of the potassium photopeak would be reduced producing a
high Si/K ratio as observed by XPS. Koppelman (1977a) later found
excellent (<5% differential) agreement for XPS measurement versus

bulk chemical composition for the Si/Al ratio in kaolinite and chlorite.

Adams (1977) evaluated XPS as a quantitative technique for surface analysis of aluminosilicate minerals. This work made determinations of experimental relative cross sections for the 1s (Li-F), 2s (Na-K) and 2p (Na-K) subshells and used these determinations to evaluate Si/metal atom ratios in ground polycrystalline minerals (kaolinite, montmorillonite and others) and freshly cleaved single crystals of lepidolite, muscovite and phlogopite. Their conclusions were that XPS is capable of providing bulk quantitative analyses of air-stable homogeneous solids (specifically aluminosilicates), accurate to 5% on the average for main group elements.

Trace metal analysis. Hercules (1973) reported that by adsorption onto treated (amino functional silylizing reagent) glass fiber, detection limits in solution of about 10 ppb were observed for lead, calcium, thallium and mercury. Czuha (1975) used the principle of ion exchange on silylized glass surfaces and thin polymer films to quantitatively detect and determine $Cu^{2+}$, $Fe^{3+}$, $Ba^{2+}$, $Cd^{2+}$, $Ca^{2+}$, $Pb^{2+}$ and $Hg^{2+}$. Czuha (1975) viewed possible applications of this adsorption technique in water treatment and pollution control.

Quantitative measurements of surface adsorbed species. Bancroft (1977) investigated the sorption of $Ba^{2+}$ ions from solution by powdered and crystalline calcite. The purpose of this study was to develop a direct XPS method of determining the amount of ions adsorbed on minerals. Direct surface analysis of sorbed $Ba^{2+}$ ions down to $\sim 10^{-9}$ g/cm$^2$ of sorbate could be accomplished. Surface quantity of $Ba^{2+}$ determined by XPS was in excellent agreement with that calculated for the powdered $CaCO_3$ using atomic absorption as the analytical probe to determine $Ba^{2+}$ adsorption. Furthermore, using an XPS calibration line for $Ba^{2+}$ surface analysis, Bancroft determined $Ba^{2+}$ sorption curves for different initial $Ba^{2+}$ concentrations as a function of time. This data also indicated that sorption (rather than precipitation) of $Ba^{2+}$ was occurring, since the amount of $Ba^{2+}$ on the surface increased both with increasing initial $Ba^{2+}$ concentration and with contact time in the solution.

Bonding applications

The information XPS provides is not only analytical, but also can give insight into the bonding nature of the element in question. The binding energy of a photoejected electron is dependent upon the chemical environment of the orbital from which the electron was removed. Oxidation state, type of bonding, (ie. ionic versus covalent) spin state, and nearest neighbor atoms are some chemical factors which can influence the binding energy of an electron.

Bonding nature of lattice components of aluminosilicate minerals.
Yin (1971) first used XPS to study the nature of oxygen atoms in olivines and pyroxenes. The oxygen 1s spectrum for olivines exhibited only one narrow oxygen 1s photopeak whereas pyroxenes contained two distinguishably different oxygen photopeaks. The intensity ratio of these two components (in pyroxene) was 2:1 with an energy separation of about 1 eV. It was suggested that the two oxygen components were the result of a difference in binding energy between bridging and non-bridging oxygen atoms within a silicate chain in the pyroxene structure.

Adams (1972) measured the core electron binding energies for Fe, Mg, Al, Si, and O in a number of well-characterized silicate minerals. Adams was unable to correlate Fe 2p binding energies with iron oxidation state in the minerals examined. It was also observed that O 1s peak widths for minerals with only one type of oxygen were generally narrower than those containing oxygen in more than one type of chemical environment. Small differences in Al 2p binding energy for aluminum in four coordination and aluminum in six coordination were reported.

Nicholls (1972) studied a series of magnesium and aluminum compounds with XPS and x-ray emission spectroscopy. He concluded that increasing coordination number (from four to six) increased the binding energy of both magnesium and aluminum electrons. It was also noted that increasing the electronegativity of the ligand (from oxygen to fluorine) further increased the magnesium and aluminum binding energy.

Anderson (1974), upon examining the minerals, Kyanite, sillimanite, and mullite with XPS, found that the Al 2p binding energy for sillimanite, with aluminum in both fourfold and sixfold coordinations, was experimentally identical with that in Kyanite, where aluminum is in only sixfold coordination. It was concluded that XPS could not be used to differentiate between aluminum atoms in different coordinations. This conclusion was strengthened by the XPS data for mullite where the Al 2p binding energy and peak shape were identical to those of Kyanite and sillimanite.

Lindsay (1973) reasoned that the significant difference in Al 2p binding energy between microcline (aluminum in fourfold coordination) and $Al_2O_3$ (aluminum in sixfold coordination) observed by Nicholls

(1972) could be explained by using ionic model concepts. He indicated that the presence of additional potassium cations in the crystal lattice of microcline had the effect of reducing the electron-attracting ability of the oxygen atoms. This would result in a decrease in Al 2p binding energy as the number of positively charged ions increased and therefore could account for the Al 2p binding energy in microcline being 1.4 eV lower than in alumina.

Urch (1974) determined the Al 2p and 2s binding energies for a series of aluminosilicate minerals which included microcline and alumina. He observed a 0.5 eV increase in Al 2p and 2s binding energy in going from microcline (Al-O bond length of 1.75 Å) to α-alumina (Al-O bond length of 1.92 Å). It was concluded that there was a correlation between bond length and orbital ionization (binding energies.

Schultz (1974), in an effort to identify silicate minerals in respirable coal dust, used XPS to measure the Si 2p binding energy in a series of aluminosilicate minerals. He observed five different silicon chemical environments in coal, and three different silicon environments in respirable coal dust. It should be noted that Schultz (1974) observed a 6.0 eV range in Si 2p binding energies for the various minerals he examined.

Huntress (1972) used the XPS technique to obtain rapid, non-destructive elemental (qualitative) analysis of selected lunar samples. Huntress was able to use the binding energy of the Fe 2p photopeak to identify iron in lunar samples as being in the ferrous oxidation state.

Koppelman (1975) observed that the binding energies for Si, Al and O, three major lattice constituents of kaolinite, chlorite and illite, varied little from mineral to mineral. The binding energy for the Si 2p electrons (average of 102.5 eV) was in good agreement with values published previously by Huntress (1972) and Adams (1972). This value has been confirmed by Carriere (1977).

XPS was also recently been employed to study the dissolution mechanism of feldspars (Petrovic, 1976). Examination of the K, Al and Si content of the surface of feldspar grains both prior to and after dissolution experiments with XPS revealed no evidence for silica or potassium depletion relative to aluminum within the outermost 10-20 Å. It was shown that the surface of the reacted feldspar had the same composition, within experimental error, as unreacted feldspar. This evidence led to the conclusion that the kinetics of feldspar dissolution (on a laboratory time scale) are not controlled by diffusion through a tightly adhering protective layer of hydrous aluminum oxide, kaolinite, or decationated feldspar, but rather through processes occuring at the fresh feldspar/solution interface.

Oxidation state of iron. The desire to establish the oxidation state of iron in both the bulk and surface regions of a mineral has long been a goal of geochemists. In this regard, the XPS examination of the binding energy of the Fe 2p level has proven to be of considerable interest.

Adams (1972) was unable to distinguish between ferric or ferrous iron species in the minerals he examined, solely by binding energy measurements. Huntress (1972) was, however, able to use binding energy results to assign a +2 oxidation state to iron in the lunar samples. Koppelman (1975, 1976), compared the Fe 2p binding energy for nontronite (determined by Mössbauer spectroscopy to contain only $Fe^{3+}$) and that of chlorite ($Fe^{2+}$ only) and observed a difference of 1.9 eV. Using these binding energies for the Fe 2p photopeaks of chlorite and nontronite, the rather broad Fe 2p photopeak of illite was deconvoluted into its ferric and ferrous components. Comparison of the $Fe^{2+}/Fe^{3+}$ ratios obtained by wet chemical analysis and Mössbauer spectroscopy (0.20) with that obtained by XPS measurements (0.35) indicated only fair agreement (Koppelman, 1976). It was suggested that this may be an indication of a difference in iron oxidation in the surface region as compared to the bulk mineral phase. Koppelman (1975) was unable to detect an Fe 2p photopeak for kaolinite, although bulk chemical analysis revealed approximately 0.5% $Fe_2O_3$. This lack of sensitivity was attributed to iron in kaolinite being located well within the bulk of the mineral.

Stucki (1976) used XPS to examine the redox reactions of nontronite and biotite. For the unaltered minerals a difference of 1.8 eV between the Fe 2p photopeak of nontronite ($Fe^{3+}$) and that of biotite ($Fe^{2+}$) was noted. Upon reduction of the nontronite sample with either hydrazine or dithionite, peak broadening of the Fe 2p photopeak was observed with a shift to lower binding energy. Oxidation of biotite in heated bromine water caused the Fe 2p photopeak to broaden and shift to higher binding energy. In a later study, Stucki (1977) was able to use XPS, in conjunction with infrared and Mössbauer spectroscopy to postulate a mechanism of iron redox in nontronite. The spectroscopic results were supportive of a two-step mechanism that involves an initial reduction of $Fe^{3+}$ to $Fe^{2+}$ with an accompanying increase in layer charge and no structural changes. In a second step, a further reduction of $Fe^{3+}$ was postulated with layer charge remaining constant through elimination of structural OH and alteration of iron coordination number.

Chemical nature of adsorbed species. Counts (1973) used XPS to examine the chemical nature of lead adsorbed on montmorillonite. Comparison of the Pb 4f photopeaks for the Pb/montmorillonite sample with those of elemental lead, lead oxide (PbO) and lead dioxide ($PbO_2$) indicated the lead adsorbed on montmorillonite was in a similar bonding state as lead in lead oxide.

The interaction of gibbsite with $Ca(H_2PO_4)_2$, $Si(OH)_4$, $CaSiO_3$ and $Ca(NO_3)_2$ has been examined (Alvarez (1976)) using the XPS technique. Gibbsite samples treated with $Si(OH)_4$ and $Ca(NO_3)_2$ revealed little or no detectable Si, Ca, or N photopeaks. However, upon treatment with $CaSiO_3$, significant calcium and silicon photopeaks were noted. Similarly, after treatment with $Ca(H_2PO_4)_2$, Ca and P photopeaks were also observed. From the absence of detectable calcium or silicon signals after respective $Ca(NO_3)_2$ and $Si(OH)_4$ treatments, and the detection of Ca and Si signals after $CaSiO_3$ treatments, it was suggested that a silicon adsorption mechanism onto gibbsite is dependent upon the availability of calcium ions.

Emerson (1978) examined the interaction of $Ba^{2+}$ with ripidolite as a function of pH. XPS measurement of these barium treated clay samples revealed that the binding energy of the adsorbed barium species was virtually unchanged throughout the pH range examined. Furthermore, from the Si/Ba XPS ratios obtained, the amount of barium adsorbed on the mineral surface was dependent upon pH, with the degrees of adsorption increasing with increasing pH.

The chemical nature of mineral adsorbed iron species was first probed by XPS in the work of Koppelman (1975). Untreated kaolinite, which contained no XPS detectable surface iron species was reacted with $Fe(NO_3)_3$ solutions at pH values low enough to prevent hydroxide precipitation. XPS examination of this iron treated kaolinite sample revealed a distinct Fe 2p photopeak at a binding energy 1.1 eV lower than lattice $Fe^{3+}$ in nontronite and 1.1 eV higher than lattice $Fe^{2+}$ in chlorite. Deconvolution of the rather broad Fe 2p photopeaks of chlorite and illite which had been subjected to similar $Fe^{3+}$ treatment revealed adsorbed iron at the same binding energy as iron adsorbed on kaolinite. The lowering of binding energy for adsorbed $Fe^{3+}$ relative to lattice $Fe^{3+}$ was interpreted to indicate that electron density in the Stern layer where ions are adsorbed is donated to the metal ion thus lowering its binding energy.

A similar variation in binding energy for chromium adsorbed on kaolinite, chlorite and illite was also observed (Koppelman (1978a)). In this study, these three minerals were reacted with $Cr(NO_3)_3$ solution at pH values of 2, 3, 4, 6, 8, 10. Si 2p and Al 2p binding energies were found to be pH invariant. Cr 2p binding energies varied only slightly between pH values 2-4, but remained constant at pH values 6, 8, 10. Above pH 6, the binding energy for chromium in the clay samples was identical with that of $Cr(OH)_3$ indicating precipitation had occurred. The binding energies of adsorbed $Cr^{3+}$ below pH 6 was significantly (1.0 eV) lower than $Cr^{3+}$ substituted in a octahedral lattice site (Kämmerite). These results were similar to those obtained from $Fe^{3+}$ adsorption, supporting Koppelman's (1975) conclusions.

Adsorption of $Co(H_2O)_6^{2+}$ on chlorite has been investigated by Koppelman (1978b) at pH values of 3 and 7 using XPS. It was observed that the binding energy of the adsorbed $Co^{2+}$ species was independent of pH, but was 0.5 eV lower than $Co^{2+}$ substituted in an octahedral site in lusakite. It was suggested that the degree of reduction in the adsorbed metal ion binding energy was dependent upon the oxidation state of the adsorbed species. Furthermore, it was noted that the binding energy for adsorbed $Co^{2+}$ was significantly different than that of $Co(OH)_2$ indicating precipitation had not occurred.

Tewari (1975 a,b) studied the adsorption of cobalt(II) on $Al_2O_3$ and $ZrO_2$ using XPS and electrophoretic mobility measurements. A comparison of binding energies for adsorbed $Co^{2+}$ with those for cobalt oxides and hydroxides revealed that cobalt adsorbed on alumina and zirconia exists as $Co(OH)_2$. However, at 200° C., for cobalt adsorbed on the alumina surface, the cobalt photopeaks were similar to $CoAl_2O_4$, suggestive of surface transformation.

The chemical nature of cobalt adsorbed on $MnO_2$ has been studied using XPS (Dillard (1978)). From the cobalt binding energies and the separation energies for the Co $2p_{1/2}$ and Co $2p_{3/2}$ energy levels, it was concluded that cobalt was adsorbed on $MnO_2$ as a Co(III) species. Examination of the binding energies for manganese and oxygen indicated that the surface region of the substrate ($MnO_2$) was characteristic of predominantly Mn(IV).

XPS was used to examine the products of the reactions of kaolinite, chlorite and illite with Cr(III) and Co (III) ammine complexes (Koppelman 1976, (1978b,c)). In the interaction of both chromium and cobalt hexammine complexes with chlorite, rapid and unanticipated rates of hydrolysis of the dissolved complexes were observed.

XPS examination of the cobalt complex treated chlorite after both short (1 day) and long (one week) interaction periods revealed that cobalt had been reduced to cobalt(II). Relative rates of clay catalyzed hydrolysis of both chromium and cobalt hexammine complexes could be related to the amount of unoxidized (ferrous) iron in the surface region. XPS atom ratio measurements for N/Cr suggest that significant loss of coordinated ammine had occurred upon adsorption. In a study of the adsorption of $Co(NH_3)_6^{3+}$ on a Y-type zeolite (Lunsford (1978)), XPS results indicated that $Co(NH_3)_6^{3+}$ is adsorbed as Co(III). However, no examination of the Co/N atomic ratios was carried out to discover whether dissociation or decomposition of the complex occurred upon adsorption and subsequent heat treatment.

Examination of the mode of bonding of the metal ions Cu(II) and Ni(II) to clay minerals using XPS has been investigated (Koppelman (1977)). Comparison of the binding energy for the adsorbed $Ni^{2+}$ species with that of Ni(II) substituted in an octahedral site in lizardite revealed a lowering (0.4 eV) of Ni 2p binding energy for the adsorbed nickel species. It was noted that this result was consistent with the results obtained for Co(II) adsorption (Koppelman, (1976, 1978b)). The binding energy of the adsorbed copper species did not, however, show the same (~0.5 eV) reduction in Cu 2p binding energy relative to dioptase. It was suggested from solution pH data during the reaction of the clays with $Cu(NO_3)_2$ that the adsorbed Cu(II) species was $Cu(OH)^+$.

Angular study applications. Grazing angle measurements of electron escape from solid samples may be used to enhance surface sensitivity in XPS studies. Grazing angle studies of highly polished metals have revealed that surface sensitivity may be enhanced by at least an order of magnitude (Fadley (1971, 1974), Fraser (1973)). More recently, grazing angle spectra of silicon and calcium adsorbed on $Al_2O_3$ (Baird (1976)) and of mammalian cells (Millard (1977)) have been published. In the present study grazing angle intensity measurements and binding energy determinations were carried out for take-off angles of 90° and 11° for Co(II) adsorbed on chlorite and on illite at pH 7. The quoted angle is the electron emergence angle measured between the sample surface and the entrance to the analyzer.

The enhancement ratio is defined in a manner similar to that used by Baird (1976) except that the Si 2s level was used as the reference level. The results are presented in Table 1, where the intensity enhancement ratio (IER) is defined as

$$\frac{(Co\ 2p_{3/2}/Si\ 2s)_{11°}}{(Co\ 2p_{3/2}/Si\ 2s)_{90°}} \tag{2}$$

The measured binding energies for cobalt are also presented in Table 1. A similar tabulation of the Al 2p/Si 2s grazing angle data are summarized in Table 1. The results indicate a significant enhancement of the cobalt signal upon reducing the emergence angle while that for aluminum is relatively unaffected. The enhancement of the cobalt signal indicates that adsorbed cobalt is predominantly a surface species. The measured binding energies for cobalt determined at 11° are equivalent to those measured at 90°. Additionally the Co $2p_{1/2}$, Co $2p_{3/2}$ energy separation (16 eV) is unchanged at 11° compared to 90°. These results are consistent with the notion that cobalt is adsorbed as an aquo Co(II) ion and not as Co(OH)$_2$ as Tewari (1975 a,b) observed for the Co(II)/alumina systems. Furthermore, there is no evidence that there has been surface initiated cobalt oxidation at adsorption sites near the clay surfaces as observed with the Co(II)/MnO$_2$ system (Dillard (1978)). The binding energy results for aluminum indicate that the chemical nature of surface aluminum is similar to that deeper in the sample. It is noted that significant surface enhancement for aluminum is not realized. This fact indicates that aluminum is homogeneously distributed in the surface region and may resemble bulk aluminum.

TABLE 1
Grazing angle measurements for chlorite and illite clays: IER and Binding Energy results

|  | Co | | Al | |
| --- | --- | --- | --- | --- |
|  | IER(11°) | BE | IER(11°) | BE |
| chlorite | 1.22 | 782.1 | 0.98 | 74.1 |
| illite | 1.31 | 782.2 | 1.01 | 74.2 |

CONCLUSIONS

The applicability of XPS to the study of the chemistry of mineral surfaces is evident. XPS provides the researcher with a spectroscopic tool that is unique in that it is able to probe the surface region directly rather than by inference. Furthermore, XPS can provide insight into bonding at mineral surfaces as well as monitoring interfacial reaction processes. The analytical implications of XPS are clear and obvious. XPS is a tool which should not be limited to use only by chemists and physicists, and the realm of its usage should be explored by all physical scientists.

REFERENCES

1 Adams, I., Thomas, J.M. and Bancroft, G.M., 1972. An ESCA study of silicate minerals. Earth Planet. Sci. Lett., 16:429-432.
2 Adams, J.M., Evans, S., Reid, P.I., Thomas, J.M. and Walters, J.M., 1977. Quantitative analysis of alumino-silicates and other solids by X-ray photoelectron spectroscopy. Anal. Chem., 49:2001-2007.
3 Alvarez, R., Fadley, C.S., Silva, J.A. and Uehara, G., 1976. A study of silicate adsorption on gibbsite ($Al(OH)_3$) by X-ray photoelectron spectroscopy (XPS). Soil Sci. Soc. Amer. J., 40:615-617.
4 Anderson, P.R. and Swartz, W.E., Jr., 1974. X-ray photoelectron spectroscopy of some aluminosilicates. Inorg. Chem., 13:2293-2294.
5 Baird, R.J., Fadley, C.S., Kawamoto, S.K., Mehta, M., Alvarez, R. and Silva, J.A., 1976. Concentration profiles for irregular surfaces from X-ray photoelectron angular distributions. Anal. Chem., 48:843-846.
6 Bancroft, G.M., Brown, J.R. and Fyfe, W.S., 1977. Quantitative X-ray photoelectron spectroscopy (ESCA): studies of $Ba^{2+}$ sorption on calcite. Chem. Geol., 19:131-144.
7 Brundle, C.R. and Roberts, M.W., 1973. Surface sensitivity of photoelectron spectroscopy. Chem. Phys. Lett., 18:380-386.
8 Carriere, B. and Deville, J.P., 1977. X-ray photoelectron study of some silicon-oxygen compounds. J. Electron Spectrosc. Relat. Phenom., 10:85-91.
9 Counts, M.E., Jen, J.S.C., and Wightman, J.P., 1973. An electron spectroscopy for chemical analysis study of lead adsorbed on montmorillonite. J. Phys. Chem., 77:1924-1925.
10 Czuha, M. and Riggs, W.M., 1975. X-ray photoelectron spectroscopy for trace metal determination by ion-exchange absorption from solution. Anal. Chem., 47:1836-1838.
11 Dillard, J.G. and Murray, J.W., 1978. An XPS study of the nature of cobalt adsorbed on $MnO_2$. In preparation.
12 Ebel, M.F. and Ebel, H., 1974. About the charging effect in X-ray photoelectron spectroscopy. J. Electron Spectrosc. Relat. Phenom., 3:169-180.
13 Emerson, A.B., 1978. An XPS investigation of the effect of pH on chromium and cobalt adsorption on clay minerals. M.S. thesis, VPI & SU, Blacksburg.
14 Fadley, C.S., 1974. Instrumentation for surface studies; XPS angular distribution. J. Electron Spectrosc. Relat. Phenom., 5:725-754.
15 Fadley, C.S. and Bergstrom, S.A.L., 1971. Angular distribution of photoelectrons from a metal single crystal. Phys. Lett. A., 35:375-376.
16 Fraser, W.A., Florio, J.V., Delgass, W.N. and Robertson, W.D., 1973. Surface sensitivity and angular dependence of X-ray photoelectron spectroscopy. Surf. Sci., 36:661-674.
17 Hercules, D.M., 1974. Electron spectroscopy...for chemical analysis. J. Electron Spectrosc. Relat. Phenom., 5:811-826.
18 Hercules, D.M., Cox, L.E., Onisick, S., Nichols, G.D. and Carver, J.C., 1973. Electron spectroscopy (ESCA): use for trace analysis. Anal. Chem., 45:1973-1975.
19 Huntress, W.T., Jr., and Wilson, L., 1972. An ESCA study of lunar and terrestrial materials. Earth Planet. Sci. Lett., 15:59-64.

20 Jenkin, J.G., Leckey, R.C.G. and Liesegang, J., 1977. The development of X-ray photoelectron spectroscopy: 1900-1960. J. Electron Spectrosc. Relat. Phenom., 12:1-35.
21 Johansson, G., Hedman, J., Berndtsson, A., Klasson, M. and Nilsson, R., 1973. Calibration of electron spectra. J. Electron Spectrosc. Relat. Phenom., 2:295-317.
22 Klasson, M., Berndtsson, A., Hedman, J., Nilsson, R., Nyholm, R. and Nordling, C., 1974. Electron escape depth in silicon. J. Electron Spectrosc. Relat. Phenom., 3:427-434.
23 Koppelman, M.H., 1976. An X-ray photoelectron spectroscopic Investigation of the adsorption of metal ions on marine clay minerals. Ph.D. thesis, VPI & SU, Blacksburg, 251 pp.
24 Koppelman, M.H. and Dillard, J.G., 1975. An ESCA study of sorbed metal ions on clay minerals. In: T.M. Church (Editor), Marine Chemistry in the Coastal Environment, ACS Symposium Ser. #18, pp. 186-201.
25 Koppelman, M.H. and Dillard, J.G., 1977 a. Unpublished data.
26 Koppelman, M.H. and Dillard, J.G., 1977 b. A study of the adsorption of Ni(II) and Cu(II) by clay minerals. Clays, Clay Miner., 25:457-462.
27 Koppelman, M.H. and Dillard, J.G., 1978 a. The adsorption of chromium (III) on clay minerals chlorite, illite and kaolinite. Submitted to Geochim. Cosmochim. Acta.
28 Koppelman, M.H. and Dillard, J.G., 1978 B. An X-ray photo-electron spectroscopic (XPS) study of cobalt adsorbed on the clay mineral chlorite. J. Colloid Interface Sci., 59:000.
29 Koppelman, M.H. and Dillard, J.G., 1978 c. XPS study of the adsorption of $Cr(NH_3)_6^{3+}$ and $Cr(en)_3^{3+}$ on clay minerals. In preparation.
30 Lindau, I. and Spicer, W.E., 1974. The probing depth in photo-emission and auger-electron spectroscopy. J. Electron Spectrosc. Relat. Phenom., 3:409-413.
31 Lindsay, J.R., Rose, H.J., Jr., Swartz, W.E., Jr., Watts, P.H., Jr., and Rayburn, K.A., 1973. X-ray photoelectron spectra of aluminum oxides: structural effects on the chemical shift. Appl. Spect., 27:1-4.
32 Lunsford, J.H., Hutta, P.J., Lin, M.J., and Whitehorst, K.A., 1978. Cobalt nitrosyl complexes in zeolites A,X, and Y. Inorganic Chemistry, 17: 606-610.
33 Millard, M.M. and Bartholomew, J.C., 1977. Surface studies of mammalian cells grown in culture by X-ray photoelectron spectroscopy. Anal. Chem., 49:1290-1296.
34 Nicholls, C.J., Urch, D.S. and Kay, A.N.L., 1972. Determination of coordination number in some compounds of magnesium and aluminum: a comparison of X-ray photoelectron (ESCA) and X-ray emission spectroscopies. J.C.S. Chem. Comm., 1972: 1198-1199.
35 Petrovic, R., Berner, R.A. and Goldhaber, M.B., 1976. Rate control in dissolution of alkali feldspars--I. Study of residual grains by X-ray photoelectron spectroscopy. Geochim. Cosmochim. Acta, 40:537-548.
36 Schultz, H.D., Vesely, C.J. and Langer, D.W., 1974. Electron binding energies for silicon minerals occurring in respirable coal dust. Appl. Spect., 28:374-375.
37 Siegbahn, M., 1931. Spektroskopie der Röntgenstrahlen, Springer, Berlin.
38 Siegbahn, K., Nordling, C., Fahlman, A., Nordberg, R., Hamrin, K., Hedman, J., Johansson, G., Bergmark, T., Karlsson, S.E., Lindgren, I. and Kindbert, B., 1967. ESCA, Atomic, Molecular and Solid State Structure Studied by Means of Electron Spectroscopy, Almqvist and Wiksells, Uppsala.

39 Stucki, J.W., Roth, C.B. and Baitinger, W.E., 1976. Analysis of iron-bearing clay minerals by electron spectroscopy for chemical analysis (ESCA), Clays, Clay Miner., 24:289-292.
40 Stucki, J.W. and Roth, C.B., 1977. Oxidation-reduction mechanism for structural iron in nontronite. Soil Sci. Soc. Amer. J., 41:808-814.
41 Szalkowski, F.J., 1977. The characterization of surfaces by electron spectroscopy. J. Colloid Interface Sci., 58:199-215.
42 Tewari, P.H. and Lee, W.J., 1975 a. Adsorption of Co(II) at the oxide-water interface. J. Colloid Interface Sci., 52:77-88.
43 Tewari, P.H. and McIntyre, N.S., 1975 b. Characterization of adsorbed cobalt at the oxide-water interface. AlChE. Symposium Sci., 71:134-137.
44 Urch, D.S. and Murphy, S., 1974. The relationship between bond lengths and orbital ionization energies for a series of aluminosilicates. J. Electron Spectrosc. Relat. Phenom., 5:167-171.
45 Yin, L.I., Ghose, S. and Adler, I., 1971. Core electron binding energy difference between bridging and non-bridging oxygen atoms in a silicate chain. Science, 173:633-635.

# EVOLUTION OF EXCHANGE PROPERTIES AND CRYSTALLOGRAPHIC CHARACTERISTICS OF BIIONIC K-Ca MONTMORILLONITE SUBMITTED TO ALTERNATE WETTING AND DRYING.

J.P. GAULTIER and J. MAMY

Institut National de la Recherche Agronomique, Route de St Cyr
78000 VERSAILLES - FRANCE

## ABSTRACT

In order to understand why an irreversible K fixation is frequently observed in soils containing a high proportion of montmorillonite clay, the case of biionic K-Ca montmorillonite submitted to alternating wetting and drying cycles was studied. When the K-Ca montmorillonite samples are submitted to this treatment a progressively increasing amount of K becomes non exchangeable, and is independant of the proportion of K. Crystallographic studies show a progressive separation of two phases :

- The first phase is entirely saturated with K and evolves according to a reorganization in the stacking of the sheets.
- The second, probably containing a K-Ca mixture, keeps a disordered stacking of the sheets (turbostratic structure).

The characteristics of the evolution of the samples are related to the proportion of Ca in the samples, and are discussed according to the demixion phenomena.

## INTRODUCTION

It has already been shown (Mamy and Gaultier, 1976) that the decrease in K exchangeability for K montmorillonite is due to a structural reorganization of the elementary layers. This reorganization concerns the stacking of the layers contained in the crystalline particles. Initially the layers are stacked randomly. But under the influence of an increasing number of wetting and drying cycles (W-D) this disorientated structure disappears : it is replaced by a stacking in which the hexagonal cavities of two adjacent layers are facing each other.

Consequently the K becomes more strongly bound and the layers do not expand in water.

Such a mechanism cannot be proposed as a complete explanation in the case of K fixation in soils. In fact, montmorillonite is never fully saturated with K in soils. In order to clarify the question the evolution of biionic K-Ca montmorillonite under the effect of W-D cycles was studied ; the percentage of Ca in the samples varying between 0 and 100.

## PREPARATION AND TREATMENT OF THE SAMPLES

The clay studied was essentially Wyoming montmorillonite. But results already published by Chaussidon (1963), and complementary crystallographic observations made with Camp-Berteaux montmorillonite have shown that the two clays have the same evolution when submitted to the W-D cycles.

The biionic samples where prepared by shaking together Ca montmorillonite and K montmorillonite suspensions in definite proportions.

The interlayer distances of the different samples, measured by X-ray diffraction in a relative humidity of 70 %, showed that initially there is no interstratification. This means that K and Ca are homogeneously distributed between the sheets, which, then, holds a mixture of the two cations.

The different samples studied were saturated with K and Ca in the following proportions, expressed as the percentage of Ca : 0, 15, 30, 40, 50, 80 and 100.

The samples were submitted to a series of alternating wetting and drying cycles obtained by shaking the clay in water and then drying it in a oven at 80° C.

After a given number of W-D cycles, a little of each sample was taken in which K and Ca were extracted by 0,5 N $NH_4Cl$ solutions. Complementary observations with X-ray diffraction and electron S.A.D. were made at the same time on these samples.

## EVOLUTION OF EXCHANGE PROPERTIES

### K exchangeability

In the case of 100 % K montmorillonite it has been already shown (Mamy and Gaultier, 1976) that W-D cycles produce a decrease in the K exchangeability. After 100 W-D cycles the amount of exchangeable K reaches a limiting value of about 30 meqt/100 g.

In the case of K-Ca montmorillonite a decrease in K exchangeability with the

W-D cycles was also observed. The figure 1 shows the variation of K exchangeability with the number of W-D cycles for samples containing various proportions of Ca. This series of curves gives evidence that whatever the Ca proportion may be, an increasing proportion of K becomes non exchangeable when the number of W-D cycles is increased. After 100 W-D cycles more than 60 % of total K is fixed. It is then possible to conclude that Ca, even in high proportion, does not disturb the K fixation process.

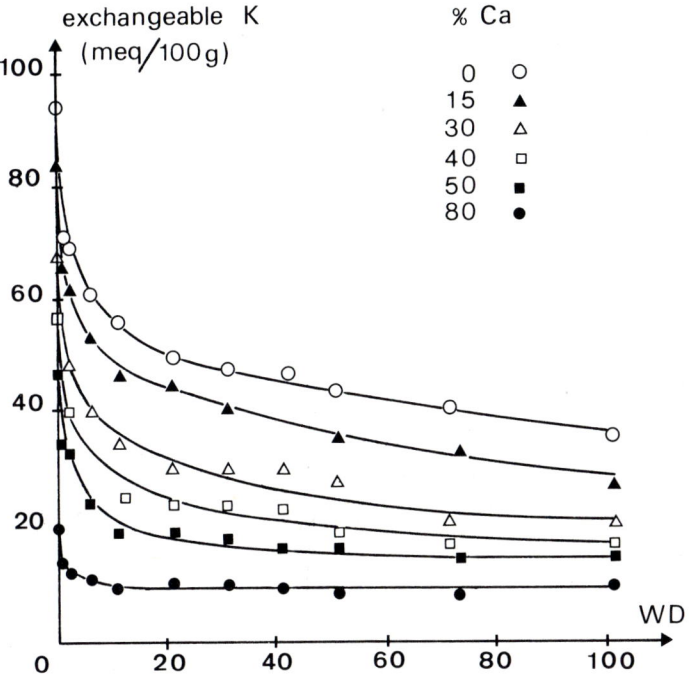

Fig. 1. Variation of K exchangeability with the number of W-D cycles for various proportions of Ca.

Ca exchangeability

Figure 2 shows that whatever the proportion of Ca and the number of W-D cycles may be, all the Ca remains exchangeable. Accordingly, it can be said that the W-D cycles have no effect on Ca exchangeability.

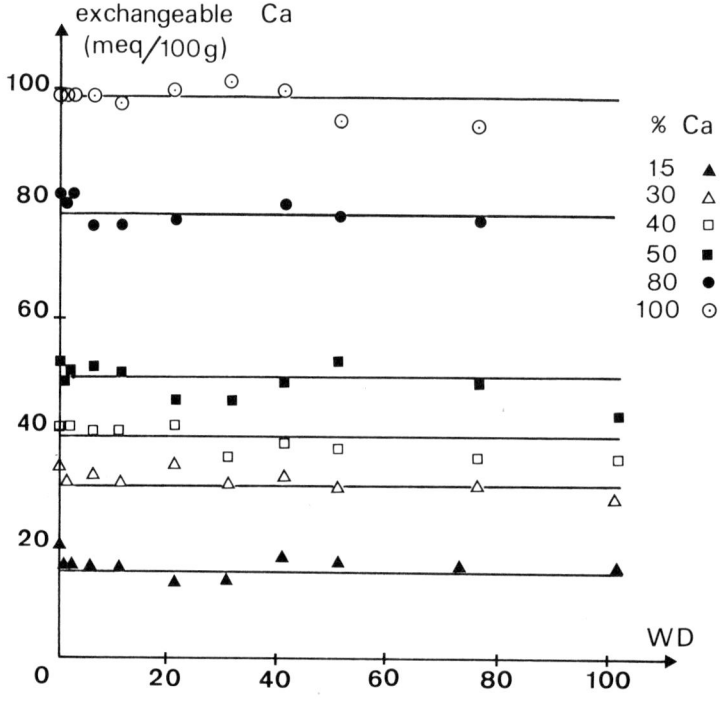

Fig.2. Variation of Ca exchangeability with the number of W-D cycles for various proportions of Ca.

Connection between K and Ca exchangeability

The above results may seem to be contradictory if one tries to explain them without considering that a redistribution of the Ca and K cations between the layers occurs. Indeed, and particularly when the proportion of Ca is equal to 50 %, there is necessarily at the beginning K and Ca cations in the interlayer spaces. The fact that a part of K becomes fixed means a collapse of the layers ; if Ca cations were between layers they would be fixed too, but this is not the case.

The only hypothesis compatible with these experimental results is that the W-D cycles play a double role :

i - The W-D cycles induce a redistribution of the Ca and K cations in separate interlayer spaces. This demixing phenomena is analogous to that revealed by Mering and Glaeser (1954) in the case of Ca-Na montmorillonite ; and already postulated by Chaussidon (1963) for the K-Ca Camp-Berteaux montmorillonite

ii - The W-D cycles produce a structural reorganization of the layers containing only K, identical to that observed for the 100 % K montmorillonite (Mamy and Gaultier, 1976).

## AN EXPERIMENTAL STUDY OF THE SWELLING OF BIIONIC K-Ca MONTMORILLONITE SUBMITTED TO ALTERNATING W-D CYCLES

The swelling of K-Ca montmorillonite was studied after isothermal equilibration of the samples at 20° C with a controlled relative water vapor pressure of 0,7. It is well known that under such conditions Ca montmorillonite exhibits a 15,2 Å interlayer space, corresponding to a two-layer water complex, and K montmorillonite a 12,4 Å interlayer space corresponding to a monolayer water complex; however this latter is eliminated after about 100 W-D cycles, after which the interlayer distance falls to 10,2 Å.

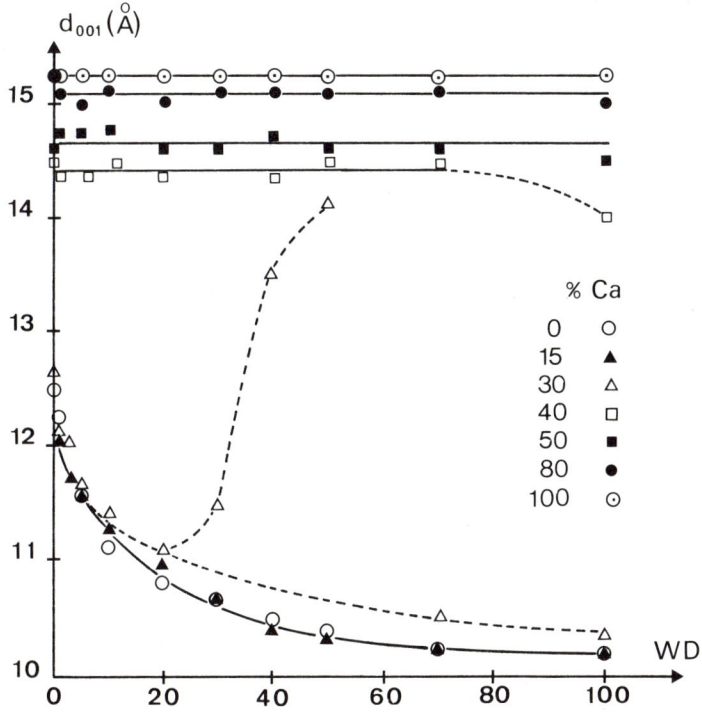

Fig. 3. Evolution of the $d_{001}$ spacing of montmorillonite with the number of W-D cycles for various proportions of Ca. At 20° C and at a relative water vapor pressure of 0,7. For all the samples, the $d_{001}$ spacing corresponds to the larger of the two maxima in the X ray diffraction band; but in the case of 30 % ca montmorillonite the relative intensities of the two maxima are inversed after 20 W-D cycles.

Figure 3 shows the evolution of the interlayer spacing of the different K-Ca montmorillonite samples with the number of W-D cycles. But it is necessary to bear in mind that these values correspond to the position of the apparent maxima of the 001 reflexions, and that considering figure 4 :

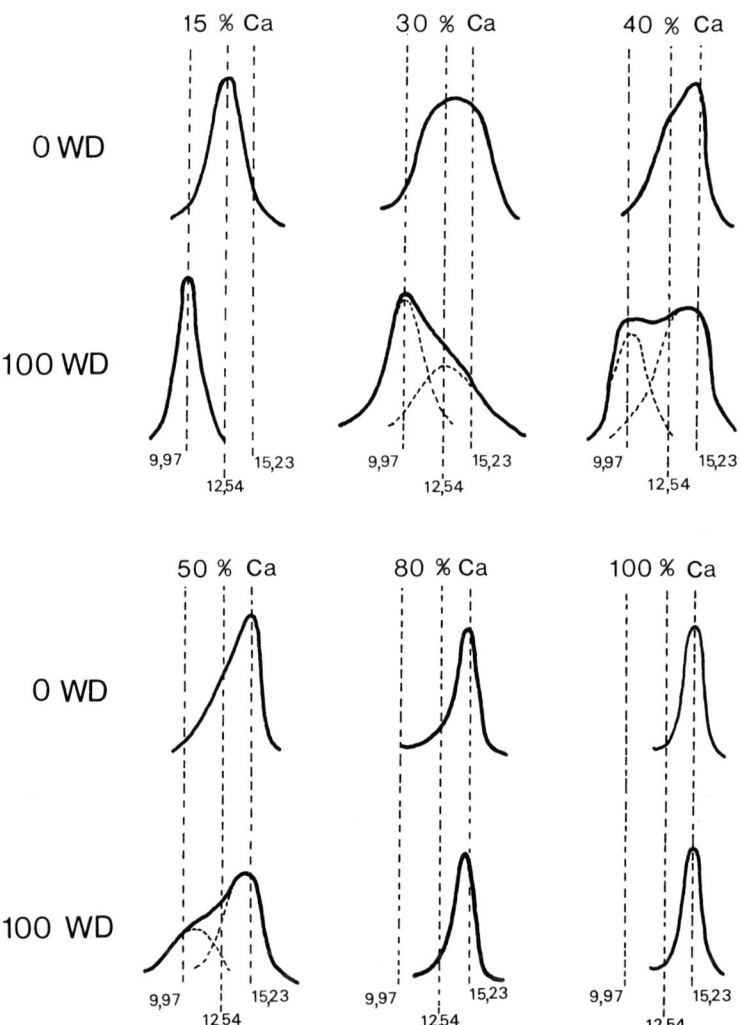

Fig. 4. Evolution of the 001 reflexion of K-Ca montmorillonite after 0 and 100 W-D cycles for various proportions of Ca. At 20° C and at a relative water vapor pressure of 0,7.

i - This reflexion is generally very nonsymetric (except for uniionic K or Ca montmorillonite).

ii - In the case where the proportion of Ca lies between 30 % and 50 % and after a high number of W-D cycles, the 001 band is resolved into two separate reflexions : one indicating 15 Å, the other about 10 Å.

After 50 W-D cycles all the 001 reflexions may be considered as being due to the superposition of the 10 Å and 15 Å reflexions.

This analysis in terms of two separate reflexions shows that the formation of the two crystalline phases is not an interstratification, but actually the formation of homogeneous crystalline domains of sufficient extent to give the X-ray diffraction effects observed.

iii - When the amount of Ca is low (15 % in the case studied) the montmorillonite samples appear to evolve as though they were 100 % K. But taking into account the external surfaces on which the Ca cations may be located, the 15,4 Å reflexion, even if it existed, would be too weak to be observed.

Likewise, in the case of Ca predominance (80 % in the case studied) for the same reason, the 10 Å reflexion due to K is not observable.

The most reasonable interpretation of the results presented here is the following : the W-D cycles produces a separation of two crystalline phases ; one containing K which becomes more and more collapsed and gives a 10 Å reflexion, and another containing Ca, or eventually a K-Ca mixture, which can swell and gives a 15 Å reflexion.

## EVOLUTION OF THE CRYSTALLOGRAPHIC CHARACTERISTICS OF BIIONIC K-Ca MONTMORILLONITE SUBMITTED TO ALTERNATE W-D CYCLES.

### X-ray diffraction observations

X-ray diffraction patterns obtained from well oriented clay films or from powder, show that the effect of W-D cycles results in an increasing modulation of the profiles of the hk bands. But these modulations, the intensity of which increases with the number W-D cycles, become more and more diffuse as the proportion of Ca in the montmorillonite sample is increased. On the other hand modulations in the 100 % Ca montmorillonite samples have never been observed.

The development of the modulations of the profile in the hk bands may be interpreted as being due to the transformation of the initially random stacking of the elementary layers. The turbostratic structure evolves into a well ordered one.

In the latter the hexagonal cavities of the adjacent layers are facing each other. This corresponds to the development of a three-dimensional organization.

This evolution is only observed when the montmorillonite samples contains K.

Electron Selected Area Diffraction observations

The elementary particles are parallel to the surface of the grid, consequently the diffraction patterns arise from the intersection of the hk reciprocal lines with the plane of the photographic film. The distribution of the diffraction spots is identical to that of the hk lines of the successive layers contained in a crystalline particle.

Observations on the K-Ca montmorillonite particles give the following informations :

i - Before any dessication, all the elementary particles of the different K-Ca montmorillonite samples are constituted of layers randomly stacked. The S.A.D. patterns form a set of spots distributed on concentric circles, without apparent symmetry.

ii - The greater the number of W-D cycles, an increasing number of elementary particles appear, for which the diffraction spots form an hexagonal array.

These patterns are due to particles in which the successive layers are regularly stacked. This implies that the crystallographic axis of two adjacent layers are probably rotated by multiples of 60° with respect to one another.

iii - For a given number of W-D cycles the number of well ordered particles is in inverse proportion to the proportion of Ca present.

iiii - For the 100 % Ca montmorillonite, whatever the number of W-D cycles, all the particles are turbostratic in structure.

All these observations are in accord with those obtained by the analysis of the X-ray diffraction patterns, but show in addition that under the effect of W-D cycles K and Ca cations are distributed in separate crystalline phases.

CONCLUSION

This investigation of the effect of a high number of W-D cycles on K-Ca montmorillonite shows that the exchangeability of K may decrease even when it is initially accompanied by Ca in the interlayer spaces. The Ca remaining fully exchangeable.

Initially the W-D cycles induce a demixing of the K and Ca cations, which progressively occupy distinctive positions on the crystalline particles or in the domains.

Secondly, the crystalline K phase is submitted to a specific structural reorganization which is due to translations and rotations of the successive layers identical to that described elsewhere in these proceedings (Plançon et al. 1978) for the 100 % K montmorillonite.

REFERENCES

Chaussidon, J, 1963. Evolution des caractéristiques chimiques et cristallographiques de montmorillonites biioniques K-Ca au cours d'alternances répétées d'humectation-dessiccation.
 in : I. Th. Rosenqvist (Editor), Proceedings of International Clay Conference Vol 1 : 195-201.

Mamy, J., Gaultier, J. P., 1976. Evolution structurale de la montmorillonite associée au phénomène de fixation irréversible du potassium.
 Ann. Agron., 27 : 1-16.

Mering, J., Glaeser, R., 1954. Sur le rôle de la valence des cations échangeables dans la montmorillonite. Bull. Soc. Franç. Miner. Crist. 77 : 519-530.

Plançon, A., Besson, G., Gaultier, J.P., Mamy, J., Tchoubar, C., 1978. Qualitative and quantitative study of a structural reorganization in montmorillonite after potassium fixation. these proceedings.

# FORMATION OF METALLIC SILVER AS RELATED TO IRON OXIDATION IN K-DEPLETED MICAS

M. SAYIN, B. BEYME, and H. GRAF VON REICHENBACH
Institut für Bodenkunde der Technischen Universität Hannover,
Hannover (G.F.R.)

ABSTRACT

Octahedral ferrous iron in K-depleted micas can be oxidized by $Ag^+$ ions penetrating from $AgNO_3$ solutions into expanded interlayers by ion exchange. Crystalline metallic silver is produced by this reaction. A blue colour develops as a result of oxidation in the low-iron micas like muscovite and phlogopite. The deposition of metallic silver within oxidized biotite samples was inspected by the use of light and electron scanning microscopy. The observed distribution pattern emphasizes the role of structural defects in promoting charge transfer processes during the reaction.

Chemical analysis, X-ray diffraction and IR spectroscopy reveal that loss of negative charge from the structure is only partly balanced by ejection of octahedral cations as well as by deprotonation of structural OH groups.

---

INTRODUCTION

Earlier investigations on the oxidation of structural ferrous iron in K-depleted biotite led to the conclusion that this reaction can be connected with and thus be dependent on ion exchange processes between aqueous solution and mica interlayers (Beyme and Graf v. Reichenbach, 1977). Ferric iron produced by the addition of hydrogen peroxide to suspensions containing K-depleted biotite and low concentrations of $FeSO_4$ displaced exchangeable cations from interlayer positions and acted as acceptors for electrons, which were released by ferrous iron in octahedral position. However, pH values of the solution had to be kept low in order to avoid interference of hydrolysis with the redox titration used for following the reaction. Low pH values at or below pH 2.8 in turn jeopardize the stability of of layer silicates. With regard to the standard reduction potential of 0.7996 V for the reaction $Ag^o = Ag^+ + e^-$ (Hunsberger, 1977) and to lack of hydrolysis under neutral conditions, silver ions were considered suitable for substituting ferric iron as an oxidizing agent in these experiments. The metallic luster of vermiculites and the blue colour of K-depleted muscovites, which were incidentally observed after treating these minerals with silver solutions, further suggested the oxidizing effect of silver ions upon structural ferrous iron.

Some results of investigations in which silver ions were used to oxidize ferrous iron in various vermiculized mica species are presented in this paper.

MATERIALS AND METHODS

Two biotites (from East Africa and from Bamle, Norway) one phlogopite (from Bamle, Norway) and one muscovite (from Schwarzenbach, Germany) obtained from F. Krantz Co., Bonn, as large, clean flakes, were wet ground and separated into different particle size fractions.

Interlayer potassium was removed from biotite (5 - 20μm) by treating 25 mg portions of the mineral with 1 liter of 0.1 N $BaCl_2$ solution at 70°C for 8 hours. More than 98 % of the original K content was exchanged by this treatment. Larger particles of phlogopite of (>50μm) and muscovite (>50μm) were also prepared, the latter at 120°C in a autoclave, but K removal remained incomplete with these minerals.

Details of the treatments with $AgNO_3$ solutions are given in the following sections.

Total amounts of K, Na, Mg, Ba, Fe and Ag were determined with a Perkin-Elmer Model 303 atomic absorption spectrophotometer after $HF/HClO_4$ treatment and $HNO_3$ dissolution. Exchangeable cations were removed for analysis by treating the minerals repeatedly with nitrate solutions in centrifuge tubes.

Ferrous iron and metallic silver were determined by redox titration. 10 - 20 mg of sample were mixed with 20 ml of 4.5 N $H_2SO_4$ and 1 ml of 0.05 N $FeSO_4$ and titrated with 0.05 N $H_2O_2$ by using a Metrohm automatic titrimeter (Multi Dosimat). This simultaneous dissolution-oxidation method is ineffective with muscovite due to incomplete dissolution of this mineral. In presence of metallic Ag structural $Fe^{2+}$ had to be determined indirectly by measuring the metallic Ag content spectrophotometrically after exchanging ionic silver and digesting the sample in $HF/HClO_4$. Subsequently the amount of metallic Ag was subtracted from the titrimetrically determined sum of metallic Ag and $Fe^{2+}$. Na-dithionite extractable iron has been determined following the method by Mehra and Jackson (1960).

X-ray diffractograms were obtained from samples dispersed on glass slides or ceramic tiles with a Seifert diffractometer using Ni filtered CuKα radiation. 060 reflections were measured according to the method given by Rich (1957).

For infrared spectroscopy, 1.5 mg of sample was mixed well with 300 mg KBr powder and pressed to 13 mm disks in an evacuable die. The disks were then heated overnight at $220°C$ to drive off water and repressed to obtain transparency. Spectra were recorded with a Beckman Model 4220 IR spectrophotometer.

For scanning electron microscopy samples were sedimented from aqueous suspensions on photograph paper fixed on the sample holders with double-stick tape. Mounted samples were coated with a gold film of 100 - 200 $Å$ thickness and micrographs taken with an ETEC Autoscan scanning electron microscope.


RESULTS AND DISCUSSION
Microscopic observations

K-depleted samples were subjected to repeated washings with 0.2 N $AgNO_3$ solution in the centrifuge. After a total reaction period of several hours the remaining Ag ions were exchanged with nitrate solutions of Ba or Mg and the minerals examined under the light microscope. An intense blue colour was observed in partly K-depleted samples of muscovite and phlogopite, which shows up as darkened areas on the particles in the photographs (Fig.1 and 2). The spatial distribution of darkened zones within distinct particles is obviously related to the distribution of K-depleted and expanded areas as can be concluded from preferential formation along mineral edges, fissures and cracks. The well established fact that oxidation of structural ferrous iron in micas in aqueous suspensions presupposes layer expansion suggests that the blue colour develops where ionic silver is reduced by oxidation of structural ferrous iron and the resulting metallic silver is deposited in the interlayers or between the layer packets.

A different picture is obtained with biotites (Fig.3 and 4) after the same treatment. No blue colour develops, but again metallic silver is formed, the distribution of which depends on experimental conditions. If silver ions in solution stay in contact with silver saturated minerals during the reaction, metallic silver is deposited in the form of large particles along the mineral edges (Fig.3). If, in contrast, the $AgNO_3$ solution is displaced after Ag saturation of the minerals, metallic silver produces dark zones preferentially in the center of mineral particles (Fig.4). These observations would indicate that $Ag^+$ ions of solution which come into contact with interlayer $Ag^+$ ions at the edges of the biotite flakes are reduced in the first case (see below) while a direct ejection of metallic silver out of the interlayers occurs in the second case. The blue colour may be a result of the existence of finely distributed silver mainly in colloidal form within the individual particles. No coloration was observed with lepidolite, a mineral from which K extraction is extremely difficult.

The higher resolution scanning electron microscopy adds further information. The regular shape of many of the particles deposited at the edges of the silicate points to the crystalline character of th silver grains (Fig.5). Special attention should be directed to differences in size and to distribution of silver crystals on the mineral surface (Fig.6). Large silver particles are accumulated mainly at the lateral faces of the minerals, where formation of silver crystals is favoured either by unrestrained delivery of silver from the adjacent interlayer space or by electron transfer to silver ions in solution. In addition, however, much smaller crystals are deposited all over the biotite surface, partly forming regular distribution patterns. The formation of these crystals may well be related to points of disorder and to small structural defects in silicate layers opening a pathway for electron transfer through a restricted number of layers.

Electron transfer may be pictured as hopping of electrons from octahedral ferrous iron to $Ag^+$ ions in interlayer position and, subsequent channelling through the interlayers by exchangeable $Ag^+$ ions. Electron transfer could also be bridged by octahedral ferric iron along the central plane of silicate layers on to $Ag^+$ ions contacting lateral layers edges. Obviously, these processes could be considerably enhanced by any forms of structural disorder which locally disturb the screening effect of tetrahedral sheets.

In both cases oxidation would result in the deposition of metallic silver at the layer faces and close to fissures or cracks as exhibited by the distribution patterns shown in Figures 3, 5 and 6, and within the particles as shown in Fig.4.

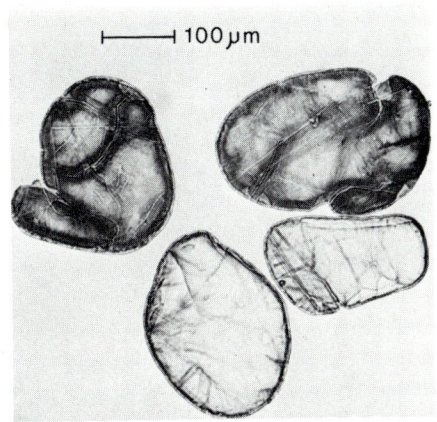

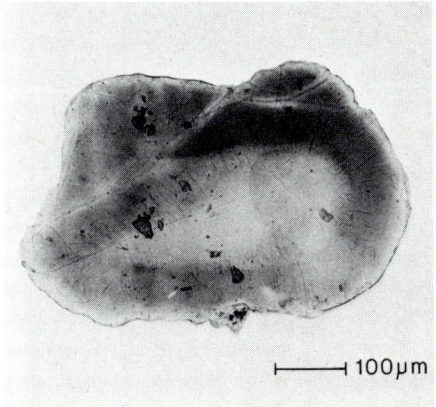

Fig.1: Partly K-depleted muscovite after oxidation with AgNO$_3$ solution.

Fig.2: Partly K-depleted phlogopite after oxidation with AgNO$_3$.

## Chemical observations

Quantitative aspects of the above observations were investigated by chemical analysis of the reaction induced by saturating K-depleted biotites with silver ions. For this purpose 600 mg Na-saturated biotite was treated for 1 minute with 100 ml 0.2 N AgNO$_3$ solution. The content of ferrous Fe, total Fe and Mg of the Na-saturated sample amounted to 13.01, 21.41, and 4.10 % of weight, respectively. Preliminary investigations revealed that unlike Ba$^{++}$ or Mg$^{++}$, Na$^+$ is almost completely exchanged by Ag$^+$ under these conditions. After 1 minute the AgNO$_3$ solution was filtered off. The minerals were suspended in distilled water and aliquots of the suspension removed for further analysis after the time intervals indicated in Table 1.

One part of these aliquots was used for total chemical analysis, another portion was transferred to centrifuge tubes, treated five times with 1 N NaNO$_3$ solution in order to determine exchangeable cations in the extracts, and after several washings with water also subjected to total analysis. The exchangeable amount of H$^+$ was

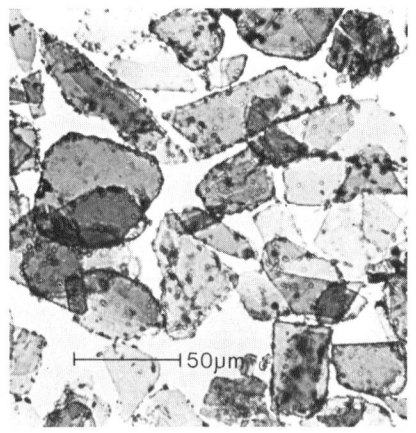
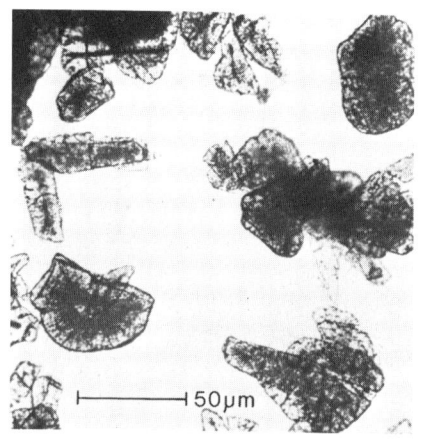

Fig.3: K-depleted biotite oxidized in contact with $AgNO_3$ solution.

Fig.4: K-depleted biotite oxidized after removal of $AgNO_3$ solution.

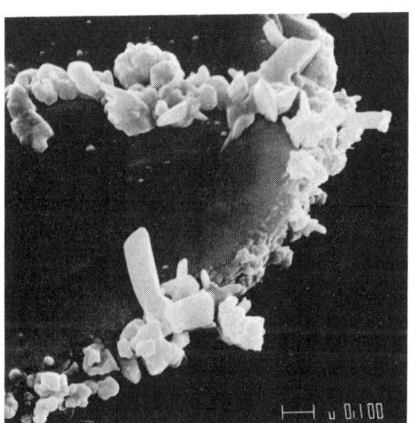
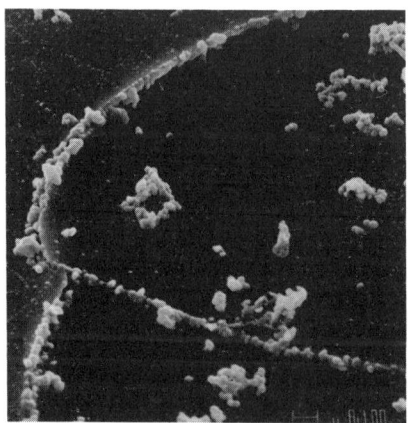

Figures 5 (left) and 6: Scanning electron micrographs exhibiting metallic silver crystals on the surfaces of K-depleted biotite after oxidation with $AgNO_3$ solution.

determined by titrating the exchange solutions as well as the Na-resaturated reaction products with NaOH to pH 5.8.

The results of these determinations are given in Tables 1 and 2.

The content of unexchangeable reduced silver increasing to 144 meq per 100 g after 6 hours of reaction indicates that the formation of metallic silver is due to an equivalent oxidation of structural ferrous iron. As a consequence, the consumption of $H_2O_2$ for the titration of oxidizable cations remains constant irrespective of the reaction time.

Oxidation of structural ferrous iron is not equivalent to the loss of cation exchange capacity (CEC). Compared with the formation of 144 meq of metallic silver, the CEC as determined by Na contents as well as by summation of exchangeable cations decreases only from 157 meq of untreated mineral to 109 meq. This difference is even more reduced, if the titratable (unexchangeable) amount of $H^+$ in the Na-resaturated sample is added to the exchangeable cations.

Ejection of octahedral cations during oxidation is demonstrated by increasing amounts of exchangeable $Fe^{2+}$ and $Mg^{2+}$ ions. In addition hydrolyzed ferric iron from the same source has been extracted with Na-dithionite. However, due to the insufficient selectivity of this extraction procedure, the data given in Table 2 are unsuitable for further quantitative interpretations. The existence of dithionite-insoluble iron compounds within the particles, if any, needs further study.

Considerable production of $H^+$ ions must have been initiated by oxidation leading to an $H^+$-saturation of about 75 % of CEC after 6 hours of reaction. Summation of exchangeable $H^+$, titrated unexchangeable $H^+$, and loss of $H^+$ to solution during the $AgNO_3$ treatment yields total amounts of $H^+$ production which progessively fall behind the amounts of reduced Ag.

The resulting charge loss indicates that the removal of electrons from structural ferrous iron is not fully compensated by an equivalent transfer of positive charges consisting either of cation ejection or deprotonation of structural hydroxyl.

On the other hand, a large difference in equivalency is found between total production of $H^+$ and ejection of octahedral iron. Thus, in spite of the remaining uncertainty concerning the exact amount of iron extractable by dithionite, $H^+$ production can only partly be attributed to hydrolysis of ejected iron. A rough estimate rather suggests that deprotonation becomes the prevailing

TABLE 1

Exchangeable cations, cation exchange capacity (CEC), loss of layer charge and silver reduction as dependent on reaction time after treatment of Na-saturated East Africa biotite with $AgNO_3$ solution. (Data expressed as meq calculated on the basis of 100 g of substance free of exchangeable cations, metallic silver and water ($300°C$)

| Reaction time (minutes) | Exchangeable cations | | | | CEC | unexch. $H^+$ | Loss of layer charge | red. Ag | $H_2O_2$ Cons. |
|---|---|---|---|---|---|---|---|---|---|
| | $Ag^+$ | $Fe^{2+}$ | $Mg^{2+}$ | $H^+$ | | | | | |
| 10 | 132 | 3.8 | 0.3 | 15.9 | 152 | 3.1 | 1.9 | 27 | 215 |
| 30 | 92 | 11.2 | 0.8 | 38.0 | 142 | 8.6 | 6.4 | 65 | 212 |
| 90 | 51 | 16.2 | 1.4 | 58.9 | 127 | 17.1 | 12.9 | 106 | 212 |
| 360 | 5 | 17.4 | 3.7 | 82.9 | 109 | 24.9 | 23.1 | 144 | 211 |

TABLE 2

Production of $H^+$ and ejection of Fe from octahedral position as related to oxidation of structural ferrous iron after different period of reaction. (Basis of calculation as indicated in Table 1)

| Reaction time (minutes) | Loss of $H^+$ to solution mmol | Total $H^+$ prod. mmol | $\dfrac{H_{prod.}}{Ag_{red.}}$ | Fe extracted (dithion.) mmol | Total Fe ejected mmol | $\dfrac{H_{prod.}}{Fe_{eject.}}$ |
|---|---|---|---|---|---|---|
| 10 | 8.7 | 27.7 | 1.02 | 4.0 | 5.9 | 4.7 |
| 30 | 8.7 | 55.3 | 0.85 | 5.0 | 10.6 | 5.2 |
| 90 | 8.7 | 84.7 | 0.80 | 6.8 | 14.9 | 5.7 |
| 360 | 8.7 | 116.5 | 0.81 | 9.0 | 17.7 | 6.6 |

mechanism of positive charge transfer during the oxidation.

The extent of oxidation strongly depends on the amount of $Ag^+$ ions introduced into interlayer position. Thus, an almost linear relationship is found by plotting log $Ag^+_{exch.}$ against reaction time.

Further evidence is obtained from complementary experiments. In correlation with the respective exchange equilibria, oxidation is diminished if the minerals are presaturated with cations like $Mg^{2+}$ or $Ba^{2+}$ which exhibit a higher affinity to exchange sites as compared to $Na^+$. The extent of oxidation also depends on the type and equivalent proportion of other cations added to $AgNO_3$ solutions.

## Structural observations on reaction products

In biotite, the existence of metallic silver as a result of the treatment can easily be substantiated by X-ray diffraction. Time-dependent formation of metallic silver can be followed by the intensity increase of $d_{111}$ reflections at 2.359 Å. Peak heights corrected for mass absorption show a linear relationship with the amount of oxidized iron. The instant crystallisation indicates that $Ag^+$ ions after being reduced by electron acceptance do not remain in the original position of exchangeable cations, but rather move to points of crystallisation.

With increasing oxidation $d_{001}$ spacings of Ag-saturated samples dried at room temperature progressively decline from 14.5 to 12.4 Å. Even after resaturation of reaction products with $Mg^{2+}$ ions a slight decrease of $d_{001}$ spacings from 14.5 to 14.0 Å is observed. Increasing contractibility may be interpreted as a result of either deprotonation of structural OH groups or inclination of O-H bonds to octahedral vacancies induced by cation ejection. These explanations are supported by the occasional occurence of regular interstratification indicated by reflections at 8 and 24 Å.

060 reflections of biotites decrease as a result of oxidation from 1.5476 Å for the unoxidized Na-saturated biotite over 1.5434, 1.5383, 1.5383 to 1.5378 Å after reaction periods of 10, 30, 90, and 360 minutes, respectively. This also points at a rearrangement of the structure as resulting from the ejection of octahedral cations.

The IR spectra exhibit concomitant increase of absorbance at 3550 cm$^{-1}$ with oxidation of structural ferrous iron (Fig.7). This band must be ascribed to hydroxyl groups associated

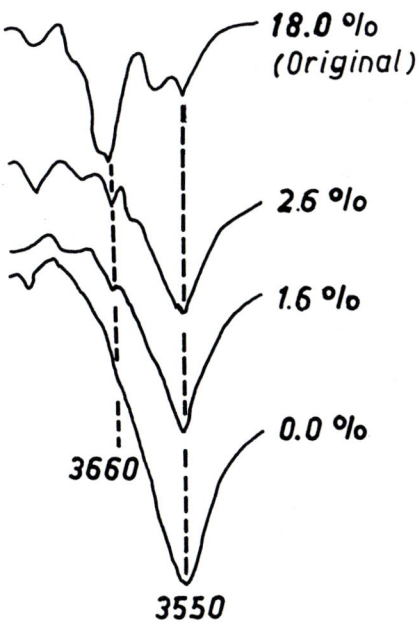

Fig.7: OH stretching vibrations (cm$^{-1}$) of East Africa biotite at different contents (%) of ferrous iron remaining after oxidation.

with $Fe^{3+}Fe^{3+}$ pairs (Vedder, 1964). Octahedral vacancies must therefore be continuously produced by the ejection of octahedral cations from the structure. The decrease and eventual disappearence of the bands at 3660 $cm^{-1}$ which are related to groups of octahedral ferrous iron (Vedder and Wilkins, 1969), are in agreement with this contention. The origin of the 3550 $cm^{-1}$ band in original biotite is unknown. Change of saturating cation from Na to Mg did not cause a variation in the vibrational frequency of the vacancy band.

Ejected metallic silver made the KBr disks more opaque and reduced the transmission of infrared radiation.

ACKNOWLEDGEMENTS

The authors are indebted to the Bundesanstalt für Geowissenschaften und Rohstoffe in Hannover for permitting the use of the scanning electron microscope. Thanks are also due to Miss A. Müller, who conducted most of the chemical analyses, and to Mr. B. Tippkötter, who prepared the photomicrographs.

REFERENCES

Beyme, B. and Graf v. Reichenbach, H., 1977. Oxidation of structural Fe in altered biotite with $H_2O_2$ under controlled redox conditions. Abstract of paper presented at the Third European Clay Conference in Oslo, Norway.

Hunsberger, J.F., 1977. Electrochemical series. In: R.C. Weast (Editor), CRC Handbook of Chemistry and Physics 1977. CRC Press Inc., Cleveland, Ohio, p. D-141.

Mehra, O.P., and Jackson, M.L., 1960. Iron oxide removal from soils and clays by a dithionite-citrate system buffered with sodium bicarbonate. Clays Clay Miner. 7, 317 - 327.

Rich, C.I., 1957. Determination of (060) reflections of clay minerals by means of counter type X-ray diffration instruments. Am. Miner. 42, 569 - 570.

Vedder, W., 1964. Correlations between infrared spectrum and chemical composition of mica. Am. Miner. 49, 736 - 768.

Vedder, W. and Wilkins, R.W.T., 1969. Dehydroxylation and rehydroxylation, oxidation and reduction of micas. Am.Miner. 54, 482 -509.

# PROTONATION OF BASES IN CLAY SUSPENSIONS

J. R. FELDKAMP and J. L. WHITE

Department of Agronomy, Purdue University, West Lafayette, Ind. 47907 (U.S.A.)

## ABSTRACT

A new mechanism is proposed to explain the observation that protonation of bases in clay suspensions often proceeds to a much greater extent than that predicted from the pH of the bulk solution phase. Thermodynamic measurements made by spectroscopic titrations show that this phenomenon has its origin in the ability of the clay to interact with and stabilize the protonated form of the base to a much greater extent than the neutral form. The degree of this stabilization depends on the apparent magnitude of the negative charge on the clay as encountered by the basic compound. This relative difference in stabilization then results in a displacement of the equilibrium toward production of the protonated form of the base. Present ideas concerning protonation of bases in clay-water systems as expressed in terms of "surface acidity" are discussed in light of the proposed mechanism.

## INTRODUCTION

When basic compounds are placed in a clay suspension, it is widely recognized that protonation quite often proceeds to a much greater extent than that expected based upon the pH of the bulk solution phase and the pK of the compound (Bailey and Karickhoff, 1973; Karickhoff and Bailey, 1976). For example, a suspension having a $pH^b = 4$ may cause considerably more protonation than an aqueous solution (clay-free) having that same pH value (the superscript b refers to the bulk solution phase; $pH^b$ might be measured by measuring the pH of the supernatant liquid after centrifugation). Examination of the system shows that the excess protonated base is in the interfacial region so that this phenomenon is strictly surface induced.

The theory commonly invoked to explain the surface effect as described above states that a variety of potential proton donors such as $H_2O$ or surface OH groups are in a highly dissociated state and that protons arising from such donors are subsequently available for protonating a basic molecule in the surface phase.

In addition, it is said that the surface is at a lower pH than the bulk solution phase. If pH is to be defined by pH = -log $a_{H^+}$ where $a_{H^+}$ is itself defined by $\mu_i = \mu_i^* + RT \ln a_{H^+}$, then pH certainly does decrease as a negatively-charged surface is approached more closely. In fact, according to double layer theory, pH will vary linearly with electrical potential. So a negatively-charged clay mineral surface will tend to accumulate protons and hence develop a zone of lower pH. The effect of the zone of lowered pH is to cause increased surface protonation. Those protons arising from the previously mentioned dissociated donor groups will be distributed in some fashion throughout the electrical double layer.

The error in the concept discussed above lies in the assumption that all surface phase protons (generated according to the above mentioned mechanisms) are equally available for protonation. It can readily be shown by using thermodynamic principles that this is not the case for suspensions. We wish to consider the reaction $BH^+ = B + H^+$ where B is a weak base. The condition that this reaction be at equilibrium is the following

$$\bar{G}_B + \bar{G}_{H^+} - \bar{G}_{BH^+} = 0 \qquad [1]$$

where the individual $\bar{G}_i$ are the partial molar Gibb's free energies for the three components. This equation states quite clearly that the $\bar{G}_i$ (we wish to focus specifically on $\bar{G}_{H^+}$ here) are the important quantities influencing the chemical equilibrium. Thus, neither the degree of dissociation of specific proton donors nor the concentration of protons in the interfacial region are relevant when considering the point of stable equilibrium for this reaction. If the interfacial region is considered to be in equilibrium with the bulk solution phase (at least with respect to proton transfer processes) then it is true that $\bar{G}_{H^+}$ has the same value throughout the system (i.e. $\bar{G}_{H^+}$ (surface) = $\bar{G}_{H^+}$ (bulk solution)). The conclusion that follows is that both the surface phase and the bulk solution phase are equally able to supply a proton to participate in the acid-base equilibrium. Hence, even though there exists a higher proton density near the surface, not all of these protons are to be regarded as available for protonation; however, this does not mean that this property will be unimportant in kinetic phenomena.

It should be pointed out that no study of surface acidity involving acid-base indicators actually provides a measure of proton availability at the mineral surface (an indicator is regarded to be any acidic or basic compound on the mineral surface, whose form can be monitored by some means.) The behavior of the indicator compound simply reflects the extent to which the surface environment allows the formation of the conjugate acid from the base. This will depend as much on the specific interactions between B, $BH^+$ and the surface environment as it does on proton availability as determined by the magnitude of $\bar{G}_{H^+}$. For the effect descri in paragraph one, only these specific interactions are important. This must be the case, since, as already pointed out, $\bar{G}_{H^+}$ has the same value throughout a suspension at equilibrium.

The explanation that we wish to provide for the surface protonation phenomenon described earlier (which will hereafter be referred to as "enhanced protonation"), is based on what is commonly known as Le Chatelier's principle. This states that if a stress acts on a thermodynamic system, the system will respond in such a way that the stress is relieved. In our case, the stress is in the form of a new environment (clay-H$_2$O interface) for the base B and is relieved by B reacting to form BH$^+$. To illustrate this in detail requires application of thermodynamic principles in the manner described below.

When the chemical reaction (BH$^+$ = B + H$^+$) reaches equilibrium, an important quantity characterizing the equilibrium is the thermodynamic equilibrium constant K, which is at most a function of temperature and pressure but independent of composition. For this reaction, K is defined in the following way

$$K = a_{H^+} \frac{a_B}{a_{BH^+}} \qquad [2]$$

with the activities defined by

$$\bar{G}_i = \bar{G}_i^* + RT \ln a_i \qquad [3]$$

The activity will be written here in terms of the average concentration $\bar{C}_i$ (Mokady and Low, 1966); $a_i = \bar{\gamma}_i \bar{C}_i$. $\bar{\gamma}_i$ is simply the activity coefficient to be associated with $\bar{C}_i$. The standard state is chosen so that $\bar{\gamma}_i \to 1$ as $\bar{C}_i \to 0$ in water, in the absence of clay and any solutes. The equilibrium constant K can then be written

$$K = a_{H^+} \frac{\bar{\gamma}_B \bar{C}_B}{\bar{\gamma}_{BH^+} \bar{C}_{BH^+}} = \frac{\bar{\gamma}_B}{\bar{\gamma}_{BH^+}} a_{H^+} \frac{\bar{C}_B}{\bar{C}_{BH^+}} = \frac{\bar{\gamma}_B}{\bar{\gamma}_{BH^+}} K_{eff} \qquad [4]$$

This then serves as a definition for $K_{eff}$, an effective or apparent equilibrium constant

$$K_{eff} = a_{H^+} \frac{\bar{C}_B}{\bar{C}_{BH^+}} \qquad [5]$$

Important to our development is the effect on $\bar{G}_i$ (i = B, BH$^+$) brought about by the addition of clay, holding all other variables constant (i.e. T, P, and composition). If $m_c$ represents the mass of clay in the system, then the variation in $\bar{G}_i$ due to varying $m_c$ can be expressed as

$$\frac{\partial \bar{G}_i}{\partial m_c} = RT \frac{\partial \ln \bar{\gamma}_i}{\partial m_c} \quad \text{(constant T, P and composition)}; \quad [6]$$

this may be integrated, keeping T, P and composition constant, between $m_c = 0$ and $m_c$ so that

$$\Delta \bar{G}_i = RT \ln \bar{\gamma}_i \quad [7]$$

where it is assumed that at $m_c = 0$, the solution is sufficiently dilute that $\bar{\gamma}_i = 1$. If $\Delta \bar{G}_i$ is not zero, component i must have undergone some net interaction (adsorption) with the clay. $\Delta \bar{G}_i$, $\bar{\gamma}_i$, and $K_{eff}$ all depend on the composition of the system, but in the limit as $\bar{C}_i \to 0$, these quantities will approach definite values that reflect an intrinsic interaction of $BH^+$ or B with the surface independent of $\bar{C}_i$ (i = B, $BH^+$), but not independent of the other variables. Hereafter, interactions occurring in the limit when $\bar{C}_i$ is vanishingly small will be referred to as intrinsic interactions.

We wish to examine the effect that such intrinsic interactions have on the equilibrium $BH^+ = B + H^+$. If $\lambda$ represents the progress variable or extent of reaction (Moore, 1972), then the change in the Gibb's free energy accompanying a small change in $\lambda$, $\delta\lambda$, is

$$\frac{\delta G}{\delta \lambda} = \bar{G}_{H^+} + \bar{G}_B - \bar{G}_{BH^+} \quad [8]$$

When the reaction is at equilibrium, $\delta G/\delta\lambda = 0$. If clay is added to a dilute aqueous solution containing an equilibrium mixture of B, $BH^+$ and $H^+$ at constant T, P and composition, the $\bar{G}_i$ in this equation will in general change and do so according to equation [7]. Moreover, they are most likely to change in such a way that $\delta G/\delta\lambda$ is no longer equal to zero. In order to restore equilibrium, the reaction must proceed in the proper direction with appropriate changes in $\bar{C}_i$ until $\delta G/\delta\lambda$ is again equal to zero. The reaction has therefore been displaced from its original equilibrium due to interactions between B, $BH^+$, $H^+$ and the clay mineral surface.

Now the changes in the $\bar{G}_i$ that are important to the production of enhanced protonation are those which occur in $\bar{G}_{BH^+}$ and $\bar{G}_B$. To see this more clearly, first note that the ratio $K_{eff}/K$ represents the ability of a clay suspension to cause enhanced protonation. If a suspension has the ability to generate excess $BH^+$, then

$$K_{eff}/K < 1 \text{ or } \Delta pK \equiv pK_{eff} - pK > 0 \quad [9]$$

On the other hand, if the suspension causes protonation only to the extent predicted on the basis of the known value for $pH^b$, then

$$K_{eff}/K = 1 \text{ or } \Delta pK = 0 \qquad [10]$$

That this is true becomes clear if the appropriate expressions are substituted into $K_{eff}/K$:

$$K_{eff}/K = \frac{a_{H^+} \overline{C}_B/\overline{C}_{BH^+}}{a_{H^+} C_B/C_{BH^+}} = \frac{\overline{C}_B/\overline{C}_{BH^+}}{C_B/C_{BH^+}} \qquad [11]$$

$C_B/C_{BH^+}$ is the ratio of concentrations for the two forms in a hypothetical ideal solution having a hydrogen ion activity equal to $a_{H^+}$. $\overline{C}_B/\overline{C}_{BH^+}$ is the ratio of average concentrations for the two forms in the clay suspension at the same hydrogen ion activity. (It could be envisioned that this hypothetical ideal solution and clay suspension, both of which contain B and $BH^+$, are equilibrated across a membrane impermeable to clay so that both environments have the same $a_{H^+}$ value).

Using equations [4] and [7], it is observed that

$$\Delta \overline{G}_{BH^+} - \Delta \overline{G}_B = RT \ln K_{eff}/K \qquad [12]$$

If $K_{eff}/K < 1$, implying enhanced protonation, then $\Delta \overline{G}_{BH^+} < \Delta \overline{G}_B$. The converse of this statement is also true and clearly indicates that in order for enhanced protonation to take place, $BH^+$ must be stabilized (i.e. its potential must be lowered) to a greater extent in the clay-water interface than B. If $\Delta \overline{G}_{BH^+} = \Delta \overline{G}_B$, implying that $K_{eff}/K = 1$, then B and $BH^+$ interact to the same extent with the clay surface so that no excess protonation will take place.

The experiments discussed in this report are intended to examine the perturbation of the reaction $BH^+ = B + H^+$, due to interactions between B, $BH^+$ and the clay, by comparing the magnitudes of $pK_{eff}$ and $pK$, and by measuring the energy changes $\Delta \overline{G}_{BH^+}$ and $\Delta \overline{G}_B$. Equations relating experimental quantities to these quantities are discussed fully in a paper by the authors (Feldkamp and White). There, the necessary equations were developed with the assumption that the system is sufficiently dilute with respect to all components (clay, $BH^+$, B and $H^+$) that the bulk solution phase may be regarded as ideal for B and $BH^+$. All experiments have been performed with the additional assumption that dilute solutions having the same pH also have the same hydrogen activity.

## MATERIALS AND METHODS

The Ca-saturated montmorillonite used in this study was prepared from < 2μ

Wyoming Bentonite (Upton, Wyoming) by washing twice with 1M $CaCl_2$ then washing free of salts. The compounds used in this study were 2,4-bis (ethylamino)-6-methoxy-s-triazine (simetone) and 2,4-bis (isopropylamino)-6-methylmercapto-s-triazine (prometryne). Simetone and prometryne are both herbicides and have water solubilities of 3200 mg/l and 48 mg/l, respectively. Stock solutions of highly purified samples were prepared and passed through an ultrafine sintered glass filter to remove particulate matter. Weber (1967) obtained pK values of 4.15 for simetone and 4.05 for prometryne, both of which were used in calculations in this report. The pK of simetone was determined in this work to be 4.10, which is in good agreement with Weber's (1967) values.

The ratio $\bar{C}_B/\bar{C}_{BH^+}$ (or $C_B/C_{BH^+}$ if clay is absent) can be measured for the alkyl-amino-s-triazines by UV spectroscopy at the appropriate wavelength (Weber, 1967). If the total amount of base present is constant, $\bar{C}_B/\bar{C}_{BH^+}$ is related to the absorbance A by

$$\bar{C}_B/\bar{C}_{BH^+} = \frac{A - A_+}{A_- - A} \quad \text{(if } A_+ < A_-\text{)} \quad [13]$$

where $A_+$ is the absorbance when all of the base is in the protonated form and $A_-$ is the absorbance when all of the base is in its neutral form. UV spectrophotometric measurements of A (and consequently $\bar{C}_B/\bar{C}_{BH^+}$) as a function of pH will then constitute a titration curve for the basic compound. The product of $a_{H^+}$ and $\bar{C}_B/\bar{C}_{BH^+}$ provides a measurement for $pK_{eff}$. $a_{H^+}$ is calculated from the measured pH value, and is regarded experimentally as an empirical parameter rather than an absolute activity. This is all that is required since the magnitude of $pK_{eff}$ relative to pK ($\Delta pK$) is the desired quantity and is independent of any absolute measurement of $a_{H^+}$. Details of the pH measurements will be discussed below. Simetone was titrated at 218.5 nm and prometryne at 223 nm. The bandwidth was set at 0.8 nm for all titrations.

Spectrophotometric titrations were carried out in suspensions of salt-free Ca-saturated montmorillonite as well as with excess salt present. Only double-distilled water was used in these experiments. The concentrations of the various components of the system during titration were as follows: 200 mg clay/l; 5 mg base/l; salt additions were at the rate of 20 mg/l. $pH^b$ measurements were made by immersing the combination electrode directly into the stirred dilute suspension and recording the pH. This value differed by less than 1% from the value obtained by first centrifuging the suspension, followed by measuring the pH of the supernatant phase, thus the suspension effect may be ignored. For additional details concerning measurements, the paper previously cited should be consulted (Feldkamp and White).

## RESULTS AND DISCUSSION

As was demonstrated in the Introduction, a basic compound B will undergo enhanced protonation if $\overline{\Delta G}_{BH^+} < \overline{\Delta G}_B$. It is our intention to show that the specific characteristic of the surface responsible for generating the condition $\overline{\Delta G}_{BH^+} < \overline{\Delta G}_B$ is its ability to lower the electrical potential for $BH^+$ relative to the bulk solution phase, which is due to its negatively-charged sites. For this reason, $BH^+$ will have a much more favorable interaction with the surface than B. Therefore, the extent to which $\overline{\Delta G}_{BH^+}$ is less than $\overline{\Delta G}_B$ (or the magnitude of $\Delta pK$) should depend on the apparent magnitude of negative surface charge as encountered by the basic molecule. This apparent negative charge will be affected by the ability of cations and water in the surface phase (and perhaps other molecules present) to screen the electrical field arising from the mineral surface.

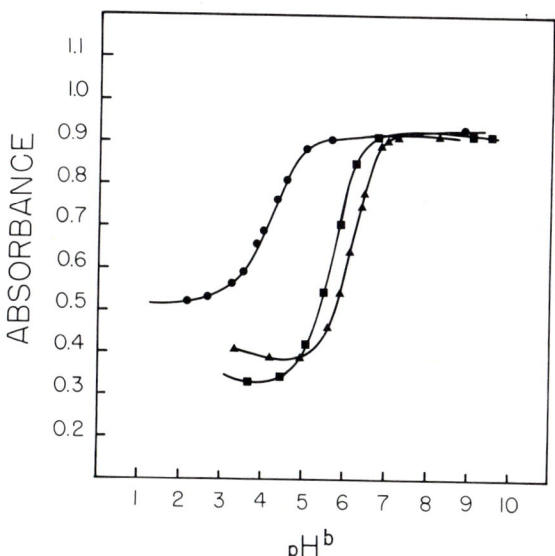

Fig. 1. Titration of simetone in dilute aqueous solution (●), Ca-montmorillonite suspension with salt added (■), and salt-free Ca-montmorillonite suspension (▲).

Fig. 1 shows the titration curves for simetone in three different environments: dilute aqueous solution, salt-free Ca-saturated montmorillonite, and Ca-montmorillonite with $CaCl_2$ added at the rate of 20 mg/l. It is clear that the $pK_{eff}$ (pH at the inflection point) is greater than pK = 4.15 with or without excess salt added. $pK_{eff}$ is largest for the salt-free case with $pK_{eff}$ = 6.15; when salt was added, the degree of displacement is reduced so that $pK_{eff}$ = 5.63. The prometryne titrations (Fig. 2) showed behavior similar to that observed for simetone.

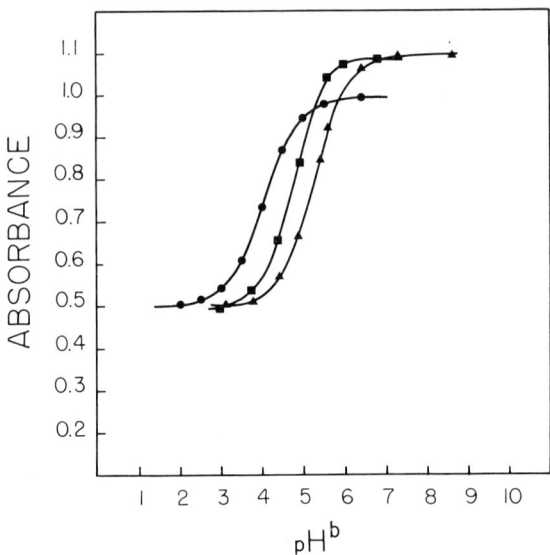

Fig. 2. Titration of prometryne in dilute aqueous solution (●), Ca-montmorillonite suspension with salt added (■), and salt-free Ca-montmorillonite suspension (▲).

In the salt-free suspension, $pK_{eff}$ = 5.32, but with $CaCl_2$ salt added to the suspension at the rate of 20 mg/l, $pK_{eff}$ = 4.78. This trend indicates that displacement of the reaction requires that $BH^+$ have access to the negatively-charged mineral surface since with a higher concentration of $Ca^{++}$ ions present, the negatively-charged surface is more effectively screened. It should be emphasized that the behavior observed here is the opposite of that predicted by Mortland and Raman (1968), i.e. that a greater ratio of $Ca^{++}/H_2O$ should promote greater protonation.

The calculated value of $\overline{\Delta G}_{BH^+}$ for simetone in the salt-free suspension is -2.8 kcal/mole and for prometryne it is -1.73 kcal/mole. It is clear that protonated simetone has a stronger interaction with the clay surface than protonated prometryne. This leads one to conclude that simetone-$H^+$ is more compatible with the surface ion-water structure than is prometryne-$H^+$. In addition, even though these two molecules have nearly the same pK (4.15 and 4.05, respectively) and similar chemical structures, their prediction of the magnitude of so-called surface acidity when used as indicators would be considerably different, as evidenced by their widely differing $pK_{eff}$ values (6.15 for simetone and 5.32 for prometryne, both in salt-free suspensions). This type of behavior is ignored by those who use acid-base indicators to measure the magnitude of surface acidity (Benesi, 1956; Frenkel, 1974; Walling, 1950).

Based upon the experiments discussed here and elsewhere (Feldkamp and White), it seems reasonable to conclude that the surface conditions most suitable for inducing a large enhanced protonation effect are a highly diffuse electrical double layer (favored by low ionic strength), coupled with the use of basic compounds that are highly compatible with the structured environment at the interface. A highly diffuse electrical double layer will minimize screening of surface fields, or viewed another way, will minimize competition with cations already present at the interface. It is expected that trivalent ions will be the most effective at screening out surface fields, followed by divalent, then monovalent ions. Those trivalent ions that are usually considered, such as $Al^{3+}$ and $Fe^{3+}$, will only retain their +3 charged state below a certain $pH^b$ value (e.g. for Al $pH^b \leq 4.2$ (Feldkamp and White; Low, 1955)). As the $pH^b$ is raised above this specific $pH^b$ value, the ion will gradually become discharged, eventually to form the hydroxide polymer. Hence, in clay systems having these ions as the saturating ions, the observed degree of protonation enhancement will be dependent on $pH^b$, with the enhanced protonation effect expected to be least in the $pH^b$ range where the ion is triply charged. As a consequence of the discharging process, greater enhanced protonation (as reflected in an increase in $\Delta pK$) is expected to occur as the $pH^b$ increases (Feldkamp and White).

In a recent publication by Karickhoff and Bailey (1976), it was observed that Ca- and Mg-saturated montmorillonite clay suspensions exhibited less surface acidity than Na- or K-saturated montmorillonite suspensions. They regarded this observation as "an exception to the general rule" since Mortland and Raman's work (1968) predicts just the opposite behavior. According to the discussion presented thus far, it is not at all surprising that such behavior should occur; indeed, the concept proposed in this paper would predict this result.

Even though our analysis has been directed toward suspensions, it seems very reasonable to expect that the mechanism described above ought to play a significant role in the protonation of bases at air-dry clay mineral surfaces. Lower water contents will force greater interactions, leading to perhaps larger perturbations of the chemical equilibrium. This could explain the experimental results observed for such systems as well as the idea that proton availability increases upon dehydration. Actually, there is no direct evidence suggesting that $\bar{G}_{H^+}$ (the true measure of proton availability) increases upon decreasing the $H_2O$ content.

Finally, our proposed concepts provide a plausible explanation for the fact that acidic compounds (e.g. $HA = H^+ + A^-$, where HA is a weak acid) do not display a surface protonation effect in dilute suspensions of negatively-charged colloidal particles as bases obviously do. At the higher water contents found in dilute suspensions, the anion $A^-$ avoids the negatively-charged surface so that $\Delta \bar{G}_{A^-} \simeq 0$

upon addition of clay to the system. Upon lowering the water content, an interaction with the negatively-charged surface will be forced upon A⁻ so that $\Delta \overline{G}_{A^-} > 0$. Since the magnitude of $\Delta \overline{G}_{HA}$ is expected to be small in comparison with $\Delta \overline{G}_{A^-}$, the reaction would be displaced toward the production of HA (i.e. $\Delta \overline{G}_{HA} < \Delta \overline{G}_{A^-}$) which is what is observed (Harter and Ahlrichs, 1967; Harter and Ahlrichs, 1969).

ACKNOWLEDGMENTS

Financial support from the Purdue Research Foundation in the form of a David Ross XR Fellowship for J. R. Feldkamp is gratefully acknowledged.

Journal Paper No.7177 of the Purdue University Agricultural Experiment Station, West Lafayette, Indiana 47907, U.S.A.

REFERENCES

Bailey, G. W., and Karickhoff, S. W. 1973. An ultraviolet method for monitoring surface acidity of clay minerals under varying water content. Clays Clay Minerals, 21:471-477.
Benesi, H. A. 1956. Acidity of catalyst surfaces. I. Acid strength from colors of adsorbed indicators. J. Amer. Chem. Soc. 78:5490-5494.
Feldkamp, J. R. and White, J. L. Acid-base equilibria in clay suspensions. (submitted for publication).
Frenkel, M. 1974. Surface acidity of montmorillonites. Clays Clay Minerals, 22:435-441.
Harter, R. D., and Ahlrichs, J. L. 1967. Determination of clay surface acidity by infrared spectroscopy. Soil Sci. Soc. Amer. Proc. 31:30-33.
Harter, R. D., and Ahlrichs, J. L. 1969. Effect of acidity on reactions of organic acids and amines with montmorillonitic clay surfaces. Soil Sci. Soc. Amer. Proc., 33:859-863.
Karickhoff, S. W., and Bailey, G. W. 1976. Protonation of organic bases in clay-water systems. Clays Clay Minerals, 24:170-176.
Low, P. F. 1955. The role of aluminum in the titration of bentonite. Soil Sci. Soc. Amer. Proc., 19:135-139.
Mokady, R. S., and Low, P. F. 1966. Electrochemical determination of diffusion coefficients. Soil Sci. Soc. Amer. Proc., 30:438-442.
Moore, W. J. 1972. Physical Chemistry, 4th edn., Prentice-Hall Inc., Englewood Cliffs, N. J.
Mortland, M. M., and Raman, K. V. 1968. Surface acidity of smectites in relation to hydration, exchangeable cation and structure. Clays Clay Minerals, 16:393-398.
Walling, C. 1950. The acid strength of surfaces. J. Amer. Chem. Soc., 72:1164-1168.
Weber, J. B. 1967. Spectrophotometric determination of ionization constants. Spectrochim. Acta, 23A:458-461.

INFRARED STUDY OF SEPIOLITE AND PALYGORSKITE SURFACES

C.J. SERNA and G.E. VANSCOYOC

Instituto de Edafología y Biología Vegetal, C.S.I.C., Madrid (6) Spain and Department of Agronomy, Purdue University, West Lafayette, Indiana 47907 U.S.A.

ABSTRACT

The interaction of water, primary alcohols and ammonia with the surfaces of sepiolite and palygorskite has been studied by IR spectroscopy. Water is adsorbed physically through hydrogen bonding to water molecules coordinated to the magnesium ions at the external surfaces and in the channels, but does not completely fill the channels. Ammonia is adsorbed by totally replacing the coordinated water. The high values reported for the specific surface by adsorption of ammonia are due to such replacement. In addition, the absence of IR absorption bands for ammonium suggests that these surfaces do not contain the acidic water found when exchangeable cations are present. Primary alcohols are adsorbed initially by hydrogen bonding through the coordinated water, but subsequently they replace up to 50% of this water by successive adsorptions and desorptions, irrespective of the length of the carbon chain, suggesting that the water replaced was mainly on the external surfaces. Only methanol and ethanol penetrated into the channels significantly.

It is concluded that, in the absence of steric hindrance, the extent of penetration of molecules into the channels of sepiolite and palygorskite is governed by the polarity and vapor pressure of the molecules.

---

INTRODUCTION

Fibrous clay minerals, sepiolite and palygorskite, have numerous industrial and pharmaceutical applications due mainly to their high surface area and intracrystalline channels. Studies of the surface of palygorskite (Barrer and Mackenzie, 1954) and sepiolite (Dandy, 1968, 1971; Fernandez Alvarez, 1970), using conventional adsorption isotherms of gases and vapors, yield information on amounts adsorbed and energies of adsorption, but do not provide direct evidence of the active sites or of the accessibility of the intracrystalline channels.  These studies indicated that only ammonia and water were adsorbed into the channels of sepiolite and palygorskite, and IR spectroscopy has confirmed this for palygorskite (VanScoyoc et al., 1978).

Evacuation at room temperature removes the physically adsorbed water, leaving two water molecules coordinated to the edge octahedral cations (Prost, 1975; Serna et al., 1975a). Therefore, the availability of these mineral surfaces can be tested by the adsorption of vapors that can interact with the coordinated water.

## METHODS AND MATERIALS

The Vallecas sepiolite and Georgia palygorskite used have been characterized by Ahlrichs et al., 1975 and Serna et al., 1977. Methanol, ethanol, n-propanol and n-butanol, all of a purity greater than 99.5%, and high purity ammonia were used as adsorbates. The quartz spiral used to weigh the water adsorbed on sepiolite at 35°C had a sensitivity of about 1.82 ± 0.01 mm/mg, which was checked between runs. Elongations read to 0.001 cm measured weight changes of 0.1 mg per gram of adsorbent.

Self-supporting films (2 mg/cm$^2$) were mounted in aluminum foil holders and placed in an evacuable infrared cell (Serna et al., 1977). All samples were evacuated at less than $10^{-5}$ mmHg prior to and after adsorption of the polar compounds. IR spectra were obtained using a Perkin-Elmer 225 or 421 spectrophotometer.

## RESULTS AND DISCUSSION

### Adsorption of water

Although absorption bands of structural hydroxyls and coordinated water overlap in palygorskite, deuteration under vacuum can produce an isolated spectrum for the two stretching bands of coordinated $D_2O$ at 2700 and 2585 cm$^{-1}$ (Fig. 1E right) or structural OH can be deuterated under vacuum leaving the two stretching bands of coordinated water at 3620 and 3505 cm$^{-1}$ (Fig. 1D left). The discussion of the interaction between adsorbed and coordinated water will be centered around the spectra in the OD region since it gives better resolution. As $D_2O$ is adsorbed by the mineral the bands at 2700 and 2585 cm$^{-1}$ shift to 2635 and 2515 cm$^{-1}$, respectively (Fig. 1D, C and B). Both of the coordinated water frequencies undergo similar shifts, indicating that both protons are equally affected as consequence of the water adsorbed.

When additional $D_2O$ is present, broad bands are formed at 2500 and 2430 cm$^{-1}$. The absorption band at 2500 cm$^{-1}$ appears to be due to the overlapping between the adsorbed water and the displaced low frequency band of coordinated water. The 2430 cm$^{-1}$ band may be a fundamental vibration or an overtone of the water bending vibration. In the region of $H_2O$ bending vibrations, (not shown here), a band at 1625 cm$^{-1}$ broadens and shifts to higher frequency (1650 cm$^{-1}$) as the stretching bands shift to lower frequency (Prost, 1975).

The stretching vibrations of the coordinated water in evacuated sepiolite lie at 3620 and 3560 cm$^{-1}$ and are not resolved as clearly as in palygorskite, 3620 and 3505 cm$^{-1}$ (Figs. 1 and 2). Since the lower frequency vibration of the coordinated water in these minerals arise from an OH directed toward the tetrahedral sheets

(Serna et al., 1977), the larger substitution of Al for Si in palygorskite (Weaver and Pollard, 1973) could explain the greater asymmetry of the coordinated water in this mineral.

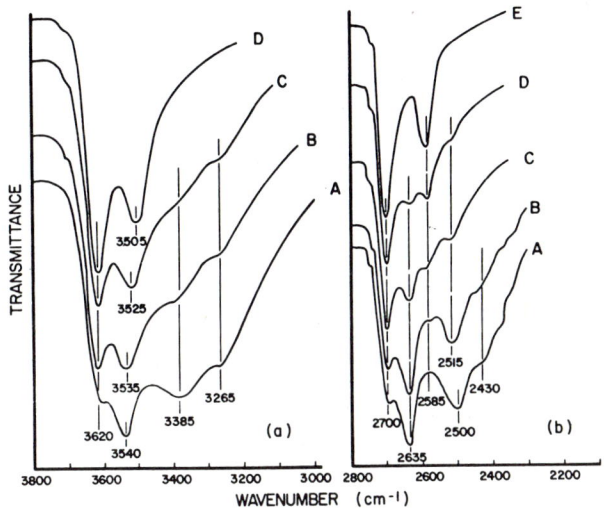

Fig. 1. Infrared spectra of palygorskite under vacuum and at different $H_2O$ (left) or $D_2O$ (right) contents.

The interaction between adsorbed and coordinated water in sepiolite is similar to that observed in palygorskite (Prost, 1975). Hydrogen bonding causes the vibrations of coordinated water molecules to shift to lower frequency, 3560 and 3370 $cm^{-1}$. As in palygorskite, the interaction by hydrogen bonding was also reflected by the shift to higher frequency and the decrease in the intensity of the bending vibration of water at 1620 $cm^{-1}$ (Prost, 1975; Serna et al., 1975b).

The important conclusion to be drawn here is that the IR spectra of the minerals under ambient conditions indicated that not all the coordinated water molecules are perturbed, as weak bands characteristic of coordinated water still persist around 3620 and 1620 $cm^{-1}$. Previous adsorption studies of water on palygorskite have also indicated that the surface is only partially covered (Barrer and Mackenzie, 1954). We have found sepiolite to have a specific surface of 275 $m^2/g$ by adsorption of water vapor at 35°C; this value is low compared with the theoretical ~900 $m^2/g$, suggesting that penetration of water into the channels is not complete.

Other absorption bands are also affected by the adsorption of water onto sepiolite and palygorskite. In sepiolite, the Si-O stretching vibration at 1195 $cm^{-1}$ shifts to 1210 $cm^{-1}$ and in palygorskite an absorption band around 1200 $cm^{-1}$ is observed only when adsorbed water is present. The adsorbed water also interacts with the lattice hydroxyls of both sepiolite and palygorskite (Prost, 1975; Serna et al., 1975b).

## Adsorption of ammonia

Repeated flushing and evacuation of sepiolite with ammonia gradually replaces the directly coordinated water, and absorption bands due to coordinated ammonia appear at 3370, 3290, 1615 and 1245 cm$^{-1}$ (Fig. 2B and C). The absence in the spectrum of $NH_4^+$ absorption around 1430 cm$^{-1}$ suggests that the mineral does not posses exchangeable cations (Mortland et al., 1963). Therefore, the isomorphic substitutions in the structure (Ahlrichs et al., 1975) must be compensated internally. Slight ammonium production was observed in ammoniated palygorskite (VanScoyoc et al., 1978), indicating greater surface acidity in palygorskite than in sepiolite, probably due to the greater amount of tetrahedral Al or to the presence of a slight montmorillonitic impurity.

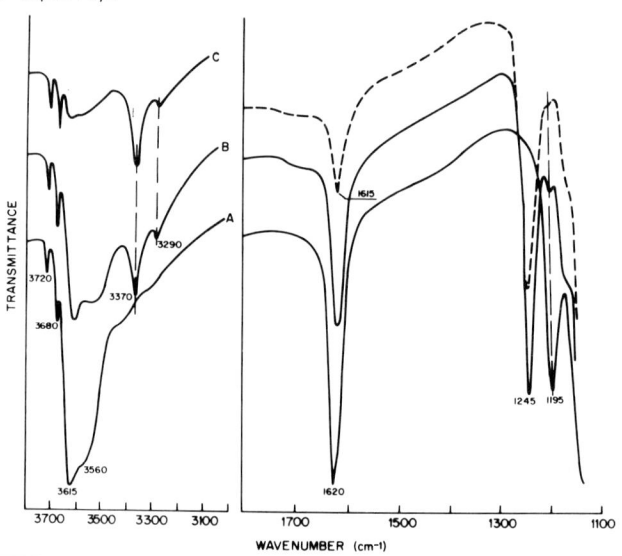

Fig. 2. Adsorption of ammonia on sepiolite evacuated at room temperature.
A.- Initial surface  B.- Partial replacement of coordinated water  C.- Total replacement of coordinated water.

Since the symmetric deformation frequency of coordinated ammonia in sepiolite at 1245 cm$^{-1}$, is almost identical to that found in palygorskite at 1248 cm$^{-1}$ (VanScoyoc et al., 1978), and since this vibration has been reported to be sensitive to the nature of the cation to which ammonia is coordinated (Russell, 1965), it would indicate that magnesium cations predominate in the external surfaces of palygorskite as previously suggested (Serna et al., 1977).

Partial replacement under vacuum of coordinated water by ammonia (and vice versa) does not alter the shape of the stretching bands of coordinated water but only their intensity. This seems to support the conclusion that, in vacuum sepiolite contains only coordinated water.

In contrast to palygorskite (VanScoyoc et al., 1978), evacuation of ammoniated sepiolite at room temperature produces no splitting of the structural hydroxyl bands, indicating that the structure of the mineral remains unfolded (Fig. 2). This observation parallels that found in the dehydration of sepiolite and palygorskite, where the weaker palygorskite structure is more easily folded under similar conditions (Serna et al., 1975a; VanScoyoc et al., 1978).

It is interesting to note that the characteristic absorption at 1195 $cm^{-1}$ is sensitive to the replacement of coordinated water by ammonia (Fig. 2). It shifts progressively to lower frequency as ammonia replaces the coordinated water until it is obscured by other Si-O vibration of the structure. This corroborates the nature of this absorption as a Si-O vibration and also illustrates that this vibration can be easily moved to lower frequency where it is obscured as has been shown in palygorskite under evacuated conditions (Prost, 1975; VanScoyoc, 1976).

Adsorption of primary alcohols

Since similar interactions between primary alcohols and the surfaces of both minerals were observed, only the results for sepiolite will be presented. Figs. 3 and 4 show spectra of sepiolite under vacuum and after being exposed to methanol and n-butanol vapors.

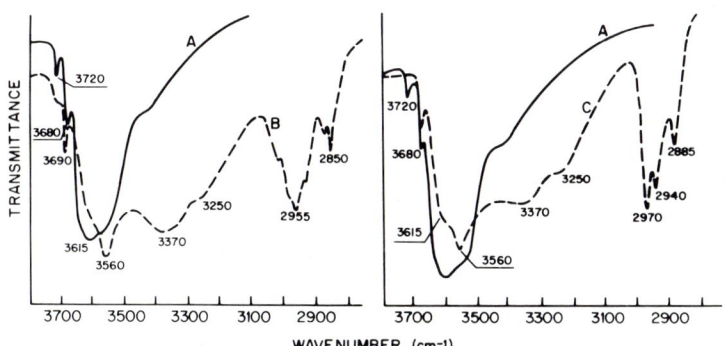

Fig. 3. Adsorption of alcohols on sepiolite evacuated at room temperature.
A.- Initial surface  B.- Exposed to an atmosphere of methanol (P/Po≅1.0)
C.- Exposed to an atmosphere of n-butanol (P/Po≅1.0)

Apart from perturbations of the structural hydroxyl groups of sepiolite at 3720 and 3680 $cm^{-1}$, the most striking effect is on the absorption bands of the coordinated water, resembling that observed in the presence of adsorbed water (Prost, 1975; Serna et al., 1975b). Therefore, hydrogen bonding is taking place between the coordinated water and the adsorbed alcohols. This is supported by the presence of absorption bands at 3370, 3250 and 1660 $cm^{-1}$ (Figs. 3 and 4). As was observed with the absorption of water, the Si-O vibration at 1195 $cm^{-1}$ is also perturbed to higher frequency (1210 $cm^{-1}$) by the adsorbed alcohols; similarly, in palygorskite a band around 1200 $cm^{-1}$ appears when methanol is adsorbed (not shown here).

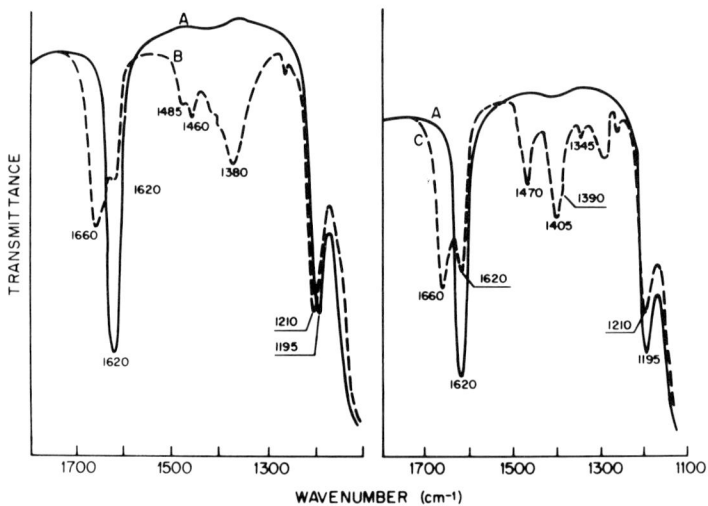

Fig. 4. Adsorption of alcohols on sepiolite evacuated at room temperature.
A.- Initial surface  B.- Exposed to an atmosphere of methanol (P/Po≃1.0)
C.- Exposed to an atmosphere of n-butanol (P/Po≃1.0)

Qualitatively, there are no important differences in the interaction of the primary alcohols on sepiolite and palygorskite surfaces. However, less water is perturbed in going from methanol to n-butanol in both fibrous minerals.

Repeated adsorptions and evacuations of alcohols were conducted in both minerals until the amount of coordinated water left in the sample remained constant (Fig. 5A and C), and the area under the bending mode of coordinated water before and after alcohol adsorption was used to determine the percentage of water replaced (Table 1): this decreases with increase in the chain length of the alcohols, but the differences between ethanol, n-propanol and n-butanol are small.

TABLE 1. Percentage of coordinated water replaced by alcohol adsorption

|  | Sepiolite | Palygorskite |
|---|---|---|
| Methanol | 52% | 60% |
| Ethanol | 48% | 35% |
| n-Propanol | 46% | 33% |
| n-Butanol | 40% | 32% |

It is interesting to note that the amount of water replaced by n-butanol is quite large if we consider that the length of this molecule (∼9Å) is very close to the width of the channels in sepiolite (∼10.6Å) and larger than those in palygorskite (∼6.4Å), suggesting that both minerals have a large external surface.

In smectites, alcohol-cation interactions through the oxygen atom raise the OH bending frequency of the alcohol from its unperturbed value in vapor (Dowdy and Mortland, 1967). Indications that the alcohol molecules are also coordinated to the exposed octahedral cations of sepiolite and palygorskite were clearly observed for adsorbed methanol: its OH bending vibration was shifted by 40 $cm^{-1}$ to higher frequencies from its vapor value, 1340 $cm^{-1}$ (Welty and Stephany, 1968).

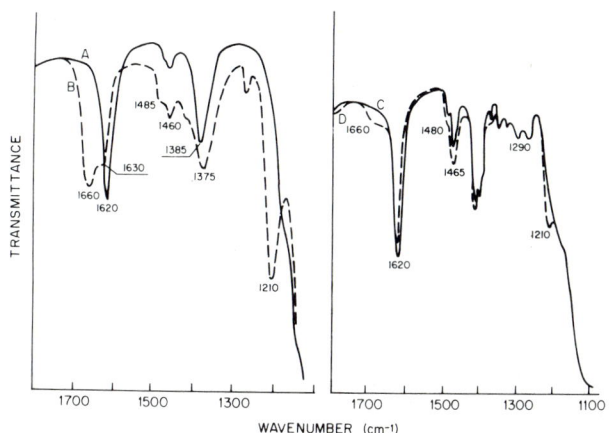

Fig. 5. Infrared spectra of sepiolite under vacuum after several adsorption-desorption of alcohols. A.- Methanol  C.- n-butanol  B.- As in A but reexposed to an atmosphere of methanol  D.- As in C but reexposed to an atmosphere of n-butanol.

Although methanol replaces only about 50% and 60% of the coordinated water in sepiolite and palygorskite, respectively, much of the remaining coordinated water seems to be accessible to this small molecule because it becomes perturbed when the minerals were reexposed to the alcohol vapor (Fig. 5B). A large perturbation of the remaining coordinated water was also observed for ethanol but no for n-propanol and n-butanol (Fig. 5D). Therefore, it can be concluded that intracrystalline penetration occurs to a larger extent for the two smaller alcohols and corroborates the results obtained by Barrer and Mackenzie (1954) from adsorption isotherms.

Surface accessibility - A model

Sepiolite and palygorskite from several sources have been previously described and shown to differ in chemical composition as well as in crystallinity (Weaver and Pollard, 1973; Martín Vivaldi and Robertson, 1971; Tchoubar et al., 1973). The degree of crystallinity, which determines the surface properties, has been related to the content of Si-OH groups on the surface of the mineral as a consequence of normal edge exposure and crystal defects (Ahlrichs et al., 1975). Therefore, the authors want to caution that the conclusions drawn here about the surface accessibility of the different molecules may vary for sepiolites and palygorskites from other sources. This is exemplified by the variations reported for the specific surface of sepiolite from different locations using nitrogen adsorption techniques. A value of 380 m$^2$/g was reported for sepiolite from Amboseli (Dandy, 1968); in contrast, 320 m$^2$/g was found for sepiolite from Vallecas (Fernandez Alvarez, 1970).

The morphological features of the sepiolite and palygorskite used in this study have been previously investigated by electron microscopy. Typical fiber dimensions of 8000 x 250 x 40 Å were reported for sepiolite (Robertson, 1957) and 10000 x 150 x 75 Å for palygorskite (Bates, 1958). Taking into account these average fiber sizes and the relative abundance of the [110] and [100] planes found in electron micrographs (Martín Vivaldi and Robertson, 1971; Rautureau and Tchoubar, 1972), fiber models for the two minerals are obtained (Fig. 6). Of course, this is a simplification since neither crystalline imperfections nor irregular channel size, which may be present, have been considered (Martín Vivaldi and Linares Gonzalez, 1962).

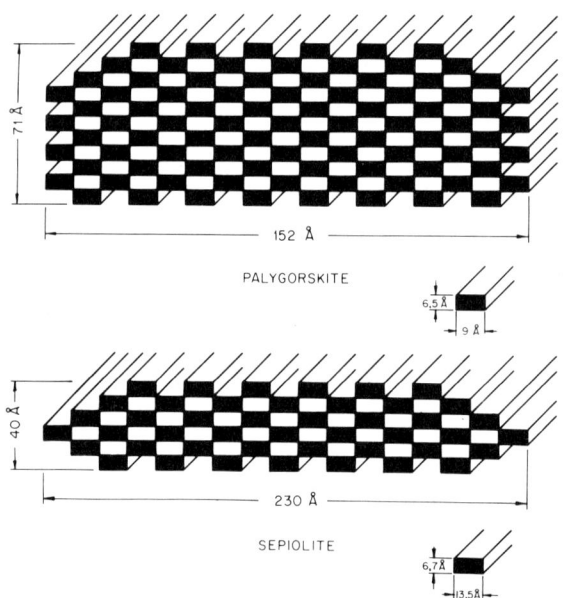

Fig. 6. Average fiber model of sepiolite and palygorskite

Some characteristics of these fibrous models are shown in Table 2. The theoretical values of the surface areas, of course, far exceed experimental determinations due to the presence of wedge-shaped capillaries and channels.

In the model, the predicted Si-OH to lattice OH ratio in palygorskite is greater than sepiolite. This is contrary to previous estimates (Ahlrichs et al., 1975; Serna et al., 1977). Two factors may be responsible for this apparent disagreement. First, the intensity for the absorption for the lattice hydroxyl is much greater in palygorskite than in sepiolite due to its dioctahedral character (Rousseaux et al., 1973). Second, more crystal imperfections may exist in sepiolite than in palygorskite. These imperfections have been shown to exist in sepiolite by electron microscopy (Tchoubar et al., 1973), but no parallel studies for palygorskite have been published.

TABLE 2. Theoretical values for average fibers according to model in figure 6.

|  | Sepiolite | Palygorskite |
|---|---|---|
| Channel dimensions (Å) | 3.7 x 10.6 | 3.7 x 6.4 |
| External surface ($m^2/g$) | 400 | 280 |
| Internal surface ($m^2/g$) | 500 | 635 |
| % of external coordinated water | 38 | 27 |
| Si-OH/Lattice OH | .20 | .36 |

The models presented give a reasonable explanation for the fact that 30% of the coordinated water is replaced by n-butanol in palygorskite and 40% in sepiolite, since this is its percentage of water on the external surface (Table 2). The slightly larger replacement of coordinated water in both minerals with decreasing alcohol size could be interpreted by steric effects due to the channel size and/or the presence of micropores (pores larger than the individual channel due to imperfections). The greater replacement of water by methanol in palygorskite than in sepiolite can not be satisfactorily explained since the channel size of sepiolite is larger. However, in connection with this observation, ammonia, which penetrated the channel of both minerals, replaced the coordinated water in palygorskite more easily than that in sepiolite.

In conclusion, despite some interaction with the structural hydroxyls, the most active adsorption site on the surface of these minerals is the coordinated water. Adsorption can occur by replacement of the coordinated water (ion-dipole interaction) and by hydrogen bonding to the water. Without steric hindrance, intracrystalline penetration appears to be strongly related to the magnitude of the dipole moment of the molecule and is apparently related to the vapor pressure.

REFERENCES

Ahlrichs, J.L., C.J. Serna and J.M. Serratosa, 1975. Structural hydroxyls in sepiolites. Clays Clay Miner. 23: 119-24
Barrer, R.M. and N. Mackenzie, 1954. Sorption by attapulgite. Part.I. Availability of intracrystalline channels. J. Phys. Chem. 58: 560-68
Bates, T.F., 1958. Selected electron micrographs of clays, 45,46. Circular n°51, Mineral Industries Experiment Station, State Univ. Pennsylvania.
Dandy, A.J., 1968. Sorption of vapors by sepiolite. J. Phys. Chem. 72: 334-39
Dandy, A.J., 1971. Zeolitic water content and adsorption capacity for ammonia of micropores sepiolite. J. Chem. Soc. A. 2383-87.
Dowdy, R.M. and M.M. Mortland, 1967. Alcohol-water interactions on montmorillonite surfaces I. Ethanol. Fifteenth Conf. Clays Clay Miner. 259-71
Fernandez Alvarez, T., 1970. Superficie específica y estructura de poro de la sepiolita calentada a diversas temperaturas. Proc. Reunión Hispano-Belga de minerales de la arcilla, Madrid. 202-09
Martín Vivaldi, J.L. and J. Linares Gonzalez, 1962. A random intergrowth of sepiolite and attapulgite. Clays Clay Miner. 9: 592-602

Martín Vivaldi, J.L. and R.H.S. Robertson, 1971. Palygorskite and sepiolite (The hormites) In The electron-optical Investigation of Clays. Gard J.A. (Editor) Mineralogical Society, London. 255-76

Mortland, M.M., J.J. Fripiat, J. Chaussidon and J.B. Uytterhoeven, 1963. Interaction between ammonia and the expanding lattice of montmorillonite and vermiculite. J. Phys. Chem. 67: 248-58

Prost, R., 1975. Etude de l'hydration des argiles. Interaction eau-mineral and mecanisme de la retention de l'eau. Ph.D. Thesis, Univ. of Paris VI, France.

Rautureau, M. and C. Tchoubar, 1972. Etude morphologique de la sepiolite par microscopie electronique. J. Microscopie. 14: 139-46

Robertson, R.H.S., 1957. Sepiolite: a versatil raw material. Chem. Ind. 1492-95

Rousseaux, J.M., C. Gomez Laverde, Y. Nathan and P.G. Rouxhet, 1973. Correlation between the hydroxyl stretching bands and the chemical composition of micas. Proc. Int. Clay. Conf. Madrid. 89-99.

Russell, J.D., 1965. Infrared study of the reaction of ammonia with montmorillonite and saponite. Trans. Faraday Soc. 61: 2284-94

Serna, C.J., J.L. Ahlrichs and J.M. Serratosa, 1975a. Folding in sepiolite crystals. Clays Clay Miner. 23: 452-57

Serna, C.J., M. Rautureau, R. Prost, C. Tchoubar and J.M. Serratosa, 1975b. Etude de la sepiolite a l'aide des donnes de la microscopie electronique, de l'analyse thermoponderale et de la spectroscopie infrarrouge. Bull. Grpe. Fr. Argiles. 26: 153-63

Serna, C.J., G.E. VanScoyoc and J.L. Ahlrichs, 1977. Hydroxyl groups and water in palygorskite. Am. Miner. 62: 784-92

Tchoubar, C., M. Rautureau, Ch. Clinard and J.P. Ragot, 1973. Technique d'inclusion appliquee a l'etude des silicates lamellaires et fibreux. J. Microscopie 18: 147-54

VanScoyoc, G.E., 1976 Surface and structural properties of palygorskite. Ph.D. Thesis, Purdue University, U.S.A.

VanScoyoc, G.E., C.J. Serna and J.L. Ahlrichs, 1978. Structural changes in palygorskite upon dehydration (Submitted to Am. Miner.)

Weaver, C.E. and L. Pollard, 1973. Attapulgite and Palygorskite. In The Chemistry of Clay Minerals. Elsevier Sci. Publication Co. 119-26

Welty, D. and R. Stephany, 1968. Some comments on the infrared spectra of vapors. Appl. Spectrosc. 22: 678-88

ION EXCHANGE OF THE POLY-AMINE COMPLEXES OF SOME TRANSITION METAL
IONS IN MONTMORILLONITE

P. PEIGNEUR, A. MAES and A. CREMERS
Centrum voor Oppervlaktescheikunde en Colloïdale Scheikunde,
Katholieke Universiteit Leuven, De Croylaan 42, B-3030 Heverlee,
(Belgium)

ABSTRACT

Using ion exchange selectivity measurements, a study is made on
the stabilities of the complexes of Cu, Ni, Zn, Cd and Hg with
ethylenediamine and of the complexes of Cu with diethylenetriamine,
triethylenetetramine and tetraethylenepentamine in montmorillonite.
It is shown that the overall stability constants of these complexes
exceed the corresponding values in solution by some three orders
of magnitude. A new method is presented for the selective removal
of heavy metals from very dilute solutions, which is based on the
very high stabilities of the metal-tetraethylenepentamine complexes.

INTRODUCTION

It is well known that transition metal-amine complexes can
easily be exchanged in the interlayer space of montmorillonite clay
(Bodenheimer et al., 1963a, 1963b; Mantin, 1969; Laura and Cloos,
1970). In a series of recent papers from this laboratory (Pleysier
and Cremers, 1975; Maes et al., 1977; Maes et al., 1978; Maes and
Cremers, 1978), it was shown that the formation of metal uncharged
ligand complexes in ion exchangers can be characterized thermodynamically, as is customary in solution chemistry. These stability
constants can be derived from (a combination of) three different
methods : ion exchange selectivity measurements of the coordinatively saturated complex, the Bjerrum complex formation function, and
the measurement of selectivity change with varying ligand concentration. Using the first of these methods, this paper presents a
systematic study on the polyamine complexes of some transition

metals and proposes a new and selective method for removing such metals from very dilute solutions.

MATERIALS AND METHODS

The clay mostly used in this study is the well known montmorillonite from Camp Berteau (CB), some experiments being also carried out on Wyoming bentonite (WB). The ions involved in this study are Cu, Ni, Zn, Cd, Hg and their complexes with (some of) the polyamines ethylenediamine (*en*), diethylenetriamine (*dien*), triethylenetetramine (*trien*) and tetraethylenepentamine (*tetren*).

The ion exchange selectivity of these complexes is studied at 25°C on CB, relying on methods, the essentials of which are described in earlier papers (Pleysier and Cremers, 1975; Maes et al., 1978). The selectivity of the Ni(*en*)$_3$ complex is measured with reference to the calcium ion by mixing 10 mls of Ca-CB clay ($\sim$ 10 g/L in 0.01 N Ca(NO$_3$)$_2$) with 20 mls of calcium-nickel nitrate solutions of varying Ni/Ca ratios at 0.01 total normality and in the presence of an appropriate excess of *en*. The selectivity of the other *en* complexes is studied with reference to the Ni(*en*)$_3$ complex by mixing 10 mls of Na-CB clay ($\sim$ 10 g/L and 0.01 N in NaNO$_3$) with 20 mls of a mixture of the two complexes studied, in various ratios and 0.01 in total normality. In all experiments, concentrations of both metals are measured on the equilibrium solution in two duplicate batches, using either radioisotope labelling ($^{45}$Ca, $^{63}$Ni, $^{65}$Zn, $^{115m}$Cd, $^{203}$Hg) or atomic absorption (Cu). Exactly the same procedure is followed in studying the copper complexes with *dien*, *trien* and *tetren* (Merck; analytical grade). In cases where the ligand distribution is measured, a third batch was run, using $^{14}$C *en*.

In measuring the exchange selectivity of such complexes, precautions must be taken to ensure high enough free ligand concentrations as to form the coordinatively saturated complex in both phases. The minimal free ligand concentrations are readily calculated (Beck, 1970) from the stepwise stability constants (Sillèn and Martell, 1964) which are shown in Table 1.

In all cases, the molar ligand-metal ratios are adjusted to 2 (or 3 or 1 depending on the complex), adding an excess of ligand, depending on complex stability, and taking account of dilution effects, pH and protonation constants of the ligand (Sillèn and Martell, 1964).

TABLE 1

Stability constants of metal polyamine complexes and minimal free ligand concentration, $L_{min}$, for complex formation ($\bar{n}$ = 0.99, 1.95 or 2.95)

| Ligand | Metal | Log $K_1$ | Log $K_2$ | Log $K_3$ | -Log $L_{min}$ (M) |
|---|---|---|---|---|---|
| $en$ | Cu | 10.71 | 9.31 | | 8 |
| | Ni | 7.51 | 6.35 | 4.42 | 3 |
| | Zn | 5.92 | 5.15 | 1.86 | 0.5 |
| | Cd | 5.63 | 4.59 | 2.07 | 0.7 |
| | Hg | | 23.2($\beta_2$) | | |
| $dien$ | Cu | 16.0 | 5.3 | | 4 |
| $trien$ | Cu | 20.1 | | | 18 |
| $tetren$ | Cu | 22.8 | | | 21 |
| | Ni | 17.5 | | | 15.5 |
| | Zn | 15.4 | | | 13.4 |
| | Cd | 14.0 | | | 12 |
| | Hg | 27.7 | | | 25.7 |

RESULTS AND DISCUSSION

Ethylenediamine-complexes of Ni, Zn, Cd, Hg

The ion exchange isotherms for the $en$-complexes of Ni, Zn, Cd and Hg are shown in Fig. 1 as the equivalent fraction of complex ions in the clay phase Z, versus its equivalent fraction in the equilibrium solution S.

Before proceeding to a thermodynamic analysis of these data, it is imperative to establish the identity of the metal complexes in the clay. In the case of Zn and Cd, the high ligand concentrations, necessary to ensure the formation of the coordinatively saturated complexes in solution, preclude ligand distribution measurements. Ligand numbers were therefore measured in separate Ca-Zn and Ca-Cd equilibria at low free ligand concentration (0.008-0.009 M) and at high Zn or Cd loading (> 1 meq/g) : the average ligand numbers $\bar{n}$ were 2.99 ± 0.08 (Zn) and 3.04 ± 0.10 (Cd). These results were confirmed by ligand distribution measurements in Ni-Zn and Ni-Cd equilibria in a ligand concentration range 0.0025-0.001 M ($\bar{n}$ = 3.01 ± 0.11). In the case of the Ni-Hg equilibria at 0.02 M in free $en$, $\bar{n}$ values were near 3 with a tendency towards lower values at high Hg-loading of the clay. These data show that complex formation on the clay is completed earlier, i.e. at lower ligand concentration, than in solution. Therefore, at the ligand concentrations reported in Fig. 1, we are confident in dealing with the coordinatively saturated complexes in both phases. These conclusions are corroborated by $d_{001}$ spacings of 1.42-1.45 nm (Ni, Zn, Cd) and

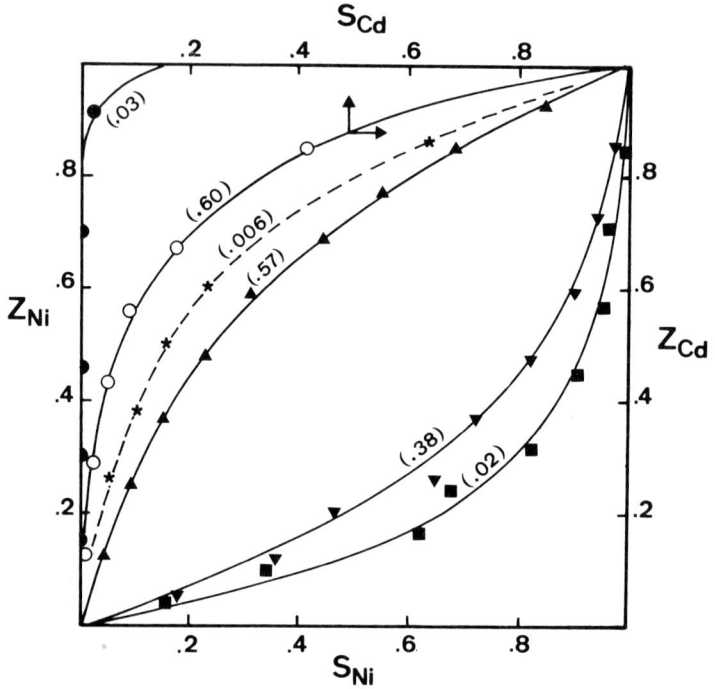

Fig. 1. Ion exchange isotherms for the *en* complexes in CB montmorillonite at 25°C : Ca-Ni (●); Ni-Zn (✱,▲); Ni-Cd (▼); Ni-Hg (■); Zn-Cd (○). Numbers in parenthesis refer to free *en* concentration.

1.36 nm (Hg) obtained on air-dry clay films, saturated with the various *en* complexes. The exchange capacity of the clay, obtained as the sum of exchange levels of the two complexes, varies in the range 1.3-1.4 meq/gram i.e., a reasonable value in view of the high pH values (10-11.9 depending on the ligand concentrations).

Selectivities are quantitatively compared in terms of selectivity coefficients, defined as

$$^A K_C = \frac{Z_A \, m_B}{Z_B \, m_A} \tag{1}$$

in which $Z_A \, Z_B$ and $m_B \, m_A$ are the equivalent fractions in the clay and the molalities in the equilibrium solutions of the two competing ions. Fig. 2 shows the variation of $\ln K_c$ for all equilibria as a function of ion exchanger composition. The values of $\ln K$, the equilibrium constant, obtained by graphical integration (Gaines and Thomas, 1953) and the corresponding standard free energy changes are summarized in Table 2A, which includes published values for the Ca-Cu and Ca-Cu(en)$_2$ equilibria on the same clay (Maes et al., 1978).

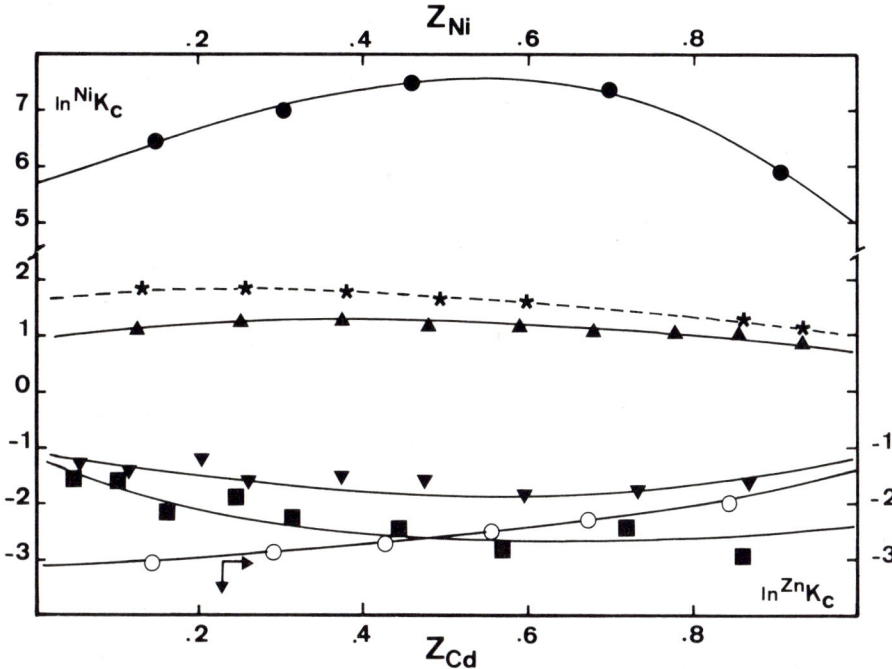

Fig. 2. ln $K_c$ versus ion exchanger composition for the en complexes in CB montmorillonite at 25°C : Ca-Ni (●), Ni-Zn (*,▲), Ni-Cd (▼), Ni-Hg (■), Zn-Cd (○).

In calculating ln K and $\Delta G_o$, we are taking for granted that these reactions are thermodynamically reversible. This assumption is validated by the finding that the $\Delta G_o$ value for the Zn(en)$_3$-Cd(en)$_3$ equilibrium corresponds nearly exactly with the prediction based on the Ni(en)$_3$-Zn(en)$_3$ and Ni(en)$_3$-Cd(en)$_3$ equilibria. Further proof is available from a study on the effect of en concentration on the Cu/Ni equilibrium. It is found that $^{Cu}K_c$ decreases asymptotically to a value 1-1.1 near 0.01 M in en, i.e. the free ligand concentration corresponding to the formation of the coordinatively saturated Ni(en)$_3$ complex : this is the result one would predict on the basis of the Ca-Cu(en)$_2$ and Ca-Ni(en)$_3$ equilibria, shown in Table 2A.

The data in Table 2A may be converted into $\Delta G_o$ values for the reversible displacement of the hydrated metal ion by its en complex,

TABLE 2

A. Equilibrium constants (ln K) and standard free energy changes $\Delta G_o$ (KJ/mole) for *en* complexes in montmorillonite clay
B. Excess standard free energy loss $\Delta G_o^{ex}$ (KJ/mole) and stabilisation factors (log $\bar{\beta}_n/\beta_n$) for the formation of *en* complexes in montmorillonite clay.

| A | | | B | | |
|---|---|---|---|---|---|
| Reaction | ln K | $-\Delta G_o$ | Complex | $-\Delta G_o^{ex}$ | log $\bar{\beta}_n/\beta_n$ |
| ZCa + Ni(en)$_3$ | 6.65 | 16.50 | Ni(en)$_3$ | 15.9 | 2.78 |
| ZCa + Cu(en)$_2$ | 7.00[a] | 17.35 | Cu(en)$_2$ | 16.7 | 2.92 |
| ZNi(en)$_3$ + Zn(en)$_3$ | -1.05 | -2.60 | Zn(en)$_3$ | 13.3 | 2.32 |
| ZNi(en)$_3$ + Cd(en)$_3$ | 1.59 | 4.00 | Cd(en)$_3$ | 19.9 | 3.47 |
| ZNi(en)$_3$ + Hg(en)$_2$ | 2.35 | 5.85 | Hg(en)$_{2-3}$ | 21.7 | 3.80 |
| ZZn(en)$_3$ + Cd(en)$_3$ | 2.55 | 6.35 | | | |
| ZCa + Cu | 0.25[a] | 0.65 | | | |

[a]Maes et al., 1978.

i.e. the standard excess free energy loss for complex formation in the ion exchanger (Maes et al., 1977). The results of these calculations, neglecting the small differences between the hydrated metal ions (Maes et al., 1976) are given in Table 2B, along with the corresponding stabilisation factors $\bar{\beta}_n/\beta_n$ (i.e. the ratio of overall stability constants). It is seen that the overall stability constants are increased by 2.5 to nearly 4 orders of magnitude. No effort was made to evaluate the contribution of the stepwise addition of the ligands. It is in any case certain that the coordination of the third ligand contributes appreciably to the stabilisation : this appears from the higher ligand numbers in the clay (Zn, Cd) and the effect of *en* concentration on selectivity (Ni, Zn) below the critical *en* level, necessary for the formation of the saturated complex.

No unambiguous interpretation is available for the high selectivity of these complexes for clay surfaces. It appears that the enhancement of the effect within the Zn, Cd, Hg series seems to validate our recently proposed hypothesis[+] of an enhanced charge delocalisation on the complexes in the interlayer space.

## Cu-complexes of *dien*, *trien* and *tetren*

The effect of complex formation with *dien*, *trien* and *tetren* on the ion exchange selectivity of copper is entirely similar to the

[+]Maes, A. and Cremers, A., 1978. Submitted for publication.

one found in the case of *en* (Maes et al., 1978). This is illustrated in Fig. 3 showing the variation of the Cu-Ca selectivity coefficients as a function of ion exchanger loading : the overall stabilisation is 16.5-17 KJ/mole.

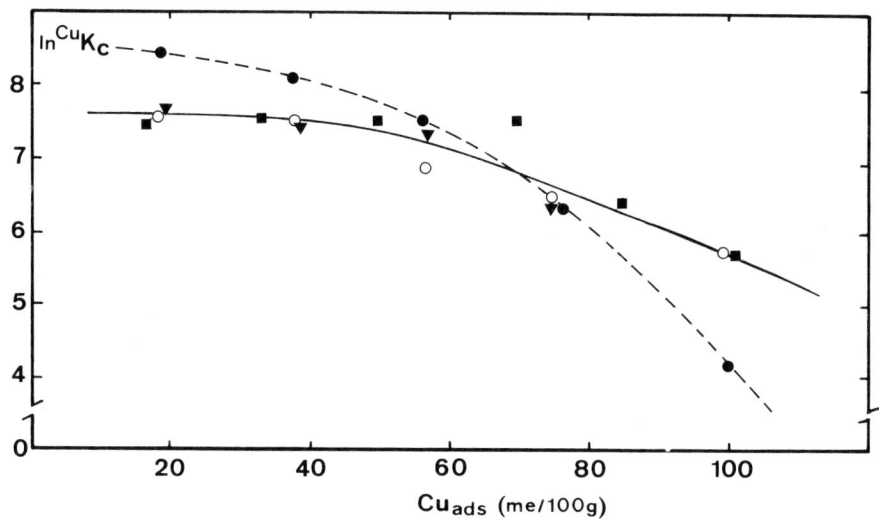

Fig. 3. ln $K_C$ versus ion exchanger composition for the Ca-Cu polyamine complexes in CB montmorillonite at 25°C : *en* (■), *dien* (●), *trien* (○), *tetren* (▼).

The $d_{001}$ spacings (rational) for the *dien* and *trien* Cu-clay complexes, as measured on air-dry films amount to 1.3-1.31 nm, a value which is slightly higher than the value found for $Cu(en)_2$ : 1.26 nm (Maes et al., 1978). These values indicate that, in these cases, the four coordinating N-atoms are oriented parallel to the clay surface. The slightly higher value for the Cu-*tetren* complex (1.39-1.41 nm) points to a pentadentate structure.

*Tetren*-clay : a selective adsorbent for heavy metals.

The pronounced synergistic effect of complex formation with polyamines on the ion exchange adsorption of transition metal ions in clays may form the basis for the development of extremely selective methods for the removal of minute traces of metals from waste waters. In fact, such efforts seem rather timely in view of the rather severe limits which are now being implemented in this area.

The very high stability of the *tetren* complexes, necessitating only a marginal excess of ligand for obtaining the desired effect, would seem to make *tetren* a most appropriate complexing agent. The fact that the *tetren* complexes are more stable than most of the complexing agents, present in waste waters of some industries (e.g. metal finishing) and which interfere seriously with current waste water treatments, makes such method even more attractive. The possibilities of using *tetren* in combination with montmorillonite (CB) for removing traces of metals are illustrated in Table 3. It is sufficiently evident that the method is quite effective, even in the presence of EDTA, one of the most powerful complexing agents for metal ions.

TABLE 3

Ion exchange adsorption of transition metal ions in CB montmorillonite at pH = 7.5 in the presence of *tetren* $10^{-4}$ M, calcium 5 . $10^{-3}$ M, and citrate or EDTA; the "free" *tetren* concentration is $\sim 10^{-9}$ M.

| Metal | Initial conc. (ppm) | Complexing agent conc. (M) | Amount of clay used (g/meq metal) | Equilibrium conc. (ppm) |
|---|---|---|---|---|
| Cu | 3.2 | $10^{-3}$ EDTA | 2 | 0.21 |
| Hg | 10 | – | 2.5 | 0.06 |
|    | 10 | $10^{-3}$ citrate | 2.5 | 0.05 |
|    | 10 | $10^{-3}$ EDTA | 2.5 | 0.05 |
| Ni | 3 | – | 2.5 | 0.10 |
|    | 3 | $10^{-3}$ citrate | 2.5 | 0.06 |
| Zn | 3.3 | – | 2.5 | 0.07 |
|    | 3.3 | $10^{-3}$ citrate | 2.5 | 0.06 |

A rather more practical alternative relies on using "*tetren*-clay" which is prepared by exchange with *tetren* at a pH of 7∿8 : under such conditions, *tetren* occurs predominantly as a trivalent cation, exhibiting a very high affinity for clays. Dispersion of a *tetren*-clay, such as *tetren*-bentonite, in a solution containing metal ions leads to a proton-metal exchange, as evidenced by the pH shift. There is some doubt about the exact nature of the resulting interlayer metal complexes and it is not unlikely that, as indicated by some recent data, trivalent complexes such as *tetren* $HM^{+3}$ may be involved. Whatever the exact stoichiometry, the process may be represented by an expression of the form

$$\text{clay } tetren \text{ H}_n^{n+} + M^{++} \longrightarrow \text{clay } tetren \text{ MH}_{n'}^{(n'+2)+} + (n-n')H^+ \quad (2)$$

It is obvious that, for optimum yield, the reaction should preferably be carried out at constant pH. Some typical results obtained with *tetren*-bentonite at pH $\sim$ 7.5 are summarized in Table 4. The efficiency in using the *tetren*-bentonite formulation is seen to be equally good. Moreover, it is apparent that the presence of trivial ions such as calcium, has a minor effect on effectiveness, even in a hundredfold excess.

TABLE 4

Ion exchange adsorption of transition metals ions in *tetren*-bentonite at pH $\sim$ 7.5; (mixing time : 1 hr).

| Metal | Initial conc. (ppm) | Ca conc. (M) | Complexing agent conc. (M) | Amount of clay used (g/me) | Equilibrium conc. (ppm) |
|---|---|---|---|---|---|
| mixture Cu | 3.2 | – | – | 2.5 | 0.02 |
| Ni | 2.9 | – | – | | 0.01 |
| Cd | 5.6 | – | – | | 0.25 |
| Zn | 3.3 | – | – | | 0.06 |
| Hg | 10 | $5 \cdot 10^{-3}$ | – | 4 | 0.05 |
|  | 10 | $5 \cdot 10^{-3}$ | $10^{-3}$ citrate | 4 | 0.08 |
|  | 10 | $5 \cdot 10^{-3}$ | $10^{-3}$ EDTA | 4 | 0.06 |
|  | 10 | – | – | 4 | 0.09 |
|  | 1 | – | – | 3 | 0.05 |
| Zn | 3.3 | – | – | 4 | 0.05 |
|  | 3.3 | $5 \cdot 10^{-3}$ | – | 4 | 0.10 |
|  | 3.3 | – | $10^{-3}$ citrate | 4 | 0.04 |
|  | 3.3 | $5 \cdot 10^{-3}$ | $10^{-3}$ citrate | 4 | 0.10 |

Perhaps it is worth emphasizing other, less obvious areas of relevance such as the immobilization of excessive quantities of heavy metals in soils. For example, it has been shown[+] that complex formation with *tetren* can restore normal plant growth in soils containing up to 450 ppm in copper.

ACKNOWLEDGEMENTS

The financial support of the Belgian Government (Programmatie van het Wetenschapsbeleid) and the "Fonds voor Kollektief Fundamenteel Onderzoek (F.K.F.O.) is gratefully acknowledged. P. Peigneur is indebted to the "Instituut tot Aanmoediging van het Wetenschappelijk Onderzoek in Nijverheid en Landbouw" (I.W.O.N.L.) for a doctoral fellowship.

---

[+] F. Smeulders, J. Sinnaeve and A. Cremers : to be presented at the 8th International Colloquium on Plant Analysis and Fertiliser problems. Auckland, N.Z., 1978.

REFERENCES

Bodenheimer, W., Heller, L., Kirson, B. and Yariv, S., 1963a. Organo-metallic clay complexes. III. Copper-polyamine clay complexes. Proc. Int. Clay Conf. 351-363.

Bodenheimer, W., Heller, L., Kirson, B., and Yariv, S., 1963b. Organo-metallic clay complexes. IV. Nickel and mercury aliphatic polyamines. Israel J. Chem. 1 : 391-403.

Beck, M.T., 1970. Chemistry of complex equilibria. Van Nostrand Reinhold, London, 285 pp.

Gaines, L.G. and Thomas, H.C., 1953. Adsorption studies on clay minerals. II. A formulation of exchange adsorption. J. Chem. Phys. 21 : 714-718.

Laura, R.D. and Cloos, P., 1970. Adsorption of ethylenediamine on montmorillonite saturated with different cations. Proc. Reunion Hispano-Belga de Minerales de la Arcilla, 76-86.

Maes, A., Peigneur, P. and Cremers, A., 1976. Thermodynamics of transition metal ion exchange in montmorillonite. Proc. Int. Clay Conf., Applied Publ. Ltd., 319-329.

Maes, A., Marijnen, P. and Cremers, A., 1977. Stability of metal-uncharged ligand complexes in ion exchangers. I. Quantitative characterization and thermodynamic basis. J.C.S. Faraday I, 73 : 1297-1301.

Maes, A., Peigneur, P. and Cremers, A., 1978. Stability of metal-uncharged ligand complexes in ion exchangers. II. The copper-ethylenediamine complex in montmorillonite and sulphonic acid resin. J.C.S. Faraday I, 74 : 182-189.

Maes, A. and Cremers, A., 1978. The stability of metal-uncharged ligand complexes in ion exchangers. III. Complex ion selectivity and stepwise stability constants. J.C.S. Faraday I (in press).

Mantin, I., 1969. Mesure de la capacité d'échange des minéraux argileux par l'éthylène diamine et les ions complexes de l'éthylène diamine. C.R. Acad. Sci. Paris 269 : 815-818.

Pleysier, J. and Cremers, A., 1975. Stability of silver-thiourea complexes in montmorillonite clay. J.C.S. Faraday I, 71 : 256-264.

Sillèn, G.L. and Martell, A.E., 1964. Stability constants of metal-ion complexes. The Chemical Society, London, 754 pp.

X.P.S. STUDY OF THE INTERACTION OF SOME PORPHYRINS
AND METALLOPORPHYRINS WITH MONTMORILLONITE

P.CANESSON, M.I.CRUZ and H.VAN DAMME
Centre de Recherche sur les Solides à Organisation Cristalline Imparfaite
Rue de la Férollerie, 45045 ORLEANS, FRANCE.

ABSTRACT

The reactivity of two simple meso-substituted porphyrins, i.e. tetraphenyl (TPP) and tetrapyridyl (TPyP) porphyrins in the interlamellar space of montmorillonite has been investigated by X.P.S. When TPP is adsorbed on the clay surface, the molecule undergoes essentially protonation whatever the exchange cation is. TPyP is adsorbed in larger amount than TPP. Considerations about the relative intensities of the various species lead to the conclusion that adsorbed TPyP is completely protonated on the pyridyl groups with $Co^{2+}$, $Mn^{2+}$, $Sn^{4+}$, $Fe^{2+}$, $Cu^{2+}$ and $Ni^{2+}$ as exchangeable cations but only partially protonated with $Na^+$. The quantity of metalloporphyrin in the interlamellar space is in the order $Co^{2+}$, $Cu^{2+}$ >> $Mn^{2+}$, $Ni^{2+}$ > $Fe^{2+}$ > $Sn^{4+}$.

INTRODUCTION

The interlayer space of montmorillonite can be swollen by a wide variety of compounds : water, alcohols, ketones, aldehydes, ethers, nitriles, fatty acids, amines, etc...(Theng, 1974). The adsorption mechanism is dependent upon the nature of the adsorbed species. Positively charged species enter into the clay by a simple exchange mechanism, polar molecules can be coordinated to the interlayer cation, and they can eventually form true organometallic complexes (Doner and Mortland, 1969).

Since self-supporting films of very large area can be obtained from these clays, they offer a possibility to make macroscopic pseudo-membranes, in which organic molecules of biological interest can be introduced.

This work deals with the characterization of the adsorbed species obtained when two mesosubstituted porphyrins of different basicity, namely meso-tetra-phenylporphyrin (TPP) and meso-tetra-pyridylporphyrin (TPyP), are in interaction with montmorillonites exchanged with various cations. X.P.S. has been chosen for this study since this technique is highly selective for surface species and it has been shown (Defosse and Canesson, 1975) that the position of the N1s level of adsorbed bases is sensitive to the surface environment. These samples have been previously studied by chemical methods as well as by U.V., visible and I.R. spectroscopy (Van Damme et al., 1978).

EXPERIMENTAL METHODS

Materials

A sample of Wyoming bentonite was purified by dispersion in water in order to remove the quartz impurities. Only the fraction < 2 µ was used. A detailed procedure for the preparation of exchanged samples and adsorption of porphyrins is given elsewhere (Van Damme et al., 1978). The cationic contents before and after the adsorption of porphyrins are given in tables I and II.

TABLE 1
Cationic contents before and after the adsorption of TPP

| cation | cationic contents $10^{-6}$ eq.g$^{-1}$ | | TPP content $10^{-6}$ mole g$^{-1}$ |
|---|---|---|---|
| | Initial | Final | |
| $Na^{+}$ | 970 | 948 | 9.5 |
| $Fe^{2+}$ | 878 | 860 | 32.4 |
| $Ni^{2+}$ | 887 | 791 | 29.6 |
| $Mn^{2+}$ | 839 | 843 | 16.2 |
| $Cu^{2+}$ | 898 | 781 | 34.4 |
| $Co^{2+}$ | 817 | 807 | 42.9 |

TABLE 2
Cationic contents before and after adsorption of TPyP

| Cation | Cationic contents $10^{-6}$ eq.g$^{-1}$ | | TPyP content $10^{-6}$ mole g$^{-1}$ |
|---|---|---|---|
| | Initial | Final | |
| $Fe^{2+}$ | 878 | 760 | 180 |
| $Ni^{2+}$ | 887 | 123 | 187 |
| $Mn^{2+}$ | 839 | 118 | 189 |
| $Cu^{2+}$ | 898 | 397 | 193 |
| $Co^{2+}$ | 817 | 279 | 182 |

X.P.S.Measurements

X.P.S. spectra have been taken using an AEI ES 200B spectrometer, equipped with a magnesium anode (12 kv, 20 mA, hν = 1253.6 eV). In order to avoid an important charging effect a self-supporting film about 10 mm long and 4 mm wide was stuck onto a conducting adhesive tape. A SEIN signal averager permitted accumulation of repetitive scans. The sweep width about the 248 channels was 12.5 eV in most of the cases. The peaks of interest namely those of the N1s and, also when possible that of the cation levels present in the sample, were sandwiched between two scans of a reference peak, namely that of Si2p . This procedure allowed monitoring of the growth of a carbon contamination overlayer. The sampling sequence was C1s, O1s, Si2p, N1s, Si2p, cation when possible, Si2p, Al2s then O1s and C1s again. Cumulated time per channel was typically 1.5s for C1s, 0.5s for O1s, 2s for Si2p, 10s for Al2s and 30s for N1s.

Peaks were smoothed and decomposed when asymmetrical, using a computerized least square fitting program.

The Si2p line has been used as reference to calibrate both binding energy scale (B.E.) and intensity. The position of the Si2p line has been arbitrarily fixed at 101.5 eV and the accuracy of the binding energy determinations is estimated to be ± 0.2 eV. All the intensity values are given as intensity ratios $R(x)$

$$R(x) = \frac{I(x)}{I(Si2p)}$$

where $I(x)$ and $I(Si2p)$ are respectively the intensity of the line to be measured and that of the Si2p level.

RESULTS

When tetraphenylporphyrin (TPP) is adsorbed in the clay interlamellar space the X.P.S. spectrum of the N1s level can be decomposed into two or three gaussian peaks, depending upon the nature of the charge balancing cation in the montmorillonite. Typical spectra are shown in figure 1 for the $Cu^{2+}$ and $Fe^{2+}$ clay treated by TPP and also for pure TPP, without clay.

The relative abundance and the position of the various species and the $R(N1s)$ ratios are summarized in table 3.

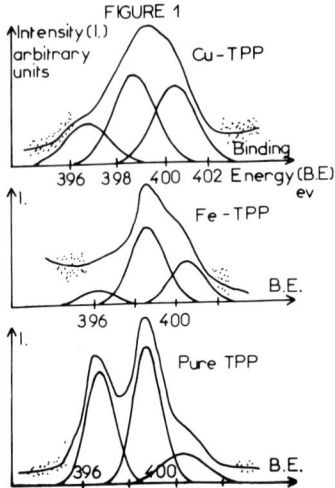

Figure 1 : Various X.P.S. profiles for TPP adsorbed on montmorillonite

TABLE 3

Position and intensity of the N1s levels of TPP adsorbed by the specified monoionic montmorillonite.

| Montmorillonite | N1s B.E. (eV) | R(X) | R(N1s) |
|---|---|---|---|
| $Cu^{2+}$ | 396.8<br>398.8<br>400.5 | 20<br>42.5<br>37.5 | 0.0410 |
| $Co^{2+}$ | 396.9<br>399.0<br>401.0 | 14<br>54.5<br>31.5 | 0.0574 |
| $Sn^{4+}$ | 396.5<br>398.1<br>399.7 | 14<br>57<br>29 | |
| $Fe^{2+}$ | 396.2<br>398.8<br>400.8 | 11<br>58<br>31 | 0.0476 |
| $Mn^{2+}$ | 395.7<br>398.6<br>400.6 | 11.5<br>55.5<br>33 | 0.0365 |
| $Ni^{2+}$ | 398.7<br>400.5 | 69.5<br>30.5 | 0.0315 |
| $Na^{+}$ | 398.1<br>400.0 | 63.5<br>36.5 | 0.0268 |
| Pure TPP | 396.7<br>398.8<br>400.3 | 38<br>46<br>16 | |

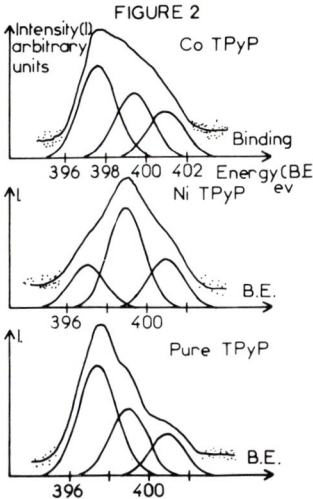

Figure 2 : Various X.P.S. profiles for TPyP adsorbed on montmorillonite

TABLE 4

Position and intensity of the N1s level of TPyP adsorbed by the specified monoionic montmorillonite.

| Montmorillonite | N1s B.E. (eV) | R(X) | R(N1s) |
|---|---|---|---|
| $Cu^{2+}$ | 397.4<br>399.3<br>400.9 | 45<br>32<br>23 | 0.159 |
| $Co^{2+}$ | 397.6<br>399.4<br>401.0 | 46<br>34<br>20 | 0.154 |
| $Mn^{2+}$ | 397.2<br>399.1<br>401.1 | 22<br>52.5<br>25.5 | 0.130 |
| $Ni^{2+}$ | 397.1<br>399.0<br>401.0 | 22.5<br>51.5<br>26 | 0.118 |
| $Fe^{2+}$ | 397.2<br>399.2<br>401.1 | 20<br>54<br>26 | 0.136 |
| $Sn^{4+}$ | * 396.5<br>* 398.2<br>* 400.6 | 16<br>65.5<br>18.5 | |
| $Na^{+}$ | 397.2<br>399.0<br>400.8 | 31.5<br>47.5<br>21 | 0.213 |
| Pure TPyP | 397.5<br>399.0<br>401.0 | 61<br>18<br>21 | |

* Shift of about 0.6 eV to lower B.E. values: instabilities of the charging effect.

From table 3 and figure 1, it is evident that the relative abundance of the various species is different from one sample to another. Some spectra for adsorbed TPyP are shown in figure 2 for $Co^{2+}$, $Ni^{2+}$ and pure TPyP whereas table 4 summarizes the observations.

DISCUSSION
=====

In acidic medium TPP acts as a basic molecule and undergoes protonation, leading to the acidic form $H_4TPP^{2+}$ :

Form I                    Form II

In form I, there are 2 nitrogen species : the pyrrole atoms with one hydrogen and the aza ones without hydrogen, whereas in form II the four nitrogen atoms are equivalent.

With TPyP, two sites are available for protonation, namely the 4 pyridyl groups and the porphin ring. Preferential protonation occurs on the pyridyl groups and a molecule of TPyP is able to fix 4 protons without modifying the porphin ring :

For pure TPP, the N1s spectrum can be easily decomposed into three gaussian peaks which is not surprising since the binding energy of the N1s level is sensitive to the environment (Perry et al., 1975). The first peak near 396.5 eV can be attributed to N atoms of the porphin ring without hydrogen (aza atoms) and the second peak near 398.6 eV to N atoms with one hydrogen (pyrrolic atoms). The intensity ratio between these two peaks is near unity, and these attributions are in good agreement with published data (Zeller and Hayes, 1973; Niwa et al., 19

Niwa et al., 1974b). As suggested by Niwa et al. (1974b), the third peak on the high B.E. side, has been attributed to a satellite line. In pure TPP, nitrogen bearing hydrogen should have a positive charge greater than that observed for the aza nitrogen.

The complete protonation of the molecule would result in a complete disappearance of the N1s level situated near 396.5 eV. This situation is actually observed when TPP is adsorbed by $Na^+$ or $Ni^{2+}$ montmorillonites. In these cases TPP is entirely protonated in the interlamellar space of the clay. On the contrary for the other homoionic clays, the peak near 396.5 eV is still present with various intensities, suggesting that the porphin ring of TPP is not entirely protonated and thus that complexation with the charge balancing cations may occur to some extent. According to the stoichiometry of the reaction, this metallation of the TPP is never complete because in that case the X.P.S. line arising from N atoms of the pyrrole rings would entirely disappear. This line remains the most important for the $Cu^{2+}$, $Co^{2+}$, $Sn^{4+}$, $Fe^{2+}$, $Mn^{2+}$ clays, showing that protonation is the main reaction that TPP undergoes in the interlamellar space. This is in striking agreement with previous results obtained on the same samples by U.V. and I.R. spectroscopy (Van Damme et al., 1978).

As the areas of the various N1s peaks are directly proportional to the number of nitrogen atoms in the various environments, to a first approximation it is possible to calculate the quantity of metallated and protonated base. Assuming that all the TPP molecules are either protonated or metallated, the ratio corresponding to the peak situated near 396.5 eV in table 3 gives the percentage of metallation : the complexation order is then $Cu^{2+} > Co^{2+}, Sn^{4+} > Fe^{2+}, Mn^{2+}$.

The third N1s peak near 400.5 eV is amazingly intense as compared to that observed for pure TPP. For the pure compound the third peak intensity is in good agreement with data published by Niwa et al. (1974a). The very high intensity value for this line in the adsorbed state suggests that there is another nitrogen atom environment responsible for it. The high value of the B.E. suggests the presence of N atoms with a positive charge greater than that observed in the protonated TPP. This would correspond to TPP molecules which have fixed more than two protons. The experimental conditions inherent to X.P.S. technique : vacuum better than $10^{-8}$ torr and X-ray bombardment of the sample would imply a desorption of adsorbed water; in turn, with clays with a very low water content it is well known that acidity is very high (Fripiat et al., 1965; Karickhoff and Bailey, 1976). One hypothesis to explain the line positioned near 400.5 eV would be that adsorbed TPP would be exposed to a super acid medium leading to a partial additional protonation of the porphin ring to $H_4TPP^{2+}$. In order to check this hypothesis some experiments involving protonation of TPP in super acids solutions are in progress.

Since TPP undergoes mainly protonation, it would be expected that the mesosubstituents of TPyP would act as protons sinks, allowing the formation of metal complexes on the montmorillonite surface. For pure TPyP, there are 6 very similar nitrogen atoms, namely the 4 N of the mesopyridyl groups and the two aza nitrogen of the porphin ring; they lead to a single X.P.S. line (figure 2) located near 397.3 eV. As in TPP, the second line corresponds to N atoms with a higher binding energy : the pyrrolic atoms. The intensity ratio between these two lines is in good agreement with the theoretical ratio (6/2). As with TPP and pure pyridine (Defosse and Canesson, 1976), a protonation of a nitrogen atom would be indicated in the X.P.S. spectrum by a shift of about 2 eV of the corresponding line towards higher binding energies; a molecule of TPyP entirely protonated on the 4 pyridyl groups would have an X.P.S. spectrum constituted of 2 lines situated near 397.3 eV and 399.1 eV, the intensity ratio between these two lines being 2/6 . This situation is actually observed for the $Sn^{4+}$-TPyP sample.

If metallation of the porphin ring was the only process resulting from adsorption, the X.P.S. spectrum would be reduced to one line only, namely that with lower binding energy. For a metallation of the ring and protonation of the pyridyl groups, two lines of equal importance should be observed. This is observed when TPyP is introduced into the interlamellar space of $Co^{2+}$ and $Cu^{2+}$ montmorillonites. Here, all the molecules of TPyP are in close interaction with the metal as metalloporphyrin and at the same time, protonation occurs on the pyridyl groups. The situation is more complicated for the other clays, but considerations about the relative intensity of the first N1s line leads to the conclusion that the quantity of metalloporphyrin in the interlamellar space is in the order $Co^{2+}$, $Cu^{2+}$ >> $Mn^{2+}$, $Ni^{2+}$ > $Fe^{2+}$ confirming the results previously published (Van Damme et al., 1978).

Particular attention must be drawn to the $Na^+$ sample. The R(N1s) ratio in table 4 indicates a high TPyP content. However, the relatively high intensity of the X.P.S. line of lower binding energy cannot be attributed to the formation of a metalloporphyrin. It is preferably interpreted assuming that the TPyP content is too high to obtain a total protonation of the pyridyl groups.

In contrast with TPP, the intensity of the satellite line is constant irrespective of the sample. This could indicate that the quantity of basic centers when TPyP is adsorbed is sufficient for a total neutralisation of the super acid centers.

REFERENCES

Defosse C. and Canesson P., 1975. Preliminary ESCA Study of Aniline Adsorption on HY Zeolites Heated at Various Temperatures. Reaction Kinetics Catalysis Letters, 3 : 161.

Defosse C. and Canesson P., 1976. Potentiality of Photoelectron Spectroscopy in the Characterization of Surface Acidity : Photoelectron and Infrared Spectroscopic Comparative Study of Pyridine Adsorption on $NH_4$-Y Zeolite Activated at Various Temperatures. J.Chem.Soc.Faraday Trans.I, 72 : 2565-2576.

Doner H. and Mortland M.M., 1969. Benzene Complexes with Copper (II) Montmorillonite. Science, 166 : 1406.

Fripiat J.J., Jelli A.N., Poncelet G. and Andre J., 1965. Thermodynamic Properties of Adsorbed Water Molecules and Electrical Conduction in Montmorillonites and Silicas. J.Phys.Chem., 69 : 2185.

Karickhoff S.W. and Bailey G.W., 1976. Protonation of Organic bases in Clay-Water Systems. Clays and Clay Minerals, 24 : 170.

King B.K.G., 1973. The Chemistry of Clay-Organic Reactions. Adam-Hilger/London.

Niwa Y., Kobayashi H. and Tsuchiya T., 1974 A. X-ray Photoelectron Spectroscopy of Tetraphenylporphin and Phtalocyanine. J.Chem.Phys., 60 : 799.

Niwa Y., Kobayashi H. and Tsuchiya T. 1974 b. X-ray Photoelectron Spectroscopy of Azaporphyrins. Inorganic Chemistry, 13 : 2891.

Perry W.B,, Schaaf T.F. and Jolly W.L., 1975. An X-ray Photoelectron Spectroscopic Study of Charge Distributions in Tetravalent Compounds of Nitrogen and Phosphorus. J.Amer.Chem.Soc., 97 : 4899.

Van Damme H., Crespin M., Obrecht F., Cruz M.I. and Fripiat J.J., 1978. Acid-Base and Complexation Behavior of Porphyrins in the Intracrystal Space of Swelling Clays : Meso-Tetraphenylporphyrin and Meso-Tetra (4-Pyridyl) porphyrin in Montmorillonites. J.Colloid Interf.Sci., in the press.

Zeller M.V. and Hayes R.G., 1973. X-ray Photoelectron Spectroscopic Studies on the Electronic Structures of Porphyrin and Phtalocyanain Compounds. J.Amer.Chem.Soc., 95 : 3855.

# ADSORPTION OF CHLORDIMEFORM BY MONTMORILLONITE

J.L. PEREZ RODRIGUEZ and M.C. HERMOSIN
Centro de Edafologia y Biologia Aplicada del Cuarto. Apartado 1052
Sevilla (ESPAÑA)

## ABSTRACT

The adsorption mechanism of the pesticide chlordimeform ($N'$-(4-chloro-2-methylphenyl)-N,N-dimethylmethanimidamide hydrochloride) by Wyoming montmorillonite containing different exchangeable cations ($Na^+$, $K^+$, $Mg^{++}$ and $Ca^{++}$) has been investigated.

The adsorption isotherms of chlordimeform in aqueous solutions on the clay have been measured and they are of "L" or Langmuir type. The corresponding values of adsorption maxima in single-layer complexes are about 1.1 mmol $g^{-1}$ in agreement with the equivalent area of the clay, obtained from the layer charge.

The adsorption of chlordimeform on clay is essentially a cation exchange reaction, increasing with temperature, in which the exchangeable inorganic cations are released into the solution.

The "L" type isotherms, X-ray diffraction and infrared spectra suggest that the chlordimeform ions lie flat between the silicate layers. Interlayered pesticide is difficult to displace with distilled water or aqueous solutions of inorganic cations.

## INTRODUCTION

In agreement with Bailey and White (1970), the phenomenon of adsorption-desorption on clay minerals appears to be one of the main factors affecting interactions between pesticides and soil colloids.

The ion exchange reaction between organic cations and clay minerals has been known for many years (e.g. Smith 1934). Thus ionic pesticides may be adsorbed by ion exchange and ultimately reside in the interlamellar region as exchange cations. This has been shown by Weed and Weber (1968) for diquat and paraquat. With regard to the behaviour and persistence of pesticides in the soil, it is very important to know the stoichiometry of the ion exchange reaction and the ease of replaceability of the adsorbed organic cation. Also of vital interest is the adstion capacity.

In this paper we study the adsorption mechanism of the pesticide chlordimeform by Wyoming montmorillonite saturated with $Na^+$, $K^+$, $Mg^{++}$ and $Ca^{++}$.

RESULTS AND DISCUSSION

Adsorption and desorption

The adsorption isotherms were determined by equilibrating, for 24 known amounts of clays in aqueous solutions of chlordimeform at 1 to 50 mmol/l. After equilibration the suspensions were centrifuged and supernatants analysed. The amount of chlordimeform adsorbed was calculated from the variation in concentration of the solution before and after equilibration, determined spectrophotometrically by UV absorption at 240nm.

The adsorption isotherms are shown in Fig. 1 and are of "L" or Langmuir type according to the classification of Giles et al. (1960). These authors and Knight and Tomlinson (1967) have suggested that this type of isotherm indicates a specific interaction between ions and adsorption sites. Also the "L" type isotherms indicate that the adsorte lie flat on the surface of the adsorbent (Giles et al. 1960).

Data fitting the Langmuir model may be linearized, as shown in Fig. 2, by plotting C/X vs C where C is the molar concentration at equilibrium and X is the number of moles adsorbed per unit adsorbent at concentration C. Table 1 summarizes adsorption maxima, $X_m$, determi-

TABLE 1

Values of adsorption maxima of chlordimeform on montmorillonite saturated with different cations

| Exchangeable cation | Temperature °C | Adsorption maximum mmol . g$^{-1}$ |
|---|---|---|
| $Na^+$ | 25 | 1.14 |
| $Na^+$ | 40 | 1.41 |
| $K^+$ | 25 | 1.12 |
| $Mg^{++}$ | 25 | 0.93 |
| $Ca^{++}$ | 25 | 1.05 |

ned by slopes of such straight lines. It is observed that the shapes of the curves (Fig. 1) and $X_m$ values are similar for all the exchange cations studied. This has been observed by others for ion-exchange reactions in clay-organic interactions (Weed and Weber, 1969; Hayes et al. 1972, 1974).

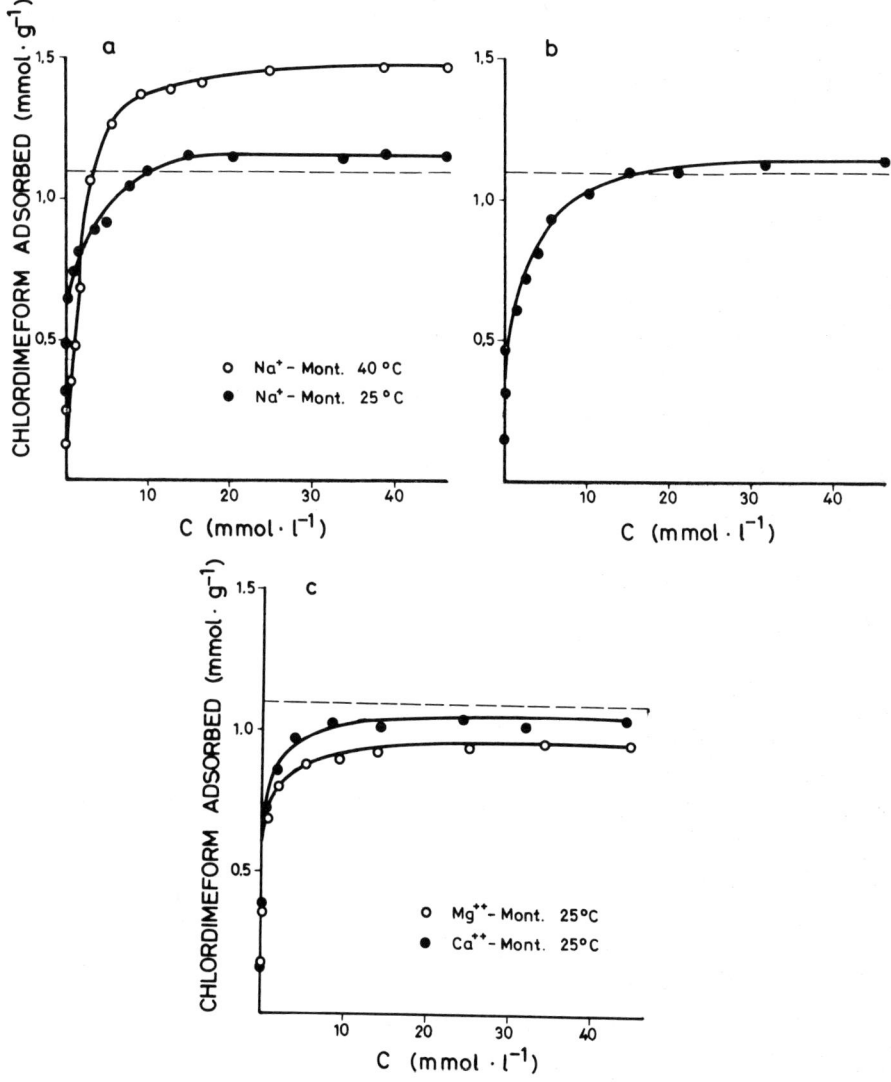

Fig. 1. The adsorption isotherms of chlordimeform on montmorillonite saturated with: a) $Na^+$, b) $K^+$ and c) $Mg^{++}$ and $Ca^{++}$. The dotted lines represent the total exchange capacity of montmorillonite.

The CEC of the montmorillonite due to interlayer cations is 0.92 meq · $g^{-1}$, as calculated from the layer charge (Lagaly et al. 1977) and the total measured CEC is about 1.1 meq·$g^{-1}$. The $X_m$ values agree with the total CEC because organic cations are essentially adsorbed on montmorillonite by cation exchange on external and internal surfa-

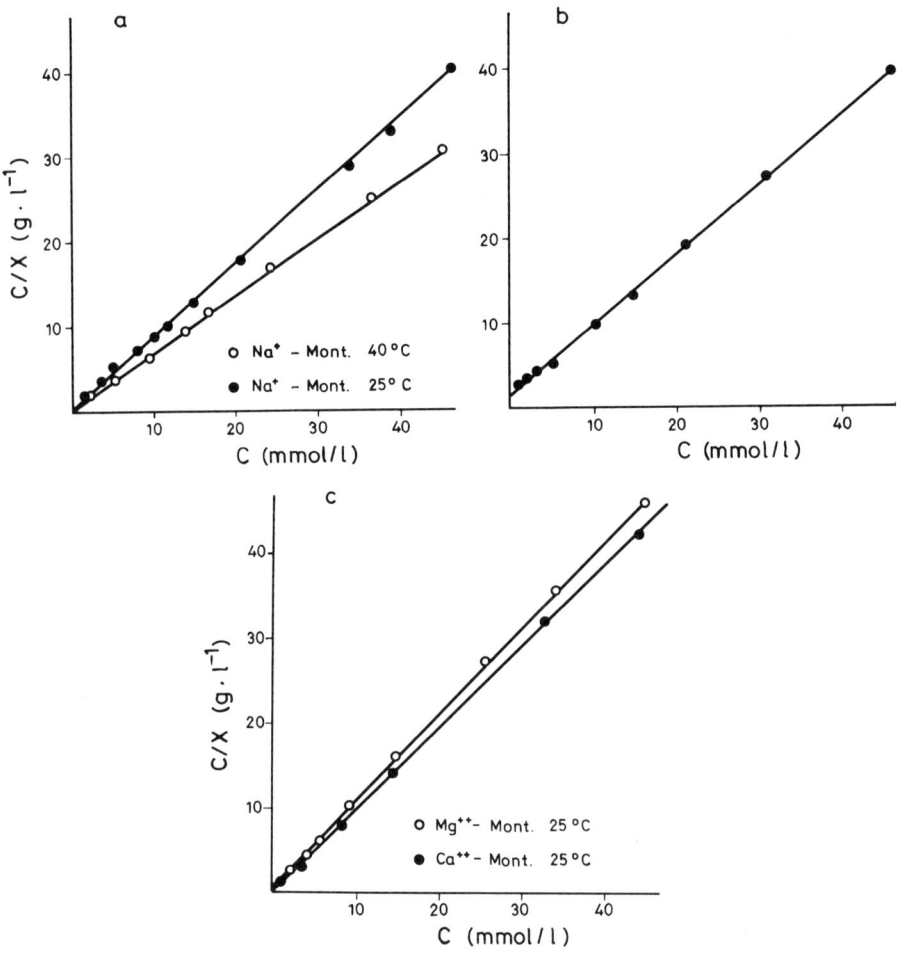

Fig. 2. Langmuir plots for adsorption isotherms of chlordimeform by montmorillonite saturated with: a) $Na^+$, b) $K^+$ and c) $Mg^{++}$ and $Ca^{++}$

ces. It was observed that chlordimeform adsorbed externally is weakly bonded and can easily be removed by washing with water, whereas chlordimeform adsorbed in the interlamellar spaces is strongly adsorbed and it is not removed by either water or aqueous solutions of inorganic cations ($Na^+$ and $Ca^{++}$). On washing sucessively with water, chlordimeform is not released after the fourth treatment (Fig. 3), when about 0.9 meq.g$^{-1}$ remains, in agreement with the interlayer CEC value. During adsorption process the exchangeable inorganic cations of montmorillonite are released to the solution, the amount of outgoing cations being slightly smaller (about 10-20%) than the adsorbed organic

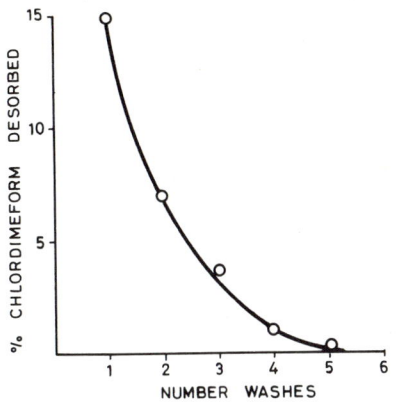

Fig. 3. Chlordimeform released by washing with water.

cations.

The $X_m$ value from the adsorption isotherm for $Na^+$-montmorillonite at 40°C is greater than those at 25°C (Table 1). This fact is explained because the ionic mobility increases with temperature. In this way the cation-exchange process is favoured. Howeber, the initial part of the isotherm at 40°C (Fig. 1a) is further from the Y axis than that at 25°C, probably indicating weaker adsorption at 40°C. On the other hand, the high adsorption plateau at 40°C (Fig. 1a) seems to indicate that some adsorption process other than ion exchange must also operate.

X-ray Data.

The interplanar spacings between the montmorillonite layers were determined by X-ray diffraction, using severals orders of the 001 reflections. The measurements were made by the Debye-Sherrer method, with samples mounted in glass capillary tubes and evacuated at $10^{-6}$ mm Hg.

The spacings $d_{001}$ of the organic-clay complexes ranged from 12.0 to 12.6 Å (Table 2). According to Greene-Kelly (1955a and b) these values indicate that organic cations are adsorbed in the interlamellar spaces with the aromatic rings lying flat on the silicate layers. This result confirms conclusions drawn above from adsorption-desorption studies.

TABLE 2.

The $d_{001}$ values of the chlordimeform-montmorillonite complexes

| Clay | Chlordimeform adsorbed $mmol.g^{-1}$ | $d_{001}$ Å |
|---|---|---|
| $Na^+$-Montmorillonite | 1.11 | 12.6 |
| $K^+$-Montmorillonite | 0.84 | 12.1 |
| $Mg^{++}$-Montmorillonite | 0.80 | 12.0 |
| $Ca^{++}$-Montmorillonite | 0.82 | 12.6 |

By means of a scale model of the chlordimeform ion, the theoretical spacing of the chlordimeform-montmorillonite complex was calculated

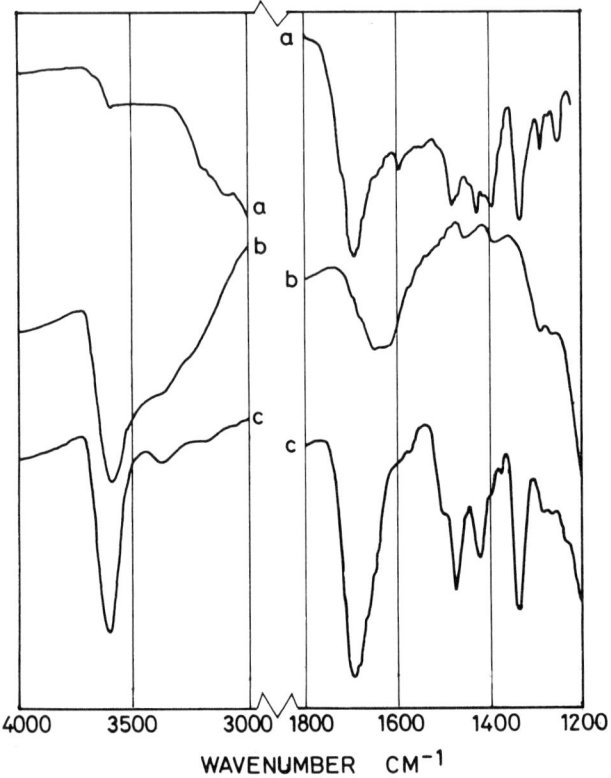

Fig. 4. Infrared spectra of: a) chlordimeform hydrochloride in a KBr disc, b) $Na^+$-montmorillonite as a self-supporting film and c) chlordimeform-montmorillonite complex as a self-supporting film

giving a value of 13.9 Å which is higher than the observed spacing (12.6 Å). Knight and Tomlinson (1967) ascribed this descrepancy to a close approach of the organic ion to the lattice surface caused by: a) Coulombic forces between the organic cations and the negatively charged layer silicate and b) Van der Waals interactions or electron-delocalization forces between the planar ion and the silicate surface.

Infrared spectra.

The infrared spectra were recorded between 4000 and 400 $cm^{-1}$. Self-supporting films, weighing about 4 $mg/cm^2$, were prepared from clay suspensions. The films were saturated in aqueous solutions of the chlordimeform. Fig. 4 shows the IR spectrum of chlordimeform hydrochloride (KBr disc), $Na^+$-montmorillonite and the chlordimeform-montmorillonite complex. The absorption bands of stretching and bending modes of water molecules associated with inorganic cations at 3420 $cm^{-1}$ and 1650 $cm^{-1}$ are not observed in the IR spectrum of chlordimeform-montmorillonite. Also the IR spectrum of the organo-clay complex shows in the region from 1700 to 1200 cm$^{-1}$ the characteristic bands of chlordimeform ions (Fig. 4). These results confirm that chlordimeform ions have displaced the inorganic cations from the interlamellar space.

Layer-charge of clay and adsorption.

The layer charge of montmorillonite determined by the exchange of n-alkylammonium ions (Lagaly and Weiss, 1969) is 0.33 eq/(Si, Al)$_4$O$_{10}$ which corresponds to an equivalent area (Ae) of 70 Å$^2$/unit charge. Using a scale model, the area occupied by a chlordimeform ion lying flat is 77 Å$^2$. The good agreement between these two area values explains the strong and high adsorption of the chlordimeform ion by montmorillonite. In the same way we have calculated 0.8 molecules of chlordimeform per 100 Å$^2$, on the basis that the total specific surface of the clay is 748 m$^2$ g$^{-1}$. This results agree with the dimensions of the chlordimeform ion and the equivalent area given above.

REFERENCES

Bailey, G.W. and White, J.L. 1970. Factors influencing the adsorption, desorption and movement of pesticides in soil. Residue Rev., 32: 29-92.

Giles, C.H., MacEwan, T.H., Nakhwa, S.N. and Smith, D., 1960. Studies in adsorption. Part. XI. A system of classification of solution adsorption isotherms, and its use in diagnosis of adsorption mechanisms and measurement of specific suface areas of solids. J. Chem. Soc., 1960: 3973-3993.

Greene-Kelly, R., 1955a. Sorption of aromatic organic compounds by montmorillonite. I. Orientation studies. Trans. Faraday Soc., 51: 412-425.

Greene-Kelly, R., 1955b. Sorption of aromatic organic compounds by montmorillonite. II. Packing studies with pyridine. Trans. Faraday Soc., 51: 425-437.

Hayes, M.H.B., Pick, M.E., Stacey, M. and Toms, B.A., 1972. Microcalorimetric investigations of the interactions between clay minerals and bipyridylium salts. Proc. Int. Clay Conf. Madrid, pp. 675-682.

Hayes, M.H.B., Pick, M.E., Stacey, M., Toms, B.A. and Quinn, C.M., 1974. The different interactions of paraquat and diquat with montmorillonite and vermiculite. Proc. Xth Int. Congr. Soil Sci. Moscow, 7: 90-93.

Knight, B.A.G. and Tomlinson, T.E., 1967. The interaction of paraquat (1:1 Dimethyl-4,4-Dipyridylium Dichloride) with minerals soils. J. Soil Sci., 18: 233-242.

Lagaly, G. and Weiss, A., 1969. Determination of the layer charge in Mica-type layer silicates. Proc. Int. Clay Conf. Tokyo, pp. 61-80.

Lagaly, G., Müller-Vonmoos, H. and Kahr, G., 1977. Layer charge and cation exchange capacity of smectites. 3th Meet. Europ. Clay Groups. Oslo 1977, pp. 97-99.

Smith, C.R., 1934. Base exchange reactions of bentonite and salts of organic bases. J. Am. Chem. Soc., 56: 1561-1568.

Weed, S.B. and Weber, J.B., 1969. The effect of cation exchange capacity on the retention of diquat $2^+$ an paraquat $2^+$ by three-layer type clay minerals. I. Adsorption and release. Soil Sci. Soc. Am. Proc., 33: 379-382.

OTHER PAPERS PRESENTED IN SECTION 2

ELECTRICAL POLARIZATION OF WATER MOLECULES ADSORBED BY SMECTITES:
AN INFRARED STUDY

C. Poinsignon[1], J.M. Cases[1] and J.J. Fripiat[2] ([1] Centre de Recherches sur la Valorisation des Minerais - ENSG BP 42 54001 - Nancy - France.
[2] Centre de Recherches sur les Solides à Organisation Cristalline Imparfaite - C.N.R.S. 1 b, Rue de la Férollerie - 45045 Orleans, France.)

CHARACTERIZATION OF INTERLAMELLAR WATER IN MAGADIITE BY IR AND NMR SPECTROSCOPY

J. Sanz and E. Ruiz-Hitzky (Instituto de Edafología y Biología Vegetal, C.S.I.C., Serrano 115 dpdo, Madrid-6, Spain.)

WATER IN HYDRATED HALLOYSITE (ENDELLITE)

M.I. Cruz, M. Letellier and J.J. Fripiat (Centre de Recherche sur les Solides à Organisation Cristalline Imparfaite, C.N.R.S., rue de la Ferollerie, 45045 Orleans, France.)

MICROORGANISATION OF WATER-SATURATED KAOLINITE, ILLITE AND MONTMORILLONITE SAMPLES: INFLUENCE OF CATION CONCENTRATION AND pF VALUE

D. Tessier and G. Pédro (I.N.R.A. Station de Science du Sol, 7800 Versailles, France.)

THE INFLUENCE OF THE EXCHANGEABLE CATIONS AND THE SAMPLE PRETREATMENT ON THE SPECIFIC SURFACE AREA OF MONTMORILLONITE

L. Van Leemput, M. Stul and J.B. Uytterhoeven (Katholieke Universiteit Leuven, Centrum voor Oppervlaktescheikunde en Colloidale Scheikunde, De Croylaan 42, B-3030 Heverlee, Belgium.)

THE INFLUENCE OF STRUCTURAL CHARACTERISTICS AND LAYER CHARGE ON POTASSIUM FIXATION BY SMECTITES

G.G. Ristori, S. Cecconi and E. Daniele (Istituto di Chimica Agraria e Forestale della Universita di Firenze, Centro di Studio C.N.R. per i Colloidi del Suolo, Piazzale Cascine 28, 50144 Firenze, Italy.)

EXCHANGE REACTION AND INTERSTRATIFICATION IN K-Ca MONTMORILLONITE

A. Inoue and H. Minato (Department of Earth Science and Astronomy, The College of General Education of University of Tokyo, Komaba, Meguro-ku, Tokyo 153, Japan.)

PHASE VARIETIES AND NOMENCLATURE OF ALCOHOL-SMECTITE COMPLEXES

Armin Weiss, G. Lagaly, S. Fitz, P. Brunner and C. Holm (Institut für Anorganische Chemie der Universität München, Meiserstrasse 1, 8000 München 2, Federal Republic of Germany.)

ADSORPTION OF ISOPROPANOL BY CHARGE-DEFICIENT MONTMORILLONITES

F. Annabi-Bergaya, M.I. Cruz, L. Gatineau and J.J. Fripiat (Centre de Recherche sur les Solides a Organisation Cristalline Imparfaite, 1B rue de la Ferollerie, 45045 Orleans, France.)

REACTIVITY OF A LONG SPACING AMMONIUM-PROPIONATE-KAOLINITE INTERCALATE TOWARDS DIOL, DIAMINES AND QUATERNARY AMMONIUM SALTS

H. Seto and M.I. Cruz-Cumplido (Centre de Recherche sur les Solides à Organisation Cristalline Imparfaite, C.N.R.S. Rue de la Férollerie, 45045 Orleans, France.)

THE INTERACTION OF BISQUATERNARY AMMONIUM IONS WITH MONTMORILLONITE

L. Van Leemput, A. Maes and J.B. Uytterhoeven (Katholieke Universiteit Leuven, Centrum voor Oppervlaktescheikunde en Colloidale Scheikunde, De Croylaan 42, B-3030 Heverlee, Belgium.)

THERMODYNAMIC PARAMETERS FOR THE ADSORPTION OF VARIOUS ALKANES ON CLAY MINERALS

William R. Almon (Cities Service Oil Company, Research Laboratory, PO Box 50408, Tulsa, Oklahoma, USA.)

ADSORPTION OF POLY(ETHYLENE GLYCOL) BY MONTMORILLONITE

Simon Burchill and Michael H.B. Hayes (Chemistry Department, University of Birmingham, Birmingham B15 2TT, U.K.)

A REARRANGEMENT REACTION OF PARATHION ON CLAYS AND OTHER SURFACES

U. Mingelgrin and S. Saltzman (Institute of Soils and Water ARO, Volcani Center, Bet-Dagan P.O.B. 6, Israel.)

INTERLAMELLAR COMPLEXES OF PHYLLOSILICATES WITH MACROCYCLIC POLYETHERS (CROWN ETHERS)

E. Ruiz-Hitzky and B. Casal (Instituto de Edafología y Biología Vegetal, C.S.I.C., Serrano 115 dpdo, Madrid-6, Spain.)

KINETICS OF GELATION OF WYOMING BENTONITE

E. Pekenc and B. Rand (Department of Ceramics, Glasses and Polymers, University of Sheffield, England.)

MICROSCOPIC CHARACTERIZATION OF THE ORGANISATION OF HYDRATED SMECTITE

D. Tessier, J. Berrier and M. Robert (Department de Science du Sol, I.N.R.A., Route de St-Cyr, 78000 Versailles, France.)

IRRADIATION OF CLAY MINERALS BY X-RAYS. EFFECT ON SPECIFIC SURFACE AREA AND PORE STRUCTURE

T. Fernandez Alvarez and H. Carbajal (Departamento de Fisica-Quimica, Instituto de Edafologia, Serrano 115 bis, Madrid 6, Spain.)

EFFECT OF EXCHANGEABLE CATION ON THE FAR INFRARED SPECTRA OF CLAY MINERALS

C.B. Roth (Department of Agronomy, Purdue University, West Lafayette, Indiana 47907, USA.)

SOME ION-EXCHANGE REACTIONS OF A VERMICULITE

R. Le Dred, D. Saehr and R. Wey (Laboratoire de Chimie Minérale Générale, Ecole Nationale Supérieure de Chimie, Rue A. Werner, 68093 Mulhouse Cedex, France.)

ALKINE AND ISONITRILE COMPLEXES OF MONTMORILLONITES

Heinrich Meyer, H. Lüddeke, R. Lehmann and Armin Weiss (Institut für Anorganische Chemie der Universität München, Meiserstrasse 1, 8000 München 2, Federal Republic of Germany.)

AN X-RAY PHOTOELECTRON SPECTROSCOPIC INVESTIGATION OF THE ADSORPTION OF COBALT COMPLEXES BY CLAY MINERALS

M.H. Koppelman[1] and J.G. Dillard[2] ([1] Georgia Kaolin Company, Elizabeth, New Jersey, USA. [2] Dept. of Chemistry, Virginia Polytechnic Institute and State University, Blacksburg, Virginia, USA.)

STABILITY OF COPPER AND SILVER ETHYLENEDIAMINE COMPLEXES ON WYOMING BENTONITE

A. Maes, E. Rasquin and A. Cremers (Katholieke Universiteit Leuven, Centrum voor Oppervlaktescheikunde en Colloidale Scheikunde, De Croylaan 42, B-3030 Heverlee, Belgium.)

TRANSITION METAL ION COMPLEXES ON LAYER SILICATES.
PART 2. THE COMPLEXES OF DIETHYLENETRIAMINE(DIEN) AND TETRAETHYLENEPENTAMINE-(TETREN) WITH Cu(II) AND Ni(II) ON HECTORITE

R.A. Schoonheydt, F. Velghe, F. Pelgrims and J.B. Uytterhoeven (Centrum voor Oppervlaktescheikunde en Colloidale Scheikunde, Katholieke Universiteit Leuven, De Croylaan 42, B-3030 Leuven (Heverlee), Belgium.)

ION EXCHANGE AND INTERSALATION REACTIONS OF HECTORITE WITH TRISBIPYRIDYL METAL COMPLEXES

Sister Mary Frances Traynor, F.S.E., M.M. Mortland and T.J. Pinnavaia (Departments of Crop and Soil Science and Chemistry, Michigan State University, East Lansing, Michigan 48824, USA.)

INTERACTION OF ORGANOPHOSPHORUS PESTICIDES WITH MONTMORILLONITE.
IV. IMIDAN [O,O-DIMETHYL-S(N-PHTHALIMIDOMETHYL)-DITHIOPHOSPHATE]

M. Sánchez Camazano and M.J. Sánchez Martín (Centro de Edafología y Biología Aplicada de Salamanca del C.S.I.C., Salamanca, Spain.)

# SECTION 3

# Geology and Sedimentology

*Chairman's Introduction*

STUDIES OF CLAY MINERALS IN SEDIMENTS-A REVIEW

T. SUDO

20-7, Miyasaka, 3-chome, Setagaya-ku, Tokyo (Japan)

ABSTRACT

A significant area of recent study is research on clay-mineral diagenesis in argillaceous sediments. The features of diagenesis have been demonstrated in detail by applying many parameters such as clay mineralogy, inorganic and organic geochemical indicators, density, and porosity. Specifically, small changes in the degree of diagenesis can be detected by following the change in crystallochemical properties of mixed-layer minerals. Today, the roles of clay minerals in reflecting the degree of diagenesis, and of montmorillonite particularly in solving the problems concerning oil formation (generation, and migration), and in assessment of the potential of rocks as source of petroleum, are all well-accepted in the study of clay-mineral diagenesis of sediments in oil fields. In this review, brief accounts are also given for some selected topics concerning the implication of clay-mineralogy for the study of sediments.

INTRODUCTION

I appreciate the opportunity to deliver this Introductory Lecture for the AIPEA 1978 International Clay Conference. Because of the limited space, it is difficult to make this review article to be comprehensive. A major topic in this review is concerned with the clay-mineral diagenesis in sediments. Brief accounts are also given for some selected topics concerning the implication of clay-mineralogy for the study of sediments. In the references cited in this review, the mica clay minerals are usually described with the term "illite", which may be defined as a term for a 10 $\overset{\circ}{A}$ mica that does not change spacing on heating and solvation. Mixed-layer minerals of illite-montmorillonite will be simply referred to hereafter as I/M minerals. The nomenclature of clay minerals in this review will follow that of the original literatures.

CLAY-MINERAL DIAGENESIS

Earlier work

Studies of the Gulf Coast pelitic sediments revealed that clay-mineral diagenesis is indicated by the occurrence of montmorillonite transforming toward illite through I/M minerals with increasing depth (Powers, 1959, Burst, 1969). Perry et al. (1970,1972) reported that discrete montmorillonite and illite phases are ambiguous in the total depth studied, and that the diagenesis is represented by the occurrence of I/M minerals with decreasing expandability with increasing depth, accompanying montmorillonite dehydration which is considered to occur in the following four stages: I, expulsion of interstitial water and interlayer water present in excess of two water interlayers from montmorillonite layers in I/M minerals with a varying but small (less than 30 percent) proportion of illite layers; II, random collapse of montmorillonite layers up to about 65 percent of the layers, and concomitant interlayer water expulsion; III, transition of random to ordered I/M minerals with collapse of montmorillonite layers up to about 80 percent of the layers; IV, occurrence of nonmixed-layered illite with the lost of interlayer water is anticipated. They suggested that the transformation resulted from redistribution of potassium from feldspar or detrital mica to montmorillonite layers within rocks, and an increasing fixation of potassium on montmorillonite layers which may be due to alteration of the montmorillonite layers, characterized by an increasing substitution of $Al^{3+}$ for $Si^{4+}$ in the tetrahedral sheet. Weaver (1960) indicated possible use of clay minerals in search for oil. Burst (1959) reported that hydrocarbon production depths are distributed in a statistically consistent relation to montmorillonite-dehydration depths, and suggested that hydrocarbon migration resulted from dewatering of montmorillonite occurring in connection with the formation of the mixed-layer minerals. Aoyagi (1969) performed an extensive study of clay minerals in argillaceous sediments distributed in the oil fields of Japan and pointed out the relation between clay minerals and source material, lithology, depositional environment, degree of diagenesis, and total production of oil. It was suggested that the average content of swelling clay minerals in each area is related in some degree to the total production of oil in the area. Shimoyama et al. (1971) and Johns et al

(1972) showed experimentally the catalytic role of montmorillonite for hydrocarbon generation involving decarboxylation and cracking reactions. Using kinetic data, these authors related the decarboxylation and cracking reactions at various depths to an increase in the dehydration of montmorillonite. Assuming the geothermal gradient as $1.1°$ C/100 ft (30 m), the boundary between the decarboxylation and cracking zones can be set at about 3,000 ft (900 m) in depth; the former corresponds closely to Stage I, and the latter to the depth-interval involving Stages II and III.

Mode of diagenetic alteration

The concept of a mechanism for the formation of the mixed-layer minerals proposed by Perry et al. (1970, 1972) has successively been supported by the structural formulas derived from precise chemical analyses (Foscolos et al.,1974), by additional data for pelitic sediments in the Gulf Coast (Hower et al.,1976), by radiogentic argon analyses (Aronson et al.,1976), and by the magnitude of the activation energy (19.6 kcal/mole) obtained from kinetic analyses (Eberl et al.,1976).

In connection with this, attention may be drawn to the study of glauconite. It is well known that opinions concerning the origin of glauconite have been divided. Shutov et al. (1973) adopted a view that the formation of glauconite resulted from a diagenetic transformation of dioctahedral iron-rich smectite. Turnaw-Morawska et al. (1975) studied the Ordivician glauconite from Poland, and pointed out that the properties of monomineralic glauconite are slightly different from earlier data for glauconite in having higher aluminium and potassium, lower amounts of divalent cations and swelling layers, and also having a well-ordered structure. These authors favoured Shutov's view and suggested that these characteristics resulted from a diagenetic transformation of weathered material in an advanced stage of diagenesis through illitization.

Hower et al. (1976) reported that the conversion of smectite into illite by diagenetic alteration may involve loss of magnesium and iron from growing mixed-layer minerals, and chlorite is then likely formed from magnesium and iron lost by smectite. The total reaction may then be expressed as : smectite + potassium feldspar (or mica) = illite + quartz + chlorite.

Perry et al, (1976) studied the sediment/pore-water system in the sediments of Hole 149 (DSDP), and reported that the volcanic materials undergo submarine weathering and early diagenetic alteration which release calcium and magnesium to the pore-water, and the resulting smectite (probably magnesium-rich sort) removed magnesium from the pore-water. The net reaction resulted in little change of the pore-water alkalinity.

Suchecki et al. (1977) gave an interesting opinion about the origin of the clay-mineral suite consisting of corrensite and I/M minerals (5-10 percent expandable) occurring in shales and volcanogenic sandstones of Ordvician age. Referring to the study of Hower et al. (1976), these authors considered that the suite has been derived from magnesium- and iron-rich volcanic detritus by a diagenetic alteration. They inferred that ready availability of much magnesium from the parent material may favour the formation of corrensite instead of chlorite, and corrensite may continue to exist with a higher proportion of expandable layers than the I/M minerals because of its trioctahedral character (Suchecki et al.,1977, p.169).

Factors governing diagenesis

Perry et al. (1970,1972), in the study of the argillaceous sediments in the Gulf Coast, demonstrated that the dependency of expandability of I/M minerals upon burial depths is quite different in the wells with different geothermal gradients. These authors initially explained the difference by suggesting that the transformation of montmorillonite into mixed-layer minerals may be represented by a temperature-dependent reaction. However, they finally suggested that the differences that still remain may be attributable to the other factors such as pressure, geological ages, and sometimes pore-water chemistry.

Hower et al. (1976) demonstrated that the compositional changes of the pelitic sediments in the Gulf Coast which vary as a function of diagenetic grades, are closely related to those of shales which vary as a function of geological age. These authors favoured a concept that the chemical variations in shales of different geological ages reflect post-depositional modification. They also pointed out that the changes of the Gulf Coast sediments are relatively short time changes (a few tens of millions of years) that have been accelerated by rapid deep burial with accompanying higher temperatures.

Eberl et al. (1976) studied an experimental conversion of potassium-beidellite glass into illite, and analysed the results by kinetics. These authors, using a value of the activation energy obtained as 19.6 kcal/mole, made a diagram showing the relationship between temperature and reaction time. According to this diagram, the starting material will convert, at $80°$ C, to a 20 percent expandable mixed-layer mineral in less than one million years. In the Gulf Coast wells, mixed-layer minerals collected at $80°$ C from Miocene sediments are usually more than 40 percent expandable. Because of undetermined factors that may slow reaction in the natural system, they stated that the kinetic data are consistent with either equilibrium or a kinetic interpretation for the data given by Perry et al. (1970,1972). Recently Lippmann (1977) argued that the clay minerals with extensive isomorphous substitutions of ions are likely to be unstable on the basis of solubility product considerations.

Venturelli et al. (1977) studied clay minerals in shaly rocks from the northern Apennines, Italy, and pointed out that the following general concept may be acceptable: the rocks having I/M minerals may be ranked in a low grade of diagenesis, and the occurrence of a illite-chlorite suite may signify a deep burial stage of diagenetic alteration.

Viczián (1975) provided a review article concerning clay mineral suites in Hungarian sedimentary rocks of different geological ages. Although it was difficult to find a consistently well-defined relation between the clay-mineral suite and geological age, he was able to find a broad generalization represented by the decline of expandable minerals and kaolinite at the expense of illite and chlorite toward older geological ages. It has been accepted that such a general tendency may indicate an advancing diagenetic grade with increasing depth. However individually speaking, it is impossible to overlook the diversity in the relationships among the factors such as geological age, diagenetic grade and clay-mineral suite as indicated by the observation that a kaolinite sometimes occurs even in the anchizone, or of subsynchronous rocks which have undergone diagenetic alteration to different degrees as indicated by the evidence of clay-mineral suites. Therefore it was stated that the diversity may be due to different efficiencies of the factors governing diagenesis itself (burial depth, temperature, tectonic effect, chemical composition of pore-water).

Inorganic and organic geochemical indicators

Foscolos et al. (1974) evaluated the degree of diagenesis in Lower Cretaceous shales (Buckinghorse Formation) by applying many indicators involving (a) some properties of illite such as the percent 2M polymorph, crystallinity, and crystallite size, (b) some properties of mixed-layer minerals such as percent illite layers, the amount of potassium and the ratio of Al/Si, and (c) cation exchange capacity and surface area. With increasing burial depth, the indicators of (a) and (b) tend to increase whereas (c) to decrease.

Aoyagi et al. (1975) studied clay minerals in many core samples from oil fields of Japan. They correlated the data for the samples (clay mineral composition, density, and geological and stratigraphical data) with burial depths for each field. Each clay mineral was determined quantitatively by X-ray analyses. The result demonstrates a well-defined correlative relationship between the depth and the other data, e.g. as indicated by a decrease of expandable minerals and density toward older geological ages and with advancing diagenetic stage.

Foscolos et al. (1976) studied clay-mineral diagenesis in Lower Cretaceous shales involving Buckinghorse Formation and Sully-Lepine Series by applying clay mineral and inorganic and organic indicators. These authors proposed a scheme for the classification of the diagenetic stages on the basis of the relation between these parameters. According to these authors, little hydrocarbon generation occurs in eodiagenesis (early diagenesis). In mesodiagenesis (middle diagenesis), both the yields of total extracts and hydrocarbon are likely higher than in the other stages. Hydrocarbon generation commences in early mesodiagenesis. In late mesodiagenesis, cracking of organic matter may proceed by a carbonium ion mechanism (Johns et al., 1972). In telodiagenesis (late diagenesis) extracts become extremely low. It was considered that mesodiagenesis corresponds to the main phase of oil genesis. According to Foscolos et al. (1974), early mesodiagenesis is represented by 6,000 ft (1,800 m) in depth prior to uplift, $94^\circ$ C in temperature, and 40-50 percent illite layers in the mixed-layer minerals. The depth, temperature, and percent illite layers assigned to late mesodiagenesis are 10,000 ft (3,000 m), $141^\circ$ C, and 70-80 percent respectively. The geothermal gradient was set there as $1.2^\circ$ C/33.33 m (surface temperature, $75^\circ$ F ($24^\circ$ C)).

Asakawa (1975) performed an organogeochemical study of the sediments distributed in the oil fields in Japan, and reported that the ratio of hydrocarbon to organic carbon abruptly increases with increasing burial depth within the depth-interval of 2,000-3,000 m and the temperature-interval of 60-105° C. The geothermal gradient was set as 2.3° C/100 m (surface temperature, 15° C).

Porosity

Inami et al. (1974) performed an experimental study of the relation between compressibility and porosity of argillaceous rocks from Japan, and demonstrated that the porosities abruptly decrease up to about 30 percent, and then decrease gradually with decreasing compressibility. Aoyagi et al. (1976) studied the compaction behaviour of sodium-montmorillonite using the triaxial hydrostatic compaction apparatus. These authors measured absolute porosities of the samples obtained at the final stages of the experiments performed under the conditions as 500-900 kg/cm$^2$ and 60-100° C, and found that the porosities remain nearly unchanged at about 30 percent. Aoyagi and co-workers measured the porosities of argillaceous rocks distributed in the oil fields of Japan. Although there is no consistent coincidence among the features of depth-dependency of the porosities, they are likely to decrease rapidly up to 25-30 percent and then to gradually decrease. These authors grouped diagenetic stages into three in regard to porosity: I(80-30 percent), II(30-10 percent), and III(less than 10 percent). Setting the geothermal gradient as 2.9° C/100 m (surface temperature, 15° C), Stage II is confined within the depth-interval of 1,400-2,800 m, and the lowest limit of the Stage II is close to the level of the inception of the mixed-layer mineral occurrence, which has been estimated to be about 3,000 m by these authors. They stressed that primary migration of oil may favourably occur in Stage II where discharge of of the interstitial water still continues, and vigorous expulsion of inter-layer water accompanying the main phase of oil generation will commence.

Concluding remarks

Diagenetic patterns of sedimentary rocks are usually diverse owing to different efficiencies of the factors governing diagenesis itself. A relatively well-defined pattern has been revealed from thick pelitic sediments since many indicators are capable of acting as useful guides for evaluating the degree of diagenesis. Specifically, changes in properties of I/M minerals with increasing burial depths may serve to evaluate the degree of diagenesis. Montmorillonite may play an important role as a promoter of petroleum migration and generation. In connection with the study of clay-mineral diagenesis of argillaceous rocks, it may be worthwhile to refer to some selected data summarized as follows. The geothermal gradients reported are involved in a range of 2.3-3.6° c/100 m (surface temperatures are set in a range of 15-24° C). Most of these values have been obtained by calibration or taking account of the method of measurement. The lowest limit of disappearance of discrete montmorillonite, which corresponds to the inception of the occurrence of mixed-layer minerals, has been estimated to be the levels of 1,000-3,000 m, and 80-100° C. The lowest limit of the occurrence of mixed-layer minerals, which corresponds to the inception of nonmixed-layered illite, has been estimated to be 4,000-5,000 m. However one's attention should be drawn to the study of Perry et al. (1970, 1972) reporting that discrete montmorillonite and illite phases are ambiguous in the total depth-range studied (3,700-5,500 m). The main phase of oil genesis has been estimated to be at the 1,000-3,700 m. depths.

## REMARKS ON THE CLAY-MINERAL IMPLICATION FOR THE STUDY OF SEDIMENTS

Detrital clay minerals are found in recent as well as ancient sediments. Each clay-mineral population in detrital clays may show variation in composition or particle size among individual flakes, and each clay mineral itself has been subjected to a slight modification in crystallochemical properties. The inhomogeneity and modification in properties as such have been inherited from provenance, and may represent weathering history in predepositional stages.

Weathered illite, particularly under acid-leaching conditions, is usually capable of potassium fixation, and tends to return to normal illite by fixing potassium when exposed to sea water or layed down in sediments in marine environment. The potassium fixation ability may be due to the development of high interlayer charges which are essentially vermiculitic interlayer sites (Perry et al., 1976). The degree of degradation in illite is observable in subtle-to moderate extent. Sometimes I/M minerals and dioctahedral vermiculite (14 Å) are regarded as degraded illites. Here it should be observed that the I/M minerals in this case are alteration products of illite to montmorillonite in contrast to those occurring in an advancing stage of diagenesis. Dioctahedral vermiculite is also regarded as a degraded illite in which most of the interlayer potassium has been removed by weathering. Degraded illite can sometimes be found under deep burial where chemical conditions of sediments have favoured the persistence of the mineral; in such a case, degraded illite may be used as an indicator of depositional environment.

Brown et al. (1977) studied the clay-mineral variation in the sediments (Middle Pennsylvanian age) represented by deltaic-interdeltaic model occurring in complicated stratigraphical sequences. These authors pointed out that the changes from marine to nonmarine conditions tend to coincide with a decrease of illite and an increase in kaolinite and degraded illites involving I/M minerals and dioctahedral vermiculite (14 Å).

Bodine et al. (1977) reported the properties of illite and chlorite occurring in Upper Silurian rock salts representing a special chemical environment. Chlorite is homogeneous in composition, which is, however, distinct from shale chlorites. The chlorite is authigenic and its formation resulted from hyperhalmyrolysis which is suggested by these authors as applying to mineral reactions occurring in the marine evaporite environment from the time of mineral deposition or precipitation in the brine to the time of its burial. Illite, showing a wide compositional variation among individual flakes, is considered to be detrital in nature, approaching an equilibrium between illite and hypersaline brine.

Finally, it is impressive that mixed-layer minerals have often been called into question in the study of sediments. It has been known that component mineral layers of mixed-layer minerals are occasionally abnormal (unusual) in crystallochemical behaviour (Sudo et al.,1977). It would be interesting to extend the study of component layers in the future addressing the question as to whether or not subtle variability in crystallochemical behaviour of component layers can be used to evaluate small changes in diagenetic stages.

ACKNOWLEDGEMENT

The writer is indebted to many colleagues from home and abroad for providing valuable information for this review article.

REFERENCES

Aoyagi, K., 1969. Mineralogical study of sedimentary rocks in the oil fields of Japan by the X-ray diffraction method, and its application to petroleum geology. Part 4: Applications of the studies of mineral compositions of the sedimentary rocks to the petroleum geology. Clay. Sci., 3:126-139.

Aoyagi, K., Kobayashi, N. and Kazama, T., 1975. Clay mineral facies in argillaceous rocks of Japan and their sedimentary petrological meanings. In: S.W. Bailey (Editor), Proc. Intern. Clay Conf., 1975, Mexico City, 101-110.

Aoyagi, K., Kazama, T., and Sudo, Y., 1976. Experimental compaction of Na-bentonite under programming temperature and pressure. J. Jap. Assoc. Petroleum, Tech., 41:125-130.

Aronson, J.L. and Hower, J., 1976. Mechanism of burial metamorphism of argillaceous sediments: 2. Radiogenic argon evidence. Geol. Soc. Am. Bull., 87:738-744.

Asakawa, T., 1975. Relationships between normal alkanes and maturation of petroleum of oilfields in Japan. J. Jap. Assoc. Petroleum, Tech., 40:117-126.

Bodine, M,W, and Standaert, R.R., 1977. Chlorite and illite compositions from Upper Silurian rock salts, Retsof, New York, Clays Clay Minerals, 25:57-71.

Brown, L.F., Bailey, S.W., Cline, L.M., and Lister, J. S., 1977. Clay mineralogy in relation to deltaic sedimentation patterns of Desmoinesian cyclothems in Iowa-Missouri. Clays Clay Minerals, 25:171-186.

Burst, J.F., 1969. Diagenesis of Gulf Coast clayey sediments and its possible relation of petroleum migration. Am. Assoc. Petroleum, Geol. Bull., 53:73-93.

Eberl, D, and Hower, J., 1976. Kinetics of illite formation . Geol. Soc. Am. Bull., 87:1326-1330.

Foscolos, A.E. and Kodama, H., 1974. Diagenesis of clay minerals from Lower Cretaceous shales of north eastern British Columbia. Clays Clay Minerals, 22:319-33

Foscolos, A. E., Powell, T.G., and Gunther, P.R., 1976. The use of clay minerals and inorganic and organic geochemical indicators for evaluating the degree of diagenesis and oil generating potential of shales.Geochim.Cosmochim.Acta,40:953

Hower, J., Eslinger, E.V.,Hower, M.E., and Perry, E.A., 1976. Mechanism of burial metamorphism of argillaceous sediments: I. Mineralogical and chemical evidence. Geol. Soc. Am. Bull., 87:725-737.
Inami, K. and Hoshino, K., 1974. Compressibility and compaction of clastic sedimentary rocks. J. Jap. Assoc. Petroleum, Tech., 39:357-374.
Johns, W.D. and Shimoyama, A., 1972. Clay minerals and petroleum forming reactions during burial and diagenesis. Am. Assoc. Petroleum, Geol. Bull., 56:2160-2167.
Lippmann, F., 1977. The solubility products of complex minerals, mixed crystals, and three-layer clay minerals. Neues Jahrb. Mineral. Abh., 130:243-263.
Perry, E.A. and Hower, J., 1970. Burial diagenesis in Gulf Coast pelitic sediments. Clays Clay Minerals, 18:165-177.
Perry, E.A. and Hower, J., 1972. Late-stage dehydration in deeply buried pelitic sediments. Am. Assoc. Petroleum, Geol. Bull., 56:2013-2021.
Perry, E.A. and Gieskes, J.M., and Lawrence , J. R., 1976. Mg, Ca and $O^{18}/O^{19}$ exchange in the sediment-pore water system, Hole 149, DSDP. Geochim. Cosmochim. Acta, 40:413-423.
Powers, M.C., 1959. Adjustment of clays to chemical change and the concept of the equivalence level. In: Ada Swineford (Editor), Clays Clay Minerals, 6:309-326.
Shimoyama, A. and Johns, W.D., 1971. Catalytic conversion of fatty acids to petroleum-like paraffins and their maturation. Nature, Phys. Sci., 232:140-144.
Shutov, V.D., Katz, M. Ya., Drits, V.A., Sokolova, A.L., and Kazakow, G.A., 1973. Crystallochemical heterogeneity of glauconite as depending on the conditions of its formation and postsedimentary changes. In: J.M. Serratosa (Editor), Proc. Intern. Clay Conf., 1972, Madrid, 269-279.
Suchecki, R.K., Perry, E.A., and Hubert, J.F., 1977. Clay petrology of Cambro-Ordvician continental margin, Cow Head Klippe, western New-Foundland. Clays Clay Minerals, 25:163-170.
Sudo, T. and Shimoda, S., 1977. Interstratified clay minerals-mode of occurrence and origin. Miner. Sci. Eng., 9:3-24.
Turnau-Morawska, M., Łacką, B., and Wiewióra, A. 1975. Charakterystyka krystalochemiczna glaukonitu ze skal ordowiku obnizenia podlaskiego na tle litologii i genezy. Kwartalnik Geologiczny, 19:829-843.
Venturelli, G. and Frey, M., 1977. Anchizone metamorphism in sedimentary sequences of the northern Apennines. Rendiconti Societa Italiana di Mineralogia e Petrologia. 33:109-123.
Viczián, I., 1975. A review of the clay mineralogy of Hungarian sedimentary rocks (with special regard to the distribution of diagenetic zones). Acta Geol. Sci. Hungaricae, 19:243-256.
Weaver, C.E., 1960. Possible uses of clay minerals in search for oil. Am. Assoc. Petroleum, Geol. Bull., 44:1505-1518.

# CORRELATION BETWEEN COAL AND CLAY DIAGENESIS IN THE CARBONIFEROUS OF THE UPPER SILESIAN COAL BASIN

J. ŚRODOŃ

Institute of Geology of Polish Academy of Sciences, Kraków (Poland)

## ABSTRACT

The range of diagenesis of the Carboniferous rocks in the profile of the Upper Silesian Basin is indicated by coal rank parameters: $R_o$ = 0.8 -2.2 % and V.M. = 38 - 12 %. Diagenetic chlorite is detectable in the lower part of the profile. Illite 1M forms from kaolinite in the entire profile, but the process intensifies in the lower part. Illite/smectite in shales evolves, with a change in structure, from 25 % smectite at the top, to 10 % smectite at the bottom. In pyroclastic materials, mixed layers are always more smectitic, at least by 10 %. Diagenesis then produces a mixture of illite/smectite with discrete illite. A new method has been devised for identification of such materials, and for their use as indicators of the degree of diagenesis. The method is compared with the widely used method of Kubler.

## INTRODUCTION

The reaction series smectite - smectite/illite - illite - muscovite is the most generally applicable and sensitive clay indicator of the degree of diagenesis and low-grade metamorphism of sedimentary rocks.

The most detailed information on the low- temperature part of this series, down to about 25% smectite mixed layer, comes from the Tertiary of the Gulf Coast (Perry and Hower, 1970; Hower et al., 1976). The illite - muscovite part of the series was studied by means of the 2M/1Md polytype ratio (Maxwell and Hower, 1967), and by measurements of the assymetry (Weaver, 1960), and broadening (Kubler, 1964) of the 001 reflections. The last method especially gave a good scale for late diagenetic and low-grade metamorphic phenomena (see Dunoyer de Segonzac, 1970), which is now well correlated with the degree of coalification measured by the vitrinite reflectance (Weber, 1972; Wolf, 1975).

Much less attention has been paid to the intermediate stage, i.e., the transformation of an ordered 25% smectite mixed layer into pure 100% illite. Several papers describing such materials (reviewed in Dunoyer de Segonzac, 1970; Drits and Kopor lin, 1973) do not give much data. The purpose of this study was to look in detail at this intermediate stage of diagenesis and to correlate changes shown by clay and coal.

MATERIALS

The materials studied were mostly core samples of shales from two wells 2000 m deep, located west of the centre of the Upper Silesian Basin, chosen to cover the complete range of coalification known in the Basin. In addition, scattered sample of shales and pyroclastic materials were collected from other wells and from coal mines.

ANALYTICAL TECHNIQUES

Bulk rock mineral composition

The method of quantitative mineral analysis employed here will be published in detail in a separate paper. It is based on the formula of Klug and Alexander (197 p.545) for the direct diffraction - absorption method. However, it makes use only of the mass absorption coefficients of standards. Those of the samples are not actually measured, the mineral composition being obtained by normalization to 100

The standard for 1M illite was separated from the rocks studied. A mineral sel ted to be the chlorite standard had intensity ratios of the basal reflections clo to these encountered in the rock samples. Where possible, non-basal reflections c clay minerals were used (Fig. 1), because their intensities seem to be more repro cible than those of the 001 reflections, and they offer the possibility of quanti ing the polytypes.

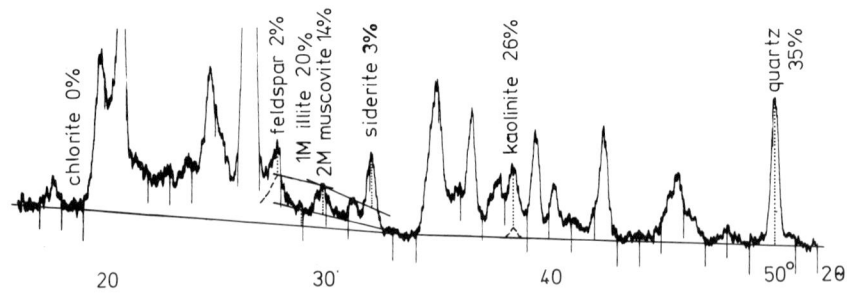

Fig. 1. An example of the intensity measurements for the quantitative X-ray diffraction analysis of shales.

Fine fraction study

4 g of a drill-powdered sample were soaked overnight in 100 ml of distilled wa sonified for 5 min with an ultrasonic probe, and the <0.2 μm fraction separated by centrifuging. For one sample, coarser fractions were also collected. About

40 mg of the dried clay was resuspended with ultrasonics and pipetted onto a glass slide to give an oriented film. Both air-dry and glycolated specimens were X-rayed in the range 50-2° 2θ CuK$_\alpha$.

RESULTS

Variability of mineral composition

The mineral composition of the bulk rock is very stable throughout entire profiles (Fig. 2). Generally, more fine-grained samples were selected from profile II, so they contain less quartz.

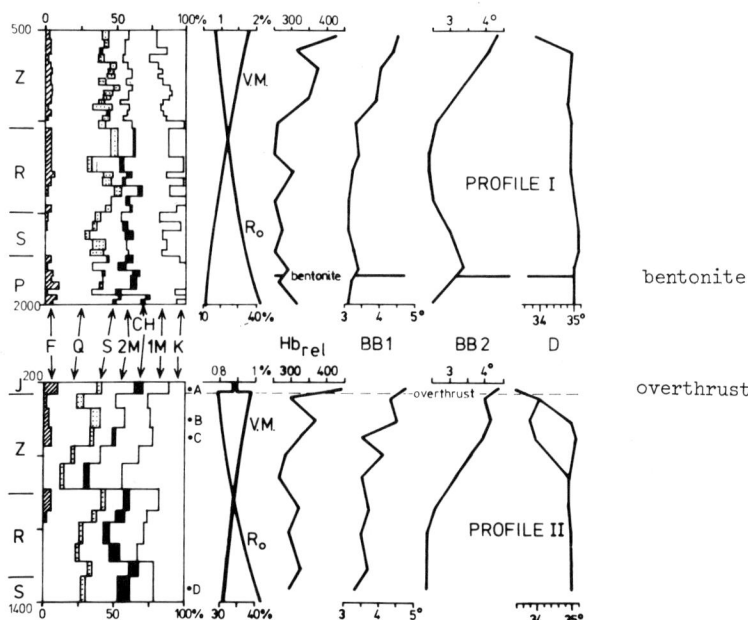

Fig. 2. Variability of mineral composition of bulk shales and indices of coal and clay diagenesis in two profiles of the Upper Silesian Basin. The vertical scale is in meters below the surface. Z - Załęskie Beds, R - Rudzkie B., S - Siodłowe B., P - Porębskie B., J - Jaklowieckie B. A,B,C,D - samples shown in Fig. 4. F - feldspars, Q - quartz, S - siderite , 2M - 2M muscovite, CH - chlorite, 1M - 1M illite, K - kaolinite, $R_o$ - vitrinite reflectance, V.M. - volatile matter. Hb$_{rel}$, BB1, BB2, and D are characteristics of the illite/smectite, defined in the text.

Chlorite seems to be both detrital and diagenetic. In the upper parts of both profiles, chlorite is stratigraphically controlled, being almost absent from the Załęskie Beds, though they represent different stages of diagenesis (Fig. 2). On the other hand, in the lowermost section of profile I, an admixture of clay-sized chlorite was found within a bentonite layer (Fig. 3). Chlorite has never been encountered in Carboniferous bentonites from shallow localities (Środoń, 1972;

Parachoniak and Środoń, 1974; this paper), so, in the author's opinion, this find gives good evidence for the diagenetic crystallization of chlorite in this deep section of the profile.

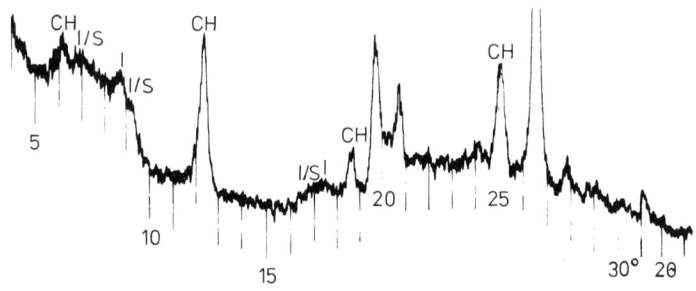

Fig. 3. X-ray diffraction pattern of a glycolated preparation of the bulk bentonit from profile I. CH - chlorite, I/S - illite/smectite, I - illite.

Additional data supporting this conclusion (Table 1) come from the paper by Dopita and Králík (1977) on the southern part of the Coal Basin.

TABLE 1

Correlation between the basal spacing of air-dry illite/smectite and the occurrenc of chlorite ( + ) in tuffaceous horizons of the Coal Basin.

| | | | |
|---|---|---|---|
| 10.03 Å − | 10.42 + | 10.83 − | 11.21 − |
| 10.09 + | 10.63 − | 11.00 − | 11.36 − |
| 10.09 − | 10.65 + | 11.05 − | 11.42 − |
| 10.10 + | 10.68 − | 11.09 − | 11.64 − |
| 10.14 + | 10.80 − | 11.20 − | |
| 10.16 + | 10.82 + | 11.21 − | |

It is evident that chlorite in pyroclastics is associated with highly illitic mixed layers ( < 25% smectite, based on comparison with the author's own materials As will be shown later, the illitization is a diagenetic process.

The absence of chlorite from the lower part of the Siodłowe Beds in profile I (Fig. 2) suggest that detectable amount of this mineral forms diagenetically at $R_o$ (mean vitrinite reflectance in oil) $> 1.7\%$.

The amount of kaolinite in profile II shows negative correlation with the amoun of quartz. In profile I this relation does not exist, and the amount of kaolinite decreases down the profile, whereas the sum of the major clay minerals, i.e. illi 1M + kaolinite, remains more or less constant. This suggests diagenetic illitizat of kaolinite, which becomes evident in a more advanced diagenetic stage ($R_o > 1.2\%$ volatile matter (dry, ash-free), V.M. $< 25\%$). It is in a good agreement with data given by Dopita and Králík (1977, Fig. 63).

The data obtained from several pyroclastic horizons support this conclusion. The common features of these rocks are barrel-shaped clay pseudomorphs after biotite and long, vermicular clay aggregates (Środoń, 1972; Parachoniak and Środoń, 1974). In samples, of a diagenetic stage comparable, according to V.M. data, with the upper part of profile II, these forms consist of kaolinite or kaolinite with narrow laths of pure, dioctahedral, aluminous illite, as identified by microprobe (Środoń, 1976, Fig. 19). In the bentonite from profile I (Fig. 3), these forms consist of illite. Evidently, the illitization of kaolinite proceeded in the whole investigated range of diagenesis and intensified with depth.

Both the crystallization of chlorite and the illitization of kaolinite are difficult to quantify and cannot be used as sensitive indicators of the degree of diagenesis. It is therefore necessary to look in detail at the component identified generally as " 1M illite ".

Nature of the " illite "

The <0.2 μm fraction of the shales consists principally of a dioctahedral 2:1 mineral component (1M polytype) here called " illite ", with usually minor quantities of kaolinite and/or chlorite (Fig. 4). Looking at the glycol patterns of the shallowest samples we can identify " illite " as a two-component mixture of a pure mineral illite and an ordered illite/smectite (hereafter designated I/S): The $d_{001}$ values of illitic peaks after glycol treatment are stable throughout the profiles (Fig. 4). The small shifts of the low angle reflections probably result from a fine crystallite size of illite (Reynolds, 1968). The separate reflection around 12.8 Å indicates the presence of an ordered I/S (Reynolds and Hower, 1970). The study of various grain fractions supports the identification of " illite " as the two-component mixture. The mixed layer tends to accumulate in the finest fractions (Fig. 5).

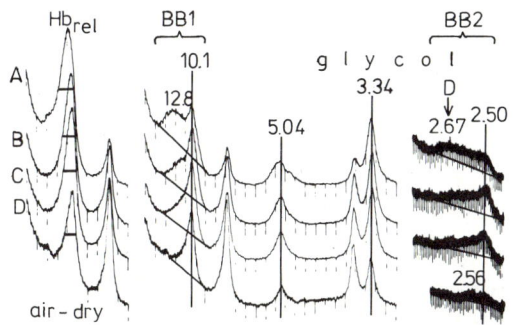

Fig. 4. X-ray traces of < 0.2 μm fractions of samples from profile II. The location is shown in Fig. 2.

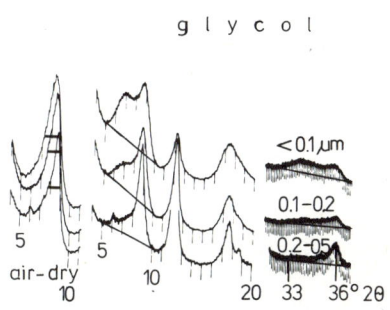

Fig. 5. X-ray traces of different size fractions.

The Kubler index of "illite", $Hb_{rel}$, measured according to the definition of Weber (1972), decreases in the coarser fractions (Fig. 5), proving that it is sensitive to the relative amount of I/S (in the fractions $>0.5$ μm, $Hb_{rel}$ is strongly influenced by the presence of 2M muscovite). On the other hand, $Hb_{rel}$ decreases systematically down the profiles (Fig. 2), indicating an advanced diagenetic stage, but still far from the beginning of anchimetamorphism (Weber, 1972). We can presume that the Kubler index is sensitive not only to the quantity but also to the nature of I/S component in a sample. The question is which parameter changes down the profiles.

After a close inspection of the computer-calculated glycol patterns of I/S given by Reynolds and Hower, three features were selected which should allow for identification of highly illitic I/S in a mixture with illite (Fig. 4):

BB1 - the joint breadth of the 10 Å and any low angle reflection, if present, measured in degrees 2θ $CuK_\alpha$ at the background line;

BB2 - the joint breadth of the I/S and illite reflections in the range $33 - 36°$ 2θ, at the background line;

D - the position of the mixed reflection between $33$ and $35°$ 2θ.

All three values are uninfluenced by the presence of discrete illite. In addition the low angle mixed reflection usually remains distinguishable down to about 20% smectite.

Reynolds and Hower described two types of ordering of illite/smectites: IS and ISII, the latter known only from >85% illite minerals. For both types, BB1 and BB2 decrease slowly with increasing illite percent, but the values are different: $>4.0$ for the IS structure and $<3.5$ for the ISII structure. Numerous additional computer simulations performed by the present author, with the use of the Reynolds and Hower program, convinced him that BB values are independent of any parameter other than the illite:smectite ratio and the type of ordering. The mixed reflection D of the IS structure moves to $34°$ for the 85% illite mineral and disappears if the illite conte increases. The distinct reflection between $34$ and $35.5°$ proves that ISII structure is present.

BB1 and BB2 are almost the same for different fractions (Fig. 5), proving that I/S does not change but only its quantity decreases in the coarser fractions. In the profiles, these values and D do change (Fig. 2,4), showing that the evolution of I/S is the major measurable diagenetic feature.

Diagenetic evolution of illite/smectites

Shales. Using two reflections, 12.8 and 2.67 Å, I/S from the shallowest sample from profile II (Fig. 4) can be identified as IS ordered, 22-23% smectite. It is the most smectitic sample in the profiles studied, and this is probably true for shales of the whole Basin. Several scattered samples from very shallow depths and the youngest coal-bearing strata were studied, and no minerals beyond 25% smectite have been found, though I/S was present in every sample.

Down the profile, the low angle reflection gradually disappears, leaving a low angle tail to the 10 Å peak. The 2.67 Å peak also becomes poorly defined, moves toward higher angles and finally there appears a distinct reflection at about 35°, with little further change to the bottom of the section. These observations and the BB values suggest a change from IS to ISII type of ordering at about 85% illite, taking place at $R_o$ = 0.85-0.90% (Fig. 2). In the profile II, the evolution of the IS structure in the range 25-15% smectite occurs within less than 500 m. Below, the ISII structure evolves slowly, not farther than to about 10% smectite composition.

According to the computer simulation studies (to be published separately), the anomalously high intensity and low d value of the D reflection of the ISII structure, as compared with the data of Reynolds and Hower, can be explained by assuming a 16.6 Å thickness for the smectite-glycol complex.

Pyroclastics. In a previous paper (Środoń, 1976), several ash-derived, i.e. bentonite and tonstein, horizons from the Upper Silesian Basin were described in detail. Five other scattered occurrences were studied later. It is possible to locate these materials in the general diagenetic profile, using the available data (V.M.) for the neighbouring coal seams. One more bentonite layer was found near the bottom of profile I. Fig. 6 shows that I/S in pyroclastics are always more smectitic than in surrounding shales, at least by 10%. The ample data from the southern part of the Basin, presented by Dopita and Králík (1977), support this conclusion. In the coalification range V.M. = 10-35% they recorded 22-42% smectite in I/S from tuffaceous horizons.

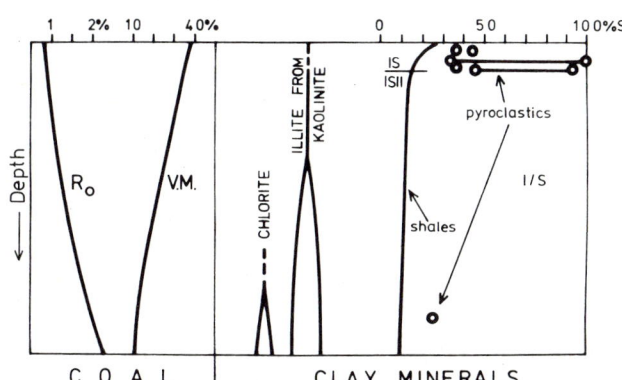

Fig. 6. The general scheme of diagenetic phenomena in the Upper Silesian Basin. The joined circles show the range of expandability within thick tuffaceous beds.

CONCLUSIONS

1. There is a decrease of about 10% in the illite content in I/S from pyroclastic as compared with normal shales, both in thin beds and in the outer parts of thick be The illite content decreases more, even down to pure smectite (Fig. 6), toward the centres of thick beds (Parachoniak and Środoń, 1974; Środoń, 1976). This kind of variation within a bed is a general phenomenon, and has been recognized in the Cretaceous of Montana (J.Hower, personal communication, 1976) and in the Ordovician of Sweden (A.M.Brusevitz, personal communication, 1977). Hower explains it by the deficiency of $K^+$ within bentonite beds. This explanation may be valid also for the basin studied here because:
a) the $K^+$ necessary to form I/S in bentonites was supplied from outside (Środoń, 197
b) the mixed layers are unstable with respect to excess $K^+$ (op.cit.);
c) the present paper shows that the illitization of smectite is temperature dependent, so that the low-temperature fresh water processes, proposed in the previous paper, are probably not essential for the degree of illitization.

The above hypothesis implies that, in nature, the process supplying $K^+$ is an independent variable, controlling, along with temperature, the rate of transformatio of smectite toward illite. On the other hand, when starting from an expanding clay and kaolinite, diagenesis produces a mixture of I/S and illite (Fig. 3), and obvious the product of illitization depends also on the nature of starting material. It is unknown, to what extent the difference between natural smectites can influence the product of illitization. This triple control (temperature, $K^+$ supply, starting material) ought to be recognized in more detail before the illite/smectite ratio of I/S can be used for the estimation of maximum burial temperatures.

2. The diagenesis of expanding clays does not reflect the overthrust, whereas coal diagenesis does (Fig. 2). It is known from independent evidence that the over thrust is synsedimentary (W. Bogacz, personal communication, 1978). The conclusion is that clay diagenesis proceeds more slowly than the coalification process, and thi should be kept in mind when comparing the indexes of diagenesis in tectonized areas. In the range of diagenesis studied, the best correlation point between coal and shal diagenesis is the change from IS to ISII structure at about $R_o$ = 0.85-0.90%. $R_o$ curves testify that:
a) The paleothermal gradient was relatively stable along a profile. It leads to t conclusion that the increase of illite content of I/S with temperature is much faste for IS than ISII structure.
b) The paleothermal gradient varied significantly between the two profiles, though they are situated only a few kilometers from each other. The lithology, and, in consequence, the thermal conductivity of the rocks are very similar, so there must h been big variations in the heat flow. This suggests that, like in many other coal basins (Damberger, 1974), diagenesis is not only a function of burial but also of telemagmatic activity. If so, it is impossible to estimate the duration of the

maximum heating period from tectonic data, and there is no ground for determination of the maximum diagenetic temperature from the degree of coalification (Epstein et. al., 1977).

3. The BB and D parameters measured in this study correspond very well with the values computed by Reynolds and Hower (1970), thus corroborating their hypothesis of the IS and ISII types of ordering of I/S. It remains unknown how the experimental data correspond with alternative structural models suggested by Tettenhorst and Grim (1975), because these authors have calculated only $G^2$, instead of complete diffraction patterns, and they have shown only the range 2-30° 2θ.

4. The Kubler index is an inconvenient measure of the degree of diagenesis as long as an expandable material is present, being sensitive both to the quantity and the nature of the I/S component. Three new indices are proposed which reflect only the latter parameter. The application of the Kubler index is well justified at higher temperature stages, i.e. for non-expandable 10 Å minerals, for which the physical meaning of this index is more restricted: it is then a measure of crystallite thickness and perfection.

ACKNOWLEDGEMENTS

The author is greatly indebted to A. Kotas and Z. Bula from the Upper Silesian Branch of the Geological Survey, who kindly supplied the sample material together with the coal diagenesis data. I thank W.D. Johns and B. Velde for critical reading of the manuscript, and V.C. Farmer for language corrections.

REFERENCES

Damberger, H.H., 1974. Coalification patterns of Pennsylvanian coal basins of the eastern United States. In: R.R. Dutcher et al. (Editors), Carbonaceous materials as indicators of metamorphism. Geol. Soc. Am. Spec. Paper, 153: 53-74.
Dopita, M. and Kralik, J., 1977. Uhelne Tonsteiny Ostravsko-Karvinskeho Reviru. Ostrava, 213 pp.
Drits, V.A. and Koporulin, V.I., 1973. K postsedimentacijonnoj transformacji montmorillonita v gidrosljudu. Litol. Polezn. Iskop., 5: 145-148 (in Russian).
Dunoyer de Segonzac, G., 1970. The transformation of clay minerals during diagenesis and low-grade metamorphism: a review. Sedimentology, 15: 281-346.
Epstein, A.G., Epstein, J.B., and Harris, L.D., 1977. Conodont color alteration - an index to organic metamorphism. Geol. Surv. Prof. Paper 995, 27 pp.
Hower, J., Eslinger, E.V., Hower, M.E., and Perry, E.A., 1976. The mechanism of burial metamorphism of argillaceous sediment: 1. Mineralogical and chemical evidence. Geol. Soc. Am. Bull., 87: 725-737.
Klug, H.P. and Alexander, L.E., 1974. X-Ray Diffraction Procedures for Polycrystalline and Amorphous Materials. Wiley, New York, 966 pp.
Kubler, B., 1964. Les Argiles, indicateurs de metamorphisme. Rev. Inst. Fr. Pet., 19: 1093-1112.
Maxwell, D.T. and Hower, J. 1967. High-grade diagenesis and low-grade metamorphism of illite in the Precambrian belt series. Am. Mineral. 52: 843-857.
Parachoniak, W. and Srodon, J., 1974. The formation of kaolinite, montmorillonite and mixed-layer montmorillonite-illites during the alteration of Carboniferous tuff (the Upper Silesian Coal Basin). Mineral. Pol., 4: 37-56.
Perry, E.A. and Hower, J., 1970. Burial diagenesis in Gulf Coast pelitic sediments. Clays Clay Miner., 18: 165-177.

Reynolds, R.C., 1968. The effect of particle size on apparent lattice spacings. Acta Crystallogr., A24: 319-320.
Reynolds, R.C. and Hower, J., 1970. The nature of interlayering in mixed-layer ill montmorillonites. Clays Clay Miner., 18: 25-36.
Srodon, J., 1972. Mineralogy of coal-tonstein and K-bentonite from coal seam no. 6. Bytom trough (Upper Silesian Coal Basin, Poland). Bull. Acad. Pol. Sci., Ser. Sci. Terre, 20: 155-164.
Srodon, J., 1976. Mixed-layer smectite/illites in the bentonites and tonsteins of the upper Silesian Coal Basin. Prace Mineral., 49, 84 pp.
Tettenhorst, R. and Grim, R.E., 1975. Interstratified clays. Am. Mineral., 60: 49-65.
Weaver, Ch.E., 1960. Possible uses of clay minerals in search for oil. Bull. Am. Assoc. Pet. Geol., 44: 1505-1518.
Weber, K., 1972. Kristallinitat des Illits in Tonschiefern und andere Kritereen Schwacher Metamorphose in nordostlichen Rheinischen Schiefergebirge. Neues Jahrb. Geol. Palaontol. Abh., 141: 333-363.
Wolf, M., 1975. Uber die Beziehungen zwischen Illit-Kristallinitat und Inkohlung. Neues Jahrb. Geol. Palaontol. Monatsh., 7: 437-447.

MINERALOGICAL AND GEOCHEMICAL TRANSFORMATION OF CLAYS DURING BURIAL-DIAGENESIS
(CATAGENESIS): RELATION TO OIL GENERATION

A.E. Foscolos and T.G. Powell
Institute of Sedimentary and Petroleum Geology, Geological Survey of Canada,
Calgary, Alberta

ABSTRACT

The thermal diagenesis (catagenesis) of clays has been investigated in samples from the Panarctic et al. North Sabine H-49 well in the Sverdrup Basin in the Canadian Northwest Territories.

Upon burial of the sediments, the concentration of expandable 2:1 layer silicate, kaolinite and amorphous inorganic material decreases whereas illite increases in concentration in the clay fractions. The first dehydration of the interstratified clays is made permanent by isomorphic substitution of $Si^{4+}$ by $Al^{3+}$ and the ensuing absorption of $K^+$ and coincides with the onset of hydrocarbon generation from the sedimentary organic matter. The clay size fraction also decreases with depth due to destruction of the hydrous clay minerals. Both the destruction of inorganic amorphous material and hydrous clay minerals provides water to the pore system but only the destruction of the latter occurs within the hydrocarbon generation zone. The second dehydration step of the interstratified clays occurs below the oil generation zone.

INTRODUCTION

During burial of sedimentary rocks, both clay minerals and organic matter undergo diagenetic reactions (Foscolos et al., 1976; Powell et al., in press). In order to study the relationships between changes in clay minerals and changes in organic matter, a detailed study of samples from six wells in the Sverdrup and Beaufort-Mackenzie Basins in the Northwest Territories of Canada has been undertaken. The results presented herein are concerned with the qualitative and quantitative changes in the crystalline and amorphous material in the 0.2-2.0 μm and <0.2 μm clay fractions and their relation to organic diagenesis in one of the wells from this study. The well studied was Panarctic et al. North Sabine H-49 (Lat. 76°48'15.1"N, Long. 108°45'11.2"W) which was drilled in the Sverdrup Basin in the Canadian Arctic Islands. The well penetrated a sequence ranging from Late Cretaceous to Late

Triassic in age. The procedures followed for the clay mineralogy and inorganic geochemical studies are identical to those used previously (Foscolos et al., 1976).

The clay mineralogy of two size fractions was examined. The clay fraction below 0.2 μm in size consists of mixed layer clays, amorphous inorganic material and kaolinite. The 2 to 0.2 μm fraction consisted of quartz, feldspars, kaolinite, chlorite; illite, 2:1 expandable layer clays and interstratified clays.

## CLAY MINERALOGY

### Expansible and interstratified clays

X-ray analysis shows that the expandable 2:1 layer silicates in the 2 to 0.2 μm fraction decrease in abundance with depth. A sample from 4900 feet is typical of the unaltered samples (Fig. 1A). Smectite is recognized from the continuous shift of the $d_{001}$ spacing from 10Å with sodium saturation and 0 per cent relative humidity (R.H. to 18.0Å with sodium saturation and glycerolation. Upon saturation with potassium, only a limited expansion of the lattice was observed at increasing levels of humidit and glycerolation indicating that smectite is of the low-swelling variety. The peak at 10Å in the diffractograms obtained at 50% R.H., 80% R.H. and under glycerolation indicates that illite is present as a discrete mineral.

A sample from 9550 feet is typical of the more deeply buried samples in which interstratified clays can be readily recognized (Fig. 1B). The first order basal reflection shifts from 10.2Å at 0% R.H. to 11.6Å at 50% R.H. At 80% R.H. the peak splits into two at 14Å and 10Å, respectively, indicating the presence of mica and expansible clays as a mixed layer. Upon glycerolation the 14Å peak separates into a broad peak at 14Å and a minor peak at 18Å. The continuous peak movement with increase in relative humidity and glycerolation indicates the presence of smectite. Upon heating at 550°C, all layers collapse to 10Å indicating the absence of a chlorite-like compound in the mixed layers. Upon $K^+$ saturation and glycerolation, only limited expansion of the basal spacing is observed; this indicates that in the interstratified clays there is an expansible component other than smectite which inhibits expansion upon $K^+$ saturation. This component is considered to be a vermiculite-like layer silicate. Thus the mixed layer comprises a ternary system of mica-vermiculite-smectite.

Previous results (Powell et al., in press) showed that all the <0.2 μm fractions, whether from shallow or greater depths, contained a ternary system of mixed layer clays when the amorphous inorganic material had been removed. Fourier analysis of the X-ray diffractograms showed a systematic change in the composition of <0.2 μm clay fraction (Table 1). The proportion of smectite in the mixed layers decreases from around 30 per cent above 4900 feet to less than 20 per cent below 6000 feet depth.

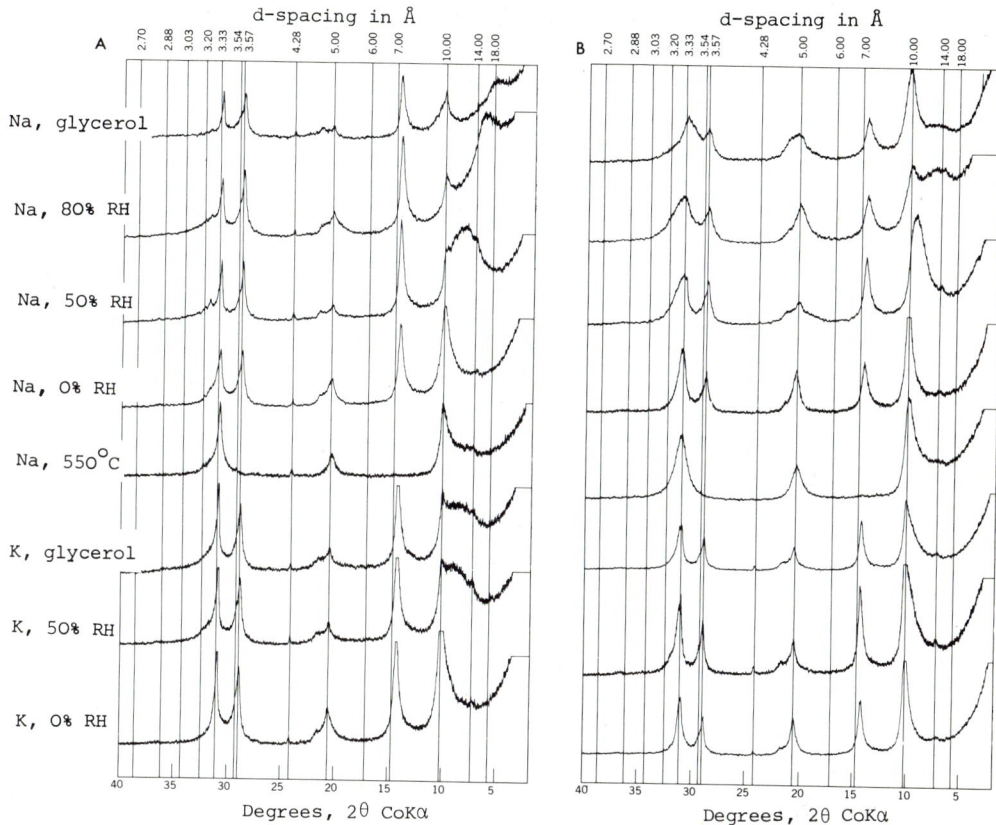

Figure 1. X-ray diffraction patterns of the oriented 2 to 0.2 μm fraction from the North Sabine well. A. 4900 feet depth B. 9550 feet depth.

TABLE 1

Variation in composition of interstratified clays in the <0.2 μm fraction with depth in the North Sabine well.

| Depth in feet | Illite | Vermiculite percent | Smectite |
|---|---|---|---|
| 1,400 | 39 | 28 | 33 |
| 2,400 | 42 | 25 | 33 |
| 3,700 | 45 | 27 | 28 |
| 4,900 | 46 | 23 | 31 |
| 6,000 | 52 | 24 | 24 |
| 7,000 | 64 | 21 | 15 |
| 8,000 | 65 | 22 | 13 |
| 9,050 | 63 | 20 | 17 |
| 10,150 | 57 | 23 | 20 |
| 11,000 | 60 | 23 | 17 |

The change in composition of the clays reflects the first dehydration of the mixed layer which occurs in the vicinity of 6000 feet. This dehydration process is indicated in the diffractograms by a shift in the $d_{001}$ spacing of the Ca saturated clays from over 14Å above 6000 feet to less than 12Å below 6000 feet (Fig. 2A). Similarly in the 2 μm to 0.2 μm fraction, the first dehydration of smectite as observed from the peak migration of the $d_{001}$ spacing was encountered between 4900 and 6000 feet (Fig. 2B).

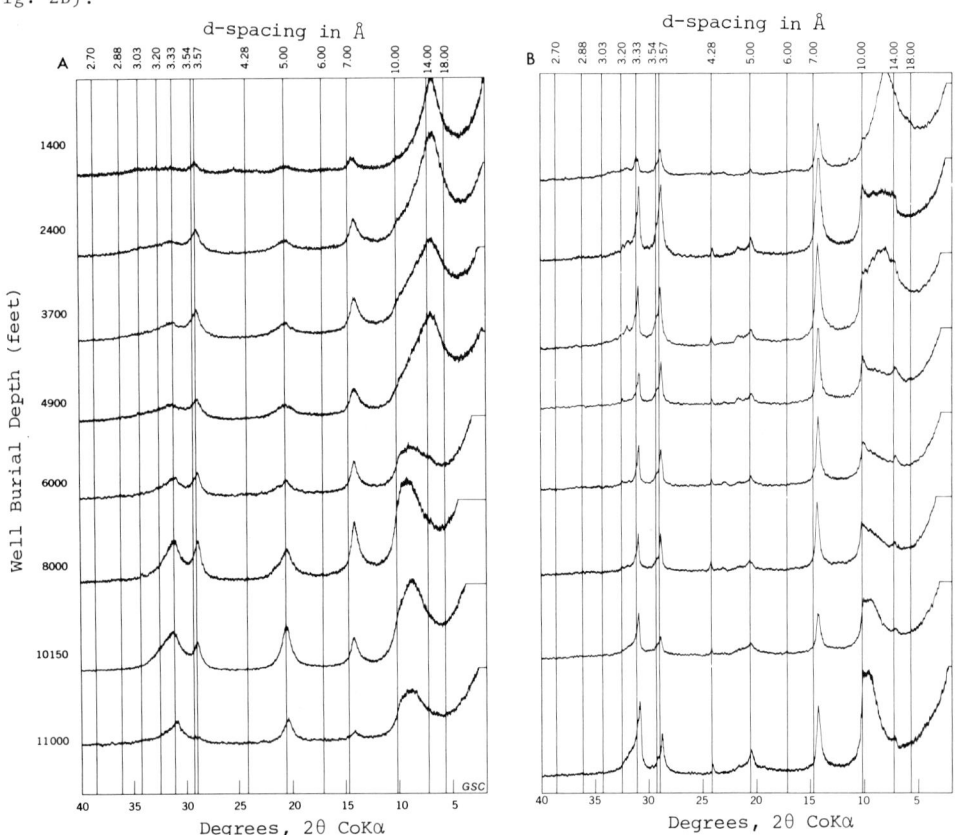

Figure 2.  X-ray diffraction patterns of oriented clay specimens at 50% R.H. from the North Sabine well. A. Ca saturated <0.2 μm fraction. B. Na saturated 0.2 to 2 μm fraction.

It seems that vermiculite is an intermediary step in the conversion of smectite to illite. Thus in the <0.2 μm fraction there is a progressive decrease in the proportion of smectite in the mixed layer and a corresponding increase in illite but the vermiculite content remains relatively constant. In the 0.2 μm to 2 μm fraction smectite is transformed to a mixed layer clay in which vermiculite is a significant component. The presence of vermiculite accounts for the absence of a second clay dehydration step even at depths of 17 000 feet in this area (Powell *et al.*, in press). Work by Kittrick (1969) indicates that $Mg^{2+}$ or $Ca^{2+}$ vermiculites undergo the second dehydration around 230°C. Since all the samples in this study contained varying amounts of calcite and dolomite, it follows that the second clay dehydration step has been retarded because of the presence of a $Ca^{2+}$ vermiculite component.

## Kaolinite and Chlorite

The presence of kaolinite was indicated by the occurrence of a peak at 7.2Å in the diffractograms of the <0.2 μm and 0.2 to 2.0 μm fractions after they had been treated with warm dilute HCl to remove chlorite. This peak decreases below 8000 feet in the North Sabine well indicating removal of kaolinite during catagenesis.

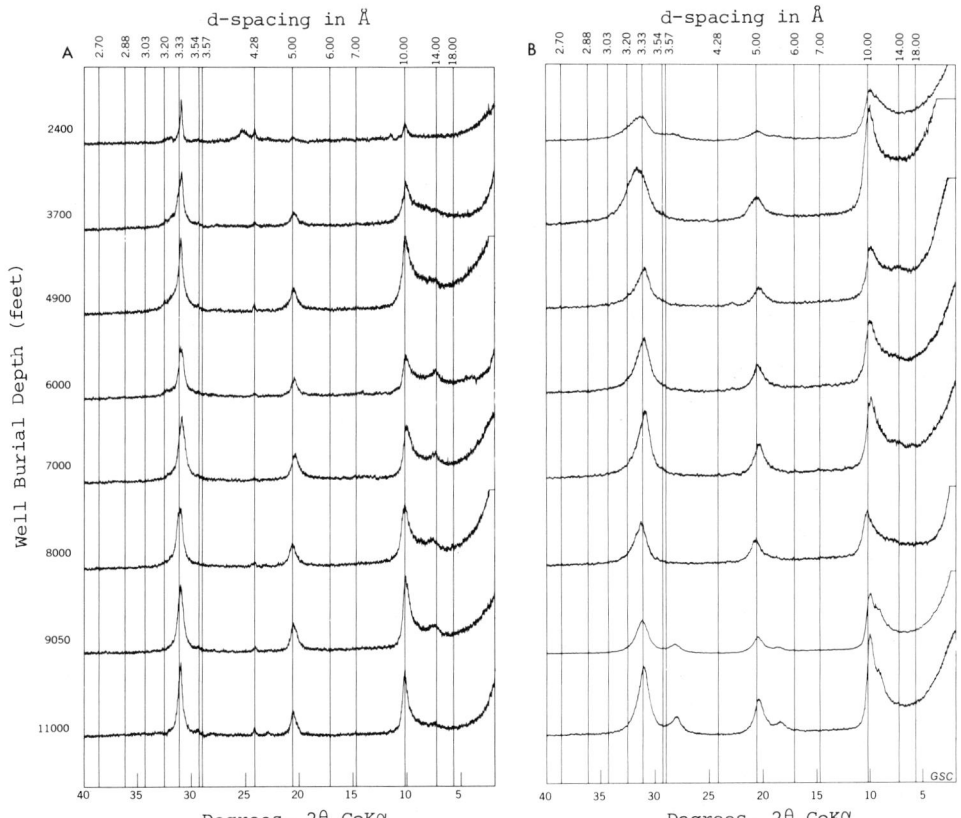

Figure 3. X-ray diffraction patterns of oriented clay specimens from the North Sabine well after heating at 550°C. A. Na saturated 0.2 to 2.0 μm fraction. B. Ca saturated <0.2 μm fraction.

Chlorite was identified from the 14Å peak in the 0.2 to 2 μm samples after they had been heated at 550°C for two hours. Samples from shallow depths (above 4900 ft. do not contain discrete chlorite. As burial depth increases, chlorite intergrades appear to be converted into discrete chlorites (Fig. 3A). No discrete chlorite was identified in the <0.2 μm fraction. However X-ray diffractograms of the whole Ca saturated <0.2 μm fraction (i.e. amorphous material present) indicates that amorphous material may participate in the structure, forming a chlorite-like component in certain samples (Fig. 3B). The presence of a chlorite-like component is indicated b $d_{001}$, $d_{002}$ and $d_{003}$ spacings at 11.1Å, 5.56Å and 3.65Å, respectively, in Ca saturate samples which had been heated to 500°C (Fig. 3B samples 10150 and 11000). These peaks are absent from the patterns obtained after the amorphous material had been removed. It seems therefore that a chlorite-like mineral is formed during burial by adsorbing amorphous $Fe_2O_3$ and/or $Al_2O_3$ in the expansible 2:1 layer silicates to form a pseudoquaternary system of interstratified clays consisting of smectite-vermiculit chlorite-illite.

## INORGANIC GEOCHEMISTRY OF CLAY FRACTIONS

Chemical analysis of the <0.2 μm and 2 to 0.2 μm fractions was carried out in order to determine changes in major elements, and to measure changes in amorphous matter content.

TABLE 2

Normalized elemental analysis of mixed layers in the <0.2 μm fraction after correcti for kaolinite and amorphous inorganic material in the North Sabine well.

| Depth in feet | Elemental composition per cent | | | | | | | | | |
|---|---|---|---|---|---|---|---|---|---|---|
| | $SiO_2$ | $Al_2O_3$ | $Fe_2O_3$ | CaO | MgO | $Na_2O$ | $K_2O$ | $P_2O_5$ +TiO | Loss on Ignition | $H_2$ |
| 2,400 | 46.46 | 8.19 | 4.04 | 3.14 | 3.08 | 1.18 | 2.23 | 4.26 | 9.18 | 18. |
| 3,700 | 47.52 | 9.95 | 8.56 | 2.86 | 3.19 | 0.49 | 2.56 | 2.86 | 7.54 | 14. |
| 4,900 | 46.70 | 13.30 | 8.10 | 2.77 | 4.24 | 0.27 | 4.02 | 1.41 | 7.92 | 11. |
| 6,000 | 38.46 | 22.18 | 10.09 | 1.91 | 3.61 | 0.38 | 5.42 | 1.56 | 8.03 | 8. |
| 7,000 | 47.03 | 22.75 | 5.15 | 1.33 | 2.97 | 0.24 | 4.86 | 1.06 | 8.76 | 5. |
| 8,000 | 45.67 | 19.34 | 5.85 | 1.43 | 3.25 | 0.32 | 5.30 | 1.95 | 9.11 | 7. |
| 9,050 | 46.74 | 22.27 | 4.13 | 1.47 | 2.51 | 0.25 | 5.91 | 1.77 | 8.11 | 6. |
| 10,150 | 47.47 | 19.98 | 4.04 | 1.39 | 2.72 | 0.29 | 6.21 | 2.50 | 8.70 | 6. |
| 11,000 | 46.93 | 23.57 | 2.64 | 1.40 | 2.35 | 0.36 | 6.43 | 1.20 | 8.57 | 6. |

The normalized chemical analysis of <0.2 μm clays (Table 2) was obtained by subtracting the contribution made by amorphous inorganic material and kaolinite from the elemental analysis of the whole fraction. The analysis indicates that aluminum and potassium increases with burial depth whereas adsorbed water decreases. The CaO content (an index of cation exchange capacity in the Ca saturated <0.2 μm fraction also decreases with burial depth. These changes reflect the transformation of expansible 2:1 layer silicates to non-expansible 2:1 layer silicates with depth. Aluminum is substituted for silicon in the clay lattice; $K^+$ is adsorbed on the surface while cavity and adsorbed $H_2O$ decreases and the cation exchange capacity decreases.

In the same manner, the elemental analysis of the 2 to 0.2 μm fraction (Table 3) of the North Sabine well shows that $Al_2O_3$ and $K_2O$ contents increase while the $H_2O$ content decreases with burial depth as the expansible 2:1 layer silicates transform to the non-expansible 2:1 layer silicates. No other systematic variations were observed due to variations in the amounts of quartz and feldspars in these fractions.

Analysis of the <0.2 μm and 0.2 to 2 μm fractions shows that amorphous $SiO_2$ and $Fe_2O_3$ decline with burial depth whereas $Al_2O_3$ remains relatively constant. The decrease in these components occurs prior to the first clay dehydration step noted above (Table 4).

TABLE 3

Elemental analysis of the 2 to 0.2 μm fractions in the North Sabine well

| Depth in feet | Elemental composition - per cent ||||||||| Loss on Ignition | Total |
|---|---|---|---|---|---|---|---|---|---|---|---|
| | $SiO_2$ | $Al_2O_3$ | $Fe_2O_3$ | $TiO_2$ | $CaO$ | $MgO$ | $K_2O$ | $Na_2O$ | $H_2O^-$ | | |
| 2,400 | 49.55 | 13.70 | 15.86 | 1.06 | 0.60 | 0.10 | 0.99 | 0.41 | 0.37 | 16.00 | 99.14 |
| 3,200 | 55.00 | 22.42 | 8.26 | 1.09 | 0.67 | 0.21 | 2.31 | 0.43 | - | - | 99.76 |
| 3,700 | 52.48 | 23.23 | 8.72 | 1.32 | 0.67 | 0.24 | 2.40 | 0.26 | 2.50 | 8.63 | 99.76 |
| 4,900 | 53.29 | 23.74 | 7.84 | 1.37 | 0.66 | 0.24 | 2.41 | 0.42 | 1.63 | 8.81 | 100.41 |
| 6,000 | 51.34 | 24.60 | 8.20 | 1.31 | 0.42 | 0.27 | 3.23 | 0.71 | 1.63 | 9.19 | 100.90 |
| 7,000 | 52.58 | 25.05 | 7.70 | 1.17 | 0.41 | 0.23 | 3.19 | 0.65 | 0.97 | 9.00 | 100.95 |
| 8,000 | 53.09 | 25.46 | 7.12 | 1.10 | 0.42 | 0.53 | 3.29 | 0.47 | 1.35 | 8.88 | 101.71 |
| 9,050 | 52.45 | 25.20 | 6.73 | 1.12 | 0.36 | 0.58 | 3.54 | 0.46 | 1.60 | 7.94 | 99.98 |
| 10,150 | 54.54 | 24.38 | 4.97 | 1.34 | 0.39 | 0.60 | 3.68 | 0.24 | 1.47 | 7.38 | 99.15 |
| 11,000 | 55.95 | 24.34 | 4.84 | 1.35 | 0.36 | 0.50 | 3.70 | 0.66 | 1.45 | 6.44 | 99.59 |

VARIATIONS IN CLAY CONTENT WITH DEPTH

Mechanical analyses of the samples shows that the <2.0 μm size fraction decreases with burial depth (Table 4). This was noted in other wells (unpublished results) and appears not to be dependent on age or formation. As a result the proportion of expandable clays and kaolinite in the clay size fraction is decreased when considered as a proportion of the total rock. In addition, although the relative concentration of illite in the 2 to 0.2 μm fraction increases with depth, the absolute concentration of illite in the rock decreases with depth. Only extremely small amounts of clay were observed in the >2.0 μm fraction. This suggests that the loss of clays from the rock occurs during burial whereas Hower et al. (1976) suggested in a study of a well from the U.S. Gulf Coast that clay was lost from the small size fractions

TABLE 4

Variations in clay size fractions and amorphous inorganic matter with depth in the North Sabine well.

| Depth in feet | Size Fractions <2.0 μm % of rock | % of <0.2 μm | | | | Amorphous Inorganic Matter % of 0.2-2.0 μm | | | |
|---|---|---|---|---|---|---|---|---|---|
| | | $SiO_2$ | $Al_2O_3$ | $Fe_2O_3$ | $H_2O$ | $SiO_2$ | $Al_2O_3$ | $Fe_2O_3$ | $H_2O$ |
| 1,400  |      | 6.79  | 1.30 | 3.22  | 1.63 | 18.60 | 1.25 | 15.02 | 1.81 |
| 2,400  | 50.0 | 12.24 | 1.40 | 12.55 | 1.52 | 4.80  | 1.47 | 3.00  | -    |
| 3,700  | 76.0 | 4.55  | 1.72 | 4.44  | 1.40 | 3.80  | 1.47 | 3.58  | -    |
| 4,900  | 58.0 | 2.82  | 1.33 | 2.95  | 1.34 | 2.80  | 1.21 | 2.07  | -    |
| 6,000  | 35.5 | 1.68  | 1.02 | 3.35  | 1.12 | 2.00  | 0.83 | 3.22  | -    |
| 7,000  | 26.5 | 2.47  | 1.50 | 3.69  | 0.81 | 2.00  | 0.98 | 3.93  | -    |
| 8,000  | 19.5 | 2.32  | 1.38 | 2.90  | 0.75 | 2.20  | 1.13 | 3.58  | -    |
| 9,050  | 19.0 | 2.33  | 1.43 | 2.21  | 0.87 | 1.90  | 0.76 | 2.00  | -    |
| 10,150 | 18.5 | 2.96  | 1.55 | 2.03  | 0.43 | 2.60  | 0.76 | 1.43  | -    |
| 11,000 | 12.0 | 3.06  | 1.40 | 1.17  | 0.93 | 2.60  | 0.87 | 0.86  | -    |

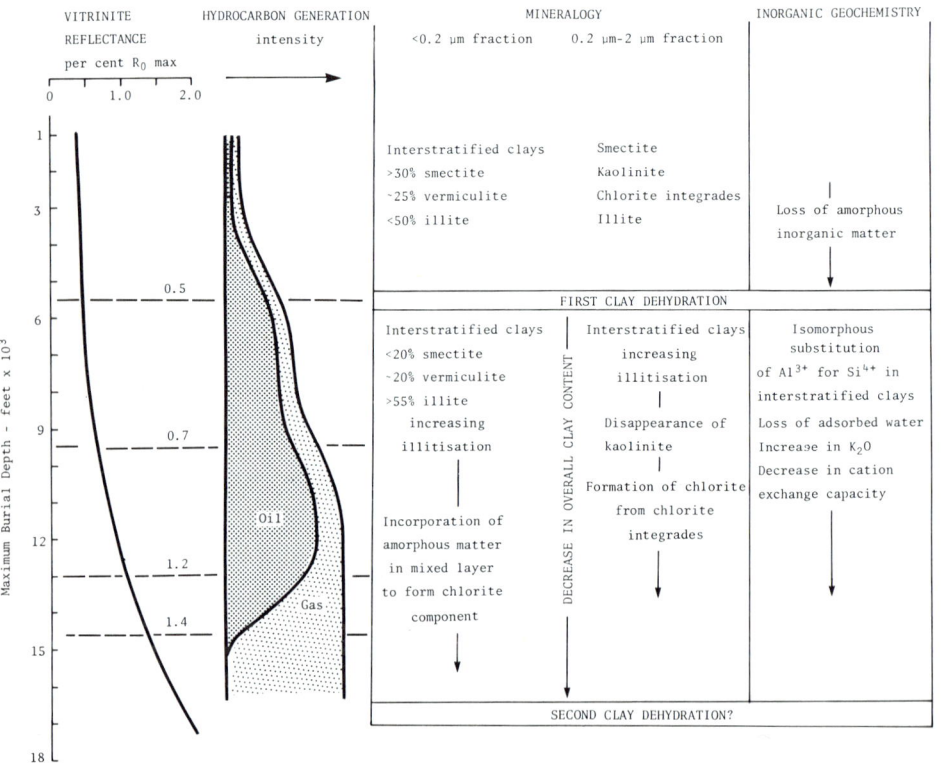

Figure 4. Relation between clay diagenesis and organic diagenesis in the vicinity of the North Sabine well. Western Sverdrup Basin.

to the large-size fractions by cementation of fine clay particles.

## DISCUSSION AND CONCLUSIONS

Powell et al. (in press) have shown that organic matter in the Sverdrup Basin began to generate petroleum at a burial depth of approximately 5000 feet. Diagenetic processes in respect of the organic matter can be conveniently measured by vitrinite reflectance. A vitrinite reflectance value of 0.5% $R_0$ (max) coincides with the onset of oil generation. The main phase of oil generation commences at a vitrinite reflectance value of 0.7% $R_0$ (max) whereas cracking of liquid hydrocarbons to gas commences at a reflectance level of 1.2% $R_0$ (max). The relationship between clay diagenesis and organic diagenesis for the North Sabine H-49 well are illustrated in Figure 4. Amorphous inorganic material loses water and decreases in amount some 2000 feet above the first dehydration step of montmorillonite. The first clay dehydration step coincides with 0.5% $R_0$ vitrinite reflectance and occurs several thousand feet above the main phase of oil generation while the second dehydration step takes place below the oil generating phase (Foscolos et al., 1976; Powell et al., in press) since vermiculite is a major component in the mixed layers. These results suggest that the depths of clay dehydration are not coincident with the depths of oil generation and that water derived by this process does not have a role to play in oil migration. However it is noted that the concentration of clay minerals in the rock diminishes as a function of depth below the first dehydration step. These latter processes occur within the oil window and hence the release of crystal lattice water due to clay destruction may assume an important role in petroleum migration.

The results of this work show that water is released not only by compaction of the sediments upon burial but also from the amorphous inorganic material, by clay dehydration as expandable 2:1 layer silicates are converted to non-expandable 2:1 layer silicates and by destruction of the hydrous layer silicates. Besides water being released from the clay destruction there are other materials released into the pore system such as $SiO_2$, $Al_2O_3$, $Fe_2O_3$, $CaO$, $MgO$ and $Na_2O$. Thus the clays supply cementing agents that might stay within the finegrained rock or migrate to reservoirs prior to or concommitantly with oil migration. Thus the mineralogical trends within the clay fraction of the North Sabine well are interpreted as due to internal readjustment under the prevailing physicochemical conditions. Hower et al. (1976) has arrived at essentially similar conclusions.

## ACKNOWLEDGMENTS

We are indebted to A. Heinrich, G. Jamro and M. Northcott for technical assistance.

REFERENCES

Foscolos, A.E., Powell, T.G. and Gunther, P.R. 1976. The use of clay minerals and inorganic and organic indicators for evaluating the degree of diagenesis and oil generating potential of shales. Geochim. Cosmochim. Acta, 40: 953-966.

Hower, J., Eslinger, E.V., Hower, M.E. and Perry, E.A. 1976. Mechanism of burial metamorphism of argillaceous sediment: I. Mineralogical and chemical evidence. Bull. Geol. Soc. Amer., 87: 725-737.

Kittrick, J.A. 1969. Quantitative evaluation of the strong force model for expansion and contraction of vermiculite. Soil Sci. Soc. Amer. Proc., 33: 223-225.

Powell, T.G., Foscolos, A.E., Gunther, P.R. and Snowdon, L.R. in press. Diagenesis of organic matter and fine clay minerals: A comparative study. Geochim. Cosmochim. Acta.

CLAY MINERALS AS INDICATORS OF THE CENOZOIC EVOLUTION OF THE NORTH ATLANTIC OCEAN.

C. LATOUCHE

Institut de Géologie du Bassin d'Aquitaine. Université de Bordeaux I (FRANCE)

ABSTRACT

Clay minerals of ancient marine deposits are sensitive indicators for the reconstruction of the history of oceans and of the main stages of their paleogeographical and paleohydrological evolution. This paper presents an example illustrating the accuracy of such tracers for this purpose. It concerns mineralogical studies of 400 samples of Cenozoic deposits collected during two legs of the INTERNATIONAL PROJECT OF DEEP SEA DRILLING (IPOD) in the North-East Atlantic (leg 47B - W. of the Iberian Peninsula ; leg 48 - Rockall region W. of Great Britain).

The mineralogical composition of the deposits makes it possible to trace, for different periods, the origin and dispersion paths of detrital materials, climatic conditions of weathering in adjacent continental areas and environmental characteristics of sedimentary basins. An important fact is the generally new aspect of the mineralogy of deposits at the Eocene/Oligocene boundary which clearly denotes the occurence of a major hydrodynamic event. It marks the establishment of the system of thermohaline circulation of the modern Atlantic Ocean, when cold dense waters derived from northern regions began to sink to the bottom of the North Atlantic Basin and bring towards the South clay fractions generated by weathering in cold climatic areas.

---

INTRODUCTION

Within the framework of the Deep Sea Drilling Project - International Phase of Ocean Drilling (DSDP-IPOD), we have carried out a mineralogical study of 400 samples from the Cenozoic deposits of 5 sites prospected during legs 47 B and 48 (sites 398, 403, 404, 405 and 406), West of Portugal and in the Rockall region (Fig. 1). Analyses were carried out by X-ray diffractometry, first on the sample as a whole, then on the portion of less than 2 µm after decarbonation.

Methods of work, results of analyses and the preliminary interpretation of mineralogical assemblages have been collected for each site in the Initial Reports of the Deep Sea Drilling (Chamley et al.., 1978 - Latouche and Maillet, 1978). In this paper we present a synthesis of these results which make it possible to assemble a body of information concerning the paleogeographic and paleohydrologic evolution of the North Atlantic during the Cenozoic.

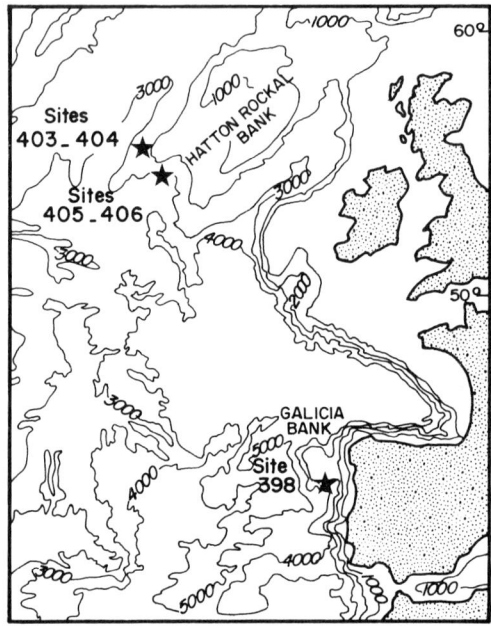

Fig. 1. Location of the I.P.O.D. Sites.

MINERALOGICAL SERIES ENCOUNTERED.

The mineralogical data (table 1) shows 3 successive sedimentary episodes, or "mineralogical units" (Fig. 2), occuring more or less contemporaneously in all the sections studied. The oldest of the episodes presents mineralogical associations which vary from site to site. Conversely, the associations for the two more recent episodes are identical in all 5 sites studied.

The possible origin of these minerals and their significance were considered in the light of the work of several authors, in particular, that relative to marine clay mineral sedimentation (Biscaye, 1965 - Griffin et al., 1968 - Millot, 1964 - Rateev et al., 1969 - Yeroshev-Shak, 1954).

Episode 1.

It includes deposits formed between the end of the Cretaceous and the middle Oligocene and is characterised by the following mineralogical associations, depending on the site :

In sites 403 and 404 at Rockall, the association designated as 1a consists of :

TABLE 1

Mean values of the composition of mineralogical units

| D.S.D.P. Sites | Mineralogical units | Depth (m) | Samples number | < 2 μm fraction ||||||| | Bulk Sediment ||||
|---|---|---|---|---|---|---|---|---|---|---|---|---|---|---|---|
| | | | | Smectites | Illite | Kaolinite | Chlorite | Attapulgite | Sépiolite | Opal C.T. | Zeolite | Quartz | Calcite | Alkali feldspars | Plagioclase feldspars |
| 403 | 3 | 5,3<br>34,7 | 6 | 35 | 37 | 12 | 16 | — | — | — | T | 5 | 54 | T | 2 |
| | 2 | 52,5<br>235 | 19 | 63 | 24 | 6 | 7 | — | — | — | P | T | 88 | — | — |
| | 1 a | 252<br>395 | 23 | 89 | 5 | — | — | — | — | *— | 6 | 4 | 2 | 3 | 10 |
| 404 | 3 | 0,7<br>30,3 | 8 | 29 | 41 | 15 | 15 | — | — | — | — | 6 | 44 | 4 | 4 |
| | 2 | 104,7<br>182,9 | 4 | 64 | 20 | 9 | 5 | — | — | — | 2 | 0 | 86 | — | — |
| | 1 a | 200,7<br>379,6 | 16 | 70 | 2 | — | — | — | — | *— | 28 | 3 | 8 | 5 | 10 |
| 405 | 3 | 2,4<br>46,6 | 10 | 34 | 38 | 14 | 14 | — | — | — | — | 8 | 44 | 3 | 5 |
| | 2 | 55,6 | 1 | 40 | 40 | 10 | 10 | — | — | — | — | — | 96 | — | — |
| | 1 b | 67,4<br>405,1 | 40 | 62 | — | — | — | — | — | *35 | 3 | T | 23 | T | T |
| 406 | 3 | 3,6<br>215,1 | 6 | 33 | 35 | 13 | 16 | — | — | 3 | — | 6 | 47 | 1 | 6 |
| | 2 | 216,5<br>613,1 | 25 | 73 | 16 | 5 | 4 | — | — | 2 | — | T | 73 | — | — |
| | 1 b | 615,5<br>793,6 | 35 | 91 | 1 | — | — | — | — | *7 | T | T | 64 | — | — |
| 398 | 3 | 9,1<br>104 | 10 | 30 | 45 | 14 | 11 | — | — | — | — | 4 | 27 | — | — |
| | 2 | 105,6<br>565,8 | 87 | 40 | 40 | 14 | 6 | — | — | — | — | 4 | 60 | — | T |
| | 1 c | 594<br>792 | 118 | 46 | 20 | 8 | 6 | 10 | 10 | — | — | 7 | 41 | 4 | 6 |

P : present  
T : trace  
* : Presence of amorphous silica

- for the fine-grained fraction (< 2 μm), predominantly a poorly crystallised smectite slow to swell, as previously described in DSDP deposits (Fan Pow-Foong and Zemmels, 1972) ; zeolite (clinoptilolite) (Mumpton, 1960), often abundant ; amorphous silica which shows up on the diffractograms as an increase in the continuous background (Greenwood, 1973) ; and a trace of illite.
- for the sediment as a whole, predominantly feldspars (especially plagioclases) with quartz (poorly represented). Carbonates are scanty to absent.

This association appears to be characteristic of a detritic type of sedimentation. The smectites could be the result of the re-working of soils under a hot climate with contrasting seasons (Paquet, 1969). It will be noted, moreover, that the sediments are rich in volcano-clastic debris. Littoral deposits presenting very similar lithological and mineralogical characteristics are encountered at the present day in the neighbourhood of the basaltic Faeroes Archipelago (Latouche, 1976). While association 1 a may thus in some cases be the result of a given climate, it seems above all to point to a basaltic environment, the alteration products of which, when modified, were apparently deposited in a littoral context.

In sites 405 and 406, again at Rockall, the deposits dating from the Eocene up to the middle Oligocene are characterised by an association designated 1 b, consisting of a fine-grained fraction containing smectites of excellent crystallinity and silic which is either amorphous or in the form of C.T. opal (Nayudu, 1964 - Jones and Segnit, 1971). In the sediment as a whole, carbonates are scanty in 405 and plentiful in 406. Other minerals (quartz, feldspars) are virtually absent.

The high concentrations of silica, the excellent crystallinity of smectites and the absence of detritics all suggest sedimentation of a chemical to biochemical type. Moreover, siliceous authigeneses are to be observed in the deposits at site 405 and siliceous organisms are everywhere present in quantity. The silica could come from solutions originating in a continent influenced by hydrolysing climates (Leclaire, 1974 - Tardy, 1969) and which have accumulated in a trench isolated from circulation. However, it might also be connected with volcanic phenomena, as the result of underwater alteration of volcanic flows or glasses (Griffin et al., 1972 - Nayudu, 1964) or of siliceous hydrothermal fluids issuing from sea-floor springs ; the presence of volcano-clastic debris at the base of 405, situated near an important fault, is indicative of volcanic activity ; the smectites might themselves be neoformations produced by volcanic matter (Bonatti, 1967).

In site 398, in the vicinity of Portugal, the Paleocene and Eocene deposits are characterised by an association designated 1 c. It consists of complex clay phases where smectites are dominant and illites, chlorites, kaolinites and fibrous minerals (attapulgites and sepiolites) are also present. Carbonates are more or less abundant while quartz and feldspars detritics are always present.

The abundance of primary minerals suggests a detritic sedimentation underlined by slumping-type sedimentary structures. At the same time, fibrous clay minerals may signify chemical sedimentation in a confined basin (Millot, 1964) or a hydrothermal volcanic environment (Hathaway and Sachs, 1965). However, the absence of any indication of volcanic action and the appearance of plankton microfauna with globigerins characteristic of a pelagic environment seems to prove that none of the conditions necessary for the neoformation of fibrous minerals exist. It must therefore be a case of detritic sedimentation in which modified terrigenous and chemical constituents are associated, having been carried towards deep-water facies by tectonic movements. The age of the deposits coincides with that of the major phase of movement in the Pyrenees (Choukroune, 1973).

Episode 2

This involves deposits of Oligocene, Miocene and Lower Pliocene age. Unlike episode 1, this episode is characteristised by an identical mineralogical association in all sites. Compared to the association found in the previous episode, the present association can always be distinguished by the appearance or the increase of three minerals : illite, chlorite and calcite and by the disappearance or decrease of terrigenous detritics - quartz and feldspars.

The transition is sharp in the Rockall sites where the deposits show numerous interruptions ; it is more gradual in the site in West Portugal where certain characteristics of episode 1 re-emerge during the Oligocene. On the other hand, an evolution towards a sedimentation of pelagic type is indicated in all cases ; the increase in biogenic carbonates, the disappearance of quartz and feldspars and the decrease in smectites at the sites where sedimentation was previously of a patently continental origin (site 398), emphasise the fact that terrigenous contributions slacken off. In the Portuguese site the increase in primary clay minerals (illites and chlorites) may be connected with an adduction of the product of rejuvenation of neighbouring reliefs. However, this interpretation is at odds with the lesser detritic nature of the sedimentation denoted by all other parameters and with the general aspect of the phenomenon ; this applies equally to those zones where illite and chlorite were entirely absent in the under-lying detritic episode (sites 403 and 404). It is therefore more logical to think that these minerals have the same origin in all cases and, in view of observations made concerning the present distribution of clays as regards latitude (Biscaye, 1965 - Rateev et al., 1969), that they are the product of cold climatic domains. We are thus led to envisage the inauguration of a modern Atlantic type circulation, that is, of a kind where high and medium latitudes come into communication through the action of ocean currents carrying biogenic (surface currents) or mineral (deep currents) pelagic elements.

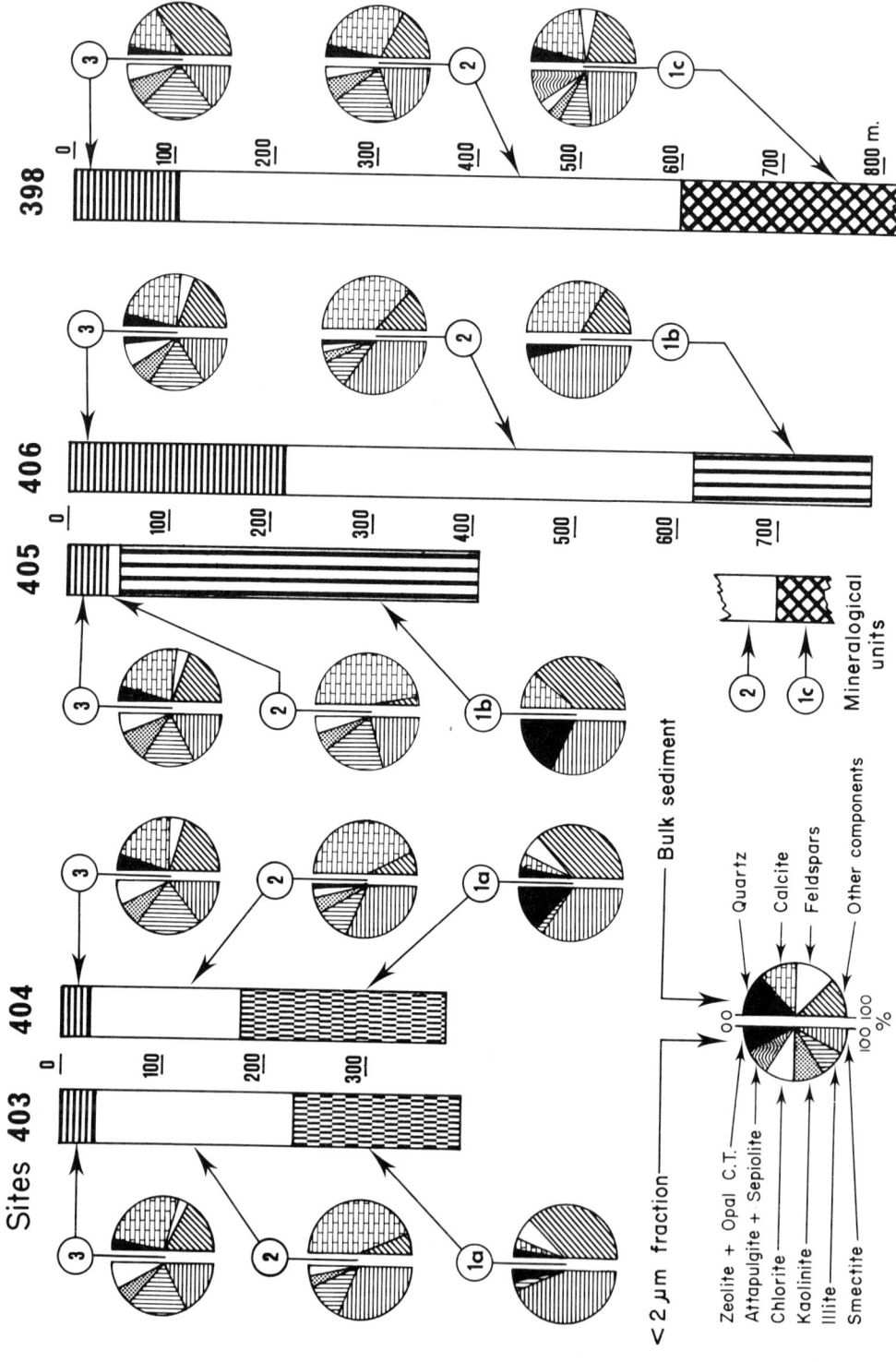

Fig. 2. Schematic representation of the mineralogical

Episode 3

The upper Pliocene and Pleistocene deposits have the following characteristics which distinguish them from the deposits of episode 2 : the recrudescence (398) or sudden reappearance (Rockall) of quartz and feldspars detritics ; a decrease in or irregular appearance of carbonates ; and the dominance of illite and chlorite in the clay phases.

This evolution points to a slackening of biological production, while detritic contributions reappear ; however, they remain relatively scarce. The decrease in smectites and the increase in illites and chlorites suggest a cooling of the climate or a greater rapidity of exchanges with northern latitudes, possibly in the form of ice-rafting phenomena associated with the northern European glaciations at the end of the Tertiary (Berggren and Hollister, 1974).

SYNTHESIS AND CONCLUSIONS.

The mineralogical characteristics thus show that in all 5 sites sedimentation evolved along identical lines during the Cenozoic (Fig. 2). Varied deposits of highly distinctive character according to the individual context of the deposit or geological environment of the region (mineralogical units 1a - 1b - 1c) are succeeded by pelagic deposits of a uniformly "Atlantic" nature belonging to a climate which is at first hot to temperate (unit 2) then cold (unit 3). The transition between 1 and 2 constitutes an important change marking the end of "regionalism" and the establishment of a generalized system of the Atlantic type. Because of the large number of hiatuses, it is difficult to put a precise date to this change at Rockall. In site 406, phase 1 is prolonged right up to the middle Oligocene, while at site 403 the middle Eocene is already part of episode 2. For sites 404 and 405 the change is marked by an important hiatus lasting from the middle Eocene to the upper Miocene. In West Portugal continuous sedimentation reveals a gradual change which nevertheless reaches a climax at the Oligocene-Eocene boundary. Generally speaking, the mineralogical renewal is thus centred on the boundary between the Eocene and the Oligocene and at most varies between the base of the middle Eocene and the centre of the middle Oligocene. This space of time corresponds exactly to the period in which important hiatuses are observed in the D.S.D.P. surveys in the North Atlantic. These hiatuses may be attributed "to increased bottom currents caused by the opening of the northeastern Atlantic and the initiation of vigorous North Atlantic deep circulation" (Berggren and Hollister, 1974). Considerable sluicing of deposits apparently occurred at this time at Rockall which seems to have constituted a zone of obstruction for the passage of water and where sedimentation would only have been possible during limited episodes and at localised points (sites 403 and 406). On the contrary, towards the South (site 398), the competence of deep currents decreases as they arrive in a widen part of the ocean and an almost continuous sedimentation, which at first combines autochthonal and Atlantic characteristics occurs.

From the upper Miocene on, the lines of communication between periartic and European areas are enlarged, either because of late tectonic adjustments or as a result of erosion. The competence of currents decreases in the Rockall zone and sedimentation is re-established. In the South, the increase in exchange is shown in the more markedly northern origin of the material involved. This type of sedimentation persists right up to the upper Pliocene, at which point the first effects of glaciation become apparent.

Thus, the mineralogical data collected during this study have given access to a body of information all pointing to the same conclusions as regards the geographical, climatic and hydrological changes which took place in the Atlantic zone in the course of the Cenozoic. In particular, they give evidence of a remarkably important boundary in the history of the Atlantic at the end of the Eocene : the transition from a system where individual basins in the old Atlantic exist side by side, to a pattern characterised by North-South circulation of a type similar to that of the Atlantic at the present day, giving rise to a homogenous sedimentation over the whole North-eastern sector.

ACKNOWLEDGEMENTS

We wish to thank W.B.F. Ryan and J.C. Sibuet co-chief scientist of leg 47 B and L. Montadert and D.G. Roberts co-chief scientist of leg 48 who made it possible for us to participate in this study.

The investigation was supported by ATP n° 2.685 from the Centre National de la Recherche Scientifique.

REFERENCES

Berggren, W.A. and Hollister, C.D., 1974. Paleogeography, paleobiogeography and the history of circulation in the Atlantic Ocean. In : W.W. Hay (Editor), Studies in Paleo-Oceanography. Society of Economic Paleontologists and Mineralogists, Sp. Publ. n° 20 : 126-186.
Biscaye, P.E., 1965. Mineralogy and sedimentation of recent deep sea clay in the Atlantic Ocean and adjacent seas and ocean. Geol. Soc. Am. Bull., 76 : 803-832.
Bonatti, E., 1967. Mechanisms of deep sea volcanism in the south Pacific. In : Abelson, P.H., Researches in Geochemistry. Wiley, New York, V.2 : 453-491.
Chamley, H., Debrabant, P., Foulon, J., Giroud d'Argoud, G., Latouche, C., Maillet, N., Maillot, H., Sommer, F., 1978. Mineralogy and geochemistry of Cretaceous and Cenozoic Atlantic sediments off the Iberian Peninsula (site 398, leg 47 B, D.S.D.P.). In : Ryan, W.B.F. and Sibuet, J.C. (Editors), Initial Reports of the Deep Sea Drilling Project. U.S. Government Printing Office, Washington. V. 47 B. To appear.
Choukroune, P., 1973. Phase tectonique d'âge variable dans les Pyrénées : évolution du domaine plissé au cours du Tertiaire. C.r. Acad. Sci., D, 276 : 909-912.
Fan Pow-Foong, F. and Zemmels, I., 1972. X-ray mineralogy studies - Leg 12. In : Berggren, W.A., Laughton, A.S. and al. (Editors). Initial Reports of the Deep Sea Drilling Project. U.S. Government Printing Office, Washington, V. 12, pp. 1127-1154.
Greenwood, R., 1973. Cristobalite : its relationship to chert formation in selected samples from the Deep Sea Drilling Project. J. Sediment. Petrol., 43 : 700-708.

Griffin, J.J., Window, H. and Goldberg, E.D., 1968. The distribution of clay minerals in the world ocean. Deep Sea Res., 15 : 433-459.

Griffin, J.J., Kroide, M., Hohndorf, A., Hawkins, J.W. and Goldberg, E.D.,1972. Sediments of the Law Basin - rapidly accumulating volcanic deposits. Deep Sea Res., 19 : 139-148.

Hathaway, J.C. and Sachs, P.L., 1965. Sepiolite and clinoptilolite from the Mid Atlantic ridge. Am. Mineral., 50 : 852-867.

Jones, J.B. and Segnit, E.R., 1971. The nature of opal. 1, nomenclature and constituent phases. J. Geol. Soc. Aust., 18 : 57-58.

Latouche, C., 1976. Les minéraux argileux des sédiments actuels de l'Atlantique Nord-oriental et du Sud de la Mer de Norvège. In : Bailey, S.W. (Editor), Proceeding of the International Clay Conference Mexico City 1975. Applied Publishing Ltd, Wilmette, Ill. : 45-53.

Latouche, C. and Maillet, N., 1978. X-ray mineralogy studies, leg 48 - Rockall region (sites 403 - 404 - 405 - 406), 1978. In : Montadert, L. and Roberts, D.G. (Editors), Initial Reports of the Deep Sea Drilling Project, U.S. Government Printing Office, Washington. V. 48 - 129. To appear.

Leclaire, L., 1974. Hypothèse sur l'origine des silicifications dans les grands bassins océaniques. Le rôle des climats hydrolysants. Bull. Soc. géol. France, 7 : 214-224.

Millot, G., 1964. Géologie des Argiles. Masson et Cie, Paris, 499 p.

Mumpton, F.A., 1960. Clinoptilolite redefined, Am. Mineral. 45 : 351-369.

Nayudu, Y.R., 1964. Palogonite tuffs (hyaloclastites) and the products of post-eruptive processes. Bull. Volcanol., 27 : 391-410.

Paquet, H., 1969. Evolution géochimique des minéraux argileux dans les altérations et les sols des climats méditerranéens et tropicaux à saisons contrastées. Mém. Serv. Carte Géol. Alsace Lorraine, 30 : 212 p.

Rateev, M.A., Gorbunova, Z.N., Lisitzyn, A.P. and Nosov, G.L., 1969. The distribution of clay minerals in the oceans. Sedimentology, 13 : 21-43.

Tardy, Y., 1969. Géochimie des altérations. Etude des arènes et des eaux de quelques massifs cristallins d'Europe et d'Afrique. Mém. Serv. Carte Géol. Alsace Lorraine, 31 : 199 p.

Yeroshev-Shak, V.A., 1954. Clay minerals of the Atlantic Ocean. Sov. Oceanogr., 30-2 : 90-106.

# CHANGES IN MINERALOGICAL COMPOSITION OF TERTIARY SEDIMENTS FROM NORTH SEA WELLS

W. Karlsson, J. Vollset, K. Bjørlykke (x) and P. Jørgensen
Department of Geology, The University of Oslo, Blindern, Oslo.
(x) Department of Geology, The University of Bergen.

ABSTRACT

Well 2/11-1 contains a Paleocene to Pliocene sedimentary sequence which is almost 3000 m thick. Quantitative clay composition was determined by combining X-ray diffraction with normative calculation based on chemical analyses. High amounts of smectite in the Lower Tertiary are interpreted as a result of halmyrolytic transformation of volcanic glass. Observed changes in clay mineralogy are interpreted as a response to a colder climate and to higher rates of erosion in the Upper Miocene. Clay mineralogy indicates two sources for the Pliocene sediments, namely material from Scandinavia containing chlorite and reworked older sediments containing kaolinite.

INTRODUCTION

While Tertiary sections in N. Europe are mostly discontinuous, the central part of the North Sea basin contains a sequence which may exceed 3,000 m and represents continuous sedimentation from the Paleocene to the Pliocene.

Wells drilled in this part of the North Sea provide very useful material for detailed stratigraphical and sedimentological studies based on seismic information, well logs and samples (mostly cuttings).

Well 2/11-1 is located in the southeastern part of the central Tertiary basin near the junction between British, Danish and Norwegian territory (Fig. 1).

The purpose of the present study is to analyse mineralogical and textural changes in well samples (mostly cuttings and some sidewall cores) in order to find out if systematic changes, particularly in clay mineral composition exist. Such changes may then be interpreted in terms of source rocks, paleoclimate, rate of deposition, and postdepositional processes, including authigenic growth as well as halmyrolytic and early diagenetic mineral changes. The data presented are taken from two university theses by Karlsson (1977) and Vollset (1977).

## METHODOLOGY

X-ray diffraction analyses: Samples, mainly cuttings from regular intervals in the well, were suspended in water and the $<2$ μm fraction was separated and sucked onto a Millipore filter (poresize 0.2 μm). The clay film was then transferred to a glass slide by gently pressing the clay on to the glass and carefully removing the filter paper. This method is fast and gives highly reproducible x-ray diffraction intensities. An alternative method of drying a suspension which has been pipetted directly onto a glass slide tends to give a thin upper film enriched in fine-grained clay minerals, notably smectite.

Clay mineral composition: The identification of clay minerals was carried out following criteria published by Brown (1961). Typical x-ray diffraction curves from 3 different stratigraphical levels are presented in Fig. 2. The lowest sample contains mainly smectite with additional small amounts of illite, kaolinite and quartz. The clay fraction from the next sample (6660-90 ft.) contains less smectite and quartz and more illite and kaolinite. The uppermost sample contains much less smectite while the amounts of kaolinite and illite have increased considerably. Chlorite is present in significant amounts and small amounts of mixed-layer minerals can be detected.

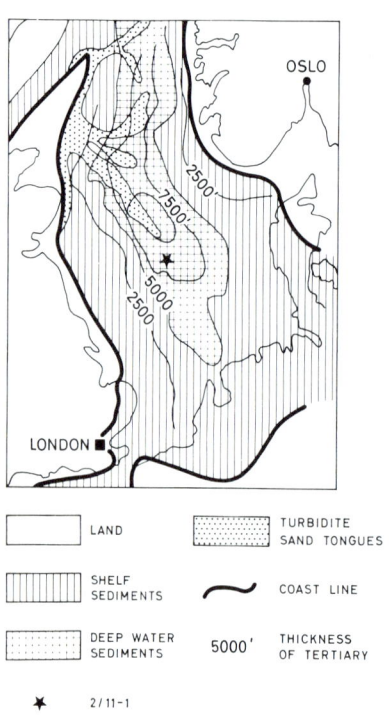

Fig. 1. Paleogeographical map (Paleogene, Ziegler 1975). Showing location of well 2/11-1.

## QUANTITATIVE DETERMINATIONS

Peak intensities were calculated by multiplying peak height by width at half peak height. In order to calculate the semiquantitative mineral composition, peak intensities of smectite (17 Å), illite (10 Å), chlorite (7 Å) and kaolinite (7 Å) were multiplied by factors of 1, 4, 2 and 2 respectively (Biscaye, 1965). In samples where chlorite was found in addition to the other phyllosilicates the kaolinite/chlorite ratios were calculated from slowscans over the 002 and 004 (3.57-3.54 Å) peaks (Fig. 3). The calculation method was the same as that used by Bjørlykke and Elverhøi (1975).

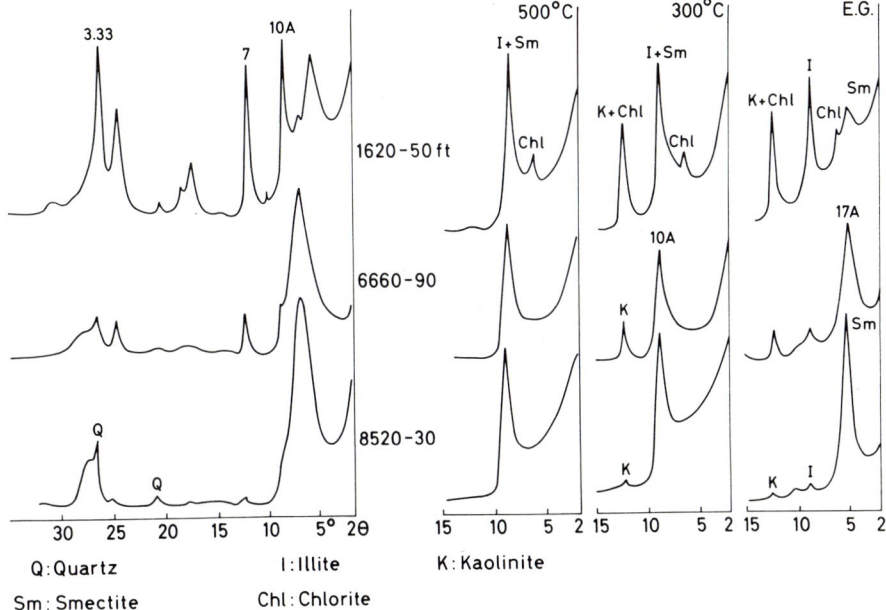

Fig. 2. X-ray diffraction curves of the clay fraction from three different depths.

Clay mineral compositions of the same samples were also calculated from chemical analyses. By assuming that potassium is essentially present in illite in these sediments (which contain minor amounts of K-feldspar) the normative amount of an illite containing 9% $K_2O$ can easily be calculated. Having determined the cationic exchange capacity (C.E.C.), the amount of smectite (in samples containing illite, kaolinite and smectite) was calculated using the following equation:

$$(C.E.C.) = \frac{80}{100}(X) + \frac{15}{100}(Y) + \frac{5}{100}(100-X-Y)$$

where X and Y are the contents of smectite and illite (%), and it is assumed that pure smectite, illite and kaolinite + quartz have C.E.C. values of 80, 15 and 5 meqv./100 g. The amount of quartz was determined from peak intensity (4.26 Å), and the sum of clay minerals was recalculated to 100%.

Fig. 4 shows that there is a good correlation between illite and smectite contents calculated from X-ray data and from chemical analyses. Consequently we believe that compositions calculated from X-ray data alone are close to the true values.

CONTAMINATION OF CUTTINGS BY DRILLING MUD

The use of cuttings for mineralogical and geochemical studies presents several

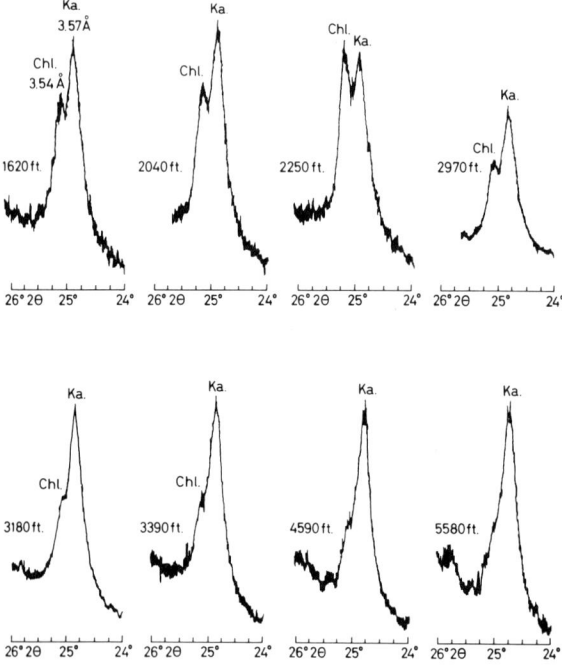

Fig. 3. Slowscan diffraction curves for the 24-26° 2θ region. Notice the difference between the two lowermost samples containing kaolinite and the samples above containing kaolinite and chlorite.

problems, the most important being "caving" and contamination by drilling mud. Cuttings represent, however, a very important source of information about the composition of sediments in the North Sea, particularly in the shaly Tertiary facies where cores are very rare. Possible contamination by drilling mud or "caving" was estimated by comparing analyses of side wall cores with analyses of cuttings at the same depths. Table 1 shows that these two types of analyses give only small differences in clay mineralogy (probably within the analytical error) suggesting that contamination is of minor importance and that the cuttings are representative of the sediments at the various depths.

A DISCUSSION OF THE STRATIGRAPHICAL VARIATIONS IN MINERALOGICAL COMPOSITION

On the basis of X-ray diffraction and chemical data the composition of the clay fraction for the whole well has been determined (Fig. 5).
Paleocene and Eocene: Paleocene shales contain mainly smectite with only minor amoun of kaolinite and illite. The clay fraction also has a high quartz content. The Eocen

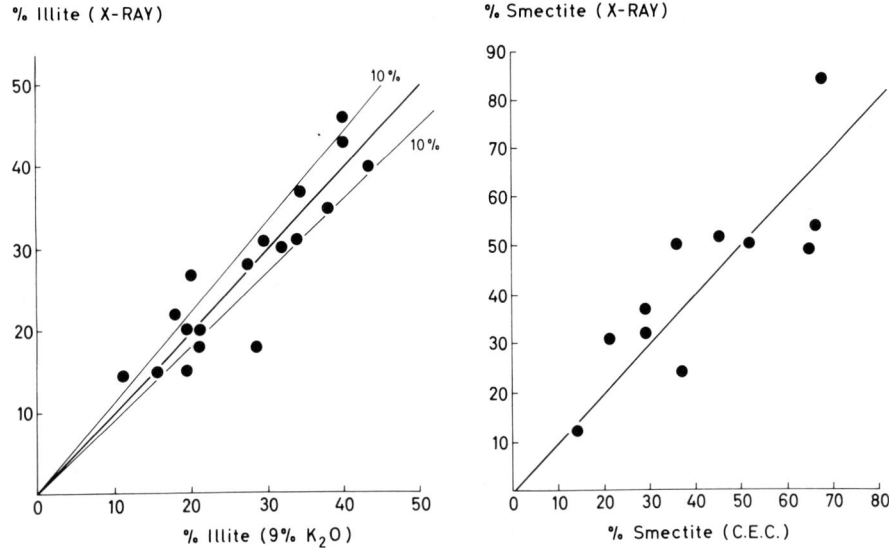

Fig 4. Correlation between illite and smectite contents calculated by X-ray methods and from chemical analyses.

TABLE I

Mineralogical composition of cuttings and sidewall cores

| Depth in feet | Material | Amounts in % | | |
|---|---|---|---|---|
| | | Smectite | Illite | Kaolinite |
| 8500-10 | Cuttings | 71 | 10 | 19 |
| 8510 | Core | 65 | 14 | 22 |
| 8520-30 | Cuttings | 82 | 14 | 4 |
| 8530 | Core | 78 | 15 | 7 |
| 8540-50 | Cuttings | 69 | 19 | 12 |
| 8546 | Core | 72 | 21 | 7 |

shales show a gradual upwards increase in illite and kaolinite.

The widespread "tuff" at the Paleocene-Eocene boundary is a record of the opening of the Atlantic (Ziegler, 1975) and Jacque and Thouvenin (1975) who described the Lower Tertiary tuffs in the North Sea basin showed that the volcanism had mainly a basic composition. The ash beds contain glass, plagioclase, anatase, zeolites and carbonates (calcite and siderite). From this it seems reasonable to assume that the very high smectite contents found in the Paleocene and Eocene are due to halmyrolytic alteration of volcanic material. The content of volcanic glass fragments and tuffaceous aggregates in these sediments support this conclusion.

Authigenic pure smectite (beidellite) has also been described from the Tertiary basin of N.W. Germany in association with albite, analcime and Na-jarosite (Buchman & Ritzkowski, 1976). High contents of smectite in Paleocene and Eocene beds have al been observed in Denmark, where they have been attributed to halmyrolytic alteratio of volcanic glass (Nielsen, 1974), and the clay mineralogy of Lower Eocene sediment from the studied well compares with Eocene tuffs from Ølst in Denmark. Since older sediments, as the Danian limestone, also contain smectite it is possible that some of the smectite was inherited from these sediments.

As pointed out by Hopkinson and Nysæther (1974) a marked regression between the Cretaceous and Tertiary favoured the reworking of older sediments.
The authigenic formation of smectite can be illustrated by the following simple equation (Müller 1967):

GLASS + $H_2O$ $\longrightarrow$ SMECTITE + SILICA + ZEOLITE + METALLIC IONS

High contents of silica in the pore water will inhibit this transformation as illustrated for the Danish Mo-clays (Tank 1963).
Since the metallic ions have higher mobilities than silica ($H_4SiO_4$) they can leave the sediment by diffusion, while silica will remain. Consequently neoformation of silica compounds is required for the process described by this equation to proceed. The high quartz contents in the tuffaceous formations are therefore explained by extensive halmyrolytic transformation of volcanic glass. The quartz may have been formed diagenetically from amorphous silica or cristobalite. The onset of middle diagenesis is characterized by smectite being stable while cristobalite converts to quartz (Mitsui, 1975).

The variations in clay mineralogy through the Tertiary can not be explained by diagenetic alterations. MacEwan et al (1961) has pointed out that variations in the amount of illitic layers in randomly ordered illite-smectite mixed-layer minerals are best studied by the migration of the 002/003 peak. Since there are no signs of systematic alteration of smectite to illite via the illite-smectite mixed-layer pha even in the deepest part of this well, it is concluded that this diagenetic alteration has not started at the base of the Tertiary in this area. The Eocene and Oligocene shales show a gradual upwards increase in illite and kaolinite which were probably present in the fine suspension load of these clastic sediments.
The lack of abrupt changes in clay mineralogy in these shales suggests a homogeneous source area and/or a good mixing of fine suspended material in the basin.

The grain size distribution of the Paleocene sediments is similar to recent dist turbidity current deposits. This is in agreement with available reconstructions of the Lower Tertiary paleogeography of the North Sea basin (Parker 1975). It is suggested that this well (2/11-1) is located in the distal part of a submarine fan with a source to the north-west. In this deep and distal part of the basin, facies

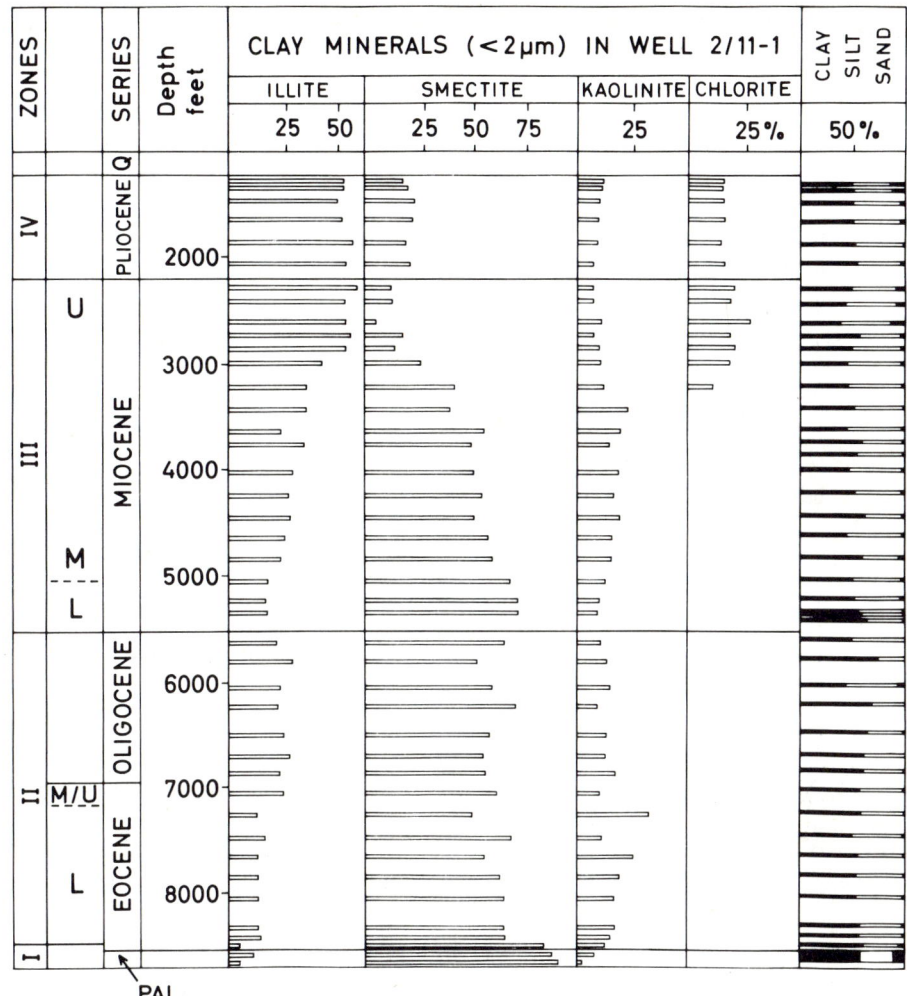

Fig. 5. Stratigraphical variation in clay mineralogy and mechanical composition for well 2/11-1.

variations resulting from transgressions and regressions during Tertiary time were not as pronounced as in the shallower deltaic or fluvial facies.

MIOCENE AND PLIOCENE

The most conspicuous variation through the Miocene is a gradual decrease in smectite content accompanied by increasing amounts of illite. This is probably caused by a gradual change in the ratio of fine suspended clastic material to volcanic glass. This is reasonable as volcanic activity in the North Sea basin terminated in the

Lower Miocene. Significant quantities of chlorite appear in the Upper Miocene beds and continue to be an important constituent in Pliocene and Quaternary sediments, while this mineral is absent in Lower Tertiary sediments in the well.

The appearance of chlorite in the upper part of the Tertiary is accompanied by a reduction of kaolinite and smectite while the illite content increases. This shift coincides with a change towards a colder climate in northern Europe. Similar changes have been observed in deep sea sediments (Jacobs and Hays, 1972 and Jacobs, 1974). Higher sedimentation rates in the Upper Miocene and Pliocene also imply more rapid erosion and consequently less weathering. We conclude that the presence of chlorite is a clear indication of reduced intensity of weathering in the source areas due to colder climate and higher relief that prevailed around the North Sea Tertiary basin in late Miocene and Pliocene times.

The fact that kaolinite persists in both Pliocene and Quaternary sediments probably reflects reworking of older kaolinite-rich sediments rather than continental weathering (Bjørlykke and Elverhøi, 1975 and Rønningsland, 1976). Compared with recent sediments, the Lower Tertiary sediments have a "Low lattitude composition" while the upper ones, from Middle Miocene, have a "High lattitude composition". The clay mineralogy thus indicates two sources for the Pliocene sediments; viz. material from Scandinavia containing chlorite and reworked older sediments containing kaolinite.

CONCLUSIONS

The present study shows that clay mineral analyses of cuttings are, at least in this case, representative of the formations, since the level of contamination of cuttings by drilling mud is relatively low.

A good correlation can be observed between quantitative clay mineral analyses using x-ray diffraction and results obtained by normative calculation of illite and smectite based on chemical analyses and cation exchange capacities (C.E.C.).

High amounts of smectite in Lower Tertiary sediments are interpreted as a result of halmyrolytic transformation of volcanic glass, due to volcanism associated with the opening of the North Atlantic.

Systematic changes in clay mineralogy can be observed from the Paleocene and Eocene sediments (dominated by smectite) to the Upper Miocene and Pliocene beds which contain illite, chlorite and smaller amounts of kaolinite and smectite.

The observed changes in clay mineralogy are interpreted as a response to climatic deterioration accompanied by higher rates of erosion in Upper Miocene time. The persistence of kaolinite and smectite in Pliocene and Pleistocene sediments is probably due to reworking of older sediments rather than continental weathering.

ACKNOWLEDGEMENTS

We are indebted to the AMOCO/NOCO group for letting us use material from their well.

REFERENCES

Biscaye, P.E., 1965. Mineralogy and Sedimentation of Recent Deep-Sea Clay in the Atlantic Ocean and adjacent Seas and Oceans. Geol. Soc. Am. Bull., 76:803-832.
Bjørlykke, K. and Elverhøi, A., 1975. Reworking of Mesozoic clayey material in the northwestern part of the Barents Sea. Marine Geol., 18: 29-34.
Brown, G. (Editor), 1961. The X-ray Identification and Crystal Structures of Clay Minerals. Mineralogical Society, London, 544 pp.
Bühmann, D. and Ritzkowski, S., 1975. Clay Minerals of the Northwest German Tertiary Basin. IGCP Project 124, The Northwest European Tertiary Basin, Rept. nr. 1:56-62.
Hopkinson, J.P. and Nysæther, E., 1974. North Sea Petroleum Geology. Exploration Geology and Geophysics. Offshore North Sea Technology Conference, Stavanger:1-33.
Jacobs, M.B., 1974. Clay mineral changes in Antarctic deep-sea sediments and Cenozoic climatic events. Journ. Sedim. Petrol., 44: 1079-1086.
Jacobs, M.B. and Hays, J.D., 1972. Paleo-Climatic Events Indicated by Mineralogical Changes in Deep-Sea Sediments. Journ. Sedim. Petrol., 42: 889-898.
Jacque, M. and Thouvenin, J., 1975. Lower Tertiary tuffs and volcanic activity on the North Sea. In: A.W. Woodland (Editor), Petroleum and the Continental Shelf of North-west Europe. Geology, 1: 455-465. Applied Science Publishers, London.
Karlsson, W., 1977. Tertiære sedimenter fra Nordsjøen. En teksturell, mineralogisk og geokjemisk undersøkelse med vekt på den leirmineralogiske utvikling. Unpubl. Thesis, Dept. of Geol., University of Oslo, 211 pp.
MacEwan, D.M.C., Ruiz Amil, A. and Brown, G., 1961. Interstratified Clay Minerals. In: G. Brown (Editor), The X-ray Identification and Crystal Structures of Clay Minerals. Mineralogical Society, London: 393-445.
Mitsui, K., 1975. Diagenetic alteration of some minerals in argillaceous sediments in Western Hokkaido, Japan. Sci. Rept. Tokohu Univ., Third Ser., XIII, No. 1: 13-65.
Müller, G., 1967. Diagenesis in argillaceous sediments. Developments in sedimentology, 8. Elsevier.
Nielsen, O.B., 1974. Sedimentation and Diagenesis of Lower Eocene Sediments at Ølst, Denmark. Sedim. Geol., 12: 25-44.
Parker, J.R., 1975. Lower Tertiary Sand Development in the Central North Sea. In: A.W. Woodland (Editor), Petroleum and the Continental Shelf of North-west Europe. Geology 1. Applied Science Publishers, London.
Rønningsland, T.M., 1976. Mineralogi og geokjemi av resente leirsedimenter i Skagerak, Kattegat og tilgrensende fjordområder. Unpubl. Thesis, Dept. of Geol., University of Oslo, 194 pp.
Tank, R.W., 1963. Clay mineralogy of some Lower Tertiary sediments from Denmark. Dan. Geol. Unders., Bd. 4, No. 9: 45 pp.
Vollset, J., 1977. Nordsjøens tertiære sedimentasjonsbasseng, en undersøkelse av materiale fra borhullene 2/11-1 og 9/12-1. Unpubl. Thesis, Dept. of Geol., University of Oslo, 228 pp.
Ziegler, W.H., 1975. Outline of the Geological History of the North Sea. In: A.W. Woodland (Editor), Petroleum and the Continental Shelf of North-west Europe. Geology 1: 165. Applied Science Publishers, London.

THE ORIGIN OF CLAY MINERALS IN CENOMANIAN LITTORAL DEPOSITS AROUND THE ARMORICAN MASSIF.

J. LOUAIL, J. ESTEOULE and J. ESTEOULE-CHOUX
Laboratoire de Géologie G. 4, Université de RENNES, 35042 RENNES CEDEX (France)
Laboratoire de Minéralogie et Géotechnique I.N.S.A. 35031 RENNES CEDEX (France)

ABSTRACT

The study of Cretaceous deposits around the Armorican Massif shows that the mineralogical composition of the detrital material entering the area during the Cenomanian was almost constant. However, the distribution of the clay mineral fraction in the sediments appears to be related to two facies: 1° a continental nearshore facies with kaolinite, smectite and mica as the main clay mineral constituents;

2° a marine facies in which the clay fraction contains smectites and glauconite, with various proportions of zeolites and cristobalite. Similar observations have been frequently described in the literature and the segregation of the clay minerals in such a basin has been accounted for by different settling velocities under different environmental conditions. However this interpretation cannot be applied to this study.

Analysis of the data suggests that the more or less stable silicate detritus originating from continental weathering has undergone modification within the basin. In the nearshore facies, under the influence of fresh water the detrital material is stabilized. In the marine facies, however, under the influence of saline waters, reorganization of disordered silicates in an iron- and silica-rich and magnesium-poor environment gives rise to ferriferous montmorillonites and glauconites; still higher silica activity favours the appearance of zeolites and cristobalite.

These reactions occur within the newly deposited sediment by a mechanism of ionic diffusion in the interstitial fluids.

---

INTRODUCTION

Sur la bordure orientale du Massif Armoricain la transgression mésocrétacée a scellé des profils d'altération qui se sont développés sur des substratum divers pendant l'émersion du Massif entre le Jurassique supérieur et le Cénomanien. Ces profils sont les témoins d'une couverture d'altération beaucoup plus étendue dont le démantèlement a alimenté la sédimentation terrigène au cours du Cénomanien. Pourtant, l'analyse

précise des faits montre qu'il est parfois difficile de relier directement la miné
logie de la fraction fine des dépôts et celle des altérations. Même dans le cas d'u
sédimentation à dominante terrigène, l'héritage seul ne suffit pas à rendre compte
faits observés.

LES FAITS

Situation des dépôts et lithostratigraphie

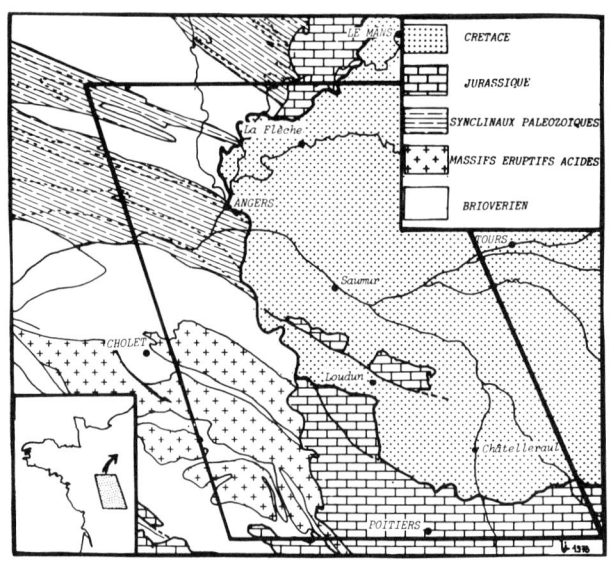

Fig. 1. Esquisse géologique et localisation de la région étudiée.

L'étude a été effectuée sur les assises cénomaniennes littorales de la marge Sud-
Est du Massif armoricain entre le Loir au Nord et le Poitou au Sud (fig. 1). Ces dépôts, toujours fortement influencés par des apports terrigènes d'origine armoricaine
(LOUAIL, 1969,), s'ordonnent suivant une mégaséquence transgressive marquée par la
diminution progressive de la taille et du pourcentage des détritiques vers le sommet
de la série. Cette mégaséquence transgressive se décompose en séquences mineures pos
tives et négatives qui ne sont pas partout synchrones ce qui provoque régionalement
certain diachronisme des faciès. Ainsi, au Cénomanien terminal, un afflux littoral c
terrigène se marque en Anjou septentrional (Sable de Bousse) ; il s'atténue progres
sivement vers la région Poitevine qui pendant la même période montre une légère tendance à l'émersion.

Les dépôts cénomaniens reposent tantôt sur le Briovérien et le Paléozoïque au
niveau du synclinorium d'Angers, tantôt sur le Jurassique au Nord et au Sud. Mais,
plus à l'Ouest les témoins cénomaniens isolés par l'érosion se trouvent en général

directement sur le socle armoricain, montrant ainsi que le Jurassique avait été en partie décapé avant la transgression crétacée : ceci est confirmé par l'étude de la fraction détritique grossière qui prouve que la sédimentation cénomanienne a été essentiellement alimentée par le démantèlement des assises paléozoïques de l'arrière pays et de leur manteau d'altération.

Certains profils d'altération se trouvent scellés sous les horizons de base de la transgression : ils sont toujours établis sur des roches sédimentaires plus ou moins métamorphisées mais jamais sur les massifs éruptifs qui forment actuellement des reliefs dépourvus de couverture crétacée.

## Les minéraux argileux des altérations

Les profils d'altération conservés sous le Cénomanien, sur schistes briovériens ou siluriens qui constituent par ailleurs la plus grande part de l'arrière pays, sont tous semblables : à partir d'une roche mère à mica dominant se développe progressivement une phase argileuse de plus en plus riche en kaolinite. Cette altération est en fait banale pour ce type de roche et elle a été retrouvée de façon quasi systématique sur tout le Massif Armoricain (ESTEOULE-CHOUX, 1967,). Par contre sur certaines roches pétrographiquement moins évoluées, telles que les schistes et grauwackes du Houiller de la terminaison orientale du synclinorium d'Ancenis, l'altération est différente : à la base des profils se trouvent du mica et des chlorites accompagnés de petites quantités de kaolinite et de smectites. Vers le sommet cet assemblage cède progressivement la place à l'association smectite-kaolinite-mica en passant par une zone où des interstratifiés chlorite-smectite apparaissent de façon fugace.

Quant aux massifs éruptifs acides, ils sont actuellement fortement arénisés mais dépourvus de toute couverture sédimentaire crétacée qui permettrait éventuellement de dater cette altération. Leur contribution à la sédimentation crétacée peut toutefois être établie par l'étude de la fraction terrigène grossière qui comprend des minéraux caractéristiques de roches éruptives acides, des biotites et des feldspaths potassiques.

Malgré l'évidence d'un déblaiement précoce constaté d'après la géométrie des affleurements, une alimentation à partir du Jurassique ne peut être totalement exclue. Dans ces dépôts essentiellement carbonatés, la kaolinite est présente à la base de la série associée à des quantités plus ou moins importantes d'argile micacée, d'interstratifiés illite-smectites, ainsi qu'à de petites quantités de chlorite. Les smectites ne deviennent prépondérantes qu'à partir du Callovien supérieur, et constituent l'essentiel de la fraction fine des marnes oxfordiennes et des calcaires portlandiens. Il faut souligner que les indices du remaniement de Jurassique et en particulier les silex, sont extrêmement rares dans la série cénomanienne ce qui laisse penser que la participation du Jurassique à la sédimentation crétacée a été très réduite dans cette région.

En résumé, la généralisation à l'ensemble de l'arrière pays des divers modes d'altération retrouvés sous les premiers dépôts cénomaniens permet de penser qu'il y avait en stock sur le continent au début de la période cénomanienne, des quantités importantes de kaolinite, d'argile micacée et vraisemblablement en proportion moindre des smectites, des chlorites, des minéraux interstratifiés et une fraction en cours d'évolution : minéraux désorganisés, fragments de charpentes silicatées, etc... Cette hypothèse est indirectement confirmée par l'existence de dépôts kaoliniques piégés dans des cavités du Jurassique sous les premiers sédiments cénomaniens et qui témoignent d'une altération puissante avant cette période (ESTEOULE-CHOUX et al, 1969,).

Les minéraux argileux des dépôts cénomaniens

L'évolution de la phase argileuse des dépôts cénomaniens est résumée dans les figures 2,3 et 4.

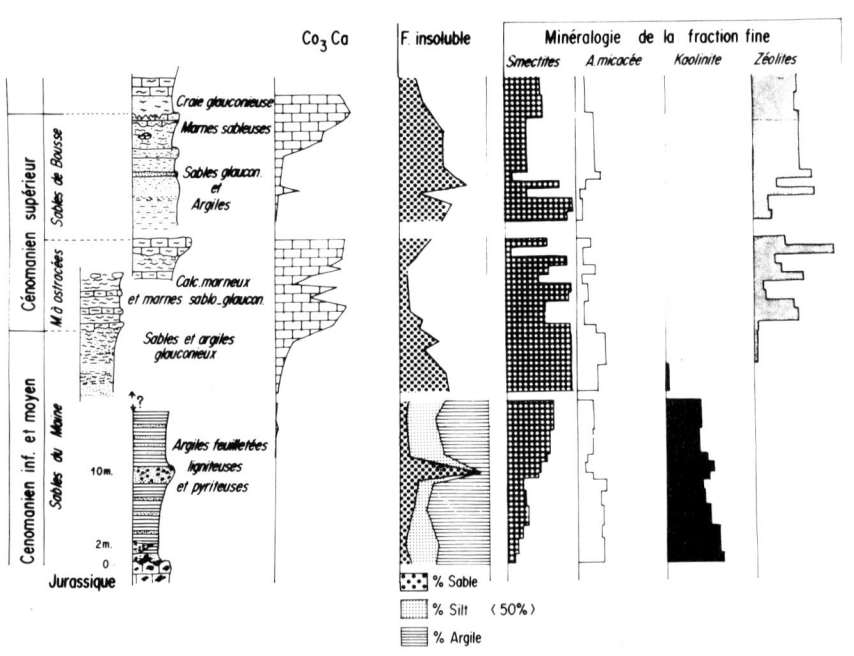

Fig. 2. Lithologie et évolution de la fraction fine des dépôts cénomaniens en Anjou septentrional (Beaugeois).

En Anjou septentrional (fig. 2) la kaolinite, abondante à la base dans les sédiments les plus littoraux où elle est associée à une argile micacée et à des smectites ainsi qu'à des quantités peu importantes d'édifices interstratifiés, perd progressivement de son importance au fur et à mesure que le faciès devient plus marin. Elle n'apparaît plus que de façon fugace dans les Marnes à Ostracées et disparaît presque totalement dans les Sables de Bousse dont la fraction fine est en outre caractérisée par la présence importante de clinoptilolite et de cristobalite.

Dans le Saumurois, (fig. 3) la kaolinite est le minéral cardinal des argiles ligniteuses et pyriteuses de la base, associée à une smectite toujours subordonnée et à une argile micacée. Puis la smectite envahit les dépôts et la kaolinite n'apparaît plus que sporadiquement en quantité minime ; l'argile micacée toujours présente n'est jamais importante. Dans le terme supérieur, c'est toujours la smectite qui caractérise la fraction fine, accompagnée d'une argile micacée, mais ici la kaolinite réapparaît en quantité notable à la base de cette séquence transgressive. A côté de ces minéraux argileux, apparaît une zéolite, la clinoptilolite, qui se rencontre en très faible quantité dans la presque totalité des échantillons étudiés et devient très importante dans un niveau particulier où elle est accompagnée de cristobalite présente sous forme de très nombreux spicules de Spongiaires.

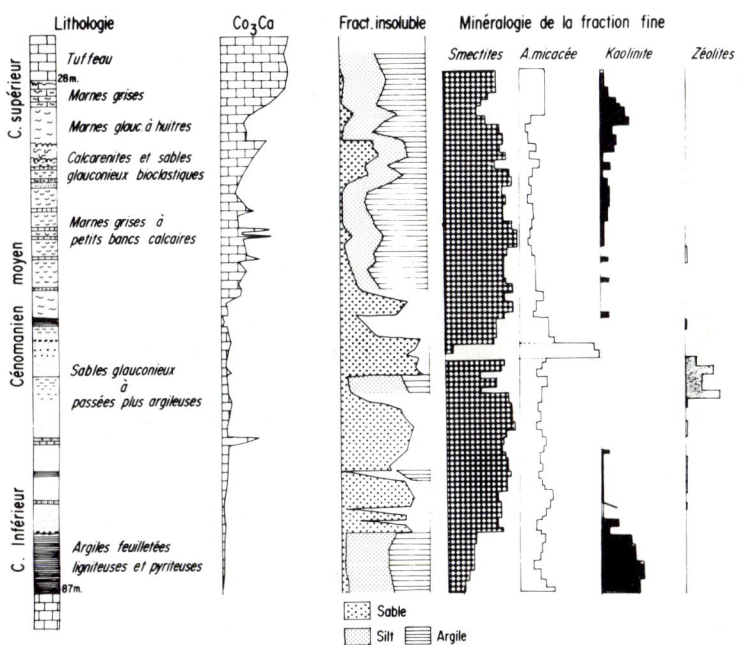

Fig. 3. Lithologie et évolution de la fraction fine des dépôts cénomaniens dans le Saumurois (sondage de Loudun).

Enfin dans la région poitevine (fig. 4), les argiles ligniteuses de base sont beaucoup moins développées et n'existent qu'en intercalations dans des sables glauconieux. La fraction fine de ces sables et argiles essentiellement smectitique comprend parfois localement des quantités notables de kaolinite. La kaolinite disparaît ensuite rapidement et les smectites deviennent le composant exclusif de la fraction fine aussi bien dans les faciès marnoglauconieux que dans les calcarénites de haute énergie où elles se trouvent accompagnées par une proportion plus élevée de mica ouvert.

Dans les termes supérieurs, marnes à Ostracées et craies glauconieuses qui ici so⟨nt⟩ très réduites, ce sont les smectites qui sont les minéraux cardinaux ; les zéolites qui apparaissent de façon sporadique dans l'ensemble de la série, deviennent alors plus fréquentes et abondantes dans certains échantillons de craie glauconieuse.

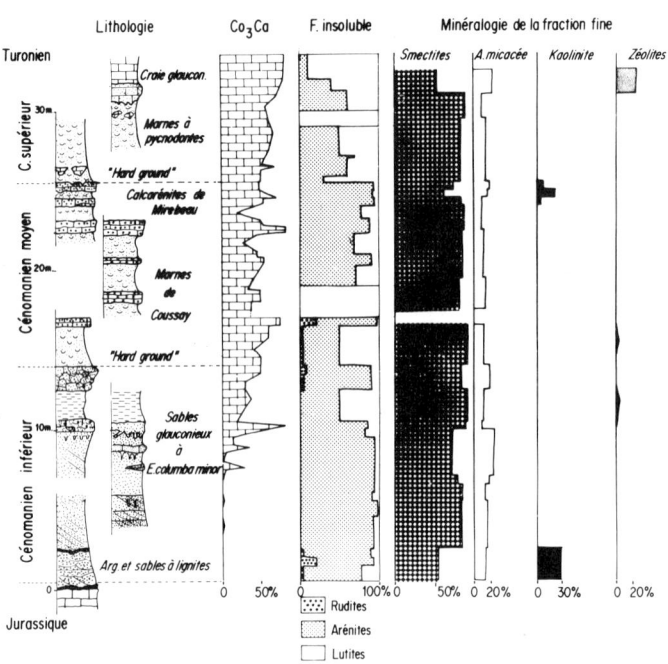

Fig. 4. Lithologie et évolution de la fraction fine des dépôts cénomaniens en Poitou (région de Chatellerault).

Dans cet inventaire, il n'a été tenu compte que de la fraction argileuse diffuse ; il existe en outre de façon quasi constante, une phase granulaire glauconieuse dont la composition minéralogique va d'une beidellite ferrifère à une illite $IM$.

En résumé, il apparaît de façon évidente que la kaolinite est plus abondante à la base de la série, surtout dans les dépôts très littoraux et qu'elle disparaît ensuite au profit des smectites. Il est à remarquer que cette disparition n'obéit pas à un moindre afflux des terrigènes et que sa réapparition locale dans des faciès variés ne peut être reliée strictement à des variations d'ordre sédimentologique.

En réalité, cette évolution que l'on constate dans l'ensemble de la série et qui conduit à des dépôts de plus en plus riches en smectites, est compliquée par le fait que les faciès ayant des cortèges argileux semblables ne sont pas partout de même âge. Les argiles riches en kaolinite se rencontrent dans le Cénomanien inférieur et moyen et latéralement passent à des sédiments souvent plus marins où la kaolinite laisse la place aux smectites. En outre la kaolinite réapparaît parfois localement en des points divers de la série alors que l'ensemble des dépôts de même âge est à smectites.

Il semble donc difficile d'expliquer cette variation de la minéralogie de la fraction argileuse par une simple variation des apports.

DISCUSSION

Un des traits caractéristiques de la sédimentation cénomanienne de la bordure Sud-Ouest du Bassin de Paris est la présence constante de smectites plus ou moins ferrifères. Cette prédominance s'observe partout dans le Bassin anglo-parisien au Crétacé supérieur quelque soit le faciès (MILLOT et al., 1957 ; WEIR et al., 1965 ; STEINBERG, 1967 ; BROWN et al., 1969 ; JEANS, 1968, 1971 ; MATHIEU et al., 1972 ; JUIGNET, 1974 ; THOREZ et al., 1974 ; MORGAN-JONES, 1977,). Toutefois dans la partie la plus septentrionale du Bassin, l'importance des smectites par rapport aux autres minéraux apparaît moins marquée et les auteurs britanniques ont souligné à plusieurs reprises cette différence avec le Bassin de Paris.

Diverses origines ont été envisagées pour les smectites du Crétacé supérieur : pour certains elles résultent d'une néoformation dans le bassin (MILLOT et al., 1957 ; MILLOT, 1964,) ou au cours d'une phase diagénétique (JEANS, 1968,) pour d'autres elles sont héritées (HEIM, 1957,) ; enfin, à un héritage évident pourrait s'ajouter soit un ajustement dans le bassin (JEANS, 1971,) soit une genèse à partir de débris pyroclastiques (VALETON, 1960 ; AUBRY et al., 1975 ; POMEROL, 1977,). Mais aucune de ces hypothèses n'explique de façon satisfaisante la prépondérance des smectites dans les dépôts essentiellement terrigènes du Cénomanien du Sud-Ouest du Bassin de Paris. Comme l'indiquent la diversité minéralogique des argiles des altérations antécénomaniennes qui ont pu être observées et l'apparition sporadique de minéraux relativement fragiles (biotites, feldspaths potassiques) tout au long de la série, l'alimentation du bassin n'était pas homogène : elle ne pouvait donc fournir uniquement des smectites. Cependant, à cette diversité des apports qui est conservée dans les faciès très littoraux et fluviatiles, la réponse sédimentaire en milieu marin est la présence quasi constante de smectites.

La ségrégation des minéraux argileux dans un bassin est souvent expliquée en invoquant la différence de leur vitesse de sédimentation. Cette explication concevable à l'échelle d'un bassin où il existe un gradient de dépôt en fonction de l'éloignement des zones d'apport ne peut être retenue dans le cas présent. En effet, les sédiments étudiés se situent en bordure de bassin dans un milieu où l'énergie des dépôts varie beaucoup d'un point à un autre : les smectites se rencontrent aussi bien dans des zones calmes à sédimentation essentiellement argileuse que dans des zones agitées où elles sont mêlées à des clastiques grossiers.

L'explication alternative de néoformation par précipitation à partir des solutions en milieu chimique basique souvent proposée ne se conçoit pas mieux dans un milieu caractérisé par l'afflux important de détritiques terrigènes. A cette explication peut se rattacher l'hypothèse selon laquelle les smectites proviendraient de l'évolution de cendres ou de poussières volcaniques en milieu marin. Aucune évidence de volcanisme n'a pu être décelée dans ces dépôts ; de plus, une telle hypothèse impliquerait que l'on observe la concentration de smectites dans des niveaux privilégiés, ce qui n'est pas le cas.

Toutes ces considérations amènent à rechercher l'origine des smectites dans un ajustement diagénétique précoce de la phase terrigène. Les observations montrent qu'elle est constituée en partie de minéraux qui ont pris naissance au cours des phénomènes d'altération (essentiellement kaolinite et smectites) et de minéraux plus ou moins désorganisés en cours d'évolution. Durant cette phase d'altération partielle en milieu continental, les charpentes silicatées résiduelles sont évidemment enrichies en alumine, en oxyde de fer ferrique et en silice. De plus, en ce qui concerne les feldspaths, les plagioclases ont été altérés les premiers tandis que les potassiques qui résistent mieux se retrouvent en plus grande quantité dans le bassin : le sodium et le calcium sont donc évacués en solution tandis que la phase terrigène est enrichie en potassium. Un schéma analogue peut être donné pour l'altération de la chlorite et de l'argile micacée qui sont les phyllosilicates principaux des schistes briovériens et paléozoïques. La chlorite s'altère en premier lieu, évacuant en solution le magnésium et une partie du fer sous forme réduite tandis que l'argile micacée subsiste plus longtemps (ESTEOULE-CHOUX, 1967,). Ceci conduit de la même manière à un appauvrissement relatif de la phase terrigène en magnésium et à son enrichissement en potassium.

Les vases argileuses déposées dans le bassin sont donc constituées pour partie de minéraux qui ont pris naissance au cours de l'altération en milieu continental et pour partie des débris de minéraux instables désorganisés en cours d'évolution : les uns comme les autres sont instables en milieu marin et vont donc subir un ajustement intrasédimentaire qui les conduira vers des termes plus stables. Le dépôt dont la majeure partie des éléments a été amenée sous forme figurée est enrichi en alumine, en oxyde de fer ferrique, en silice et de manière relative en potassium. Une partie du calcium évacué en solution se trouve réintroduit dans cette boue sous forme d'allochems. De plus, une fois le dépôt mis en place, les solutions vont diffuser plus difficilement et les échanges tendant à équilibrer les concentrations avec le milieu franchement marin vont être considérablement freinés. Ce sédiment fraîchement déposé constitue donc en milieu éminemment favorable à la formation rapide, par ajustement, de minéraux stables :
- 1° l'instabilité du matériel d'apport va permettre l'établissement de concentrations en solution élevées du fait de l'importance du potentiel chimique des éléments cons-

titutifs des réseaux cristallins plus ou moins désorganisés,
- 2° les réseaux phylliteux plus ou moins désorganisés jouent le rôle de germe de cristallisation et par abaissement de l'énergie d'activation, ils vont accélérer la formation des nouvelles phases minérales auxquelles ils servent d'ébauche,
- 3° enfin, ces résidus en voie de dissolution sont capables de fournir des quantités importantes de matière et maintiennent de façon constante les solutions intrasédimentaires à des concentrations élevées.

Si le passage en solution de la majorité des éléments constituant les nouvelles phases minérales est indéniable, il est à noter que l'eau dans ces processus ne joue que le rôle d'un vecteur à petite distance qui permet aux ions de se réagencer d'une particule à l'autre. La quasi totalité de la matière minérale mise en jeu dans ces transformations a été transportée sous forme figurée. Seuls les ultimes ajustements se sont effectués par l'intermédiaire d'ions en milieu aqueux.

La coexistence de grains de glauconie avec la phase smectitique diffuse pose le problème de leurs relations génétiques. Si l'on écarte le cas où les grains de glauconie apparaissent remaniés, les observations montrent que ces deux minéraux prennent naissance concurremment dans un même milieu (LOUAIL et al., 1977,).

Il n'est donc pas possible de relier la formation préférentielle de glauconite à des enrichissements locaux de solutions en fer ou en potassium qui amèneraient la fermeture de certains feuillets ; par contre, il semble préférable de considérer la constitution cristallochimique de ces deux minéraux. En effet, les beidellites ferrifères, évidemment riches en alumine, présentent une teneur en magnésium relativement basse qui varie peu autour de 1,5% tandis que les glauconites, outre qu'elles ont une teneur élevée en fer, présentent une teneur en magnésium plus grande qui se situe en moyenne de 2,5 à 3%. Cette localisation préférentielle du magnésium dans le réseau des glauconites trouve son explication dans le fait que les cations octaédriques de ce minéral ont un encombrement moyen plus grand que celui des cations octaédriques de la beidellite ferrifère : les paramètres b mesurés sur les glauconites sont supérieurs à 9 Å tandis que ceux des beidellites sont légèrement inférieurs. Ces deux minéraux argileux dioctaédriques de composition cristallochimique stable constituent un couple capable de traduire par la variation de leurs proportions relatives, les fluctuations de l'alimentation du bassin.

Localement un enrichissement élevé en silice qui se manifeste également par l'apparition de cristobalite diagénétique, entraîne la formation de clinoptilolite dans les pores. Si, compte tenu des observations dont on dispose actuellement, il n'est pas possible d'envisager précisément les mécanismes qui en sont responsables, l'apparition de zéolites traduit de la même marnière que la présence de glauconite, un ajustement diagénétique capable de répondre aux fluctuations du chimisme du milieu.

CONCLUSION

Cette étude des dépôts Cénomaniens de la bordure Sud-Ouest du Bassin de Paris mont que la phase argileuse constituée de beidellites ferrifères et de glauconite est le résultat d'une transformation diagénétique précoce d'éléments terrigène, la contribution directe des apports d'éléments en solution paraissant négligeable.

REFERENCES

AUBRY, M. P. et POMEROL, B., 1975. La pétrogénèse des craies du Bassin de Paris est-elle une conséquence de l'expansion océanique ? C. R. Acad. Sci. Paris, 280 : 2081 2084.
BROWN, G., CATT, J. A. and WEIR, A. H., 1969. Zeolites of the clinoptilolite - heulandite type in sediments of south-east England. Miner. Mag., 37 : 480-488.
ESTEOULE-CHOUX, J., 1967. Contribution à l'étude des Argiles du Massif Armoricain. Argiles des altérations et argiles des bassins sédimentaires tertiaires. Thèse Sci Rennes, 319 pp.
ESTEOULE-CHOUX, J., ESTEOULE, J. et LOUAIL, J., 1969. Sur la présence d'un dépôt à kaolinite et à gibbsite entre le Bajocien et le Cénomanien en Maine-et-Loire. C. R Acad. Sc. Paris, 268 : 891-893.
HEIM, D., 1957. Uber die mineralischen, nicht karbonitischen Bestandteile des Cenoman und Turon der mitteldeutschen Kreidemulden und ihre Verteilung. Heidelberg. Beitr., 5 : 302-330.
JEANS, C. V., 1968. The origin of the montmorillonite of the European chalk with special reference to the lower chalk of England. Clay Min., 7 : 311-329.
JEANS, C. V., 1971. The neoformation of clay minerals in the brackish and marine env ronments. Clay. Min., 9 : 209-217.
JUIGNET, P., 1974. La transgression crétacée sur la bordure orientale du Massif armoricain. Thèse Sci. Caen, 806 pp., 174 fig., 28 pl.
LOUAIL, J., 1969. Etude sédimentologique des sables et graviers de Jumelles (Maine-et-Loire). Thèse 3ème cycle, Rennes, 113 pp.
LOUAIL, J., ESTEOULE, J. et ESTEOULE-CHOUX, J., 1977. La glauconitisation, transformation ou néogénèse : conditions de formation des glauconies du Cénomanien du Poit 5ème Réun. An. Sc. de la Terre, Rennes : 322.
MATHIEU, C. et CONINCK (de), F., 1972. Caractérisations physico-chimiques et sédimentologiques des craies turoniennes et coniaciennes de Thiérache et du Marlois (Nord-Est du Bassin de Paris). Bull. Inf. Géol. Bass. Paris, 34 : 3-14.
MILLOT, G., CAMEZ, T. et BONTE, A., 1957. Sur la montmorillonite dans les craies. Bull. Serv. Carte géol. Als. Lor., 10, n° 2 : 25-26.
MILLOT, G., 1964. Géologie des Argiles. Masson, Paris, 499pp.
MORGAN-JONES, M., 1977. Mineralogy of the non carbonate material from the chalk of Berkshire and Oxfordshire, England. Clay Min., 12 : 331-343.
POMEROL, B., 1977. La connaissance des paléo-océans : données fournies par l'étude des formations sédimentaires des marges continentales. C. R. Acad. Sci. Paris, 284 341-344.
STEINBERG, M., 1967. Contribution à l'étude des formations continentales du Poitou (Sidérolithique des auteurs). Thèse Sci., Paris : 415 pp.
THOREZ, J. et MONTJOIE, A., 1974. Lithologie et assemblages argileux de la smectite de Herve dans les craies campaniennes et maestrichtiennes dans le Nord-Est de la Belgique. An. Soc. Géol. Belgique, 96 : 651-670.
VALETON, I., 1960. Vulkanische Tuffiteinlagerung in der nordwestdeutschen Oberkreide. Mitt. Geol. Staatsinst. Hamburg, 29 : 26-41.
WEIR, A. H. and CATT. J. A., 1965. The mineralogy of some upper chalk samples from the Arundel Area, Sussex. Clay. Min., 6 : 97-110

A MONTMORILLONITE, KAOLINITE ASSOCIATION IN THE LOWER CRETACEOUS OF SOUTH-EAST ENGLAND

D.J. MORGAN, D.E. HIGHLEY and D.J. BLAND
Institute of Geological Sciences, London.

ABSTRACT

A clay bed occurring in an Upper Aptian-Lower Albian sandstone sequence at Godstone, Surrey, UK, contains montmorillonite and well-ordered kaolinite, the proportions of which vary both laterally and vertically. Electron microscopy shows that kaolinite is forming at the expense of montmorillonite and also suggests that there is an epitaxial relationship between the original montmorillonite flakes and the newly-formed kaolinite crystals. Levels of Zr, Nb, Y, La and Ce indicate that the original montmorillonite of the bed is the alteration product of alkaline volcanic ash of similar composition to that which gave rise to the slightly older fuller's earth deposits of the area.

INTRODUCTION

Economically important deposits of Ca-montmorillonite (fuller's earth) are found in Lower Greensand sediments between Redhill and Godstone, Surrey (Dines and Edmunds, 1933). The fuller's earth beds, which show convincing evidence of derivation from volcanic ash (Jeans et al., 1977), are restricted to the Sandgate Beds which are of Upper Aptian age (Casey, 1961). Recently, Mr N G Ward of British Industrial Sand Ltd drew the authors' attention to a clay bed resembling a fuller's earth occurring much higher in the Lower Greensand succession in an abandoned quarry in the Folkestone Beds (Upper Aptian-Lower Albian) at Godstone. This clay bed proved to contain both montmorillonite and well-ordered kaolinite, vertical and lateral relationships between these two minerals strongly suggesting that kaolinite was secondary after montmorillonite. This preliminary report describes the mineralogy of the bed, the morphology of the clay constituents as observed by both scanning and transmission electron microscopy, and changes in major and minor element chemistry accompanying the transition from montmorillonite to kaolinite.

GEOLOGICAL SETTING AND GENERAL MINERALOGY OF BED

The Folkestone Beds exposed in the worked-out Forge glass sand quarry, Godstone (National Grid Reference: TQ 3452 5182), consist of massive and cross-bedded, white to

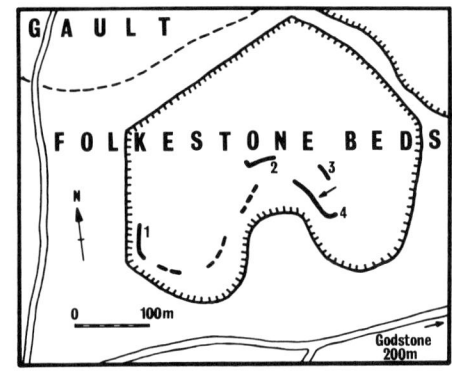

Fig.1. Sketch map of Forge sand quarry, Godstone, showing outcrop (thick lines) of clay bed.

buff sandstones above and reddish-brown sandstones beneath. The montmorillonite, kaolinite clay bed, with a silty sandstone immediately below, marks the junction between these two lithologies. It occurs some 30m below the base of the Gault and m represent the 'Silt Band' recorded by Gossling (Kirkaldy, 1947) and used as a local marker horizon.

Quarrying was confined mainly to the cleaner sands and the clay bed is visible on in the western and central part of the quarry where some working of the underlying sands took place. Much of this area now lies below the water table. Fig.1. is a sketch map of the quarry showing exposed sections of the bed. It varies in thicknes from 10 to 40cm and dips at a shallow angle to the NNW, disappearing under water in this direction. The intermittent nature of the exposure elsewhere is due to regradi and slumping of the quarry sides.

In all the exposed sections the bed has a very sharp base but an irregular upper surface, due mainly to the presence of loadcast structures in the overlying sandston Sand-filled burrows are also present in the top few cm of the clay. The bed is thickest in sections 1 and 2 where it consists of three lithologically distinct zone (Fig.2). X-ray diffraction traces of representative material from these zones

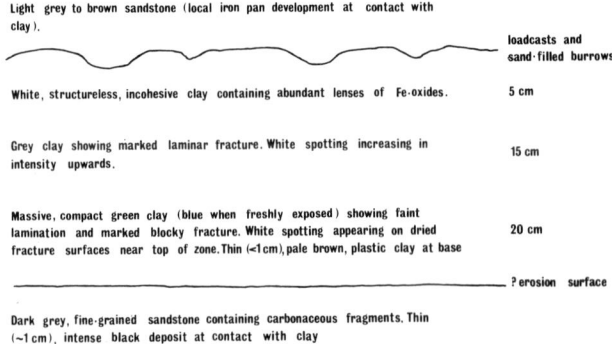

Fig.2. Lithological description of clay bed exposed in sections 1 and 2.

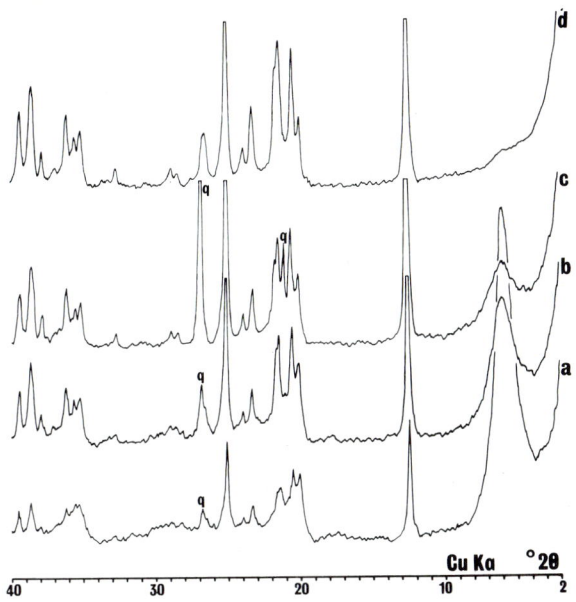

Fig.3. X-ray diffraction traces: (a) green clay zone, (b) grey clay zone, (c) white clay zone, section 1; (d) buff clay, section 4. All original samples.

are shown in Fig.3. Montmorillonite (dioctahedral, $d(060) = 1.498$ Å) predominates over kaolinite in the green clay comprising the lower half of the bed. This material has an ethylene glycol monoethyl ether surface area (Carter et al., 1965) of 546 $m^2/g$ which indicates a bulk composition near 65% montmorillonite and 35% kaolinite (assuming values of 800 and 40 $m^2/g$ for the surface areas of the pure minerals). No more than 2-3% of non-clay minerals, mainly quartz and pyrite, are present. Local concentrations of kaolinite are common towards the top of the zone, imparting a faint white spotted effect to dried fracture surfaces of the clay. The green clay passes upwards into a greyish clay (Fig.3, trace b) which has a surface area of 312 $m^2/g$, equivalent to a bulk composition of approximately 35% montmorillonite and 60% kaolinite (most of the remainder being quartz). About 80% kaolinite is present in the white structureless clay comprising the top of the bed. 'Hardpan' seams and lenses of Fe-oxides/hydroxides are common near the contact with the overlying sandstone and much quartz is present (Fig.3, trace c).

In the remaining exposures in the SW corner of the quarry (un-numbered sections in Fig.1) the bed shows a similar upward increase in amount of kaolinite relative to montmorillonite. In no instance, however, is the lithological zoning so well-defined as in sections 1 and 2.

In section 3 the bed is 37cm thick and is composed throughout of white, incohesive, structureless clay, faintly mottled in shades of pink. From the base to within 5cm of the top the bed has a uniform composition near 30% montmorillonite and 70% kaolinite. The kaolinite content of the top 5cm is about 80%.

At the NW end of section 4 the bed is only 20cm thick and consists of mottled white and buff friable clay with a composition again near 30% montmorillonite and 70% kaolinite. Traced SE-wards along the section the bed gradually thins and increases in overall kaolinite content. Halfway along the section, at the point marked by an arrow in Fig.1, the kaolinite content reaches 90%. An X-ray diffraction trace of this clay (surface area $60m^2/g$) is given in Fig.3 (trace d). The well-ordered nature of the kaolinite is evident from the extent of peak development between 19 and $22°2\theta$; traces run at slower scanning speeds consistently gave Hinckley 'crystallinity' indexes >1.6. A sharp infrared spectrum is also given by this kaolinite (Fig.4). Towards the end of the section, small-scale laminated and cross-bedded structures are picked out in the upper portion of the bed by reddish-brown secondary Fe-oxides/hydroxides.

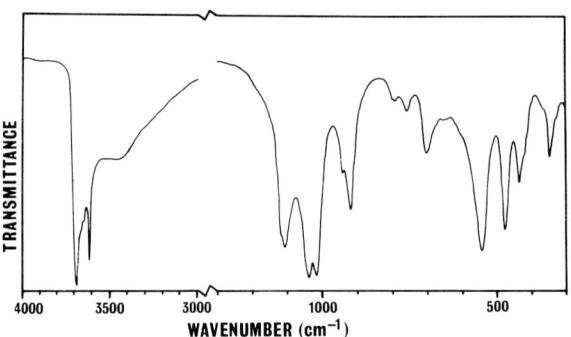

Fig.4. Infrared spectrum of buff clay, section 4.

ELECTRON MICROSCOPY

Samples for electron microscopic examination were taken from the bed exposed in sections 1 and 4. Transmission electron micrographs were made using a Philips 201 electron microscope; specimens were prepared by dropping dilute aqueous dispersions on to a carbon-coated support grid. Scanning electron microscopy (using a Cambridge Stereoscan IIA) was carried out on fresh fracture surfaces of the specimens after coating first with carbon then with gold/palladium.

Bed exposed in section 1

A transmission electron micrograph of the basal green clay from this section is given in Fig.5a. Kaolinite is present as thick, rounded to subhedral crystals, mainly 0.3 to 0.4μm in diameter. Montmorillonite mostly occurs as thin laths about 0.3μm long and 0.1μm wide, these being in various stages of aggregation. To the left of the micrograph is a large thin undispersed montmorillonite flake. This has an unusually regular outline with many straight edges meeting at 120° angles, suggesting that conversion to kaolinite may be taking place by lateral epitaxial growth.

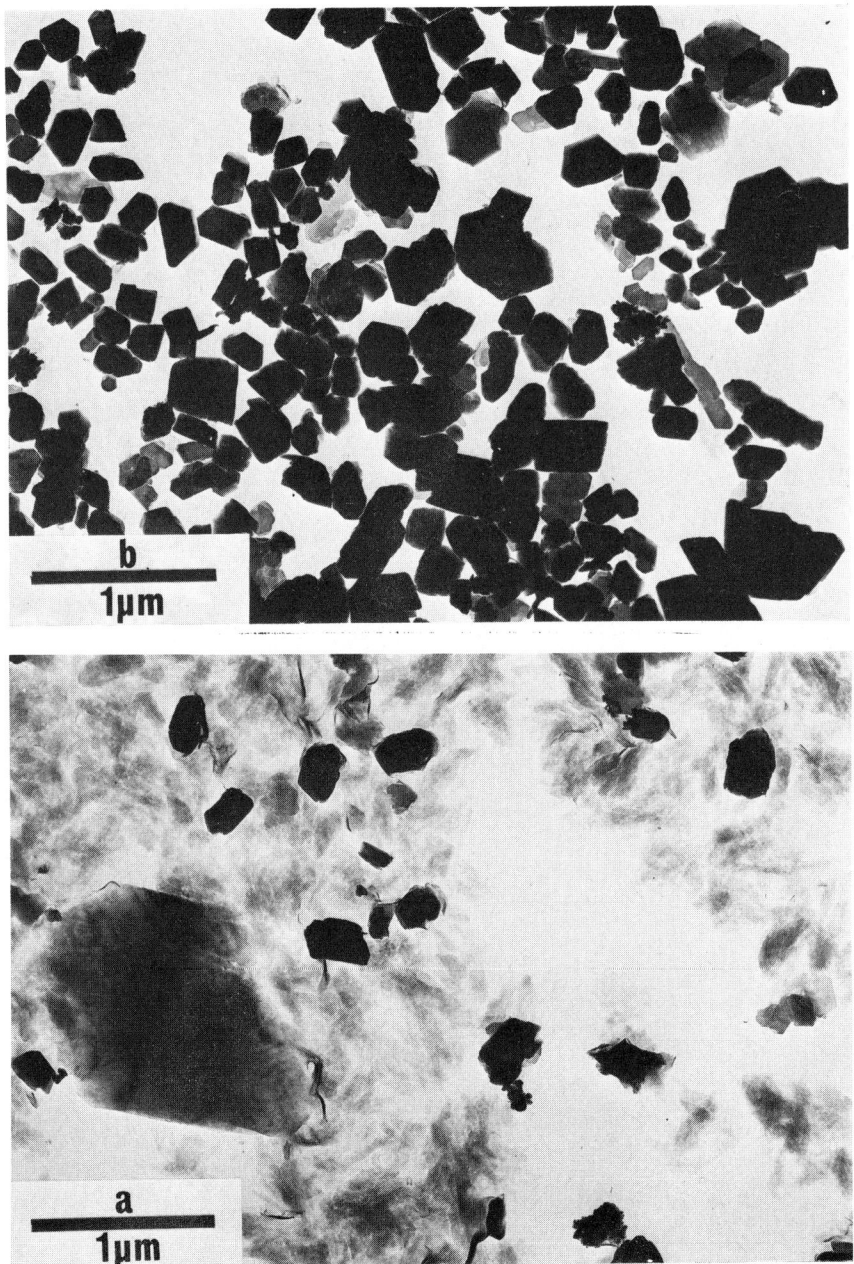

Fig.5. Transmission electron micrographs: (a) green clay zone, section 1; (b) buff clay, section 4.

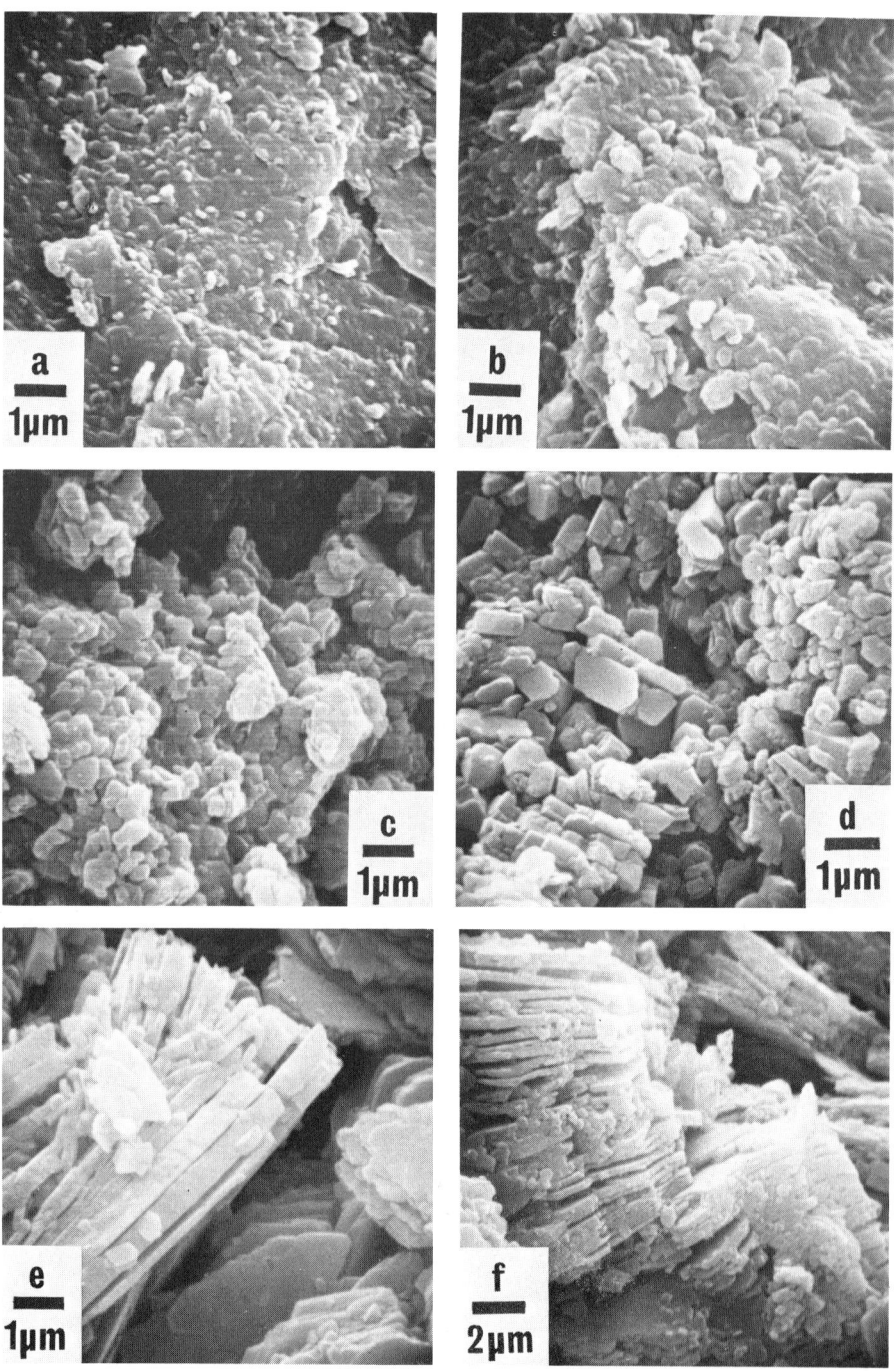

(a)-(b) green clay zone, (c) grey clay zone, section 1; (d)-(f) buff clay, section 4.

Scanning electron microscopic examination of samples from this section confirms that there is a morphological relationship between the original montmorillonite flakes and the newly-formed kaolinite crystals. Fig. 6a is a view looking down on the bedding plane surface of the basal green clay. Margins of the clay flakes are generally rounded but occasional straight edges intersecting at ~120° angles can be distinguished. In Fig. 6b, which is of an area in the green clay locally richer in kaolinite (corresponding to one of the 'white spots' noted in Fig.2), the larger clay flakes appear to be breaking down into smaller flakes of definite hexagonal outline. In the overlying grey clay where kaolinization is further advanced, most of the original montmorillonite flakes are replaced by a mosaic of hexagonal kaolinite crystals (Fig. 6c). Individual crystals in the mosaic are between 0.1 and 0.5µm in diameter and all lie with their a-b planes parallel to the bedding plane (by now also a well-defined laminar fracture plane) of the clay.

Bed exposed in section 4

A transmission electron micrograph of a sample from the completely kaolinized portion of this section is given in Fig. 5b. Although most of the kaolinite crystals have sharply-defined hexagonal outlines, a significant proportion show square to rectangular cross-sections. These must be lying with their c-axes parallel to the substrate, indicating that they have high thickness : diameter ratios. The blocky nature of the kaolinite crystals is apparent in a scanning electron micrograph (Fig. 6d) of the clay. This has an open, porous texture; individual crystals are randomly oriented and have diameters generally between 0.5 and 1.5µm and thicknesses of up to 0.5µm. Occasional pockets containing thick books of large kaolinite plates (Fig. 6e) and vermicular growths (Fig. 6f) are present, small euhedral crystals of kaolinite commonly occurring on the edges of the plates.

These textural features are characteristic of kaolinite formed by an in situ alteration process (Keller, 1978). It would appear that once the montmorillonite structure was completely destroyed, conditions prevailed that were either at or near equilibrium with respect to kaolinite crystallization. Kaolinite crystals originally formed in parallel mosaic growth as shown in Fig. 6c dissolved and then recrystallized as larger blocky crystals; large plates developed where there was ample space for growth, although it is possible that these represent a later dissolution/recrystallization phenomenon. The small euhedral crystals coating the edges of the large kaolinite plates obviously formed later still and may have been precipitated from solutions percolating through the bed.

CHEMISTRY

Changes in chemical composition of the bed with increase in extent of kaolinization are illustrated by the analyses in Table 1. Molar $Al_2O_3$ : $SiO_2$ ratios increase from approximately 1:2 in the basal green clay from section 1 to almost 1:1 (the theoretical kaolinite value) in the buff clay from section 4. Apart from Si, the

TABLE 1

Chemical analyses

| % | section 1. | | section 3. | section 4 |
|---|---|---|---|---|
| | green clay | grey clay | pink clay | buff clay |
| $SiO_2$ (reactive) | 45.0 | 43.0 | 42.5 | 42.3 |
| $SiO_2$ (free) | 0.3 | 3.0 | 2.0 | 1.4 |
| $Al_2O_3$ | 24.0 | 29.8 | 30.3 | 36.4 |
| $TiO_2$ | 0.90 | 1.49 | 1.22 | 1.49 |
| $Fe_2O_3$ | 5.11 | 2.71 | 4.31 | 2.18 |
| FeO | 0.35 | 0.06 | 0.10 | 0.10 |
| CaO | 1.34 | 0.65 | 0.78 | 0.17 |
| MgO | 1.64 | 0.63 | 0.65 | 0.14 |
| $Na_2O$ | 0.10 | 0.10 | 0.10 | 0.20 |
| $K_2O$ | 0.20 | 0.15 | 0.15 | 0.15 |
| MnO | 0.01 | 0.01 | 0.01 | 0.01 |
| $P_2O_5$ | 0.16 | 0.16 | 0.20 | 0.27 |
| loss on ignition (1000°C) | 21.0 | 17.4 | 17.7 | 14.8 |
| (loss at 105°C) | (12.5) | (6.5) | (6.8) | (1.6) |
| | 100.11 | 99.16 | 100.02 | 99.61 |
| ppm Zr | 2120 | 3910 | 2650 | 3330 |
| Nb | 260 | 370 | 350 | 380 |
| Y | 110 | 250 | 150 | 230 |
| La | 220 | 350 | 260 | 470 |
| Ce | 570 | 930 | 710 | 1310 |

main elements removed are Mg, present in the octahedral layer of the original montmorillonite, and Ca, mainly occupying exchange sites on the surfaces of this mineral. Not all the Fe resulting from breakdown of the montmorillonite structure is removed from the bed, a large proportion remaining dispersed within it in the form of secondary Fe-oxides/hydroxides.

The bed contains unusually high amounts of Zr, Nb, Y, La and Ce. In general, levels of these elements increase with increasing degree of alteration. As kaolinite itself can be dismissed as a possible host phase on crystallochemical grounds, the most likely explanation for this behaviour is that the elements are present in constituents of a resistate mineral suite which is being progressively upgraded with loss of Si, Mg, Ca etc. from the bed. Amounts of Zr, Nb and Y were found to decrease only slightly in <2μm products whilst La and Ce values actually increased, indicating that the possible resistate host minerals must be extremely fine-grained. In an attempt to isolate these, a 25g sample of the <2μm product from the grey clay zone in section 1 was treated with an excess of HF for 6 hours. Apart from a complex magnesium aluminium fluoride commonly precipitated during prolonged HF digestion of clay minerals (Campbell et al., 1972), the only phases detected by X-ray powder photography of the ~200mg residue were zircon and anatase. X-ray fluorescence analysis of the residue, however, indicated major amounts of all five elements. Zircon, the bulk of which occurs below 2um, is thus the source of the Zr in the bed and the strong correlation shown between Zr and Y values in Table 1 suggests that the latter element is present in isomorphous substitution in the zircon structure. Anatase is the probable source of the Nb, the geochemical association of Ti and Nb being well-documented (Vlasov, 1966). Neither of these minerals can be responsible for more than a small proportion of the La and Ce detected in the HF residue, these elements most probably having been co-precipitated by the complex magnesium aluminium fluoride following dissolution of the original La, Ce-bearing phase. Evidence bearing on the nature of this was obtained from electron microprobe analysis of an araldite-impregnated block of the buff clay from section 4. Small areas approximately 2um long and 1um across were located which, using a standard X-ray imaging technique, proved to consist of the element association Ca-Ce-P. Both morphological and chemical data would be compatible with a rare earth-bearing apatite.

## DISCUSSION AND CONCLUSIONS

According to Jeans et al. (1977), the Upper Aptian fuller's earths of the Redhill-Godstone area represent the alteration products of volcanic ash carried into a shallow, nearshore marine environment by rivers draining the London - Brabant Platform to the north. High levels of Zr, Nb, Y, La and Ce shown by the fuller's earths were considered by Morgan (1974) to be a reflection of the alkaline nature of the precursor volcanic ash. The original montmorillonite of the present bed was derived from ash of similar composition and source. Fine-grained zircon, and possibly a rare earth-bearing apatite, were crystalline components of the ash. The Folkestone Beds consist characteristically of well-sorted, cross-bedded sands which were deposited in shallow water under the influence of strong current activity. However, the presence of abundant carbonaceous matter in the fine-grained sandstones immediately associated with the bed suggests that relatively quiet conditions prevailed about the time of deposition of the ash. Alteration to montmorillonite occurred either during or soon after deposition. Kaolinization of the bed occurred much later and was probably effected by acidic ground water circulating in the overlying sandstone. The extent of kaolinization appears to have been controlled mainly by the thickness of the bed.

Direct intracrystalline transformation of montmorillonite to kaolinite has been postulated on the basis of electron microscopic examination. The feasibility of this transformation has not been generally accepted although Altschuler et al. (1963) have suggested a provisional mechanism involving an intermediate regular 1:1 mixed-layer montmorillonite-kaolinite phase. Detailed X-ray examination of samples from the present bed failed to detect such a phase, however, and work is currently in progress aimed at devising an alternative mechanism.

ACKNOWLEDGEMENTS

We should like to thank Miss A M Shilston for help with the laboratory investigations, N Cogger for major element chemical analyses and Dr A Weir for supplying transmission electron micrographs. We should also like to thank The East Surrey Water Company, the present owners of the quarry, for allowing us access. This paper is published by permission of the Director, Institute of Geological Sciences.

REFERENCES

Altschuler, Z.S., Dwornik, E.J. and Kramer, H., 1963. Transformation of montmorillonite to kaolinite during weathering. Science, 141, 148-152.

Campbell, A.S., Adams, J.A. and Howarth, D.T., 1972. Some problems encountered in the identification of plumbogummite minerals in soils. Clay Miner., 9, 415-423.

Carter, D.L., Heilman, M.D. and Gonzalez, C.L., 1965. Ethylene glycol monoethyl ether for determining surface area of silicate minerals. Soil Sci., 100, 356-360.

Casey, R., 1961. The stratigraphical palaeontology of the Lower Greensand. Palaeontology, 3, 487-621.

Dines, H.G. and Edmunds, E.H., 1933. The Geology of the Country around Reigate and Dorking. Mem. Geol. Surv. Engl. Wales, HMSO, London, 204 pp.

Jeans, C.V., Merriman, R.J. and Mitchell, J.G., 1977. Origin of Middle Jurassic and Lower Cretaceous fuller's earths in England. Clay Miner., 12, 11-44.

Keller, W.D., 1978. Classification of kaolins exemplified by their textures in scan electron micrographs. Clays and Clay Miner., 26, 1-20.

Kirkaldy, J.F., 1947. The work of the late Mr Frank Gossling on the stratigraphy of the Lower Greensand between Brockham (Surrey) and Westerham (Kent). Proc. Geol Ass., 58, 178-192.

Morgan, D.J., 1974. Calcium montmorillonite and related mixed-layer illite/ montmorillonite clays from the UK: the interdependence of physical properties with mineralogy, chemistry and mode of occurrence. Unpublished Ph.D. Thesis, University of London.

Vlasov, K.A., 1966. Geochemistry of Rare Elements. IPST, Jerusalem, 688 pp.

MINERAL DISTRIBUTIONS IN SEDIMENTS ASSOCIATED WITH THE ALTON MARINE BAND NEAR PENISTONE, SOUTH YORKSHIRE

D.A. ASHBY* and M.J. PEARSON[+]
*Hepworth Iron Company, Hazlehead, Stocksbridge, Sheffield, S30 5HG.
[+]Department of Geology and Mineralogy, The University, Marischal College, Aberdeen, AB9 1AS.

ABSTRACT

Three stratigraphically equivalent mudstone/shale sequences associated with the Alton (G. listeri) Marine Band have been freshly exposed in quarry sections near Penistone, South Yorkshire. The succession comprises a seat earth, coal and sideritic black shale lithology that is common in the paralic sedimentary facies of the Lower Westphalian A in the Pennine Province.

The mineralogy of these predominantly fine grained sediments is complex. The detrital minerals are quartz, kaolinite, mica ($2M_1$ and illite-smectite) and subsidiary chlorite, while the principal diagenetic minerals are siderite and pyrite. Mineral estimates have been derived using a new technique for interpreting the geochemical data obtained from some 58 samples taken from the sequence.

The mineral distributions correlate well with faunal and lithological changes and provide evidence for the nature of the depositional environment, the origin of the source material and the lateral effects of synsedimentary tectonics.

---

INTRODUCTION

The use of quantitative clay mineral estimates to highlight trends in the vertical and lateral distribution of minerals in the sedimentary environment has seen only limited application. Two reasons for this are the lack of a sufficiently accurate technique capable of rapidly determining clay mineral abundances in a wide range of sediment types and also the large sample populations generated by such studies (Griffon et al., 1968; Porrenga, 1966). In this paper, a computer based technique for mineralogical analysis has been applied to a range of commercially important mudrocks as an aid to mineral facies interpretation.

Carboniferous coal measure sediments are used by the Hepworth Iron Company for the manufacture of vitrified drainageware near Penistone, South Yorkshire. One particular sequence that has been extensively worked is the black shales and underclays associated with the Halifax Hard Bed Coal and the Alton (Gastrioceras listeri) Marine Band. This has provided a unique opportunity to study freshly exposed material in three quarry sections over a distance of some 4 kilometres.

GEOLOGICAL SETTING

Lower Westphalian A sediments in the Pennine Province are characterized by rhythmic units up to 30m in thickness. The predominant lithology is black shale which frequently grades upwards into a siltstone or sandstone. These coarser clastics are often overlain by poorly developed coals and show seat earth modification. The overall impression is one of a paralic sedimentary environment and this view is supported by the occurrence of eight separate marine incursions (Calver, 1968) and by the lack of true non-marine fauna.

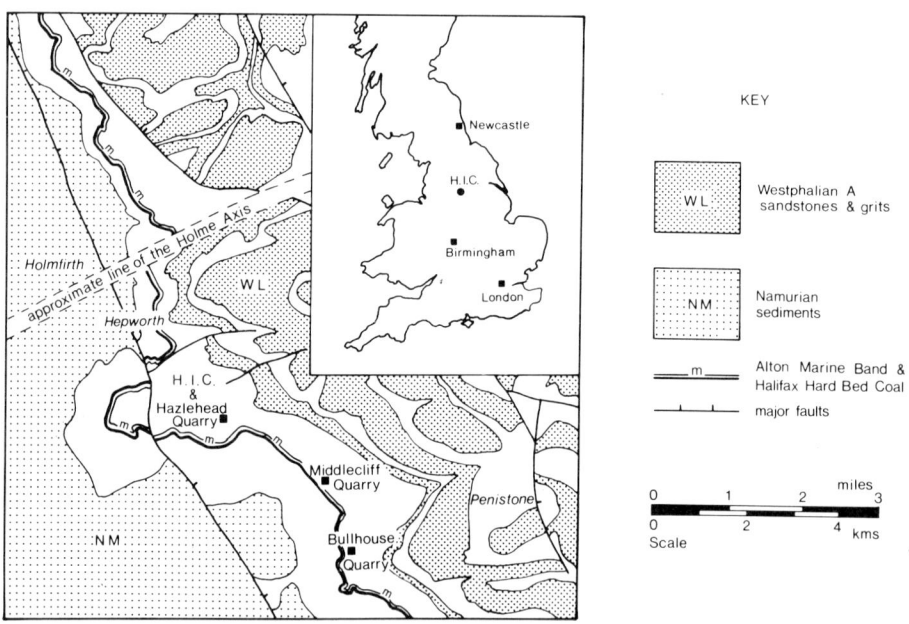

Fig.1. Regional geology and quarry location map.

Typical of this type of sediment sequence are the strata associated with the Alton Marine Band which are currently worked at Bullhouse (Grid Ref. SE209023) and Middlecliff (SE119038) and formerly at Hazlehead (SE182043). The lateral relationship of these quarries to the regional geology and the Holme Axis in particular is shown in Fig.1. This axis is a line of contemporaneous uplift that has caused attenuation of the strata in the Holmfirth and Hepworth areas. Although of minor economic significance in this locality, to the east it causes thinning of important coal seams and forms the boundary between the West and South Yorkshire Coalfields.

THE SEDIMENTS

The lithology and palaeontology described in Fig.2 is representative of the succession in this area. The sedimentary environment and the field relationships between the quarries however require further comment.

The basal sediments are considered to have been black shales coarsening upwards into sheet sandstones prior to modification by groundwater movement and seat earth development in coal swamp conditions. The underclay sequence including the mottled "blue" clays and ganister are the end product of this process. The thickness of these basal measures varies within, as well as between, each quarry, but there is no evidence to suggest that the Holme Axis affected or controlled the development of the seat earth.

The sediments above the Hard Bed Coal are largely sideritic black shales. The Alton Marine Band immediately overlying the coal contains abundant pyritized marine fauna in a very fine-grained carbonaceous shale. The lack of true bethonic fauna suggests that the bottom waters were either euxinic or that the sediment was insufficiently consolidated to support the weight of bottom dwelling species. The total absence of bioturbation in the sediment indicates that very early reducing conditions prevailed.

The shales coarsen upwards into a silty shale or mudstone with mica flakes evident on bedding planes. Here the ostracod Geisina arcuata and the small lamellibranch Curvirimula subovata are found. In the most silty shale these occur in association with Carbonicola prisca. This large lamellibranch is found in situ, indicating that the sediment was now capable of supporting benthonic species, although the faunal association is thought to indicate brackish water rather than truly non-marine conditions (Ameron, 1975). This environment was sharply terminated by the onset of a marine incursion marked by a fine-grained pyritic shale containing siliceous foraminifera and designated the Parkhouse Marine Band (Calver, 1968). Above this horizon the sediments are at first fine-grained, but begin to coarsen towards the top of the quarry sections.

SAMPLING AND ANALYSIS

Hand specimens were collected from the Hazlehead Quarry during the course of an earlier study. The geochemistry carried out on these samples formed the guidelines for the present work and is described in detail elsewhere (Pearson, 1978b).

In subsequent work a series of 50kg channel samples was taken over one metre

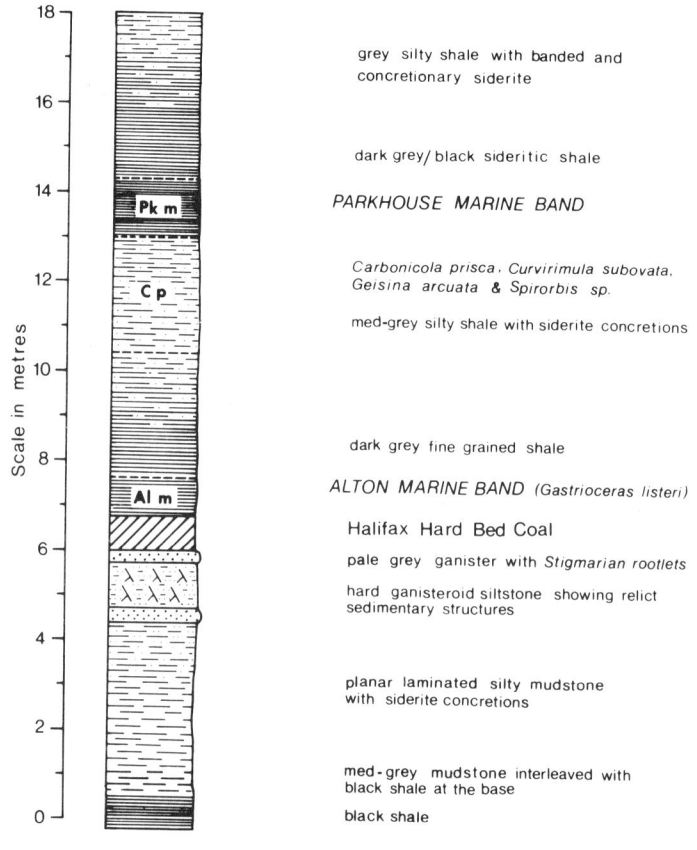

Fig.2. The geological succession (Middlecliff quarry)

intervals from the Middlecliff and Bullhouse Quarries. The bulk samples were dried and ground to pass a 2.5mm sieve and subsamples were "Tema" ground prior to analysis. Major elements were determined by routine X.R.F. spectrometry, whilst carbonate $CO_2$, organic carbon and FeO were analysed by standard wet chemical methods. Quantitative X-ray diffraction was employed to determine quartz (Till and Spears, 1969).

MINERALOGY

The principal diagenetic minerals are siderite and pyrite; calcite is confined largely to the Alton Marine Band and the Hard Bed Coal. Cone-in-cone calcite is rare but can occur throughout the sequence apart from the ganister and the coal.

The major detrital minerals are quartz, $2M_1$ mica, kaolinite and, to a lesser extent, chlorite.

Particle size fractionation was undertaken on three representative samples, firstly to highlight trends in mineral composition with respect to grain size, and secondly to check for minerals not found during the course of routine X-ray diffraction of whole rock (powder) specimens. Samples selected were an underclay, a shale from the Alton Marine Band and a silty shale from the C. prisca band. The fractions separated were <0.5μm, 1-2μm and 53-106μm.

Interstratified illite-smectite and disordered kaolinite were by far the most abundant minerals in the <0.5μm fractions. The only other mineral found was non-expandable chlorite in the underclay and the C. prisca shale but this was totally absent in the Alton shale. No attempt was made to quantify the degree of illite-smectite mixed-layering, although the smectite component was greatest in the underclay. In the 1-2μm fractions there was a marked increase in the discrete 10Å component of the illite-smectite concomitant with a decrease in total expandable material. Quartz was detected in all three samples. Small quantities of feldspar were found in the 53-106μm fractions of both the C. prisca shale and the underclay but not in the Alton shale.

A particle size distribution was determined for each of the three representative samples. In no case did the coarse fraction exceed 7% of the whole rock. Feldspar is therefore unlikely to exceed more than one percent of the total mineral content. Illite-smectite is confined to the finest fractions (<2μm) which together constitute 17-25% of the disaggregated samples. This suggests an approximate whole rock abundance for the smectite layers of 5-10%. Little error is therefore likely to result from neglecting feldspar in the rational analysis whilst the small smectite content can be accommodated in calculated illite.

MINERALOGICAL CALCULATIONS

Deduction of the non-clay phases from the bulk data assumed a composition for pyrite of $FeS_2$, anatase ($TiO_2$), quartz ($SiO_2$) and hydroxyapatite ($Ca_{10}(PO_4)_6(OH)_2$). Detailed studies of the Hepworth carbonate geochemistry have shown that siderite has an average composition of $(Fe, Mn)_{0.82}Mg_{0.11}Ca_{0.07}CO_3$ (Pearson, 1978b). The calculation procedure is illustrated in the flow chart (Fig.3) using sample M48 (Bullhouse black shale) as an example. The residual data are assigned to the clay minerals. The selection of and justification for using the clay standards in matrix $A_{44}$ is discussed in detail elsewhere (Pearson, 1978a).

Assumptions made in the calculation include a fixed composition for kaolinite and the partitioning of mica between muscovite and "Fithian" type illite end members. The iron-rich chlorite standard is justified on the basis of microprobe studies (unpublished) and the work of Collins (1976) which both suggest that such varieties predominate in Carboniferous mudrocks. The standard clay values for

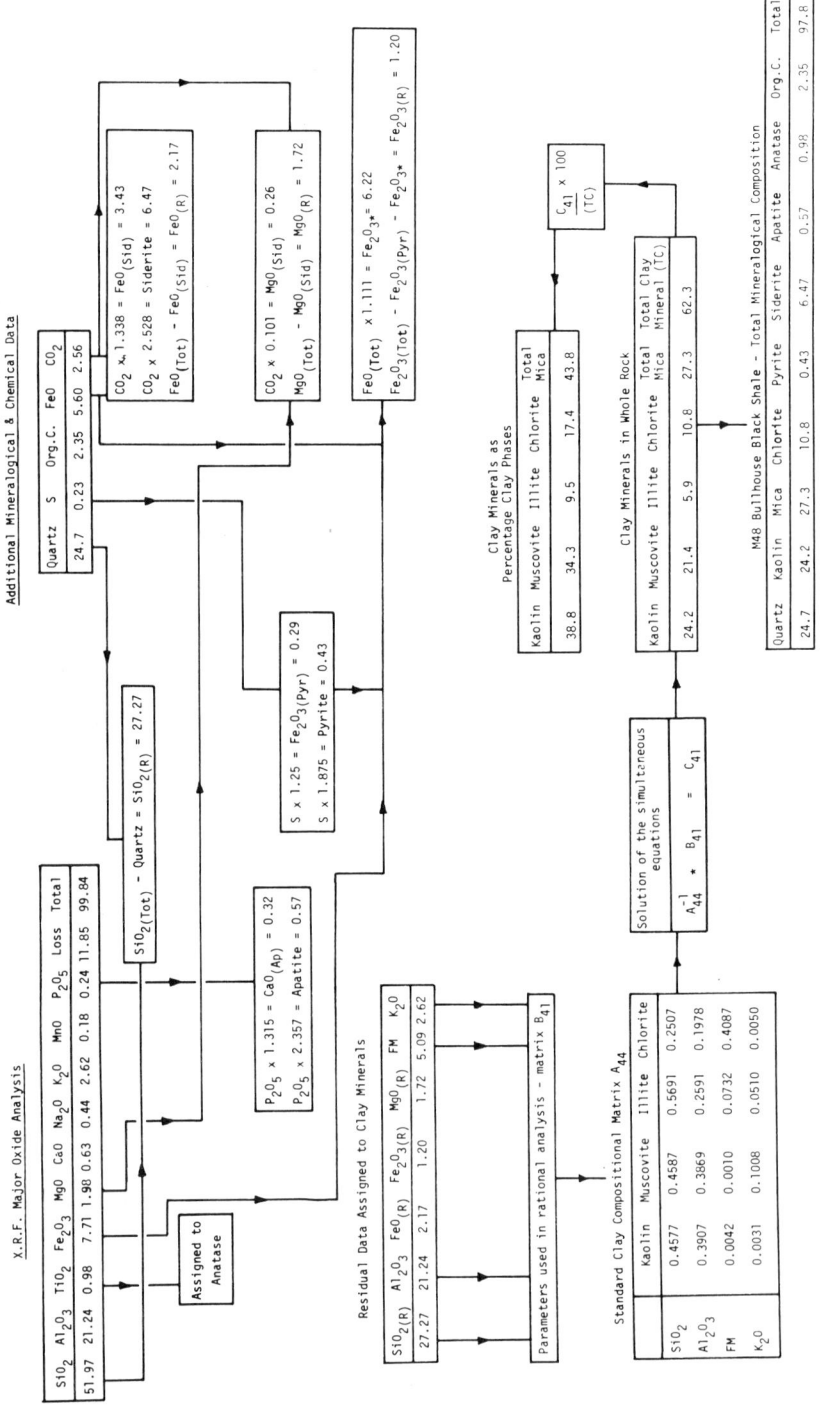

Fig.3. Flow diagram of mineralogical calculations.

the variables $SiO_2$, $Al_2O_3$, $K_2O$ and FM (total of iron and magnesium oxides) are shown in the matrix $A_{44}$. The bulk sample values of these variables are assigned from the residual data of each sample to the matrix $B_{41}$ and computer methods are used to solve the set of simultaneous equations.

## THE MINERAL DISTRIBUTIONS

Distribution diagrams have been plotted for quartz, pyrite and siderite (Fig.4) and for the clay minerals (Fig.5). Point distributions are plotted for the Hazlehead quarry where small hand samples were taken whereas block distributions for the other quarries reflect the continuous channel sampling employed at these sites.

Increases in quartz reflect increases in sediment grain size observed in the quarry sections. Quartz content may therefore be taken as an indicator of the energy of the sedimentary environment. The correspondence of minima or near-minima at all three localities with the Alton Marine Band accords with the quiet conditions of deposition commonly assumed for this lithology whilst the general increase upwards indicates a return to more active sedimentation which reaches a maximum at the C. prisca horizon (closest approach to freshwater conditions) before dropping sharply to a new minimum somewhat above the Parkhouse Marine Band. Settling velocity was therefore closely related to salinity. The generally higher quartz contents in the underclays are partly a reflection of upward coarsening into the ganisteroid siltstone and partly due to silica cementation which probably developed during attack of clays under coal swamp conditions.

Pyrite develops by bacterial reduction of sulphate ions either in anoxic bottom waters or in the immediately underlying sediment. Pyrite distribution clearly reflects the availability of sulphate ions with a sharp increase at each of the Alton and Parkhouse marine incursions whilst a further increase is seen at the top of the sequence at Middlecliff and Bullhouse. This latter may represent the Upper Parkhouse Marine Band, although no confirmatory faunal evidence was found.

Siderite occurs generally through the succession in disseminated and concretionary form but is notably absent from the Alton Marine Band, the Hard Bed Coal and underlying ganister. Siderite precipitates in pore space during early diagenesis (Curtis, 1967), the concretionary form lying on channels of pore fluid expulsion whilst the disseminated form probably appears when permeability is low (Pearson, 1978b). Its distribution is thus complex but does show minima in the marine bands where pyrite development has exhausted available iron.

Clay mineral composition correlates excellently with quartz distribution: mica abundance increases with energy of the depositional environment whilst kaolinite decreases. Two very important conclusions may be drawn. Firstly, that the dominant control on the distribution of clay mineral types is sedimentational:

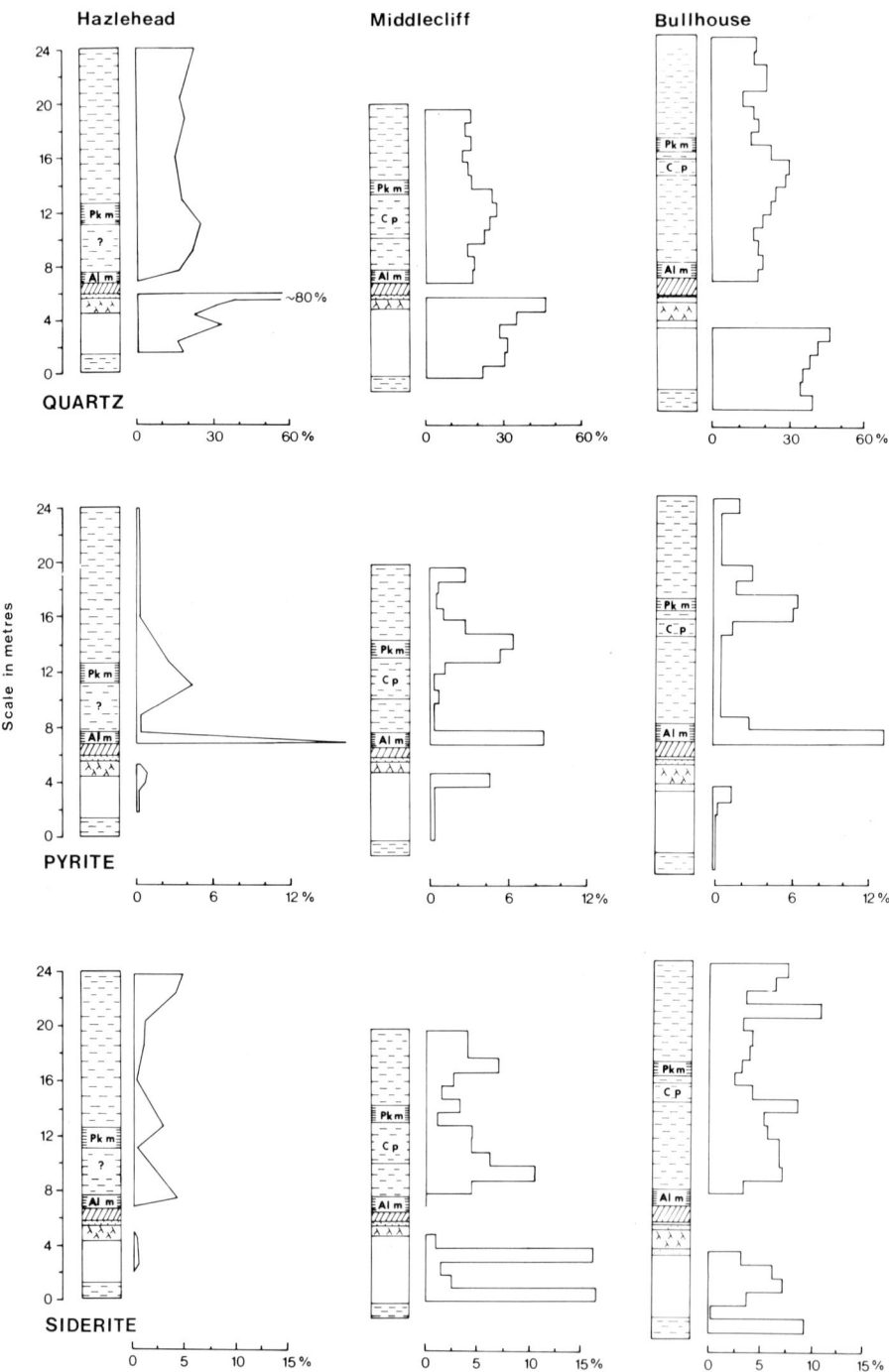

Fig.4. Quartz and diagenetic mineral distributions.

319

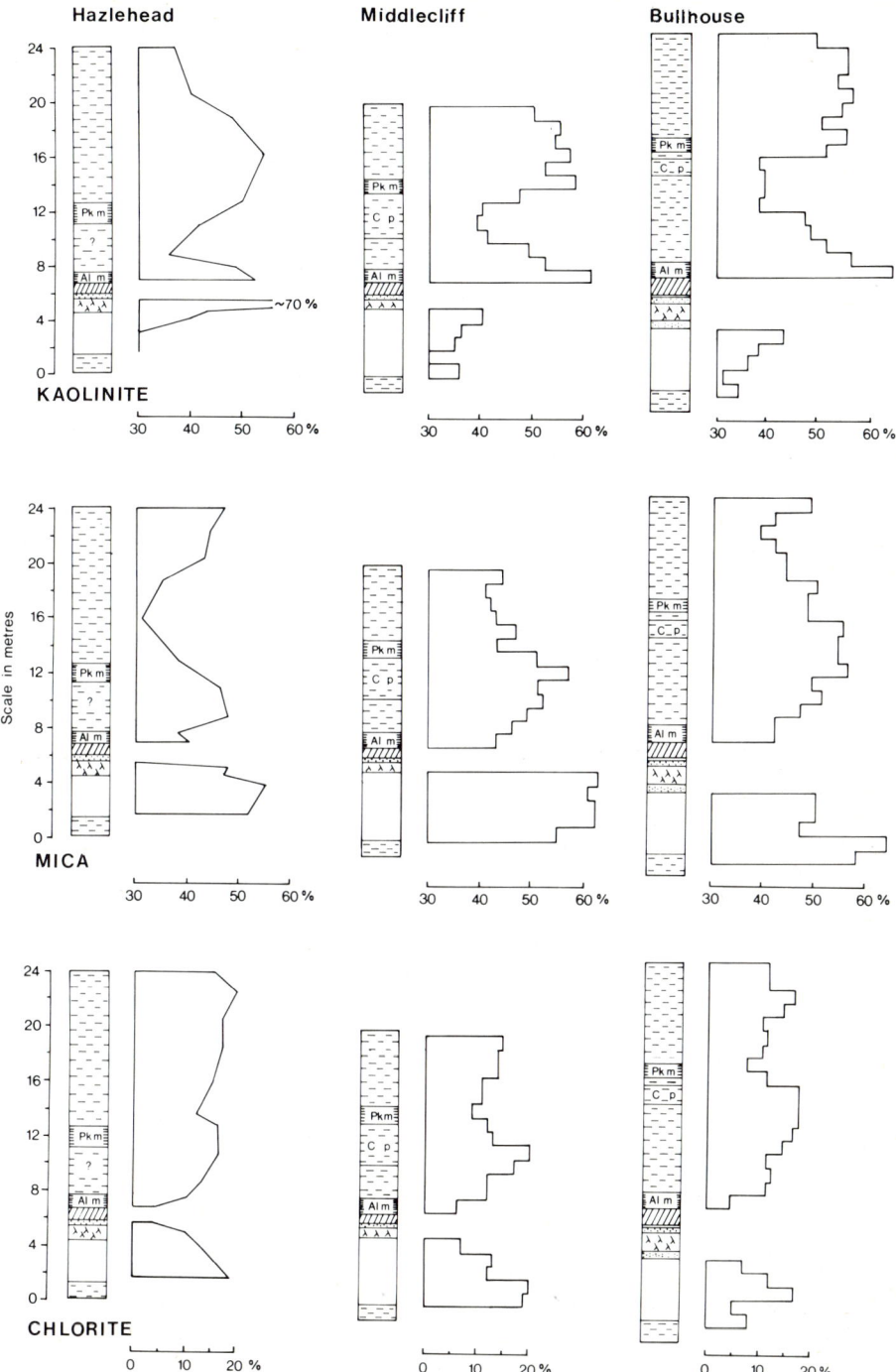

Fig.5. Clay mineral distributions (as % clay phases).

they are detrital clays derived from the weathering of soils. Secondly, that their differentiation may reasonably be interpreted as a grain size effect: kaolinite was the finest material and was therefore probably neoformed in the strong leaching régime of interior tropical Westphalian soils; primary micas, probably degraded, survived weathering and were deposited in relatively coarser fractions of the sediment.

Chlorite, which follows similar trends to quartz and mica, is also of detrital origin but is strongly depleted in the marine bands. This is partly due to the fine-grained nature of these sediments but may indicate further that high sulphide activity accompanied by low pH mobilised both Fe and Mg in the sediment and thereby degraded available chlorite. Mn/Fe ratios suggest that the dominant source of iron for diagenetic mineral development was an oxidate phase such as goethite or ferric hydroxide. The larger part of the sediment was probably derived from intensely weathered soils similar to a modern latosol but the presence of detrital chlorite and coarse-grained mica suggests that other less mature sources contributed significantly.

The underclays contrast with the major part of the succession in that kaolinite increases upwards into the ganister with increasing quartz whilst chlorite decreases to a minimum. Kaolinite was almost certainly neoformed in these sediments in response to increasing leaching intensity in the seat earth whilst chlorite was simultaneously degraded.

LATERAL TRENDS

From palaeontological study it can be shown that the sediments thin northwards towards Hazlehead where C. prisca is absent and the facies is represented only by G. arcuata and C. subovata suggesting that conditions were then too stagnant or bottom sediments insufficiently consolidated for C. prisca to thrive. The interval between the two marine bands indicates most clearly this attenuation of the strata against the Holme Axis.

The most striking aspect of the mineral distributions at the three localities is their general similarity. Quartz, pyrite and the clay minerals follow very similar trends relative to the positions of the marine incursions in all three cases, confirming that their distributions are sedimentologically controlled. Siderite is more variable but consistently low abundances in the marine shales reflect the loss of available iron to pyrite.

The southward trend towards a fuller establishment of open circulation freshwater conditions in the C. prisca band is reflected in modest mineralogical differences. Quartz and more obviously mica reach higher maxima in this horizon at Middlecliff and Bullhouse than at Hazlehead where the sediments are finer in the proximity of the Holme Axis. Sulphide levels in both the Alton and Parkhouse Marine Bands are laterally uniform, an indication that these transgressions were widespread and controlled by factors outwith the depositional basin. The considerable difference in the relative developments of ganister and seat earth on the other hand reflect local topographic and/or vegetation differences during the period of emergence.

CONCLUSIONS

The geochemistry of three stratigraphically equivalent mudstone sequences has been studied and mineral distributions computed. These distributions correlate well with faunal and lithological changes and indicate that quartz and the clay minerals are in general of detrital origin and have abundances related to the energy and salinity of the depositional environment. Pyrite distribution is related to salinity through the latter's control on the availability of sulphate ions whilst siderite distribution, though dominantly controlled by pore water migration channels, shows an inverse relationship to pyrite through their mutual dependence on available iron. Both intensely weathered tropical soils and immature profiles are considered to have contributed to the sediment supply.

The regional geology and palaeontology indicate that the overall sedimentary environment was paralic. The regional tectonic effect of the Holme Axis is seen in the northward thinning of the sediments and the fining and poorer circulation during C. prisca times at Hazlehead concurrent with an increase in mica and coarse clastics to the south.

Mineral distributions in the seat earth and underclays are modified by neoformation of kaolinite during an emergent soil phase and subsequent silica cementation under coal swamp conditions.

REFERENCES

Ameron, H.W.J., Van, 1975. Biostratigrafie van het carboon in Nederland. In: Zagwijn and Staalduinen (Editors), Geologische Overzichskaarten van Nederland. Rijks Geologische Dienst.
Calver, M.A., 1968. Distribution of Westphalian marine faunas in Northern England and adjoining areas. Proc.Yorks.Geol.Soc., 37: 1-72.
Collins, R.J., 1976. A method for measuring the mineralogical variations of the spoils from British collieries. Clay Miner., 11: 31-50.
Curtis, C.D., 1967. Diagenetic iron minerals in some British Carboniferous sediments. Geochim.Cosmochim.Acta, 31: 2109-2123.
Griffon, J.J., Windom, H. and Goldberg, E.D., 1968. The distribution of clay minerals in the world ocean. Deep Sea Res., 15: 433-459.
Pearson, M.J., 1978a. Quantitative clay mineralogical analyses from the bulk chemistry of sedimentary rocks. Clays Clay Miner., In press.
Pearson, M.J., 1978b. Geochemistry of the Hepworth Carboniferous sediment sequence and origin of the diagenetic iron minerals and concretions. Geochim.Cosmochim. Acta. In press.
Porrenga, D.H., 1966. Clay minerals in recent sediments in the Niger delta. Clays Clay Miner.,14: 221-233.
Till, G.A.R. and Spears, D.A., 1969. The determination of quartz in sedimentary rocks using an X-ray diffraction method. Clays Clay Miner.,17: 323-327.

PETROLOGY OF K-BENTONITE BEDS IN THE CARBONATE SERIES OF THE VISEAN AND TOURNAISIAN STAGES OF BELGIUM

J.THOREZ and H.PIRLET
Mineralogical and Geological Depts., Liège University (Belgium)

ABSTRACT

The occurrence of several thin but widespread K-bentonites and subsidiary Tonstein were noticed in the Namur and Dinant synclines of Belgium as intercalated beds in the carbonate Visean and Tournaisian stages ( Lower Carboniferous). Microscopically the texture varies from a dense cryptocrystalline clay to, locally, a highly recrystallized phyllitic matrix resulting from compaction.K-bentonite layers exhibit various types of coarse particles and phenocrysts which are differently associated according to the stratigraphic level or interval: pseudomorphed biotite and feldspar, quartz shards, idiomorphic apatite and zircon. The minus 2 microns fraction was investigated using X-ray diffraction. The clay composition is highly polymineralic but comprises predominantly illite-smectite or illite-vermiculite mixed-layer minerals. These are associated with other clay components: tri- or dioctahedral chlorites, kandites, smectite, illite-chlorite mixed-layer. The composition and texture indicate a volcanic origin. The beds are used as marker-horizons in conjunction with microfossils.

---

INTRODUCTION

During a study of the sedimentology and stratigraphy of the Visean and Tournaisian succession in the Namur and Dinant synclines by Pirlet (1966), several thin but widespread argillaceous beds were noticed to be intercalated in the carbonate sedimentation.

Macroscopically these beds exhibit several specific features which differ readily from those of the occasional shales occurring in the Lower Tournaisian: a conchoidal fracture, wavy lustres, a non-fissile nature, a massive structure. The colours vary between dark grey and different shades of brown and yellow. These beds - a few centimetres in thickness - sometimes show internal darker streaks, close-set vertical jointings, and have sharp contacts with the enclosing rock.

These characters serve to distinguish them and make them suitable for marker-beds as confirmed by rhythmic analyses ( Pirlet,1968) and biostratigraphic correlations ( Conil et al.,1969; Groessens,1974).

Because of the lively interest of such laterally persistent deposi comprehensive petrographic, X-ray diffraction and chemical analyses were carried out on about a hundred samples from the Visean and Tournaisian succession in different localities. A preliminary study demonstrates that these peculiar layers correspond to various K-bentonites and, consequently, bear a volcanic origin.

STRATIGRAPHY

The stratigraphic framework of Tournaisian and Visean stages in Belgium is now sufficiently well established to locate accurately the analysed samples ( Table 1). However only a preliminary spot-sampling has been carried out here.

TABLE 1

Stratigraphic positions and localities from which K-bentonites have been sampled and analysed

| Stratigraphy | | Localities (Dinant and Namur synclines) |
|---|---|---|
| VISEAN | V3c | Warnant,Bioul,Trooz |
| | V3bα-V3bγ | Gesves |
| | V3bβ | Gesves,Thon Samson,Yvoir |
| | V3bα-V3bβ | Gesves,Andenne |
| | V3bα | Samson valley,Frasnière,Yvoir,Gesves,Flemalle |
| | V3b-V3a | Thon Samson, Boulonnais (Northern France) |
| | V3aβ | Yvoir,Hastière,Andenne |
| | V3aα | Boulonnais (Northern France) |
| | V3a-V2b | Andenne,Tramaka,Forge Thiry,Dinant |
| | V2b | Fonds de Leffe |
| | V2b γ | Dinant,Theux,Hastière |
| | V2b β | Ronde Hage,Lovegnée,Awirs,Trooz,Engihoul, Landelies |
| | V2b α | cf:"Banc d'Or":Molignée valley,Ben Ahin, Trooz, Aachen (Germany) |
| VISEAN-TOURNAISIAN | | Nattoye,Walcourt |
| TOURNAISIAN | Tn3c | Tournai (borehole) |
| | Tn2c | Martinrive,Maredsous,Peruwelz,Moha, Tournai |
| | Tn2a | Feluy,Moha,Pernillière |
| | Tn1 | Méan |

DEFINITION OF BENTONITE AND K-BENTONITE

As defined by Ross and Shannon(1926) "a bentonite is a rock composed essentially of a crystalline clay-like mineral formed by the devitification and accompanying alteration of a glassy igneous material, usually a tuff or volcanic ash". The implication is thus made that a bentonite is formed in situ.These authors stated that the characteristic clay mineral is montmorillonite, or less often,beidellite. Consequently the term bentonite carries definite mineralogic, textural and genetic connotations.Weaver and Bates(1952) have assigned the name K-bentonite to material of the Ordovician from Pennsylvania, indicating that its high content distinguishes Ordovician bentonite from others such as those found in the Cretaceous of the United States (Roberson,1964).The inability to identify in the field the fine-grained particles which compose a bentonite has led to a certain practice of using and extending the term bentonite to any relatively pure clay deposit which is thought to be both predominantly of montmorillonite composition and a product of the chemical alteration of an ashfall material. Schultz(1962) has pointed out that just as all montmorillonite is not found in bentonites neither are all bentonites made up solely of smectites. A group of non-smectitic bentonites is just the potassium-rich one (Weaver,1953).

In Belgium the Tournaisian and Visean highly fossiliferous carbonates have accumulated in a shallow to moderately deep marine environment.Usually no fossils occur in the intercalated K-bentonites whose clay compositions are far from being uniform and simple in the entire stratigraphic interval,polymineralic associations being the rule even within a specific narrow range.

## PETROLOGIC STUDY OF VISEAN AND TOURNAISIAN K-BENTONITES
### Procedure

For the X-ray diffraction analysis the fraction below two microns was dispersed in distilled water, and prepared as oriented aggregates and random powdered preparations. Various post-treatments were applied to refine the identification of the clay components particularly: $KCl$,$MgCl_2$ and $LiCl$ saturations with subsequent heating and ethylene glycol or glycerol solvation,hydrazine saturation,4N HCl attack and treatment in KOH (Thorez,1975 and 1976).Because of the general polymineralic compositions, no attempt has been made up to now to complete the investigation with DTA.Electron microscopic analyses failed also to provide conclusive data.About sixty chemical analyses have been carried out on the bulk material, but detailed analyses related to the minus two microns fractions will be later published in an extended paper.

Several thin sections have been studied to enable a characterization of the mineralogy and texture of the coarser material which comprises aggregates and phenocrysts. Although it is rather difficult to study the fine-grained matrix of any bentonite under a petrographic microscope, it remains possible to differentiate the peculiarities of the coarse particles which normally escape the X-ray analysis of the finer fractions.

Results of the microscopic study

For all the investigated material a clay matrix was generally noted which bears a dense, crypto- to microcrystalline texture or, locally, a highly recrystallized and reoriented phyllitic composition resulting from compaction. Such features provide at first glance a rather basic uniformity. However a clear differentiation is further obtained according to the occurrence, nature and granulometry of the coarser particles and phenocrysts embedded in the clay matrix.

The heavy mineral suites, even when not systematically analysed because of the difficulties of their extraction from the bulk material provide an unusual high concentration of perfectly euhedral, unrounded and well-sorted zircon and/or apatite associated with minute grains of opaque minerals.

Typically some layers exhibit an upwards-fining of phenocrysts, particularly of biotite flakes. Such a feature indicates the natural fractionation of the aggregates during the accumulation at the bottom of the sedimentary basin. On the other hand, reworking of such a material seldom occurs and when present is typified by small "nests" of clayey or microscopic vermicular aggregates embedded in the fossiliferous micrite to microsparite.

As a rule the quartz content remains very low (less than 1%). When occurring as scattered grains, it exhibits elongated acicular shapes (splinters). Also in some compacted phyllitic layers there appears intercalated microlayers made up of isotropic material which can be interpreted as former glassy clasts and shards.

Unaltered feldspars do not exist apart from traces of sanidine in some Tournaisian layers. However some perfect euhedral "square"-shaped phenocrysts 30 to 50 microns in size are present in some stratigraphic levels. They are interpreted as pseudomorphs of feldspars now entirely replaced by kaolinite whose grain-size is larger than the enclosing clay matrix.

Biotite as well-sorted, sometimes euhedral plates occurs scattered or concentrated into microlayers. It is unaltered or more or less pseudomorphed. "Fresh" biotite withholds its parent but weaker pleochroism, and shows very fine black streaks parallel to elongation. Altered plates typically bear a "barrel" habit with jagged edges and interlayered colourless (very low birefringent) zones.The latter correspond to the partial kaolinitization of the biotite with subsequent swelling and widening.Electron microprobe analysis (Srodon,1976) could reveal an iron-rich composition of the kaolinite.

Other barrel-like pseudomorphed grains are composed of tri- or dioctahedral chlorite while the clay matrix yields an illite-smectite composition.Such a transformation of biotite into chlorite was noted by Byström(1954) and Schultz(1962) for K-bentonites occurring in marine sediments.

"Graupen" composed of a brown isotropic clay, and "Krystal-tonstein" with a typical "rouleaux".or a vermicular habit are restricted to certain stratigraphic intervals (Table 2).Some quartz or glass shards and microfragments of volcanic material have also been entirely pseudomorphed into a low-birefringent clay material whose crystallites size is larger than the enclosing cryptocrystalline clay matrix.

TABLE 2
Main results of the microscopic investigation

| Stratigraphy | Clay matrix | Phenocrysts main composition |
| --- | --- | --- |
| V3c | cryptocrystalline | microscopic vesicles |
| V3b$\gamma$ | cryptocrystalline | graupen,Krystal-tonstein,quartz splinter |
| V3b$\beta$ | phyllitic | isotropic splinters and flatttened volcanic glass particles |
| V3b$\alpha$ | cryptocrystalline or phyllitic | Krystal-tonstein,graupen,vermicules, splinters,vesicles,apatite |
| V3a$\alpha$-V3a$\beta$ | cryptocrystalline | vermicular chlorite;locally calcitized volcanic glass |
| V3a$\beta$ | cryptocrystalline | apatite |
| V2b$\beta$ | cryptocrystalline | graupen,biotite (well-sorted,unaltered or partially kaolinitized; upwards-fining plates), pseudomorphed feldspar phenocrysts and minute volcanic debris |
| V2b$\alpha$ | cryptocrystalline or phyllitic | abundant,well-sorted and idiomorphic apatite and zircon |
| Tn | cryptocrystalline | undiscernible |

Summing up the results of the petrographic investigations,it is clear that the mineralogic and textural variations of layers are due to volcanic material laid down in the marine environment and further altered in situ.The composition of the parent ash certainly differed in time according to the present heterogeneity of both the clay assemblages and phenocrysts in the whole stratigraphic succession from Visean to Tournaisian.

## $K_2O$ content of the K-bentonites

By considering the lack of unaltered feldspars - apart from some sanidine in some rare beds - the $K_2O$ content was attributed to clay minerals. The combined results of sixty analysed samples are listed in Table 3 to indicate the range of variation within the different stratigraphic levels.

TABLE 3

$K_2O$ content of the different K-bentonites (mean values)

| Stratigraphy | %$K_2O$ |
|---|---|
| V3c | 0.41-1.11 |
| V3bβ-V3bα | 1.15 |
| V3bα | 0.22-3.43 |
| V3aβ | 0.27-4.77 |
| V3a | 0.80 |
| V3b -V2bδ | 4.57 |
| V2bδ | 6.62 |
| V2bβ | 0.20-2.40 |
| V2bα | 2.43-5.74 |
| Tn3c | 4.81 |
| Tn2c | 2.23 |

## Results of the X-ray diffraction analysis of the clay fractions

The results of the X-ray diffraction analyses are listed in Table 4 wherein the clay-assemblages are grouped according to their importance expressed for the predominant components as a "relative frequency" in a range of 1-10 units.

The qualitative and semi-quantitative composition was obtained from the analysis of the oriented aggregates before and after the different identification treatments. Complementary data were provided from the powdered preparations which permitted recognition of the illite polymorphs, the di- or trioctahedral character of chlorite, and the crystallinity of kandites (kaolinite,halloysite). For the mixed-layer illite-smectite, the nature of the swelling component was completed by Greene-Kelly's 1953 Li-test,whereas the degree of ordering and the amount of swelling (inter)layers were calculated after Reynolds and Hower(1970) by using a quantitative chart published by Srodon(1976

Saturation of the material in KOH for 15 hours, followed by heating in 1N KOH for 1 hour at 90°C, only partially collapsed the expansible (inter)layers of the illite-smectite mixed-layers and the 10.4Å "illite" to about 10.2-10.3Å ( the unsolvated position being from 10.4 to 10.9Å).This incomplete collapse suggests that the mixed-layers as well the "illite" was derived from a non-micaceous material (Weaver,1958).

TABLE 4

Clay compositions and clay-assemblages (fraction below 2 microns)

| Stratigraphy | Clay minerals Predominant | Minor | Accessory |
|---|---|---|---|
| V3c | (I-M)(2.5-7.5) | - $C_D$ | - (Bio-C); 1M I |
| V3b β | (I-M) or $C_D$ (5) | - (10-14c) | |
| | (I-M) (5) | - (Bio-C) | |
| V3b α | (I-V) (6) | - (Bio-C) | - C; (14c-14v) |
| | (I-M) (6) | - C | - Sm,(10-14c) |
| | K (5-6) | - (Bio-C) | - Sm,C,(10-14m) |
| | (I-M) (5-8) | - C; $C_D$ | |
| | 1M I (8) | - (10-14m) | - C |
| | $C_D$ (5-8) | - (I-M) | - 1M I;K |
| V3b-V3a | (I-M) + $C_D$ (8) | | |
| V3a β | 1M I (9) | | - (10-14c) |
| | Sm (10) | | |
| | $C_D$ (7) | - (I-M) | |
| | (I-M) (7-9) | | - (Bio-C) |
| V2b-V3a | (I-M) (6) | - $C_D$ | |
| | 1M I (7-9) | | - (I-M),(10-14c),C |
| V2b δ | (Bio-V)(10) | | |
| V2b γ | (I-M) (6) | - $C_D$ | |
| | (10-14m) (5-8) | - (10-14v) | - C |
| | C (6) | - (Bio-V) | |
| V2b α | 1M I (9) | | - (10-14c) |
| | (I-M) (9) | | - (10-14c) |
| Visean-Tourn. | (I-M) (8) | - Sm,C | |
| | K (9) | | - (10-14c) |
| | 1M I (6) | - (I-M) | - (10-14c) |
| Tn3c-Tn3b | (I-M) (8-9) | | - Sm; (10-14c) |
| Tn2c | Halloysite (6) | - (10-14c) | |
| | Halloysite (5) | | - 1M I, (10-14v) |
| | 1M I (8) | | - (10-14c),C |
| | Sm (10) | | |
| Tn2a | 1M I (6-8) | - (I-V);C | |
| | 1M I (6-8) | - (Bio-C) | |
| Tn1 | 1M I (8-9) | | - (10-14c);C |
| | (10-14m) (5) | - Sm | - (10-14c) |
| | (I-M) (6) | - Sm;(10-14c) | |

I=illite (1M polymorph);C=chlorite;$C_D$= dioctahedral chlorite;Sm=smectite; K=kaolinite;V= vermiculite;Bio= biotite;(I-M)=illite-montmorillonite;(I-V)=illite-vermiculite;(Bio-C)=biotite-chlorite;(10-14m)= illite-montmorillonite;(10-14v)=illite-vermiculite;(10-14c)=illite-chlorite;(14c-14v)=chlorite-vermiculite mixed-layers.Figures between parentheses refer to the relative frequency in the range 1-10 units.

In random powdered preparation both illite-smectite mixed-layers -referred as (I-M) when regular, and as (10-14m) when randomly interstratified (Thorez,1975 and 1976) - and the "illite" itself show characteristic reflections of a 1M mica. A value of d(060)=1.50Å indicates that the component layers of the interstratification and the illite are dioctahedral. By contrast the illite particles extracted from the enclosing carbonates and pelitic beds bear the 2M polymorph, and are thus a variety formed at high temperatures as indicated by Yoder and Eugster(1955). The mica must therefore be considered as detrital in origin and resulting from diagenetic evolution of the sediment. The %Sm in the illite-smectite mixed-layers varies from 15 to 55.

Besides the (I-M) or (10-14m) illite-smectite mixed-layers there also exists other varieties of mixed-layered minerals, regularly or randomly interstratified: (I-V),(10-14v),(Bio-C),(10-14c). Chlorite is either a Mg- or Fe-rich trioctahedral mineral, or a dioctahedral one, the latter being characterized by: the intense (003) peak at 4.7Å, a d(060)=1.503Å, and a resistance when boiled in 4N HCl. Kandites are either well-crystallized kaolinites or halloysite. Smectites, as montmorillonite or beidellite, are badly crystallized.

ORIGIN OF THE BELGIAN TOURNAISIAN AND VISEAN K-BENTONITES

From this preliminary study it seems that Visean and Tournaisian K-bentonites were originally windborne volcanic ashes laid down in a marine environment before becoming altered to illite-smectite or illite-vermiculite mixed-layers. These deposits are named K-bentonites according to Weaver's(1953) accepted concept which assumes halmyrolitic weathering of volcanic ashes and subsequent illitization of smectite under the influence of sea water and later through some diagenesis

Monomineralic composition seldom occurs, polymineralic clay-assemblages being the rule and including predominantly mixed-layers illite-smectite (I-M), less often illite-vermiculite. These minerals are also associated with chlorite specifically the dioctahedral (aluminous) variety. The coarser fractions vary in several ways according to their composition, associated natures, and texture, and show usually an upwards-fining distribution in a single layer. As a consequence sensitive differences do occur in compositions of both the clay fraction and coarser aggregates and phenocrysts as well as in the texture of all the K-bentonite deposits comprised in the Tournaisian-Visean interval. Such differences may be considered as the results of chemical variations within the parent volcanic ash laid down in a sedimentary basin whose environmental conditions have not varied too drastically during the whole Visean to Tournaisian interval. On the other hand mineralogic and petrographic differentiation noted in a single layer -depending on whether the analysis was made at the base or top- certainly corresponds to an eolian differentiation. This last process could become responsible for the grain-size distribution and fining-up of the coarse particles, and for variations of the composition and textures of the K-bentonite layers. However these variations may also be connected with the chemistry of the original ashes and their later alteration in sea waters. A genetic reconstitution of these variables is actually in progress by considering that the volcanic vents lie outside the investigated area.

Despite the restrictions and the preliminary aspect of this study, and the fact that all the layers have not yet been comprehensively analysed but already recognised in the field, the succession of these beds can be used as marker-horizons in close conjunction with micropalaeontologic and rhythmic methods of correlation. These layers extend as far as the Boulonnais area (Northern France) and the Aachen area (Federal Republic of Germany) near the Belgian border.

What remains unclear, and has still to be established, is whether clay composition and textural characters of a specific layer change from one locality to another through the entire area when it is comprised in the same stratigraphic level. If such a non-uniformity becomes materialized within long distance, it would be fallacious to attempt lateral correlations by referring exclusively to the clay composition and petrographic characters of the layer.

CONCLUSION

K-bentonite beds in several sets of thin layers have been recognised in marine Tournaisian to Visean (Lower Carboniferous) succession of the Dinant and Namur synclines in Belgium. Their clay composition is analogous with specific characters of well-known K-bentonites in Europe and North America. This first recognised occurrence in non-coal Lower Carboniferous but marine sediments is emphasized. The clay composition is highly polymineralic, and the clay-assemblages vary from one interval to another providing already, at the present state of the study, some sensitive tools for the vertical stratigraphic recognition of the whole Tournaisian-Visean succession in the investigated area. The numerous layers are readily discernible in the field from the enclosing carbonates and accessorily pelitic sediments. Their lateral persistence is used for detailed correlations. Vertical variations of composition and texture do not prevent their usefulness as key-horizons during biostratigraphic and sedimentological researches.

The parent volcanic material has been laid down as ashfalls during the whole Tournaisian to Visean interval. Variations of composition and texture are related to changes in the chemistry of the ashes during halmyrolitic and diagenetic alteration in the marine environment.

REFERENCES

Byström,A.M.,1954.Mineralogy of the Ordovician bentonite at Kinnekulle,Sweden.Sver.Geol.Undersök,48 (1954),5:62 pp.
Conil,R.,Pirlet,H.,Legrand,R.,Streel,M.,Bouckaert,J.,Thorez,J.,1969. Traits dominants de l'échelle biostratigraphique du Dinantien de la Belgique.C.R.6è Congr.Intern.Strat.Géol.Carbon.,Sheffield(1967) I:45-50.
Greene-Kelly,R.,1953. The identification of montmorillonids in clays. J.Soil Sci.,4:233-237.
Groessens,E.,1974. Distribution de Conodontes dans le Dinantien de la Belgique.Intern.Symposium on Belgian Limits,Namur(1974),Pub.17: 192 pp.
Pirlet,H.,1966. Présence d'un Tonstein dans le Viséen supérieur des synclinoriums de Namur et de Dinant.Ann.Soc.Géol.de Belgique, 89:27-32.
Pirlet,H.,1968. La sédimentation rythmique et la stratigraphie du Viséen V3b,V3c inférieur dans le synclinorium de Namur et Dinant. Acad.Royale de Belgique,Mém.4,17:98 pp.
Reynolds,R.C., and Hower,J.,1970. The nature of interlayering in mixed-layers illite-montmorillonite.Clays and Clay Min.,18:25-36.
Ross,C.S. and Shannon,E.V.,1926. Minerals of bentonite and related clays and their physical properties.Am.Ceram.Soc.J.,9:77-96.
Schultz,L.G.,1962. Non-montmorillonite composition of some bentonite beds.Clays and Clay Min.,11th Conf.:169-177.
Srodon,J.,1976. Mixed-layer smectite/illites in the Bentonites and Tonsteins of the Upper Silesian coal basin.Polska Akad.Nauk., Kom.Nauk.Mineral.,49:73 pp.
Thorez,J.,1975.Phyllosilicates and Clay Minerals.A Laboratory Handbook for their X-ray Diffraction Analysis.G.Lelotte (ed.),604pp.
Thorez,J.,1976. Practical Identification of Clay Minerals. G.Lelotte (ed.),90 pp.
Weaver,C.E. and Bates,T.F.,1952. Mineralogy and petrography of the Ordovician "metabentonites" and related limestones.Clay Min.Bull., 1:258-261.
Weaver,C.E.,1953.Mineralogy and petrology of some Ordovician K-bentonites and related limestones.Bull.Geol.Soc.Am.,64:921-943.
Weaver,C.E.,1958.Effects and geological significance of potassium "fixation" by expandable clay minerals derived from muscovite, biotite,chlorite and volcanic material.Am.Min.,43:839-861.
Yoder,H.S. and Eugster,H.P.,1955.Synthetic and natural muscovites. Geochim. et Cosmochim.Acta,8:225-280.

OTHER PAPERS PRESENTED IN SECTION 3

CLAY MINERALS IN THE SEDIMENTS OF A SALINE LAKE

A.H. Weir[1] and Blair F. Jones[2] ([1] Rothamsted Experimental Station, Harpenden, England. [2] U.S. Geological Survey, Reston, Virginia, USA.)

DIFFERENCES IN CLAY MINERAL COMPOSITION OF SALT, BRACKISH AND FRESH WATER SEDIMENTS IN THE RHINE ESTUARY

A. Breeuwsma (Soil Survey Institute, Marijkeweg 11, Wageningen, The Netherlands.)

CLAY MINERALS IN SEDIMENT FROM THE N.EQUATORIAL PACIFIC Mn-NODULE BELT

James R. Hein, Elaine Alexander, and C. Robin Ross (Pacific-Arctic Branch of Marine Geology, U.S. Geological Survey, 345 Middlefield Road, Menlo Park, California 94025, USA.)

EVIDENCE FOR METASOMATIC DIAGENESIS AND NON-METASOMATIC METAMORPHISM OF ORDOVICIAN META-BENTONITES IN SWEDEN

B. Velde[1] and A.M. Brusewitz[2] ([1] Laboratoire de Pétrographie, Université P. et M. Curie, 4 Place Jussieu, 75230 Paris Cedex 05, France. [2] S.G.U., 10405 Stockholm, Sweden.)

THERMODYNAMIC PARAMETERS OF ORGANIC-ACID/CLAY MINERAL COMPLEXES AND THEIR POSSIBLE ROLE IN ORGANIC DIAGENESIS

William R. Almon (Cities Service Oil Company, Research Laboratory, PO Box 50408, Tulsa, Oklahoma, USA.)

TRANSFORMATION OF CLAY MINERALS AND ASSOCIATED CHANGES IN THE PROPERTIES OF OIL- AND GAS-BEARING ROCKS

S.G. Sarkisyan and I.D. Zkhus (Institute of Geology and Exploitation of Combustible Fuels, Moscow, USSR.)

ELEMENTAL VARIATION IN ORDOVICIAN K-BENTONITES

A. Gunal and W.D. Huff (Department of Geology, University of Cincinnati Cincinnati, Ohio USA 45221.)

DISTRIBUTION OF MINOR ELEMENTS IN CLAYS IN RELATION TO THEIR GENESIS

I. Kraus (Faculty of Natural Science of the Comenius University, Bratislava, Czechoslovakia.)

THE DIFFERENT TYPES OF TERTIARY CLAYS IN EASTERN BAVARIA

H. Kromer (Staatliches Forschungsinstitut für angewandte Mineralogie bei der TU München, Kumpfmühlerstrasse 2, 84 Regensburg, Federal Republic of Germany.)

LOWER JURASSIC REDBEDS FROM CENTRAL ARABIA

A.M. Abed (Department of Geology and Mineralogy, University of Jordan, Amman, Jordan.)

QUANTATIVE CLAY MINERALOGICAL ANALYSES FROM THE BULK CHEMISTRY OF SEDIMENTARY ROCKS

M.J. Pearson (Department of Geology and Mineralogy, The University, Marischal College, Aberdeen, AB9 1AS, Scotland.)

ORIGIN AND COMPOSITION OF RECENT CLAYS IN THE KATTEGAT, SKAGERRAK AND ADJACENT SCANDINAVIAN COASTAL WATERS

T.M. Rønningsland[1], E. Roaldset[2] and I.Th. Rosenqvist[2] ([1] Continental Shelf Institute, P.O. Box 1883, 7001 Trondheim, Norway. [2] Institute of Geology, University of Oslo, P.O. Box 1047, Blindern, Oslo 3, Norway.)

DIAGENETIC IMPLICATIONS OF POST-CARBONIFEROUS - PRE-PERMIAN BURIAL IN SAND BODIES

William R. Almon[1] and Brian M. Abbott[2] ([1] Cities Service Company, Tulsa, Oklahoma, USA. [2] Cities Service Europe-Africa Pet. Corp., 197 Knightsbridge, London, UK.)

MINERALOGY OF VARIEGATED CLAYS IN SOUTHERN ITALY

R. Belviso[1], C. Cherubini[1], V. Cotecchia[1], M. Del Prete[1], A. Federico[1], F. Soggetti[2] and F. Veniale[2] ([1] Istituto di Geologia Applicata e Geotecnica, Facoltà di Ingegneria, Università di Bari, Italy. [2] Istituto di Mineralogia e Petrografia, Università di Pavia, Italy.)

DIAGENESIS AND EPIMETAMORPHISM OF CLAY MINERALS IN THE VERBICARO UNIT, CALABRIA, SOUTHERN ITALY

A. Pozzuoli[1], D. Dietrich[2], P. Scandone[2], F. Huertas[3] and J. Linares[3] ([1] Istituto di Mineralogia, Universita di Napoli, Italy. [2] Istituto di Geologia e Geofisica, Universita di Napoli, Italy. [3] Estacion Experimental del Zaidin, C.S.I.C., Granada, Spain.)

SERPENTINE AND TREMOLITE ASBESTOS DEPOSITS IN CALERA DE LEON, BADAJOZ, SPAIN

A. Valero-Saez, B. García Ramos, J.L. Pérez-Rodríguez and F. González-García (Centro de Edafología y Biología Aplicada del Cuarto, C.S.I.C., Apartado 1052, Sevilla, Spain.)

MIXED-LAYER ILLITE-MONTMORILLONITE CLAY BEDS (K-BENTONITES) IN THE SILURIAN OF THE WELSH BORDERLANDS

D.J. Morgan and I.R. Basham (Institute of Geological Sciences, 64-78 Gray's Inn Road, London WC1X 8NG, England.)

A GEOCHEMICAL AND MINERALOGICAL INVESTIGATION OF SOME BRITISH AND OTHER EUROPEAN TONSTEINS

D.A. Spears and R. Kanaris-Sotiriou (Department of Geology University of Sheffield, Mappin Street, Sheffield S13JD, England.)

# SECTION 4

# Genesis and Synthesis

*Chairman's Introduction*

GENESIS AND SYNTHESIS OF CLAYS AND CLAY MINERALS : RECENT DEVELOPMENTS AND FUTURE PROSPECTS.

Bernard SIFFERT

Centre de Recherches sur la Physico-Chimie des Surfaces Solides
24, Avenue du Président Kennedy, 68200 MULHOUSE - France

INTRODUCTION

First of all, my thanks are due to the Board of Organizers for the honour of inviting me to address this conference. It is of course a formidable task to deliver a survey on the genesis and synthesis of clays and clay minerals.

The problem of clay mineral synthesis has long been a concern of geologists, pedologists, mineralogists and research chemists. The natural formation of clays occurs at the expense of crystalline rocks. Clays are generated by the disaggregation and weathering of mother rocks, under the influence of climate and drainage. MILLOT (1945) was the first to state precisely as well as to correlate the conditions of the geological deposit with the resulting type of clay mineral. It should be remembered that clays are formed under normal conditions of temperature and pressure and in the hydrothermal medium. In a first survey, the major researches can be listed under two headings :

- hydrothermal systems
- synthesis in hydrosphere conditions

To these principal domains must be added the petrogenetical aspects of clay formation which include rock alteration processes and gel crystallization.

Owing to the vast literature available on these subjects, it is almost an impossible task to report all the studies performed so far in these different fields. Therefore, in my oral presentation at the Conference, I confined myself to the papers published in the last three years, which still amount to 450 papers. Even an adequate review of this latest research work could not be done. I have therefore limited myself to a list of the most significant innovations and findings.

In this report, I shall, however, confine myself to the results published over the two last decades on clay synthesis at room temperature. I shall also attempt to draw a plausible guideline for the research work which may be performed in this field in the near future.

## CLAY MINERAL SYNTHESIS IN HYDROSPHERE CONDITIONS

The problem of preparing clay minerals at low temperature has not been entirely settled, especially the preparation of purely aluminous clays. Only a few investigations have been performed on the synthesis of clay minerals under earth-surface conditions (STRESE and HOFFMANN, 1941 ; HENIN and CAILLERE, 1961 ; SIFFERT and WEY, 1962 ; HARDER, 1965, 1977 ; POLZER et al., 1967 ; WOLLAST et al., 1968 ; KITTRICK, 1970 ; HEM et al., 1973, 1974 ; IGLESIA and MARTIN-VIVALDI, 1975). The synthesis of trioctahedral clays at low temperature is comparatively easy, owing to the presence of the magnesium mineralizing cation (HENIN and CAILLERE, 1961 ; ESTEOULE, 1969). Even in a low proportion, magnesium induces the swift formation of a brucite layer and the crystallization of layer silicates belonging to the smectite group.

Thus, HARDER (1976) has newly synthesized nontronite in various redox conditions. In particular, he pointed to the fact that the presence of a divalent iron - magnesium mixture is required to obtain three layer silicates containing aluminium and trivalent iron. Illitic mixed layer minerals have also been obtained at surface temperature by HARDER (1974). The formation of these minerals is only possible when the precipitates contain at least 6 % of MgO. A higher potassium content favors the illite mineral formation, but there is a competition between K and Mg.

The preparation of dioctahedral clay (kaolinite) performed in the absence of magnesium presents numerous difficulties. Several reasons have been put foward to account for the failure of the kaolinite synthesis :
- aluminium displays two coordination numbers (4 or 6) (GASTUCHE et al., 1962)
- the simultaneous formation of various oxides and hydroxides (HENIN and CAILLERE, 1961)
- the marked hydration shell of the aluminium cations in solution together with the formation of insoluble Al hydroxopolycations (SIFFERT and WEY, 1972)
- the insolubility of aluminium ion in the pH range of natural medium which gives rise to a disequilibrium between silica and alumina concentrations (SIFFERT and WEY, 1972)
- the formation of a surface silica deposit on the aluminous compounds (CLOOS et al., 1969)

Let us now have a closer look at those various factors and see how the researchers attempted to overcome them.

The disequilibrium between the soluble forms of silica and alumina prompted the use of highly diluted media (HENIN and CAILLERE, 1961 ; SIFFERT, 1962 ; DE KIMPE et al., 1964 ; HARDER, 1965 ; IGLESIA and MARTIN-VIVALDI, 1975). Under such conditions, the silica concentration draws nearer to that of alumina. Furthermore, the most conclusive results have been obtained under such conditions (SIFFERT and WEY, 1962 ; HARDER, 1965 ; POLZER et al., 1967 ; KITTRICK, 1970 ; LINARES and HUERTAS, 1971). The silica proportion is thus decreased with respect to alumina, as well as the probability of a possible silica polymerization.

This one and only way which provides means for obtaining low amounts of kaolinite, warrants the statement that aluminous phyllites are formed not by recrystallization of mixed gels, but following a dissolution (or dispersion) and recombination of the dissolved elements (neoformation). This is the reason why long aging times of the gels are required for the appearance of a few clay crystals. As a matter of fact, it is fair to assume that in a mixed gel, silica and alumina are distributed according to a tridimensional structure analogous to framwork silicates, such as quartz and feldpars. In order to accelerate the phenomenon, that is to say, to make more alumina available, SIFFERT and WEY (1962), SIFFERT and DENNEFELD (1971), LINARES and HUERTAS (1971), LA IGLESIA (1975) involved inorganic acids or better still aluminium complexing acids (oxalic, citric, tartric, salicylic, fulvic acids). These agents also prevent a too high hydration of the $Al^{3+}$ ion on one hand (SIFFERT, 1962), and retain the sixfold coordination number of the aluminium required for kaolinite formation on the other (GASTUCHE et al.,1962 ; SIFFERT and WEY, 1972).

Another method consists of precipitating very slowly the aluminosilicates, so as to maintain a comparatively constant balance between the silica and alumina concentrations. This technique which borders on natural medium conditions, has been employed by LA IGLESIA and MARTIN-VIVALDI (1974 ; 1975), using agents which alter very slowly the pH of the medium, such as (i) anionic resins in acidic solution (ii) cationic resins in alkaline solution and/or (iii), potassium-bearing feldspars in acidic medium.

This process complies thoroughly with the conditions of crystalline growth and nucleation, as stated by VAN OOSTERWYCK-GASTUCHE and LA IGLESIA (1977) :

- precipitation from a solution slightly supersaturated in elements with respect to the crystal
- slow and regular crystallization.

Yet, although these conditions are essential to any formation of well-ordered crystals, they do not particularly account for the difficulties in the formation of kaolinite.

The dependence on starting conditions of the crystallinity of the materials obtained points to the existence and the operation of soluble complex polyions prior to the crystalline phase. The presence of alumino-silicate complexes in the solution has been suggested as early as 1962 by SIFFERT. Experimental evidence has been provided by SCHENK (1969) who studied alumina precipitation, as well as by GUTH et al. (1974) who investigated the mechanism of zeolite formation. DE KIMPE and FRIPIAT (1968) ; RODRIGUE et al. (1973) concluded also that O-Si-O-Al-OH rows are the elements required to form phyllosilicates.

The operation of these complex polyions is likely to account for the difficulties encountered during the synthesis of aluminous phyllites. The formation of these polyions during the preparation of magnesium clays appears to occur spontaneously and readily. In the case of aluminium cation, their range of appearance must be very restricted. The understanding of the conditions in which aluminous polyions are formed is still too shallow. For all that, it seems that HARDER (1970) revealed them through obtaining a fire-clay at 20°C, by allowing a solution $3.3 \times 10^{-4}$ molar in silica and $1.76 \times 10^{-4}$ molar in aluminium to mature around pH 4.5 to 5.

Furthermore, several studies of synthesis in the presence of organic matter carried out in recent years (IGLESIA and GALAN, 1975 ; GUITAN OJEA and COLADAS, 1975 ; TAN and MC CREERY, 1975 ; TAN, 1976 ; EBERL and HOWER, 1976 ; MOINEREAU, 1977) warrant the speculation that organic reagents do not promote merely aluminium solubility by preventing its polymerization into polynuclear hydroxide species (HEM and LIND, 1974), but contribute to the creation of silico-aluminous complexes which are more or less soluble.

It is most probable that they contribute to change the charges carried by the particles dispersed in the solution, wether they are siliceous or aluminous. It appears, therefore, that a fuller understanding of the phenomena will require an investigation into the properties of the slightly polymerized silico-aluminous polycations and complexes dispersed in the solution which make up the structural elements in the constitution of phyllites. Future research work should, therefore, be directed towards the investigation of surface charge characteristics of alumino-silica particles and the stability of silica-alumina aqueous systems.

It is well known that most ions undergo hydrolysis reactions and that the hydrolysis products are frequently polymeric. This is more particularly the case for aluminium cation in neutral and slightly acidic medium. Under such pH conditions which, the fact should be stressed, are those governing kaolinite formation, we are no more in the presence of an ion, but of a dispersed alumina particle, a tiny one to be sure, which may adsorb or react with silica molecules. The problem may then be reduced to studying the properties of a freshly precipitated hydrous oxide in presence of silica. Important information on the subject may be derived from electrokinetic measurements. This method provides an opportunity for characterizing chemical reactions at the surface of insoluble oxides (OTTEWILL and HOLLOWAY, 1975). JOHANSEN and BUCHANAN (1957), as well as FUERSTENAU (1970) investigated a number of precipitated aluminosilicates with various compositions, the properties of which revealed the transition from a predominantly acidic surface ($SiO_2$) to a predominantly basic one ($\alpha-Al_2O_3$). With the exception of pure silica, all simple oxides show positive surface charges in acid medium and negative charges in alkaline solution, the main differences occurring at the pH of the isoelectric point in which case the net surface charge is zero. Silica shows no isoelectric point, as its surface displays an acidic behaviour down to at least pH 2.

Electrophoretic mobility versus pH curves for silica and α-alumina particles are shown in figure 1.

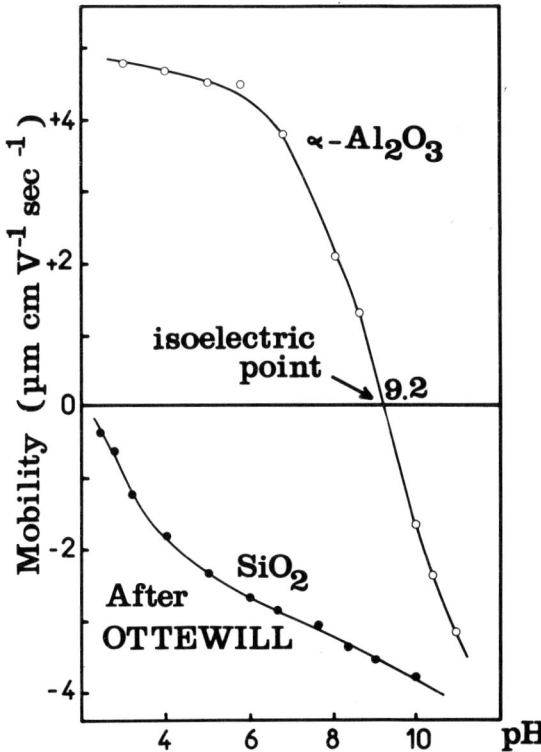

Fig. 1. Electrophoretic mobility against pH for silica and alumina particles ($10^{-3}$ mole $dm^{-3}$)

This figure reveals that the surface charge of silica is negative at any pH above 2, owing to the loss of surface protons. Conversely the surface of α-alumina particles are positively charged below pH 9. Assuming the interactions between silica and alumina to require a preliminary electrostatic attraction and the existence of an overlap in the double layer surrounding positive and negative particles (SUMNER, 1963), it appears from these data that alumina and silica can only react at pH below 9. In fact, the isoelectric point of alumina shifts to lower pH values for synthetic aluminium oxides and hydroxides (table 1), and in presence of salts (PARKS, 1965 ; OTTEWILL and HOLLOWAY, 1975).

| Products | Investigators | I.E.P. |
|---|---|---|
| $\alpha$ Al$_2$O$_3$ | Yopps and Fuerstenau (1964) | 7.7 |
| Synth. $\alpha$ AlOOH (Boehmite) | Koz'mina et al. (1963) | 7.7 |
| Synth. $\gamma$ AlOOH (Diaspore) | Schuylenborgh (1951) | 7.5 |
| Synth. $\alpha$ Al(OH)$_3$ (Gibbsite) | Koz'mina et al. (1963) | 3.8 - 5.0 |
| Natural Gibbsite | Schuylenborgh and Sänger (1949) | 5.0 - 5.2 |
| Synth. $\gamma$ Al(OH)$_3$ (Bayerite) from AlCl$_3$ + NH$_4$OH | Schuylenborgh (1951) | 7.5 |
| Synth. $\gamma$ Al(OH)$_3$ (Bayerite) from Na$_3$AlO$_3$ + CO$_2$ | Schuylenborgh (1951) | 5.4 |

Table 1. Isoelectric points (I.E.P.) of some aluminium oxides and hydroxides (after Parks).

In that case, the surface reactivity of alumina with regard to silica is depressed at pH values approximating neutrality. This interpretation allows one to understand why the aluminous phyllites cannot be obtained above pH 7 to 7.5. As soon as these pH values are exceeded, the surface of the aluminous particles become negative and repel the silica molecules or polymers. Assuming this starting hypothesis to be accurate, silica uptake or its interaction with alumina could not take place. This has been observed in attempting the synthesis of kaolinite. Yet, if any interaction occurs, the comparative silica proportion with regard to alumina is a key factor. It governs the equivalence point of mutual neutralization. The following curves, taken from MATIJEVIC et al. (1971) are a perfect illustration of this fact (fig. 2). They plotted the mobility values of a silica sol (2g.dm$^{-3}$) in presence of aluminium ions at various concentration as a function of the pH. The two curves marked with filled points show the mobility versus the pH for silica and alumina respectively. The most suitable interaction seems to be obtained when the mobility is reduced to zero near pH 5.

PARKS (1965) determined the points of zero charge for various coprecipitates of silica-alumina mixed gels (fig. 3). These points do not really tally with the actual points of equilibrium or mutual neutralization of the charges at the time of particle interaction. As a matter of fact, some authors showed (CLOOS et al., 1969 ; PERROTT, 1977) that the phenomena are intricate and that an isomorphous substitution of tetrahedrally co-ordinated Al for Si takes place concurrently. Yet, it is apparent in figure 3 that the points lie in the

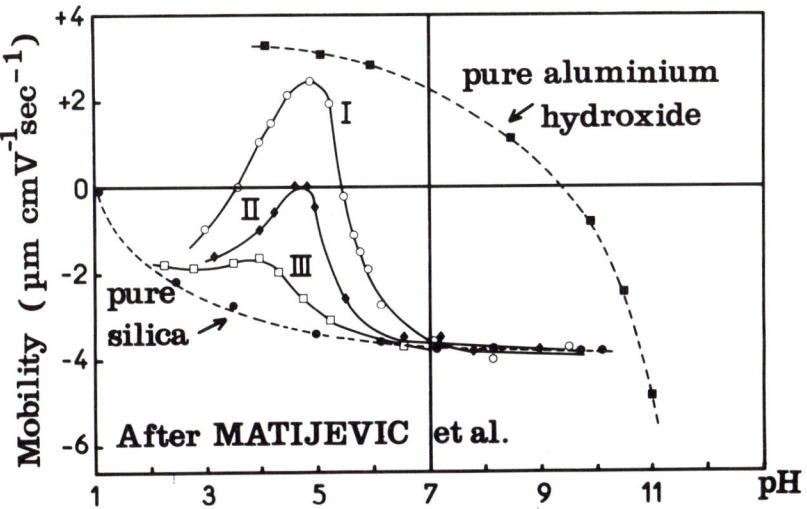

Fig. 2. Mobilities of a silica sol (2g.dm$^{-3}$) in the presence of aluminium ions as a function of pH :
I : 1.05.10$^{-3}$ iong.dm$^{-3}$ Al$^{+3}$
II : 5.25.10$^{-4}$ iong.dm$^{-3}$ Al$^{+3}$
III : 1.0 .10$^{-4}$ iong.dm$^{-3}$ Al$^{+3}$

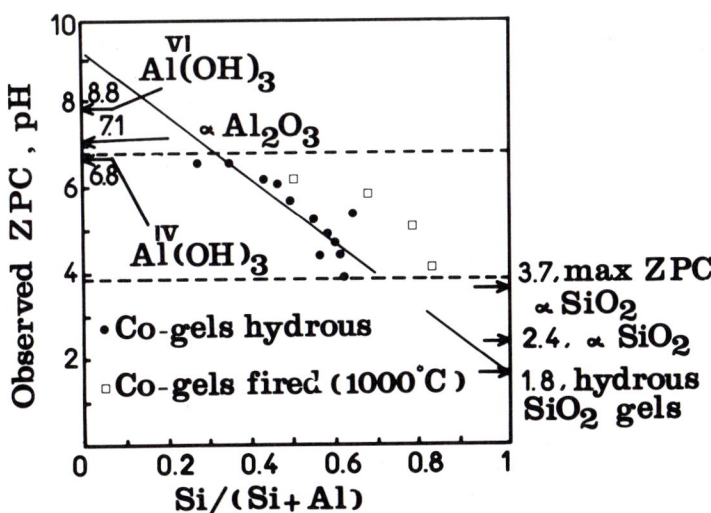

Fig. 3. Zero point charge (Z.P.C.) of coprecipitates in the Al$_2$O$_3$-SiO$_2$-H$_2$O system.

pH values ranging from 4 to 7, favouring the kaolinite formation.

The hypothesis of an electrostatic attraction previous to any phyllite formation allows one also to understand why montmorillonite is formed above pH 7.5. The aluminous particles must contain a given magnesium percentage in order to retain their positive charge above pH 7.5. As a matter of fact, the charge carried by particles of magnesium oxide (brucite) is always positive, irrespective of the pH (NEY, 1973) (Fig. 4). Trace amounts of magnesium oxide in the aluminium oxide (mixed oxide) allow the alumina particle to keep the charge positive beyond pH 7.5.

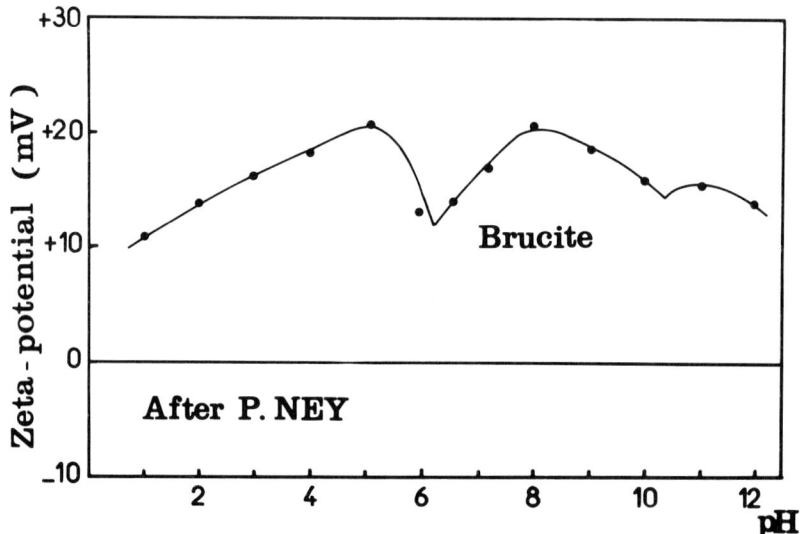

Fig. 4. Zeta potentiel against pH for brucite particles.

These few data taken from the literature show the interest which is attached to electrophoresis techniques when investigating the factors governing clay formation. In my opinion, if applied to perfectly defined systems, they may provide valuable information affording a further insight into the nature of the phenomena.

The hypothesis of oppositely charged particles interacting electrostatically in solution is supported by the experiments and synthesis data obtained in the presen of organic matter (SIFFERT, 1962 ; LINARES and HUERTOS, 1971 ; IGLESIA and MARTIN-VIVALDI, 1972 ; HEM and LIND, 1974 ; GUITAN and CALADOS, 1974 ; TAN, 1975 ; TAN and Mc CREERY, 1975). The organic ions, chiefly the anions, affect the electrical double layer by adsorption onto the elementary particles and alter the magnitude of surface ionization (PARKS, 1965 ; BUCHANAN and JAMES, 1966 ; KAVANAGH et al., 1975 ; OTTEWILL and HOLLOWAY, 1975). They also shift the pH and the conditions of interaction between the aluminous particles and the silica. Electrokinetic measurements carried out systematically on suspensions of aluminium oxides and hydroxides in the presence of organic reagents and silica, may permit solution of the problem and lead to more precise information regarding the conditions of clay formation. In my opinion, the thrust of our endeavours will have to be directed along these guidelines.

CONCLUDING REMARKS

At the end of a review which is far from complete, a few conclusions may be drawn. Regarding the methods, the research workers merely applied available techniques. A new approach has been appearing through the increasingly frequent use of organic compounds in synthesis experiments. The role played by organic matter in the processes, however, is generally not mentioned. In my opinion, organic substances are likely to be involved in the lowering of the activation energies and in reaction kinetics. They are also likely to affect the interaction of particles in solution. Research in the next few years may be guided towards a better understanding of these problems.

No significant interest has been shown so far in electrophoretic methods affording an insight into the interaction of colloïdal and subcolloïdal particles in aqueous solution. These techniques, however, have been applied with positive results for several years by physical chemists in interpreting surface precipitation, hydrolysis and adsorption reactions. More efforts may be directed to the use of these techniques which are likely to provide further insight into the formation of clay, notably kaolinite.

As for the ideas, there seems to be a renewed emphasis on the understanding and refinement of the relationships of the stability of the clay minerals. One also cannot help wondering that in recent years, synthesis experiments using isotopes of elements involved in the phyllite structure have pratically been completely neglected. Research work in this domain is likely to lead to a better understanding of the mechanisms of phyllite formation.

REFERENCES

BUCHANAN, A.S. and JAMES, S.D., 1966. Electrochemistry of the Interface between aluminosilicate crystals and salt solutions. II Electric charge. J. Phys. Chem., 70(11) : 3454-3457.

CLOOS, P., LEONARD, A.J., MOREAU, J.P., HERBILLON, A. and FRIPIAT, J.J., 1969. Structural Organization in amorphous silico-aluminas. Clays Clay Min., 17 : 279-287.

DE KIMPE, C.R., GASTUCHE, M.C. and BRINDLEY, G.W., 1964. Low temperature syntheses of kaolin minerals. Am. Miner., 49 : 1-16.

DE KIMPE, C.R. and FRIPIAT, J.J., 1968. Kaolinite crystallization from

H - exchanged zeolites. Am. Miner., 53 : 216-230.

EBERL, D. and HOWER, J., 1975. Kaolinite synthesis : The role of the Si/Al and (alkali)/(H$^+$) ratio in hydrothermal systems. Clays Clay Miner., 23 : 301-309.

ESTEOULE, J., 1969. Contribution à la genèse des argiles dioctaédriques dans les conditions de surface. Thèse Sc. Phys. Rennes, AO 3521 Paris.

FUERSTENAU, D.W., 1970. Interfacial processes in mineral/water systems. Pure Appl. Chem., 24 : 135-164.

GASTUCHE, M.C., FRIPIAT, J.J. and DE KIMPE, C., 1962. La genèse des minéraux argileux de la famille du kaolin. I Aspect colloïdal. Coll. C.N.R.S. n° 105 : 57-65.

GUITAN, F. and COLADAS, V., 1975. Inhibicion de la sintesis de caolinita por diversos extractos acuosos de restos vegetales. An. Edafol.Agrobiol. 33 : 979-989.

GUTH, J.L., CAULLET, Ph. and WEY, R., 1974. Contribution à l'étude du mécanisme de formation des zeolites. Bull. Soc. Chim. Fr., 9-10 : 1758-1762.

HARDER, H., 1965. Experimente zur "Ausfällung" der Kieselsäure. Geochem. Cosmochem. Acta. 29 : 429-442.

    1970. Kaolinit Synthese bei niedrigen Temperaturen. Naturwis. 57 : 193-198.
    1974. Illite Mineral synthesis at surface temperatures. Chem. Geol., 14 : 241-253.
    1976. Nontronite synthesis at low temperatures. Chem. Geol. 18 : 169-180.
    1977. Clay mineral formation under lateritic weathering conditions. Clay Min., 12 : 281-288.

HEM, J.D., ROBERSON, C.E., LIND, C.J. and POLZER, W.L., 1973. Chemical Interactions of aluminium aqueous silica at 25°C. Geol. Survey Water-Supply Pap. 1827-E : 1-54.

HEM, J.D. and LIND, C.J., 1974. Kaolinite synthesis at 25°C. Science, 184 : 1171-1173.

HENIN, S. and CAILLERE, S., 1961. Vue d'ensemble sur la synthèse des minéraux phylliteux à basse température. Colloque Int. C.N.R.S. 105 : 31-43.

JOHANSEN, P.G. and BUCHANAN, A.S., 1957. An application of the micro-electrophoresis method to the study of the surface properties of insoluble oxides. Aust. J. Chem., 10 : 398-403.

KAVANAGH, B.V., POSNER, A.M. and QUIRK, J.P., 1975. Effect of polymer adsorption on the properties of the electrical double layer. Faraday Discuss. Chem. Soc., 59 : 242-249.

KITTRICK, J.A., 1970. Precipitation of kaolinite at 25°C and 1 atm. Clays Clay Min., 18 : 261-268.

LA IGLESIA, A. and GALAN, E., 1975. Halloysite-kaolinite transformation at room temperature. Clays Clay Min., 23 : 109-113.

LA IGLESIA, A., MARTIN CABALLERO, J.L. and MARTIN-VIVALDI, J.L., 1974. Formation de kaolinite par précipitation homogène à température ambiante. Emploi de feldspaths potassiques. C.R. Acad. Sc. Paris, 279 : 1143-1145.

LA IGLESIA, A. and MARTIN-VIVALDI, J.L., 1975. Synthesis of kaolinite by homogeneous precipitation at room temperatures. I. Use of anionic resins in (OH) form. Clay Min., 10 : 399-405.

LINARES, J. and HUERTAS, F. 1971. Sintesis de minerales a temperature ordinaria. I. Estudio preliminar. Bol. Inst. Geol. Min., 82 : 77-86.

MATIJEVIC, E., MANGRAVITE, F.J. and CASSELL, E.A., 1971. Stability of colloïdal Silica. IV. The silica-alumina system. J. Coll. Int. Sci., 35 : 560-568.

MILLOT, G., 1949. Relations entre la constitution et la genèse des roches sédimentaires argileuses. Geol. Appl.Prosp. Min., 11 : 2, 3 et 4.

MOINEREAU, J., 1977. Absorption de composés humiques par une montmorillonite $H^+$, $Al^{3+}$. Présence de smectites à couches interfoliaires organo-minérales. Clay Min., 12 : 75-82.

NEY, P., 1973. Zeta Potentiale und Flotierbarkeit von Mineralen. Springer Verlag, Wien, New-York, 214 pp.

OTTEWILL, R.H. and HOLLOWAY, L.R., 1975. Electrokinetic properties of particles. Phys. Chem. Sci. Res. Rep., 1 : 599-621.

PARKS, A.G., 1965. The isoelectric points of solid oxides, solid hydroxides and aqueous hydroxo complex systems. Chem. Rev., 65 : 177-198.

PERROTT, K.W., 1977. Surface charge characteristics of amorphous aluminosilicates. Clays Clay Min., 25 : 417-421.

POLZER, W.L., HEM, J.D. and GABE, H.J., 1967. Formation of crystalline hydrous aluminosilicates in aqueous solutions at room temperature. U.S. Geol. Survey Prof. Paper, 575-B : 128-132.

RODRIGUE, L., PONCELET, G. and HERBILLON, A., 1972. Importance of the silica substraction process during the hydrothermal kaolinization of amorphous silico-aluminas. Proc. Int. Clay Conf., Madrid : 187-198.

SCHENK, J.E., 1969. Interaction of monomeric silica with iron, manganese and aluminium in aqueous solution. Ph. D. Thesis, Michigan State University.

SIFFERT, B., 1962. Quelques réactions de la silice en solution : la formation des argiles. Mem. Serv. Carte Geol. Als. Lor., 21 : 1-84.

SIFFERT, B. and DENNEFELD, F., 1971. Etude de l'influence des acides minéraux forts sur la synthèse de la kaolinite à partir du mélange gibbsite silice amorphe. Bull. Gr. Fr. Arg., 23 : 91-106.

SIFFERT, B. and WEY, R., 1961. Sur la synthèse de la kaolinite à température ordinaire. C.R. Acad. Sci. Paris, 253 : 142-144.
   1972. Contribution à la connaissance de la synthèse des kaolins. Proc.
   Int. Clay Conf., Madrid, 159-172.

STRESE, H. and HOFFMANN, U., 1941. Synthese von Magnesiumsilikate-Gelen mit zweidimensional regelmässriger Struktur. Z. Anorg. Allg. Chem., 247 : 65-95.

SUMNER, M.E., 1963. Effect of alcohol washing and pH value of leaching solution on positive and negative charges in ferroginous soils. Nature, 198 : 1018-1019.

TAN, K.H. and Mc CREERY, R.A., 1975. Humic acid complex formation and intermicellar adsorption by bentonite. Proc. Int. Clay Conf., Mexico City, 629-641.

TAN, K.H., 1976. Complex formation between humic acid and clays as revealed by gel filtration and infrared spectroscopy. Soil, Biol. Biochem., 8 : 235-239.

VAN OOSTERWYCK-GASTUCHE, M.C. and LA IGLESIA, A., 1977. Kaolinite synthesis. A discussion on the factors influencing the rate process. Proc. Third Meeting Eur. Clay Groups, 141-143.

WOLLAST, R., Mac KENZIE, F.T. and BRICKER, O.P., 1968. Experimental precipitation and genesis of sepiolite at earth-surface conditions. Am. Min., 53 : 1645-1662.

CLAY MINERAL COMPOSITION AND POTASSIUM STATUS OF SOME TYPICAL
HUNGARIAN SOILS

E.M. VARJU and P. STEFANOVITS
Central Research Institute for Chemistry, Budapest and
University of Agricultural Sciences, Gödöllő (Hungary)

ABSTRACT

The clay mineralogy of six profiles including two Calcaric Chernozems, an Orthic Luvisol, a Haplic Phaeozem and two Humic Gleys has been investigated by X-ray diffraction. The soils were developed on loess or loess-like sediments. The clay fraction is of polymineral character consisting of a mixture of illite, smectites and their intermediates accompanied by a relatively constant amount of chlorite. Due to soil forming processes and potassium fertilization, illitization, the transformation of inherited smectites to illite, can be observed. A correlation exists between clay mineral composition and K-fixation as indicated by electro-ultrafiltration method.

INTRODUCTION

The aim of the present investigation was to determine the clay mineral composition of soils formed on loess or loess-like sediments and to evaluate the effects of inheritance and pedogenesis on the clay minerals in the soils.

This paper presents the results of studies on clay minerals in profiles representing important maize production areas in Hungary with particular respect to the relationship between clay mineral composition and potassium status.

GENERAL GEOLOGY AND THE SOILS

The majority of the soil profiles investigated are formed on loess or loess-like sediment. According to the genesis of the parent material, two main groups of loess can be distinguished:
1. Transdanubian loess which is mainly of eolian origin and was

deposited on dry surfaces in the late Pleistocene. It is well drained. Weathering processes are slow as shown by the high proportion (25-35%) of the 0.02-0.05 mm coarse silt fraction, characteristic of eolian loess, and by the relatively small amount (15-20%) of the clay fraction ( < 0.002 mm). The fine silt fraction (0.02-0.002 mm) is about 20-35%.

2. <u>Lowland loess</u> was developed in two steps. First by eolian deposition on wet sites in the late Pleistocene, later in the early Holocene by fluviatile transport of the Pleistocene loess eroded partly by water. As a result of the more intensive weathering, this parent material is more finely textured, consisting of a smaller proportion of coarse silt (15-25%) and a greater amount of fine silt (30-35%) and clay fraction (25-30%).

A forest-steppe vegetation was established on loess formed in dry conditions and the formation of Chernozems became predominant. Soil <u>profile No.1</u> is a <u>Calcaric Chernozem on Transdanubian loess</u> from Mezőföld. <u>Profile No.2</u> is a <u>Calcaric Chernozem</u> formed on the <u>Lowland loess</u> deposits of the river Maros from the loess ridge of the south part of the Hungarian Great Plain. The soil reaction is mildly alkaline. Carbonates gradually increase downward from 2-3% in the Ap up to 20-30% in the C1 horizon. The appearance of a pseudo-micelial carbonate layer in the A12- A13- A-4 horizons is a characteristic feature of this soil type. The humus layer is thick with relatively high humus content, decreasing gradually from about 3% in Ap to 0.9% in C1. This indicates that humus formation played a predominant role in the pedogenesis. The soil texture is silt loam with 50-60% silt and 18-25% clay.

On the same loess covered with forest vegetation Luvisols were formed. Soil <u>profile No.3</u> is an <u>Orthic Luvisol</u> (lessivated brown forest soil) developed on <u>Transdanubian loess</u> from Zala Hills. Soil reaction varies between pH 5.9-8.2 as a result of the leaching of carbonates from A and B horizons. The amount of carbonates reaches 20% in the C horizon. The humus content is 1.4% in the Ap horizon decreasing to 0.2% in the C horizon. The soil texture is silty loam with 60-70% silt and with 8-15% clay fraction. The argillic horizon formed by lessivage is most characteristic of this soil.

Under hydromorphic conditions, however, Phaeozems were developed. <u>Profile No.4</u> is a <u>Haplic Phaeozem</u> formed on <u>Transdanubian loess</u> mixed with Pannonian marine deposits from Mezőföld. Soil reaction is neutral or mildly alkaline. Carbonates increase gradually downward the profile from 0 to about 30%. The humus content is 2.5% in Ap, decreasing to 0.4% in the C1 horizon. The soil texture is loam with 35-40% silt and 20-25% clay.

The low-lying areas adjoining loess ridges are covered with Holocene fluviatile sediments. This parent material is strongly weathered, containing large amounts of clay. Under hydromorphic conditions and hydrophyte vegetation, Humic Gley soils were formed. Soil profile No.5 developed on the clayey deposits in the valley of the river Körös; Profile No.6 formed on the alluvial deposits of the river Rába. The soil reaction is slightly acidic or neutral. Carbonates appear occasionally in the C horizon below 100 cm. The humus percentage is about 2.5% in Ap, decreasing to 0.6% in the C horizon. The high silt (40-50%) and clay fractions (40-55%) grade them by texture in the silty clay group. The ground water level can be at 1-2 m.

METHODS

Primary and clay minerals were studied first by X-ray diffraction in about 1600 soil samples in order to select typical profiles.

The routine physical and chemical analyses (particle size analysis, pH, humus, exchangeable cations, etc.) were carried out by usual soil survey laboratory methods. The soils were subjected to gentle wet grinding and the clay fraction ( < 1.4 μm) was dispersed by repeated washing with water and separated by sedimentation. Only ammonium hydroxide was used as dispersing agent, the aim being to characterize as far as possible the clay fraction as it exists in the soil. In parallel with original samples, Mg-saturated specimens were examined. Oriented aggregates were prepared by sedimenting on glass slides. The different clay minerals were identified by observing the response to ethylene glycol and heat (335 and 550°C) treatments. The effect of K-saturation was also studied. X-ray diffraction patterns were recorded through the 3-25° 2θ range on a Philips PW 1050 powder diffractometer using Ni-filtered $CuK_\alpha$ radiation ( $\lambda$ = 1.5418 Å) with a scanning speed of 1/2° (2θ)/min. K, Mg, Ca and Na desorption curves were obtained by electro-ultrafiltration (EUF) as proposed by Németh (1976).

RESULTS AND DISCUSSION

The clay fraction of soils developed on loess is of polymineral character and despite the variety of depositional conditions the

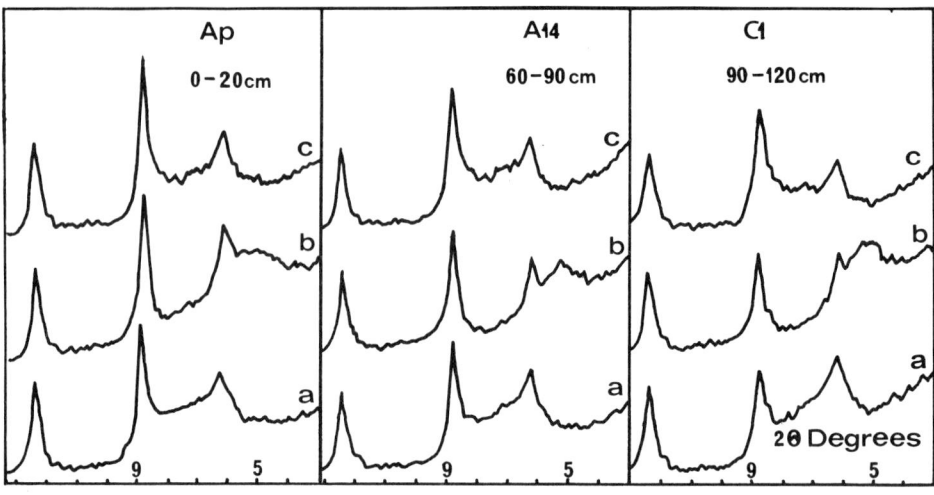

Fig. 1. X-ray diffraction patterns of the clay fractions from Ap, A14, and C1 horizons of profile No.1, a Chernozem on Transdanubian loess. (a) Mg-saturated, (b) Mg-saturated and treated with ethylene glycol, (c) K-saturated.

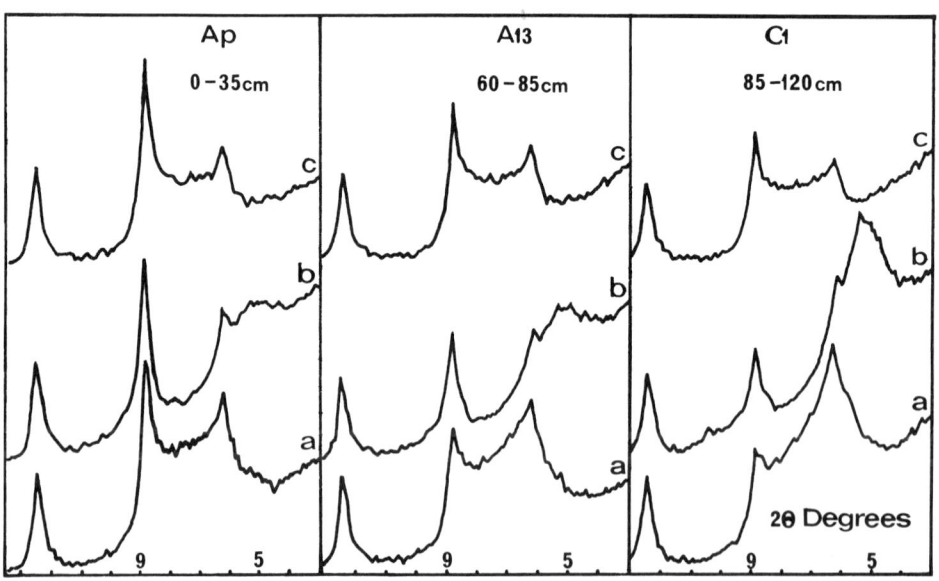

Fig. 2. X-ray diffraction patterns of the clay fractions from Ap, A13, and C1 horizons of profile No.2, a Chernozem on Lowland loess. See (a) (b) (c) in Fig. 1.

clays tend to be uniform in composition, being usually a mixture of illite, smectites and their intermediates in various proportions, accompanied by a relatively constant amount of chlorite (Fig.1, 2, 3 and 4).

Smectites predominate <u>in the parent material</u> of the profiles, showing a strong reflection at 14Å, accompanied by a series of intermediate type clay minerals which yield a broadened reflection ranging between 10.5 and 14Å. They all expand to 15-20Å spacing or sometimes even above, with a maximum intensity at about 17Å. Heating at 335 and 550°C weakens the 14Å reflection and intensifies that at 10Å. The non-expanding phase contributing to the 14Å reflection was identified as chlorite. The illite content is always smaller in the loess, than in upper horizons. The asymmetry of the 10Å illite reflection toward low angles is often decreased by ethylene glycol indicating interstratification. <u>Saturation with potassium</u> ions results in a considerable intensity decrease of 14Å smectite, a partial change of the broadened reflection between 10.5 and 14Å and an intensity increase in the 10Å illite reflection. Similar X-ray diffraction patterns were obtained for the different profiles showing the appearance of a predominant quantity of illite. The adjoining broadened reflection is of different intensity. Sometimes the existence of certain intermediates can also be recognized in the form of separate small peaks.

These results can be explained by complex interstratification, i.e. by the presence of mixed layer illite-smectite. On the basis of the X-ray patterns, smectites of 11-14Å can be considered as a more or less continuous series. On weathering of loess, micaceous minerals were presumably converted to <u>smectite(-like) minerals with open layers</u>. These minerals are classified in the literature as degraded illite, mica intermediate, soil montmorillonite, expanding illite, beidellitic smectite. As shown above, they are particularly liable to hold potassium by the closure of interlayer spaces.

These <u>smectites</u> are presumably inherited components of the parent material, the greatest amount of which can be found in the C horizon of the profiles studied. However, the effects of the conditions of formation before sedimentation or thereafter during pedogenesis can be observed to some extent in the clay mineral composition of loess. The parent material of Chernozem profile No.1, i.e. Transdanubian loess, contains a smaller amount of smectite than that of profile No.2, i.e. Lowland loess, in which smectites predominate, possibly

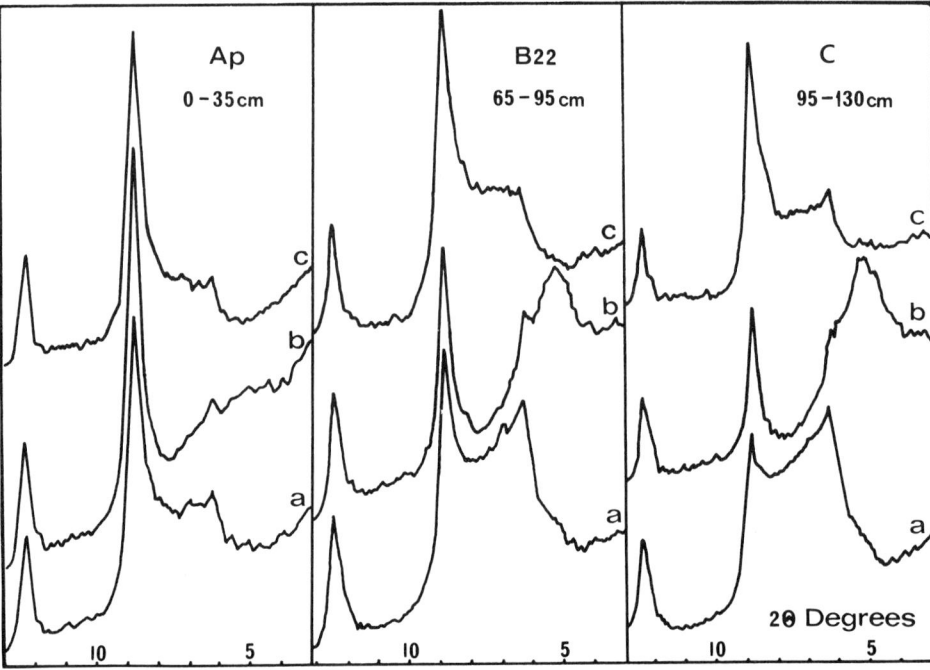

Fig. 3. X-ray diffraction patterns of the clay fractions from Ap, B22, and C horizons of profile No.3, a Luvisol on Transdanubian loess. See (a) (b) (c) in Fig. 1.

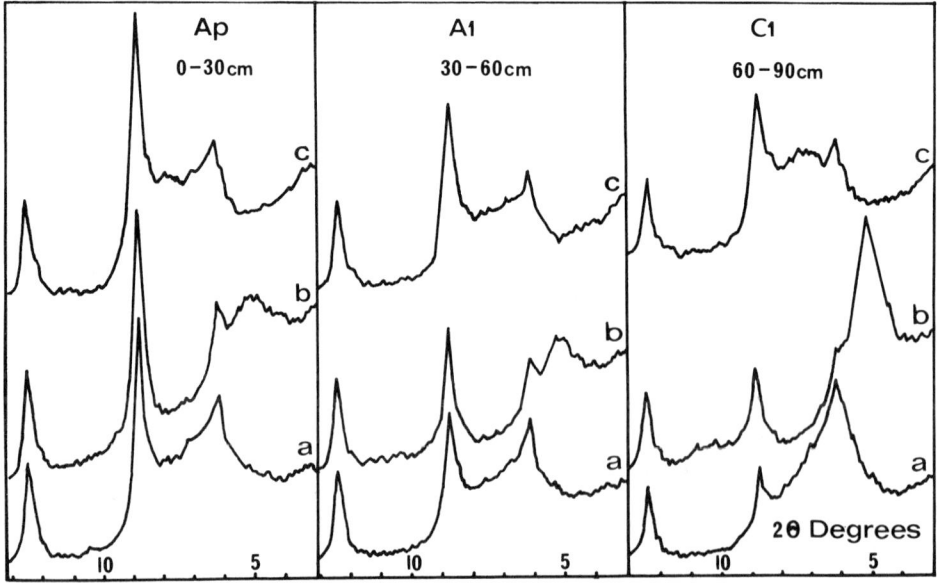

Fig. 4. X-ray diffraction patterns of the clay fractions from Ap, A1 and C1 horizons of profile No.4, a Phaeozem on loess mixed with Pannonian clay. See (a) (b) (c) in Fig. 1.

due to weathering under hydromorphic conditions and in magnesium--rich environment (See Figs. 1 and 2, C1 horizons). The great proportion of smectites in the parent material of Luvisol (profile No.4) can be explained by the admixing of Pannonian marine deposits rich in expansible minerals.

Smectites dominate most frequently in the parent material of soils formed on Pleistocene loess although a decrease of their amount can always be observed in the upper horizons. This decrease seems to be connected with an increase of the amount of illite. It is difficult to arrive at any firm conclusion in this matter. Nevertheless, in agreement with Paquet and Millot (1973) and Niederbudde (1973 and 1975), this phenomenon can perhaps be attributed to the transformation of smectites to illite, i.e. the partial reversion of the weathering cycle. Weathering is used in a specific sense suggested by Nemecz (1973) as the removal of potassium ions and the increase of the degree of interstratification.

The soils formed on loess still contain carbonates at the surface. Therefore the selective translocation of smectites with a concentration of illite related to that in the upper horizons is not probable. Micas resist weathering in a calcareous medium at pH $\gtrsim$ 7.0 and the quantity of micaceous minerals and feldspar does not change appreciably with depth. Thus these minerals did not serve as fundamental sources for the formation of illite during pedogenesis.

The transformation of smectites to illite ("illitization") involves a pedogenetic process, resulting in the incorporation (fixation) of potassium in the interlayer spaces. As the identified smectites contract remarkably on potassium saturation, the probability of the transformation depends mainly on the concentration of potassium ions produced in pedogenetic processes. Potassium originates mostly from humified plant material accumulated in the top soil and to a smaller extent from soluble weathering products of feldspars and micas. Its migration into the deeper horizons can be considerably influenced by drainage conditions in the soil.

The progress of illitization depends apparently on soil-forming processes. Owing to intensive humus formation, illitization is more advanced down the profiles in Chernozems (see Figs. 1 and 2) than in other soils. Due to hydromorphic conditions, illitization was slowed down in the Phaeozem profile (see Fig.4), whereas in the Luvisol the progress of the transformation was not appreciably influenced by lessivage and the more intense leaching.

The clay fraction of Humic Gley soils (profiles No.5 and 6) formed on Holocene fluviatile sediments contains a predominant quantity of smectites, associated with illite and chlorite (Fig.5). As in the loess soil clays, the presence of intermediates can be observed, which yield a broadened reflection between the 10Å

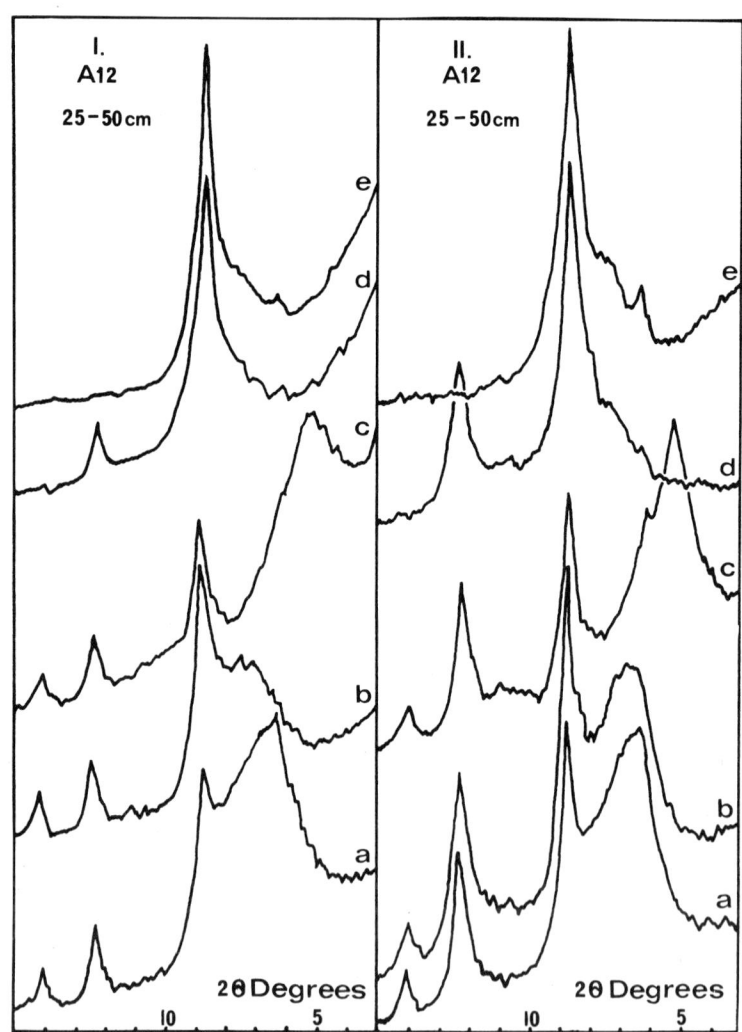

Fig. 5. X-ray diffraction patterns of the clay fractions from A12 horizons of profile No.5 (I.) and profile No.6 (II.), Humic Gleys on Holocene sediment. (a) Mg-saturated, (b) K-saturated, (c) treated with ethylene glycol (d) heated at 335°C (e) heated at 550° C.

illite and the 14Å smectite peak. The ethylene glycol treatment results in the appearance of an intensive 17Å peak, broadening toward low angles. On heating at 335 and 550°C, the smectites collapse to a great extent. The amount of chlorite is different in the profiles. The parent material of profile No.5 contains the detrital products from the Carpathian Mountains, where interstratified illite-smectite is common. Profile No.6 consists of weathering products of the chlorite-rich Alps and its foothills.

<u>Saturation with potassium ions</u> results in different responses. In profile No.5 the 14Å smectite reflection disappears, a broad reflection at 11-14Å with small peaks at 11.8 and 12.4Å appears, accompanied by a marked increase of the 10Å illite reflection. On the other hand, in profile No.6, K-treatment caused only a small alteration of the 14Å reflection and a slight increase of the 10Å peak. These results can be explained by the presence of mixed layer illite--smectite derived from parent material which contain smectites of different weathering grade.

The mineral composition of soil clays shows no significant change throughout the profiles. These soils were formed later than the soils on Pleistocene loess. Transformation of smectites to illite in profiles No.5, and No.6 was not observed, possibly due to the

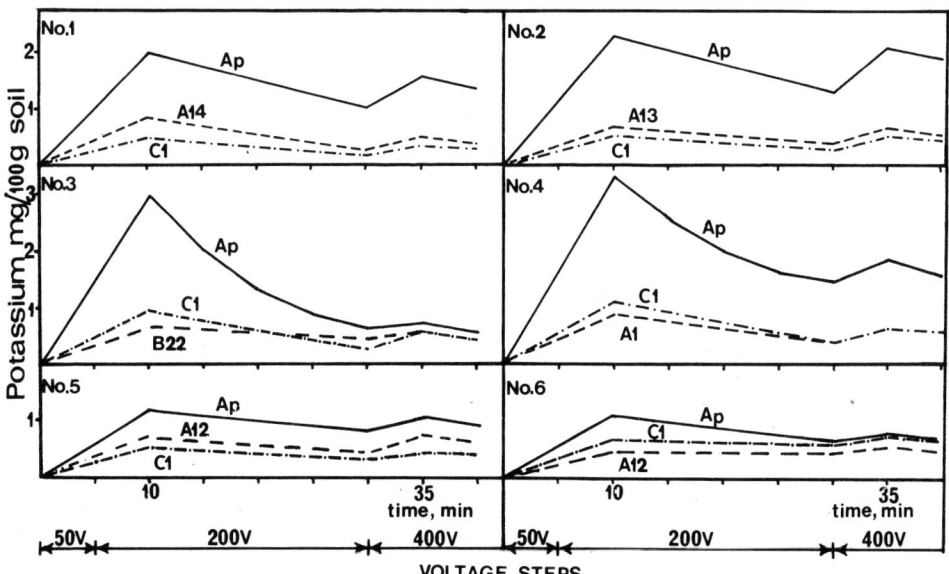

Fig. 6. Potassium-desorption curves of soil profiles No.1-6 prepared at varied voltage.

heavy texture, the decreased water permeability and the magnesium-
-rich medium.

Illitization may be the result of human activities. The effect of
regular fertilization, i.e. the addition of large amounts of potassium (300-500 kg $K_2O$/ha in the last few years), seems to be reflected
in the clay mineral composition of soils on loess. There are significant differences between the illite content of the Ap horizons and
the horizons underneath. The top horizons are relatively well saturated with potassium as proved by the small differences in X-ray
patterns of untreated and K-saturated samples. This is clearly indicated by the results of electro-ultrafiltration studies. The
potassium-desorption curves (Fig.6) show large amounts of available
potassium in the top horizons. There are great differences in K-fixation between the soils formed on loess and the soils developed on
fluviatile sediments. Correlation can be found between the clay
mineral composition and potassium-desorption curves, with increasing
smectite content the potassium desorption is considerably decreased.

ACKNOWLEDGEMENT

The authors thank Mrs. J.Matkó for technical assistance, Drs.
J.Tóth and J.Kárpáti for financial support and kind interest.
Thanks are due to Prof. A.Kálmán for stimulating discussions.

REFERENCES

Nemecz, E., 1973. Agyagásványok (Clay minerals). Akadémiai Kiadó,
    Budapest, pp. 464-467.
Németh, K., 1976. The determination of effective and potential
    availability of nutrients in the soil by electro-ultrafiltration.
    Applied Sciences and Development, 8: 89-111.
Niederbudde, E.A., 1973. Beziehungen zwischen K-Fixierungsvermögen
    und Dreischicht-Tonmineralien in Bodenprofilen aus Löss.
    Z. Pflanzenernähr. Bodenkunde, 135: 196-208.
Niederbudde, E.A., 1975. Veränderungen von Dreischicht-Tonmineralien durch natives K in holozänen Lössböden Mitteldeutschlands
    und Niederbayerns. Z. Pflanzenernähr. Bodenkunde, Heft 2:
    217-234.
Paquet, H. and Millot, G., 1973. Geochemical evolution of clay
    minerals in the weathered products and soil of Mediterranean
    climates. Proc. Intern. Clay Conf. 1972, 199-206.

ALTERATION OF BASALTIC ROCKS BY HYDROTHERMAL ACTIVITY AT 100-300°C

H. KRISTMANNSDOTTIR

National Energy Authority, Laugavegur 116, Reykjavik, Iceland

ABSTRACT

Six high-temperature geothermal areas in Iceland have been investigated by deep drilling. The underground rocks down to 2 km depth are basaltic hyaloclastites and lavas. Olivine-tholeiites are dominant in some areas, but tholeiites in others. Hydrothermal alteration has taken place at 100-300°C.

In all the fields the same groups of alteration minerals are formed at the same rock temperature. However, minor differences are observed due to the varying composition of the basaltic rocks.

Smectites (iron-rich saponites), zeolites, calcium silicates, calcite, pyrite and quartz are formed at rock temperatures below 200°C. The smectites have transformed into mixed-layer clay minerals and swelling chlorites at 200-230°C. Most zeolites and the calcium silicates disappear in this temperature interval. Chlorites become the dominant sheet-silicates when the rock temperature exceeds 240°C. Epidote and prehnite are formed at slightly higher temperature. Actinolite appears near to 300°C.

The clay minerals are quantitatively the most significant alteration minerals. Further they respond quickly to temperature changes in the geothermal system. An examination of the clay minerals and their relation to other alteration minerals is very useful for interpretation of the thermal history of a geothermal field.

INTRODUCTION

Drillhole data from active geothermal areas are useful for the study of hydrothermal crystallization of minerals under natural conditions. Most of the factors which influence the alteration process such as temperature, pressure, fluid and rock composition can be measured directly or calculated. Further, the location of the drillhole in the convecting geothermal system is determined.

In the early stages of the study of the Icelandic geothermal areas a certain relation between the mineral assemblages and the rock temperature was noticed. This has been confirmed by subsequent studies. At specified conditions, stability ranges have been obtained for some mineral assemblanges.

The main centers of geothermal activity in Iceland are shown in Fig. 1.

Geothermal areas in Iceland are divided into two groups. First, high - temperature areas where the rock temperature exceeds 200°C at 1 km depth. Second, low - temperature areas where the rock temperature is below 150°C at 1 km depth.

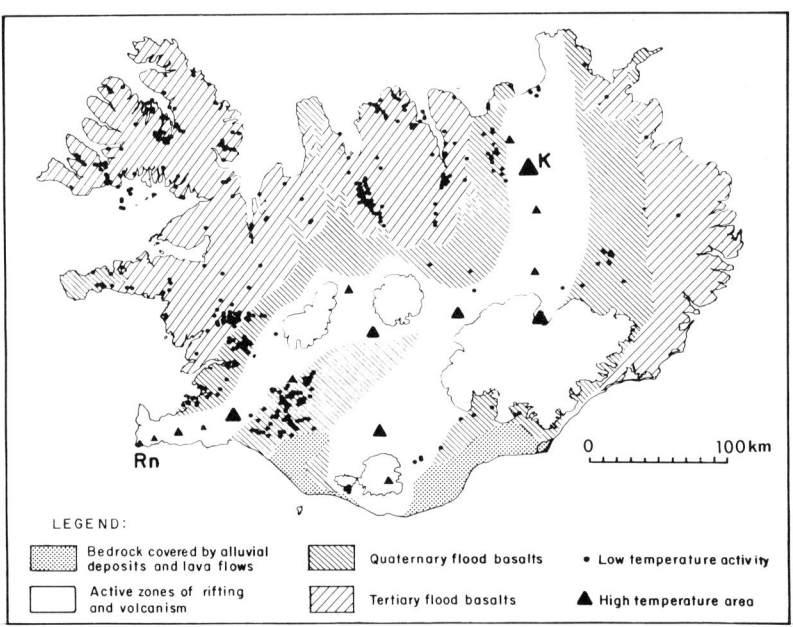

Fig. 1. The main centers of geothermal activity in Iceland. Compiled by Sæmundsson, K. Rn is the high - temperature geothermal area on Reykjanes. K is the Krafla geothermal area.

High temperature geothermal areas are located within the zones of rifting and volcanism. The rocks there are very young and have not been altered previously at higher temperatures. On the other hand the low - temperature geothermal areas are located in the older Quaternary and Tertiary rock formations, in which the rocks may have undergone previous metamorphic events. Therefore, the correlation of alteration effects to recent geothermal activity can be complicated. However, an overprint of low - temperature alteration can sometimes be distinguished in the Quaternary rock formation, where the permeability is still fairly high. A connection between mineral stability and rock temperature has been established at 40-100°C by correlation of data from many low-temperature areas (Kristmannsdóttir and Tómasson, 1976).

The results for the rock temperature interval 100-300°C, which are reported in the present paper are based on studies of the six high-temperature areas where deep drillings have been carried out. In two of the areas (Reykjanes, Svartsengi) the geothermal fluid is highly saline, but in the others (Krafla, Námafjall, Nesjavellir, Hveragerði) the fluid is of meteoric origin. The fluids of meteoric origin are slightly alkaline and rather low in dissolved solids. According to Arnórsson et. al. (1978) the composition of the geothermal fluids is largely fixed by the temperature and salinity. The saline high-temperature geothermal fluids are originally sea-water. The fluids are nearly neutral, and relative to sea-water, depleted in $Mg^{2+}$, $SO_4^{2-}$ and enriched in $K^+$, $Ca^{2+}$ and $SiO_2$. This modification of sea-water already starts by reaction with rocks at temperatures below 50°C (Tómasson et. al., 1977).

The underground rocks in all the six areas are predominantly basaltic hyaloclastites and lavas. Intrusive rocks are common at greater depths. A typical example of a stratigrapic section through a high-temperature geothermal area is given in Fig. 2.

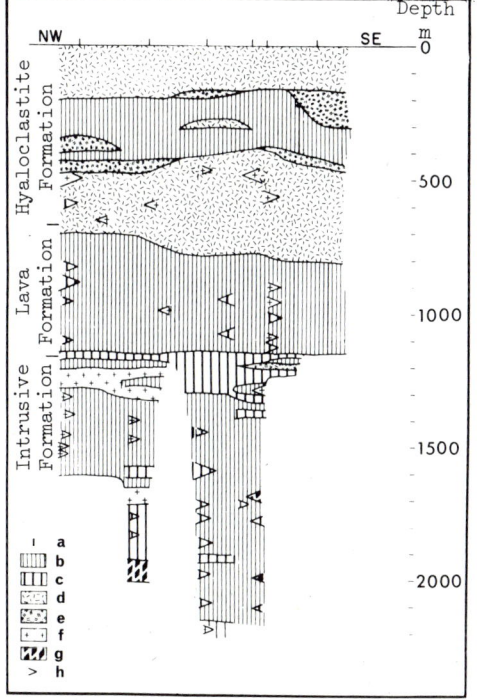

a   Drillholes
b   Basalt lava altered
c   Basalt lava fresh
d   Hyaloclastite
e   Breccia, basalt rich
f   Dolerite
g   Granophyre
h   Dykes

Fig. 2. A stratigraphic section through the high-temperature geothermal area in Krafla NE Iceland. The horizontal scale is the same as the vertical scale. The surface is 480 m above sea level.

Olivine-tholeiite basalt is dominant in two of the areas and tholeiite in the others.

The pressure in the uppermost 2 km is below 0.5 kbar. The rock temperature is determined from inhole temperature logging by selecting points which lie near to undisturbed conditions. The maximum recorded temperature in a high-temperature area is about 340°C, but maximum temperatures of 280-300°C are the most common.

Temperature is the dominant factor in the alteration process in areas where the original rocks are similar, the geothermal fluid is of the same origin, the permeability is high, and the effects of the overburden pressure are insignificant.

GENERAL ASPECTS OF HYDROTHERMAL ALTERATION

Due to the seepage of geothermal fluids through the rocks original glass and minerals are replaced by secondary minerals which also fill the pores and cracks. Hydrothermal alteration is characterised by its localized distribution whereby zones of completely recrystallized rocks are found close to rocks which are very slightly affected by alteration.

The rate of this alteration process is dependent on many factors such as temperature, chemistry, lithology and structure of the rocks. For a given rock temperature the lithology and permeability of the rocks are the most critical factors for the progress of alteration. The glassy rocks are most easily altered, whereas holocrystalline dense basalts are much more resistant. The flow through the rocks is highly governed by the tectonic pattern.

Mineralogical changes in the rocks show that most elements must have been mobilized during the hydrothermal alteration.

The rock samples studied are mostly drill cuttings and represent average samples from 5-10 m of penetration. Chemical exchanges on a small scale will therefore not be seen, but a reliable measure of the large scale changes can be obtained by a careful choice of samples. Hydration is the most prominent change in rock composition due to hydrothermal alteration (Kristmannsdóttir, 1975 (a), Kristmannsdóttir, in preparation). An increase in sulphur and carbonate content is also noticable, especially at upper levels in the geothermal areas.

The relative changes in contents of major cations are not found to be extensive. As an example of variation of major elements with depth are shown in Fig. 3 results from a drillhole in the Krafla area. The geothermal fluid in Krafla is of meteoric origin. The hyaloclastites in the upper part of the section are highly hydrated. Silica shows local depletion or increase, but is on the average similar to that of the fresh rocks. Total iron and calcium show some variations but no definite trends of enrichment or depletion. Magnesium shows a slight relative increase at upper levels. Sodium is leached out of the rocks at upper levels.

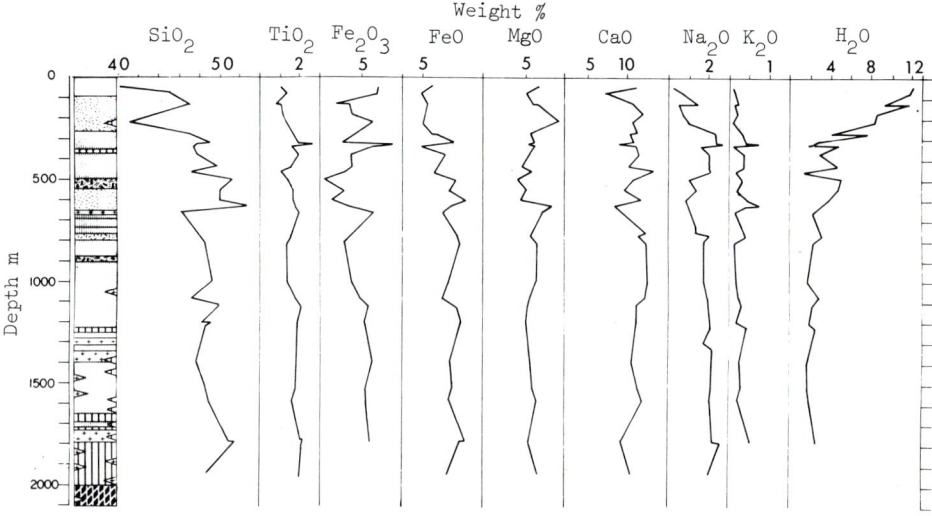

Fig. 3. Compositional variation of some major elements by depth in drill cuttings from a drillhole in Krafla. A simplified geological section is also shown. Legend as in Fig. 2.

The alkalies are the most mobile cations. Potassium increases in the zone of upflow and boiling in the geothermal areas, although not shown in this example.

The chemical exchanges between saline geothermal water and basaltic rocks are somewhat different (Tómasson and Kristmannsdóttir, 1972, Kristmannsdóttir, in preparation). The same sequence for mobility of the elements is found. The enrichment of potassium is however more distinct in the zones of upflow and boiling and an increase of calcium with depth is indicated.

ALTERATION MINERALS

The study of altered basaltic rocks in the geothermal areas shows a sequence of mineral assemblages related to increased temperature and depth. The clay minerals are the most voluminous alteration minerals. They replace basalt glass and olivine, pyroxene and partly plagioclase in the basalts. Also they are precipitated in pores and cracks. The clay minerals are mostly finegrained and poorly crystalline. Smectites form at the lower rock temperatures. Montmorillonite is found in the near surface acid leaching zone of geothermal areas and in mudpots at the surface. The smectites formed below by the action of the geothermal fluids (saline & fresh) are iron-rich saponites (Kristmannsdóttir,1975 (b), 1976).Their compositional variation within one area or from one area to another is not known. Data from XRD, DTA, IR spectra and chemical analyses of clay fractions suggest a considerable variation between areas, but a small variation within a single area. An interlayering of the smectites starts at rock temperature below 200°C and a gradual transformation to chlorite proceeds at 200-240°C. Smectites are seldom recorded at rock temperature exceeding 200°C. Mixed - layer clay minerals of smectite and chlorite are dominant at 200-240°C. The study by XRD of the clay minerals recorded at 200-240°C indicates a variety of different mixed-layer clay minerals and swelling chlorite. DTA analyses and studies of the IR spectra show a close structural relationship between all these minerals (Kristmannsdóttir 1976). Chlorite has become the dominant sheet silicate at rock temperatures of 230-250°C: In a geothermal area where the maximum temperature does not exceed 240°C chlorite is only found sporadically. All chlorites are iron-rich, but their composition varies from area to area. Table 1 shows the composition of a typical chlorite from Krafla.

TABLE 1

Microprobe analyses of some typical alteration minerals from the Krafla area

| Weight % | Heulandite | Laumontite | Wairakite | Prehnite | Epidote | Chlorite | Actinolite |
|---|---|---|---|---|---|---|---|
| $SiO_2$ | 49.60 | 55.11 | 55.98 | 42.44 | 39.92 | 28.17 | 49.52 |
| $TiO_2$ | - | - | - | - | - | 0.12 | 0.00 |
| $Al_2O_3$ | 21.16 | 19.32 | 20.82 | 22.10 | 25.17 | 17.96 | 3.32 |
| FeO (total) | 0.03 | 0.05 | 0.00 | 2.37 | 8.80 | 26.19 | 10.85 |
| MnO | - | - | - | - | - | 0.80 | 0.37 |
| MgO | - | - | - | - | - | 15.22 | 16.46 |
| CaO | 12.23 | 11.17 | 12.41 | 27.38 | 25.38 | 0.12 | 11.15 |
| $Na_2O$ | 0.15 | 0.27 | 0.30 | 0.04 | 0.02 | 0.07 | 0.36 |
| $K_2O$ | 0.00 | 0.13 | 0.01 | 0.03 | 0.01 | 0.02 | 0.20 |
| Total | 83.17 | 86.05 | 89.52 | 94.36 | 99.30 | 88.67 | 92.23 |

Zeolites (all Ca-rich varieties. Table 1) and calcium silicates are precipitated mostly in pores and cracks but are also found dispersed in replaced glass. A variety of zeolites is recorded at rock temperature below 100°C. Zeolites are less common in areas with saline geothermal fluids. Below 100°C two groups of zeolites occur i.e. mordenite, heulandite, stilbite and epistilbite which are common in tholeiite basalts, and chabazite, thomsonite and mesolite-scolectie which are more common in olivine tholeiite basalts. Laumontite replaces other zeolites at 100-120°C. In the high-temperature areas the thermal gradient is very steep and therefore a sequence of zeolite zones is seldom distinct (Kristmannsdóttir and Tómasson, 1976). Wairakite appears at 180°C and all other zeolites have disappeared at about 200°C. Wairakite has been recorded at rock temperatures as high as 300°C.

The calcium silicates, gyrolite, reyerite, zeophyllite and truscottite are found at shallow levels where the rock temperature is below 200°C.

Calcite is common at all depths at rock temperatures up to 270°C.

Chalcedony is precipitated at temperatures below 100°C, whereas quartz forms at higher temperatures and persists up to the highest recorded rock temperature. Locally quartz amounts to over 20% of the samples.

Epidote may occur sporadically at 200-260°C, but appears not to be precipitated in any quantity until the rock temperature reaches about 260°C. It is mostly precipitated in veins and cracks, but also replaces plagioclase. The highest contents of epidote recorded in the hydrothermally altered rocks reach 15-20% by volume. Epidote is common as a relict alteration mineral in low-temperature areas.

Prehnite is found in most of the areas at similar depths to epidote. No reliable correlation with rock temperature has been established for prehnite.

Actinolite is recorded from two areas at rock temperature exceeding 280°C. It crystallizes in finegrained aggregates together with chlorite and epidote and is also found as a replacement of pyroxene. Few chemical analyses have been obtained of this mineral (Table 1), but it appears to vary considerably in composition.

Albite is found to replace plagioclase, but no extensive albitization is observed in the rocks. Albite is observed at rock temperatures exceeding about 220°C, but is more common at rock temperatures near to 300°C.

Wollastonite is mostly recorded as a contact metamorphic replacement of calcite and quartz near to dykes intruding the geothermal system. It appears to form hydrothermally at rock temperatures near to and above 300°C.

Pyrite is precipitated over the whole temperature range. It is found in veins with other precipitates and also dispersed in the groundmass of the rocks.

Pyrrhotite is formed together with pyrite, but is less common and is not found in areas with saline fluids. Gunnlaugsson (1977) has shown that all high-temperature geothermal fluids in Icelandic areas, except for the saline ones, are in equilibrium with pyrite and pyrrhotite.

Leucoxene and sphene have been found occasionally replacing ilmenite. No correlation with rock temperature has been established.

Anhydrite is found dispersed in the areas with non saline fluids and is common in the areas with highly saline fluids. Anhydrite is precipitated in the highest quantities, during the inflow and rise of temperature of the water (Tómasson and Kristmannsdóttir 1972).

The data show that the clay minerals respond quickly to temperature changes within the range 200-240°C. The zeolites also appear to attain near equilibrium conditions fairly easily. Epidote, on the other hand, responds slowly to changes in the system.

SUMMARY

The same main mineral groups are formed at the same rock temperature in the

Icelandic geothermal areas, whatever is the quantitative rate of alteration. This applies to the basaltic hyaloclastites and lavas, but not always to the intrusive rocks of young age and very low permeability. In detail, the alteration mineral assemblages differ due to the variations of the original basaltic rocks. Alteration mineral assemblages, formed by the action of saline geothermal fluids, are in all main aspects similar to those formed by fluids of meteoric origin. One can distinguish between four significant alteration zones in the geothermal areas. The zones are defined by certain index minerals. Fig. 4 shows the zones and summarizes the mineralogical changes, and the characteristics of each zone.

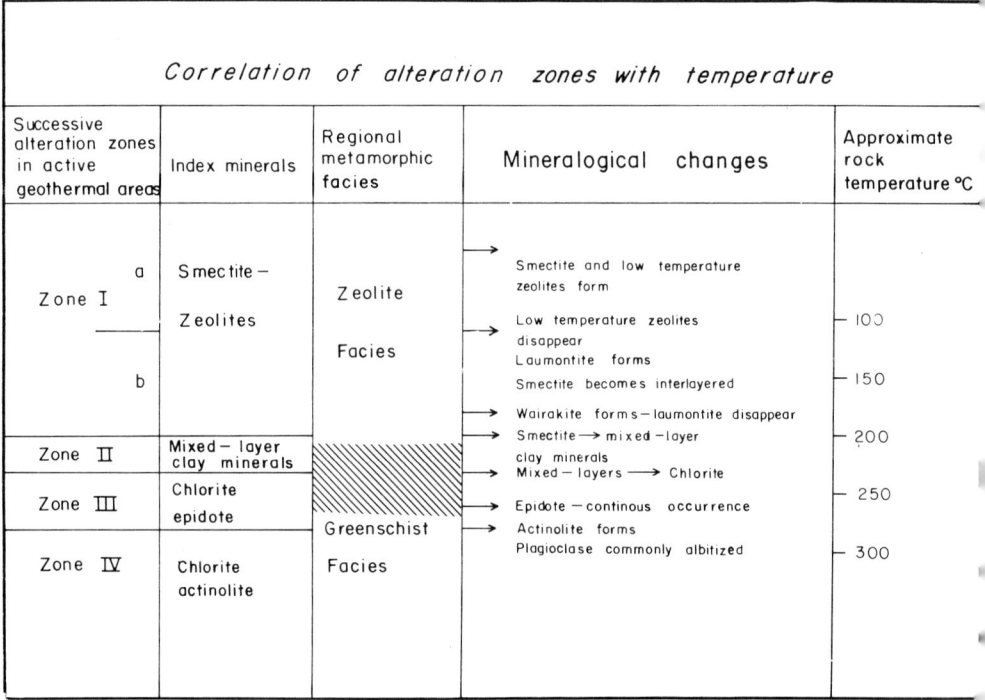

Fig. 4. Alteration zones in the high-temperature geothermal areas correlated to rock temperature and metamorphic facies.

The correlation with rock temperature is shown, and a relationship to facies in regional metamorphic rocks is suggested.

Interpretation of the thermal history of altered rocks from their mineralogy is a potential use of the results. A fossil geothermal gradient can be determined and the development of a geothermal area can be reconstructed.

The stability ranges for the minerals, established by the present study, are at distinctly lower temperatures than previously reported (see for example Miyashiro,

1973). Experimental data do not exist for most of the mineral transformations which are considered above. However, Liou (1971, 1970) reports stability ranges of laumontite and wairakite at temperatures which are a few tens of degrees higher than observed at high-temperature areas in Iceland.

In general, some discrepancy is to be expected between laboratory experiments and processes in nature. This is due to different conditions, such as composition of fluids and starting materials, fluid pressures, and different time scales for the processes.

REFERENCES

Arnórsson, S., Sigurðsson S. and Grönwold, K., 1978. Aquifer chemistry of four geothermal areas in Iceland. Geochem., Cosmocim., Acta. (In press).

Gunnlaugsson, E., 1977. The origin and distribution of sulphur in fresh and geothermally altered rocks in Iceland. A Ph.D. thesis, Univ. Leeds, October 1977.

Kristmannsdóttir, H., 1975(a)Hydrothermal alteration of basaltic rocks in Icelandic geothermal areas. Proceedings, Second U.N. Symposium on the development and use of Geothermal Resources, San Francisco, 441-445.

Kristmannsdóttir, H., 1975(b)Clay minerals formed by hydrothermal alteration of basaltic rocks in Icelandic geothermal fields. GFF. The Transactions of the Geological Society of Sweden. 97, 289-292.

Kristmannsdóttir, H., 1976. Types of clay minerals in hydrothermally altered basaltic rocks, Reykjanes Iceland. Jökull, 26, 30-39.

Kristmannsdóttir, H. and Tómasson, J., 1976. Zeolite zones in geothermal areas in Iceland. Proceedings, "Zeolite '76". Tucson, Arizona, in Press.

Liou, J.G., 1970. Synthesis and stability relations of vairakite. $CaAl_2Si_4O_{12}2H_2O$. Contr. Mineral. Petrol. 27, 259-282.

Liou, J.G. 1971. Stilbite-laumonite equilibrium. Contr. Mineral. Petrol. 31, 171-177.

Miyashiro, A., 1973. Metamorphism and metamorphic belts. George Allen & Unwin Ltd. 492 pp.

Tómasson. J., Kristmannsdóttir, H., Arnórsson, M., 1977. The interaction of Sea-Water with basaltic volcanic rocks on the Reykjanes penisula. Proceedings, The Second International Symposium on Water-Rock Interaction Strasbourg, I, 327-333.

Tómasson, J., and Kristmannsdóttir, H., 1972. High temperature alteration minerals and thermal brines, Reykjanes, Iceland. Contr. Miner. Petrol. 36, 123-134.

CLAYS AND CLAY MINERALS OF HYDROTHERMAL ORIGIN IN HAWAII[1]

Pow-foong Fan
Hawaii Institute of Geophysics, University of Hawaii, Honolulu,
Hawaii 96822 USA

ABSTRACT

Hydrothermal clays and clay minerals of Hawaii include actinolite, albite, alunite, anhydrite, calcite, cristobalite, hematite, heulandite, laumontite, mordenite, pyrite, quartz, chlorite, kaolinite, montmorillonite, mixed-layered chlorite-montmorillonite and mixed-layered chlorite-vermiculite. They can be grouped into five zones: (1) actinolite, (2) chlorite, (3) montmorillonite, (4) kaolinite, and (5) alunite. Four examples of hydrothermally altered Hawaiian basalts of Hawaii are discussed. They are (1) cores from Hawaii geothermal project well-A, Kilauea, Hawaii, (2) Waiahewahewa Gulch near Hoolehua, Molokai, (3) Koolau Caldera, Oahu, and (4) Waianae Caldera, Oahu. The zonal distribution of hydrothermal minerals is related to temperature, permeability and composition of fluid.

INTRODUCTION

Hydrothermal alteration of Kilauea basalts has been described by Macdonald (1944) and Stearns and Macdonald (1946). Fujishima and Fan (1977) investigated the older hydrothermally altered basalts of the Kailua Volcanic Series, Oahu, and described both vertical and horizontal distribution of clay minerals.

Four examples of clays and clay minerals derived from hydrothermally altered basalts of Hawaii are discussed in this study (Fig. 1). They are (1) cores from Hawaii geothermal project well-A, Kilauea, Hawaii, (2) southwestern Molokai, (3) Koolau Caldera, Oahu, and (4) Waianae Caldera, Oahu.

---

[1] Hawaii Institute of Geophysics Contribution No. 877.

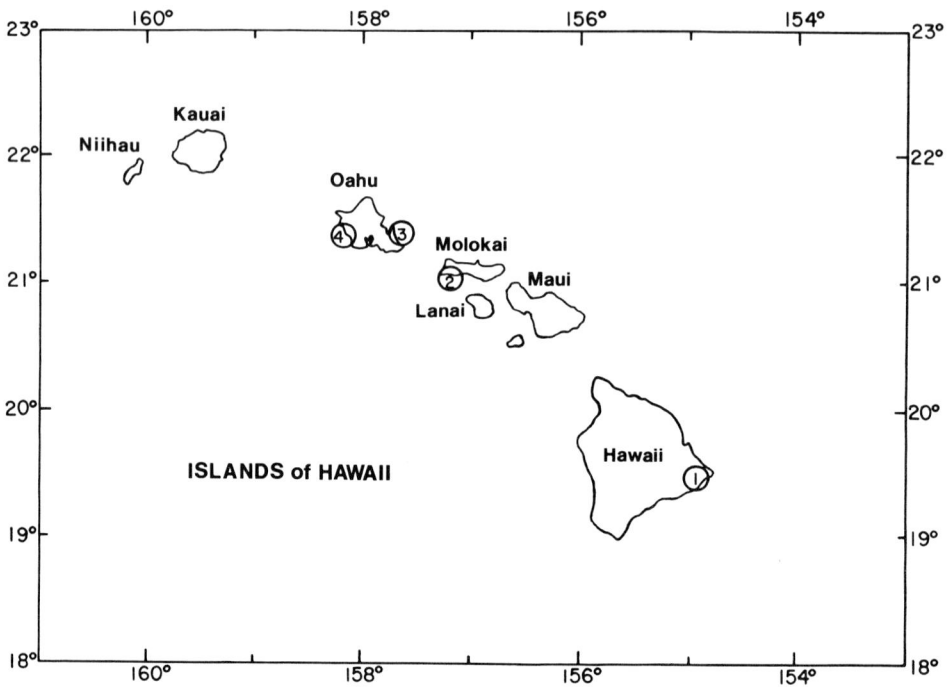

Fig. 1. Map of Hawaii, showing the location of (1) Hawaii geothermal test well-A; (2) Waiahewahewa Gulch, Molokai; (3) Koolau Caldera, Oahu; and (4) Waianae Caldera, Oahu.

RESULTS

X-ray diffraction analysis of cores obtained during 1976 successful drilling of a 1962-meter geothermal well in the east rift zone of Kilauea Volcano, Hawaii (Fig. 2) reveals three zones of hydrothermal alteration beneath a zone of unaltered lavas (Stone and Fan, 1978). The first zone, 675-1300 m, is characterized by the presence of montmorillonite with minor amounts of pyrite, quartz, calcite, chlorite, anhydrite, and heulandite. The temperature ranges from 290-325°C because the well has not returned to thermal equilibrium (Stone and Fan, 1978). Montmorillonite occurs as pseudomorphs after olivine. The second zone, 1300-1896 m, is characterized by the presence of chlorite as the dominant mineral, with minor amounts of quartz, actinolite, montmorillonite, albite, hematite, calcite, anhydrite, and pyrite. The temperature of the zone ranges from 307 to 340°C. On the basis of the X-ray intensity ratio of [(002/004)] 7Å/14Å peaks (Carroll, 1969), Mg- and

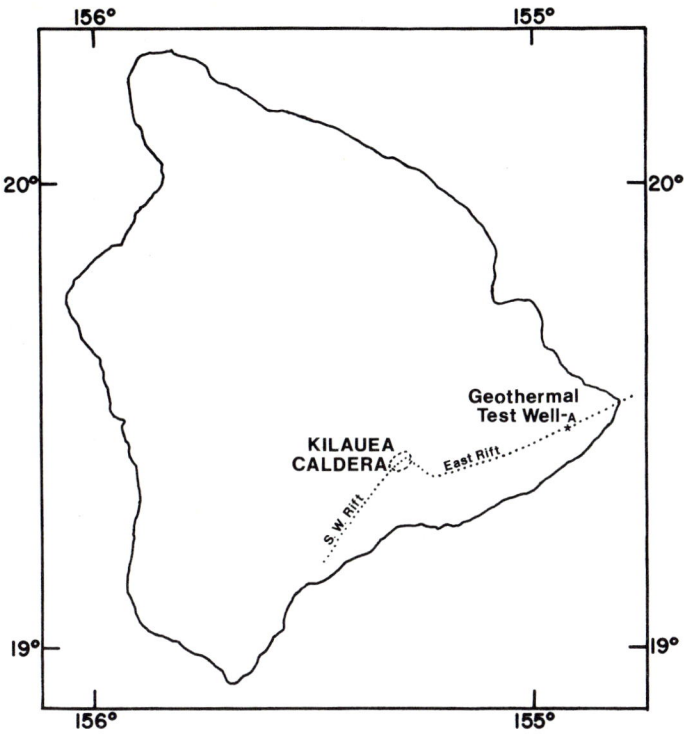

Fig. 2. Map of Hawaii, showing the location of Kilauea Caldera and geothermal test well-A.

Fe-chlorite are present. The Fe-chlorite content increases with depth (Stone and Fan, 1978). Chlorite is observed replacing actinolite, augite, and olivine. Montmorillonite is not present below 1896 m. The third zone, 1896-1959 m, is characterized by the presence of actinolite, with small amounts of chlorite, quartz, hematite, pyrite, calcite, and anhydrite. The temperature range is higher than 307-340°C. Actinolite is sparsely distributed in Zone 2, between 1300 to 1896 m, but becomes the dominant mineral below 1896.

Hydrothermally altered samples were collected along Waiahewahewa Gulch, about 3 km southwest of Molokai Airport, Hoolehua, Molokai (Fig. 3). X-ray diffraction analysis shows two types of mineral assemblages: (1) kaolinite as the major mineral, with small amounts of alunite, quartz and cristobalite; and (2) alunite as the dominant mineral, with minor amounts of quartz and cristobalite. Opal is also present in both assemblages.

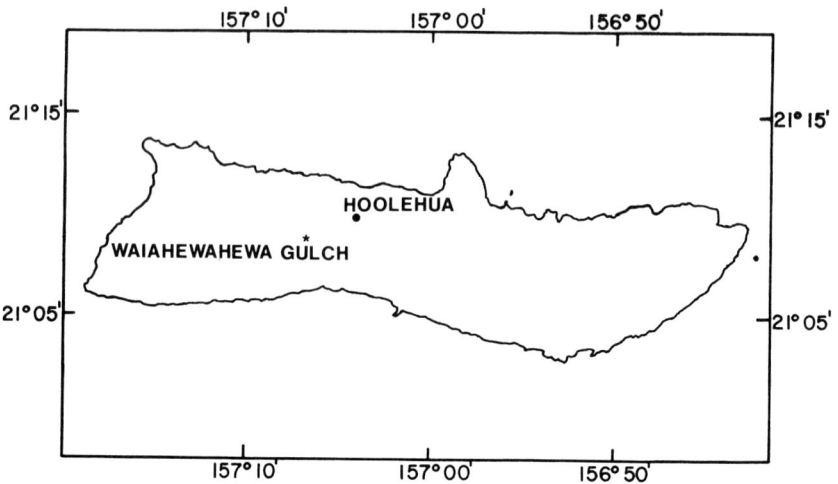

Fig. 3. Map of Molokai, showing the location of Waiahewahewa Gulch.

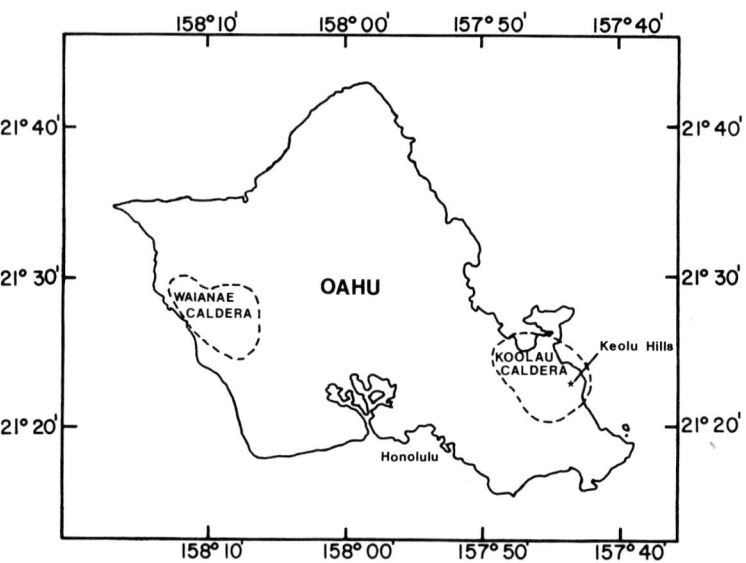

Fig. 4. Map of Oahu, showing the location of Koolau and Waianae Calderas.

The Kailua Volcanic Series is located in the deeply eroded Koolau Caldera on northeastern Oahu, Hawaii (Fig. 4). The mineralogy of the hydrothermally altered basalts examined by X-ray diffraction analysis showed two types of assemblages: (1) a vertical zonal boundary between chlorite (lower) and mixed-layered chlorite-vermiculite (higher), and (2) a horizontal zonation between mixed-layered chlorite-montmorillonite and chlorite or mixed-layered chlorite-vermiculite (Fujishima and Fan, 1977). The other minerals found in Koolau Caldera are quartz, laumontite, cristobalite, and mordenite.

The deeply eroded Waianae Caldera is located at the northwestern part of Oahu (Fig. 4). The samples for this study have been collected at the eastern edge of the caldera. X-ray diffraction analysis of the hydrothermally altered basalts shows the presence of only montmorillonite and hematite.

DISCUSSION

The alteration mineral zones (actinolite, chlorite, and montmorillonite) observed in the Hawaii geothermal project well-A cores, Kilauea, Hawaii and their temperature measurements are very similar to the high-temperature geothermal wells of Iceland, especially in the Nesjavellir area (Kristmannsdóttir, 1975; Kristmannsdóttir and Tómasson, 1975). Epidote is present in the Iceland geothermal field but has not been found in Hawaii geothermal project well-A cores.

Hydrothermal clays and clay minerals of Hawaii can be grouped into five zones according to decreasing temperature: (1) actinolite (>307-340°C), (2) chlorite (290-325°C), (3) montmorillonite (<290-325°C), (4) kaolinite, and (5) alunite. Each alteration zone is characterized by the dominant mineral (Table 1). With the exception of the actinolite zone, these zones are similar to the hydrothermal zones of Matsukawa geothermal area in Japan (Sumi, 1969). Actinolite, chlorite and montmorillonite zones are present in Hawaiian geothermal project well-A, Kilauea, Hawaii. Kaolinite and alunite zones are present in Waiahewahewa Gulch, near Hoolehua, Molokai. The chlorite zone is present in Koolau Caldera, Oahu, and the montmorillonite zone in Waianae Caldera, Oahu. All five hydrothermal mineral zones have not been located within a single location.

ACKNOWLEDGMENTS

I thank G.A. Macdonald, who provided the samples from Molokai. I thank K.A. Pankiwskyj and J.M. Sinton for reviewing this manuscript. This study was supported by a grant from the Energy Research and Development Administration, ERDA E(04-4)-1093.

TABLE 1

Mineral associations in the hydrothermal zones of Hawaii

| Minerals | Actinolite Zone | Chlorite Zone | Montmorillonite Zone | Kaolinite Zone | Alunite Zone |
|---|---|---|---|---|---|
| Actinolite | — | — — — — | | | |
| Albite | — | — — — — | | | |
| Alunite | | | | | — |
| Anhydrite | — | — — — — | | — — — — | |
| Calcite | — | — — — — | | — — — — | |
| Chlorite | | — | | | |
| Cristobalite | | | — | | — |
| Hematite | — | | | | |
| Heulandite | — — | — — — — | — — — — | | |
| Kaolinite | | | | — | |
| Laumontite | | — | | | |
| Montmorillonite | | | — | | |
| Mordenite | | | — | | |
| Opal | | | | — | — |
| Pyrite | — | | | | |
| Quartz | — | | | | — |
| Mixed-layered Chlorite-mont. | | — — | | | |
| Chlorite-vermicu. | | — — | | | |

REFERENCES

Carroll, D., 1969. Chlorite in central north Pacific Ocean sediments. In L. Heller (Editor), Proc. Int. Clay Conf., 1:335-338.
Fujishima, K.Y. and Fan, P.F., 1977. Hydrothermal mineralogy of Keolu Hills, Oahu, Hawaii. Am. Mineralogist, 62: 574-582.
Kristmannsdóttir, H., 1975. Hydrothermal alteration of basaltic rocks in Icelandic geothermal areas. In Second UN Symposium on the Development and Use of Geothermal Resources, San Francisco Proceedings. Lawrence Lab., Univ. Calif., 441-445
Kristmannsdóttir, H. and Tómasson, J., 1975. Hydrothermal alteration in Icelandic geothermal fields. Soc. Sci. Isl., 167-176.
Macdonald, G.A., 1944. Solfataric alteration of rocks at Kilauea Volcano. Am. Jour. Sci., 242: 495-505.
Stearns, H.T. and Macdonald, G.A., 1946. Geology and groundwater resources of the island of Hawaii. Hawaii Division of Hydrography, Bull., 9, 363 pp.
Stone, C. and Fan, P.F., 1978. Hydrothermal alteration of basalts from Hawaii geothermal project well-A, Kilauea, Hawaii. Submitted to Geology.
Sumi, K., 1969. Zonal distribution of clay minerals in the Matsukawa geothermal area, Japan. In L. Heller, Editor, Proceedings of the Int. Clay Conference, 1: 501-512/

REACTION SERIES FOR DIOCTAHEDRAL SMECTITE: THE SYNTHESIS OF MIXED-LAYER PYROPHYLLITE/SMECTITE

DENNIS EBERL

Geology Department, University of Illinois, Urbana, Illinois 61801 (U.S.A.)

ABSTRACT

Mixed-layer pyrophyllite/smectite, a member of the Al-smectite reaction series, has been synthesized from Ca- and Na-saturated Wyoming montmorillonite by treating the clay hydrothermally in a run solution containing $Al^{3+}$. Temperatures for synthesis ranged between 320°C and 400°C at autoclave pressures, and run times ranged between 3 and 30 days. Reaction towards pyrophyllite was favored by increasing run time, temperature and concentration of $Al^{3+}$. Calculated diffraction profiles for random and ordered pyrophyllite/smectite are presented.

INTRODUCTION

Many dioctahedral clay minerals can be classified into one of four structural trends: the mica trend, the chlorite trend, the kaolinite trend, and the pyrophyllite trend. Each trend is marked by the completely expandable smectite structure on the one extreme, and the completely non-expandable mica, chlorite, kaolinite or pyrophyllite structure on the other. Between these end-members are mixed-layer clays which may range in expandability, including mixed-layer mica/smectite, chlorite/smectite, kaolinite/smectite and pyrophyllite/smectite.

These structural trends can be produced hydrothermally from dioctahedral smectite starting material (Eberl, 1978). The trend which evolves depends on the interlayer cation and on the concentration of that cation in the run solution. Generally, alkali and alkaline earth interlayer cations yield the mica trend with increasing reaction time and temperature: smectite → mica/smectite → mica (the second step has not yet been accomplished experimentally for some cations). The exception to this pattern is Li-smectite which yields the chlorite trend: Li-smectite → chlorite/smectite → chlorite. It does yield the mica trend, however, when run in a LiCl solution. Mg-montmorillonite yields the chlorite trend when run in a $MgCl_2$ solution. The kaolinite and pyrophyllite trends are produced when the run solution contains $Al^{3+}$.

The family of clays which forms from the association of smectite with a particular cation in a hydrothermal system is known as a reaction series and is named for the cation. The K-smectite series includes randomly interstratified illite/smectite, K-rectorite and illite; the Na-smectite series includes randomly interstratified paragonite/smectite, Na-rectorite and paragonite; the Li-smectite series includes Li-tosudite, Li-rectorite, and cookeite; the Ca-smectite series includes Ca-rectorite and theoretically margarite; the Mg-smectite series includes Mg-rectorite, tosudite and possibly sudoite; and the Al-smectite series includes kaolinite/smectite, kaolinite, pyrophyllite/smectite and pyrophyllite.

The present experiments further explore the Al-series, and focus on its unusual higher-temperature member, mixed-layer pyrophyllite/smectite. This phase has not yet been found in nature. A report of ordered mixed-layer pyrophyllite/smectite from Japan (Kodama, 1958) is probably rectorite; the author seems to be following the usage of Bradley (1950) who misidentified rectorite as ordered pyrophyllite/smectite.

TECHNIQUES

The hydrothermal techniques used in this study have been described elsewhere (Eberl, 1978; Eberl and Hower, 1976). Briefly, both Ca and Na-saturated Wyoming montmorillonite were used as starting materials. They were welded into gold capsules with a known weight of $AlCl_3 \cdot 6H_2O$ and a solution of known composition (Table 1). Capsules were heated in water-filled autoclaves at constant temperatures that ranged between 245°C and 400°C. Capsules were weighed before and after treatment to detect fluid loss or gain.

Run products were identified from oriented, ethylene glycol-treated glass slide mounts by X-ray diffraction using a Norelco diffractometer and Ni-filtered Cu Kα radiation. Expandabilities (per cent smectite in the mixed-layer phase) were determined by comparing the real X-ray patterns with computer-simulated patterns using a modified version of the computer program of Reynolds and Hower (1970). For these computations, 9.2 Å was taken to be the thickness of a pyrophyllite layer, and 16.9 Å for a smectite layer. Crystallites were assumed to range in thickness between 10 and 15 layers. Both random and ordered structures were computed. In the random structures there is no pattern to the interlayering between pyrophyllite (Py) and smectite (S) layers, whereas in ordered structures there is preference for "PyS" pairs. Lattice coordinates for atoms in the 2:1 layers of illite were retained from the original program, rather than substituting those for pyrophyllite (Wardle and Brindley, 1972), because the former gave simulated X-ray patterns which more closely resembled those of the run products. KBr disks of the run products were studied with a Perkin Elmer model 467 grating infrared spectrophotometer.

TABLE 1

Hydrothermal runs

| Run no. | Starting clay | Clay/AlCl$_3$·6H$_2$O (by weight) | Solution composition | Time (days) | Temp (°C) | Run products |
|---|---|---|---|---|---|---|
| 1 | Ca-mont | 3.0 | 40 μℓ 1.0N CaCl$_2$ | 5 | 245 | Smectite, Q |
| 2 | " | " | 40 μℓ 0.5N CaCl$_2$ | 12 | 260 | Kao, Q, minor mixed layers |
| 3 | " | " | " | 6 | 280 | Smectite (slightly mixed-layered), Kao, Q |
| 4 | " | 1.5 | 50 μℓ 0.5N CaCl$_2$ | " | " | Kao, Q |
| 5 | " | 0.77 | 69 μℓ 0.5N CaCl$_2$ | 11 | 320 | Py/S(35) partly ord, Kao, Q |
| 6 | " | 0.79 | 68 μℓ H$_2$O | " | " | Py/S(10), Kao, Q |
| 7* | " | 3.0 | 50 μℓ 0.5N CaCl$_2$ | " | " | Py/S(20), Kao, Q, An |
| 8 | " | 3.33 | 50 μℓ 0.5N CaCl$_2$ | 9 | " | Py/S(20), Kao, Q |
| 9* | " | 5.0 | " | 3 | " | Py/S(40), Kao, Q |
| 10 | " | 10.0 | " | 9 | " | Smectite, Q |
| 11* | " | 3.33 | 50 μℓ 1.0N CaCl$_2$ | 3 | " | Py/S(25) slightly ord, Kao, Q |
| 12 | " | " | " | 9 | " | Py/S(25), Kao, Q |
| 13* | " | 5.0 | " | " | " | Py/S(85), Q |
| 14* | " | 3.33 | 50 μℓ H$_2$O | 3 | " | Py/S(25) slightly ord, Kao, Q |
| 15* | " | " | " | 9 | " | Py/S(10), Kao, Q |
| 16 | " | 2.5 | 30 μℓ 1.0N CaCl$_2$ | 30 | 400 | Py/S(30), Kao, Q |
| 17* | Na-mont | 3.33 | 50 μℓ 1.0N NaCl | 3 | 320 | Py/S(30), Kao, Q |
| 18 | " | " | " | 9 | " | Py/S(25), Kao, Q |
| 19* | " | " | 50 μℓ H$_2$O | 3 | " | Py/S(35), Kao, Q |
| 20 | " | " | " | 9 | " | Py/S(30), Kao, Q |

All runs made at autoclave pressures. Py/S(25) = 25% expandable, randomly interstratified (unless noted otherwise) pyrophyllite/smectite; Kao = kaolinite; Q = quartz; An = anorthite; Ca-mont = Ca-saturated Wyoming montmorillonite; Na-mont = Na-saturated Wyoming montmorillonite; ord = ordered. Runs 1 through 7 had 30 mg of starting clay, and the rest had 50 mg.
*Runs did not change weight during hydrothermal treatment.

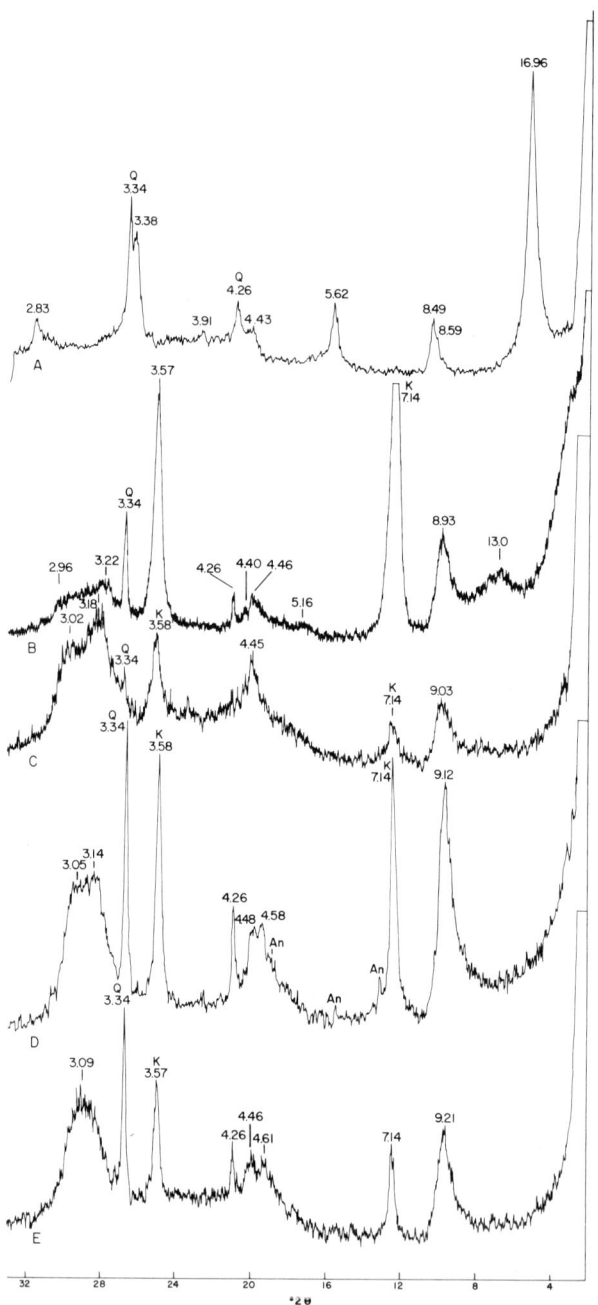

Fig. 1. Diffractograms of synthetic pyrophyllite/smectite. A = 85% expandable (run 13); B = 35% (run 5); C = 30% (run 20); D = 20% (run 7); E = 10% (run 15). The peak at 4.46 Å to 4.48 Å is the 020 reflection.

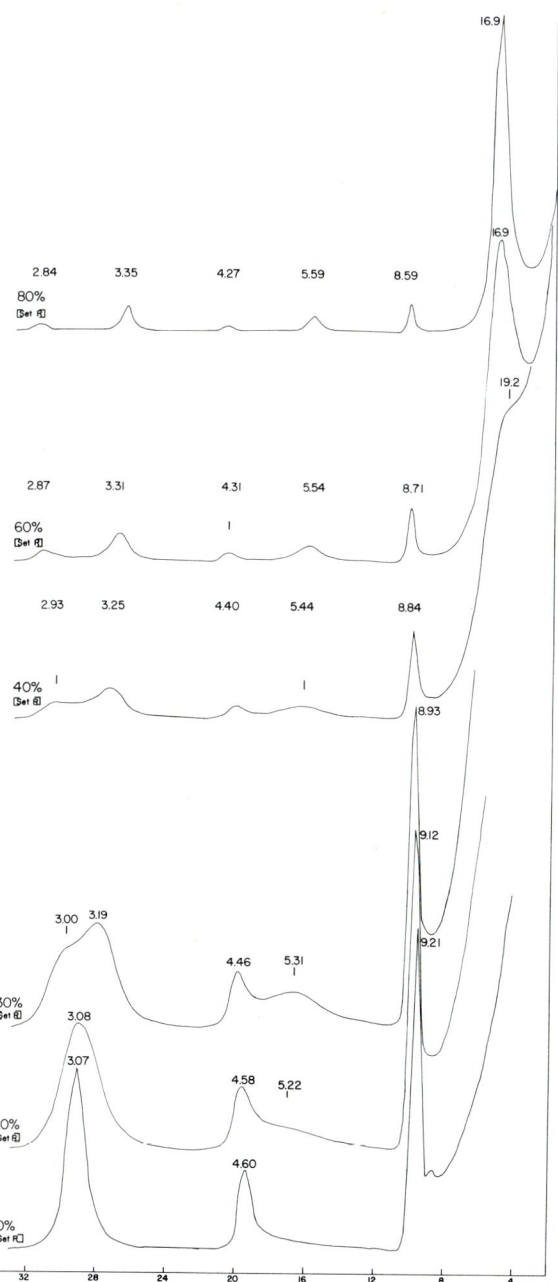

Fig. 2. Calculated diffraction profiles for random pyrophyllite/smectite.

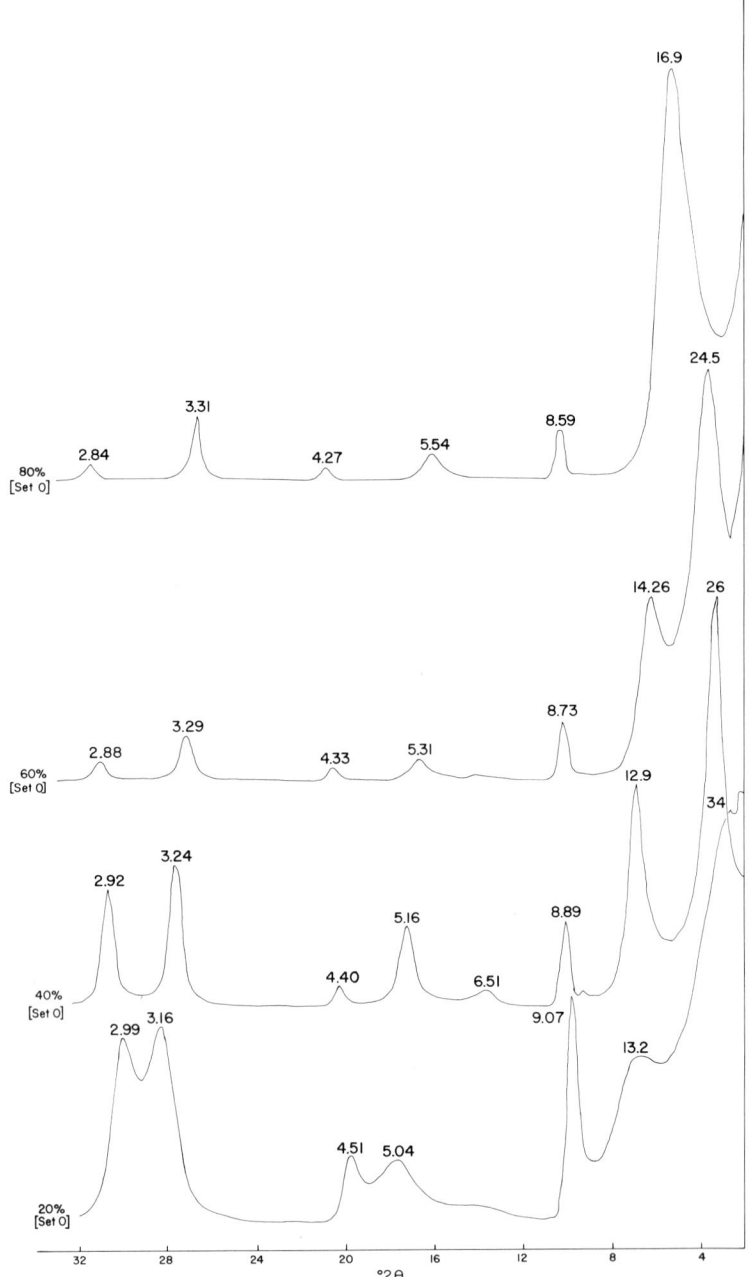

Fig. 3. Calculated diffraction profiles for ordered pyrophyllite/smectite.

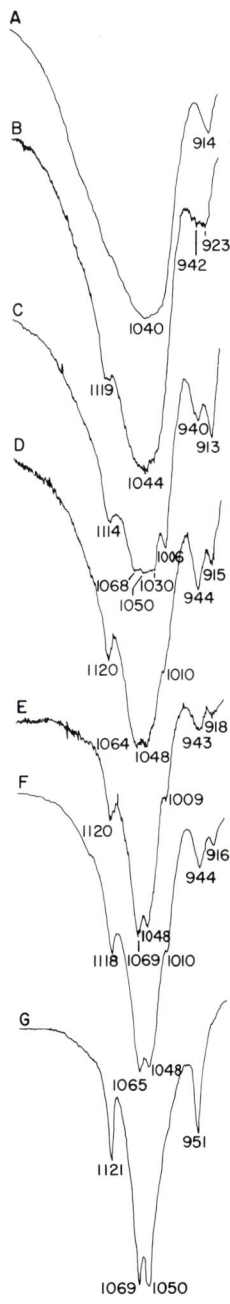

## RESULTS

Run results are given in Table 1. Sample X-ray patterns of synthetic mixed-layer pyrophyllite/smectites which cover a range of expandabilities are given in Figure 1. These patterns can be compared with calculated patterns of similar expandability shown in Figure 2. The calculated patterns in Figures 2 and 3 are presented so that random or PyS-ordered pyrophyllite/smectite can be identified in the future without a computer.

Corroborative evidence for the existence of mixed-layer pyrophyllite/smectite, in addition to the close match between the real and the calculated X-ray patterns, is given by infrared patterns in Figure 4. The figure shows the gradual development of a pyrophyllite-like IR pattern with an increasing percentage of pyrophyllite layers in the mixed-layer phase. Pyrophyllite-like Si-O stretching absorptions develop at about 1120 cm$^{-1}$, 1069 cm$^{-1}$ and 1048 cm$^{-1}$ and an OH bending absorption forms at about 944 cm$^{-1}$ (Farmer and Russell, 1964). The appearance of an absorption at about 915 cm$^{-1}$ is attributed to smectite and kaolinite, and the shoulder at 1010 cm$^{-1}$ is unidentified.

Some broad generalizations are possible concerning the effect of run chemistry on the formation of pyrophyllite/smectite: (1) The ratio of the weights of clay to $AlCl_3 \cdot 6H_2O$ is crucial in determining the extent of reaction. Increasing the amount of aluminum relative to montmorillonite increases the amount of pyrophyllite layers that form during treatment (compare runs 8 and 10, 12 and 13, Table 1). (2) Pyrophyllite/smectite forms in these experiments only at 320°C and above. Between 245°C and 280°C montmorillonite either remains unreactive, or reacts to form discrete kaolinite (runs 1-4). At still lower

Fig. 4. Infrared patterns of synthetic pyrophyllite/smectite. A = Ca-saturated Wyoming montmorillonite starting material (100% expandable). B = 85% expandable Py/S (run 13); C = 35% (run 5); D = 30% (run 20). E = 20% (run 7); F = 10% (run 15); G = natural pyrophyllite (Shoko, Japan). See Fig. 1 for X-ray patterns of these phases.

temperatures (150°C), Środoń (in preparation) has shown that hydrothermal runs with similar chemistries will react to form mixed-layer kaolinite/smectite. (3) Increasing the run time from 3 to 9 days favors the formation of pyrophyllite layers (compare runs 14 and 15, 17 and 18, 19 and 20), showing that expandabilities are a function of reaction time as well as run chemistry. (4) The concentration of alkalis in the run solution seemed to have only a minor effect on the reaction in the two runs (runs 17 and 19) that can be rigorously compared (i.e., did not leak and had everything else held constant). These runs lasted only 3 days, however, and perhaps a larger effect would be seen in longer runs.

More detailed conclusions regarding the effect of run chemistry can be gathered only from runs which neither gained nor lost weight during treatment (runs marked by an asterisk in Table 1). Unfortunately, many runs did show a weight change after treatment, indicating that there was some exchange between the solution in the capsule and water in the autoclave. From a geological viewpoint, a more rigorous and thorough investigation of the reaction of smectite to pyrophyllite/smectite does not seem to be warranted at this time. The phase has not yet been recognized in nature, and the synthetic variety has no known economic importance.

SUMMARY

The synthesis of mixed-layer pyrophyllite/smectite, a member of the Al-smectite reaction series, has been accomplished by running Ca- and Na-saturated Wyoming montmorillonite in hydrothermal solutions containing aluminum trichloride. The existence of this unusual phase was confirmed by comparing X-ray diffraction patterns of the phase with calculated diffraction profiles, and by comparing the development of a pyrophyllite-like infrared pattern with the development of pyrophyllite layers in the mixed-layer phase. The reaction of montmorillonite towards pyrophyllite is favored by increasing the amount of aluminum in the system, and by increasing run time and temperature.

ACKNOWLEDGEMENTS

The author thanks T. Anderson and G. Whitney for their comments on the original manuscript, and T. Nishiyama for the sample of natural pyrophyllite. This research was supported by the Earth Sciences Section, National Science Foundation, NSF Grant EAR 76-13368.

REFERENCES

Bradley, W. F., 1950. Alternating sequence of rectorite. Am. Mineralogist, 35: 590-595.
Eberl, D., 1978. Reaction series for dioctahedral smectite. Clays Clay Minerals. In press.
Eberl, D. and Hower, J., 1976. Kinetics of illite formation. Geol. Soc. Am. Bull., 87:1326-1330.
Farmer, V. C. and Russell, J. D., 1964. The infra-red spectra of layer silicates. Spectrochim. Acta, 20:1149-1173.
Kodama, H., 1958. Mineralogical study on some pyrophyllites in Japan. Min. J. (Japan), 2, 236.
Reynolds, R. C., Jr. and Hower, J., 1970. The nature of interlayering in mixed-layer illite-montmorillonites. Clays Clay Minerals, 18:25-36.
Wardle, R. and Brindley, G. W., 1972. The crystal structures of pyrophyllite, 1Tc, and of its dehydroxylate. Am. Mineralogist, 57:732-750.

STABILITE DES MINERAUX PHYLLITEUX 2/1 EN CONDITIONS ACIDES
ROLE DE LA COMPOSITION OCTAEDRIQUE.

M. Robert et G. Veneau
Département de Science du Sol, I.N.R.A., Versailles-France

ABSTRACT

This study concerns the stability of 2/1 clay minerals in the pH range 3-5 which occurs in acid soils. Dioctahedral (aluminous and ferric) and trioctahedral 2/1 phyllosilicates and the corresponding hydroxides have been employed. The studies aim to distinguish the role of mineralogical factors from those which depend on the weathering conditions.

The crystal chemical characteristics and particularly the di or trioctahedral composition are the most important factors affecting stability. Thus the dioctahedral minerals even if they are ferric are very stable.

Therefore it seems necessary to distinguish stability as a function of determined conditions of the alteration medium, considering the percolating conditions (open or closed cystem) as well as the physicochemical conditions (acid or acid and complexing).

INTRODUCTION

Il est classique depuis Goldich (1938) de présenter l'ordre d'altération des minéraux selon des séquences. La plus connue, donnée par cet auteur, qui va de l'olivine (le moins stable) jusqu'au quartz, intéresse essentiellement les minéraux primaires. Elle est logique dans la mesure où elle correspond à des conditions de formation (ordre de cristallisation à partir du refroidissement d'un magna), et où elle est en relation étroite à la fois avec une compostion chimique et un certain type de structure.

Etablir de telles séquences pour les minéraux secondaires (argiles - oxydes - hydroxydes), qui se trouvent en équilibre à la surface du globe, apparaît beaucoup plus difficile dans la mesure où de multiples facteurs sont à prendre en compte : composition chimique et structures trés variées, faible taille des particules, ou au contraire grandes surfaces développées.

Il existe cependant un grand nombre de tentatives pour classer le comportement des minéraux. Les unes peuvent être considérées comme thermodynamiques dans la mesure où elles utilisent soit une énergis réticulaire de formation des minéraux

(Fairbairn 1943, Gruner 1950, soit des énergies libres standards de formation
(Garrels and Christ, 1965 , Tardy et Gac, 1978) qui caractérisent une stabilité intrinsèque du minéral d'une façon indépendante du milieu.

D'autres approches (Jackson and Sherman, 1953) sont au contraire basées sur l'étude de la stabilité des minéraux vis à vis de l'altération dans le milieu naturel et une séquence en a été déduite valable pour les minéraux secondaires (tableau 3). Il s'agit déjà là d'une stabilité extrinsèque dépendante des conditions générales du milieu d'altération (de type hydrolytique).

Notre démarche comme celle de Huang et Keller (1972) est expérimentale et elle cherche à caractériser ici le comportement de différents minéraux 2/1 dans des conditions déterminées du milieu, qui sont celles que l'on peut trouver dans les sols acides (pH 3-5).

Les résultats obtenus permettront une première discussion des séquences classiques proposées par différents auteurs sur la stabilité des minéraux.

## Minéraux étudiés et conditions expérimentales utilisées

Les expériences intéressent ici les principaux minéraux susceptibles d'être présents dans la fraction argile des sols. Il s'agit essentiellement des phyllosilicates 2/1 à 10 Å ou à 14 Å (tableau 1) qui diffèrent par leur degré d'occupation octaédrique (Mg, $Al^{3+}$, $Fe^{3+}$), et des oxydes et hydroxydes correspondant à ces 3 cations (brucite, gibbsite, goethite et hématite).

TABLEAU 1

Formules structurales des minéraux sur base $O_{10}(OH)_2$

P  $\{Si_{2,69}Al_{1,31}\}\{Al_{0,13}Ti_{0,06}Fe^{3+}_{0,12}Fe^{2+}_{0,18}Mg_{2,47}\}\ Ca_{0,03}K_{0,88}$

V  $\{Si_{2,69}Al_{1,31}\}\{Al_{0,14}Ti_{0,02}Fe^{3+}_{0,22}Fe^{2+}_{0,02}Mg_{2,58}\}\ Ca_{0,46}$

M  $\{Si_{3,92}Al_{0,08}\}\{Al_{1,55}Ti_{0,02}Fe^{3+}_{0,04}Fe^{2+}_{0,05}Mg_{0,38}\}\ Ca_{0,18}$

I  $\{Si_{3,48}Al_{0,52}\}\{Al_{1,24}Ti_{0,04}Fe^{3+}_{0,32}Fe^{2+}_{0,02}Mg_{0,38}\}\ Ca_{0,07}K_{0,66}$

N  $\{Si_{3,46}Al_{0,54}\}\{Al_{0,01}Fe^{3+}_{1,92}Fe^{2+}_{0,02}Mg_{0,31}\}\ Ca_{0,23}$

Gl  $\{Si_{3,74}Al_{0,26}\}\{Al_{0,39}Ti_{0,02}Fe^{3+}_{0,95}Fe^{2+}_{0,13}Mg_{0,49}\}\ Ca_{0,07}K_{0,75}$

Hec  $\{Si_4\}\ \{Mg_{2,71}\ Li_{0,29}\}\ Ca_{0,15}\ (F^-\ OH^-)_2 O_{10}$

P = Phlogopite ; V = Vermiculite Santa Olalla ; M = Montmorillonite grecque ;
I = Illite Du Puy ; N = Nontronite Garfield ; Hec = Hectorite d'Hector ;
Gl = Glauconite Gl ; O (La Roche et al, 1976)

La vermiculite a été utilisée soit en framents de 500 μ à 1 mm, soit après séparation < 2 μ. Cette même séparation dans l'eau a été effectuée pour tous les autres minéraux, parfois après broyage préalable. Les surfaces externes ont été mesurées par méthode BET (tableau 2).

TABLEAU 2

Valeurs des surfaces externes des minéraux étudiés (en $m^2/g$)

| M | I | N | Gl | V | Hec | B | Goe | Hem | Gib |
|---|---|---|---|---|---|---|---|---|---|
| 58 | 112 | 94 | 74 | 3 | 72 | 25 | 25 | 7,7 | 16 |

B = Brucite ; Goe = Goethite (Sénégal) ; Hem = Hématite (Brésil) ; Gib = Gibbsite

Les minéraux possédant une capacité d'échange ont été saturés par du calcium et tous ceux qui ont une taille < 2 μ ont été placés en membrane de dialyse. Tous les minéraux ont été traités à une température analogue (60°C) dans 2 types de systèmes correspondant à des conditions hydrodynamiques différentes :

- en fiole de plastique où les solutions sont renouvelées tous les 4 jours lorsqu'un équilibre est pratiquement atteint.
- en cellules de téflon où les solutions percolent continuellement sur les minéraux à raison de 1 1/24 heures à l'aide d'une pompe péristaltique.

Les études ont été effectuées dans des conditions acides pas trop éloignées de celles se développant dans les sols acides (pH 3 à 5). Si pour certaines études 7 acides différents ont été utilisés, le choix a souvent porté uniquement sur un acide complexant (oxalique ou citrique) et un acide peu complexant (HCl ou lactique).

Tous les éléments chimiques majeurs ont été dosés dans les solutions et à la fin des expériences des minéraux ont été traités par KCl N pour obtenir les cations échangeables, puis par une solution d'oxalate d'ammonium à pH 4,2 pour obtenir les éléments libres non échangeables.

RESULTATS

Le comportement des minéraux en conditions acides a été caractérisé en se basant sur la sortie de l'élément chimique constitutif qui est dominant en couche octaédrique. Les résultats, qui ont été exprimés en % de l'élément présent dans le minéral montrent que d'une façon générale le pourcentage de libération de cet élément octaédrique reflète le pourcentage global de destruction du minéral. Il s'agit donc là en quelque sorte d'une appréciation quantitative de la stabilité du minéral.

Expériences avec percolation

Dans ce milieu les solutions venant en contact du minéral sont constamment renouvelées, et il n'existe pas de blocage de l'évolution. Bien que les expériences aient portées sur 100 litres, les résultats présentés dans la fig.1 intéressent ici

20 litres de solution (4 x 5 litres). Le premier élément intéressant est la différence qui existe entre les minéraux trioctaédriques (Mg) et les minéraux dioctaédriques (contenant $Al^{3+}$ ou $Fe^{3+}$). Il a été en effet nécessaire d'utiliser des échelles qui diffèrent d'un facteur de 10 pour représenter la sortie des élémen dans des conditions semblables. Si l'on étudie d'abord les <u>minéraux trioctaédriques</u> il faut tout d'abord distinguer le cas de la vermiculite où le magnésium occupe la place prépondérante en couche octaédrique mais où l'aluminium entre pour un tiers dans l'occupation tétraédrique. Les 2 milieux I (complexant) et II (acide peu complexant) conduisent à des résultats très différents : les pourcentages de sortie des éléments du réseau, sont pour le milieu I $Al > Fe^{3+} > Mg > Si$ et pour le milieu II $Mg > Si > Al-Fe^{3+}$ ; les quantités libérées varient dans un rapport de 10 pour Al et de 3 à 4 pour Mg entre les 2 milieux (Fig.1).

Ceci ne se retrouve pas dans le cas de l'hectorite et de la brucite, et les 2 types d'acides conduisent alors aux mêmes résultats : ceci s'explique très bien dans la mesure où le magnésium est peu complexable et où les 2 minéraux concernés ne contiennent ni aluminium ni fer.

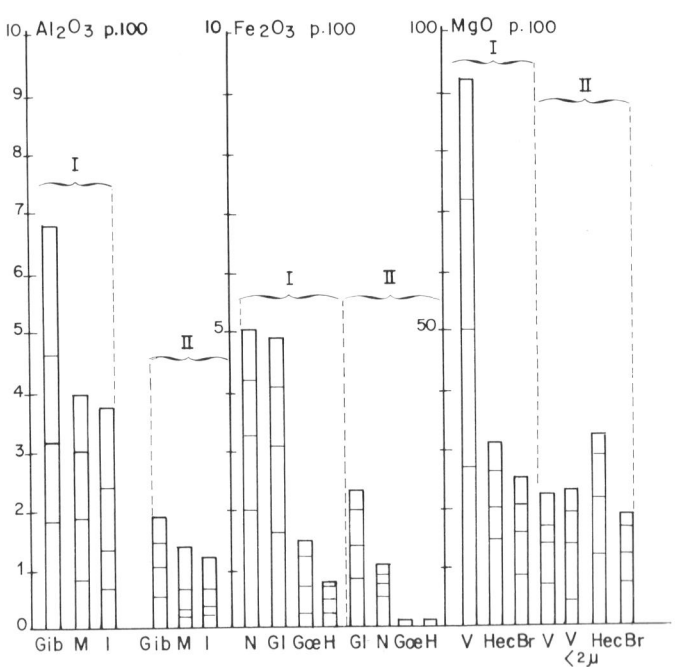

Fig. 1 - Evolution des différents minéraux en conditions acides ($10^{-3}$N) et en système avec percolation (pour 20 litres)
I milieu acide et complexant - II milieu acide

La vermiculite a dans la plupart de nos expériences une taille de 500 µ à 1 mm.
D'une façon assez curieuse, si on fait varier la taille d'un facteur de 500 en
prenant des particules < 2 µ, la libération de Mg obtenue est sensiblement la
même, ce qui indiquerait que l'attaque se fait sur toute la surface du minéral
(Miller, 1965). Ce résultat permet d'établir directement une comparaison entre les
minéraux riches en Mg : en système percolant et en milieu acide le comportement des
minéraux 2/1 riches en Mg est très similaire, et il se rapproche de celui de
l'hydroxyde correspondant (la brucite). Dès qu'un acide complexant entre en jeu, la
vermiculite est par contre 3 à 4 fois plus instable que les autres minéraux trioctaédriques.

Pour le cas des minéraux dioctaédriques (Fig.1) et d'une manière générale, on
peut immédiatement noter que les libérations de $Fe_2O_3$ des minéraux ferrifères et
de $Al_2O_3$ des minéraux alumineux étant représentées avec la même échelle, sont très
voisines. La stabilité des minéraux dioctaédriques considérés ici est donc quelle
que soit leur composition, de 10 à 20 fois plus forte que celle des minéraux
trioctaédriques.

La différence entre système complexant et non complexant est toujours accusée et
le rapport des éléments extraits peut aller de 2 à 5. Si maintenant on considère
l'ordre relatif de stabilité, on peut noter que la gibbsite, qui est la plus stable,
devient particulièrement fragile en milieu complexant. En comparaison la goethite
et l'hématite sont au moins 5 fois plus stables. On est frappé par contre par la
similitude de comportement de la montmorillonite et de l'illite d'une part, et de la
nontronite et de la glauconite d'autre part. Ces minéraux qui ont à peu près la
même surface externe, diffèrent pourtant beaucoup par leur surface interne et
l'accessibilité de leur zone interfoliaire.

Expériences en système "clos"

Seuls ont été retenus ici les comparaisons vermiculite - mica pour les minéraux
trioctaédriques, et glauconite - nontronite pour les minéraux dioctaédriques.
La fig.2 rassemble les résultats obtenus par 5 traitements de 4 jours effectués
avec 7 acides en concentration $10^{-3}N$ sur la vermiculite préalablement saturée par
$Ca^{2+}$. Pour chaque acide, les résultats sont comparés avec ceux obtenus par Razzaghe
(1976) avec une phlogopite dont la composition est très proche de celle de la
vermiculite. Seuls 2 éléments occupant la couche octaédrique ont été retenus ici.
En ce qui concerne la libération du fer, on trouve un certain parallélisme entre le
comportement des 2 minéraux, aussi bien en ce qui concerne les quantités d'éléments
extraites, que pour l'ordre d'efficacité des acides dans la destruction. Cet ordre
est logique et correspond au pouvoir complexant des acides et à leur efficacité dans
l'altération (Robert et Razzaghe, 1977). Ainsi les acides les plus complexants
(oxalique, citrique) sont en général 2 à 3 fois plus actifs que HCl ou lactique qui
constituaient le milieu II (acide non complexant) dans les expériences précédentes.

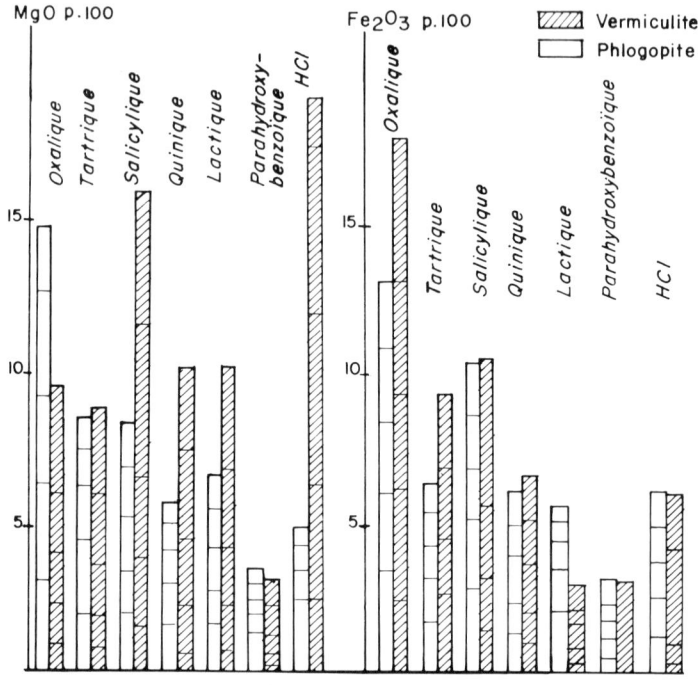

Fig. 2 - Evolution des minéraux trioctaédriques en conditions acides ($10^{-3}$N) et en système clos - rapport minéral/solution = 200 mg/200 ml avec 5 renouvellements.

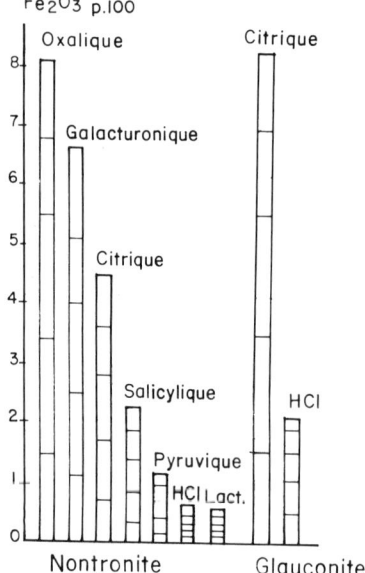

Fig. 3 - Evolution des minéraux ferrifères en conditions acides ($10^{-3}$N) et en système clos (200 mg/500 ml) avec 5 renouvellements.

Cette opposition est d'ailleurs plus nette dans le cas de la vermiculite et surtout de la nontronite où le fer est entièrement à l'état ferrique dans la structure (Fig.3). Pour ce dernier minéral le rapport entre le fer extrait par l'acide oxalique et celui extrait par HCl est de 13 alors qu'il était seulement de 5 en milieu ouvert (Fig.1). Le système confiné semble ainsi favoriser la réaction de complexation du fer (Razzaghe et Robert, 1977).

En ce qui concerne la sortie de Mg de la vermiculite et du mica, les résultats ne suivent plus les mêmes règles logiques que pour le fer. Tout d'abord l'ordre d'efficacité des acides est inversé et HCl est de loin le plus efficace ; d'autre part avec les acides les plus actifs, la vermiculite devient de 3 à 5 fois plus instable que le mica. Les phénomènes de dissolution apparaissent eux-mêmes différents dans la mesure où la quantité de MgO extraite est constante ou tend à diminuer au cours des traitements pour le mica, alors qu'elle augmente pour la vermiculite. Il serait possible de montrer que cette augmentation coïncide avec la libération de 80 à 100 % du Ca échangeable ; au contraire les faibles valeurs obtenues pour MgO avec les acides oxalique, tartrique et parahydroxybenzoïque coïncident avec une vitesse de désaturation plus lente.

L'analyse de la réactivité des minéraux en système clos est donc particulièrement complexe, et cette complexité apparaît liée à l'existence d'une couche interfoliaire et d'une capacité d'échange importante. Plusieurs phénomènes peuvent d'ailleurs jouer : - d'une part la libération de certains éléments (Mg mais aussi Al) est très ralentie tant que la désaturation du minéral n'est pas complète. - D'autre part, il se trouve que ces mêmes éléments sont les seuls qui peuvent venir remplacer $Ca^{2+}$ lors de l'attaque acide du réseau. Pour la vermiculite, en milieu complexant, $Mg^{2+}$ sera le cation qui s'accumulera préférentiellement, alors qu'en milieu non complexant, ce sera l'aluminium (Robert et Veneau, 1977). Dans le cas des minéraux dioctaédriques (nontronite, montmorillonite) seul l'aluminium s'accumule entre les feuillets.

La sortie de Al et Mg, éléments qui peuvent transiter en couche interfoliaire est affectée, mais pas celle de $Fe^{3+}$ et Si : on peut donc penser que les mécanismes de mise en solution de ces éléments sont de nature différente.

Ces phénomènes expliquent en grande partie le comportement relatif des minéraux qui pourra d'ailleurs être différent selon le stade d'évolution : au début, une vermiculite est ainsi plus stable qu'un mica de même composition et ceci jusqu'à complète désaturation. A partir de ce moment-là, elle se révèle au contraire de 2 à 3 fois plus altérable qu'un mica. Il faut noter aussi qu'en milieu non complexant la formation d'une couche hydroxyalumineuse entre les feuillets se traduit ensuite par une diminution de l'altérabilité.

DISCUSSION

La discussion de ces résultats peut porter sur 2 points : d'une part sur la natur[e] des phénomènes d'altération en milieu acide, d'autre part sur les séquences de stabi[-]lité qui peuvent être déduites.

Un certain nombre d'auteurs (Brindley et Youell, 1951 ; Gastuche, 1963 ; Arvieu e[t] Chaussidon, 1964 ; Miller, 1965), ont pu considérer que les phénomènes d'attaque acide des minéraux 2/1 étaient une réaction assimilable à une réaction du premier ordre, et qui intéresse toute la surface du minéral ; une telle réaction est ainsi différente de celle intervenant pour la kaolinite qui est une réaction de diffusion limitée à l'interface des seules surfaces latérales. Un certain nombre de nos résultats vont dans le sens indiqué par ces auteurs, en particulier le fait que la vermiculite, quelle que soit sa taille, ait le même comportement.

Cependant les résultats obtenus en milieu clos montrent qu'en réalité, au sein d'une même structure de type 2/1, la libération des divers éléments semble elle aussi obéir à des lois différentes. Ainsi, Al et Mg pourraient suivre une réaction d[u] premier ordre, alors que la libération de Fe et Si ne serait limitée que par une réaction de type diffusion (Huang et Keller, 1972). Il nous est de plus possible de préciser que si la libération des premiers éléments (Al et Mg) peut se faire par la surface interne du minéral (lorsqu'elle existe) où ils peuvent d'ailleurs s'accumule[r] la libération des autres éléments ne se fait que par la surface externe.

Au niveau du comportement relatif des différents minéraux au cours de l'altératio[n] il est possible de comparer la séquence expérimentale obtenue en milieu percolant aéré (fig.1) avec celle, classique pour les minéraux secondaires, présentée par Jackson et Sherman (1953). Un certain nombre de divergences apparaissent pour les minéraux 2/1, (tableau 3) qu'il est possible d'expliquer par les remarques suivantes

1) En milieu percolant, le comportement des minéraux micacés à 10 Å et de ceux à 14 Å, expansibles ou pas, est très semblable. Le facteur essentiel qui joue dans ces conditions n'est donc pas la surface totale (surface interne comprise), mais la surface externe du minéral. Ceci explique très bien que les couples de minéraux phlogopite - vermiculite, nontronite - glauconite, montmorillonite - illite, aient un comportement très proche.

2) La constitution cristallochimique du minéral importe donc beaucoup plus que son type de structure, et la principale distinction est celle qui existe entre les minéraux trioctaédriques et dioctaédriques. La différence de stabilité intervient alors pour un facteur de plus de 10 et elle n'apparaît pourtant pas dans la séquence de Jackson (tableau 3). Ainsi on peut penser que l'hectorite, la phlogopite ont la même stabilité que la biotite et la saponite, stabilité qui sera également proche de celle de l'hydroxyde correspondant (la brucite). Si la stabilité des minéraux alumineux est bien connue, nos résultats attirent l'attention sur la grande stabilité des minéraux 2/1 ferrifères. Ce sont eux aussi des minéraux dioctaédriques

TABLEAU 3

Séquences d'altération pour les minéraux présents dans la fraction argile des sols et des sédiments (par ordre de stabilité croissante)

(A) Jackson and Sherman (1953) (milieu hydrolytique)

(1) Gypse (2) Calcite (3) Olivine-hornblende (4) Biotite (+ glauconite nontronite - Mg chlorite) (5) Albite (6) Quartz (7) Muscovite (+ séricite - illite) (8) Interstratifiés 2/1 et vermiculite (9) Montmorillonite (beidellite, saponite) (10) Kaolinite (11) Gibbsite (+ boehmite) (12) Hématite (goethite) (13) Anatase (Zircon, rutile, ilménite corindon)

(B) Séquence obtenue expérimentalement en milieu acide non complexant système ouvert

(4) Vermiculite (saponite phlogopite biotite) (5) Hectorite - Brucite
(8) Glauconite (9) Gibbsite (10) Montmorillonite - Illite - Nontronite
(11) Goethite (12) Hématite.

dont le comportement sera proche de celui des minéraux alumineux.

Mais les expériences présentées montrent que déjà selon les acides utilisés, et en particulier leurs propriétés plus ou moins complexantes, un certain nombre de modifications peuvent intervenir dans la stabilité relative des minéraux. Ainsi en milieu complexant, les minéraux ferrifères, la gibbsite ou les minéraux alumineux deviennent beaucoup plus instables. Cela vaut également pour la vermiculite (et certainement la saponite) dans la mesure où le présence d'aluminium tétraédrique devient alors un facteur d'instabilité, ceci par rapport à l'hectorite ou la brucite qui n'en contiennent pas.

Les modifications sont elles aussi importantes si on considère un système clos où les réactions d'échange utilisant la couche interfoliaire, et les réactions de complexation, peuvent se manifester plus aisément. Il est probable aussi que des phénomènes de réduction très importants pour les minéraux ferrifères pourront aussi se développer.

CONCLUSION

Les séquences de stabilité des minéraux ont le plus souvent été obtenues lors d'études cinétiques en milieu acide minéral concentré ou de la réalisation d'équilibre thermodynamique en milieu aqueux. De même celles établies dns le milieu naturel sont basées en général sur une résistance à l'altération hydrolytique. Il semble nécessaire dorénavant de soumettre les minéraux à la fois à des conditions plus variées, et plus proches de celles du milieu naturel. En conditions acides, la composition octaédrique d'un phyllosilicate apparaît comme un des facteurs essentiels régissant sa stabilité. Mais la stabilité d'un minéral est relative et dépend étroitement de la nature du système d'agression (Hénin et al, 1968). Pour les seules conditions acides, il nous apparaît ainsi nécessaire de distinguer à la fois 2 types de système hydrodynamique, percolant (et aéré) ou clos, et 2 types de milieux physicochimiques, acide et complexant ou acide non complexant.

REFERENCES

Arvieu, J.C. et Chaussidon, J., 1964. Etude de la solubilisation acide d'une illite extraction du potassium et évolution du résidu. Ann. Agron.,15, 3, 207-229.

Brindley, G.W. and Youell, R.F., 1951. A chemical determination of tetrahedral and octahedral aluminum ions in a silicate. Acta Crystallogr., 4, 495-496.

Fairbairn, H.W., 1943. Packing in ionic minerals Bull. Geol. Soc. Am., 54 : 1305-137.

Garrels, R.M. and Christ, C.L. 1965. Harper and Row, New York 450 pp.

Gastuche, M.C., 1963. Kinetics of acid dissolution of biotite. Proc. Int. Clay Conf. Stockholm Pergamon Press London p.67-83.

Goldich, S.S., 1938. A study of rock weathering. J. Geol., 46 : 17-58.

Gruner, J.W., 1950. An attempt to arrange silicates in the order of reaction energies at relative low temperatures Am. Mineral. 35 : 137-148.

Hénin, S., Pédro, G. et Robert, M., 1968. Considérations sur les notions de stabilité et d'instabilité des minéraux en fonction des conditions du milieu ; essai de classification des systèmes d'agression. 9th Int. Cong. Soil Sci. Trans. Adelaïde, III, 79-90.

Huang, W.H. and Keller, W.D., 1972. Kinetics and mechanisms of dissolution of Fithian Illite in two complexing organic acids. Proc. Int. Clay Conf., Madrid, 321-331.

Jackson, M.L. and Sherman, G.D., 1953. Chemical weathering of minerals in soils. Adv. Agr. 5, : 219-318.

Laroche, H., Govindaraju, K. et Odin, G.S., 1976. Préparation d'un étalon analytique de glauconite. Analusis, 4 : 385-397.

Miller, R.J., 1965. Mechanisms for hydrogen to aluminum transformations in clays. Soil Sci. Soc. Am. Proc., 29 ; 36-39.

Razzaghe-Karimi, M., 1976. Thèse Univ. Paris VI, 222 pp. Contribution à l'étude expérimentale des phénomènes d'altération en milieu organique acide.

Razzaghe-Karimi, M. et Robert, M., 1977. A distinction between the main mechanisms governing weathering by organic acid compounds. Proc. 2nd Int. Symp. on water-rock interaction, Strasbourg, IV, 250-253.

Robert, M. et Veneau, G., 1977. Ion exchange and dissolution reactions in clay minerals in acidic conditions. Proc. 3rd Eur. Clay Conf. Oslo, 160-162.

Tardy, Y. et Gac, J., 1978. Contrôle de la composition chimique des solutions par la précipitation des minéraux dans les sols. Ebauche d'un modèle thermodynamique pour la formation des argiles. Séminaire altération des roches en milieu superficiel ; Science du Sol, Versailles (sous presse).

SYNTHETIC ILLITE IN THE CHEMICAL SYSTEM $K_2O$-$Al_2O_3$-$SiO_2$-$H_2O$ AT 300°C AND 2Kb

B. VELDE, Laboratoire de Pétrographie, Université P. et M. Curie, Paris, France.
A.H. WEIR, Rothamsted Experimental Station, Harpenden, England.

ABSTRACT

A limited degree of solid solution in a partially-ordered 1M micaceous mineral has been inferred from results of XRD, TG, IRA and TEM on samples of bulk composition varying between muscovite ($KAl_3Si_3$) and pyrophyllite ($Al_2Si_4$). Illite is the term that may be used, in a restricted sense, for the end-member of the series with a composition of about $K_{0.8}Al_2(Si_{3.2}Al_{0.8})O_{10}(OH)_2 \cdot xH_2O$, although it is not intended to limit the use of the term illite to aluminous micaceous minerals only. It has a $\underline{b}$ cell dimension of 8.97Å, which is more similar to that of K-saturated beidellite (8.95Å) than to that of synthetic muscovite (9.00Å). Compositions more pyrophyllite-rich than 80% muscovite contain illite-beidellite interstratifications.

We think that the similarity of cell dimensions between illite and beidellite may be necessary for interlayering to occur between them and that it is the higher Si and lower K content of illite compared to muscovite that controls this small but important change in cell dimensions.

---

INTRODUCTION

Mineral phase relationships in the muscovite-pyrophyllite system have been described by Velde (1973). He showed that micaceous minerals are formed at 2 Kb pressure at temperatures between 225 and 420°C and compositions between 100 and 75% muscovite. We have repeated this work in more detail to see if a distinction can be made between mica and illite - named by Grim, Bray and Bradley (1937) - and to find the relationship between the micaceous minerals and expanding mixed-layer clays that form at lower K contents.

RESULTS

Synthesis

Table 1 presents data for the hydrothermal runs. Most were for three months at 2 Kb and 320°C, but there were a few exceptions to this and, in an attempt to make a 1M muscovite, one sample was heated to 603°C for 12 days. The sample numbers indicate approximate mol.% muscovite in the compositions. The starting gels (Velde, 1969) were ignited at 950°C overnight and then crushed and sieved to fine-sand grade. The products were dried powders that contained all the elements of the starting gels with or without additional water. They varied in appearance from friable powders to massive aggregates; one sample (70b) gave evidence of inhomogeneity in that some parts of it contained traces of kaolinite. The only other impurity recognised was in sample 65, which gave a single diffraction reflection of a 4.06Å spacing, representing a small amount of opal-CT. The alkali dissolution test of Hashimoto and Jackson (1960) was used to test for unreacted gel. It dissolved almost 100% of one of the starting gels, but only 2% of Sample 100-3 and 4% of 65. As fine-grained aluminosilicates are somewhat soluble in this test, these results are taken as evidence that crystallisation of the samples was essentially complete.

TABLE 1

Duration, pressure, temperature and initial gel composition of the hydrothermal runs

| Sample | Days | Pressure Kb | Temperature °C | Composition, weight %, ignited basis | | |
|---|---|---|---|---|---|---|
| | | | | $SiO_2$ | $Al_2O_3$ | $K_2O$ |
| 100-6 | 12 | 2 | 603 | 46.5 | 41.1 | 12.4 |
| 100-3 | 31 | 2 | 320 | 46.5 | 41.1 | 12.4 |
| 95 | 97 | 2 | 320 | 49.6 | 38.3 | 12.1 |
| 90a | 97 | 2 | 320 | 49.8 | 38.9 | 11.3 |
| 90b | 97 | 2 | 320 | 40.0 | 39.9 | 11.1 |
| 85 | 96 | 2 | 320 | 51.7 | 37.7 | 10.6 |
| 80 | 97 | 2 | 380 | 50.8 | 39.0 | 10.2 |
| 70a | 97 | 2 | 320 | 53.4 | 37.1 | 9.5 |
| 70b | 97 | 2 | 320 | 53.7 | 37.6 | 8.7 |
| 65 | 191 | 2 | 330 | 55.8 | 35.7 | 8.5 |

Particle size and shape

Examination of fracture surfaces of blocks and small natural aggregates by scanning electron microscope showed that they consist of masses of intergrown small crystals with platelets protruding around the edges to give a hedgehog effect. Diffractometer specimens made of crushed, lightly tamped powders showed no preferred orientation. A proportion of crystals in each specimen had apparently formed in more open conditions, however, because hand shaking in suspension at pH 9 dispersed sufficient flakes for transmission electron microscopy (TEM) and a few seconds with an ultrasonic probe sufficient for oriented aggregates for X-ray diffractometry (XRD). The dispersed flakes were representative of the whole sample, however; they did not give a concentration of expanding phases at the expense of non-expanding. TEM of unshadowed particles showed that sample 100-6 contains euhedral, tabular flakes two to three times larger than 100-3. Plate 1a shows that sample 100-3 contains a proportion of euhedral to subhedral hexagonal plates, 0.5-1 μm in large dimension and exceeding

200Å in thickness. With an increase in Si and reduction in K and Al the crystals become smaller and thinner but still platy - this is particularly noticeable for sample 95. Plate 1b illustrates the masses of thin platelets, some more than 1 μm in large dimension, together with undispersed aggregates that make up sample 85. With the occurrence of expanding interlayers the morphology of the samples changes and samples 70a and b and 65 contain a mixture of plate- and lath-shaped particles (Plate 1c).

 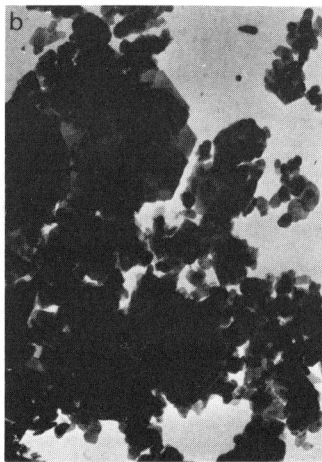 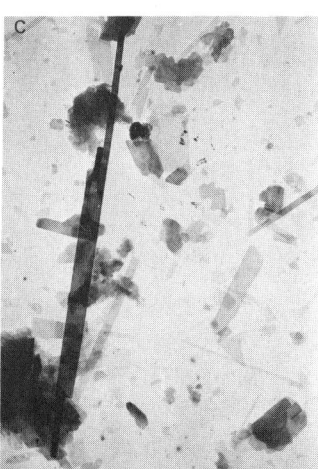

Plate 1. Transmission electron micrographs of dispersed particles of (a) sample 100-3, (b) 85 and (c) 65. Unshadowed mounts, magnification 20,000X.

Chemical composition

The compositions of the starting gels were confirmed by X-ray fluorescence analysis using the method of Norrish and Hutton (1969). These were taken to be the compositions of the crystalline products, since no material was lost. The data in Table 1 was supplemented by exchange capacity measurements and TG data. Table 2 gives results for exchangeable K and Ca obtained by replacing exchangeable K with 0.1M $CaCl_2$ solutions, washing salt-free, and then replacing the exchangeable Ca with 0.1M $MgCl_2$ solutions.

TABLE 2

Exchangeable and non-exchangeable cations, me per 100g, 100°C weight basis

| Sample | 100-3 | 95 | 90a | 90b | 85 | 80 | 70a | 70b | 65 |
|---|---|---|---|---|---|---|---|---|---|
| Total K | 251 | 245 | 229 | 225 | 215 | 217 | 192 | 176 | 172 |
| Exchangeable K | 11 | 12 | 6 | 10 | 13 | 9 | 27 | 21 | 19 |
| Exchangeable Ca | 9 | 13 | 10 | 13 | 14 | 11 | 28 | 26 | 24 |
| Non-exchangeable K | 240 | 233 | 223 | 215 | 202 | 198 | 165 | 155 | 153 |

All the samples from 100-3 to 80 have similar exchange capacities of about 10 me per 100g, whereas those with expanding interlayers, 70a, 70b and 65, have larger capacities.

The water contents of the samples were measured on 10 mg, K-saturated samples with a Stanton TG 770 thermobalance operated in air at a heating rate of $5°$ min.$^{-1}$. Results are given in Fig. 1 and Table 3.

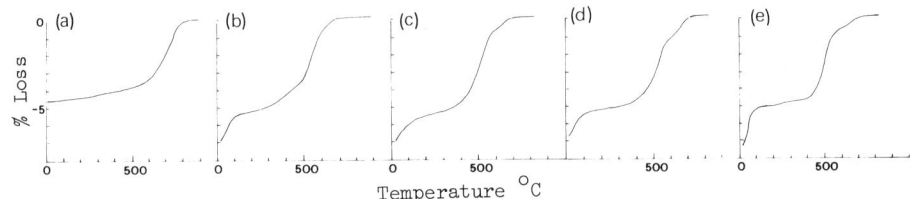

Fig.1 TG curves for 10 mg K-saturated samples heated in air at $5°$ minute$^{-1}$, (a) sample 100-6, (b) 100-3, (c) 90b, (d) 80 and (e) 70b.

The curves in Fig. 1 show that only the high-temperature muscovite, sample 100-6, contains little or no hygroscopic moisture and loses, approximately, its theoretical OH content above $150°C$. All the other samples contain appreciable hygroscopic moisture, those with expanding interlayers the most, and excess OH water, although for some samples the excess is small. All samples other than those of 100 composition have a double dehydroxylation reaction with an inflexion point on the TG curve between 575 and $675°C$. Both the temperature of inflexion and the amount of water lost above it are quite variable. The cause is not known.

TABLE 3

TG data for K-saturated samples heated in air; % weight losses are on an ignited basis

| Sample | 100-6 | 100-3 | 95 | 90a | 90b | 85 | 80 | 70a | 70b | 65 |
|---|---|---|---|---|---|---|---|---|---|---|
| Loss, 20-150°C | 0.11 | 1.19 | 1.34 | 0.72 | 1.68 | 1.75 | 1.46 | 3.07 | 2.14 | 2.33 |
| " 150-900°C | 4.43 | 5.38 | 4.97 | 5.02 | 5.64 | 5.90 | 5.38 | 5.50 | 5.04 | 5.15 |
| Inflexion temp. °C | - | - | 620 | 675 | 625 | 590 | 615 | 575 | 625 | 575 |
| Loss, IT-900°C | - | - | 0.60 | 0.35 | 0.60 | 0.85 | 1.25 | 0.60 | 0.60 | 1.75 |

It appears from the data presented that excess water, both hygroscopic and high-temperature, is a feature of micaceous clays formed at low temperatures, not only of K-deficient illites but of muscovites as well.

Table 4 lists unit formulae for the samples calculated on the basis of the half unit cell. The total K has been sub-divided into exchangeable and non-exchangeable categories to emphasize that exchangeable cations balance part of the permanent negative charge of non-expanding micaceous clays as well as expanding clays. The difference between exchangeable K on the illites, average $0.04K/Si_4O_{10}$, and the interstratifications, $0.07-0.11K/Si_4O_{10}$, represents the cations associated with expanding layers. The unit formulae in Table 4 list water in excess of that needed for structural OH as essential $H_2O$. It is not known whether this is present as trapped liquid water or as positive $H_3O^+$ or negative $OH^-$ ions attached to broken bonds on the clay surfaces. The variability of the amount suggests it is possibly a function of crystal growth and the abundance of more very fine particles in some specimens than others. Including it as $H_2O$ records that it is ubiquitous in these samples, but does not arbitrarily distort the balance of charge, as attributing it to either $H_3O^+$ or $OH^-$ ions would do (Gaudette et al., 1966).

TABLE 4

Unit formulae

| Sample | K | | $Al^{VI}$ | $Si^{IV}$ | $Al^{IV}$ | O | (OH) | $H_2O$ |
|---|---|---|---|---|---|---|---|---|
| | Exch. | Non-ex. | | | | | | |
| 100-3 | 0.04 | 0.96 | 2.02 | 2.95 | 1.05 | 10 | 2 | 0.27 |
| 95    | 0.05 | 0.92 | 1.97 | 3.12 | 0.88 | 10 | 2 | 0.09 |
| 90a   | 0.02 | 0.88 | 1.99 | 3.12 | 0.88 | 10 | 2 | 0.10 |
| 90b   | 0.04 | 0.85 | 2.01 | 3.07 | 0.93 | 10 | 2 | 0.36 |
| 85    | 0.05 | 0.79 | 1.98 | 3.22 | 0.78 | 10 | 2 | 0.45 |
| 80    | 0.04 | 0.77 | 2.01 | 3.16 | 0.84 | 10 | 2 | 0.23 |
| 70a   | 0.11 | 0.61 | 1.99 | 3.29 | 0.71 | 10 | 2 | 0.26 |
| 70b   | 0.08 | 0.60 | 2.01 | 3.29 | 0.71 | 10 | 2 | 0.06 |
| 65    | 0.07 | 0.59 | 1.98 | 3.41 | 0.59 | 10 | 2 | 0.10 |

## Structure

X-ray powder diffractometry. Fig. 2 compares the powder diffraction of the synthetic micaceous minerals with a trace computed for 1M muscovite with the POWD-5 computer program (Clark et al., 1973) using cell dimensions given by Yoder and Eugster (1955) and atom coordinates for 2M1 muscovite by Guven (1971) modified to fit a 1M cell. An alternative calculation using the slightly different atom coordinates of a phlogopite gave intensities that agree less well with those of sample 100-6. Examination of the 100-6 trace shows that all the reflections of 1M muscovite can be accounted for, but there are in addition a number of weak reflections of the 2M1 polytype. The intensities of the hkl reflections, particularly the 112 and $\overline{1}12$, although much more intense than for the other specimens examined, are less intense than in the computed trace. It is thought, therefore, that 100-6, although mainly well-crystallised 1M muscovite, is not fully ordered. Moreover, it is in the process of being replaced by the 2M1 polytype, which is the equilibrium phase. Comparison of the trace of 100-3 (Fig. 2c) with those in (a) and (b) shows that the prism

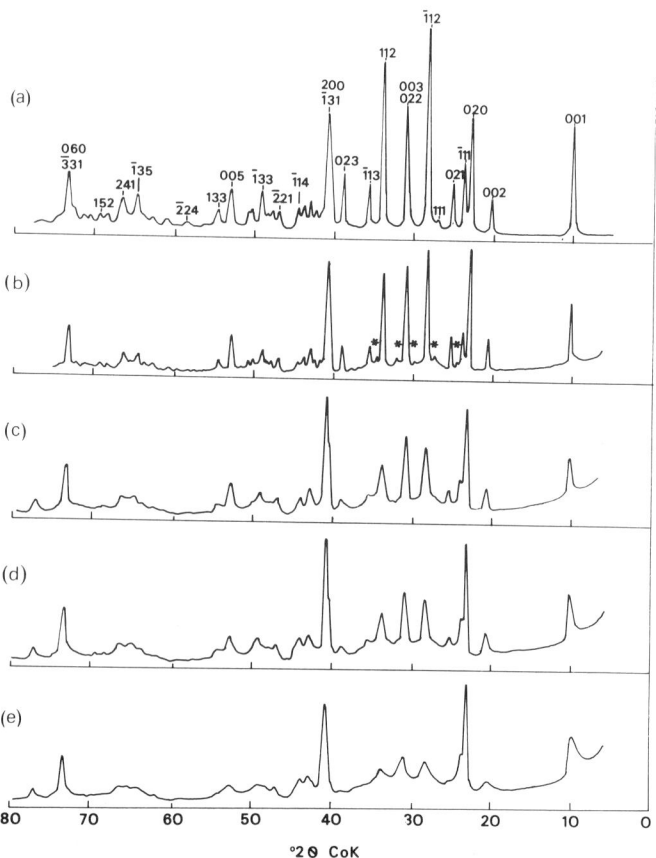

Fig.2 XRD traces of powder samples with Fe-filtered Co radiation, (a) computed trace from POWD-5 program with 0.5° peak broadening of 1M muscovite, (b) sample 100-6, * = reflections of 2M1 polytype, (c) 100-3, (d) 90b and (e) 70a, all samples K-saturated, in air at 20°C.

reflections, 020, 200 and 060, are sharp and intense, but that all the others, including the basal reflections, are less intense and broader. This is particularly noticeable for the 112 and $\bar{1}12$ reflections. Samples 90b (Fig. 2d), 95, 90a, 85 and 80 all resemble 100-3, whereas 70a and 65 resemble 70b with considerable broadening of the basal reflections due to interlamellar expansion and weakening of these and the general reflections. There are small but significant changes in cell dimensions in passing from sample 100-3 to 80 (Table 5), but otherwise they form a solid-solution series of very similar, partially-ordered 1M polytypes. Even sample 70a contains <u>hkl</u> reflections and not the <u>hk</u> bands that would be given by a turbostratic arrangement of

successive silicate layers. Sample 70a therefore has a partially ordered 1M
rather than the fully disordered 1Md structure described by Yoder and Eugster
(1955) for some samples they synthesized at 200°C and 1Kb.

TABLE 5

Powder data and derived parameters for 060 and 005 reflections

| Sample | 100-6 | 100-3 | 95 | 90a | 90b | 85 | 80 | 70a | 70b | 65 |
|---|---|---|---|---|---|---|---|---|---|---|
| **060** | | | | | | | | | | |
| Max. in °$2\theta$ CoK | 73.26 | 73.20 | 73.46 | 73.40 | 73.42 | 73.40 | 73.50 | 73.52 | 73.48 | 73.50 |
| Spacing in Å | 1.499 | 1.500 | 1.496 | 1.497 | 1.496 | 1.497 | 1.495 | 1.495 | 1.495 | 1.495 |
| b in Å | 8.994 | 9.000 | 8.976 | 8.982 | 8.976 | 8.982 | 8.970 | 8.970 | 8.970 | 8.970 |
| **005** | | | | | | | | | | |
| Max. in °$2\theta$ CoK | 52.78 | 52.76 | 52.90 | 52.88 | 52.84 | 52.94 | 52.88 | 52.94 | 52.90 | 53.00 |
| Spacing in Å | 2.012 | 2.013 | 2.008 | 2.009 | 2.010 | 2.007 | 2.009 | 2.007 | 2.008 | 2.005 |
| Breadth in °$2\theta$ | 0.40 | 0.45 | 0.61 | 0.50 | 0.65 | 0.75 | 0.58 | | 0.75 | |
| Breadth in Å | 630 | 470 | 250 | 360 | 240 | 180 | 270 | | 180 | |

Table 5 gives selected powder data and derived constants for the samples obtained
under standard conditions. Half peak breadths of the 005 reflections (that for
70b was for a sample heated to collapse its expanded layers) were corrected for
0.27° instrumental broadening (Jones b correction) and converted to particle
breadths using the expression given by Klug and Alexander (1954, p.503). The
results show that the thickness of the diffracting units decreases with temperature
of formation and with composition away from that of muscovite, thus reinforcing
the observation on crystal thickness made by electron microscopy. The other
results in Table 5 show that there are small changes in the size of the unit cell
with composition. For simplicity, changes in the complex 060, $\bar{3}31$ reflection
will be treated as if it is a simple 060 reflection and interpreted in terms of
the b cell dimension. As the composition changes from that of muscovite,
particularly in terms of the Si content, the b dimension decreases from 9.00
to 8.97Å. The change is towards that of K-beidellite, since separate measurements
of K-saturated Black Jack Mine beidellite gave $d_{06}$ = 1.491Å, b = 8.95Å.

Interstratification. Diffractometry of K-saturated specimens of samples 65 and
70a and b showed that they all collapse to give sharp 10Å reflections when heated
to 60°C or above. They thus contain no chlorite or chlorite-intergrade layers.
Fig. 3 shows the diffractometer traces obtained for Mg- (or Ca-) saturated 70b.
An intense 11.2Å reflection for the air-dry specimen divides into a sharp 10 and
broad 13Å reflection on glycol-solvation, the other basal reflections making
small shifts about the positions of the higher order 10Å reflections. Computer
modelling indicates a structure in which 15-20% smectite (S) layers are ordered
in IS units with some segregation of the remaining 10Å illite (I) layers. This
is similar to structures described by Reynolds and Hower (1970) and Eberl and

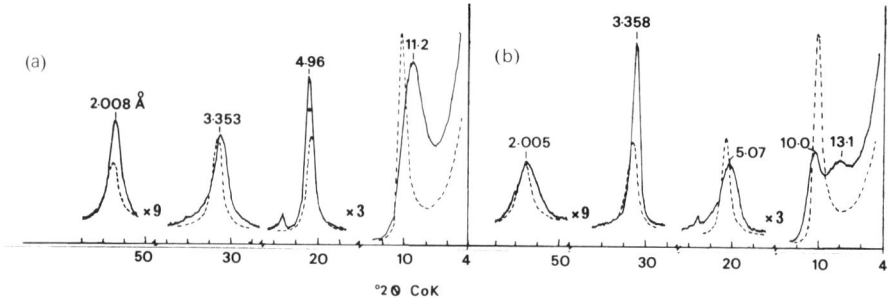

Fig. 3. X-ray diffractometer traces of Mg-saturated, oriented aggregate specimens of sample 70b, (a) in air at 20°C, (b) glycol-solvated. The broken lines are for a (less well-oriented) specimen heated to 250°C.

Hower (1977) for micaceous minerals of similar composition. The 70 compositions mark the end of the micaceous mineral series and the start of illite-smectite interstratification.

Infrared absorption. Spectra of Mg-saturated samples show little change between specimens in the major absorption bands of the 1000-350 cm$^{-1}$ region. Compositional changes in the specimens are mirrored in the 800-805 cm$^{-1}$ band, which is most distinct for the 100 sample, but becomes indiscernible between 80 and 70a. The band is due to an Al-O vibration (Farmer, 1974) and its intensity relative to the 820-830 cm$^{-1}$ OH libration band is sensitive to the mica polymorph (Velde, 1978). A similar but less distinct effect is seen for the 620-605 cm$^{-1}$ band, which diminishes in intensity between samples 70 and 85.

K-saturated samples dried at 300°C to remove adsorbed water show the presence of absorption bands at 3660 and 3637 cm$^{-1}$ in samples 65 and 70a and b. These bands are typical of beidellite, smectites showing a double OH stretch instead of the singlet given by muscovite (Farmer, 1974). Thus the presence of a smectite component is apparent in the infrared spectra of samples beyond the 80 composition.

DISCUSSION

The data given here for synthetic micaceous minerals formed at 2Kb and 300°C show too much variability to tie changes of expandability and interlayer charge exactly to overall composition. We don't think this is entirely due to experimental error, but that these fine-grained, low-temperature clays are inherently variable and many repeat preparations would be needed to define their properties exactly. Nevertheless several important points emerge. For starting gels of muscovite composition a partially ordered 1M polytype is first formed, the degree of order increasing with the temperature of formation. At 300°C this mica is composed of small, euhedral to subhedral platelets and has an exchange capacity of about 10 me per 100g and associated hygroscopic and excess high temperature water. Minerals with a similar structure form a compositional series up to $0.8K/Si_4O_{10}$, but show a reduction in particle size and $b$ cell dimension as their K and Si contents decrease. We note that the $b$ cell dimension of K-saturated beidellite hydrated with one layer of water molecules is also less than that of muscovite (muscovite 9.00Å, "synthetic" illite 8.97Å, K-beidellite 8.95Å). This reduction in cell size must favour the interlayering of smectite- and mica-like units in mixed-layer minerals.

We can use the samples studied here to characterise the compositional series of fine-grained minerals that can be termed illite, as named by Grim et al. (1937). For the K-Al-Si system a phase exists with common properties (grain size, polytype, CEC and water content) that covers the compositional range from mica (1.0K) to illite (0.8K). Such characteristics are those generally attributed to natural illites to distinguish them from muscovites of high temperature origin (Yoder and Eugster, 1955; Velde and Hower, 1963; Velde, 1965). We predict that a further series of low temperature minerals exists that can contain iron and magnesium with, possibly, different K- and Si-contents. Together with the aluminous minerals these represent illites in the general sense of the term.

The illite unit found in interstratified minerals should be that with the smallest K-content for a given bulk composition. Since reduction of K and increase of Si in the aluminous system reduces the $b$ cell dimension and since the beidellite $b$ dimension is smaller than that of muscovite it is predictable that illite should interlayer with beidellite. The effect of introducing $Fe^{3+}$, $Fe^{2+}$ or Mg into the lattice is not easy to predict, but experience of natural clays suggests that illite-smectite interstratifications form when Mg is introduced.

We suggest that the term illite should continue to be used to designate low-temperature micaceous minerals of variable composition. In this general sense it represents a solid solution series originating at the muscovite composition and varying towards lower K contents. Minerals at the low-K limit of the series (near 0.8K in the aluminous system) can be termed illite in the strict sense.

ACKNOWLEDGEMENTS

We thank J.H. Rayner for the computer traces, R.D. Woods for the electron micrographs, Jean Devonshire for the silicate analyses and D.J. Morgan of the Institute of Geological Sciences for the TG results.

REFERENCES

Clark, C.M., Smith, D.K. and Johnson, G.G., 1973. A FORTRAN IV Program for Calculating X-ray Powder Diffraction Patterns - Version 5. The Pennsylvania State University, U.S.A., 152 pages.

Eberl, D. and Hower, J., 1977. The hydrothermal transformation of sodium and potassium smectite into mixed-layer clays. Clays and Clay Minerals, 25: 215-227.

Farmer, V.C., 1974. The layer silicates. In: V.C. Farmer (Editor), The Infrared Spectra of Minerals, The Mineralogical Society, London, pp 331-363.

Gaudette, H.E., Eades, J.L. and Grim, R.E., 1966. The nature of illite. Clays and Clay Minerals, 13: 33-48.

Grim, R.E., Bray, R.H. and Bradley, W.F., 1937. Mica in argillaceous sediments. Amer. Mineral., 22: 813-829.

Guven, N., 1971. The crystal structures of 2M1 phengite and 2M1 muscovite. Z. Kristallogr., 134: 196-212.

Hashimoto, I. and Jackson, M.L., 1960. Rapid dissolution of allophane and kaolinite-halloysite after dehydration. Clays and Clay Minerals, 7: 102-113.

Klug, H.P. and Alexander, L.E., 1954. X-ray Diffraction Procedures. John Wiley & Sons, New York, p. 509 and p. 531.

Norrish, K. and Hutton, J.T., 1969. An accurate X-ray spectrographic method for the analysis of a wide range of geological samples. Geochim. et Cosmochim. Acta, 33: 431-453.

Reynolds, R.C. and Hower, J., 1970. The nature of interlayering in mixed-layer illite-montmorillonites. Clays and Clay Minerals, 18: 25-36.

Velde, B., 1965. Experimental determination of muscovite polymorph stability. Amer. Mineral., 50: 436-449.

Velde, B., 1969. The compositional join muscovite-pyrophyllite at moderate pressures and temperatures. Bull. Soc. franc. Min. Crist., 92: 360-368.

Velde, B., 1973. Phase equilibra for dioctahedral expandable phases in sediments and sedimentary rocks. In: J.M. Serratosa (Editor), Proc. Int. Clay Conf., Madrid 1972, C.S.I.C., Madrid, pp 235-248.

Velde, B., 1978. Infrared spectra of synthetic micas in the series muscovite-MgAl celadonite. Amer. Mineral., In Press.

Velde, B. and Hower, J., 1963. Petrological significance of illite polymorphism in Palaeozoic sedimentary rocks. Amer. Mineral., 79: 1239-1254.

Yoder, H.S. and Eugster, H.P., 1955. Synthetic and natural muscovites. Geochim. et Cosmochim. Acta, 8: 225-280.

BIOTITE WEATHERING IN GRANITES OF WESTERN FRANCE

A. MEUNIER
Laboratoire de PEDOLOGIE, Faculté des Sciences, Université de Poitiers
40, avenue du Recteur Pineau    86022 POITIERS CEDEX - France

B. VELDE
Laboratoire de PETROGRAPHIE, Université de Paris VI
4, Place Jussieu    75230 PARIS CEDEX 05    -    France

ABSTRACT

Electron microprobe analyses of fresh biotites, altered biotites and newly formed vermiculitic phases formed in two weathering profiles of the same calc-alkaline granite indicate that the composition of the altering fluid can greatly influence the composition of the newly formed minerals. In one profile, calcic carbonate solutions produce a regularly interstratified mineral (hydrobiotite) which concentrates Fe and Si relative to Al and Mg while another profile gives a vermiculite-intergrade mineral with a higher Al content. Hydrobiotite (interstratified mineral) persists into later stages of weathering than does vermiculite in these profiles. In both cases the final assemblage is kaolinite + oxides which also begins to form in the early stages of weathering. Biotite breaks down to a multiphase assemblage in both of the profiles.

INTRODUCTION

During the course of many studies of weathering (Coleman et al., 1963 ; Wilson, 1966 ; Ojanuga, 1973 ; Rimsaite, 1975 ; Stoch et al., 1976, for example) it has been noted that biotites are particularly vulnerable to the attack of rain water. They alter progressively in a given profile and, at times, can be used as indicators of the degree of alteration (Seddoh, 1973). We have undertaken here to determine the chemical evolution of biotites of nearly the same initial composition which destabilize, in one case, into a vermiculite-kaolinite-oxide assemblage and, in the other, a regular mixed layered mineral (hydrobiotite) kaolinite-oxide assemblage. Both sequences terminate with the kaolinite-oxide mineral pair in the profile zones of most intense weathering under conditions of a temperate "Atlantic-type" climate. The treatment which follows is based upon electron microprobe and X-ray diffraction determinations of the weathering products (extracted by ultrasonic vibration) of minerals from two profiles in a calc-alkaline granite, the "*granite de Parthenay*" (Deux-Sèvres, France). Details of the experimental method are found in Meunier (1977).

DESCRIPTION

Two profiles were studied - *La Pagerie* and *La Rayrie* - both local place names. In the first instance, the profile is terminated by a thin cover of irregular calcareous rocks which appear to give a calcic character to the altering solutions. In the second, there is no sedimentary cover and one can assume that the granite and its weathering products form the totality of the profile. One can divide the profile by textural and mineralogical methods into the following petrographic types.

a - <u>Fresh rock</u> : plagioclase (An 25 - 30) 14-16 %, quartz 32-38 %, potassic feldspar 38-41 %, muscovite 5 %, biotite 2-5 % and sericite due to retromorphic effects 5 % at *La Rayrie*.

b - <u>Compact, altered granite</u> : the rock is little altered ; alteration reaction zones are seen at grain contacts, while internal destabilization of primary minerals just begins.

c - <u>Friable rock zone</u> : as the name indicates, the initial mineral grains are easily dissociated. In thin section, many minerals show signs of internal destabilization. The plasma thus formed contains multiphase weathering mineral assemblages. This zone assumes a reddish tone in aspect in the outcrop, but the structure of the rock still maintained.

d - <u>Whitened friable rock zone</u> : sectors adjacent to diaclase fractures are notably whitened. One can still identify the original petrofabric of the rock but significant pores are developed.

e - <u>Fissured zones</u> : here one sees the effects of illuviation ; clay cutans deposit on fissure walls 10 to 1500 µ thick. The petrofabric in these zones is completely destroyed. Kaolinite plus oxides is the exclusive mineral assemblage of this zone.

Each horizon in the profiles contains several of the above petrographic types depending upon the extent of weathering. However, the chemical system active in the alteration of the biotites depends upon the local weathering stage and the speed with which water circulates more than its level in the profile. Thus one finds small fissured zones at the bottom of the profile where only kaolinite and oxides are stable or one finds micas still persisting in isolated blocks of relatively little altered rock which are found in the upper portions of the profile.

MICROMORPHOLOGIC EVOLUTION OF BIOTITES

In the earliest stages of weathering, the biotites appear little altered but show a tendency for iron to migrate to grain edges where it accumulates as amorphous oxides. The most important reactions are between biotites and potassic feldspar (Photo 1). These micas have somewhat variable compositions (Meunier and Velde, 1976).

Internal destabilization of biotites commences only in the red friable rock zone showing greater extent in smaller crystals. The morphologic evolution appears to include : sheet splitting, destruction of the structure, crystallization of secondary phases ; all of which seem to manifest themselves at the same time. (Photos 2 and 3)

The composition of the biotites from the two outcrops studied are very similar (Table 1), however the micas from *La Rayrie* have undergone a slight metamorphism during tectonization. This phase caused small pockets of kaolinite to form between mica sheets. It is probable also that this metamorphism caused these biotites to lose some potassium and possibly aluminium from their bulk composition which distinguishes them from the *La Pagerie* micas. However the difference is slight. They do, nevertheless, appear to alter more completely and at an earlier stage than their counterparts in the *La Pagerie* profile. One remarkable feature in early biotite alteration is the accumulation of iron oxides at grain cleavages and edges while titanium minerals (anatase) form throughout the relict grains. Thus iron is obviously more mobile than titanium in these systems.

|  | La Rayrie | | | | La Pagerie | | | | | |
|---|---|---|---|---|---|---|---|---|---|---|
|  | 1 | 2 | 3 | 4 | 1 | 2 | 3 | 4 | 5 | 6 |
| SiO | 34,33 | 35,35 | 35,93 | 34,85 | 33,48 | 34,48 | 33,55 | 33,80 | 34,41 | 32,99 |
| $Al_2O_3$ | 17,22 | 17,42 | 18,40 | 17,10 | 19,52 | 19,09 | 18,56 | 17,96 | 19,44 | 18,66 |
| FeO | 23,86 | 23,04 | 26,33 | 21,01 | 22,76 | 24,05 | 22,37 | 23,55 | 23,18 | 22,76 |
| $TiO_2$ | 1,57 | 1,74 | 1,78 | 1,43 | 1,96 | 1,65 | 1,67 | 1,78 | 1,78 | 1,98 |
| MgO | 5,83 | 6,17 | 5,90 | 5,63 | 5,94 | 5,64 | 5,55 | 4,62 | 5,59 | 5,45 |
| CaO | 0,72 | 0,10 | 0,06 | 0,14 | 0,12 | 0,04 | 0,14 | 0 | 0,04 | 0,13 |
| $Na_2O$ | 0,47 | 0,55 | 0,49 | 0,49 | 0,32 | 0,43 | 0,41 | 0,13 | 0,37 | 0,39 |
| $K_2O$ | 7,45 | 7,86 | 8,25 | 7,48 | 8,30 | 8,51 | 8,39 | 8,28 | 8,40 | 8,26 |
| Total | 91,45 | 92,22 | 97,14 | 88,13 | 92,40 | 93,90 | 90,65 | 90,13 | 93,23 | 90,63 |

Table 1 - *Microprobe analyses of biotite grains from samples of the least altered rock in each profile studied.*

ALTERATION PRODUCTS

The *La Rayrie* biotites alter to an intergrade vermiculite + kaolinite + oxide assemblage. The *La Pagerie* biotites alter to a hydrobiotite (regularly interstratified mica-intergrade mineral) plus minor amounts of intergrade + kaolinite + oxides.

X-ray diffraction characteristics are as follows :

a - Vermiculite intergrade potassium saturated 13.80 Å, untreated and glycollated 14.1 Å. Untreated, heated to 110°C, 13.54 Å ; 220°C, 11.7 Å ; 400°C mostly collapsed to 10 Å with some 11.7 Å material.

b - Hydrobiotite, regular interstratified : untreated, 11.65 + 24.12 Å ; glycollated, 11.6 + 24.0 Å ; potassium saturated, 11.6 + 24.0 Å ; untreated heated to 110°C, 11.5 + 24 Å ; 220°C, 10 Å + minor 12 Å.

The intergrade vermiculitic minerals were trioctahedral in type. The sequence of mineral stability is outlined in figure 1.

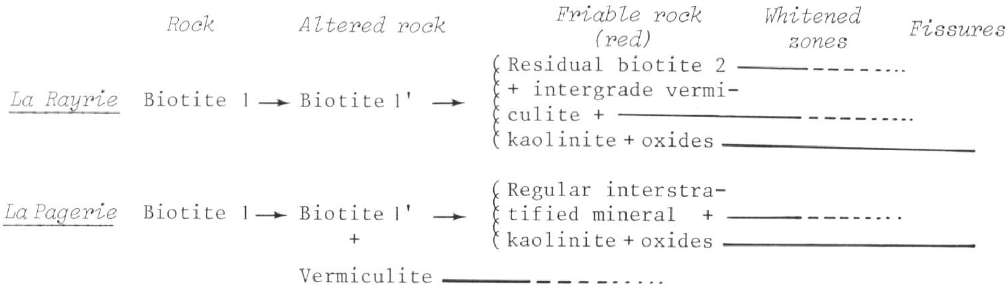

|  | Rock | Altered rock | Friable rock (red) | Whitened zones | Fissures |
|---|---|---|---|---|---|
| La Rayrie | Biotite 1 → | Biotite 1' → | Residual biotite 2 ———————<br>+ intergrade vermi-<br>culite + ————————————— ———————<br>kaolinite + oxides ——————————— |  |  |
| La Pagerie | Biotite 1 → | Biotite 1' →<br>+<br>Vermiculite ——————— ——————— | Regular interstra-<br>tified mineral + ————————— ———————<br>kaolinite + oxides ———————————— |  |  |

Figure 1

## CHEMICAL EVOLUTION OF THE BIOTITES AND THEIR WEATHERING PRODUCTS

Microprobe analyses of the biotites are presented in table 1. All iron is considered as FeO. Table 2 presents microprobe results for biotites from altered rock samples (see previous section for classification). These data show :

a - <u>Biotites, unaltered</u> : Iron/magnesium and Si/Al ratio's are quite similar for the micas of the two profiles.

b - <u>Biotites, altered</u> (<u>maintaining biotite optical properties</u>) : Figure 2 indicates that the chemical evolution of the biotites is quite different in the two outcrops. Plotted in Si-Al-K coordinates, one can see that the dispersion of compositions is greater in La Rayrie than in La Pagerie where hydrobiotite, a regular mixed layered intergrade-mica mineral is found. The La Pagerie biotites do not lose more than 50 % of $K_2O$ whereas the residual pleochroic layers in La Rayrie biotites can lose more than 80 %. They evolve toward low charge phyllosilicates and one finds associated with it a mixture of vermiculite intergrade mineral plus kaolinite plus oxides.

c - <u>Secondary alteration products found between cleavage flakes</u> : Table 3 shows the composition of the newly crystallized mineral within the old mica flakes. Generally the tendency is for the La Pagerie samples to increase the proportion of iron with respect to magnesium as $K_2O$ content decreases. In the La Rayrie sample the reverse trend occurs. In this later instance microscopic observation indicates that the iron is deposited at the edges of the biotite flakes as iron oxides. It appears then that the sample where a regular mixed layered hydrobiotite is formed keeps a larger proportion of iron in the newly crystallized phyllosilicate structures than that where an intergrade vermiculite is formed. Note that the calcium content of the intergrade mineral is higher for the La Pagerie samples where the overlying soil is slightly calcareous.

## DISCUSSION

The biotites studied here develop in two different directions both chemically and mineralogically as weathering becomes more intense. The final "facies" for both profiles developed in the former biotite grains is that of kaolinite plus oxides (Fe and Ti). However the intermediate stages differ markedly and thus liberate material into the altering solution at different rates. The *La Pagerie* profile, which is capped by a thin layer of calcareous material, develops a hydrobiotite, *i.e.* a regularly interstratified biotite-vermiculitic intergrade mineral phase. Since the bulk composition of the initial mica is about 30 % Mg (Mg/Mg + Fe atoms), the two chemical factors find confirmation in the experiments of Hoda and Hood (1972) where dilute calcic solutions acting upon ferrous biotites produce interlayered minerals similar to those found in our study. In the *La Rayrie* profile, the only major difference is the "lack" of calcic solutions. The plagioclase which furnishes the great mass of alkalis in solutions is essentially sodic (An 25-30) in both profiles.

Although the major mineral formed in both profiles is initially a trioctahedral vermiculite with intergrade characteristics, the chemical evolution is quite different, in the case of the interstratified mineral Al, K and Mg are lost relative to Fe and Si while the reverse is true when a vermiculite-kaolinite assemblage is formed in the case of the *La Rayrie* profile. In both cases K is lost and Na does not appear to increase in the newly crystallized phyllosilicates. Ca increases in the profile with calcic altering solutions.

A second important point to note is the multiphase character of the newly formed weathering minerals. In each profile, biotite destabilizes to an assemblage of the Fe and Ti oxides plus a new vermiculite phase, separate or interlayered, plus kaolinite. As weathering becomes more intense the proportion of kaolinite + oxides increases until it is predominant in the bleached zones of the profiles. The composition of the new vermiculitic mineral thus depends upon whether it is combined in an interlayered structure or is formed as an entirely new mineral in the recrystallized material.

It is interesting to note also that the mixed layered mineral, with a biotite component, persists into the latter stages of alteration by weathering. However the ultimate mineral assemblage in both profiles is one where only kaolinite plus oxides persist.

## LA RAYRIE

| | Altered rock | | | | | | | Friable red zones | | | | | Bleached zones | | | | | |
|---|---|---|---|---|---|---|---|---|---|---|---|---|---|---|---|---|---|---|
| | 1 | 2 | 3 | 4 | 5 | 6 | 7 | 1 | 2 | 3 | 4 | 5 | 1 | 2 | 3 | 4 | 5 | 6 |
| SiO$_2$ | 35,25 | 35,76 | 37,24 | 36,13 | 33,77 | 33,78 | 35,12 | 34,80 | 36,95 | 34,75 | 38,70 | 30,95 | 39,29 | 29,63 | 39,84 | 37,13 | 36,33 | 49,03 |
| Al$_2$O$_3$ | 19,93 | 20,70 | 25,12 | 22,30 | 21,64 | 20,91 | 20,19 | 27,60 | 27,75 | 26,20 | 28,62 | 24,95 | 19,51 | 21,68 | 22,67 | 22,62 | 21,56 | 35,71 |
| Fe$_2$O$_3$ | 27,67 | 27,82 | 17,26 | 24,77 | 20,60 | 14,56 | 20,93 | 17,20 | 13,02 | 17,84 | 10,30 | 19,22 | 16,42 | 20,89 | 16,31 | 28,38 | 23,37 | 2,35 |
| TiO$_2$ | 0 | 0 | 0,65 | 1,74 | 1,63 | 2,04 | 2,46 | 1,56 | 0,10 | 0,10 | -0,20 | - | - | - | - | - | - | - |
| MgO | 4,24 | 4,51 | 3,08 | 1,74 | 2,47 | 4,12 | 4,16 | 4,16 | 3,34 | 3,94 | 1,86 | 4,14 | 3,41 | 3,01 | 3,58 | 3,53 | 3,69 | 0,15 |
| CaO | 0,05 | 0,12 | 0,08 | 0,06 | 0,07 | 0,05 | 0,05 | 0,28 | 0,12 | 0,20 | 0,48 | 0,35 | 0,07 | 0,06 | 0,09 | 0 | 0,02 | 0 |
| Na$_2$O | 0,34 | 0 | 0,01 | 0,02 | 0,02 | 0 | 0,10 | 0,31 | 1,07 | 1,35 | 0,78 | 1,25 | - | - | - | - | - | - |
| K$_2$O | 9,05 | 9,49 | 6,30 | 6,28 | 7,52 | 10,00 | 9,68 | 1,14 | 1,88 | 1,67 | 0,52 | 1,35 | 9,26 | 7,10 | 5,01 | 6,26 | 8,06 | 8,98 |
| Total | 96,54 | 98,40 | 89,73 | 93,04 | 87,71 | 84,45 | 92,70 | 87,05 | 84,23 | 86,05 | 81,46 | 82,21 | 87,96 | 82,38 | 87,50 | 97,91 | 93,03 | 96,23 |

## LA PACERIE

| | Friable red zones | | | | | | Bleached zones | | |
|---|---|---|---|---|---|---|---|---|---|
| | 1 | 2 | 3 | 4 | 5 | 6 | 1 | 2 | 3 |
| SiO$_2$ | 39,30 | 39,66 | 37,10 | 35,56 | 34,24 | 34,12 | 39,09 | 34,92 | 32,33 |
| Al$_2$O$_3$ | 23,79 | 25,14 | 24,58 | 19,33 | 23,95 | 19,86 | 21,73 | 20,88 | 19,39 |
| Fe$_2$O$_3$ | 19,01 | 19,12 | 18,97 | 21,48 | 22,30 | 18,60 | 20,17 | 18,70 | 23,21 |
| TiO$_2$ | 1,97 | 1,89 | 1,95 | 1,64 | 1,57 | 1,55 | - | - | 1,65 |
| MgO | 4,43 | 4,47 | 4,80 | 4,23 | 3,91 | 2,99 | 4,49 | 4,07 | 4,16 |
| CaO | 0,29 | 0,37 | 0,47 | 0,30 | 0,37 | 0,33 | 0,52 | 0,55 | 0,70 |
| Na$_2$O | 0,13 | 0,09 | 0,04 | 0,03 | 0,07 | 0,08 | 0 | 0 | 0,67 |
| K$_2$O | 3,90 | 4,67 | 3,79 | 3,14 | 3,99 | 4,29 | 5,14 | 5,18 | 4,67 |
| Total | 92,81 | 95,43 | 91,69 | 85,72 | 90,40 | 81,81 | 91,15 | 84,29 | 86,77 |

Table 2 — *Electron microprobe analyses for altered biotites (pleochroic layers)*

## LA RAYRIE

| | Friable red zones | | | | Bleached zones | | | | | | | |
|---|---|---|---|---|---|---|---|---|---|---|---|---|
| | 1 | 2 | 3 | 4 | 1 | 2 | 3 | 4 | 5 | 6 | 7 | 8 | 9 | 10 | 11 |
| SiO$_2$ | 49,35 | 39,94 | 48,62 | 47,25 | 38,02 | 44,15 | 43,47 | 44,69 | 50,51 | 38,01 | 48,67 | 31,37 | 35,73 | 46,97 | 46,26 |
| Al$_2$O$_3$ | 33,08 | 32,21 | 29,53 | 29,38 | 34,31 | 25,79 | 20,58 | 37,81 | 35,48 | 18,83 | 28,21 | 41,73 | 23,72 | 28,08 | 27,91 |
| Fe$_2$O$_3$ | 0,64 | 1,58 | 2,80 | 2,67 | 2,75 | 2,11 | 9,28 | 2,31 | 1,90 | 16,38 | 7,00 | 4,49 | 18,58 | 2,07 | 1,72 |
| TiO$_2$ | 0,05 | 0,19 | 0,19 | 0,15 | – | – | – | – | – | – | – | – | – | – | – |
| MgO | 3,51 | 0,60 | 1,26 | 1,14 | 0,70 | 0,69 | 1,96 | 1,28 | 0,83 | 3,33 | 0,16 | 0,15 | 3,21 | 0,52 | 0,33 |
| CaO | 0,07 | 0,04 | 0,04 | 0,05 | 0,12 | 0,08 | 0,07 | 0,07 | 0,13 | 0,09 | 0,21 | 0,16 | 0,10 | 0,15 | 0,03 |
| Na$_2$O | 0,02 | 0,12 | 0,05 | 0,11 | – | – | – | – | – | – | – | – | 0,19 | 0,06 | 0,03 |
| K$_2$O | 5,87 | 4,50 | 0,65 | 5,18 | 1,56 | 4,25 | 3,86 | 0,49 | 2,11 | 4,81 | 2,20 | 1,38 | 2,36 | 1,76 | 5,18 |
| Total | 92,58 | 79,18 | 83,13 | 85,93 | 77,46 | 77,09 | 79,21 | 86,66 | 90,97 | 81,46 | 86,45 | 79,28 | 83,91 | 79,61 | 81,46 |

## LA PACERIE

| | Friable red zones | | | Bleached zones : | | | | |
|---|---|---|---|---|---|---|---|---|
| | 1 | 2 | 3 | 1 | 2 | 3 | 4 | 5 |
| SiO$_2$ | 38,68 | 40,30 | 35,20 | 30,27 | 34,68 | 36,96 | 34,85 | 46,20 |
| Al$_2$O$_3$ | 24,36 | 33,00 | 18,48 | 21,40 | 23,49 | 29,43 | 22,41 | 35,44 |
| Fe$_2$O$_3$ | 18,98 | 1,16 | 31,52 | 19,44 | 25,42 | 9,29 | 21,18 | 3,79 |
| TiO$_2$ | 1,98 | 0,39 | 1,18 | 1,37 | 1,20 | – | – | – |
| MgO | 4,65 | 0,59 | 2,76 | 2,96 | 2,95 | 0,82 | 3,07 | 1,01 |
| CaO | 0,37 | 0,06 | 0,19 | 0,33 | 0,31 | 0,60 | 0,73 | 0,02 |
| Na$_2$O | 0,08 | 0,29 | 0,13 | 0,12 | 0,14 | 0,16 | 0,09 | 0,08 |
| K$_2$O | 2,78 | 4,31 | 2,52 | 4,07 | 3,27 | 0,17 | 1,94 | 0,76 |
| Total | 91,88 | 80,10 | 92,00 | 79,96 | 91,46 | 77,43 | 87,99 | 87,29 |

Table 3 – *Electron microprobe analyses for newly crystallized minerals found between the cleavage flakes of the biotites.*

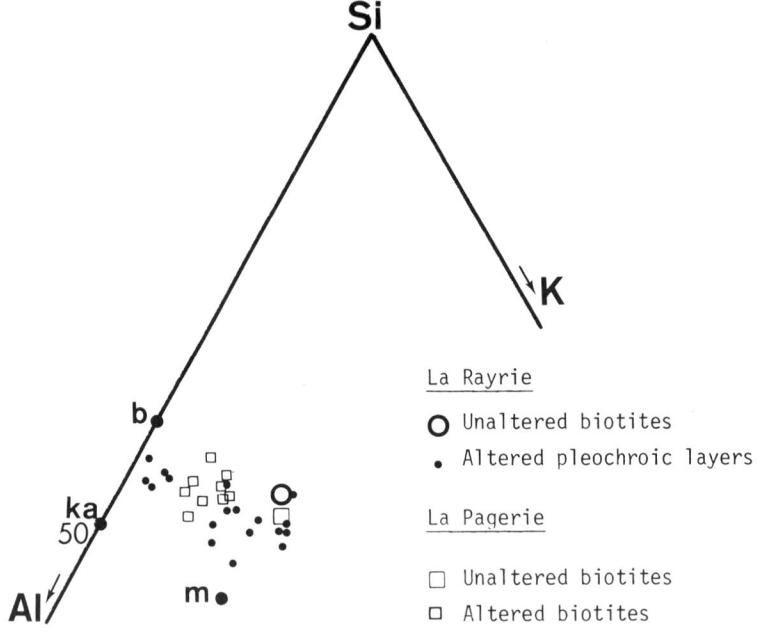

Figure 2 - *Microprobe analysis of altered biotite grains in Si - Al - K coordinates*

REFERENCES

COLEMAN N.T., LE ROUX F.H. and CADY I.G. (1963) - Biotite-hydrobiotite-vermiculite in soils. *Nature*, 198 : 409-410.

HODA S.N. and HOOD W.C. (1972) - Laboratory alteration of trioctahedral micas. *Clays and Clay min.*, 20 : 343-358.

MEUNIER A. and VELDE B. (1976) - Mineral reactions at grains contacts in early stages of granite weathering. *Clay min.*, 11 : 235-240.

MEUNIER A. (1977) - Les mécanismes de l'altération des granites et le rôle des microsystèmes. Etudes des arènes du massif granitique de Parthenay (Deux-Sèvres). *Thèse, Univ. Poitiers*, 248 pp.

OJANUGA A.G. (1973) - Weathering of biotite in soils of a humid tropical climate. *Soil Sci. Soc. Amer. Proc.*, 37, 4 : 644-646.

RIMSAITE J.H.Y. (1975) - Natural alteration of mica and reaction between released ions in mineral deposit. *Clays and Clay min.*, 23, 3 : 247-255.

SEDDOH F.K. (1973) - Altération des roches cristallines du Morvan : granites, granophyres, rhyolites. Etude minéralogique, géochimique et micromorphologique. *Thèse, Fac. Sci. Dijon - Mém. Géol. Univ. Dijon*, Doin ed., 377 pp.

STOCH L. and SIKORA W. (1976) Transformation of micas in the process of kaolinitization of granites and gneisses. *Clays and Clay min.*, 24, 4 : 156-162.

WILSON M.J. (1966) - The weathering of biotite in some Aberdeenshire soils. *Min. Mag.*, 35 : 1080-1093.

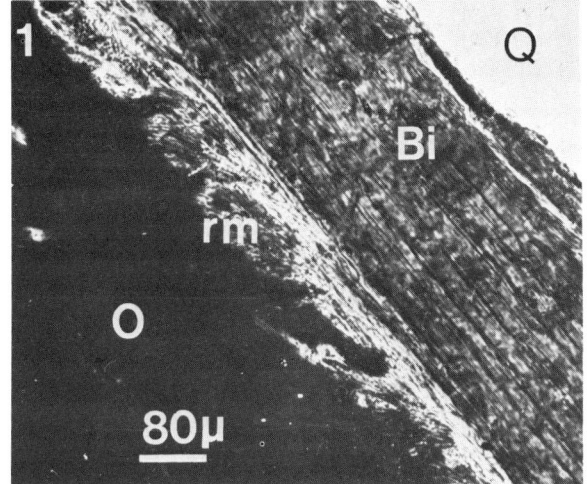

Photo 1

Reactional micas (rm) between biotite (Bi) and orthoclase (O) The biotite-quartz (Q) contact remains neutral

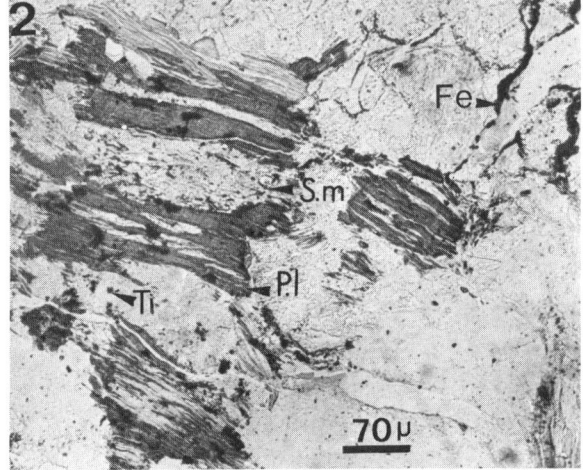

Photo 2

Altered biotite from the red friable zone of *La Rayrie*.

P.L. : pleochroic layers
S.m  : secondary minerals
Ti   : Ti-oxides
Fe   : Fe-oxides-hydroxides

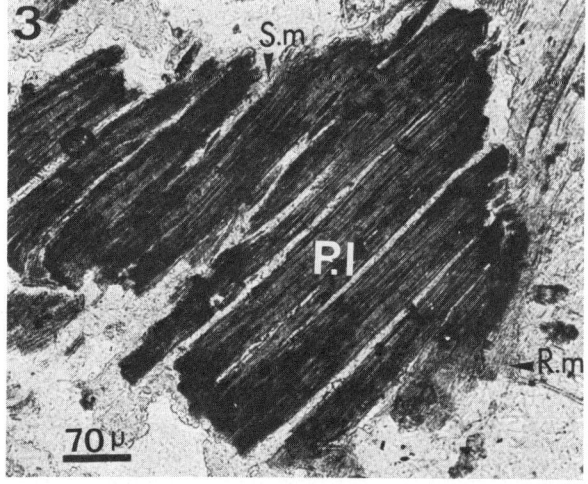

Photo 3

Altered biotite from the red friable zone of *La Pagerie*.

P.L : pleochroic layers
S.m : secondary minerals
R.m : reactional micas

MICROMORPHOLOGY OF HALLOYSITE PRODUCED BY WEATHERING OF PLAGIOCLASE IN VOLCANIC ASH

KAZUE TAZAKI
Institute for Thermal Spring Research, Okayama University, Misasa, Tottori-ken, 682-02 (JAPAN)

ABSTRACT

Halloysite derived from weathered plagioclase in volcanic ash soils in the San-in district, Japan, exhibits various micromorphological forms, namely, of spherical, walnut-meat-shaped, acicular, crinkly, platy, tubular and square-tube types. Spherical, walnut-meat-shaped and crinkly ones are allophane-halloysite aggregates, whereas the ones of other types are normal halloysite with no allophane. The crinkly type ones consist of rolling spheroidal particles of halloysite and fine allophane granules.

The weathering process can be traced morphologically from fresh plagioclase to various types of halloysite. The possible mechanisms of the crystal growth are discussed.

---

INTRODUCTION

Wide variations in morphology in halloysites are well known. Especially, tubular and spheroidal halloysite have wide occurrences in weathered rocks and soils from Hongkong (Parham, 1969), Texas (Kunze and Bradley, 1964, Askenasy et al., 1973), Japan (Okada, 1973) and New Mexico (Berner and Holdren, 1977).

In the present study, the weathering process of plagioclase in volcanic ash soils is investigated by scanning electron microscopy (SEM), transmission electron microscopy (TEM) and with the aid of other conventional methods such as X-ray diffraction and differential thermal analysis. The materials are quantitatively analyzed by electron-probe microanalyzer and electron microscopic microanalyzer (EMMA). Plagioclase samples investigated were collected from the volcanic ash soils and pyroclastic sediments of Mt. Daisen and Mt. Sambe.

Mt. Daisen and Mt. Sambe, located on the southwestern coast of the Japan Sea, are Quarternary complex volcanoes, composed of calc-alkali two-pyroxene andesite and biotite-hornblende dacite. Thick pyroclastic flows and volcanic ash soils from these volcanoes widely cover the San-in district.

A mechanism of feldspar weathering has been proposed by Berner and Holdren (1977). Acoording to them, the weathering of feldspar is controlled by chemical reaction at the feldspar-solution interface, but not by diffusion, either through aqueous solution or through a protective surface layers. Nevertheless, the present author has revealed an important role of amorphous surface layers for the formation of clays and its related minerals (Tazaki, 1976, 1977, 1978). In the present paper, the weathering process from fresh plagioclase to halloysite of various types will be demonstrated.

METHODS AND MATERIALS

Halloysite derived from plagioclase was isolated by hand-picking from the volcan ash soils of Mt. Daisen and Mt. Sambe. Weathered plagioclase was lightly broken to expose the fresh surface and the resultant chips were placed on brass stubs with wet silver cement. The mounted samples were then coated with carbon and gold films with a thickness of several hundred Å. Morphology of the mineral grains was examined by a JEOL 5A scanning electron microscope. The composition of the near-subsurface of these grains was determined by non-dispersive X-ray spectrometry, using an ORTEC Si(Li) X-ray detector and JEOL EM-ASID-4D scanning image display device.

Fresh plagioclase contained 35 to 39 mole per cent An and were identified to be andesine. The difference in d-spacings between ($1\bar{3}1$) and (131) reflections ranged from 1.90 to 1.95 suggesting that the plagioclase was of a high temperature type.

Halloysite derived from plagioclase could be easily detected by means of X-ray powder diffraction and electron microscopic observation. The X-ray diffraction peak corresponding to the d-spacing of 10Å is characteristic of halloysite and mica-group minerals, and the former is easily distinguishable from the latter by expansion of the 10Å spacing to 11Å after ethylene glycol treatment. The differential thermal analysis curve of the halloysite showed two endothermic peaks at about 120 and 550°C and a clear exothermic peak at 990°C.

RESULTS

The SEM studies indicate that dissolution of plagioclase does not initially occur over the entire surface of the mineral uniformly. Instead, the dissolution occurs at sites of chemical erosion on the smooth surface, so that oval- or square shaped etch pits and conical hollows develop on the plagioclase surface (Fig. 1,2 and 3).

As erosion proceeds, the whole surface becomes rugged and, further, changes partial into an amorphous state. The reaction product left by the dissolution of plagioclase forms as amorphous thin layer, of which the thickness is less than 0.5μm.

Fig. 4 shows the thin layer with irregular-shaped caves which suggest proceeding erosion. Subsurface fresh plagioclase could be seen below the layer (not shown).

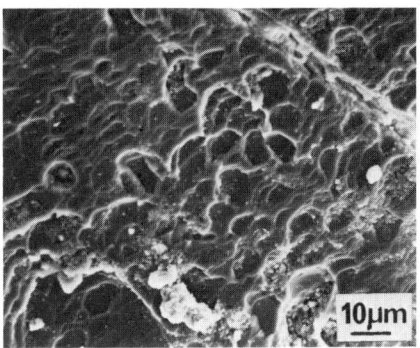

Fig. 1. Oval shaped etch pits develop on the plagioclase surface.

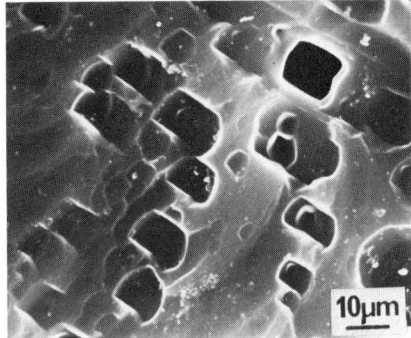

Fig. 2. Parallel development of square shaped etch pits.

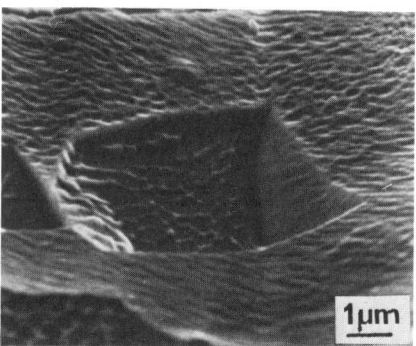

Fig. 3. Square shaped etch pits formed on the lamellae.

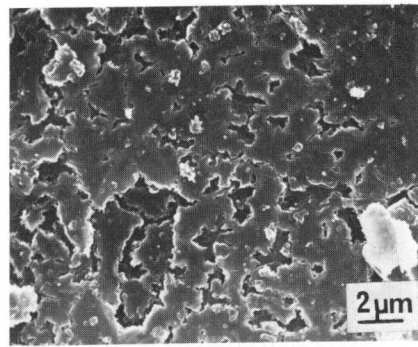

Fig. 4. The amorphous thin layer with irregular-shaped caves which suggest preceding erosion.

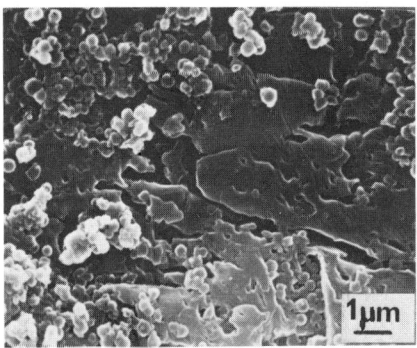

Fig. 5. Spherical grains and etched plagioclase surface.

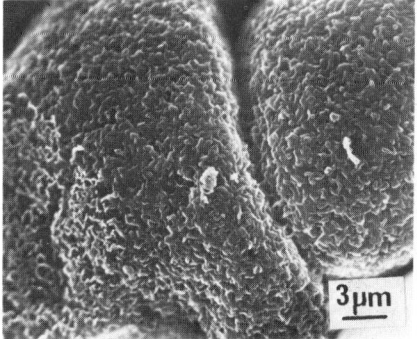

Fig. 6. Walnut-meat-shaped halloysite occurs on the swollen surface.

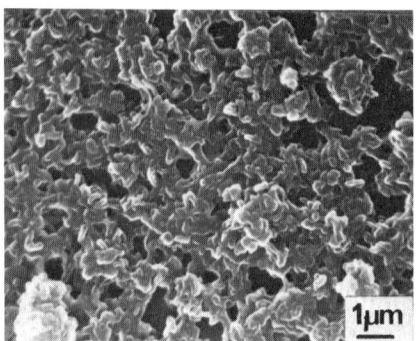

Fig. 7. Walnut-meat-shaped halloysite occurs on the smooth surface.

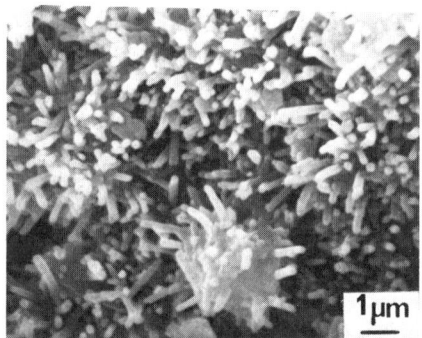

Fig. 8. Rounded aggregate of the tubular halloysite looks like a flower of chrysanthemum.

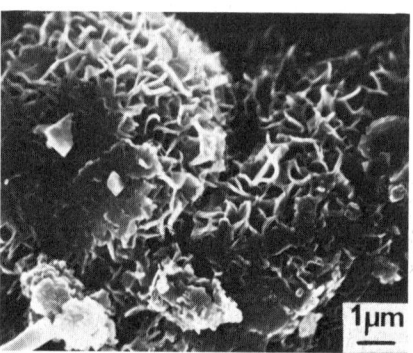

Fig. 9. The crinkly halloysite forms fringed globules with a honeycomb-like surface.

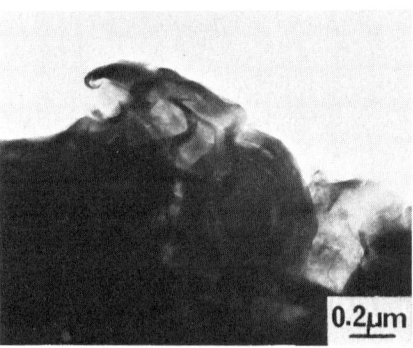

Fig. 10. The crinkly halloysite consists of spheroidal particles of halloysite and fine granules of allophane.

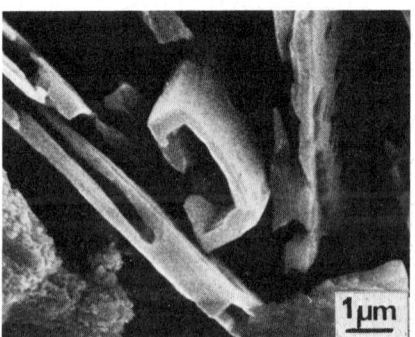

Fig. 11. The square-tube halloysite has a rectangular cross section.

Fig. 12. The platy halloysite with a fringed edge.

The X-ray diffraction of the thin layer, which was peeled off from the eroded plagioclase by ultrasonic treatment, showed an amorphous pattern with weak peaks of gibbsite Parham (1969) has shown the similar early corrosion pitting and alteration of feldspar along crystal dislocation and/or microjoints, resulting from both natural and artificial weathering.

During the subsequent stage, allophane-halloysite aggregates, of spherical, walnut-meat-shaped or crinkly types, grow at the bottom of hollows or pits. The spherical type product is conceivably a primitive stage of halloysite. Walnut-meat-shaped and crinkly type products are weakly crystallized halloysite containing allophane. In the following, various types of alteration products are described.

*Spherical type*: Spherical grains are 0.02-0.2µm in diameter and are smaller than the other forms. Figure 5 shows formation of spherical grains in the left-hand side and etched plagioclase surface in the right. Spherical halloysite often coexists with a scaly mica mineral.

*Walnut-meat-shaped type*: The walnut-meat-shaped type product consists of irregularly bent flakes and tiny tubes with a diameter less than 0.1µm (Fig. 6 and 7). Walnut-meat-shaped halloysite generally occurs on the smooth surface (Fig. 7), and sometimes on the swollen surface (Fig. 6).

*Tubular type*: The tubular crystal is morphologically well-known and is characteristic of halloysite (Sudo and Takahashi, 1956; Kurahayashi and Tsuchiya, 1960; Okada, 1973). The SEM observation of the tubular halloysite was also reported by Borst and Keller (1969) and Eswaran (1972). The tubular halloysite with various lengths was produced from plagioclase in volcanic ash soils. The size ranges from 0.2 to 2.0µm long and is about 0.1µm wide on the average (Fig. 8). The rounded aggregate of the tubular halloysite looks like a flower of chrysanthemum. Those formed by weathering of feldspars in granites are long and tubular shaped. Nagasawa and Miyazaki (1975) reported the long halloysite tubes up to 8µm.

*Acicular type*: Acicular halloysite appears spinous under low magnification (×3000). Under high magnification (×40,000) however, it is clear that the needle crystals form brush-like bundles. X-ray diffraction of this halloysite shows a relatively strong 10Å peak in spite of its primitive morphology. The cluster of elongated kaolin minerals in Sparta granite shown by Keller (1977) is similar to the present acicular-type halloysite.

*Crinkly type*: The crinkly halloysite forms fringed globules 1-5µm in diameter with a honeycomb-like surface (Fig. 9). This is the first observation of this form in halloysite. A morphology similar to this was discovered for smectite (Bohor and Hughes, 1971; Borst and Keller, 1969) of the A.P.I. standard samples. This crinkly halloysite was also investigated by X-ray powder diffraction and TEM. The TEM observation revealed that the crinkly halloysite consists of spheroidal particles of halloysite and fine granules of allophane (Fig. 10). Extremely fine granules of allophane coagulate into round grains. These grains morphologically resemble the spherical halloysite described in the foregoing section.

Observation under very high magnification (up to ×400,000) shows a rolling end of the outer layer from the spheroidal halloysite as has been observed by Sudo and Yotsumoto (1977). The morphology of partly rolled laths or ribbons may suggest a growth process in the spherical halloysite. Similar TEM morphology has been shown by Nagasawa and Miyazaki (1977) on halloysite formed by alteration of volcanic glass having shapes of balls and/or scrolls sometimes associated with tubes.

*Square-tube and platy types*: Square-tube and platy crystals coexist occasionally with crinkly halloysite. The square-tube ranges from 1 to 15μm long and 0.2 to 7μm wide in size and has a rectangular cross section (Fig. 11). The square-tube is probably fragile and easily broken down into platy pieces by grinding. The morphology which is similar to the present halloysite can be found in the TEM replica photographs of the tubular halloysite from Wagon Wheel, Colorado, taken by Dixon and Mckee (1974), although they did not mention anything about the square tube morphology. The platy halloysites may be divided morphologically into the following two groups: The one with a shape of fringed plate of the size about 5μm long and 0.2μm thick on average (Fig. 12), and the other which is characterized by a straight edge with about 0.2μm thickness on average.

EMMA analyses of platy crystals show the presence of Ca and Na as well as Si and Al. Presence of considerable amounts of CaO and $Na_2O$ may suggest preservation of plagioclase components in square-tube halloysite. As noted above, the square tube is easily broken down into platy chips by grinding for one minute. This is a distinct difference from most of the halloysites with other morphologies which do not break down either by grinding for few minutes or by heating at temperature from 200 to 700°C for one hour.

DISCUSSION

Various morphologies of halloysite can be divided into two groups, viz., primitive halloysite and normal halloysite. The primitive halloysite usually coexists with allophane, whereas the normal halloysite is free from allophane.

Spherical, walnut-meat-shaped and crinkly type halloysites contain allophane and therefore they are regarded as primitive, whereas halloysites of other types are normal because of absence of allophane. The primitive type halloysites are characterized by the presence of the allophane-halloysite balls discovered by Sudo and Takahashi (1956).

Sequences of development of halloysite from weathered plagioclase is schematically shown in Fig. 13. Among them, it is apparent that, from the morphological point of view, the main sequence starts with the formation of a rugged surface on plagioclase (Ht 2-1), passing through the spherical type (Ht 2-2), the walnut-meat-shaped type (Ht 2-3) and the short tubular type (Ht 2-4), to the chrysanthemum-flower like aggregates (Ht 2-5) which consist of the long tubular crystals, with occasional growth of well-developed halloysite tubes (Ht 2-6). Other subsidiary sequences can be recognized: the one from Ht 2-1 in Fig. 13 to the acicular (Ht 1-1) and further to the brush-like bundles (Ht 1-2), both of which may sometimes grow into the members of the main sequence as indicated in Fig. 13. Another sequence procedes from Ht 2-1 in Fig. 13 to the crinkly type (Ht 3-1), which may change into the walnut-meat-shaped type (Ht 2-3) of the main sequence or into a complex aggregate of various morphologies (Ht 3-2) composed of the platy, square-tube and crinkly type halloysites.

Dissolved silica in the permeating water may contribute to the formation of halloysite within a relatively lower stratigraphic horizon. This may be one of the processes of formation of halloysite in the heavily weathered plagioclase. A variable degree of crystallinity can be observed in weathered plagioclase within the same horizon. For example, tubular halloysite, the typical well-crystallized halloysite, occurs only in the lowest Daisen pyroclastic sediment ($D_1$). The occurrences of these halloysites, however, do not extend to the nearest outcrop even in the same stratum. In this context, no definite relation is recognized between the halloysite morphology and the stratigraphic succession. Alternate occurrences of primitive and normal halloysites are also found in the vertical succession of pyroclastic sediment ($D_1$). Several factors may be responsible for these fluctuations in the formation of clay minerals. Among numerous factors, local differences in permeability, pH of permeating water, porosity of the soil, grain size of primary minerals and micro-topography of the localites may be of importance. Thus more detailed investigation is required to clarify the decisive factors out of these possibilities.

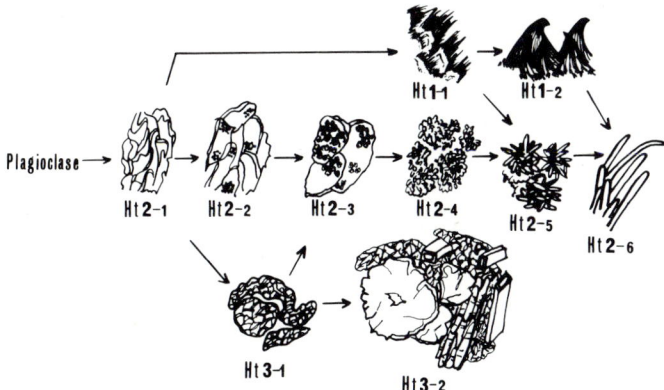

Fig. 13. The sequences of development of halloysite from weathered plagioclase. Ht1-1; acicular, Ht1-2; brush-like bundles, Ht2-1; rugged surface, Ht2-2; spherical, Ht2-3; walnut-meat-shaped, Ht2-4; short tubular, Ht2-5; chrysanthemum masses, Ht2-6; long tubular, Ht3-1; crinkly, Ht3-2; crinkly, square-tube and platy.

ACKNOWLEDGMENTS

The writer would like to express her cordial thanks to Dr. Susumu Shimoda of the Tsukuba University for his valuable suggestions, and many instructions. Thanks are also due to Professor Yoshito Matsui of the same Institute for his advice and critical reading of the manuscript. The writer also thanks Dr. Koichi Tazaki of the same Institute for many suggestions and instructions in the course of the study.

REFERENCES

Askenasy,P.E.,Dixon,J.B. and Mckee,T.R.,1973. Spheroidal halloysite in a Guatemalan Soil. Soil Sci.Soc.Am.Proc.,37:399-803.
Berner,R.A. and Holdren,Jr.,G.R.,1977. Mechanism of feldspar weathering: some observation evidence.Geology,5:369-372.
Bohor,B.F. and Hughes,R.E.,1971. Scanning electron microscopy of clays and clay minerals. Clays Clay Miner.,19:49-54.
Borst,R.L. and Keller,W.D.,1969. Scanning electron micrographs of API reference clay minerals and other selected samples. Proc. Int. Clay Conf. 1969, 871-901.
Dixon,J.B. and McKee,T.R.,1974. Internal and external morphology of tubular and spheroidal halloysite particles. Clays Clay Miner.,22:127-137.
Eswaran,H.,1972. Morphology of allophane, imogolite and halloysite. Clay Miner., 9:281-285.
Keller,W.D.,1977. Scan electron micrographs of kaolins collected from diverse environments of origin. Clays Clay Miner.,25:311-345.
Kunze,G.W. and Bradlly,W.F.,1964. Occurrence of a tabular halloysite in a Texas soil. Clays Clay Miner.,12:523-527.
Kurahayashi,S. and Tsuchiya,T.,1960. On the clay minerals of the Kanto Loam(3). Geol. Soc. Japan J.,66:586-593. (in Japanese, with English abstract)
Nagasawa,K. and Miyazaki,S.,1975. Mineralogical properties of halloysite as related to its genesis. Proc. Int. Clay Conf. 1975, 257-265.
Okada,S.,1973. Clay minerals in the Shikotsu pumice fall deposits. Geol. Soc. Japan J.,79:363-375. (in Japanese, with English abstract)
Parham,W.E.,1969. Formation of halloysite from feldspar: Low temperature, artifical weathering versus natural weathering. Clays Clay Miner.,17:13-22.
Sudo,T. and Takahashi,H.,1956. Shapes of halloysite particles in Japanese clays. Clays Clay Miner.,4:67-79.
Sudo,T. and Yotsumoto,H.,1977. The formation of halloysite tubes from spherulitic halloysite. Clays Clay Miner.,25:155-159.
Tazaki,K.,1976. Scanning electron microscopic study of formation of gibbsite from plagioclase. Pap. of the Institute for Themal Spring Research, Okayama Univ., 45:11-24.
Tazaki,K.,1977. Scanning electron microscopic study of clay minerals and non-clay minerals produced from plagioclase in volcanic ashes with regard to weathering process. (Unpublished Doctor thesis, Tokyo University of Education).
Tazaki,K.,1978. Micromorphology of plagioclase surface at incipient stage of weathering. Earth Sci. (Chikyū Kagaku), 32:8-12. (in Japanese, with English abstract)

OTHER PAPERS PRESENTED IN SECTION 4

ELECTROLYTIC WEATHERING EFFECTS IN MINERALS

R.M. Taylor and R.D. Bond (C.S.I.R.O. Division of Soils, Private Bag 2, Glen Osmond 5064, South Australia.)

CLAY MINERAL STUDIES OF A VERTISOL CHRONOSEQUENCE IN SOUTHERN TURKEY

N. Güzel and M.J. Wilson[1] (Faculty of Agriculture, University of Adana, Turkey. [1] Department of Pedology, Macaulay Institute for Soil Research, Aberdeen, Scotland.)

VARIATIONS IN CLAY MINERALOGY OF ULTISOLS FORMED ON DIFFERENT GEOMORPHIC PHASES

K.H. Tan and R.A. McCreery (Department of Agronomy, University of Georgia, Athens, GA., USA.)

A MINERALOGICAL AND PETROGRAPHICAL STUDY OF THE ALTERATION OF SOME TUFFLAVAS FROM THE LOWER CARBONIFEROUS OF SCOTLAND

D.J. Morgan and I.R. Basham (Institute of Geological Sciences, 64-78 Gray's Inn Road, London WC1X 8NG, UK.)

SOME ASPECTS OF THE CRYSTALLIZATION OF VOLCANIC GLASS TO CLAY MINERALS

N. Matsui[1] and I. Kayama[2] ([1] Tokyo University of Foreign Studies, Sumiyoshi, Fuchu, Tokyo, Japan. [2] Faculty of Science, Tokai University, Hiratsuka, Kanagawa, Japan.)

CLAY-SILICIC ACID INTERACTION IN SEAWATER

R. Dayal (Marine Sciences Research Center, State University of New York, Stony Brook, New York, USA.)

PREPARATION AND PROPERTIES OF HYDROXY-Ni,-Al,-Cr,-Zr MONTMORILLONITES

S. Yamanaka[1] and G.W. Brindley (Materials Research Laboratory, The Pennsylvania State University, University Park, Pennsylvania 16802, USA. [1] Present address: Dept. of Applied Chemistry, University of Osaka Prefecture, Sakai, Osaka 591, Japan.)

PHYSICO-CHEMICAL COMPARISON OF INTERMEDIATES FORMED IN THE SYNTHESIS OF KAOLINITES FROM SMECTITES WITH A NATURAL INTERSTRATIFIED SMECTITE-KAOLINITE SEQUENCE

M.M. Mestdagh and A.J. Herbillon (Université Catholique de Louvain, Groupe de Physico-Chimie Minérale et de Catalyse, Place Croix du Sud 1, B-1348 Louvain-La-Neuve, Belgium.)

SYNTHESIS OF TRIOCTAHEDRAL PHYLLOSILICATES ON THE TALC-POTASSIUM PHLOGOPITE JOIN

G. Whitney and D. Eberl (Department of Geology, University of Illinois Urbana, Illinois 61801, USA.)

GLAUCONITE SYNTHESIS

H. Harder (Sedimentpetrographisches Institut der Universität, Göttingen, Federal Republic of Germany.)

ANOMALOUS POTASSIUM DISTRIBUTION IN WEATHERED BIOTITE FROM SOME SCOTTISH SOILS

W.J. McHardy and M.J. Wilson (Department of Pedology, Macaulay Institute for Soil Research, Aberdeen, Scotland.)

SURFICIAL COVER ASSOCIATED WITH BASALTIC ROCKS OF THE PARANA BASIN: NATURE AND ORIGIN OF THE CLAY FRACTION AND ITS DISTRIBUTION IN THE LANDSCAPE

N.M.M. Goncalves[1], A. Carvalho[2] and A.J. Melfi[2] ([1] Department of Geology - F.F.C.L. Ribeirão Preto, USP, Ribeirão Preto, SP, Brazil. [2] Department of General Geology - I.G., USP, São Paulo, Brazil.)

THE CLAY MINERALOGY OF SOME SOILS FROM QASIM, SAUDI ARABIA

A.S. Mashhady[1], M. Reda[1], M.J. Wilson[2] and R.C. Mackenzie[2] ([1] Soil Department, Faculty of Agriculture, University of Riyad, Riyadh, Saudi Arabia. [2] The Macaulay Institute for Soil Research, Craigiebuckler, Aberdeen, Scotland.)

FIBROUS AND MONTMORILLONITE-TYPE CLAY MINERALS IN VERTISOLS AND VERTIC SOILS OF THE CANARY ISLANDS, SPAIN

C. Rodriguez-Pascual, E. Fernandez-Caldas and C.M. Rodriguez Hernandez (Instituto de Edafología y Biología Vegetal, C.S.I.C., Madrid, Spain; Centro de Edafología y Biología Aplicada de Tenerife, C.S.I.C., Santa Cruz de Tenerife, Canary Islands, Spain.)

ALTERATION OF VOLCANIC ROCK TO MONTMORILLONITE

I. Kayama, M. Okuma, Y. Arakawa and Y. Suzuki (Faculty of Science, Tokai University, Hiratsuka, Kanagawa, Japan.)

GENETIC SEQUENCE OF MONTMORILLONITE AND ZEOLITES IN THE BENTONITE DEPOSITS OF THE TSUKINUNO MINE, YAMAGATA PREFECTURE, JAPAN

S. Honda (Mining College, Akita University, Akita-shi, Japan 010.)

A NOTE ON THE FORMATION OF TALC FROM MICA THROUGH THE BREAKDOWN OF PYRRHOTITE

W.F. Cole and C.J. Lancucki (Division of Building Research, CSIRO, Graham Road, Highett, Vic. 3190, Australia.)

STABILITY DIAGRAMS INVOLVING CLAY MINERALS AT 25°C

F. Lippmann (Mineralogisches Institut der Universität, Wilhelmstrasse 56, D-7400 Tübingen, Federal Republic of Germany.)

# SECTION 5

# Applied Clay Mineralogy

*Chairman's Introduction*

RECENT DEVELOPMENTS IN APPLIED CLAY MINERALOGY

PÉRSIO DE SOUZA SANTOS
Department of Chemical Engineering, Escola Politécnica, University of São Paulo and Instituto de Pesquisas Tecnológicas, São Paulo, SP, Brasil.


LADIES AND GENTLEMEN

I am deeply honoured by AIPEA's invitation to present the Chairman's Review on the recent developments in applied clay mineralogy; some topics, like alumina from clays, soil mechanics, wastes and oil shale, are not presented due to limitations in space. Clays and clay minerals were reviewed by Patterson and Murray (1975) and by Brindley (1976).

RECENT PRODUCTION, PRICES AND FORECAST; TESTING

The world primary mineral demand in 1974, with projected and cumulated demands for 1985 and 2000, for asbestos, clays, mica, talc and vermiculite, has been evaluated by the U.S. Bureau of Mines (1976).

| World demand for clays: | 1974 | 1985 | 2000 |
|---|---|---|---|
| Asbestos (Thousand short tons) | 4 535 | 6 524 | 10 544 |
| Clays (Million short tons) | 545 | 761 | 1 061 |
| Mica (Thousand short tons) | 263 | 318 | 379 |
| Talc (Thousand short tons) | 5 064 | 7 900 | 13 650 |
| Vermiculite (Thousand short tons) | 557 | 950 | 2 010 |

Breakdown of industrial clays consumption in United States, for the years 1965-1974 (kaolin, ball clay, fireclay, bentonite, fuller's earth and other), for the production of building materials, refractories, paper products, as well similar data for asbestos, mica, talc and vermiculite exists in another publication (Bureau of Mines, 1977). Prices and production of clays in several countries are periodically reviewed and summed up every 5 years with information on uses and technology (Ampian, 1976b). Patterson and Murray (1975) published a comprehensive review on clays in which descriptions are made of several technologies for preparation and of uses of industrial clays. Liles and Heystek (1977) published a monograph on testing and evaluation of clays for industrial uses. The Institute of Geological Sciences of Great Britain has made excellent commodity reviews of ball clay, fuller's earth and talc (Highley, 1975).

## RECENT TECHNOLOGIES FOR CLAY BENEFICIATION; KAOLINS

There is a continuous development of systems of beneficiation to clays, particularly kaolins, but increasingly of interest to ball clays:purified shredded ball clay for sanitaryware is an example (Mitchell and Stentiford, 1978). This is important as inferior reserves are developed and each new deposit represents a new research problem (Clark,1978).

Crude kaolin of variable quality can be wet processed to produce kaolin products of uniform and predetermined physical and chemical properties. The kaolin slurry from the mines is pumped and stored in large tanks; kaolins of various qualities are blended to prepare "standardised" or "taylor-made" clay grades that meet given specifications; this can be controlled by analogue computers (Anon.1975c); as an example, up to five kaolins can be blended within controlled properties (maximum and minimum values) of % $K_2O$; % $Fe_2O_3$; % minus 2 microns;transverse strength and fired brightness (Anon.,1975a).

The kaolin slurries, before blending, are separated into fine and coarse fractions, by using continuous centrifuges and/or hydrocyclones (Trawinski, 1976);the complexable iron is dissolved by sulfuric acid + sodium or zinc hydrosulfite; the dewatering is made by thickening (Brociner and Vollans, 1973) or filtering (Gwilliam 1971); the kaolin cake is dried in several ways: hot pans, rotary dryers, flash or spray dryers, fluidized beds (Mortesen and Hovmand, 1976), or redispersed to be shipped as a 70% slip for industrial uses. Several methods can be used to improve the properties of kaolin; one is froth-flotation or ultra flotation, first patented by Greene et al (1961) to remove $TiO_2$, as anatase, for + 90% brightness kaolin; in the direct fine particle flotation, special designs of flotation cells are essential (Cundy, 1969); the flotation processes are used extensively in kaolin industries in Georgia (Anon., 1971b), UK and Germany.

Another is selective flocculation (Bundy et Berberich, 1969) or differential coagulation or "over flocculation", commercially exploited in Georgia: coagulation is induced in kaolin in the presence of the impurity which is not coagulated, so that a substantial differentiation of particle size can be achieved even when the true ultimate particle sizes of both impurity and product are similar; this is obtained by the use of polyacrylamides; the most important process engineering problem is the difficulty of avoiding entrainment of the unflocculated particle by the coagulum or floc of the kaolin. Electrokinetic methods for flocculation have been patented (Kunkle, 1976). The process of two-liquid separation has become of very real commercial importance in the remarkably effective separation of minerals of small dimensions in kaolin and tin-slimes (Shergold and Softhouse, 1977).

The technology of fine particle size grinding media, called attrition grinding or sand milling, has been extensively used in the recent years, either for the preparation of special pigments for paints and/or fillers composed of particles of lamellar shapes, like talcs(Ashton and Russel,1972) and kaolins (Van Buren,1974).

This type of grinder is essentially a stirred vessel containing sand at high concentration (Davis et al, 1973); design of the mill is critical as considerable quantities of energy are involved and the greatest economy of energy is essential for entering the kaolin slip and removing the product on a continuous system. Attrition grinding has been used to improve low-quality clays. The liquid mixing of high solid concentration clay suspensions in stirred liquid vessels may be considered as a form of autogenous grinding with very low impact energies with grinding media of approximately the same particle size of the material being ground . The attrition ground kaolin is commercially called "delaminated clay" (Anon.1971a) and is used for lightweight paper coatings due to the small thickness of kaolinite plates, for paint films and special rubber applications. Splitting of kaolitite plates in kaolins occurs in liquid mixing, kneading and milling, as well as in the mechanical treatment of ceramic bodies, with improvement of casting rate, plasticity, and green strength; ball clays are not affected as much as kaolins.

Impurities, such as iron and titanium minerals, may be separated by high-intensity wet magnetic separation (Iannicelli, 1976; Ashton et al, 1972; McKyes et al 1974): the main feature of the separator system is a cannister filled with a special type of fine steel wool which provides a uniform magnetic field (under 18 to 22 kilogauss), through which the kaolin slurry is pumped and the magnetic particles retained in the matrix; periodically the magnet is deenergized and a high volume of water under pressure is back flushed through the matrix to flush out the impurities.

The essential requirement in most effective beneficiation systems for fine particle minerals is that the process shall be carried out wet under the appropriate physico-chemical state: industrially, this implies a technology of dewatering. The tube press for kaolin is a development made by Gwilliam (1971): this filter is capable of operating at 100-120 atmospheres: kaolin handled in this way is now shipped in the European area at a rate approaching 100,000 tons per year. A very high pressure plate filter is now being installed in Europe, operating at about 80 atm. (Clark, 1978). Electrical dewatering of dilute clay slurries has been known for a long time; with the ever-increasing cost of thermal energy, some companies have commercial incentives to dewater using an electrokinetic technology which has a great deal of commercial promise (Kunkle,1977).

An interesting process for removing the effect of smectites in reducing the permeability and casting rate of sanitaryware slip bodies has appeared in the patent literature: the expanding properties of smectite are changed into those of a mica--like non-expanding clay mineral; this reduces the viscosity of paper clays and may improve the permeability of ceramic clays in plaster-casting caused by smectites; the reduction of the viscosity of clays may be obtained with an aluminum hydroxy-salt polymer (Kunkle and Kellman, 1977).

The mining, processing and industrial applications of kaolins were reviewed by Patterson and Murray (1975) and by Dimanche et al (1974). Hectorite and some U.S. Southern bentonites are beneficiated by a hydro-classification process (to separate non-clay and other undesirable materials); the slurry is centrifuged and dried by means of drum and spray-driers; temperatures under 150°C must be used for drying (Patterson and Murray, 1975).

## TALC AND PYROPHYLLITE

"Wonderstone" is a term applied to a massive block pyrophyllite from the Republic of South Africa: this rock is necessary for manufacturing synthetic diamond; it is a compact cryptocrystalline pyrophyllite, capable of transmitting pressures like liquid. Platy talc ore of cosmetic grade is delaminated to give a low density, high smooth product exceptionaly suitable for body powders (Ashton and Russel, 1972). Talc, graphite, crysotile asbestos and finely divided polytetrafluoroethylene are mixed with a major proportion of lubricating oil for making a high-pressure lubricating grease (Curtis, 1972). A disposable thin-walled ceramic dish can be made by firing a glazed plate, made of paper coated with a slip of pyrophyllite (Siemmen, 1972).Surface grafted talc is being used for microencapsulation.

## ADSORPTION

Adsorption by clays is a permanent field of research; two examples of recent industrial uses of adsorption by clays are given. Griffin et al (1977) investigated the potential of clay minerals for attenuating the various chemical constituents of landfill leachate; Ca-saturated clays strongly attenuated lead, cadmium mercury, zinc and chromium in the aqueous effluents. Enzymes can be immobilized or attached to water-insoluble support materials without loss of their capacities. These "immobilized enzymes" can then be repeatedly utilized in continuous reaction systems or in batch reaction systems; this permits the user to design continuous processing systems which usually require less capital expense, greater product control and, in the end, lower production cost; clays, sands and glasses can be used for that purpose (Weetall, 1976).

## SURFACE MODIFIED CLAY-MINERALS

Organic clad kaolinites and smectites have found substantial uses in industry for a long time; aluminum ions have significative effects on the surface properties of kaolinite. Crysotile asbestos, the magnesium analogue of kaolinite, has been intensively studied for obtaining derivates in which organic "grafts" have been implanted on the surface of the microfibrils (Rodrique and Brotelle-Pacco , 1975). Polymer encapsulation of "opened" chrysotile fibrils was obtained by Xanthos and Woodhams (1972); they have shown that chrysotile fibrils, surrounded by a rubbery layer, can impart greater toughness to reinforced composites as compared with composites in which the rubbery phase is dispersed at random or isolated from the interface. Similar experiments on "grafted" sepiolite were reviewed by Dr. Serratosa in this Conference. Microencapsulation of dye solutions with natural or synthetic polymers is fundamental for the NCR paper; a review on encapsulation was made by Sliwka (1975). These microcapsules must be covered and protected with a material which absorbs light pressure to avoid crushing during handling of the NCR paper; clay (kaolinite) coated with polymer is used for that purpose.

A common engineering problem in large areas of Quebec and Ontario, Canadá, is the very high mechanical sensitivy of clay soils: it is very likely that the presence of amorphous material as particle coatings has a large effect on the mechanics of these clays: the coatings may be well a major factor in the extreme sensitivy of these soils (McKyes et al, 1974).

Petroleum heavy ends (specially resins and asphaltenes) may be absorbed into clays in petroleum reservoirs and lead to alterations in rock properties: this fact may contribute to the explanation of how similar clay minerals can behave in entirely different ways in different rocks (Clementz, 1976).

## CATALYSTS

Kaolinite and halloysite can easily be transformed into zeolites. Engelhard developed a technology to use its own kaolin clay from Georgia to synthesize zeolites. The first sieve catalyst was offered to the oil refining industry for using in moving bed units in 1965; another major advance was the introduction of microspheroidal catalysts in 1970; these fine, unsupported zeolite catalysts are designed for fluid-bed oil crackers: their activity, attrition resistance and selectivity represented a significant advance over other products. The first fluid catalyst was developed during the 1970's; further informations can be found in trade literature (Anon., 1976a).

## KAOLIN AND BALL CLAYS

Improvements in kaolin beneficiation have already been discussed in this review. The Kaolin Symposia under Prof.Kuzvarth's coordination have contributed significantly to the knowledge of the kaolin deposits of the world and to their industrial uses. Production of coating kaolin, from Jari River, Brasil, has started (Anon., 1977); another kaolin clay, from Capim River, near Belém, Pará, Brasil and used for sanitaryware, is being considered for production of 280,000 tons per year of coating and filler clay.

Ball clays, like kaolins, are having a continuous development of systems of beneficiation. Efforts have been made in order to prepare synthetic mixtures of clays to simulate English ball clays: Phelps and McLaren (1977) developed criteria for the reformulation of bodies containing ball clays, using clays from several parts of the world; ball clays may be prepared by blending a non-plastic kaolin with sepiolite or with very white bentonite, both from Amargosa Valley, Nevada,USA.

## PALYGORSKITE AND SEPIOLITE

Patterson (1975) published a description of the fuller's earth deposits of Georgia and Florida, USA, and reviewed the increasing uses of palygorskite: to clarify and desulfurize lubricating oil; absorbent granules for animal bedding and floors; anticaking agent for fertilizers; suspending solid media for liquid fertilizers and liquid animal feeds; dry gelling clay; diluent for insecticides and fungicides; drilling mud for "salt formations"; pharmaceuticals designed to absorb toxins, bacteria and alkaloids in the treatment of disentery; purification of water; dry-cleaning fluids; dry-cleaning powders and granules; extenders or fillers for plastic , paints, adhesives, sealants, mastics and putties; manufacture of wall paper and of NCR (no carbon required multiple copy paper). In these districts, local uses of palygorskite containing clays exist for lightweight aggregates and cement; it is a component of phosphate slimes. Trade literature provides excellent description of the products above listed.

The industrial uses of sepiolite were reviewed by Robertson (1957) some years ago. Palygorskite and sepiolite deposits in Spain were described recently (Anon. 1976b) . The high purity sepiolite from the deposits of the Amargosa Valley, Nevada, USA, is being used in place of chrysotile asbestos, in deep drilling in which high temperatures exist (Bannerman and Davis, 1978).

## CHRYSOTILE ASBESTOS

All uses of chrysotile are as processed fibers grouped according to the length of fiber in 7 groups; chrysotile is adaptable to more than 2000 uses: 70% of the world's asbestos consumption is connected with products used in building industry; the remaining 30% is divided for the other thousand uses. In late 1976, the U.S. Bureau of Mines established the Particulate Mineralogy Unit at College Park, Md, to help clarify the existing confusion in regard to the properties of asbestos (Ampian, 1976a). The present shortages and projected demands of chrysotile indicate serious depletion of reserves by the end of the century. An excellent review of asbestos problems and technology was made by Clifton (1977). International Conferences on properties of asbestos, similar to AIPEA's, have been held: the last was in Quebec City in 1975, and the next will be in Italy in 1979.

## CERAMICS

Lightweight ceramic bodies of uniform porosity can be made with pyrophyllite after foaming and firing (Rieger, 1977). A "new taylor-made" china clay was developed with more consistent rheological properties to simplify control of casting slips to be used as the sole kaolin in vitreous china sanitaryware bodies (Anon., 1973). High quality refractory chamottes are exported by several countries.

A new process of fabrication of lightweight building block was developed out of coated pellets of expanded clay or shale (Briem and Schaefer, 1978). The differences between primary and secondary mullite from the firing of kaolins and porcelains were discussed by Kromer and Schüller (1976).

Lightweight ceramic fiber materials were developed for high-temperature insulation, as weight-saving refractories, as resilient packings and gaskets and as a filter media. The raw material is kaolin, with small additions of sodium, zirconium and boron (Chemical analysis of the ceramic fiber: $Al_2O_3$ - 45,1%; $SiO_2$ - 51,9%; $Fe_2O_3$ - 1,3%; $TiO_2$ - 1,7%; MgO - trace; CaO - 0,1%; $Na_2O$ - 0,2%; $B_2O_3$ - 0,08%); it is melted in an electric furnace and poured; the molten stream is blasted with a jet of steam; this produces a fluffy, white, cotton-like, bulky ceramic fiber. Lower-priced fibrous materials such as asbestos, glass fibers and mineral wools do a good insulating job in the lower temperature ranges: kaolin fibers may be used from 540°C to 1,260°C.

## PHOSPHOGYPSUM; PORTLAND CEMENT AND POTASSIUM

Anhydrite ($CaSO_4$) is used for the manufacture of sulphuric acid and Portland cement clinker by the well established Müller-Kühne process (Epshtein, 1969): the process, which consists of roasting ground anhydrite, clay (kaolinitic) and pulverized coke or coal at 1,200°C, was reviewed by Gutt and Smith (1973). A pollutant by-product, available in large quantities and which is pumped to sea or lakes, is "phosphogypsum", from production of fertilizer phosphoric acid. The phosphate, present in this gypsum, is deleterious for Portland cement manufacture. Processes were developed, based on the counteracting effect of fluoride, which required that the optimum limits for phosphate and fluoride must be established for each assemblage of raw materials (phosphogypsum, kaolinitic clay and coke). Since 1973, a sulphuric acid plant from phosphogypsum has been producing 350 t/day of $H_2SO_4$ and 350 t/day of clinker in Phalaborwa, Republic of South Africa, by the OSW-Krupp process (Anon. 1975b).

## VERMICULITE

Until 1950, the major demand for pyroexpanded vermiculite occurred in the building industry, for insulation purposes, and as lightweight aggregate for plaster and concrete. About 1950, there developed an awakening to vermiculite's unique chemical and physical properties as they might be applied to other areas: as a significant component in agricultural chemical formulations (weed killers, pesticides and fertilizers); in the food and pharmaceutical industries and in the feed industries. The U.S. use pattern for expanded vermiculite in 1974 is: lightweight aggregates (concrete; plasters; premixes) - 50%; thermal insulation (loosefill; block) - 20%; agriculture (horticulture and soil conditioning; fertilizer carrier) - 18% ;

miscellaneous - 2%. Crude and expanded vermiculite has found increasing uses in steelworks and foundries for its good thermal insulation properties, low density, refractoriness and ease of application with resulting low labor costs: it is generally used as a loosefill, sometimes in conjunction with an exothermic material to save heat and maintain the temperature: in ingot moulds, crude vermiculite is poured straight from the bag on to it, pyroexpanding immediately on contact. Expanded vermiculite in the highly refined form (exfoliated hydrobio - tite), micron size, is a free-flowing, soft absorbent outstanding support material for carrying a wide variety of nutrients: for instance, a free flowing fat concentrate can be made consisting of 70% fat and 30% vermiculite by weight. The use of vermiculite with appropriate fertilizer for hydroponic cultivation has been known and practised for many years in "dry" climates (Anon.,1971c ;1974).

## BENTONITES

Existing laboratory methods for testing and evaluating bentonites are listed by Patterson and Murray (1975). Commercial synthetic mica-montmorillonite is produced by Baroid Industries, in Houston, Texas, USA (Barasym SMM) and synthetic hectorite by Laporte Industries, (Laponites), Bedfordshire, UK (Mining Magazine, 1974).

Bentonite brick was used as liner in deep underground caverns for storing spent atomic waste (White, 1978). Hofmann (1975) studied the problems of recirculating foundry sands when structural changes have occured in the burned part of it. Uses of bentonite were reviewed more recently by Hofstadt and Fahn (1976); a book on bentonite by Grim and Güven is supposed to appear any time this year. The 1st International Congress on Bentonites is scheduled for October 1978, in Sassari, Sardinia, Italy.

## ACKNOWLEDGEMENTS

Due to shortage of the space allowed for this review, it is not possible to make here a proper acknowledgements and a listing of the names of people and companies who kindly supplied the requested information; but, I am extremely thankfull to all of them.

REFERENCES

Ampian,S.G., 1976a. Asbestos minerals and their non asbestos analogs. Review of Mineral Fibers Session of Electron Microscopy of Microfibers, Pennsylvania State University, Univ.Park, Pa., Aug.23.
Ampian,S.G., 1976b. Clays. Mineral Facts and Problems. 1975 Edition, p.253, Bureau of Mines Bull. 667, U.S.Dept.of Interior, Washington, 1976.
Anonymous, 1971a. Astraplate: delaminated coating clay; Kaopaques: delaminated aluminum silicates for paint systems. U.S.Patent 2,904,267, Georgia Kaolin Co. Elizabeth, New Jersey.
Anonymous, 1971b. Ultraflotation. Engelhard Minerals, Chemicals Div., Menlo Park, New Jersey.
Anonymous, 1971c. Zonolite vermiculites, properties and uses. W.R.Grace & Co., Cambridge, Massachussetts.
Anonymous, 1973. NCS New Sanitaryware clay. English China Clays, St.Austell.
Anonymous, 1974. Vermiculite. Mandoval Ltd., London SW1Y4LD.
Anonymous, 1975. ECC Western Area Laboratory Treviscoe. English China Clays,St. Austell.
Anonymous, 1975b. Sulphuric acid and cement from gypsum. Chemical Process Engineering, p.55.
Anonymous, 1975c. Western Area Analogue Computers. English China Clays, St.Austell.
Anonymous, 1976a. Petroleum cracking catalysts. Engelhard Minerals and Chemicals Division, Menlo Park, New Jersey.
Anonymous, 1976b. The industrial minerals of Spain. Industrial Minerals, April, pp.15-51.
Anonymous, 1977. Amazon 88; Coating clay: Amazon 88; $Fe_2O_3$ content of Amazon 88; Dispersion of Amazon 88. Euroclay, Hirschau, West Germany.
Ashton,W.H. and Russel,R.S., 1972. Talc beneficiation. U.S.Patent 3,684,197.
Bannerman,J.K. and Davis,N., 1978. Sepiolite muds used for hot wells, deep drilling. Oil Gas Journal, February 27, pp.142-150.
Briem,K. and Schaefer, G.F., 1978. Zytan, a new building material. Zytam Thermochemische Verfahrenstechnik, Braunschweig, West Germany.
Brindley, G.W., 1976. Current and future trends in clay mineralogy - a review . Clay Minerals 11:257-268.
Brociner,R.E. and Vollans,E.C., 1973. Thickening flocculated kaolinite slurries in the nozzle discharge multiple bowl centrifuge. Clay Min. 10:99-112.
Browning,J.S.; McVay,T.L. and Johnson,A.B., (1975). Continuous flotation of high-clay potash ores. Bureau of Mines R.I. 8081, U.S.Dept.of Interior, Washington.
Bundy,W.M. and Berberich,J.P., 1969. Kaolin treatment. U.S.Patent 3,477,809.
Bureau of Mines, 1977. Minerals in the U.S.Economy. U.S.Dept.of Interior,Washington.
Bureau of Mines, 1976. Mineral trends and forecasts. U.S.Dept. of Interior,Washingto
Clark,N.O., 1978. Future preparation technologies for ceramic raw materials. Paper presented at 22th.Brazilian Ceramic Congress, Rio de Janeiro, Brasil.
Clementz,D.M., 1976. Interaction of petroleum heavy ends with montmorillonite. Clays and Clay Minerals 24:312-319.
Clifton,R.A., 1977. Asbestos. Mineral Commodity Profiles MCP-6. Bureau of Mines, Pittsburgh, Pa., September.
Cundy,E.K., 1969. Clay flotation cells. U.S.Patent 3,450,257.
Curtis,C.E. (Assigned to Esso Research and Engineering Co.),1972. Extreme pressure grease. U.S.Patent 3,639,257.
Davis,E.G.; Collins,E.W. and Feld,I.L., 1973. Large scale continuous attrition grinding of coarse kaolin. Bureau of Mines, R.I. 7771, U.S.Dept.of Interior, Washington.
Dimanche,F.; Rassel,A; Tarte,P. and Thores,J., 1974. The kaolins: mineralogy, deposits, uses. Min.Sci.Eng.6:184-205.
Epshtein,D., 1969. Fundamentos de Tecnologia Química. Editorial MIR, Moscow,368 pp.
Fletcher, A.W.; Pearson,D. and Tron,A.R., 1966. Potash extraction from Cambrian sediments of Northwest Scotland. Miner.Proc.Extr.Metal 75:296-306.
Gutt,W. and Smith,M.A.,1973. Utilization of by-product calcium sulphate. Chemistry and Industry, July 7, pp.610-619.
Griffin,R.A.;Frost,A.K.Au.;Robinson,G.D. and Shimp,N.F., 1977. Attenuation of pollutants in municipal landfill leachate by clay minerals - Part 2: Heavy metal adsorption.Illinois State Geol.Survey,Environmental Geol.Notes nº79,Urbana,Illinois,April.

Gwilliam,R.D.,1971. The ECC filter press. Filtration and Separation,March/April,pp.1-
Greene,E.N.;Duke,J.B. and Hunter,K.,1961. Selective froth-flotation of ultrafine
minerals or slimes. Trans.SME/AIME 223:389-393(1962);U.S.Patent 2,980,958.
Highley,D.E.,1975. Ball clay. Mineral Dossier n? 11, Institute of Geological Sciences
Mineral Resources Div., London SW7.
Hofmann,F.,1975. Structural changes of circulating foundry systems sands. George
Fischer Ltd., Schaffhausen, Switzerland, March.
Hofstadt, C.E. and Fahn,R.,1976. Bentonite: a valuable and versatile mineral and raw
material. "Industrial Minerals"Intern.Congress, München, pp.1-7.
Iannicelli,J.,1976. High extraction magnetic filtration of kaolin clay. Clays and
Clay Minerals 24:64-68.
Kromer,H. and Schüller,K.H., 1975. Primärer und sekundärer Mullit: Verzuch einer
Abgrenzung. Keramische Z.27:625-627.
Kunkle,A.C.(assigned to J.M.Huber Corp.),1976. Electroflocculation cell. U.S.Patent
3,980,547.
Kunkle,A.C. (assigned to J.M.Huber Corp.),1977. Electrokinetic separation of solid
particles from aqueous suspensions thereof. U.S.Patent 4,003,811.
Kunkle,A.C. and Kellman,Jr.,C.E. (assigned to J.M.Huber Corp.),1977. Methods of re-
ducing the viscosity and for refining kaolin clays. U.S.Patent 4,030,941.
Liles,K.J. and Heystek,H.,1977. The Bureau of Mines Test Program for Clay and Ceramic
raw materials. Bureau of Mines,I.C.8729. - U.S.Dept.of Interior, Washington.
McKyes,E.; Sethi,A. and Yong,R.N., 1974. Amorphous coating of particles of sensitive
clay soils. Clays and Clay Min.22:427-434.
Mining Magazine (London),1974. Industrial Mineral Abstracts 130:324.
Mitchell,D. and Stentiford,M.J.,1978. Sanblend 75. Watts Blake Bearne & Co.,Newton
Abbott, Devon.
Mortesen,S. and Hovmand,S.,1976. Particle formation and aglomeration in the spray gra
nulator. In:Keairns;D.L.(Editor). Fluidization Technology. Hemisphere Publ.,Washingto
Vol.II, pp.519-544.
Patterson,S.H. and Murray,H.H.,1975. Clays. In:Industrial Minerals and Rocks,4th.Ed
AIMMPE, New York, pp.519-585.
Phelps,G.W. and McLaren,M.G., 1977. Control parameters of casting slips for sanitary-
ware. Cerâmica (Brasil) 23:64-78.
Robertson,R.H.S.,1957. Sepiolite, a versatile raw material. Chemistry and Industry
pp.1492-1495.
Rieger,K.C.,1977. New Process for porous ceramic bodies. Vanderbilt Exp.Corp.,
Norwalk, Conn.
Rodrique,L. and Brotelle-Pacco,F.,1975. Étude morphologique et texturale de l'asbeste
chrysotile Cassiar AK et des derivés hydrolisés et organiques de ce phyllosilicate.
J.Microscopie 24:217-230.
Shergold,H.L. and Softhouse,C.H.,1977. The purification of kaolin by two-liquid sepa-
ration process. 12th.Intern.Mineral Processing Congress, São Paulo,Brasil, September.
Siemmen,F.G. (assigned to Hall China Co.),1972. Manufacture of disposable ceramic
dishes from high-alkali pyrophyllite. U.S.Patent 3,655,843.
Sliwka,W.,1975. Mikroverkapselung - Angew.Chem. 87:556-567.
Trawinski,H., 1976. Erzielung gewünschter Eigenschaften keramicher Rohstoffe (insbeso
dere Kaoline) durch geeignete Aufbereitungverfahren. Ber.Deut.Keram.Ges.53:117-121.
Van Buren,R., 1974. Kaolin slurries:properties, handling economics. Amer.Paint J.58:
54-58.
Weetall,H.H.,1976. Immobilized enzyme technology. Cereal Food World 21:581-586.
White,W.A., 1978. Personal Communication. Ill.StateGeol.Survey, Urbana, Illinois.
Xanthos,M. and Woodhams,R.T., 1972. Polymer encapsulation of colloidal asbestos fi-
brils. J.Applied Polymer Sci. 16:381-394.

# REGIONAL APPRAISAL OF CLAY RESOURCES - A CHALLENGE TO THE CLAY MINERALOGIST

J.A. BAIN and D.E. HIGHLEY
Applied Mineralogy Unit and Mineral Intelligence Unit, Institute of Geological Sciences, London.

## ABSTRACT

A regional approach to clay resource appraisal demands a ready means of identifying and assessing the basic characteristics of clays which promote their use as industrial raw materials. The principal features and commercial uses of clays in the United Kingdom are discussed in this context and proposals made for laboratory evaluation procedures to support a broad-based reconnaissance programme.

## INTRODUCTION

Regional studies of clays are usually geologically orientated, particular emphasis being placed on the stratigraphical relationships, depositional characteristics and paragenesis of clay deposits. These investigations, unfortunately, provide little information on applied mineralogy of value in assessing the likely economic potential of clay formations as industrial raw materials, although geotechnical studies can provide useful data.

In recent years, the need for basic information to support land-use planning policies, and concern about the future availability of certain industrial minerals, has involved the Institute of Geological Sciences in the regional resource assessment of a number of important commodities. The principal effort has been devoted to sand and gravel resources and, on a more local scale, to limestone. A rapid coverage is possible because the economic parameters, such as particle grading or chemical purity, can easily be identified and the analytical results reduced to numerical factors for statistical treatment.

There is, as yet, no proposal for clay resource evaluation on a regional scale and, because of the wide variety of argillaceous strata encountered and the enormous range of potential applications for clay minerals, any blanket coverage is unlikely to be embarked upon. Nevertheless, the development of laboratory evaluation procedures appropriate to operations at a "reconnaissance" level, and the use of a standard format for recording and analysing results, would seem to be a highly desirable objective. For long-term strategy the latter would also serve as a data-banking facility for information derived from ad hoc investigations arising out of the systematic geological survey mapping programme.

## PRODUCTION AND APPLICATIONS OF CLAYS

Before considering what properties need to be evaluated for clay resource assessment it is instructive to examine, even in briefest outline, the production, uses and characteristics of the important groups of clays currently worked in the United Kingdom. A primary distinction is made between low-value clays, used mainly for captive consumption in the manufacture of structural ceramics, cement, and lightweight aggregates, in which their value is increased severalfold, and those higher-value clays such as china clay and fuller's earth which are sold in the open market and, depending on the end-use, may have been subjected to various degrees of beneficiation or modification.

## CLAYS FOR CAPTIVE CONSUMPTION

Great Britain consumption of these clays and shales, by end-use, in 1976 was as follows:

|  | Thousand tonnes |
|---|---|
| Structural clay products (bricks, tiles, pipes, etc) | 17,045 |
| Cementitious products (Portland cement) | 3,425 |
| Lightweight aggregates (bloated clays) | 561 |
| Constructional uses (civil engineering - bulk fill) | 4,247 |
| Other uses (low-grade fillers, carriers, etc.) | 484 |
| Total | 25,762 |

### Structural clay products

Although there are British Standard specifications for the products included under this heading, to cover dimensional tolerances, strength, water absorption, etc., no such specifications exist for the type of raw material required. They are all formed however, by the application of heat so that the suitability of a clay for this purpose rests primarily on its behaviour on firing. The mineralogical nature of the raw material may provide some clues in this respect but only in a general way and the use of such data may be limited to an appreciation of the adverse effect of impurities such as carbonate, sulphate and sulphide minerals as the likely cause of bloating during firing or the source of lime blowing or efflorescence behaviour in the end product. Carbonaceous matter usually presents difficulties in firing, e.g. in the appearance of black coring if not fully oxidised, but finely-divided carbonaceous matter in the Lower Oxford Clay acts as a natural fuel and is of considerable economic importance to the British fletton brick industry.

The need for fluxing constituents to promote the phase change reactions on firing or induce vitrification is supplied by clays containing alkalies or, to a lesser extent, alkaline earths and ferrous iron, and is usually satisfied by the presence of illitic types. However, it is difficult to predict the precise nature of the colour, texture, and strength of a clay body on firing and, in practice, there is no substitute for actual firing trials, particularly as vitrification range and, perhaps, deformation-under-load may be important characteristics for determining firing schedules.

It is also necessary to consider how the ware may be shaped and formed in the first place, i.e. whether the clay will behave plastically with water or require pressing in the dry state. The presence of particular clay species, or their structural modifications, may suggest a greater or lesser degree of plasticity or indicate possible abnormal behaviour, eg. the high water requirements and subsequent high drying shrinkages associated with montmorillonite, but the characterisation of argillaceous strata solely in mineralogical terms gives no indication of the indurated nature of the rock. In general, this increases with geological age and depth of burial but it is useful, in practice, to be able to quantify plastic behaviour and assess the interrelated properties of drying shrinkage and dry strength. This will indicate whether the shaping of clay ware should, or can, be accomplished in the presence of water and whether adverse behaviour can be ameliorated by mixing or blending.

### Cement manufacture

Except for the manufacture of white cement, where a pure kaolin (low iron) would be used, a wide range of clays can be used as the source of the aluminosilicate in Portland cement clinker. The principal factor is their bulk chemistry and assessment is based on total $SiO_2/Al_2O_3+Fe_2O_3$ and $Al_2O_3/Fe_2O_3$ ratios. Consideration may also be given to the possible deleterious effects of MgO, alkalies and sulphur on the properties of the finished cement.

At a lower technology level the pozzolanic properties of "burnt clay" are of interest as an alternative cementitious material but assessment is difficult on compositional grounds; short of full scale curing trials the principal evaluation factor would probably be a test of lime reactivity of the clay heated to just above its dehydroxylation temperature.

### Lightweight aggregates

Some clays and shales have the natural ability to bloat on rapid heating to temperatures in excess of $1000^\circ C$ to produce an expanded lightweight aggregate. The raw material must have sufficient fluxes to soften at moderately high temperature to a viscous glass capable of trapping gases ($CO_2$, $SO_2$ or $H_2O$) evolved from volatile-bearing constituents. Increasing demand for the product is likely as a replacement for natural aggregates and for its insulating properties, allowing considerable scope for local assessment of raw materials, firing programmes, and possibly the addition of gasifiers.

CLAYS FOR THE OPEN MARKET

The production, exports, and export value of higher-value clays worked in the United Kingdom during the year 1976 were as follows:

|  | Production (thousand tonnes) | Exports (thousand tonnes) | Value f.o.b. (£ thousand) |
|---|---|---|---|
| China clay [1] | 2,533 | 2,062 | 59,489 |
| Ball clay [1] | 670 | 375 | 5,224 |
| Fuller's earth [1] | 163 | 27.5 [2] | 2,491 |
| Fireclay | 1,513 | 16.5 | 1,017 |

(1) saleable product.  (2) includes sodium-exchanged bentonite, natural and acid-activated earth.

## China clay

Although originally valued largely for its use in whiteware ceramics, china clay is now utilised mainly for paper making (80% of U.K. output); subsidiary amounts are employed as a filler in rubber, plastics and paint. Brightness is still a predominant factor in its commercial assessment but except in the case of ceramics (and in the preparation of refractory mullite grog), firing behaviour is less important. Evaluation is based rather on purity and physical properties.

The quantity and nature of impurities determine the amount of processing required to upgrade the clay and the extent to which abrasiveness or excessive viscosity in suspension might be a deleterious factor in the end product. Data on particle size distribution are required for determining processing routes and to assess the relative amounts of "filling" and "coating" grades that could be extracted from the clay for use in paper making. Particular applications place special emphasis on other properties such as oil absorption, ink retention, and high aspect ratio for good coating power.

## Ball clays

Ball clays are used largely in the ceramic industries (75% of U.K. output) where plasticity, high body strength or cohesion, and white-firing characteristics are sought. Although essentially kaolinitic sedimentary clays they contain other constituents which markedly affect physico-chemical properties. Quartz acts largely as a diluent, mica imparts vitrification characteristics, montmorillonite affects water requirements, body strength and casting properties, and organic matter produces significant changes in water absorption, dry strength and texture on firing.

As the raw material is used in bulk, and interactions between component phases are difficult to predict, it is advisable to measure basic properties which will at least assess plasticity and behaviour on firing. Specific marketable grades are prepared selection and blending. Kaolinite-rich material is, in itself, refractory and has special uses in bonding other refractory materials prior to firing.

Fireclay and refractory clays

In the UK the term "fireclay" is used in a generic sense and is largely restricted to seatearths associated with coal-bearing strata. Although kaolinitic and widely used as a refractory material they may also contain varying amounts of quartz and mica, the latter inducing much lower vitrification temperatures and maintaining important applications in the manufacture of vitrified pipes, sanitaryware and facing bricks (which accounted for some 40% of U.K. output in 1976). For refractory purposes, appraisal is based on a determination of "melting point" (pyrometric cone equivalent) or, for initial selection, on $Al_2O_3$ content. In this context, a knowledge of the distribution of the higher-alumina clays may be of future importance in indicating possible alternative sources of $Al_2O_3$ to bauxite.

Fuller's earth and bentonite

In the U.K., deposits of smectite clays are represented only by the calcium-montmorillonite form, fuller's earth. Its pronounced absorptive capacity is utilised in such products as pet litter granules while high surface activity - enhanced by acid treatment if appropriate - is used in refining (bleaching) of glyceride oils and in various catalyst reactions. An important commercial feature, however, is its amenability to sodium exchange and the production and performance of the more highly-prized bentonite analogue. The latter's high viscosity, thixotropy and wall-sealing properties in suspension find important applications in oil-well drilling, foundation engineering and various manufacturing fields. The high bonding strength of montmorillonite is exploited in the mineral's major markets as a bonding agent for foundry moulding sands and for pelletising iron ore concentrates.

Attapulgite and sepiolite

These clays are not found in commercial quantities in the United Kingdom; although structurally and morphologically quite distinct from the smectites they have similar applications. Demand is currently dominated by their high absorption capacity for oil, grease and animal waste products, but they have important suspending and gel-forming properties in water, showing stability even at high electrolyte concentration.

LABORATORY EVALUATION OF CLAYS

Laboratory evaluation would commence with a preliminary inspection of hand specimens for colour and texture (Fig.1) but thereafter it is necessary to establish composition in greater or less detail, not least as a check on the likely existence of non-clay constituents of potential industrial value-gibbsite, alunite, mineral pigments, etc. Clay identity is usually established by X-ray diffraction but it is normally sufficient to determine whether only one clay phase is present or whether two or more clay species are involved.

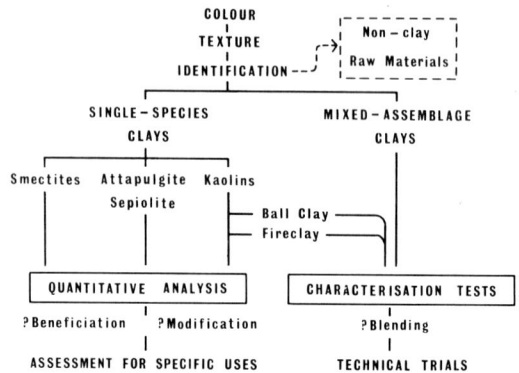

Fig. 1. Laboratory evaluation of clays

A primary subdivision into single-species and mixed-assemblage clays is a useful one in that commercial interest in the former places considerable emphasis on "grad in terms of clay content, and on the identity of the associated minerals. Composit and texture may be exploited to purify the crude clay, or a range of marketable qualities prepared by modifying textural, physical or chemical characteristics. Illitic and chloritic species are of commercial interest only as constituents of mixed-assemblage clays. In this case, however, mineral identity is less important than the properties of the material in bulk. Blending may be undertaken but the clays are not amenable to beneficiation on a commercial scale.

This two-part subdivision of industrially-useful clays is further justified by t earlier distinction made between clays sold on the "open market" and those worked f "captive consumption", where similar compositions were involved. Ball clays and fireclays, however, are more properly considered to be mixed-assemblage clays rathe than single-species clays containing kaolinite only.

APPRAISAL AT "RECONNAISSANCE LEVEL"

Exploration for a single commodity or type of clay is conducted with particular qualities or applications in mind so that laboratory evaluation can be tailored to suit and specific technical testing introduced at an early stage in the investigatic A regional survey of clay resources is executed largely as a reconnaissance operatic A supporting laboratory must be able to define relatively broad categories of potentially-useful clays, find simple cost-effective methods of determining quantitative composition and supply relevant basic data for assessment of applied properties. High throughput, in terms of sample handling capacity, may be an important consideration.

Quantitative mineralogical analysis

Although unsurpassed for identification purposes, X-ray diffraction is not sufficiently reliable for quantitative analysis of clays. For routine treatment of the single-species clays it is found more convenient to rely on rapid "assay" techniques, utilising a property specific to the mineral for assessing its concentration. Isothermal heating (to determine hydroxyl content) is used for estimation of kaolinite and associated minerals such as gibbsite, adsorption of methylene blue (as a measure of cation exchange capacity) or of a monolayer of a polar molecule (to determine surface area) for montmorillonite and body porosity or heat of wetting for attapulgite and sepiolite.

Discussion of these techniques is beyond the scope of this paper but, in essence, they provide a practical means of "screening" large numbers of prospecting samples in order to eliminate those below cut-off grade or select material suitable for more detailed study of quality or amenability to beneficiation. They are less useful for the mixed-assemblage clays although variations in the triple assemblage of quartz, kaolinite and mica typical of ball clays may be monitored solely by rapid chemical analysis for total $SiO_2$ and $K_2O$.

Physico-chemical properties

The properties which determine end-use or market value of a single-species clay, eg. the whiteness of kaolin or the drilling mud yield for bentonite, are usually fairly well circumscribed and may often be measured against specifications. The use which may be made of a mixed-assemblage clay may only emerge after extensive technical trials or manufacturing tests. At the reconnaissance level this can only be tackled by confining investigations to a few basic "characterisation" tests. From the earlier discussions on utilisation of clays it is obvious that these should be based on clay/water relationships and on firing behaviour.

SMALL-SCALE TECHNICAL TESTING

The procedures currently used within the Institute to provide basic physico-chemical data for clays are shown in diagrammatic form in Fig.2. The important property of plasticity is assessed by determination of the Atterberg "plasticity index" - the numerical difference between the minimum moisture contents required for liquid flow (Liquid Limit) and plastic moulding (Plastic Limit). The plastic limit itself is related to drying shrinkage but the latter requires a measurement of body porosity for its precise evaluation. This is done by taking a small hand-moulded specimen from the clay at its plastic limit consistency and measuring its volume by liquid displacement in kerosene and mercury before and after drying at $105°C$ (right-hand side of the diagram).

For preliminary investigations, or where sample size is limited, the same specimen may be used for a single firing at a standard temperature, conveniently taken to be $1050°C$, in order to assess colour and obtain a figure for ignition loss. A further mercury displacement measurement determines volume, for calculation of firing shrinkage, and a water absorption measurement (under vacuum) determines body porosity. Although simple in concept these measurements are consistently reproducible, may be carried out in batches by non-technical staff, and are very sensitive to changes in composition and texture in the raw material.

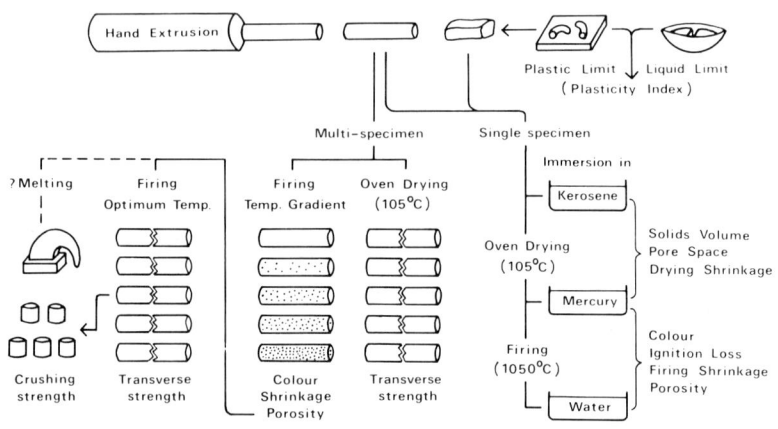

Fig. 2. Small-scale technical testing of clays

More detailed appraisal, particularly in the assessment of bonding strengths and the progressive development of properties with firing temperature, requires a facility for multi-specimen handling. To ensure consistency in specimen preparation standard rods are made by hand-extruding the clay at its plastic limit moisture content through a ½" (1.25cm) diameter nylon die and dividing the extruded column in 2" (5cm) lengths with a double wire cutter. After drying, a number of specimens are used for determining cross-breaking strength with a tensometer in order to obtain a modulus of rupture value as a measure of dry strength.

As the development of properties on firing is principally influenced by fluxing behaviour, the effect of high-temperature transformation is most easily monitored by assessing the progress of vitrification - in practice the gradual elimination of pore space in the body. Specimens fired at 50°C intervals from 950°C or 1000°C upwards, accomplished most effectively in a temperature gradient kiln, are used for porosity measurements by water absorption. The same test specimens define colour changes and if mercury displacement measurements are made before and after firing, they provide values for firing shrinkage.

If porosity data are used to determine an optimum firing temperature for production of clay ware, further investigation may be limited to a single firing trial to provide specimens for strength measurement. This may be determined as modulus of rupture but crushing strength is a more meaningful parameter for fired ware and may be determined on small 1cm thick slices cut from the fired rods with a rock saw. Provision is also made for determining refractoriness by a pyrometric cone squatting test.

RECORDING AND INTERPRETATION OF RESULTS

To cope effectively with the recording and interpretation of data from clay resource appraisal, it has been found necessary to design a data sheet with provision for both tabulation and graphical display. One half deals with location and geological association, colour and texture, mineralogical composition, and chemical analyses (as appropriate). The other is devoted to physico-chemical properties, with two charts, reproduced as Figs. 3 and 4, to deal with "wetting-and-drying" behaviour and firing behaviour. The data illustrated were obtained from a raw material of typical ball-clay composition but containing sufficient detrital feldspar and a mixed-layer clay phase to effect significant changes in bulk properties.

Clay/water relationships

Use is made of plasticity index as a guide to the workability of a clay (Fig.3), past experience having defined an optimum range for good moulding properties (centre rectangle); rather lower values are acceptable for stiff extrusion while higher values are needed for soft-mud working or hand throwing. The other parameter, plastic limit, defines a position on the chart which indicates the likely extent of shrinkage on drying; for moulding of ceramic ware excessive shrinkage would have to be modified or the clay rejected as a raw material.

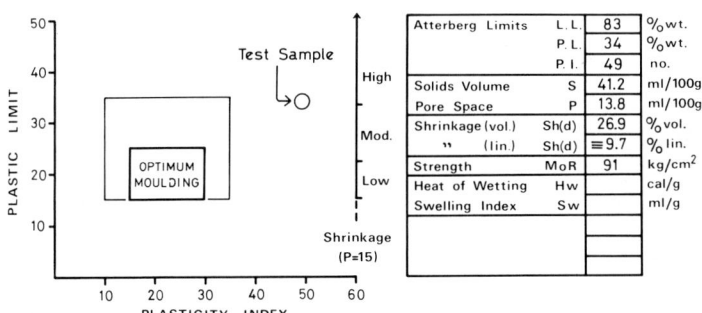

Fig. 3. Wetting and drying behaviour of clays

For a given clay mineral assemblage a good correlation is usually obtained between plasticity index and dry strength but the relationship is not sufficiently consistent to be assessed quantitatively on the chart. The example illustrated is highly plastic (almost sticky in consistency) and from its position on the chart should exhibit high dry strength and moderately-high drying shrinkage. Values actually obtained for these properties are recorded in the boxes to the right.

For characterisation purposes Atterberg Limits have the additional advantage of being important geotechnical parameters. They also have uses in evaluating the single-species clays, eg. in assessing absorptive capacities and (with the addition of $Na_2CO_3$) amenability to sodium exchange. Swelling index and heat of wetting are

occasionally useful for the same reason. Space is provided elsewhere on the form f
recording the results of viscosity, thixotropy and filtration tests when these are
required for assessing drilling mud yields, casting behaviour, etc.

Firing behaviour

The chart in Fig. 4 utilises porosity measurements to follow vitrification
behaviour during firing and to assess the extent of densification at particular kil
temperatures. The sample illustrated can be made to fire to a highly-vitrified, de
impermeable body at 1150-1200°C although shrinking appreciably in the process.
Overfiring produces slight bloating. A steep gradient on the curve signifies rapid
vitrification which may induce loss of shape in the ware and require very close
control of kiln temperatures. A shallow slope indicates a fairly long safe firing
range, as is the case in the example shown. A melting point determination assesses
the refractory nature of the material.

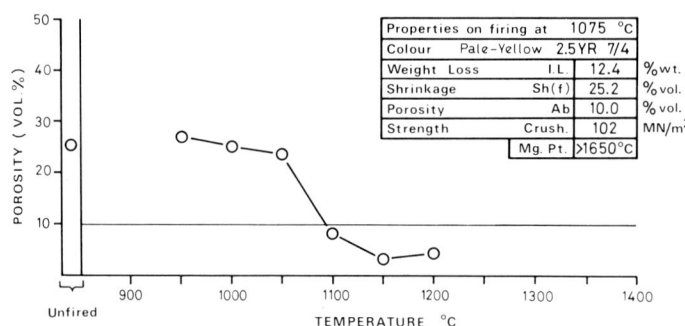

Fig. 4. Firing behaviour of clays

The boxes on the right are useful for recording preliminary results from a singl
heating experiment at 1050°C. In this case, they record test results from a follow
up firing at 1075°C to measure properties of a 10% porosity body.

The provision of even a limited amount of applied data for clays enables the
mineralogist to follow trends in behavioural characteristics with lateral
distribution and depth and to assess the possible effect of compositional changes o
uses.

ACKNOWLEDGEMENTS

The technical support of Mr F.R. Stacey is acknowledged for ceramic studies. Th
paper is published by permission of the Director, Institute of Geological Sciences.

# THE QUANTITATIVE DETERMINATION OF QUARTZ IN CLAY MIXTURES BY INFRA-RED SPECTROSCOPY

R.H. ANDREWS, J.A. GIBSON and I.M. SHAW
Hepworth Iron Company,
Ceramic Research Department,
Hazlehead, Stocksbridge, Sheffield, S30 5HG, England.

## ABSTRACT

A quantitative infra-red technique for the routine determination of quartz in clay minerals has been developed. The method relies on careful sample preparation (such as grinding and calcining), observation of the quartz absorbance at 696 cm$^{-1}$ and relating its intensity to that of a mull oil peak at 732 cm$^{-1}$. The method is accurate over a wide range of quartz contents and for a wide range of clay types and sources. Calibration curves are obtained from synthetic mixtures of calcined china clay and high purity quartz. Data is handled and processed by digital computing techniques.

## INTRODUCTION

Knowledge of the quartz content of heavy clay materials is essential for the quality control and blending of clay stocks. Quartz content affects the making, drying, and firing characteristics of clayware, and is thus an important control parameter.

Infra-red spectroscopy is a powerful laboratory tool for the identification of many mineral compositions. (Farmer, 1974; Van der Marel et al., 1976). The technique may be used on very small sample weights, and airborne dust analyses have been made on samples of 1 to 2 mg containing 10 to 100 $\mu$g of quartz (Larsen et al., 1972). There are, however, several significant disadvantages to infra-red spectroscopy. For example overlapping of vibrational modes due to different mineral constituents may occur and the natural bandwidths may not permit resolution of individual bands. Whereas bandwidths of gases can be narrow and rotational fine structure observable, the bandwidths of solids are often broad and relatively structureless.

Analyses of solids are difficult to establish on a quantitative basis and results are subject to a number of sources of error. For example the

Table 1. The estimated Mineralogical composition of some European clays

| Source | Kaolinite | Mica | Chlorite | Montmorillonite | Siderite | Pyrite | Org.C | Anatase | Qtz | Total |
|---|---|---|---|---|---|---|---|---|---|---|
| Westphalian A Mudstone [a] | 17 | 44 | 3 | n/d | 2.5 | 0.1 | 0.4 | 1.0 | 32 | 100 |
| Westphalian A Seatearth [a] | 24 | 42 | 2 | n/d | 7.6 | 0.2 | 0.5 | 1.0 | 23 | 100.3 |
| Westphalian B/C Seatearth [a] | 57 | 29 | 0 | n/d | 0.2 | 1.0 | 0.9 | 1.2 | 10 | 99.3 |
| Oligocene (Koblenz W. Germany) [a] | 34 | 25 | 4 | n/d | 0 | 0.1 | 0.1 | 1.4 | 34 | 98.6 |
| Tertiary 1 Belgium [a] | 34 | 23 | 1 | n/d | 0 | 0.1 | 0.5 | 1.2 | 35 | 94.8 |
| Tertiary 2 Belgium [a] | 16 | 19 | 2 | n/d | 0 | 0.1 | 0.2 | 0.8 | 56 | 94.1 |
| Lower Pleistocene 1 [b] | 10 | 10 | 2 | 15 | n/d | n/d | n/d | 0.7 | 60 | – |
| Lower Pleistocene 2 [b] | 15 | 15 | 2 | 20 | n/d | n/d | 0.2 | 1.0 | 43 | – |

[a] clay mineral % determined by rational analysis of chemical data. after I.R. + X.R.D. qualitative assessments of the mineral phases present.

[b] clay mineral estimates based on whole rock X.R.D. data.

n/d not detected.

Beer Lambert law is only obeyed as the particle sizes become very small (<10μm) and should ideally be smaller than the wavelength of the vibrational band under study (Van der Marel, 1966). There is considerable difficulty in the determination of the baseline in a double beam instrument. Since true absorbance measurements must be referred to this baseline large sources of error may be generated. Finally where determination of pathlengths is necessary this can be difficult since, for solid samples mulled in oil, these are typically very small, <100μm, and difficult to measure accurately.

Many of the above problems have been considered by other workers who have determined quartz quantitatively by infra-red spectroscopy. Larsen et al., (1972) used the KBr disc method with quantitative measurement of the quartz absorbance at 800 $cm^{-1}$. Sample preparations involved the use of a low temperature filter asher followed by mixing of the small quantity of ash with KBr and pressing of a pellet. An internal standard was not present and quartz figures were determined directly from absorbances by the use of a standard reference curve. Standard deviation was $\pm$ 0.38 on a quartz content of 2.54% (i.e. $\pm$ 15% absolute). A non-dispersive technique using polyvinylidene chloride windows (actually the material used as dust filters) enabled Gillieson and Farrel (1971) to measure airborne quartz but the precision was low.

Radulescu (1976) derived a series of correction equations, based on measurements of artificial mixtures, which could be used to estimate quartz in the presence of kaolin and orthoclase. Correction coefficients were also necessary when Flehming and Kurze (1973) determined quartz in the presence of illite, chlorite, kaolinite and calcite.

The requirements of routine clay control in this laboratory necessitated the development of a fast, accurate, and reproducible infra-red quartz determination. Mineralogical analyses and quartz contents of some typical clays which have application in the heavy clayware industry are shown in Table 1.

TECHNIQUE

Initial trials using KBr pressed disc methods indicated un-acceptable precision and long sample preparation times. Difficulties in baseline assignment necessitated the use of spiking techniques or inclusion of internal standards which further increased the sample preparation times.

Since satisfactory infra-red spectra can be obtained by the mull oil technique, this method was investigated, particularly with regard to the use of the mull oil as a quantitative internal standard. The infra-red spectra of quartz (99.6% purity, fine ground, British Chemical Standards) and mulling oil (commercial liquid paraffin, B.P.) are reproduced in Figure 1.

After considerable work it was decided to adopt the 732 cm$^{-1}$ mull oil peak and the 696 cm$^{-1}$ quartz peak for quantitative measurements.

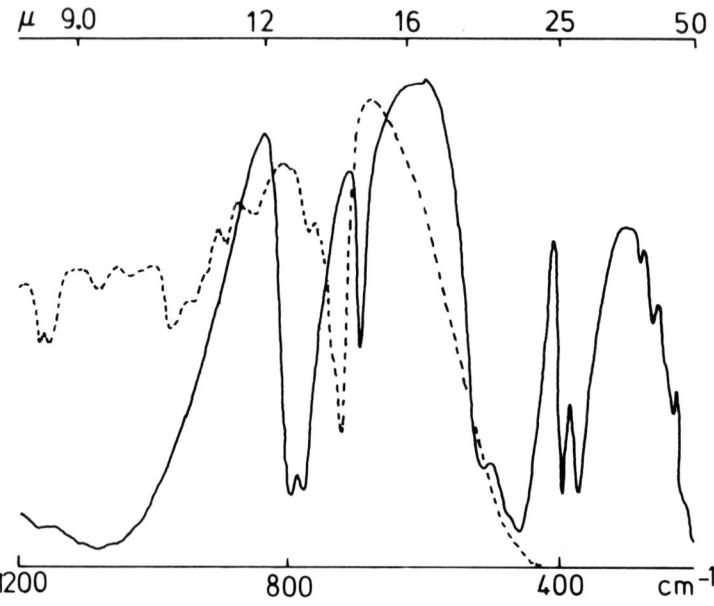

Figure 1. The Infra Red Spectra of Quartz (KBr disc) and Mulling Oil. (Broken Line, NaCl plates).

From work carried out in this laboratory and examination of work by Farmer (1974) and Van der Marel et al., (1976), it is found that remarkably few clay minerals absorb strongly in the 670 to 740 cm$^{-1}$ region. Materials such as kaolinite have broad peaks in this region which may be removed by calcining. Feldspars have very weak absorptions in this region which are not removed by calcining and consequently a check must be made that the particular clay mix under investigation does not contain large quantities of feldspars. Feldspar contents of <4% do not significantly affect the results and few of the materials common to the heavy clay industry contain feldspars at higher concentrations.

A suitable sample preparation thus involves grinding and calcining steps and quartz is determined as a % weight of calcined clay. The infra-red spectrum of a fine ground (<10 μm) calcined blended clay, as a mull between KBr plates, is reproduced in Figure 2. The quartz band 696 cm$^{-1}$ is clearly discernible and is of convenient intensity in relation to the mull oil band at 732 cm$^{-1}$. The detailed sample preparations and method for calculation of quartz contents are given below.

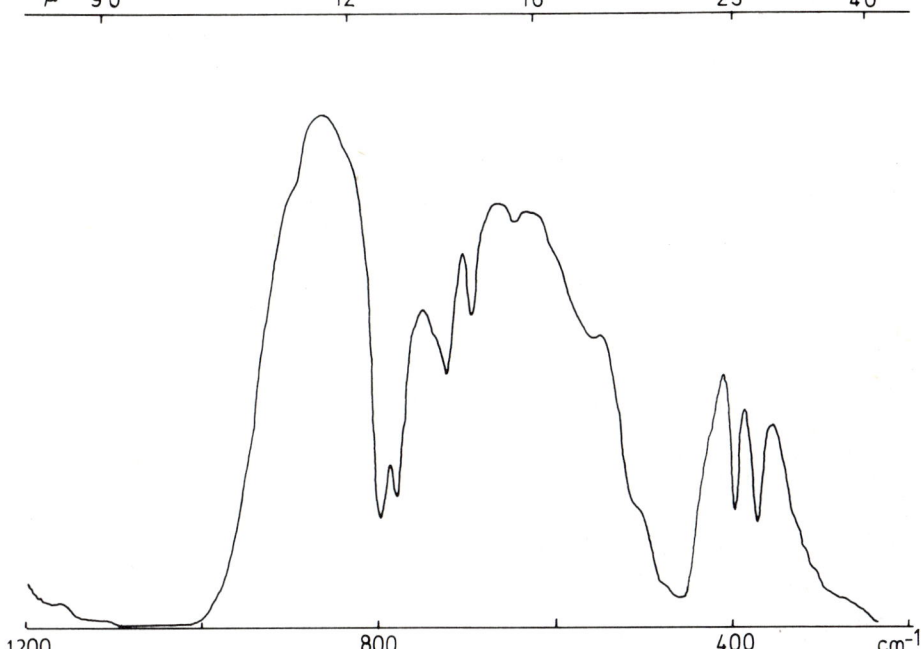

Figure 2. The Infra Red Spectrum of a Fine Ground Calcined Blended Clay in a Mull Oil Dispersion (KBr plates).


Infra-red Absorption by Dispersed Solids

For solutions of solutes in solvents the Beer Lambert law predicts the absorbance due to a given band in a spectrum.

$$I = I_o\, e^{-\varepsilon c l} \tag{1}$$

Where I = Transmitted light intensity
Io = Incident light intensity
$\varepsilon$ = extinction coefficient
c = concentration of solute
I = pathlength
$\therefore$ $A = \varepsilon c l$ (2)

Where $A = \log I_o/I$ (3)
A = absorbance

However, if a solid dispersed in a liquid media (e.g. mulling oils) is considered then it may be shown (Van der Marel, 1966) that the absorbance is a complex function of the particle diameter.

At very small particle diameters a less complex relationship exists which closely models the Beer Lambert law.

In practice to avoid such particle size factors as the Christiansen effect and wavelength selective Tyndall scattering (Van der Marel, 1966) it is necessary to use particle sizes of 2 to 10 $\mu$m. A good Beer law fit is produced and unpredictable and variable particle size effects are eliminated.

## Sample Preparation

Samples ground to c.a. 150 $\mu$m were calcined at 950° for 15 minutes prior to fine grinding. Sub 10 $\mu$m levels were achieved with a McCrone Associates Micronising Mill using Agate grinding elements and an isopropanol grinding medium. Particle sizes of 10 $\mu$m were achieved on 3g. samples with grinding times of 10 minutes. Samples were then oven dried (110°) to remove isopropanol and traces of moisture. Calcined clays prepared in this manner gave good quality reproducible spectra and no deviations from Beer law or other samples size effects were found.

Approximately 50 mg of the samples, accurately weighed, were mixed with 200 mg of oil, accurately weighed, and carefully mulled to give a fine consistent dispersion. Sufficient of the mull was then placed between two KBr plates fitted with a 0.05mm lead gasket. The spectra were scanned over the region 745 - 680 cm$^{-1}$ using a Perkin Elmer 577 Grating spectrometer at a sweep rate of 1 cm$^{-1}$/second and a time constant of 2 seconds. The quartz peak at 696 cm$^{-1}$ was clearly resolved from the mull peak down to quartz contents as low as 5% when only a shoulder was discernible. Transmission scale expansions of X5 were normally utilised.

## Calculation of Quartz Content

Since the absorbances of the mull oil at 732 cm$^{-1}$ and the quartz at 696 cm$^{-1}$ can be ratioed as a function of their weights it is immediately obvious that accurate determination of pathlength is not necessary. Using a Beer Lambert expression equation (4) may be readily derived.

$$\% Q = \frac{A_{696}}{A_{732}} \cdot \frac{W_m}{W_s} \cdot K \tag{4}$$

Where $A_{696}$ = absorbance of quartz at 696 cm$^{-1}$

$A_{732}$ = absorbance of mulling oil at 732 cm$^{-1}$
Wm = weight of mulling oil in prepared sample
Ws = weight of calcined clay mix in prepared sample
K = constant derived from extinction coefficients of mulling oil and quartz

Equation (4) would only be directly applicable if absorbances were accurately determinable from a known baseline and the absorbances at 732 cm$^{-1}$ and 696 cm$^{-1}$ were solely due to the mulling agent and quartz, respectively. In respect of the first criterion it is very difficult to decide with reasonable accuracy where a baseline should be drawn and errors of $\pm$ 15 - 20% would not be unexpected. Secondly since the peaks at 732 and 696 cm$^{-1}$ partially overlap, equation (4) would not strictly hold, i.e. the intensity of both absorbances is a combined function of the quartz and mull concentrations.

An empirical approach was adopted and an arbitrary minimum was taken as the saddle-point, B, between the mull and quartz peaks (A and C) shown in Figure 3. Intensity parameters AB and BC were measured in linear transmission units and a factor, F, was derived from equation (5).

$$F = \frac{Wm}{Ws} \times \frac{BC}{AB} \qquad (5)$$

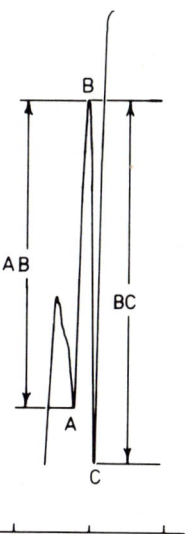

Fig. 3. Mull Oil Absorbance at 732 cm$^{-1}$ and Quartz Absorbance at 696 cm$^{-1}$.

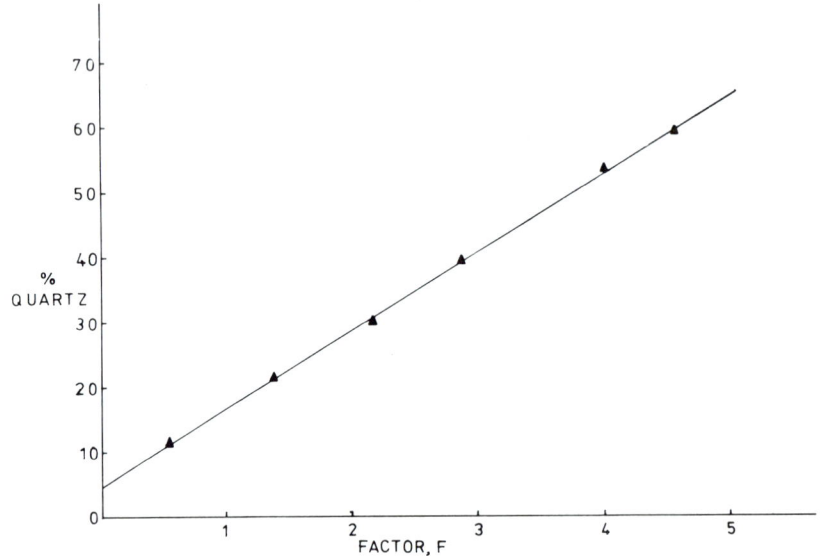

Figure 4. Plot of Factor, F, Versus Quartz Content for Synthetic Calcined China Clay/Quartz Standards.

Least squares linear regression gave equation (6), relating factor, F to the quartz content. An expected non-zero intercept at ~5% quartz was found. Although measurement of quartz below 5% is not possible by this technique (c.f. the work of Larsen et al., 1972) the method is applicable over the entire range 5 - 100% quartz.

% Q = 4.443 + 12.061 F  (6)

Quartz contents could thus be determined from this standard line on calcined clay samples with a precision of $\pm$ 4%, absolute. For example typical blended clays with quartz contents 30% are determined with a precision of $\pm$ 1.25% (one standard deviation). This technique appears to be a marked improvement on other published methods and compares favourably with lengthy wet chemical techniques. Equation (6) only applies to a given mull oil as various commercially available mulling agents show different components in their paraffinic make-up.

Computer Analysis of Results

A further refinement has been provided by the computer analysis of the output from the infra-red spectrometer converted from analogue to digital form. For rapid data collection a Dataforce Datalogger is used which records information onto cassette tapes for replay through a peripheral into the memory of a Wang 2200 system. The intensity of the peaks is logged every 0.4 seconds as a voltage (0-1000 m.v.) and one run over the 745 - 680 $cm^{-1}$ region involves collection of c.a. 160 data points. The data, which also includes sample name codes and weights of mull oil and sample, is fed to a programme written in BASIC which sorts the sample code, scans the digitised spectrum to find A, determines C and B and calculates and outputs the result directly as % Quartz content in table form.

CONCLUSIONS

A method for the precise and rapid determination of quartz in the range 5 - 100% in clay materials has been presented. There is a considerable improvement in precision compared with other published work and standard deviations of $\pm$ 4% (absolute) are obtainable on a routine basis.

REFERENCES

Farmer, V.C., Ed., 1974. The Infra-red Spectra of Minerals. Mineralogical Society, London.
Flehmig, W. and Kurze, R., 1973. Die quantitative infrarot-spektroskopische Phasenanalyse von Mineralmengen. Neues Jahrb. Mineral., 119: 101-112.
Gillieson, A.H. and Farrell, D.M., 1971. The determination of quartz in dusts by infra-red spectroscopy. Canad. Spectrosc., 16: 21-26.
Larsen, D.J., Von Doenhoff, L.J. and Crable, J.V., 1972. The quantitative determination of quartz in coal dust by infra-red spectroscopy. AM. Ind. Hyg. Assoc. J., 32: 367-372.
van der Marel, H.W., 1966. Quantitative analysis of clay minerals and their admixtures. Contr. Mineral, Petrol., 12: 96-138, and references therein.
van der Marel, H.W., Beutelspacher, H., 1976. Atlas of Infra-red Spectroscopy of Clay Minerals and their Admixtures. Elsevier, Amsterdam.
Radulescu, N., 1976. Nachweis and quantitative Bestimmung von Quartz in Vielkomponenten - Mineralgemischen durch Infrarot - Spektroscopie (Quartz, Kaolin, Orthoklas), Staub-Reinhalt. Luft, 36: 195-201.

# SORPTION PROPERTIES OF CONSOLIDATED AND COMPRESSED CLAYS

E.T. STĘPKOWSKA

Instytut Budownictwa Wodnego PAN, Gdańsk, Poland

## ABSTRACT

Studies of water sorption (WS), water retention (WR), and thermal weight loss were performed on bentonite, kaolin and two natural clays consolidated isotropically and compressed axially. WS and WR at $p/p_o$ = 0.95 and 1.0 are influenced by the values of consolidation pressure and of axial pressure, indicating changes in clay structure e.g. preferred orientation of crystallites and interaggregate water content. WS and WR at $p/p_o$ = 0.5 indicate increase in crystallite thickness, $\delta$, at low pressure ($\sigma_c < 0.05$ MPa) and decrease in $\delta$ at higher pressure ($\sigma_c > 0.2$ MPa). Pressure exceeding 0.4 MPa results in changes of interlayer structure, so that less thermal energy is necessary for escape of both sorbed and combined water. Prolonged storage in the wet state decreases the temperature of thermal weight loss.

## INTRODUCTION

The consolidated clay (smectite) is considered to consist of clay crystallites of about 100 Å to 200 Å thickness, which form parallel arrays within domains of about 1000 Å to 2000 Å thickness. These domains are in random mutual orientation within aggregates.

Thus we can distinguish three types of water: 1/water sorbed on internal and external crystallite surfaces, 2/water between the parallel crystallites which contains the diffuse layer cations, 3/interaggregate (macropore) water, which may or may not be present depending on the clay mineral and the preparation.

Comparison of calculated crystallite interactions with the measured strength of bentonite (Stępkowska, 1977a and b) indicates that during the deformation process crystallite thickness may be altered.

Clay crystallites are composed of several layers held together by ionic lattice attraction. The force between layers in clay crystallites has zero component parallel to the layer surface. The distance between layers depends on the relative water vapour pressure in the system. It is highly probable that pressure of a certain value and a certain stress distribution in the clay structure may result in layer separation and change of crystallite thickness. Change in grain size has been observed in sand or granite samples subject to pressures exceeding 5 MPa ( Feda, 1977 ).

Water sorption and water retention at $p/p_o$ = 0.5 is proportional to crystallite external specific surface, $\bar{S}$, and inversely proportional to crystallite thickness, $\delta$ (Stępkowska, 1977c).

Water sorbed at higher relative vapour pressures, $p/p_o$ = 0.95, is a measure of interlayer water within the crystallites and on their surface whereas water retention depends also on clay structure and especially on macropore (interaggregate) water content, $W_{mac}$.

The complete sorption test consists in storing the undisturbed sample at certain relative water vapour pressures followed by heating it within various temperature ranges. The same procedure was applied here: previous study has indicated that thermal weight loss depends on the sample preparation method.

## MATERIALS AND METHODS

Two predominantly monomineral clays, bentonite and kaolin, and two natural clays were subjected to strength and sorption study. Their properties are presented in Table 1.

Air-dried, powdered and sieved samples (except undried O.clay) were mixed with distilled water to obtain water contents close to the liquid limit, $W_l$, and they were stored at least four weeks ( bentonite and kaolin were stored for over one year ).

Cylindrical samples were formed and two slices were cut from the bottom part. Samples ( h = 8 to 9 cm, $\phi$ = 3,6 cm ), protected by a rubber membrane, were subjected to a hydrostatic consolidation pressure, $\sigma_c$, which was increased in a stepwise manner from 0.025 to 0.5 MPa ( 0.25 to 5.0 kG/cm$^2$ ) and continued until the termination of water escape.

Samples were unloaded, preventing the water uptake, while measuring the negative pore water pressure, $u < 0$. They were compressed axially, at atmospheric lateral pressure and at constant strain

TABLE 1. PROPERTIES OF CLAY SAMPLE DISCUSSED.

| Clay (locality) | Specific gravity $\gamma$ g/cm³ | Shrinkage limit $W_s$ % | Hygroscopic w.c. $W_h$ % | Plastic limit $W_p$ | Liquid limit $W_L$ | Grain size [%] areometer analysis | | | Content of [%] | | | | Crystallite external surface $\bar{S}$ m²/g | Crystallite thickness $\delta$ Å | C.E.C meq/g |
|---|---|---|---|---|---|---|---|---|---|---|---|---|---|---|---|
| | | | | | | >0.05 | 0.05–0.002 | <0.002 | Mo (smectite) | III (mica) | Kl | other | | | |
| 1 | 2 | 3 | 4 | 5 | 6 | 7 | 8 | 9 | 10 | 11 | 12 | 13 | 14 | 15 | 16 |
| BENTONITE (Z.M.) | 2.63± 0.14 | 30.1 | 33.5 | 77.6 | 96.5 | 35.5 | 44.5 | 20 | 94.7±2*/ | 3 | | 1.5% CaCO₃ | 127 +11 -6 | WST: 123 +2 -5 XRD: 100-160 | 1.29 1.30 (Ba²⁺) 1.08 (NH₄⁺) |
| KAOLIN SEDLEC | 2.66± 0.03 | 28.2 | 10.75 | 44.0 | 69.0 | 1.2 | 48.8 | 50 | 6.5 | 13 | 73 | | 10.0±0.6 | WST: 82±4.5 XRD: 150±250 | 0.101 0.078(0.06–pH=3.4) (0.148–pH=10) |
| Natural clay B. (BEŁCHATÓW) | 2.71± 0.08 | 14.0 | 3.46 | 14.6 | 19.5÷22.5 | 74 | 13 | 13 | 7.5±1 | 45 | <20 | 10% felsp. 8% CaCO₃ 7% Q | 11.6±1.5 | WRT: 720±50 XRD: 650-1200(III) 630-900(Kl) | 0.118 |
| Natural clay O. (OPALENIE) Nr.19 | 2.85± 0.03 | | 14.6 | 39.7 | 64.0 | 0 | 40 | 60 | 24.2 +1.5 -2.5 | ca.~25 | 27 | | 35.9 +5.4 -3.5 | WRT: 274±20 XRD: 250-320 | 0.364 |
| Nr.50 | 2.90± 0.09 | | 14.0 | 30.8 | 64.5 | 21 | 25 | 54 | 25.4±2.5 | ca.~25 | 32 | | 38.2 +6.3 -3.4 | WRT: 262±25 | 0.387 |

*/ Z.M. bentonite contains about 5% vermiculite in mixed layers with montmorillonite

rate ( 1.6 mm/hour ). They failed either by bulging or a distinct shear plane was formed at elevated axial pressures.

After unloading three pairs of slices, each of 1.5 – 3 g, were cut from the deformed cylindrical sample: from the top, bottom and from the shear zone. One sample from each pair was subject to a water retention test ( WRT ) and the other to a water sorption test ( WST ).

WRT: Samples in flat ceramic containers were stored successively: 1) for 3 weeks in a desiccator over distilled water at $p/p_o = 1.0$, 2) for 2 weeks over 10% $H_2SO_4$ solution at $p/p_o = 0.95$ and 3) for 10 days over saturated $Mg(NO_3)_2$ solution at $p/p_o = 0.5$.

WST: Samples were dried for one day in 105°C and stored successively for the above mentioned times in the opposite sequence of $p/p_o$ conditions.

The weight change at every test step was measured in both tests.

After the termination of the sorption or retention process all the samples were heated successively at 105°C ( one day ), 220°C ( one day ), 400°C ( 6 hours ), and 800°C ( one hour ), measuring the weight loss, $\Delta G$.

WST and WRT results are presented in Fig. 1a,b and 2a,b. $\Delta G$ results are presented in Fig.3a,b,c,d. WST results of sieved powder samples and oriented samples ( obtained by drying a dispersed suspension ) of Ca bentonite and kaolin are also presented for comparison.

From the water sorption at $p/p_o = 0.5$ related to sample weight at 220°C the crystallite external specific surface was calculated ( Stępkowska, 1977c ).

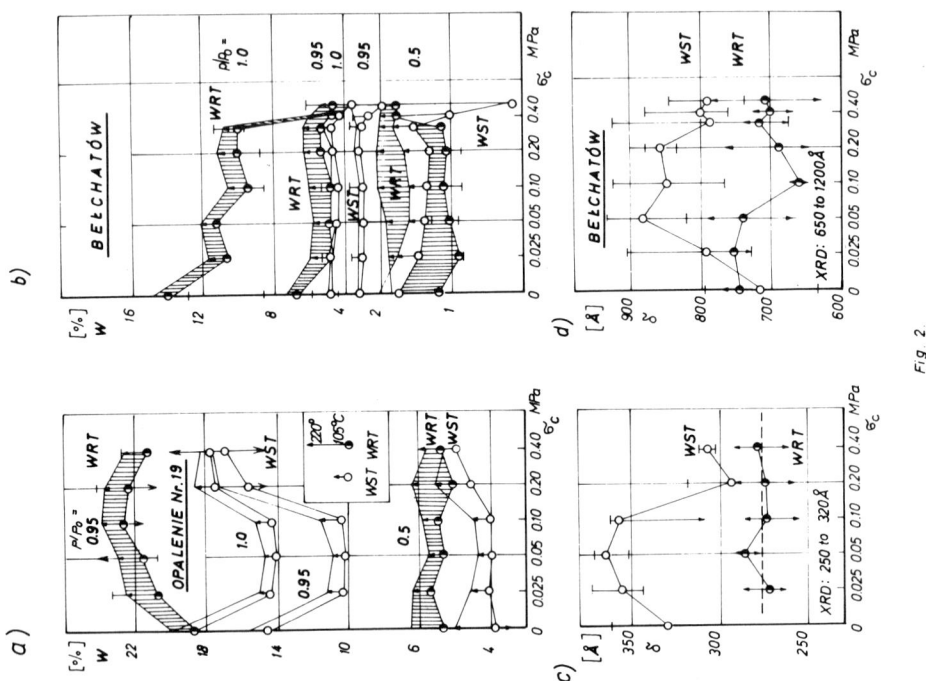

Fig. 1.

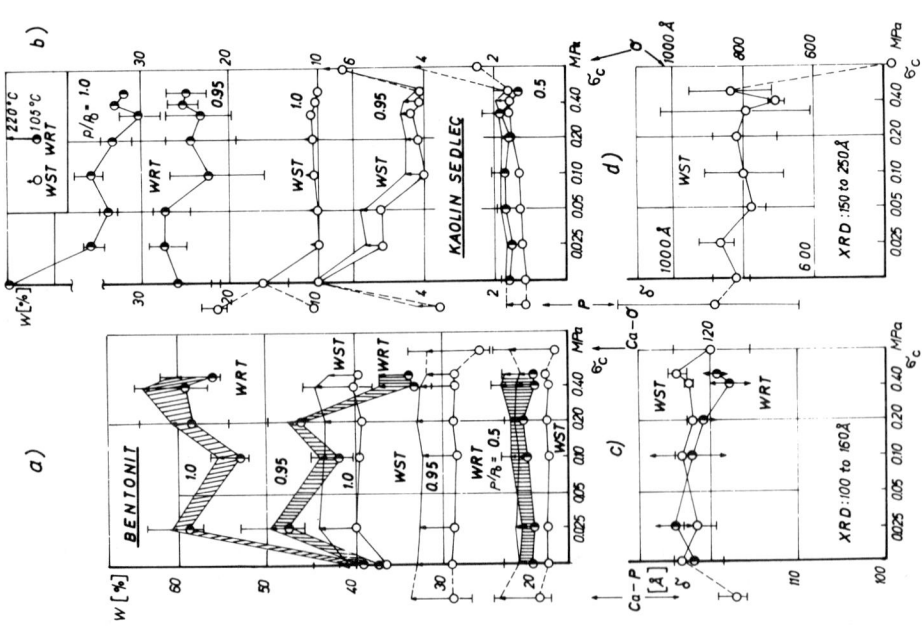

Fig. 2

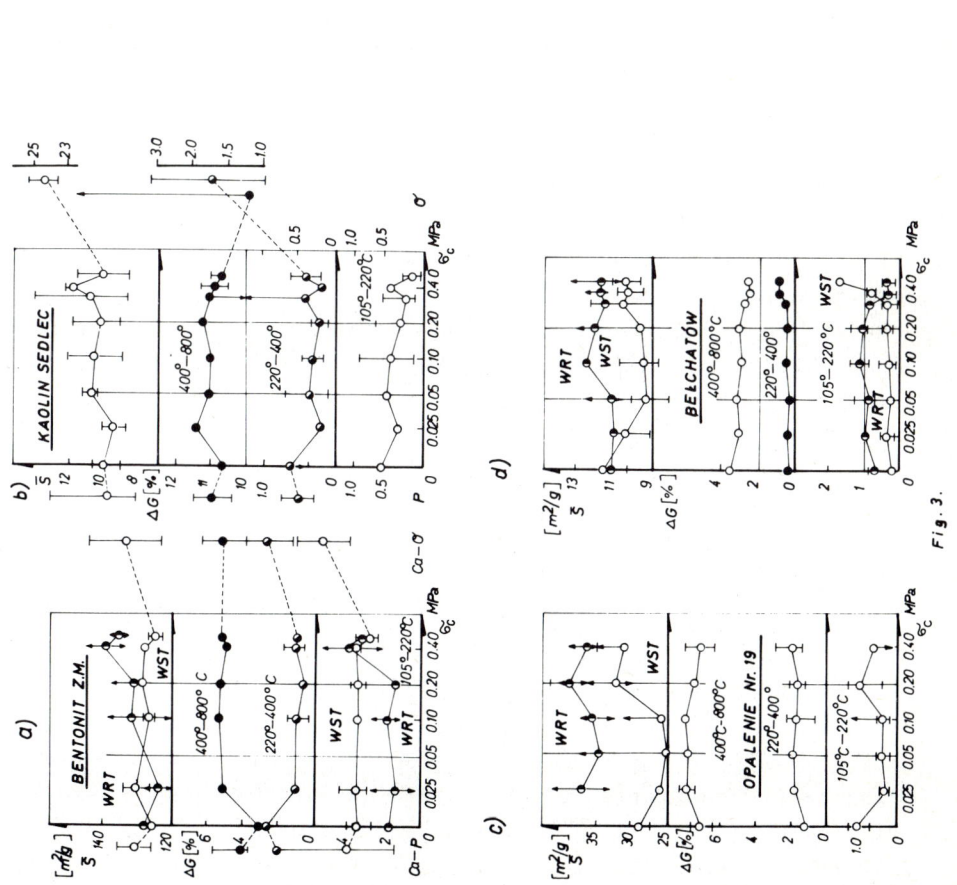

Fig. 4.

Fig. 3.

$$\bar{S} = W_m \left[ p/p_o = 0.5 \right] \times 585 \text{ m}^2/\text{g} \tag{1}$$

Average crystallite thickness was determined from geometrical consideration as:

$$\delta = \frac{7576 \text{ Å m}^2/\text{g}}{\bar{S}} + 63 \text{ Å} \tag{2}$$

Both equations were derived for natural clays. They give approximate values for kaolins containing some smectite, but they are incorrect for pure kaolinite.

Results obtained are presented in Fig. 1c,d, 2c,d, ($\delta$) and 3($\bar{S}$). Crystallite thickness measured by WST indicates similar values to those measured by XRD and observed in the scanning electron microscop (see Table 1 and Stępkowska et al., 1978).

An exception is Sedlec kaolin, where XRD indicates much lower $\delta$ values possibly due to low degree of crystallinity. Strength values (maximum axial pressure as a function of consolidation pressure) are presented in Fig. 4 to assist interpretation.

DISCUSSION

1. Water sorption and water retention at $p/p_o = 0.5$ does not depend strongly upon the consolidation pressure, retention being somewhat higher than sorption, except for B.clay. Dispersed and oriented kaoli samples indicate increased sorption properties, about 2.5 times greater than powdered samples.

2. At $p/p_o = 0.95$ water sorption changes with $\sigma_c$ in kaolin and O.clay. Little influence was detected in bentonite and B.clay. Water retention is influenced by pressure applied, decreasing at $\sigma_c \geqslant 0.4$ M (4 kG/cm$^2$) in bentonite and B.clay.

There is a reversible increase in water sorption for kaolin and O.clay stored in the wet state ($\sigma_c = 0$) and an increase in water retention of B.clay. In bentonite paste sorption and retention properties decreased reversibly.

Five powdered kaolin samples indicated two different values of water sorption at $p/p_o = 1.0$; i.e. $10.5 \pm 0.4\%$ and $21.5 \pm 1.4\%$ in spite of the fact, that all samples were stored in identical conditions (Fig. 1b).

3. Crystallite thickness, $\delta$, is influenced by consolidation and/or axial compression pressure. For $\sigma_c \leqslant 0.05$ MPa, a small increase in $\delta$ is usually observed, whereas for $\sigma_c \geqslant 0.2$ MPa, $\delta$ tends to decrease. This is about one tenth of the pressure value necessary to decrease the grain size of granite (Feda, 1977).

At a certain stress value a pronounced scatter of test results may be observed: in bentonite, B.clay and O.clay at $\sigma'_c = 0.1$ MPa (Fig.1c, 2c and 2d) in kaolin at $\sigma'_c = 0.3$ MPa (Fig. 1d). At these values inflections in axial pressure curves are observed (Fig. 4). These features may be due to contact bond formation between crystallites when attraction exceeds repulsion.
WRT indicates mostly lower crystallite thickness than WST, an exception being B.clay paste ($\sigma'_c = 0$).

4. Crystallite external specific surface, by definition inversely proportional to crystallite thickness, is influenced by the applied pressure in an opposite way to that described above.

5. The influence of the applied pressure on the thermal weight loss $\Delta G$ is surprising (Fig. 3). Within the temperature range 105°C to 220°C, $\Delta G$ decreases with $\sigma'_c$ in bentonite and kaolin after WST, whereas high $\sigma'_c$ causes an increase in $\Delta G$ after WRT. B.clay indicates an opposite change in $\Delta G$: an increase with $\sigma'_c$ after WST, and a decrease after WRT. This phenomenon may be due to reversible decrease in $\delta$, and to disturbance of interlayer structure due to compression or shearing pressure, thus enabling sorbed water to escape at a lower temperature in WRT. Drying at 105°C of samples subjected to an elevated pressure may generate a new interlayer structure resulting in stronger bonding and this increases the temperature at which interlayer sorbed water escapes.

6. B.clay indicates an influence of $\sigma'_c$ on $\Delta G$ even at more elevated temperatures (220°C to 400°C and 400°C to 800°C) :weight loss is moved towards lower temperatures (Fig.3d). The same is observed in the other samples.

Thus after the sample has been subject to elevated pressure a part of the structural (combined) water needs less kinetic energy (lower temperature) to escape from the crystal lattice. This may be connected with the decrease in crystallite thickness. It should be mentioned that powder samples of natural clays give a higher $\Delta G$, between 400°C and 800°C, than samples of undisturbed structure.

CONCLUSIONS

1. Crystallite reorientation due to applied pressure (isotropic or anisotropic) may influence sorption properties of both monomineral and natural clays.

2. Isotropic consolidation and/or axial compression causes a change in crystallite thickness, $\delta$, that may be detected by water

sorption and/or retention. Low pressure ( $\sigma_c \leq 0.05$ MPa ) may cause an increase of $\delta$, high pressure ( $\sigma_c \geq 0.1$ MPa, 0.2 MPa or 0.4 MPa depending on the tested sample ) may cause a decrease of $\delta$. Inflections on sorption and retention curves are correlated to inflections on maximum axial pressure vs. $\sigma_c$ curves.

3. The applied pressure may change the interlayer structure causing a change in temperature of sorbed water escape.

4. The changes in sample structure due to applied pressure may decrease the kinetic energy necessary for the escape of structural (combined) water.

5. Water sorption and thermal weight loss measurement is a sensitive test of structure changes in a natural clay subjected to externally applied pressure.

6. If several clay samples are stored in identical conditions a stepwise difference in sorption properties is sometimes observed.

7. Further study is needed to give appropriate qualitative and quantitative interpretation of these phenomena.

Sorption tests and strength measurements were carried out, respectively, by Mrs Ewa Qaunt and Mr Andrzej Wilczuk.

REFERENCES

J.Feda,1977. High-pressure triaxial test of a highly decomposed granite. International Symposium of Structurally Complex Formation Associazione Geotechnica Italiana, Capri 1977, V.I. p.239-244.

E.T.Stępkowska,1977a. Physics of the Shearing Process of Saturated Clays.Proc.IX International Conference on Soil Mechanics and Foundation Engineering, Tokyo, V.I.p.311-314.

E.T.Stępkowska, 1977b. Physics of shearing process in remolded bentonite with various exchangeable cations. Archiwum Hydrotechniki, V.24. Nr.2.p.215-235.

E.T.Stępkowska, 1977c. Test sorpcyjny i możliwość jego stosowania w różnych badaniach. Archiwum Hydrotechniki, V.24, N° 3, p.411-421.

E.T.Stępkowska, M.Nesteruk, 1978. Badania mineralogiczne i sorpcyjne iłu z Bełchatowa. Archiwum Hydrotechniki, V.25, N° 2.

ELECTROPHORETIC PHENOMENA AS APPLIED TO THE INVESTIGATION OF INTERACTION BETWEEN CLAYS AND ANIONIC POLYELECTROLYTES.

D. RIOCHE and B. SIFFERT
Centre de Recherches sur la Physico-Chimie des Surfaces Solides
24, Avenue du Président Kennedy, 68200 MULHOUSE - France.
(with the collaboration of Elf-Aquitaine laboratories).

ABSTRACT

The Interaction between various anionic polyelectrolytes (ferrochromelignosulfonate, carboxymethylcellulose, chromium lignite) and different clay minerals (montmorillonite and kaolinite) has been investigated, using an electrophoretic mass-transport analyzer.

The data confirm the existence of two electric double layers at the surface of the clay micelle. The results corroborate, in particular, the existence of a strong positive double layer around the kaolinite particles dependant upon the pH, and located on the Al-faces and on the edge surfaces of the clay layer. The same double positive layer located solely on the edges of montmorillonite particles displays a very low intensity.

INTRODUCTION

Drilling fluids are generally made up of aqueous bentonite suspensions with diverse added products which allow the different properties, in particular the rheological properties to be controlled.

The use of these chemicals is often empirical. There is a lack of understanding of the mechanism involved therein. Hence, an investigation was undertaken to study the interaction between two model clays (kaolinite and montmorillonite) and the three polyelectrolytes, commonly used in drilling muds :

- a ferrochrome lignosulfonate (F.C.L.) used for its fluidizing properties and its filtrate reducing ability

- a chromium lignite (L.C.) used for its synergic action towards lignosulfonate, notably at high temperature

- a carboxymethylcellulose (CMC). The CMC are used to enhance the viscosity and to reduce the filtrate of drilling muds.

Electrokinetic measurements provide convenient means for characterizing the materials and the adsorption processes of surface active minerals. Chemical reactions occuring at the surface of compounds such as oxides and clays can be traced (OTTEWILL and HOLLOWAY, 1975). With this end in view, the electrophoretic mobilities of clay suspensions have been measured in the presence of the above mentioned polyelectrolytes, using an electrophoretic mass-transport analyzer.

THE ELECTROPHORETIC MASS-TRANSPORT ANALYZER

The measurements on concentrated clay suspensions have been performed using an electrophoretic mass-transport analyzer of the type 'Micromeritics" model 1202 (fig. 1 A).

The apparatus consists of a reservoir which may contain 100 ml of suspension. The measuring cell proper contains approximately 6 ml. Before any determination, the cell containing the experimental suspension is accurately weighed ($\pm$ 0,1 mg) after level adjustement with the filling - tube. The measuring cell, fastened to the reservoir, is isolated from the later by a shutter during reservoir filling, closing of the shutter prevents particle sedimentation or diffusion before and after each measurement.

While the electrophoretic mobility is being determined, the sedimentation effects are eliminated by continuously rotating the measuring cell (about 30 r.p.m.).

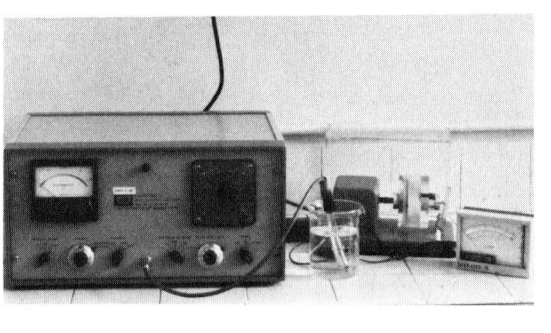
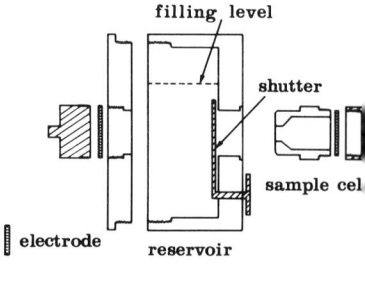

(A)                                             (B)

Fig. 1. The electrophoretic mass-transport analyzer (A), sketch of the measuring cell (B).

Reservoir and cell are made of plexiglass ; the electrodes of zinc. The apparatus with its chamber of electrophoretic mass-transport is attached to a steady intensity generator.

Under the action of the electric field, the particles migrate from the reservoir to the cell or in the opposite direction depending on the polarity of the field applied. Provision should be made to ensure a weight increase after performance of the various operations. Upon completion of a test - the experimental time being known to within a hundredth of a second - the measuring cell was once again weighed after adjusting the level with the filling-tube. The weight increase is directly related to the electrophoretic mobility of the particles. The theory of mass-transport has been described by OLIVIER and SENNET (1965). The electrophoretic mobility ($V_\varepsilon$) is given in terms of the different parameters by the general formula :

$$V_\varepsilon = \frac{\Delta W . \kappa}{R.i.t. \phi(1-\phi) (\rho_s - \rho_e)}$$

$\Delta W$ = weight increase
$R$ = resistance of the suspension
$i$ = current
$t$ = experimental time
$\phi$ = weight fraction of dispersed solid
$\rho_s$ = volumic mass of dispersed solid
$\rho_e$ = volumic mass of liquid
$\kappa$ = conductivity constant of the apparatus

The conductivity constant ($\kappa$) of the apparatus is determinated by the following equation using a N/100 solution of potassium chloride, whose specific conductance ($\lambda$) is well known.

$$\kappa = R_{KCl} \times \lambda_{KCl}$$

DESCRIPTION OF THE PRODUCTS AND THE MEASURING TECHNIQUE.

Two clay minerals have been used :
- a smectite : MILBEN bentonite
- a kaolinite from Ploemeur (Morbihan, France)

In order to work on well defined minerals, both minerals were purified and transformed into sodium minerals according to the well-known method of ion exchange.

Before performing the measurements, the suspensions were allowed to stay so that electrical and physico-chemical equilibria were reached. All concentrations were stated in grams of dry matter per liter of suspension. $C_s$ is the mineral concentration ; $C_{FCL}$, $C_{CMC}$, $C_{LC}$ are the concentrations of lignosulfonate (FCL), carboxymethylcellulose (CMC) and chromium lignite (LC) respectively.

In all measurements, $C_s$ is fixed at 50 g/l for montmorillonite and 100 g/l for kaolinite ; the ratios $C_s/C_{FCL}$, $C_{MC}$, $_{LC}$ vary with the contents of added polyelectrolytes. The mobilities have been measured in the alkaline pH range (this is always the case in drilling muds), the pH of the suspensions being adjusted with concentrated sodium hydroxide.

EXPERIMENTAL RESULTS

The electrophoretic mobilities of sodium montmorillonite and sodium kaolinite suspensions are represented in Figure 2 as a function of pH.

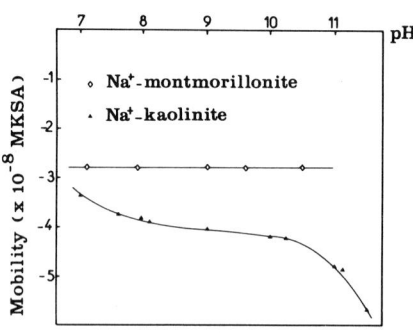

Fig. 2. Electrophoretic mobilities of montmorillonite and kaolinite suspensions as a function of pH.

The electrophoretic mobility of the sodium montmorillonite is constant in the alkaline pH range. On the other hand, the mobility of the sodium kaolinite varies with pH. This phenomena has already been reported by OTTEWILL and HOLLOWAY (1975).

The electrophoretic mobility of kaolinite is always higher in absolute value, than that of montmorillonite.

The action of CMC on clay minerals seems to be considerably dependant on the pH. Irrespective of the CMC concentration, all the electrophoretic mobility curves versus pH pass through a minimum for montmorillonite and a maximum for kaolinite at a same pH value of approximately 9 (Fig. 3).

In the presence of lignosulfonate, the mobility of sodium bentonite suspension is not subject to significant variations as a function of pH, except when the FCL concentration is high (fig. 4). All the mobility curves fall into a mobility domain ranging from - 1,6 to - 3,6 x $10^{-8} m^2 s^{-1} V^{-1}$.

On the other hand, the action of FCL on kaolinite is much more significant (Fig. 4). All the mobility curves fall into a range between - 1,9 to 6,5 $10^{-8} m^2 s^{-1} V^{-1}$, i.e. three times broader than the previous case.

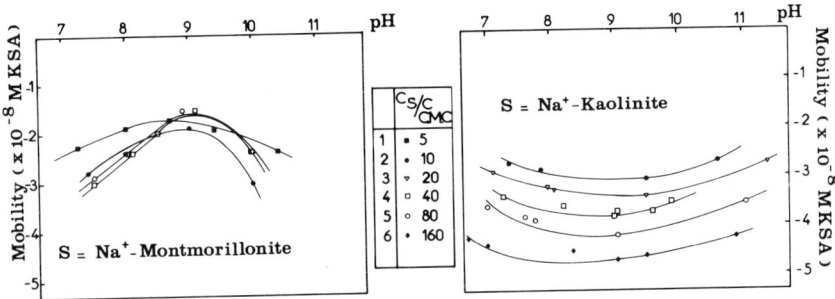

Fig. 3. Electrophoretic mobility of a sodium montmorillonite and a sodium kaolinite suspensions as a function of pH, in presence of carboxymethylcellulose (CMC).

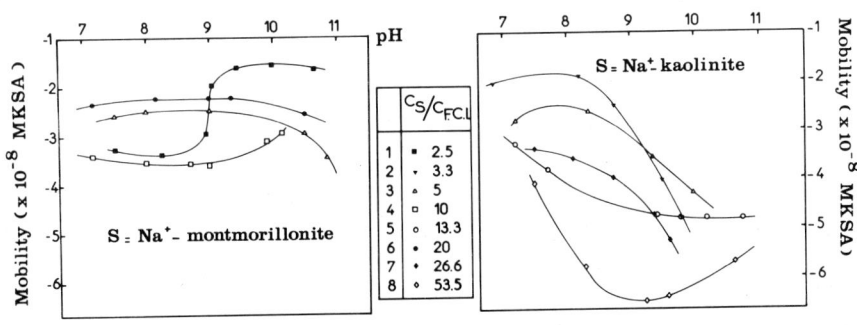

Fig. 4. Electrophoretic mobility of sodium montmorillonite and sodium kaolinite suspensions as a function of pH, in presence of ferrochrome lignosulfonate (FCL).

The action of chromium lignite (LC) on the two minerals differs significantly :
- with montmorillonite, the mobility decreases smoothly as a function of pH for a given LC concentration (Fig. 5).
- for kaolinite, one again finds a similar behaviour for high LC concentration and a sufficiently alkaline pH ; for lower L.C. concentration, however, the mobility increases as a function of pH (Fig. 5).

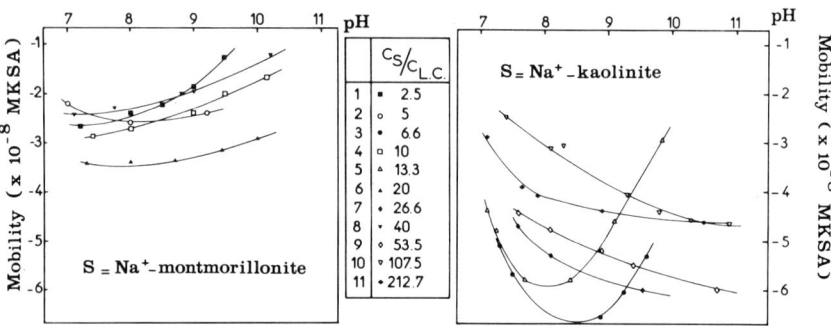

Fig. 5. Electrophoretic mobility of sodium montmorillonite and sodium kaolinite suspensions as a function of pH, in the presence of chromium lignite (LC).

INTERPRETATION OF THE PHENOMENA

It is well known that the double layers formed at the surface of pure alumina and silica particles are widely different. They have been studied extensively by JOHANSEN and BUCHANAN (1957), FUERSTENAU (1970) OTTEWILL and HOLLOWAY (1975). For alumina particles, the double layer is positive in an acidic medium ; it becomes negative in sufficiently alkaline conditions. The position of the point of zero charge depends on the nature and the structure of the aluminous surface. For pure alumina ($\alpha$ $Al_2O_3$), the isoelectric point is situated at a pH approaching 8.5. For a surface composed only of silicon atoms, the double layer is always negative, except for very acidic conditions (pH < 2). Lastly, for silico alumi-nates, the double layer is generally positive in acidic medium and negative in alkaline medium. The position of the isoelectric point varies with the $Al_2O_3/SiO_2$ ratio of the alumino-silicate (MATIJEVIC and al., 1971).

For montmorillonite, the double layer depends mostly on the intrinsic charge of the mineral layer and the silica sheet of the basal planes ; the positive charges on the layer edges (lateral surface) being negligible. The mobility of montmorillonite suspensions is then constant and negative (Fig. 2).

In case of kaolinite, on the other hand, half the basal surface of the mineral layer displays a double layer related to an aluminous surface (-Al-OH) : the mobility therefore varies according to the pH (Fig. 2).

## Action of carboxymethylcellulose (CMC)

The comparison of electrophoretic mobility curves of clay suspensions versus pH, in presence of CMC is represented in Fig. 6.

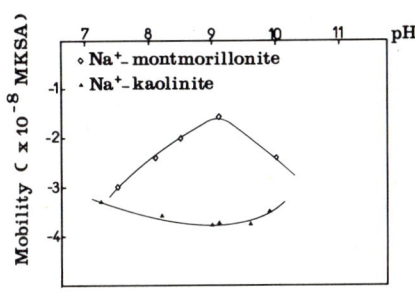

Fig. 6. Comparison of the evolution of electrophoretic mobility of clay suspensions versus pH, in the presence of carboxymethylcellulose (CMC) ($C_S/C_{FCL}$ ratio = 40).

The differences in variation of the mobility between the montmorillonite and kaolinite suspensions may be explained by taking into account the following :
- the existence of two types of electric double layers at the surface of the clay particles (Van OLPHEN, 1963)
- the hydrolysis reaction of CMC according to the pH :

$$R - CO_2^- \ Na^+ + H_2O \underset{(2)}{\overset{(1)}{\rightleftharpoons}} R - CO_2H + OH^- + Na^+$$

When the $OH^-$ ion concentration reaches a given limit, the equilibrium shifts in the direction (2) : CMC is adsorbed in anionic form in a concentrated alkaline medium.

Hence, for montmorillonite at a pH below 9, CMC is fixed predominantly in its neutral molecular form. The net negative charge of the clay particle decreases through the screening effect on the neutral molecules (protective colloid effect of the molecules) and the mobility diminishes. Around pH = 9, the CMC dissociation equilibrium reverses. The $RCO_2^-$ anions are not repelled by negative clay micelles because of the "protective" neutral layer already formed : the net negative charge will thus increase. For pH values above 9, the mobility increases.

At a pH below 9, the positive adsorption sites are numerous at the surface of kaolinite (aluminous double layer). The CMC dissociation equilibrium shifts towards the formation of anions which are fixed onto the positive sites. Hence, the net negative charge and the mobility of the clay particles increase. Above pH = 9, the anionic polymer molecules are more abundant, the repulsion between the polymer molecules and the negative siliceous surface of the clay micelles continues to exist : CMC will only be fixed in its neutral form. The screening effect of the intrinsic negative charge begins to appear, resulting in a decrease of mobility.

Action of the lignosulfonate (FCL)

The action of lignosulfonate on the dispersed clay particles of montmorillonite and kaolinite seems to be identical for $C_s/C_{FCL}$ ratios below 20. For lower FCL concentration, the action depends on the nature of the clay mineral (Fig. 7).

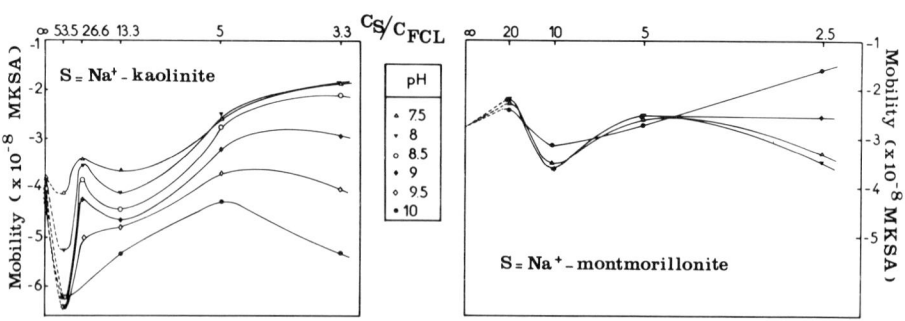

Fig. 7. Comparison of the electrophoretic mobilities of clay suspensions in the presence of lignosulfonate (FCL) at various pH values.

In order to explain the shape of the curves obtained, the existence of the two electric double layers at the kaolinite particle must always be taken into account. It is worth noting that the lignosulfonates retain their anionic form and react with the aluminium atoms irrespective of the pH. In this way, at slightly alkaline pH values (up to 8.0), the aluminous double layer is positive.

The anionic form of FCL molecules react with the alumina sites and neutralize them : the mobility decreases slightly because of a weak screening effect. For a pH above 8, the aluminous double layer itself becomes negative, but the reactivity towards the aluminium atoms continues to exist (SIFFERT and FERRAND, 1973). The polymeric anions then contribute to an increase of the negative charge of the micelle. Hence a significant mobility increase is observed.

For higher lignosulfonate concentration, the behaviour is identical for kaolinite and montmorillonite. It may be explained as follows :
- Neutralization of the remaining positive and poorly localized charges takes place on the layer edges and the bulky FCL molecule produces a partial screening of the negative charge ; hence the mobility decreases.
- For even higher FCL concentration a physical adsorption of the anionic polyelectrolyte onto the negative clay surface is observed. The mobility again

begins to increase, owing to the additional supply of negative charges.

Lastly, starting from a sufficiently high FCL concentration ($C_s/C_{FCL}$ ratio approaching 10), the clay micelle is entirely surrounded by a film of lignosulfonate molecules. The thickness of this film grows to such an extent that the lignosulfonate acts as a protective colloid and the electrophoretic mobility decreases resulting from a screening of the intrinsic charge of the clay particle.

Action of chromium lignite

The electrophoretic mobilities of montmorillonite and kaolinite suspension in the presence of chromium lignite are compared as a function of pH in Fig. 8.

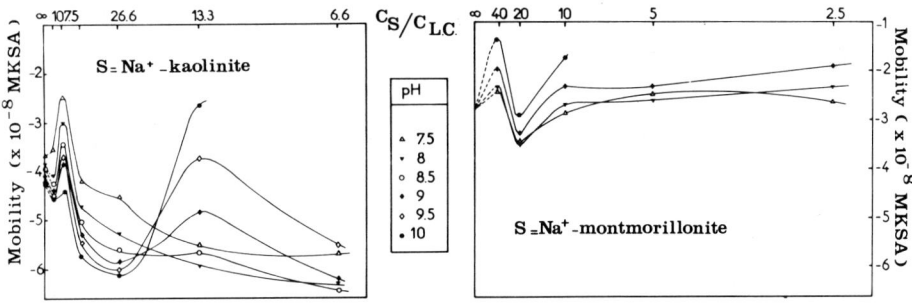

Fig. 8. Comparison of the electrophoretic mobilities of clay suspensions in presence of chromium lignite (LC) at various pH values.

The curves of both minerals are identical to those obtained in presence of lignosulfonate. The phenomena may be interpreted in the same way. It must however be appreciated that the maxima and minima of the mobilities occur for $C_s/C_{LC}$ ratios different from the $C_s/C_{FCL}$ ratios. For an identical interaction, the $C_s/C_{FCL}$ ratio is generally twice the $C_s/C_{LC}$ ratio, i.e. half the quantity of chromium lignite is required to bring about the same effect. This result may be interpreted by assuming either steric hindrances or different charges for each kind of molecule.

CONCLUSIONS

These results may account for the "change-over" (SCHOFIELD and SAMSON, 1953) in the positive electric double layer charge carried by the clay micelles, as a function of pH.

The results discussed above support the existence both of a strong negative double layer independant of pH, located on the siliceous surfaces and of a positive double layer, dependant on pH, on the kaolinite micelles, located on the Al-faces and on the edge surfaces of the layers. They also demonstrate the weak intensity of the positive double layer, located only on the edge surfaces of the layers in the case of montmorillonite particles.

ACKNOWLEDGEMENTS

The authors are grateful to ELF-AQUITAINE Society for its laboratory and financial assistance.

REFERENCES

Fuerstenau, D.W., 1970. Interfacial processes in mineral water systems. Pure and Appl. Chem., 24 : 135-164.

Johansen, P.G. and Buchanan, A.S., 1957. An Application of the microelectrophoresis method to the study of the surface properties of insoluble oxides. Aust. J. Chem. 10 : 398-403.

Matijevic, E., Mangravite, F.J. and Cassell, E.A., 1971. Stability of Colloïdal silica, IV : silica-alumina system. J. Colloïd and Interface Sci. 35 : 560-568.

Olivier, J.P. and Sennett, P., 1965. Electrokinetic effects in kaolin-water systems : the measurement of electrophoretic mobility. Fifteenth Conf. Clay Minerals, Pittsburgh.

Ottewill, R.H. and Holloway, L.R., 1975. Electrokinetic properties of particles. Phys. Chem. Sci. Res. Rep. 1 : 599-621.

Schofield, R.K. and Samson, H.R., 1953. The defloculation of kaolinite suspensions and the accompanying change-over from positive to negative chloride adsorption. Clay Min. Bull. 2 (9) : 45-51.

Siffert, B. and Ferrand, C., 1973. Contribution à l'étude du mécanisme d'interaction des argiles et des lignosulfonates. Bull. Gr. Fr. Arg. 25 : 135-148.

Van Olphen, H., 1963. An Introduction to Clay Colloïd Chemistry. Interscience Publishers, John Wiley, New-York, 301 pp.

# THE CLAY DEPOSITS OF MEXICO

LIBERTO DE PABLO-GALAN
Consejo de Recursos Minerales and Instituto de Geología, U.N.A.M.
México, D. F., México

## ABSTRACT

The kaolins mined in Mexico resulted from the primary hydrothermal alteration of Tertiary acidic rocks (deposits at Sombrerete, Ahualulco, Huayacocotla, Pathé, Sierra de San Andrés, Nayarit, and Xilosintla) or of Triassic schists (Noche Buena). Montmorillonites are sedimentary, like the swelling type from Cuencamé, in the state of Durango, or the non-swelling type from the state of Tlaxcala. The geology of the country warrants the study of the clays associated with the Triassic (San Marcial), Jurassic (Tezoatlán-El Consuelo) and Late Cretaceous (Cohauila) coal beds, as well as with the Mesozoic sediments included in the Huayacocotla, Todos Santos, Huizachal, Las Vigas, Carbonera, Cabullona, and Guayabal formations. The Paleozoic - Tertiary intrusive rocks along the Pacific coast equally represent favorable targets for finding clay deposits.

## INTRODUCTION

The kaolin deposits of Mexico, with few exceptions, are of primary hydrothermal origin and montmorillonites are of sedimentary origin. Only ceramic and refractory kaolins, sodium and calcium bentonites, attapulgite, vermiculite, and zeolites are produced, while sedimentary kaolinites, ball clays, and processed kaolins are imported.

## GEOLOGY

A brief review of the geology of the country, with reference to the Geologic Map of Mexico (Lopez-Ramos, 1976), indicates the presence of Precambrian gneiss and schist in the state of Sonora, in a few localities in Puebla, and more extensively in Oaxaca, Guerrero, and Chiapas. During the Paleozoic, transgressive seas covered the Precambrian basement with sediments (Fig. 1), deposited in a moderately cold climate during the early

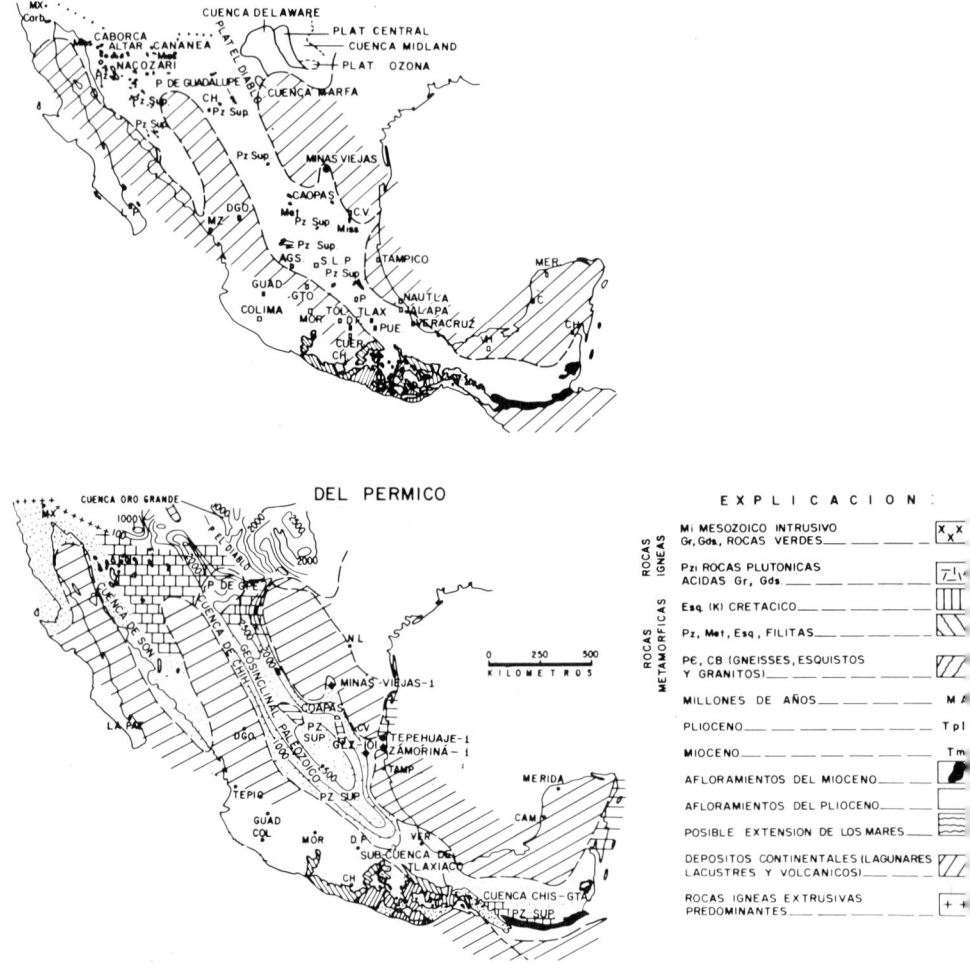

Fig. 1. Paleogeography of the Mexican Mississippian and Permian (Lopez-Ramos 1974).

Paleozoic, warm and arid during the Devonian, warm during the Carboniferous, and desertic throughout the Permian (Lopez-Ramos, 1974). Towards the end of the Paleozoic, the geosyncline emerged in the central and northern areas, developing flysch-type sedimentation. Metamorphic schist at Acatlán, Puebla, and Taxco, Guerrero suggest a weathering profile now largely obliterated by the effects of metamorphism. Paleozoic and Mesozoic intrusions have been reported by Lopez-Ramos (1974), Salas et al. (1974), and others. Guerra-Peña (1976) interpreted, from satellite imagery, extensive faulting in Paleozoic and younger formations and pointed to zones prone to hydrotherlism.

During the Mesozoic, transgressive seas covered large areas of the country (Fig. 2). The Triassic climate was warm and semi-arid (Lopez-Ramos, 1974) as evidenced by the coal-bearing clastic sequence of San Marcial, in the state of Sonora, and which extends westward into the Lower Jurassic (Alencaster, 1961), the red beds of the Nazas Formation, and the shales and argillites west of Zacatecas. In north-central Zacatecas, metavolcanics of Triassic age (Caopas Schist; Fries et al., 1965) are in juxtaposition with non-metamorphosed sediments of the same age, which calls for large scale thrusting during the Middle Jurassic (de Cserna, 1970). In south-central Mexico, the Middle Jurassic coal-bearing sequence becomes progressively marine by the beginning of the Late Jurassic (Erben, 1956a, 1956b). The presence of kandites is known from the shales of various Lower and Middle Jurassic stratigraphic units of south-central Mexico. In north- and southeastern Mexico, the Upper Jurassic contains important salt accumulations (Wall et al., 1961; Viniegra-Osorio, 1971) while, elsewhere, it consists of limestone, shale (Imlay, 1943), and some marine phosphorite (Rogers et al., 1961).

At the beginning of the Early Cretaceous, marine sedimentation was practically coextensive with that during the Late Jurassic. However, starting during the Aptian (Fig. 3), the seas gradually invaded the existing positive areas which, during the Albian and early Cenomanian acted as platforms. Along the edges of these shelves rudistid banks proliferated, whereas on top of them gypsum precipitated which gave away to bank limestone during the remainder of the Early Cretaceous (Muir, 1936; Imlay, 1944; Carrillo, 1971). In western Mexico an eugeosynclinal regime existed during the Late Jurassic and Early Cretaceous (Beal, 1948; King, 1939; Silver et al., 1956). The base of the Lower Cretaceous is represented by the clastic sediments of Las Vigas and San Marcos formations, whereas the upper part consists chiefly of limestones.

About 100 m. y. ago, in Baja California the eugeosynclinal sequence underwent metamorphism and granitic batholiths were emplaced (Böse et al., 1912; Silver et al., 1956). The process initiated a gradual rising in western Mexico that shifted progressively with time towards the east. As a result, the Upper Cretaceous-Paleocene (Fig. 4) sequence is a flysch wedge consisting of alternating shales and sandstones that became coarser upwards (de Cserna, 1960). Only at a few localities in Mexico which were protected from the influx of clastics, like eastern San Luis Potosí (Cardenas) or Chiapas (Coban), the limestone sedimentation continued well into the Late Cretaceous (Myers, 1968; Alencaster, 1971). It should be mentioned that the Upper Cretaceous of

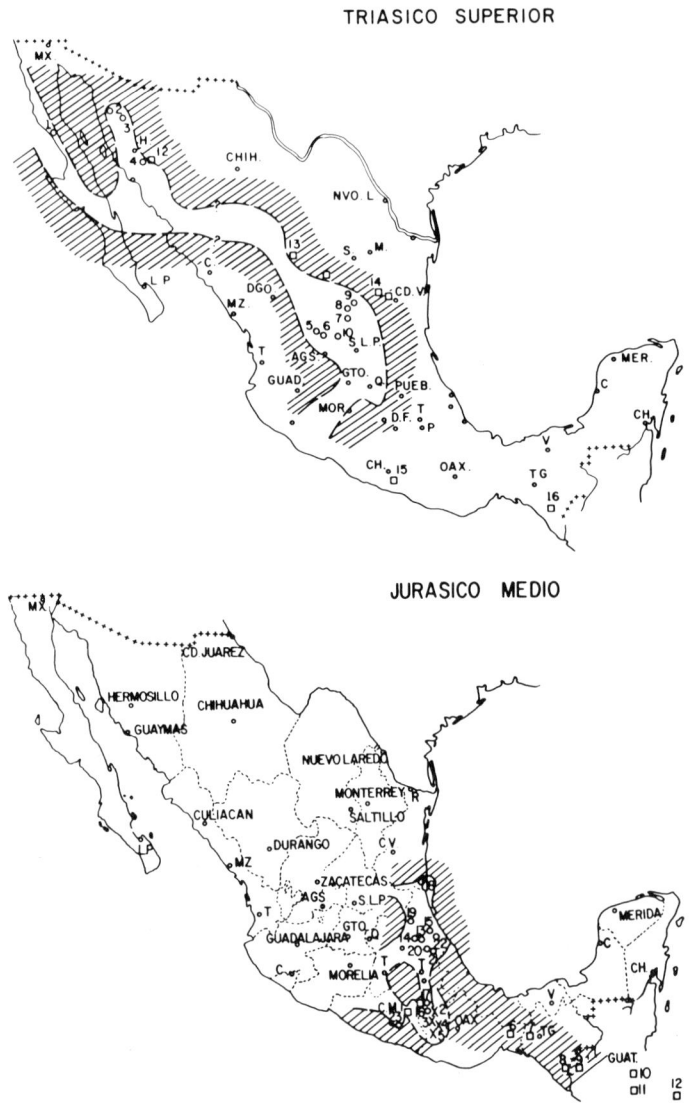

Fig. 2. Paleogeography of the Mexican Triassic and Middle Jurassic (Lopez-Ramos. 1974)

Mexico contains tuffs in many localities at several horizons (Taliaferro, 1933; Imlay, 1937; de Cserna et al., 1978) as well as bituminous coal (Maestrichtian) in northeastern Mexico (Robeck et al., 1956). The Upper Cretaceous (Cenomanian-Turonian) Coban limestone (Imlay, 1944) of southern Mexico and Central America includes the "white limestone", associated with the bauxite developments in the Caribbean area.

During the early Eocene (Fig. 4), severe folding affected the country

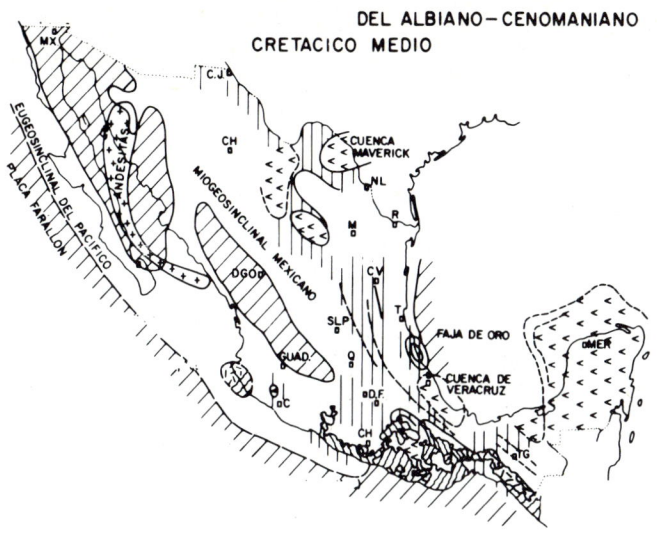

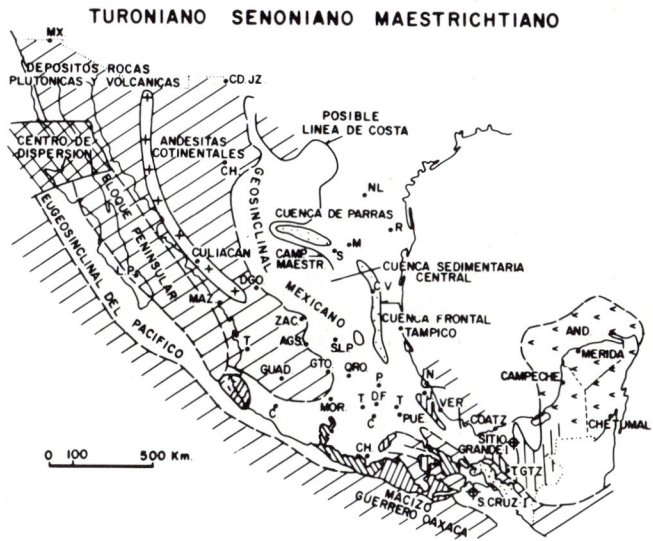

Fig. 3. Paleogeography of the Mexican Cretaceous (Lopez-Ramos, 1974)

producing the Sierra Madre Oriental, the Mexican part of the early Tertiary foreland fold and thrust belt of the North American Cordillera (de Cserna, 1971). The deformation triggered a large scale erosion of the fold structures and the debris was carried eastward into the Gulf of Mexico depositing an important molasse sequence (Muir, 1936). West of the Sierra Madre Oriental, the region experienced block-faulting, drainage blocking, the accumulation of continental clastics (Guanajuato, Balsas, Ahuichila, El Bosque and other formations), and the emplacement of numerous granitic and granodioritic plutons along the Pacific states of Jalisco, Colima, Na-

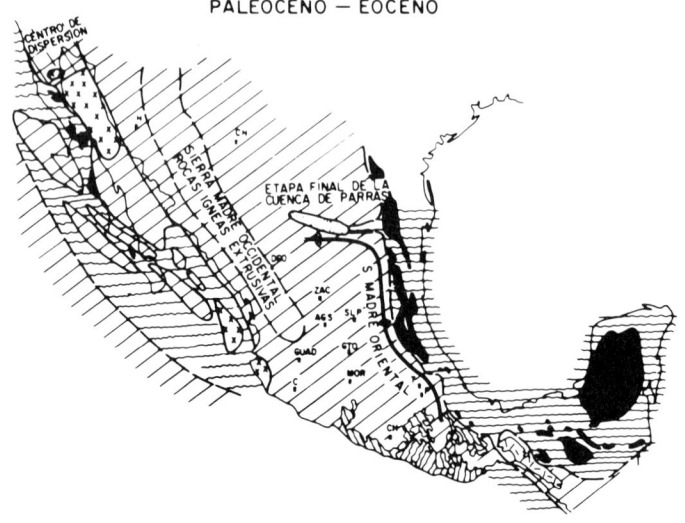

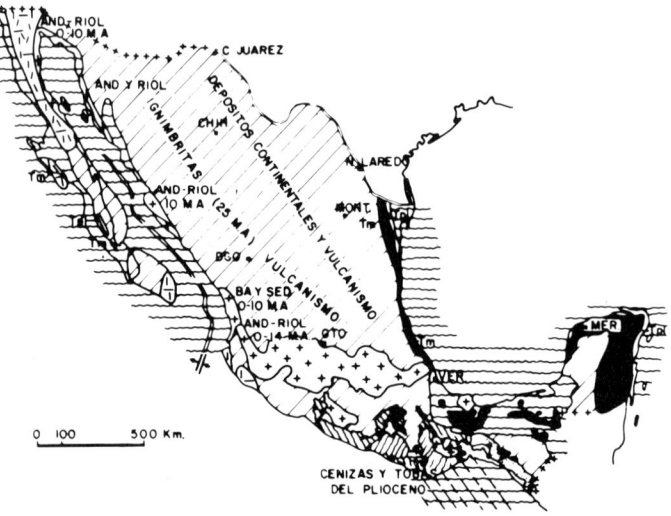

Fig. 4. Paleogeography of the Mexican Cenozoic (Lopez-Ramos, 1974)

yarit, Michoacan, and Guerrero (Fries et al., 1955; Salas et al., 1974). Prominent volcanic activity during the Oligocene-Miocene developed a huge volcanic plateau in western Mexico, the Sierra Madre Occidental, largely from rhyodacitic pyroclastic flows (McDowell et al., 1977). From the late Miocene onward, a chain of volcanoes formed across Mexico the Neo-Volcanic Belt (Flores, 1946; Blasquez, 1956; Mooser, 1956; de Cserna, 1975).

CLAY RESOURCES

The kaolin produced in Mexico is almost totally from primary hydrothermal deposits. Some of the most important ones are located in the Neo-Volcanic Belt. Many geothermal localities are associated with kaolin deposits (Anda, 1956; Blasquez et al., 1961; Alonso et al., 1964; Mooser, 1964).

Kaolin deposits in the Neo-Volcanic Belt have been commented on by several authors (Blasquez et al., 1946; Esquivel et al., 1958; Pesquera et al., 1968. In the extreme NW of the province, north of Tepic, Nayarit (Fig. 5)

Fig. 5. The clay deposits of Mexico.

refractory kaolin is mined from a relatively unknown location from where geothermal activity and intrusions have been reported. In the state of Michoacan, hydrothermal activity is evident in the deposits of Sierra de San

Andrés, where hot acidic solutions altered porphyritic and vitrophyric Cenozoic rhyolites into endellite and cristobalite (Esquivel et al., 1958; Blasquez et al., 1946; Kesler, 1970; Keller et al., 1971; Hanson, 1975).

Southwest of the city of Guanajuato, at Noche Buena, minor mining of kaolin still continues. The country rock is a chlorite schist, probably of Triassic age, invaded during the Laramide orogeny by granitic, dioritic bodies (Edwards, 1955; Schulze, 1956b; Hanson et al., 1966; de Cserna, 1975; Hanson, 1975). The alteration by rising solutions produced plastic kaolinite and quartz. At Pathé and Tecozautla, Hidalgo, where geothermal steam was tapped for energy purposes, montmorillonite from basalt and kaolinite from rhyolite are mined side by side (Arellano, 1958; de Pablo, 1962; Pesquera, 1975). Both clays are used for dinnerware and porcelain manufacture.

To the north of the Neo-Volcanic Belt, in the area of Apulco, Agua Blanca, and Huayacocotla, the alteration of Tertiary rhyolites, tuffs, and pyroclastic breccias produced a high grade kaolinite used for ceramics, in cement, and as a filler.

West of Toluca, in the state of Mexico, there is a small refractory halloysite deposit originated from volcanic ash deposited in a lake (Hanson, 1975). An unsurfaced intrusion was here indicated in the tectonic interpretation of Guerra-Peña (1976). Other deposits within the Neo-Volcanic Belt are at Vito, Tula de Allende, in the state of Guanajuato; Comonfort, Neutla, San Felipe, and Silao, in Guanajuato; Zacualco, Magdalena, Zapopan, Lagos, and Comanja in Jalisco. Their production for ceramic purposes is minor.

In the Central Plateau, NW of San Luis Potosí, the Santa Teresa mine, near Ahualulco, is the largest underground clay mine in Mexico. Here Jurassic shales, overlain by Lower Tertiary rhyolitic flow breccias high in sanidine, are altered through montmorillonite into kaolinite, halloysite, cristobalite and minor alunite (Hanson, 1975). Farther to the northwest, in the state of Zacatecas, there are a small refractory dickite deposit at Guadalupe (Hanson, 1975) and, at Cerro de La Bufa, kaolinite associated with a metaliferous vein (Schulze, 1956a). At the NW limit of the Central Plateau with the Sierra Madre Occidental, the San José de Ranchos deposit near Sombrerete is mined for refractory kaolin within Tertiary rhyolite in Cretaceous limestones (Keller and Hanson, 1969; Hanson, 1975). Other localities in the state of Zacatecas are Valdecañas, Jerez, and Villa Garcia; Peñon Blanco and Yerbaniz, Durango on the Transverse Ranges of the Sierra Madre Occidental; Jimenez and Ojinaga in Chihuahua; Ixtacamaxtitlan, Petlalcingo, Tecomatlan, and Zapotitlan, in Puebla; and Magdalena Peñasco, San Mateo Mixtepec, San Martin Zacatepec, and Mariscala de Iturbide, in Oaxaca. In north-central Guerrero, in Xilosintla and Iguala, kaolin is produced for ceramic and refractory uses.

Montmorillonites, in contrast to kaolinites, are sedimentary. The main producing area for swelling bentonites is Cuencamé, in Durango while the non-swelling type is almost totally mined near the town of Panzacola, Tlaxcala, and acid activated at a nearby plant. Numerous minor deposits have been reported throughout the country. A high magnesium montmorillonite, of primary hydrothermal alteration, is mined at Pathé for ceramic purposes.

In the Yucatan peninsula, sedimentary attapulgite is used by the local potters as it was by the ancient mayas. Localities are reported at Maxcanú, Sacalúm, and Ticúl to the south of the Sierra de Ticúl, a minor fault scarp limiting shallow lagoons where dolomite and attapulgite precipitated together (Butterlin et al., 1963; Folan, 1969; Ipshording, 1973; de Pablo, 1973; Bohor, 1975).

Sedimentary plastic kaolinites (ball clays) are not produced in Mexico. There is one small operation near Villa Juarez, Puebla, where a reddish burning kaolinite-hydrous mica mixture of low PCE value is mined for some ceramic applications. South of lake Chapala, de Pablo (1965) reported a plastic disordered kaolinite.

The Eocene siderolitic weathering front, particularly along the high humidity areas of Chignahuapan and Villa Juarez, Puebla, and in the southern states of Tabasco and Chiapas has been explored for laterites. The common association of kaolinite, gibsite, and iron hydroxides has been recorded. Quiñones (1975a) reported kaolinitic acrysols in Tabasco and, in Yucatan, anomalous kaolinite with minor montmorillonite (Quiñones, 1975b).

Zeolitic tuffs have been mined since colonial times at Etla and within the Oaxaca valley as building stone. They represent the Etla ignimbrite member of the Suchilquitongo Formation (Wilson and Clabaugh, 1970; Mumpton, 1973, 1975). The minerals are clinoptilolite and mordenite. Hydrothermal zeolites have been recorded in drill cores from Ixtlán de los Hervores, Michoacan.

CONCLUSIONS

Primary hydrothermal kaolins, sedimentary montmorillonites, continental and marine sediments are mined for their typical uses. However, the geology of the country suggests the study of the underclays associated with the Triassic San Marcial, the Tezoatlan-El Consuelo Jurassic sediments, and the Upper Cretaceous coal beds of Coahuila. Of no less interest are: the clays of the Mesozoic sediments included in the Tototapa, Huayacocotla, Divisadero, Todos Santos, Paltoltecoya, Huizachal, Consuelo, La Barranca, and La Joya formations; also, the Lower Cretaceous Las Vigas and Carbonera formations and the Upper Cretaceous Parras, Mendez, and Tamesi shales and the Cabullona group; from the Cenozoic, the Eocene El Bosque, Guanajuato, Balsas, and Chapopote continental clastics and the weathering fronts and alterations of Paleozoic and Tertiary intrusions along the Pacific coast. Tertiary volcanics in NW and W Mexico and in the Neo-Volcanic Belt include some of the most important clay localities. In general. it would be expected that economically interesting clay minerals may be more common in the country than hitherto expected but a vast amount of work remains to be done.

ACKNOWLEDGEMENTS

The author is indebted to Z. de Cserna and R. Pesquera for their revision of the manuscript and helpful comments on the geology of Mexico.

REFERENCES

Alencaster, G., 1971. Rudistas del Cretácico Superior de Chiapas. Parte I. Univ. Nac. Aut. México, Inst. de Geología, Paleontología Mexicana 34.
Alonso, H., Anda, L., and Mooser, F., 1964. Focos térmales de la Republica Mexicana. Bol. Soc. Mex. Geol. Petroleros, 16: 145-152.
Anda, L., 1956. El Campo de Energia Geotérmica de Pathé, Estado de Hidalgo, México. In: A. Garcia (Editor), Vulcanología del Cenozoico, XX Congreso Geol. Int., Instituto de Geología, México, pp. 257-283.
Arellano, A. R. V., 1958. Las minas de Santa Rosa y anexas. Publicaciones Cerámicas, 1:45-50.
Beal, C. H., 1948. Reconnaissance of the geology and oil possibilities of Baja California, México. Geol. Soc. America Mem., 31, 138 pp.
Blasquez, l., 1956. Bosquejo Fisiográfico y Vulcanologico del Occidente de México. In: Inst. de Geología (Editor), Libre Guia de la Excursión A-15, XX Congreso Geol. Int., Univ. de México, México, pp. 9-55.
Blasquez, L. and Lozano, R., 1946. Hidrogeología y minerales no metálicos de la zona norte del estado de Michoacan. Anales del Inst. Geol. de México IX.
Blasquez, L., Mooser, F., Reyes, A., and Lorenzo, J., 1961. Fenómenos Geologicos de Algunos Volcanes Mexicanos. Inst. de Geol., Univ. de México, México, 108 pp.
Bohor, B. F., 1975. Attapulgite in Yucatan. In: L. de Pablo (Editor), Guide book FT-4, 1975 Int. Clay Conference, Inst. Geol., Univ. de México, pp. 95-125.
Böse, E. and Wittich, E., 1912. Informe relativo a la exploración de la región norte de la costa occidental de la Baja California. Inst. Geol. Mexico Parargones, 4: 307-529.
Butterlin, J. and Bonet, F., 1963. Mapas geológicos de la Peninsula de Yucatan. I. Las formaciones Cenózoicas de la parte mexicana de la Peninsula de Yucatan. Ingeniería Hidraulica, 17:63-71.
Carrillo-Bravo, J., 1971. La plataforma Valles-San Luis Potosí. Bol. Asoc. Mex. Geol. Petroleros, 23:1-102.
de Cserna, Z., 1960. Orogenesis in time and space in Mexico. Geol. Rundschau, 50: 595-605.
de Cserna, Z., 1970. Mesozoic sedimentation, magmatic activity and deformation in northern Mexico. In: K. Seewald and D. Sundeen (Editors), The Geologic Framework of the Chihuahua Tectonic Belt, Midland, West Texas Geol. Soc., pp. 99-117.
de Cserna, Z., 1971. Development and structure of the Sierra Madre Oriental of Mexico. Geol. Soc. America Abs. with Programs, 3: 377-378.
de Cserna, Z., 1975. On the Geology of Parts of the Trans-Mexico Volcanic Belt and of the Mexican Central Plateau. In: L. de Pablo (Editor), Field Trip Guidebook FT-1, 1975 International Clay Conf., Inst. de Geología, Univ. de México, México, pp. 1-72.

de Cserna, Z., Palacios-Nieto, M., and Pantoja, J., 1978. Relaciones de facies de las rocas Cretácicas en el noroeste de Guerrero y en areas colindantes de México y Michoacan. Univ. Nac. A. de México, Inst. de Geología, Revista, 2: 55-64.
de Pablo, L., 1962. Investigación mineralógica de la arcilla Pathé. Minería y Metalurgia, 22: 39-64.
de Pablo, L., 1965. A Disordered Kaolinite from Concepción de Buenos Aires, Jalisco, México. In: Clays and Clay Minerals, Proc. 13th Nat. Conf., Madison. Pergamon Press, Oxford, pp. 143-150.
de Pablo, L., 1973. Attapulgite de Ticúl, Yucatan. Umpublis. report. Inst. de Geol., Univ. Nac. A. México, México, 15 pp.
Edwards, J. D., 1955. Studies of some Early Tertiary Red Conglomerates of Central Mexico. U. S. Geol. Survey Prof. Paper 264-H.
Erben, H. K., 1956a. El Jurásico Inferior de México y sus Amonitas. XX Con. Geol. Internacional. Univ. Nac. A. de México, Inst. de Geol., México, 293 pp.
Erben, H. K., 1956b. El Jurásico Medio y el Calloviano de México, XX Cong. Geol. Internacional, Univ. Nac. A. de México, Inst. de Geol., 140 pp.
Esquivel, J. and Zamora, A., 1958. Informe sobre Minerales No Metálicos. Consejo de Recursos Naturales No Renovables, México, Bol. 44, 236 pp.
Flores, T., 1946. Geología Minera de la Región Noroeste del Estado de Michoacan, Univ. Nac. A. de México, Inst. de Geol., México, 106 pp.
Folan, W., 1969. Sacalum, Yucatan: A pre-hispanic and contemporary source of attapulgite. Am. Antiquity, 34: 182-183.
Fries, C. and Rincón-Orta, C., 1965. Nuevas aportaciones y técnicas empleadas en el Laboratorio de Geocronometría. Contrb. Lab. Geocron. Univ. Nac. A. de México, Inst. de Geol., 73: 57-133.
Fries, C., Hibbart, C. W., and Dunkle, D. H., 1955. Early Cenozoic vertebrates in the red conglomerate of Guanajuato, México. Smithsonian Misc. Coll., 123: 1-25.
Guerra-Peña, F., 1976. Interpretación de la Tectónica Mexicana en las Imagenes del Satelite Artificial Landsat-1. Comisión Estudios del Territorio Nacional, México, 19 pp.
Hanson, R. F., 1975. The Geology of some Clay Deposits in the Trans-Mexico Volcanic Belt and the Mexican Central Plateau. In: L. de Pablo (Editor), Guidebook to Field Trip FT-1, 1975 International Clay Conf. Inst. de Geol. Univ. Nac. A. de México, pp. 73-87.
Hanson, R. F. and Keller, W. D., 1966. Genesis of Refractory Clay near Guanajuato, Mexico. Clays and Clay Minerals, Proc. 14th Conference. Pergamon Press, New York, pp. 259-267.
Imlay, R. W., 1937. Geology of the middle part of the Sierra de Parras. Geol. Soc. America Bull., 48: 587-630.
Imlay, R. W., 1944. Cretaceous formations in Central America and Mexico. Bull. Amer. Asoc. Petro. Geol. 28: 1077-1195.
Ipshording, W., 1973. Discussion on the occurrence and origin of sedimentary palygorskite-sepiolite deposits. Clays and Clay Minerals, 21: 391-401.
Keller, W. D. and Hanson, R. F., 1969. Classification and Problems of Hydrothermal Refractory Clay Deposits in Mexico. Proc. International Clay Conf. Tokyo. Israel Univ. Press, pp. 305-312.
Keller, W. D., Hanson, R. F., Huang, W. H., and Cervantes, A., 1971. Sequented active alteration of rhyolitic volcanic rock to endellite and a precursor phase of it at a spring in Michoacan, Mexico. Clays and Clay Minerals, 19: 121-127.
Kesler, T. L., 1970. Hydrothermal kaolinization in Michoacan, Mexico. Clays and Clay Minerals, 18: 121-124.
Lopez-Ramos, E., 1974. Geología General y de México. México, 509 pp.
Lopez-Ramos, E., 1976. Carta Geológica de la Republica Mexicana. In: S. Hernandez (Editor). Comité de la Carta Geologica de México, Inst. de Geol., Univ. Nac. A. México, México.
McDowell, W. F. and Keizer, P. R., 1977. Timing of mid-Tertiary vulcanism in the Sierra Madre Occidental between Durango City and Mazatlan, Mexico.

Geol. Soc. America Bull., 88: 1479-1487.
Mooser, F., 1956. Los Ciclos de Vulcanismo que Formaron la Cuenca de México. XX Cong. Geol. Intern., Univ. Nac. A. México, Inst. de Geol. México, pp.285-315.
Mooser, F., 1964. Las provincias geotérmicas de México. Bol. Asoc. Mex. Geol. Petr., 16: 153-161.
Mumpton, F. A., 1973. First reported occurrence of zeolites in sedimentary rocks of Mexico. Amer. Mineral., 58:287-290.
Mumpton, F. A., 1975. Zeolitic tuffs in the vicinity of Oaxaca, Mexico. In: L. de Pablo (Editor), Guidebook to Field Trip FT-4, 1975 Int. Clay Conf., Univ. Nac. A. México, Inst. de Geol., México, pp, 45-51.
Muir, J. M., 1936. Geology of the Tampico region, Mexico. Amer. Assoc. Petr. Geol., Tulsa, 280 pp.
Myers, R. L., 1968. Biostratigraphy of the Cardenas Formation (Upper Cretaceous), San Luis Potosi, Mexico. Univ. Nac. A. de México, Inst. de Geol., Paleontología Mexicana, 24.
Pesquera, R., 1975. On the Geology of the Pachuca-Real del Monte District and the Clay Deposits of Huayacocotla, Veracruz, and Pathé, Hidalgo. In: L. de Pablo (Editor), Guidebook Field Trip FT-2, 1975 Int. Clay Conf., Inst. de Geol., Univ. Nac. A. México. México, pp. 1-29.
Pesquera, R., de Pablo, L., and Carbonell, M., 1968. Kaolin Deposits of Mexico. Proc. XXIII Int. Geol. Congress, Symposium I. Kaolin Deposits. Academia, Prague, pp. 105-110.
Quiñones, H., 1975a. Soil Study Area 2 and 3. Teapa, Tabasco-Pichucalco, Chiapas and Palenque, Emiliano Zapata, Chiapas. In: L. de Pablo (Editor), Guidebook to Field Trip FT-4, 1975 Int. Clay Conf., Inst. de Geol., Univ. Nac. A. México, México, pp. 52-68.
Quiñones, H., 1975b. Soil Study Area 4. Intrazonal Soils of Northern Yucatan Peninsula. In: L. de Pablo (Editor), Guidebook to Field Trip FT-4, 1975 Int. Clay Conf., Instituto de Geología, Univ. Nac. A. México, México, pp. 70-94.
Rogers, C. A., de Cserna, Z., Vhoten, R. van, Tavera, E., and Ojeda, J., 1961. Reconocimiento geológico y depósitos de fosfatos del oeste de Zacatecas y areas adyacentes en Coahuila, Nuevo León y San Luis Potosí. Consejo de Recursos Naturales No Renovables, México, Bol. 56, 322 pp.
Salas, G. P., Cordoba, D. A., and Lopez, J., 1974. General aspects of batholits and intrusive rocks of western Mexico. Pacific Geolgy, 8: 67-72.
Schulze, G., 1956a. Estudio sobre la Naturaleza de las Formaciones Rioliticas al Este de la Ciudad de Zacatecas. In: A. Garcia (Editor), Vulcanología del Cenózoico, XX Congr. Geol. Int.. Inst. de Geol., Univ. Nac. A. México, México, pp. 285-315.
Schulze, G., 1956b. Replacement of Tertiary Bedded Red Conglomerate by Rhyolitic Formations. In: A. Garcia (Editor), Vulcanología del Cenozoico, XX Cong. Geol. Int.. Inst. de Geol., Univ. Nac. A. México, México, pp. 3 7-336.
Silver, L. T., Stahl, F. G., and Allen, C. R., 1956. Lower Cretaceous prebatholithic rocks of southern Baja California, Mexico. Amer. Assoc. Petr. Geologists Bull., 47: 2054-2059.
Taliaferro, N. L., 1933. An occurrence of Upper Cretaceous sediments in northern Sonora, Mexico. Jour. Geology, 41: 12-37.
Viniegra-Osorio, F., 1971. Age and evolution of salt basins of southeastern Mexico. Am. Assoc. Petroleum Geologists Bull., 55: 478-494.
Wall, J. R., Murray, G. E., and Diaz, T., 1961. Geologic occurrence of intrusive gypsum and its effect on structural forms in Coahuila marginal folded province on northeasthern Mexico. Am. Assoc. Petr. Geol. Bull., 45: 1504-1522.
Wilson, J. A. and Clabaugh, S. A., 1970. A new Miocene Formation and a description of Volcanic Rocks, Northern Valley of Oaxaca. Libreto Guia de la Excursión México-Oaxaca. Soc. Geol. Mexicana, pp. 120-128.

OTHER PAPERS PRESENTED IN SECTION 5

STONEWARE AND LOW DUTY REFRACTORY CLAYS ASSOCIATED WITH THE ATHABASCA OIL SANDS

D. Scafe (Alberta Research Council, 11315 - 87 Avenue, Edmonton, Alberta Canada T6G 2C2.)

USE OF NATURAL GARNIERITE-TYPE CLAY MINERALS AS CATALYSTS FOR THE METHANATION OF CARBON MONOXIDE

P.A. Jacobs, G. Poncelet[1] and J.B. Uytterhoeven (Katholieke Universiteit Leuven, Centrum voor Oppervlaktescheikunde en Colloidale Scheikunde, De Croylaan 42, B-3030 Heverlee, Belgium. [1] Université Catholique de Louvain, Groupe de Physicochimie Minérale et de Catalyse, Place Croix du Sud, 1, B-1348 Louvain-la-Neuve, Belgium.

SEPARATION OF ALUMINA FROM KAOLIN

Liberto de Pablo Galán (Consejo de Recursos Minerales, Niños Heroes y Dr. Navarro, México 7, D.F., México.)

STRENGTH, PLASTICITY AND SEDIMENT-VOLUME OF CLAYS

M. Müller-Vonmoos and G. Kahr (Institute for Foundation Engineering and Soil Mechanics, Laboratory for Clay Mineralogy, Swiss Federal Institute of Technology, Zurich, Switzerland.)

A MODERN APPRAISAL OF THE WEDGWOOD "CLAY-SHRINKAGE" PYROMETER

J.A. Bain and F.R. Stacey (Institute of Geological Sciences, 64-78 Gray's Inn Road, London, UK.)

OCCURRENCE OF PYROPHYLLITE IN LA SERENA AREA, BADAJOZ, SPAIN

J. Mesa López-Colmenar, G. García Ramos, J.L. Pérez Rodríguez and F. González García (Centro de Edafología y Biología Aplicada del Cuarto, C.S.I.C., Apartado 1052, Sevilla, Spain.)

CERAMIC PROPERTIES AND PHYSICOCHEMICAL STUDIES OF CLAYS FROM PAPIOL, BARCELONA, SPAIN

E. Tauler, C. de la Fuente and S. Martinez (Departamento de Cristalografía y Mineralogía de la Universidad de Barcelona, Avda. José Antonio 585, Barcelona, Spain.)

# SECTION 6

# Non-crystalline and Accessory Minerals

CHAIRMAN'S INTRODUCTION

NON-CRYSTALLINE AND ACCESSORY MINERALS[1]

U. SCHWERTMANN

Institut für Bodenkunde der Techn. Universität München, 805 Freising, FRG.

INTRODUCTORY REMARKS

In the past our knowledge of non-crystalline and accessory minerals has not developed to the same extent as that of clay silicates. The reasons are the difficulty of identification and further description of non-crystalline minerals and for accessory minerals, their low concentration and their frequent removal before clay mineral identification.

For the secondary accessory minerals this situation is surprising since these minerals quite often markedly influence the properties of clays and soils (aggregation, ion adsorption). Also, in contrast to clay silicates, most of these minerals can be synthesized under ambient conditions - and therefore their formation in nature can easily be studied in the laboratory.

ALLOPHANE AND IMOGOLITE

Considerable progress particularly in Japan was made recently with regard to the morphology and structure of these two hitherto called non-crystalline Al silicates (see also the paper by WADA in this volume). Short range (allophane) or long range order at least in one crystallographic direction (imogolite) was discovered. High magnification electron microscopy has shown allophane to consist of hollow spheres 35-55 Å in external diameter with walls 7-10 Å thick made up of one Al octahedral and one Si tetrahedral layer with strong Al substitution. The excess charge could be neutralized by a positive charge in the octahedral layer or by a hydroxylation of apical oxygens of the tetrahedrons. In view of

---

[1] The following is a condensed version of the oral presentation which relies mainly on recent reviews published in "Minerals in Soil Environments" (editors J.B. Dixon and S.B. Weed. Soil Sci. Soc. Am., Madison, 1977). For brevity only those references are cited here, which were not included in those reviews. The relevant chapters from this book are the following: WADA, K: Allophane and imogolite; HSU, P.H: Aluminium oxides and oxyhydroxides; SCHWERTMANN, U. and TAYLOR, R.M: Iron oxides; MACKENZIE, R.M: Manganese oxides and HUTTON, J.T: Titanium and zirconium minerals.

these findings it becomes questionable if these minerals can still be considered as non-crystalline.

Efforts have been made to clarify the role of Fe in the allophane structure by Mößbauer spectroscopy (HORIKAWA and FUIJO 1977) so far without unambiguous results. The existence of hisingerite as a separate phase has been questioned and a sample called hisingerite was later identified as poorly crystalline nontronite (KOHYAMA and SUDO 1975).

Among the methods of differential dissolution of allophane from soils and weathered ash deposits, NaOH has been replaced by acid oxalate (HIGHASHI and IKEDA 1974, HENMI and WADA 1976, FEY and LE ROUX 1977) because the latter is much less likely to dissolve finely dispersed layer silicates.

## ALUMINIUM OXIDES

The mineralogy of natural Al-oxides is fairly simple. Gibbsite is by far the most frequently occurring polymorph of $Al(OH)_3$. It is usually taken as a typical product of tropical, high leaching i.e. low Si concentration weathering environment. Therefore, gibbsite found recently in non-tropical soils trace back the existence of a warmer climate in the past (WILSON 1969, BASHAM 1974, TORRENT and BENAYAS 1977, WILKE and SCHWERTMANN 1977). Gibbsite is often a late product of weathering (see e.g. fig. 11 in SCHEFFER and SCHACHTSCHABEL 1976) but need not to be so, since it is frequently associated with silicates of medium weathering resistance such feldspars, muscovites and even biotites (BASHAM 1974, MÜLLER 1976). Therefore, gibbsite can form at any stage of weathering as long as an environment with < 1 ppm $SiO_2$ in solution exists to prevent Al-silicate formation. This appears to be the case in podsolic soils of southern France (De Coninck pers. comm.) and Spain (LANCHO et al. 1976) where Al-fulvic acid complexes are hydrolyzed to gibbsite in the B-horizon.

Nordstrandite, a much less frequent $Al(OH)_3$ polymorph seems to be a common constituent of Jamaica bauxites (HILL 1977) formed from limestone under alkaline conditions. On acidification nordstrandite converts to gibbsite in agreement with synthesis experiments. In these types of experiments, nordstrandite and bayerite preferably form under alkaline conditions and gibbsite in the acid range. An interesting explanation for this was based by SCHOEN and ROBERSON (1970) on the pH dependent polarization of OH by Al not only existing in the soluble species ($[Al(OH)_{3-n}]^{n+}$ versus $Al(OH)_4^-$) but also involved in governing the structure of the different solid polymorphs.

Although, not preparable under ambient conditions boehmite and diaspore frequently occur in bauxites. Diaspore is commonly weakly Fe substituted (3-4 mole %) (BIAIS et al. 1973, GOUT 1971) and it is likely that this substitution facilitates the formation of diaspore at lower temperature.

Amorphous Al hydroxide sometimes called cliachite was detected in volcanic
ash soils of Hawaii (WADA and WADA 1976) on the basis of selective dissolution
results and electron microscopic observations. Its Fe content might indicate some
Al-Fe substitution analogous to aluminous ferrihydrite.

IRON OXIDES

In this field two lines of research have been followed recently:
(1) Fe-oxides as characteristic and ubiquitous minerals in the weathering environment, their crystallographic properties, occurrence and genesis.
(2) Fe-oxides as finely dispersed compounds with an active surface reacting as a strong sorbent for a large variety of ions and molecules.

New minerals. Among new iron oxide minerals recently found in nature ferrihydrite (composition $5 Fe_2O_3 \cdot 9 H_2O$), formerly called amorphous ferric hydroxide, has been accepted as a mineral (CHUKHROV et al. 1973). It consists of 3-10 nm spheres of high surface area (100-300 m²/g) and has a hematite-like structure with a hexagonal unit cell ($a_o = 5.08$ Å, $c_o = 9.4$ Å) in which part of the Fe sites are vacant and O is replaced by $OH_2$. The structure is based on five diffraction lines at 2.5, 2.2, 1.9, 1.7 and 1.5 Å. A formula proposed is $Fe_5HO_8 \cdot 4H_2O$. Correspondingly, no Fe-OH stretching or bending modes are observed in the IR spectrogram.

The $\delta$-FeOOH known for long as a synthetic phase was recently found in marine iron concretions and accepted by IMA as feroxyhite (CHUKHROV et al. 1976). An intersting compound being an important precursor of Fe(III) oxides when a Fe(II) system is oxidized is the so called green rust. Structurally, it is probably a member of the pyroaurite group (BRINDLEY and BISH 1976) with the general formula $(Me^{2+})_6(Me^{3+})_2(OH)_{16}CO_3 \cdot 4 H_2O$ in which Me is FeII and FeIII and $CO_3$ can be replaced by other anions. Other formulas with varying FeII/FeIII were proposed as well. Green rust forms at a lower pH than $Fe(OH)_2$ possibly through a reaction of $Fe^{2+}$ ions with ferrihydrite. Green rust might be responsible for the green colors often observed in reduced soils and sediments, highly sensitive to aerial oxidation. Its natural occurrence has not as yet been described.

Occurrence and Genesis of Fe-oxides. One attractive aspect in studying the occurrence of the various mineral species of Fe-oxides, is to use them as indicators for particular genetic environments. The investigation in this field is facilitated because (1) all iron oxides can be easily synthesized under near natural conditions and (2) except for time, all other parameters can be duplicated in synthesis experiments.

In an aqueous weathering environment, Fe(III)oxides generally form via solution and not through solid state transformations (except possibly the magnetite⟶maghemite transformation). Solid precursors supply low molecular weight species to the solution which then nucleate and feed the growing crystal. Recent evidence for formation via solution come from (1) a relationship between monomer concentration and rate of goethite formation, (2) an initial increase in the rate during the lepidocrocite⟶ goethite and ferrihydrite⟶ hematite transformation indicating a nucleation phase, (3) XRD spectrograms and high magnification electron microscopy supplying no evidence for intermediate phases during the ferrihydrite ⟶ goethite (MURPHY et al. 1976), the ferrihydrite⟶ hematite, and the lepidocrocite ⟶ goethite transformation and (4) Al incorporation from solution into aluminous goethite.

From this it follows, that the structure of any solid precursor such as Fe silicate, carbonate, sulphide, etc. has hardly any influence on the formation of Fe(III)oxides (except through the rate of Fe release), whereas environmental factors such as temperature, pH, composition of the aqueous phase etc. are much more important.

The occurrence of certain oxide associations in soils and sediments can not be fully explained on thermodynamic reasoning. Kinetic factors must be considered as well. Metastable forms such as lepidocrocite and ferrihydrite appear kinetically favored (such as opal versus quartz) or may be stable for long periods because of sluggish transformation kinetics.

The possible pathways during the formation and transformation of the common mineral species can be briefly summarized as follows:

Under surface conditions, hematite seems to form only through a dehydration-recrystallization process from ferrihydrite but not through a dehydration of goethite. The tendency for ferrihydrite to form and induce hematite is higher under a subtropical and tropical climate because of a higher probability under these conditions that the relatively higher solubility product of ferrihydrite (vis-a-vis goethite) will be exceeded. The reasons for this are (1) higher Fe release on weathering and (2) lower concentration of Fe complexing organic ligands because of a faster turn over of organic matter at a higher temperature. The transformation of ferrihydrite to hematite is preceeded by an aggregation of the 3-5 nm spheres of ferrihydrite and involves a nucleation phase. It is again favored (over the transformation via solution to goethite) by higher temperature because a dehydration step is involved and by Al in the system (SCHWERTMANN et al. 1978).

Thermodynamically most stable goethite is ubiquitous in soils and sediments of any climatic environment. It forms either through a slow hydrolysis of low molecular weight Fe(III) cations (supplied by any Fe(III) source such as ferrihydrite) or through an oxidation of Fe(II)ions via an intermediate phase possibly related to the green rust. In the Fe(II)system goethite is favored over lepidocrocite by carbonate ions and by Al. Hematite formed under a warmer climate but later exposed to a cooler one, transforms to goethite after dissolution through reduction and/or complexation of Fe and its reprecipitation.

Lepidocrocite, very commonly associated with goethite but not with hematite, generally forms through an oxidation of FeII and is therefore typically occurring in hydromorphic, anaerobic soils and sediments. Local micro variations in pH, carbonate and Al concentration, and oxidation rate govern the proportion of lepidocrocite and goethite which are most likely formed from a common precursor (green rust) but not through a transformation. Reasonably well crystalline lepidocrocite reflects the slow oxidation in poorly aerated soils. On rapid oxidation however, as in ferriferous springs, poorly crystalline lepidocrocite with 5-10 nm layers per crystal forms with a consistently higher (020) basal spacing (analogous to pseudoboehmite) (TOWE and RÜTZLER 1968) and strong differential line broadening. Under surface conditions the transformation of lepidocrocite to the more stable goethite appears to be inhibited by slow dissolution of lepidocrocite and/or interference of goethite nucleation by solution compounds such as silicate, organics, etc.

Rapid oxidation, partly microbiological, of ferriferous waters may lead to ferrihydrite. The transformation of this rather unstable compound ($pK_s$ = 38-39) is again blocked by those compounds with a high affinity for the ferrihydrite surface. Maghemite, the ferromagnetic cubic form of $Fe_2O_3$, commonly results from the oxidation of magnetite (for the mechanism see SIDHU 1978). However, because it can be easily synthesized under ambient conditions by oxidizing green rust, it remains to be seen if magnetite is a necessary precursor for the wide spread occurrence of maghemite in tropical soils.

Isomorphous substitution. This phenomenon, very common in clay silicates, is most widespread in Fe oxides as well. The best known example is the replacement of Fe by Al for goethite known since 1941 (CORRENS and v. ENGELHARDT 1941). Other work has been recently published in this area.

Goethite appears to incorporate Al to an extent of 1/3 of the octadral positions, hematites to an extent of 1/6, whereas Al substituted lepidocrocite has not as yet been reported in the literature. The shrinking of the unit cell may not follow the Vegard rule. In soils and bauxites the substituted oxides seem to occur more commonly than the pure forms.

Al substitution strongly affects the formation and properties of synthetic and natural Fe oxides. In particular it (1) favors hematite over goethite in a ferrihydrite system and goethite over lepidocrocite in a FeII system, (2) it improves crystallinity at low substitution and retards crystal growth at high substitution in the needle direction with goethite and perpendicular to the plate with hematite (SCHWERTMANN et al. 1978) (3) it decreases dissolution rate in acids

(4) it increases phosphate adsorption per unit surface area but lowers the bonding energy (GOLDEN 1978) and (5) it produces a double peak in d.t.a.

Surface reactions. Fe oxides have constant potential surfaces with $H^+$ and $OH^-$ being the potential determining ions. Their zpc values range between pH 7 and 9 without indicating much of a mineral specifity probably because the surface of all the different forms are hydroxylated in the presence of water through a reaction between surface groups and the water molecule. All mineral species are generally strong sorbents for a large range of different anions and cations through so called specific adsorption. The amount adsorbed is again more a function of surface area than of mineral species. The same holds true for their aggregation effect on silt particles.

The surface of iron oxides can, however, not be considered homogenous. Detailed IR studies have shown recently that various anions prefer to react with certain surface groups of goethite: bicarbonate, sulphate and phosphate ions with single coordinated Fe-OH groups to form bidentate surface complexes, $CO_3^{2-}$ ions also with surface oxo groups (RUSSELL et al. 1975, PARFITT et al. 1975, 1977). Analogous observations were made with platy gibbsite crystals where only edge faces were active in absorbing phosphate and various organic anions (PARFITT et al. 1977a). Functional group specific reactions might also be the reason for crystal face specific rates of initial dissolution of goethite in strong acids: the rate was much higher for the (001) and (010) than for the (100) face which correlates with the density of single coordinated Fe-OH groups on the faces (CORNELL et al. 1974). Logically, the isotropic magnetite and maghemite dissolved homogenously (SIDHU 1978).

## MANGANESE OXIDES

The Mn oxides exhibit much greater variety than Al and Fe oxides. Birnessite and todorokite, but also ramsdellite, hollandite and lithiophorite were identified more or less frequently in soils and sediments including deep sea nodules. They vary structurally by the way $MnO_6$ octahedrons are arranged and by the type and amount of cations such as K, Li, Ca, Ba, and Al incorporated to balance for lower valent Mn. An amorphous Mn oxide, sometimes called wad, was detected recently in soils (ROSS et al. 1976).

The way the various minerals form in nature is almost completely unknown. Synthesis experiments are usually done under conditions far away from those in the weathering environment (permanganate, alkaline conditions). The main problem is to oxidize MnII to MnIV under natural conditions. Catalytic effects of oxide surfaces, foreign constitutional cations or, even more likely, plant and microbial enzymes may facilitate the oxidation (SCHWEISSFURTH a. GATTOW 1966).

Mn-oxides are of special interest as heavy metal binders, particularly for Ni, Co, Cu and Zn (review see JENNE 1977). Deep sea Mn oxide nodules are therefore of increasing interest as potential ores. In soils they can control the Co nutrition of plants. The mechanism of these reactions are not fully understood yet. Although stoichiometric amounts of protons are involved, structural Mn is also replaced by these metal cations leading to their incorporation into the structure. High concentration ratios of these metals between Mn-oxide nodules and the soil matrix are the result.

## TITANIUM OXIDES

The fate of Ti in the weathering cycle is much less understood than any of the other metals mentioned. Among the $TiO_2$ phases <u>anatase</u> appears to be a surface weathering product whereas rutile is usually detrital. Anatase was recently detected in kaolin (WEAVER 1976), in soil clay and in silcretes.

Besides the pure $TiO_2$ a series of mixed Ti-Fe-oxides also occur as weathering products, <u>pseudorutile</u> being one with a known composition ($Fe_2Ti_3O_9$) and structure.

The Ti released by weathering from minerals containing Ti as a major constituent (sphene, ilmenite, titanomagnetite) but possibly more so (due to their lower weathering resistance) from mafic minerals such as biotite and Ti-augite may recrystallize to form anatase or pseudorutile or form an amorphous oxide, called <u>leucoxene</u>. FITZPATRICK et al. at this conference have demonstrated that crystalline Ti- and Ti-Fe-oxides can easily be synthesized from an amorphous precursor at near pedogenic conditions. Their results indicate a range of mixed Ti-Fe-oxides between anatase, ferriferous anatase and pseudorutile of varying Ti/Fe-ratio.

## FUTURE WORK

The genesis of many of the non-crystalline and accessory minerals is still widely unknown. Some cannot be synthesized at all under ambient conditions although they must have formed under these conditions in nature. For others, the mode of formation likely to have taken place in nature has not yet been successfully copied in vitro. Constituents of the environment interfering in some way or another with the formation should be more and more included.

The investigation of the properties of different minerals usually makes use of mineral specimens which are as pure as possible in order to be able to explain the results. This is fully justified. However, the same gross species formed in nature might behave quite differently. Therefore, the study of natural samples should be intensified and this knowledge should govern the type of synthetic minerals to be studied.

Among the properties to be studied are crystal size and morphology, structural disorder, ion substitution and crystal face specific surface reactions. More refined spectrographic methods such as Mössbauer spectroscopy, expanded IR spectroscopy, NMR, ESR, ESCA will increasingly come into use.

Among the minerals presently called non-crystalline, the trend to discover a certain order, short or even long range, will continue and poorly ordered members of the Al, Mn and Ti-oxide groups will be discovered in nature. Procedures should be developed to enable the investigation of minerals, such as green rust which are sensitive to aerial oxidation. To facilitate the study of accessory minerals in soils and clays, methods should be worked out to concentrate them by removing the clay silicates or by high gradient magnetic separation.

REFERENCES

Basham, I.R., 1974. Mineralogical changes associated with deep weathering of gabbro in Aberdeenshire. Clay Miner. 10:189-202.
Biais, R., Gramont, X. de, Janot, C., and Charrier, J., 1973. Contribution à l'étude des substitutions Fe-Al dans des roches latéritiques ainsi-que dans des hydroxydes et oxydes de synthèse. ICSOBA troisième congrès international. Nice 1973.
Brindley, G.W. and Bish, D.L., 1976. Green Rust: a pyroaurite type structure. Nature, 263:353.
Chukhrov, F.V., Ermilova, L.P., Zvyagin, B.B. and Gorshkov, A.I., 1973. New data on iron oxides in the weathering zone. Proc. Int. Clay Conf. 1972 (Madrid) 1:397-404.
Chukhrov, F.V., Zvyagin, B.B., Yermilova, L.P. and Gorshkov, A.I., 1976. Mineralogical criteria in the origin of marine iron-manganese nodules. Mineral. Deposita (Berl.) 11:24-32.
Correns, C.W. and v. Engelhardt, W., 1941. Röntgenographische Untersuchungen über den Mineralbestand sedimentärer Eisenerze. Nachr. Akad. Wiss. Göttingen. Math.-Phys. Kl. 213:131-137.
Cornell, R.M., Posner, A.M. and Quirk, J.P., 1974. Crystal morphology and the dissolution of goethite. J. Inorg. Nucl. Chem. 36:1937-1943.
Fey, M.V. and Le Roux, J., 1977. Properties and quantitative estimation of poorly crystalline components in sesquioxidic soil clays. Clays and Clay Miner. 25:285-294.
Golden, D.C., 1978. Physical and chemical properties of aluminum-substituted goethite. Ph.D. Thesis Univ. of North Carolina.
Gout, R., 1971. Sur les ions ferriques présents dans les réseau des diaspores. 96$^e$ congrès national des sociétés savantes, Toulouse.
Henmi, T. and Wada, K., 1976. Morphology and composition of allophane. Am. Mineral. 61:379-390.
Highashi, T. and Ikeda, H., 1974. Dissolution of allophane by acid oxalate solution. Clay Sci. 4:205-212.
Hill, V.G., 1977. Syngenetic and diagenetic changes in Jamaica bauxite deposits. ICSOBA Jamaica, in press.
Horikawa, Y. and Fuijo, Y., 1977. State analysis of iron in allophanic clays. I. Mössbauer effect analysis of iron in allophanic clays. Clay Sci. 5:67-77.
Jenne, E.A., 1977. Trace element sorption by sediments and soils - sites and progresses. Chapter 5 in Chappel, W. and Peterson, K. (Eds.) Symp. on Molybdenum in the environment. M. Dekker, Inc. N.Y.
Kohyama, N. and Sudo, T., 1975. Hisingerite occurring as a weathering product or iron-rich saponite. Clays and Clay Min. 23:215-218.

Lancho, J.G., Sanchez Camanzano, M., Alonso, J.S. and Sanchez, A.G., 1976. Influencia de la materia organica en la genesis de gibsita y caolinita en suelos graniticos del centro-oeste de Espana. Clay Miner. 11:241.

Müller, H., 1976. Mineralogische und chemische Untersuchungen von lateritischen Rotlehmen aus Nepal. Sitzungsber. Österr. Akad. Wiss. Math.-naturw. Kl. Abt. I 185:43-53.

Murphy, P.J., Posner, A.M. and Quirk, J.P., 1976. Characterization of partially neutralized ferric perchlorate solutions. J. Colloid Interf. Sci. 56:289-311.

Parfitt, R.L., Atkinson, R.J. and Smart, R.St.C., 1975. The mechanism of phosphate fixation by iron oxides. Soil Sci. Soc. Amer. Proc. 39:837-841.

Parfitt, R.L. and Smart, R.St.C., 1977. Infrared spectra from binuclear bridging complexes of sulphate adsorbed on goethite. J. Chem. Soc. Faraday I. 73:796-802. See also Soil Sci. Soc. Amer. J. 1978, 42:48-50.

Parfitt, R.L., Fraser, A.R., Russell, J.D. and Farmer, V.C., 1977a. Adsorption on hydrous oxides. II. Oxalate, benzoate and phosphate on gibbsite. J. Soil Sci. 28:40-47.

Ross, S.J., Jr., Franzmeier, D.P. and Roth, C.B., 1976. Mineralogy and chemistry of manganese oxides in some Indiana soils. Soil Sci. Soc. Amer. J. 40:137-143.

Russell, J.D., Paterson, E., Fraser, A.R. and Farmer, V.C., 1975. Adsorption of carbon dioxide on goethite ($\alpha$-FeOOH) surfaces and its implications for anion adsorption. J. Chem. Soc. Faraday Trans. 71:1623-1630.

Scheffer, F. and Schachtschabel, P., 1976. Lehrbuch der Bodenkunde. 9. Aufl., neubearbeitet von P. Schachtschabel, H.-P. Blume, K.H. Hartge und U. Schwertmann Ferd. Enke Verlag, Stuttgart.

Schoen, R. and Roberson, E.C., 1970. Structures of aluminium hydroxide and geochemical implication. Am. Mineral. 55:43-77.

Schweissfurth, R. and Gattow, G., 1966. Untersuchungen über die Struktur und Zusammensetzung mikrobiell gebildeter Braunsteine. Z. allg. Mikrobiol. 6:303-308.

Schwertmann, U., Fitzpatrick, R.W., Taylor, R.M. and Lewis, D.G., 1978. The influence of Al on Fe oxide formation. Part II. Clays and Clay Miner. (submitted)

Sidhu, P.S., 1978. Some properties of trace element substituted magnetites, maghemites and hematites. Ph.D. Thesis. Univ. of Western Australia.

Torrent, J. and Benayas, J., 1977. Origin of gibbsite in a weathering profile from granite in west-central Spain. Geoderma 19:37-51.

Towe, K.M. and Rützler, K., 1968. Lepidocrocite iron mineralization in keratose sponge granules. Science 162:268-269.

Wada, K. and Wada, S., 1976. Clay mineralogy of the B-horizon of two hydrandepts, a torrox and a humitropept in Hawaii. Geoderma 16:139-157.

Weaver, Ch.E., 1976. The nature of $TiO_2$ in kaolinite. Clays and Clay Miner. 24:215-218.

Wilke, B.-M. and Schwertmann, U., 1977. Gibbsite and halloysite decomposition in strongly acid podzolic soils developed from grantic saprolite of the Bayerischer Wald. Geoderma 19:51-61.

Wilson, M.J., 1969. A gibbsitic soil derived from the weathering of an ultrabasic rock on the island of Rhum. J.Geol. 5:81-89.

REVERSIBILITY OF LATTICE COLLAPSE IN SYNTHETIC BUSERITE

M. I. TEJEDOR-TEJEDOR* and E. PATERSON
Department of Pedology, Macaulay Institute for Soil Research, Craigiebuckler, Aberdeen AB9 2QJ, Scotland
* Permanent address: Department of Analytical Chemistry, CSIC, Madrid, Spain.

ABSTRACT

The basal spacing of buserite, a hydrous manganese oxide with a layer structure, has been studied as a function of the exchangeable cation present and the relative humidity at which the sample is equilibrated. With the cations $Na^+$, $Mg^{2+}$, $Ca^{2+}$ and $Sr^{2+}$ either a 10Å or a 7Å reflection can be observed, whereas, with the cations $Li^+$, $K^+$, $Cs^+$ and $Ba^{2+}$ only a 7Å phase is detected. These results and those obtained from determination of the water content of the various ion forms suggest that the "disordered" layer, thought to be present in this compound, consists of a discrete layer of hydrated cations.

INTRODUCTION

The hydrous oxides of manganese play an important role in the sorption and release of a variety of cations in soils and sediments (Jenne, 1967); indeed, the availability of the essential micronutrient, cobalt, is thought to be completely controlled by their presence (Taylor and McKenzie, 1966). These observations have led to a number of studies on the sorption properties of manganese oxides extracted from soil nodules and concretions (McKenzie, 1967) and of synthetic hydrous oxides. The synthetic oxide most relevant to soil systems is $\delta-MnO_2$ which is known to occur as the mineral birnessite in a wide range of soils (Taylor et al., 1964).

The work of Loganathan and Burau (1973) suggests that two types of interaction occur between $\delta-MnO_2$ and a range of cations. With cations of the first transition series, manganese is released into solution on sorption, suggesting that manganese within the structure is being replaced. However, with main group cations such as sodium and calcium no such release occurs and although sorption at low concentrations shows a marked pH-dependence, consistent with adsorption into the diffuse double layer, at higher concentrations the amount adsorbed is independent of pH. This is thought to be due to penetration of the exchangeable cations into the interlayer region of the structure (Murray et al., 1968).

One of the major difficulties in interpreting the results of these studies is the uncertainty relating to specific aspects of the structure of the synthetic product used. Although it is known that the layer structured manganates, of which buserite can be considered the parent compound (Giovanoli and Burki, 1975), have as their basic structural unit a layer of edge-sharing $[MnO_6]$ octahedra, there is some doubt concerning the contents and structure of the interlayer region (Burns and Burns, 1975).

In this study the structural changes occurring on treatment of synthetic buserite with cations from Groups I and IIa of the Periodic Table have been assessed using X-ray diffraction. In addition, water vapour sorption isotherms have been used along with thermogravimetric data to interpret the changes in basal spacings observed for the different cation forms.

EXPERIMENTAL

a) Materials

Sodium buserite was synthesised using the method described by Giovanoli et al. (1970). An aqueous solution of sodium hydroxide (0.25 $dm^3$, 5.2M) was added rapidly to 0.2 $dm^3$ of 0.5M manganous sulphate solution in a 1 $dm^3$ measuring cylinder. Oxygen was immediately passed through the suspension of manganous hydroxide at a flow rate of 2 $dm^3$ $min^{-1}$ using a fritted glass tube to produce a stream of small bubbles and to ensure oxidizing conditions at the outset. After 5 hours the black precipitate was allowed to settle and the supernatant liquid poured off. The precipitate was then washed free of sulphate with distilled water and stored in aqueous suspension until required.

In preparing the other cation forms an aliquot of the sodium buserite suspension was treated with an aqueous 1M solution of the appropriate chloride. After standing overnight the buserite was washed four times with an 0.1M chloride solution, then with de-ionised water until the supernatant liquid was chloride-free. The samples were again stored as an aqueous suspension (solid concentration 5mg/ml).

b) Methods

X-ray diffraction was carried out using a Philips 2kW diffractometer with Fe-filtered Co K$\alpha$ radiation ($\lambda$ = 1.790A). The samples, prepared as oriented aggregates by sedimentation on glass slides, were dried in a desiccator over saturated salt solutions chosen to give the required relative humidity (R.H.). After equilibration, the slide was wrapped in a thin self-adhesive plastic film which maintained the water content reasonably constant until X-ray examination was complete.

Water vapour sorption isotherms were obtained by weighing samples on a semi-micro balance, with an accuracy of 0.01mg, after equilibration in a vacuum desiccator at the required humidity. The samples were dried at room humidity (approximately 40% R.H.), lightly ground to -100 mesh, then subjected to an adsorption cycle up to 96% R.H. followed by desorption.

Thermogravimetry was carried out using an electrobalance assembly, similar in many respects to the Stanton Redcroft TG 750. A sample weight of 5-10 mg was used with a heating rate of 10K min$^{-1}$ in an atmosphere of nitrogen flowing around the sample at a rate of 30 cm$^3$ min$^{-1}$.

CHARACTERISATION OF SYNTHETIC PRODUCT

Electron micrographs of the product showed that, in addition to particles with a platy morphology of the type ascribed to buserite (Giovanoli et al., 1970), a small amount of unidentified thread-like material also occurred. However, the amount was so small that it was considered negligible for the purposes of this study.

It has been suggested by Giovanoli et al. (1970) that buserite contains di-, tri- and tetravalent manganese. Since a suitable chemical method to test this was not available, the average oxidation number of the manganese was determined using a iodometric/complexometric method (Gattow and Wendlandt, 1960). A value of +3.60 was obtained - the same as that given by Giovanoli et al., (1970).

The cation-exchange capacity (Mackenzie, 1952) of the sample dried at 56% R.H. was found to be 240 meq/100g.

RESULTS AND DISCUSSION

The major peak on the X-ray diffractogram of synthetic buserite is a 002 reflection at about 10Å, consistent with a layer structure in which the sheets of octahedrally co-ordinated manganese ions occur every 10Å. For Na-buserite this spacing is known to reduce to 7Å on drying (Giovanoli et al., 1970), but the basal spacing varies with both the exchangeable cation present and the relative humidity at which the sample is examined (Table 1).

TABLE 1

Basal spacings ($d_{002}$) of various cation forms at different relative humidities.

| Cation | Radius Å | Relative Humidity | | | | |
|---|---|---|---|---|---|---|
| | | 3% | 33% | 56% | 86% | moist |
| $Li^+$ | 0.60 | 7.0 | – | 7.1 | 7.1 | 7.1 |
| $Na^+$ | 0.95 | 7.1 | – | 7.2 | 7.2 | 10.1 |
| $K^+$ | 1.33 | 7.1 | – | 7.1 | 7.1 | 7.0 |
| $Cs^+$ | 1.69 | 7.3 | – | 7.3 | 7.3 | 7.3 |
| $Mg^{2+}$ | 0.65 | 7.1 | 9.7 | 9.7 | 9.7 | 9.5 |
| $Ca^{2+}$ | 0.99 | 7.1 | 7.1 | 10.0 | 10.1 | 9.9 |
| $Sr^{2+}$ | 1.13 | 7.0 | – | 7.1 | 7.2 | 9.8 |
| $Ba^{2+}$ | 1.35 | 7.0 | – | 7.0 | 7.0 | 7.0 |

The results in Table 1 enable the various cation forms to be divided into two groups - (a) one containing Li-, K-, Cs- and Ba-buserite which gives a 7Å basal spacing at all relative humidities and (b) one containing Na-, Mg-, Ca- and Sr-buserite which gives a basal spacing that is dependent on relative humidity. The latter group can be further subdivided; thus, the Na- and Sr- saturated samples give 10Å spacings when wet but collapse irreversibly to 7Å on drying at 86% R.H., the Ca-form, with a 10Å spacing at 40% R.H., collapses reversibly to 7Å at 33% R.H., and Mg-buserite collapses irreversibly to the 7Å phase at about 10% R.H.

It is interesting to note, that despite the wide variation of cation radii, the $d_{002}$ reflections for the collapsed lattice of all samples are similar, only the very large $Cs^+$ ion causing any significant increase. Thus, the basal spacing is not controlled by the saturating cation and it seems likely that a species, such as water, intermediate in size between $Ba^{2+}$ and $Cs^+$ is responsible. From the difference between the observed spacing and the thickness of the $[MnO_6]$ layer, 4.4Å, the interlayer distance can be calculated as 2.6Å. This value is similar to the increase in basal spacing on formation of a one-layer hydrate in vermiculite and suggests that the 7Å and 10Å spacings for buserite may reflect a one-layer and a two-layer hydrate, respectively.

Both water-vapour desorption isotherms (Fig. 1) and thermogravimetric data (Table 2) were used to obtain quantitative information on the amount of water lost by the samples during dehydration.

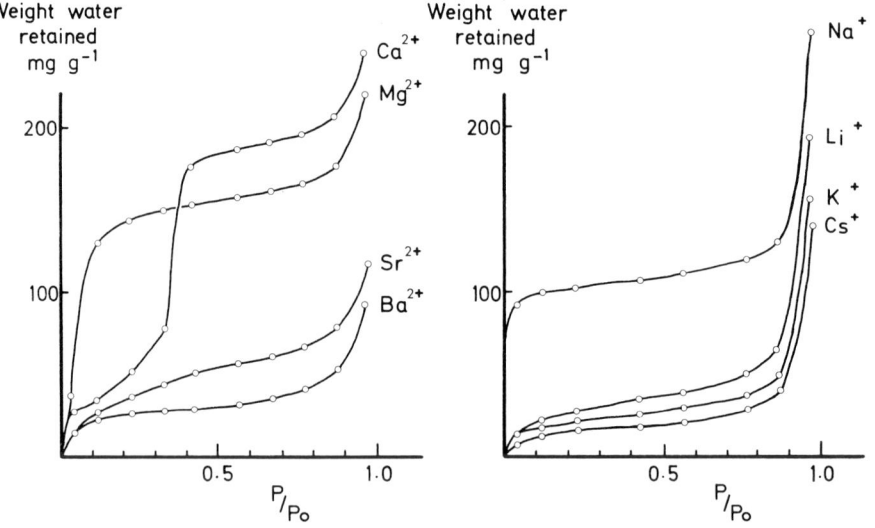

Fig. 1. Water vapour desorption isotherms for buserite saturated with various catio

The fact that buserite saturated with the monovalent cations $Li^+$, $K^+$ and $Cs^+$ gave a sigmoidal desorption curve that was almost completely reversible is consistent with an adsorption/desorption process involving external surfaces only. Although a similar shape of isotherm is obtained for Na-buserite, a considerable amount of weight is lost on drying over $P_2O_5$. Subsequent adsorption measurements show that, at any given relative humidity, only about 60% of the water content on the desorption cycle is attained, indicating the occurrence of an irreversible change on drying.

For the divalent cations, the desorption curve for Ba-buserite is very similar to those obtained for the Li-, K- and Cs-saturated samples and, again, adsorption on external surfaces is indicated. Sr-buserite exhibits a slightly different behaviour, a greater amount of water being retained at intermediate relative humidities. This is probably due to desorption from the interlayer since, throughout this range of relative humidities, there is a gradual decrease in the basal spacing from 7.2Å at 86% R.H. to 7.0Å at 3% R.H. The most striking desorption curves were obtained with the Mg- and Ca-buserites. For the Ca-saturated sample a large decrease in water content occurs between 40 and 30% R.H., corresponding to the collapse from the 10Å to the 7Å phase; a similar decrease was observed for the Mg-buserite at a much lower relative humidity. After correcting for the water lost by desorption from external surfaces the interlayer water contents of Mg- and Ca-buserite were found to be similar (110 and 120 mg/g dry sample, respectively). If a close-packed monolayer is assumed to exist then each water molecule will occupy an area of 0.105 nm² and from the water contents the interlayer area of buserite would be 800 m²g⁻¹ - a value of the same order as, but slightly larger than, the calculated area of 720 m²g⁻¹ for $\delta$-$MnO_2$ (Burns and Burns, 1975).

TABLE 2

Weight losses from TG curves for buserite saturated with various cations.

| Cation Form | % Weight Loss on Dry Basis | | |
|---|---|---|---|
| | 1st weight loss | 2nd weight loss | 3rd weight loss |
| $Li^+$ | 2.5 | 11.0 | - |
| $Na^+$ | 2.4 | 9.6 | - |
| $K^+$ | 2.6 | 6.6 | - |
| $Cs^+$ | 2.8 | 4.6 | - |
| $Mg^{2+}$ | 16.4 | 9.0 | 3.0 |
| $Ca^{2+}$ | 17.3 | 7.3 | 4.0 |
| $Sr^{2+}$ | 3.4 | 6.1 | 3.4 |
| $Ba^{2+}$ | 2.0 | 6.5 | 2.0 |

While the water vapour desorption curves for Li-, K- and Cs-buserite indicate loss of water from external surfaces only, the TG curves for these samples show that, in addition to this physically adsorbed water lost at relatively low temperatures (<100°C), a rapid weight loss occurs in the region of 160°C. The Na-sample exhibits a similar weight loss but at a somewhat lower temperature (120°C). The weight lost during this stage is markedly reduced if the sample has been evacuated over $P_2O$, thus explaining the considerable water loss observed at low relative humidities on the desorption curve for Na-buserite.

For the divalent cations, the first weight loss on the TG curves represents desorption of water from external surfaces for Ba-buserite, but possibly includes some interlayer water for the Sr-saturated sample. Mg- and Ca-buserite lose much more than the other ion forms, as would be expected from the stability of the 10Å phase. The second weight loss is associated with the removal of water from the 7Å phase and the third weight loss, which occurs at much higher temperatures, is probably due to removal of oxygen. Although the evolved gas was not analysed, TG curves obtained in an oxygen atmosphere showed a significant change in this region.

The presence of water in all of these samples when collapsed to 7Å is consistent with a structure in which the basal spacing is controlled by the largest species in the interlayer, namely water. Only for $Cs^+$, the largest cation, does the basal spacing increase to 7.3Å. In addition, the decrease in the amount of water held as the size of the interlayer cation increases suggests that a complete monolayer of water is not present but that a layer consisting of both water molecules and exchangeable cations is formed.

Heat treatment of the alkali-saturated buserites shows that a further lattice collapse occurs at temperatures similar to those at which the rapid weight loss is observed on the TG curves. The XRD curve for Li-buserite heated to 200°C shows three very weak reflections at ca 4.3, 4.82 and 6.77Å, that at 4.82Å being the most intense. Na-buserite, when heated to the same temperature, shows only two peaks at 5.64 and 6.88Å and further heating to 350°C causes the 5.64Å peak to disappear leaving only a broad reflection at 6.86Å. For K-buserite, heating to 300°C produces a small peak at 5.88Å, the main peak at 7.11 Å becoming markedly asymmetrical; further heating increases the intensity of the peak at 5.80Å and the asymmetry of the 7.07Å peak. Finally, no significant change occurs in the basal spacing of 7.29Å for Cs-buserite even on heating to temperatures far in excess of those required to remove interlayer water.

Thus, dehydration of the alkali buserites by heating causes a progressive change in the XRD trace with the appearance of a lower spacing, the position of which depends on the size of the saturating cation (Fig. 2). The calculated values have been plotted assuming that (a) the thickness of the [$MnO_6$] layer is 4.4Å and (b) there is no penetration of the cation into the [$MnO_6$] layer. The observed curvilinear relationship for the experimental values and its displacement relative to the theoretical curve suggests that penetration of the saturating cation does, in fact, occur. In addition, the presence of a small peak at 4.3-4.4Å for Li-buserite, which may represent the thickness of the octahedral layer, and the low thermal stability of the fully collapsed spacing for Li- and Na-buserite may reflect the ease with which these cations diffuse into the structure. It is not clear, however, whether the remarkable stability of the 7Å reflection is due to difficulty in removing the last traces of interlayer water or to the formation of some other interlayer species on dehydration.

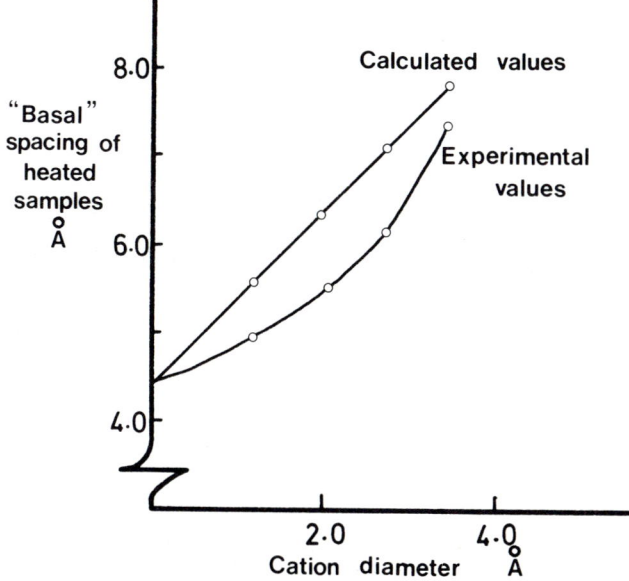

Fig. 2. Variation of "basal" spacings of heated alkali buserites as a function of the diameter of the saturating cation.

CONCLUSIONS

The stability of the 10Å phase in synthetic buserite is determined by the nature of the saturating cation. With the exception of $Li^+$, which occupies an anomalous position, the more highly hydrated cations stabilise the 10Å basal spacing.

Although some of the properties of buserite are similar to those of clay minerals such as vermiculite there are important differences, the principal one being a lack of reversibility in the dehydration reaction. Only with Ca-buserite is the collapse of the 10Å phase reversible. The reasons for this are not yet clear but there is little doubt that the high exchange capacity and its extremely localised nature may well play an important part. Despite this, the variation in lattice spacing for the different cation forms indicates a mechanism whereby large cations such as $Ba^{2+}$ can be fixed by layer-structured manganates.

ACKNOWLEDGEMENTS

The authors wish to thank Mr D. R. Clark for help in obtaining the XRD results and one of us (M. I. T-T.) would like to thank the Royal Society for financial support under the European Science Exchange Programme.

REFERENCES

Burns, R. G. and Burns, V. M., 1975. "Structural Relationships between the Manganese (IV) Oxides". Proceedings of a Manganese Dioxide Symposium, 306-327, ed. A. Kozana and R. J. Brodd, The Electrochem., Cleveland Section.
Gattow, G. and Wendlandt, H. G., 1960. "Zur analytischen Untersuchung von Braunsteinen". Z. Anal. Chem., 174: 15-23.
Giovanoli, R. and Bürki, P., 1975. "Comparison of X-ray evidence of Marine Manganese Nodules and Non-marine Manganese Ore Deposits". Chimia, 29: 266-269.
Giovanoli, R., Stähli, E. and Feitknecht, N., 1970. "Über Oxidhydroxide des vierwertigen Mangans mit Schichtengitter".
Part I "Natriummangan (II, III) manganat (IV)". Helv. Chim. Acta, 53: 209-220.
Part II "Mangan (III) manganat (IV)". Helv. Chim. Acta, 53: 453-464.
Jenne, E. A., 1967. "Controls on Mn, Fe, Co, Ni, Cu and Zn Concentrations in Soils and Water: The Significant Role of Hydrous Mn and Fe Oxides". Adv. Chem. Ser., 73: 337-387.
Loganathan, P. and Burau, R. G., 1973. "Sorption of Heavy Metal Ions by a Hydrous Manganese Oxide". Geochim. et Cosmochim. Acta, 37: 1277-1293.
Mackenzie, R. C., 1952. "A Micro Method for Determination of Cation Exchange Capacity of Clay". Clay Min. Bull., 1: 203-205.
McKenzie, R. M., 1967. "The Sorption of Cobalt by Manganese Minerals in Soils". Aust. J. Soil Res., 5: 235-246.
Murray, D. J., Healy, T. W. and Fuerstenau, D. W., 1968. "The Adsorption of Aqueous Metal on Colloidal Hydrous Manganese Oxide". Adv. Chem. Ser., 79: 74-81.
Taylor, R. M. and McKenzie, R. M., 1966. "The Association of Trace Elements with Manganese Minerals in Australian Soils". Aust. J. Soil Res., 429-39.
Taylor, R. M., McKenzie, R. M. and Norrish, K., 1964. "The Mineralogy and Chemistry of Manganese in Some Australian Soils". Aust. J. Soil Res., 2: 235-248.

INTERCALATION COMPOUNDS OF $KHSi_2O_5$ AND $H_2Si_2O_5$ WITH ALKYLAMMONIUM IONS AND ALKYLAMINES

A. KALT, B. PERATI and R. WEY
Ecole Nationale Supérieure de Chimie de Mulhouse (France)

ABSTRACT

By acid leaching, the layer silicate, $KHSi_2O_5$, can be converted into a phyllodisilicic acid, $H_2Si_2O_5$, which is a crystalline hydrated silica. Both products react with aqueous solutions of n-alkylammonium salts, or n-alkylamines, to form organic intercalation derivatives with larger basal spacings than the original materials. The basal spacings of the compounds obtained with the silicic acid increase with length of the alkyl chain. With the silicate, however, the increase is nearly the same (about 2.5 - 3.0 Å) for all short-chain ammonium ions (number of carbon atoms $n_c<7$). Longer chains give values of the same order of magnitude as with the silica. Treatment of the short-chain alkylammonium derivatives with the corresponding amines increases their spacings markedly, nearly to those of the $H_2Si_2O_5$-amine derivatives.

INTRODUCTION

Up to the middle of the sixties it was widely assumed that all hydrated silicas were amorphous. The very few papers reporting the formation of crystalline hydrated silicas did not find a large audience perhaps because they did not meet this general view (Schwarz and Menner, 1924 ; Schwarz and Richter, 1927 ; Pabst, 1943 ; Pabst, 1958). But during the last fifteen years several crystalline hydrated silicas were obtained. All were prepared from synthetic (Iler, 1964 ; Wodtke and Liebau, 1965 ; Wey and Kalt, 1967a ; Lagaly et al., 1973 ; Hubert et al., 1974) or natural (Eugster, 1967 ; Johan and Maglione, 1972) alkaline (essentially sodium) silicates with layer structures, by treatment with acids. A natural form called silhydrite has even been discovered (Gudde and Sheppard, 1972). Very recently new crystalline hydrated silicas were obtained from copper silicates (Guth et al., 1977 ; Guth et al., 1978).

Cation exchange experiments and X-ray investigations strongly suggest that these silicas keep a layer structure closely related to that of the parent silicates (Liebau, 1964 ; Brindley, 1969 ; Le Bihan et al., 1971 ; Hubert et al., 1976). The acid treatment simply replaces the interlayer cations by protons, the silicic sheets withstanding the attack. This was established by structure determinations (Liebau, 1964 ; Le Bihan et al., 1971) for the phyllodisilicic acids, $H_2Si_2O_5$, prepared from $\alpha-Na_2Si_2O_5$ and $KHSi_2O_5$.

We report below on some intercalation derivatives obtained with $KHSi_2O_5$ and the corresponding $H_2Si_2O_5$.

## SOME CHARACTERISTICS OF $KHSi_2O_5$ AND $H_2Si_2O_5$

$KHSi_2O_5$ can be prepared by hydrothermal synthesis from a mixture of KOH, amorphous silica and water (Wey and Kalt, 1967b). A treatment for a few hours, at room temperature, with an excess of 2N HCl removes quantitatively the interlayer potassium ions, giving a crystalline phyllodisilicic acid of formula $H_2Si_2O_5$ (Wey and Kalt, 1967a).

Both $KHSi_2O_5$ and $H_2Si_2O_5$ are orthorhombic. The $SiO_4$ tetrahedra are linked to form puckered six-membered rings in such a way that corrugated layers are obtained, but, unlike the tetrahedral layers in clay minerals, the unshared tetrahedral apices alternate in direction (Fig. 1).

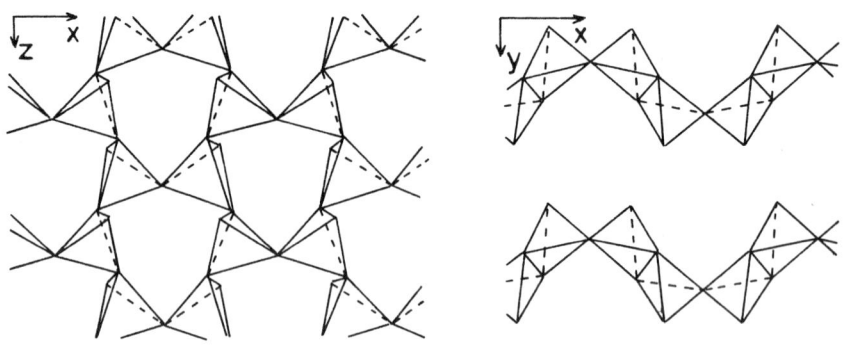

Fig. 1 : Sketch of the silica layers in $KHSi_2O_5$ and $H_2Si_2O_5$

In alkaline solutions, the protons of $H_2Si_2O_5$ can be easily exchanged by cations (Kalt and Wey, 1968).

The compound formed with $K^+$ displays the diffraction pattern of $KHSi_2O_5$ but has a lower potassium content (K = 0.44). With $Na^+$ one obtains a hydrous sodium silicate which has been found in nature and called Kanemite (Johan and Maglione, 1972). $H_2Si_2O_5$ also readily reacts with ammonia solutions to give a derivative isotypic with $KHSi_2O_5$.

In all these products the layer structure of $H_2Si_2O_5$ is retained. However, in addition to changes in the basal spacing $b$ with the interlayer cations, there are also significant variations of $a$ and $c$, indicating that the structure of the silicic layer is somewhat influenced by these cations (Table 1).

TABLE 1
Unit cell lengths (Å) of $H_2Si_2O_5$ and its derivatives

|  | $a$ | $b$ | $c$ |
| --- | --- | --- | --- |
| $H_2Si_2O_5$ | 7.47 | 11.94 | 4.91 |
| $KHSi_2O_5$ | 8.15 | 12.54 | 4.76 |
| $NH_4HSi_2O_5$ | 8.28 | 12.78 | 4.86 |
| Kanemite | 7.28 | 20.51 | 4.96 |

Remark : basal spacings are equal to $b/2$.

INTERCALATION DERIVATIVES OF $H_2Si_2O_5$ WITH n-ALKYLAMINES

We already reported briefly that $H_2Si_2O_5$ reacts with primary, secondary and tertiary alkylamines, tetramethylammonium hydroxide, ethylene diamine, etc, in aqueous solution (Kalt and Wey, 1968). These organic molecules are incorporated between the silicic layers with expansion of the basal spacing.

The derivatives obtained with n-alkylamines (1 to 10 C atoms in the alkyl chain) have now been studied in more detail.

Experimental

$H_2Si_2O_5$ was allowed to react with a large excess of aqueous solutions of n-alkylamines, at room temperature. In every case a single treatment for 24 hours resulted in the appearance of a new diffraction pattern with a sharp basal spacing. The presence of water is mandatory and anhydrous amines do not react. Alkylamines up to pentylamine are miscible with water in all proportions. For these, the concentration of the solution was not critical and the same X-ray diagrams were observed for concentrations ranging from

N/10 to almost pure amine.

For amines with longer chains the solubility decreases and the experiments were carried out with $5 \times 10^{-2}$ M solutions.

Results and discussion

The reaction products were filtered and X-rayed under mother liquid, giving the basal spacings in Table 2. These products have a regular structure and several high-order reflexions can be observed. On drying in air the basal spacings usually decrease by 1 to 2 Å, but in most cases the results are not reproducible and the disappearance of high-order reflexions suggests a less regular structure for air-dried products.

Prolonged evacuation ($p \sim 10^{-3}$ Torr) often leads to decomposition of the organic compounds and to formation of a nearly amorphous phase, except for methylamine and ethylamine where one merely observes the same decrease as on air drying.

TABLE 2

Basal spacings of organic derivatives (in Å)

| Number of C-atoms in the chain | $H_2Si_2O_5$ + n-alkylamine | | $KHSi_2O_5$ + n-alkylammonium ions |
|---|---|---|---|
| | under mother liquid | air-dried | |
| 1 | 8.84 | 7.89 | 8.18 |
| 2 | 10.3 | 8.54 | 8.84 |
| 3 | 15.4 | 13.6 | 8.78 |
| 4 | 17.6 | 15.5 | 8.92 |
| 5 | 20.3 | 19.2 | 9.11 |
| 6 | 23.4 | 22.0 | 9.01 |
| 7 | 25.8 | 24.5 | 22.6 |
| 8 | 28.2 | 26.3 | 27.2 |
| 10 | 33.3 | 32.4 | 31.5 |

Because of their small size, methyl- and ethylamine are probably trapped in the holes of the folded layers. This is not possible for amines with longer chains and one observes a striking difference between the spacings of propylamine and ethylamine derivatives (Table 2). For $n_c > 2$ the basal spacings increase nearly linearly with chain length (Fig. 2). From propylamine up to the decylamine compound the average increase is close to 2.5 Å/$CH_2$, a value which indicates a bimolecular arrangement of the alkylamine molecules in the interlayer spaces, with their long axes standing perpendicular to the silicate layers.

Similar results have been observed with other crystalline hydrated

silicas (Lagaly et al., 1974 ; Lagaly et al., 1975). However some
differences appear. H-magadiite does not react with alkylamines.
But if it is first treated with DMSO and then with alkylamines,
large basal spacings are observed for $n_c > 3$. Another form of
$H_2Si_2O_5$, obtained from $\alpha-Na_2Si_2O_5$ reacts only with methyl- and
ethylamine. Derivatives of longer chain amines can be prepared
indirectly by reaction with derivatives of shorter chain amines.

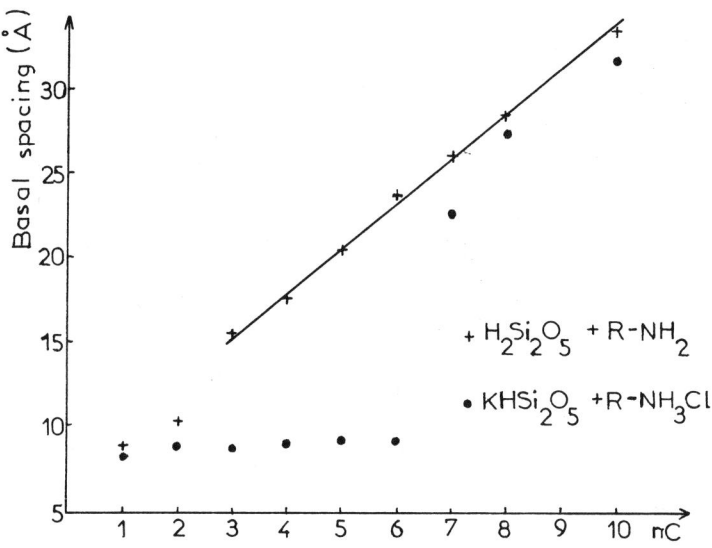

Fig. 2 : Basal spacings of intercalation compounds of $H_2Si_2O_5$ and
$KHSi_2O_5$ with n-alkylamines resp. n-alkylammonium chlorides.

## INTERCALATION DERIVATIVES OF $KHSi_2O_5$ WITH n-ALKYLAMMONIUM IONS

Experimental

$KHSi_2O_5$ was treated with aqueous solutions of n-alkylammonium
chlorides (1 to 10 C-atoms in the alkyl chain) under various
conditions with respect to reaction time, temperature and concentration. As the last factor has apparently no influence on the
basal spacings of the compounds obtained, we used 2N solutions up
to $C_7$, but only 1N octylammonium and 0.1N decylammonium chloride
solutions because of their lower solubility.

1 g $KHSi_2O_5$ was treated with 50 cm$^3$ of chloride solution for
24 hours. According to previous results obtained with magadiite
and kanemite (Lagaly et al., 1975 ; Beneke and Lagaly, 1977) first

attemps were made at 65°C in the pH range 8-9. But it appeared that the reaction proceeds as well at room temperature. As for $H_2Si_2O_5$ derivatives the reaction products were X-rayed under mother solution. The X-ray pattern of $KHSi_2O_5$ predominated in the spectrum, but a sharp line of a higher spacing appeared. Lengthening of the time of treatment (up to two weeks) had no influence, but repeated treatments for periods of 24 hours led to X-ray diagrams where all $KHSi_2O_5$ lines had disappeared. Usually four treatments were required. The complete extraction of potassium was verified by chemical analysis of the product.

The importance of pH has been emphasized (Lagaly et al., 1972) and special attention was paid to it. In every case the pH rose during the reaction. In the first experiments, the initial pH was adjusted to 8. The final values were about 9 and only a small part of $KHSi_2O_5$ had been transformed. The reaction proceeded better with an initial pH value of about 6. The final value again approached 9. The basal spacings of the derivatives so obtained do not differ from those of products prepared at higher initial pH.

Results and discussion

There is a striking difference between the derivatives obtained with short chain (up to 6 C-atoms) and long chain n-alkylammonium chlorides.

The basal spacings of the first group range between 8 and 9 Å (Table 2). The expansion is less than 3 Å with respect to $KHSi_2O_5$ and does not depend on the length of the alkyl chain. With longer alkylammonium chlorides (from $C_7$) the basal spacings increase with the length of the alkyl chain and are near those of corresponding $H_2Si_2O_5$-alkylamine derivatives.

A treatment of short-chain derivatives with aqueous alkylamine solutions strongly enlarges the basal spacings and one finds again the spacings of the $H_2Si_2O_5$-alkylamine derivatives.

Curiously the spacings of air-dried compounds are usually a little higher (0.2 - 1.0 Å) than those of the wet products, listed in Table 2. This behaviour has also been observed with derivatives of magadiite and interpreted by conformational changes (Lagaly et al., 1972).

The analytical determination of the amount of intercalated organic derivatives is uncertain, since the $H_2Si_2O_5$-amine compounds released at least a fraction of the incorporated amine on drying in air or vacuum. $KHSi_2O_5$ derivatives seemed more stable. Numerous analyses were carried out on specimen isolated under various conditions i.e. just filtered and still slightly wet, dried in air, dried in vacuum. In some cases the products were washed with water before drying. Considering only the molar ration N/Si we can express the amount x of incorporated amine (or ammonium ion) according to the formula $(RNH_3^+)_x H_{2-x} Si_2O_5, y H_2O$.

TABLE 3
Values x in compounds $|RNH_3^+|_x H_{2-x} Si_2O_5, y H_2O$

| R | from $H_2Si_2O_5$ | from $KHSi_2O_5$ |
|---|---|---|
| $CH_3$ | 0.56 - 0.70 | 0.35 - 0.60 |
| $C_2H_5$ | 0.28 - 0.31 | 0.25 - 0.45 |
| $C_4H_9$ | 0.25 - 0.30 | 0.21 - 0.33 |
| $C_{10}H_{21}$ | 0.38 - 0.50 | 0.29 - 0.42 |

Although interpretation of these rather scattered results is difficult, one can notice that the uptake of organic compounds is always far below the cation exchange capacity of $KHSi_2O_5$. Obviously the potassium ions are replaced not only by alkylammonium ions but also by protons. Two reactions must be considered.

(1) The exchange of organic cations for potassium ions,

$$K^+-HSi_2O_5 + RNH_3^+ \longrightarrow RNH_3^+-HSi_2O_5 + K^+,$$

cannot account for the increase in pH which has been observed in all exchange experiments.

(2) The proton concentration is controlled by the ammonium - amine equilibrium

$$RNH_3^+ + H_2O \rightleftharpoons RNH_2 + H_3O^+$$

For each $H^+$ replacing a potassium ion in $KHSi_2O_5$ a molecule of free amine appears and the pH increases.

But both reactions occur with short-chain as well as with long-chain alkylammonium ions, and in order to explain their different behaviour the interaction between $KHSi_2O_5$ and alkylammonium chloride solutions requires further detailed study.

REFERENCES

Beneke, K. and Lagaly, G., 1977. Kanemite - intercrystalline reactivity and relations to other sodium silicates. Am. Mineral., 62: 763-771.

Brindley, G.W.., 1969. Unit cell of magadiite in air, in vacuo and under other conditions. Am. Mineral., 54: 1583-1591.

Eugster, H.P., 1967. Hydrous sodium silicate from Lake Magadi, Kenya : precursors of bedded chert. Science, 157: 1177-1180.

Gude, A.J. and Sheppard, R.A., 1972. Silhydrite, $3SiO_2.H_2O$, a new mineral from Trinity County, California. Am. Mineral., 57: 1053-1065.

Guth, J.L., Hubert, Y., Jordan, D., Kalt, A., Perati, B. and Wey, R., 1977. Un nouveau type de silice hydratée cristallisée, de formule $H_2Si_3O_7$. C.R. Acad. Sci., 285D: 1367-1370.

Guth, J.L., Hubert, Y., Kalt, A., Perati, B. and Wey, R., 1978. Un nouveau type de silice hydratée cristallisée. C.R. Acad. Sci., 286D: 5-8.

Hubert, Y., Guth, J.L. and Wey, R., 1974. Attaque acide de différentes formes de $Na_2Si_2O_5$. Préparation d'une nouvelle silice hydratée cristallisée. C.R. Acad. Sci., 278D: 1453-1455.

Hubert, Y., Kalt, A., Guth, J.L. and Wey, R., 1976. Acide silicique cristallisée $\beta-H_2Si_2O_5$ : étude radiocristallographique et quelques propriétés. C.R. Acad. Sci., 282D: 405-408.

Iler, R.K., 1964. Ion exchange properties of a crystalline hydrated silica. J. Colloid Sci., 19: 648-657.

Johan, Z. and Maglione, G.F., 1972. La Kanemite, nouveau silicate de sodium hydraté de néoformation. Bull. Soc. Fr. Mineral. Cristallogr 95: 371-382.

Kalt, A. and Wey, R., 1968. Composés interfoliaires d'une silice hydratée cristallisée. Bull. Groupe Fr. Argiles, 20: 205-214.

Lagaly, G., Beneke, K. and Weiss, A., 1972. Organic complexes of synthetic magadiite. Proc. Int. Clay Conf. Madrid, 1972. Div. Ciencias C.S.I.C., Madrid, 1973: 663-673.

Lagaly, G., Beneke, K. and Weiss, A., 1973. Über eine neue kristalline Kieselsäure der Zusammensetzung $H_2Si_{14}O_{29}.5H_2O$ mit Schichtstruktur und Befähigung zur Bildung von Intercalationsverbindungen. Z. Naturforsch., 28b: 234-238.

Lagaly, G., Beneke, K., Dietz, P. and Weiss, A., 1974. Intercristallin reactivity of phyllodisilicic acid $H_2Si_2O_5$. Angew. Chem. Internat. Edit., 13: 819-821.

Lagaly; G., Beneke, K. and Weiss, A., 1975. Magadiite and H-magadiite. I. Sodium magadiite and some of its derivatives. Am. Mineral., 60: 642-649.

Le Bihan, M.T., Kalt, A. and Wey, R., 1971. Etude structurale de $KHSi_2O_5$ et $H_2Si_2O_5$. Bull. Soc. Fr. Mineral. Cristallogr., 94: 15-23

Liebau, F., 1964. Über Kristallstrukturen zweier Phyllokieselsäuren, $H_2Si_2O_5$. Z. Kristallogr., 120: 427-429.

Pabst, A., 1943. Crystal structure of gillespite $BaFeSi_4O_{10}$. Am. Mineral., 28: 372-390.

Pabst, A., 1958. The structure of leached gillespite, a sheet silicate Am. Mineral., 43: 970-980.

Schwarz, R. and Menner, E., 1924. Zur Kenntniss der Kieselsäuren. Ber. Dtsch. Chem. Ges., 57: 1477-1481.

Schwarz, R. and Richter, H., 1927. Zur Kenntnis der Kieselsäuren III. Ber. Dtsch. Chem. Ges., 60: 1111-1116.

Wey, R. and Kalt, A., 1967a. Synthèse d'une silice hydratée cristallisée. C.R. Acad. Sci., 265D: 1437-1440.

Wey, R. and Kalt, A., 1967b. Contribution à l'étude de l'hydrogénodisilicate de potassium. Obtention et synthèse. C.R. Acad. Sci., 265D: 1-3.

Wodtke, F. and Liebau, F., 1965. Über die Darstellung zweier Modifikationen der Phyllokieselsäure $H_2Si_2O_5$. Z. Anorg. Allg. Chem., 335: 178-188.

CRYSTALLIZATION OF NORDSTRANDITE IN CITRATE SYSTEMS AND IN THE PRESENCE OF MONTMORILLONITE

A. VIOLANTE[1] and M.L. JACKSON[2]
[1]Istituto di Chimica Agraria, Università di Napoli, 80055 Portici, Napoli, Italy.
[2]Department of Soil Science, University of Wisconsin, Madison, Wisconsin 53706, U.S.A.

ABSTRACT

Nordstrandite, one of three polymorphs of $Al(OH)_3$, is always accompanied by bayerite and/or gibbsite when samples are crystallized in the presence of chloride, nitrate, perchlorate or sulfate in the range of pH from slightly acid to highly alkaline. Pure nordstrandite (or a very high percentage of it) is obtained by precipitation of $Al(OH)_3$ in citrate systems with an Al/citrate molar ratio of 10 or more in the presence or absence of montmorillonite in the range of pH 9 to 11.

Montmorillonite at high pH values in the presence of citrate, favors the crystallization of nordstrandite by inhibiting the formation of bayerite and pseudoboehmite. In $Al(OH)_x$-montmorillonite complexes at higher than pH 9 and at different Al/citrate molar ratios, pure nordstrandite forms. Plots of Al/citrate ratio against pH show areas of stability of the various aluminous species, including pseudoboehmite, amorphous alumina, and interlayered montmorillonite. Electron micrographs of those complexes show many ovoidal particles or characteristic flamboyant clusters when the complexes are made at elevated temperatures. In citrate systems without the presence of clay, pure nordstrandite occurs mainly as rectangular particles.

INTRODUCTION

The purpose of this paper is to present the conditions for synthesis of pure nordstrandite and other aluminous species and to show the different crystal habits of this $Al(OH)_3$ polymorph as observed by electron microscopy in clay-free and in $Al(OH)_x$-montmorillonite systems. Citrate anions have a great influence not only on the rate of crystallization but also in the species of $Al(OH)_3$ polymorphs formed (A. Violante and M.L. Jackson, in manuscript). In alkaline pH (at which pure bayerite usually is formed), the presence of citrate favors the formation of nordstrandite. Montmorillonite at high pH values partially inhibits formation of bayerite and, particularly in citrate systems, has a favorable influence on the crystallization of nordstrandite and gibbsite.

Nordstrandite was found as a synthetic product mixed with bayerite and gibbsite (Van Nordstrand et al., 1956). Later it was found in natural environments by many authors (Wall et al., 1962; Hathaway and Schlanger, 1962, 1965; Davis and Hill, 1974; Milton et al., 1975). Its structure was visualized as alternating layers of gibbsite (4.85 Å) and bayerite (4.72 Å) (Lippens, 1961), but has since been presented as a (4.79Å) structure with displaced dioctahedral layers (Bosmans, 1970). None of the samples synthesized by others were pure nordstrandite and there is great confusion about optimal conditions of synthesis. Optimum pH has been given as slightly acid to neutral (Barnhisel and Rich, 1965) to as high as 13 (Papée et al., 1958). Aging aluminum hydroxide gels in aqueous solutions of ethylendiamine (Hauschild, 1963) was advocated. Crystallization of $Al(OH)_3$ polymorp at different pH values led to a conclusion that nordstrandite is an "enigma" and that "bayerite is a metastable phase that eventually recrystallizes in alkaline solutions (and in natural environments) to the stable phase nordstrandite" (Schoen and Roberson, 1970). Bayerite changed to a new unidentified species in the absence of citrate (Ng Kee Kwong and Huang, 1975, 1977; their Fig. 2); $Al(OH)_3$ remained nearly amorphous to X-rays in the presence of citrate in their experiments.

METHODS

$Al(OH)_3$ was precipitated between pH 7 and 11 by addition of 0.1 N NaOH, at 1 ml per min with stirring, to a mixture of $AlCl_3$ and citric acid in absence or presence of montmorillonite (16 meq Al/g clay) at 25°C. The final concentration of Al was $10^{-2}$M and that of citrate was varied to give an Al/citrate molar ratio of 10 or less to 1000. The samples were aged in polyethylene bottles, the pH values being kept constant by addition of 0.1 N NaOH or HCl. Oriented aggregate specimens for X-ray diffractograms were obtained by drying washed aliquots of suspension, variously aged, on glass slides. A Rigaku diffractometer, with $CuK_\alpha$ radiation generated at 30kV and 15 mA, with a scanning rate of 0.5 or 1.0° 2θ/min, was used to provide high precision 2θ values necessary to distinguish between the different $Al(OH)_3$ polymorphs. Electron micrographs (TEM) were obtained with a JEOL model JEM100B.

RESULTS

Crystallization of nordstrandite in clay-free systems

Citrate anions at low concentrations not only delay or inhibit completely the crystallization of aluminum hydroxide but have an influence on which of the $Al(OH)_3$ polymorphs crystallizes. Samples aged 120 days at pH 10, at an Al/citrate molar ratio of 100, yielded mainly bayerite (4.72 Å), with low quantities of gibbsite and nordstrandite (Fig. 1a). With increasing concentration of citrate, bayerite decreased tremendously and nordstrandite increased. Finally a pure nordstrandite

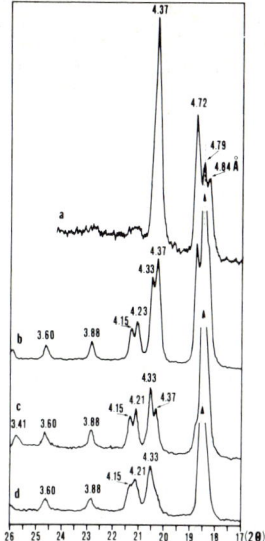

Fig. 1. X-ray diffractograms of Al(OH)$_3$ aged 120 days at pH 10 with a molar ratio Al/citrate of: (a) 100, (b) 80, (c) 50 and (d) 20, with shift from bayerite to nordstrandite.

crystallized (Fig. 1d). The strong bayerite peak at 4.37Å at higher than pH 9 (Schoen and Roberson, 1970; Violante and Violante, 1976), is easily distinguishable from that of nordstrandite at 4.33Å only when a high amount of the latter is present (curves b and c compared to curve a). However, the peaks at 4.21, 4.15, 3.88 and 3.60Å are diagnostic of nordstrandite. Pure nordstrandite or a very high percentage of it is formed at different Al/citrate molar ratios in the range of pH from 8.5 to 11.0, while bayerite crystallized only at high pH values and in the presence of very low citrate concentration. Pure synthetic nordstrandite synthesized at pH 9.5 and Al/citrate molar ratio 80 (Fig. 2) gives XRD data comparable to other reports for synthetic and natural nordstrandite (Table 1).

Rectangular parallelepipeds (up to 2.5µm) formed at pH 10.0 and Al/citrate = 20 (Fig. 3a), and twin crystals (0.4-0.7µm), at pH 9.5 and Al/citrate = 80 (Fig. 3b). At lower than pH 9.5 the crystals rarely exceed 0.5µm in length and are almost rectangular in outline.

The Al/citrate molar ratio and the pH control the formation of individual Al(OH)$_3$ polymorphs (Fig. 4a). Below pH 8.5 pure gibbsite crystallizes easily but only a small quantity of it is formed at higher pH, when the Al/citrate molar ratio is lower than 100. A large area in Fig. 4a is characterized by formation of an amorphous phase or stable pseudoboehmite after 2 months of aging at 25°C. Above pH 9, amorphous material and/or pseudoboehmite are found particularly at Al/citrate molar ratios lower than 100; at lower pH values, amorphous material is formed; at higher

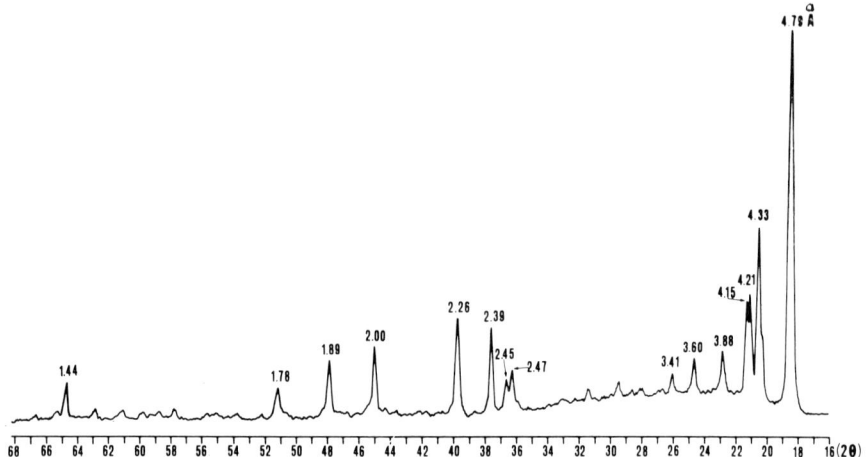

Fig. 2. X-ray diffractogram of nordstrandite synthesized at pH 9.5 and Al/citrate molar ratio = 80.

pH, pseudoboehmite and Al(OH)$_3$ form. The broad peak of pseudoboehmite at about 6.3Å, higher than those of other Al(OH)$_3$ polymorphs, often occurs after 3 or 4 months of aging (Fig. 5 La). Usually its crystallization is inhibited at a given Al/citrate ratio below certain critical ranges of pH. For example, at an Al/citrate ratio of 20 very stable pseudoboehmite is formed at pH 9.4 but pure nordstrandite crystallizes at pH 10. Analogous influence of citrate on the formation of iron oxides was reported (Schwertmann et al., 1968).

Crystallization of nordstrandite in the Al(OH)$_x$-montmorillonite complexes

In the Al(OH)$_x$-montmorillonite complexes with the same pH values and Al/citrate molar ratios as in the clay-free samples, nordstrandite and gibbsite crystallize over a wider range of conditions (Fig. 4b). In the complexes, pure nordstrandite crystallizes most readily at pH 10 or higher, but also with less reproducibility and purity, at pH 9 to 10 with Al/citrate molar ratios between 250 and 10. The areas of the amorphous material and pure bayerite (Fig. 4b) are remarkably more confined than in clay-free samples (Fig. 4a). Moreover, pure pseudoboehmite is almost completely absent. Strong peaks of nordstrandite and gibbsite occur at pH values and Al/citrate ratios at which stable pseudoboehmite is formed in clay-free systems (Fig. 5). The evidence is that montmorillonite catalyses the crystallization of nordstrandite (and often gibbsite) and inhibits the formation of pseudoboehmite and bayerite (Fig. 5).

In the presence of montmorillonite, ovoidal nordstrandite (0.3-0.5-μm) particles occur (Fig. 3c). A few twin crystals are evident. Similar ovoidal particles or nearly rectangular plates were observed in many complexes at different pH values and Al/citrate ratios. It is of interest that "rounded pellets" of nordstrandite

TABLE 1

X-ray powder diffraction data for nordstrandite

| Natural Sarawak (Wall et al., 1962) | | Synthetic (Hauschild, 1963) | | This work | |
|---|---|---|---|---|---|
| d (Å) | I | d (Å) | I | d (Å) | I |
| 4.78 | vs[a] | 4.791 | 100 | 4.791 | vs |
| 4.33 | s | 4.318 | 28 | 4.328 | s/vs |
| 4.206 | s | 4.207 | 26 | 4.207 | s |
| 4.153 | s | 4.158 | 21 | 4.148 | s |
| 3.886 | m | 3.888 | 13 | 3.880 | m |
| 3.600 | m | 3.604 | 11 | 3.604 | m |
| 3.425 | m | 3.430 | 7 | 3.411 | w/m |
| 3.140 | vw | -- | | 3.18–3.12 | vw |
| 3.023 | w | 3.027 | 4 | 3.025 | w |
| 2.849 | w | 2.848 | 4 | 2.846 | w |
| 2.704 | Bw | 2.704 | 1 | 2.712 | Bw |
| 2.637 | w | 2.627 | 1 | 2.634 | vw |
| -- | | 2.500 | 2 | 2.493 | w |
| 2.479 | Bm | 2.482 | 13 | 2.473 | m |
| 2.445 | w | 2.453 | 7 | 2.447 | m |
| 2.392 | s | 2.393 | 29 | 2.390 | s |
| 2.330 | vw | 2.330 | 2 | 2.330 | vw |
| 2.261 | s | 2.265 | 33 | 2.263 | s |
| 2.225 | w | 2.212–2.144 | 1 | 2.220 | vw |
| 2.148 | vw | -- | | 2.159 | vw |
| 2.097 | vw | 2.115 | 1 | 2.135 | vw |
| 2.073 | vw | 2.075 | 1 | 2.075 | vw |
| 2.016 | s | 2.017 | 27 | 2.008 | s |
| 1.982 | vw | 1.994 | 1 | -- | |
| 1.959 | vw | 1.947 | 1 | 1.955 | vw |
| 1.939 | vw | -- | | -- | |
| 1.899 | s | 1.901 | 22 | 1.893 | s |
| -- | | 1.866–1.802 | 1 | -- | |
| 1.779 | m/s | 1.783 | 13 | 1.779 | m |
| 1.715 | vw | 1.715 | 1 | 1.747 | vw |
| 1.704 | w | 1.703 | 2 | 1.702 | vw |
| 1.672 | w | 1.671 | 4 | 1.685 | vw |
| 1.653 | vw | 1.654 | 2 | 1.654 | vw |
| -- | | 1.633–1.619 | 1 | -- | |
| 1.593 | w/m | 1.597 | 5 | 1.591 | w |
| 1.574 | w | 1.574–1.556 | 3 | 1.574–1.556 | vw |
| 1.549 | w/m | 1.541 | 5 | 1.545 | vw |
| 1.513 | m | 1.516 | 7 | 1.517 | w |
| 1.477 | m | 1.479 | 7 | 1.477 | w |
| 1.465 | vw | 1.468 | 2 | -- | |
| 1.440 | m/s | 1.441 | 14 | 1.437 | m |

[a] vs = very strong; s = strong; m = medium; w = weak; vw = very weak; B = broad.

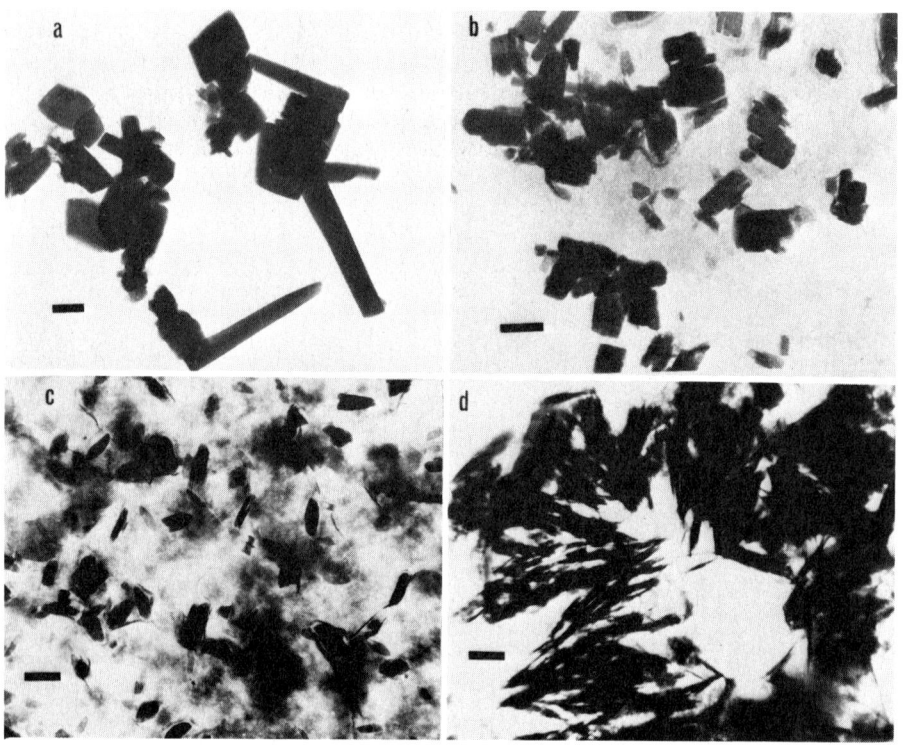

Fig. 3. Transmission electron micrographs showing nordstrandite formed in clay-free systems at: (a) pH 10 and Al/citrate molar ratio = 20, (b) pH 9.5 and Al/citrate = 80; and in Al(OH)$_x$-montmorillonite complexes, (c) at pH 9 and Al/citrate = 100, aged at 25°C, and (d) at pH 9.5 and Al/citrate = 50 aged, at 35°C. Bar = 0.5μm.

were found in soils from West Sarawak, Borneo (Wall et al., 1962) on the edge of a sinkhole in limestone where clays may be present.

In the presence of clay, particularly at pH 10 and 25°C or 35°C, or at < pH 10 and at 35°C, flamboyant clusters of nordstrandite crystals form (Fig. 3d), from 0.4 to 0.8μm in length and 0.07 to 0.1μm in width. Such radiating crystal clusters were also observed with mixtures of nordstrandite and gibbsite as ascertained by X-ray diffraction (A. Violante and M. L. Jackson, in manuscript; their Plate 1e, f and g). Perhaps the crystals of nordstrandite and gibbsite form face to face aggregates or clusters of acicular or lath-shaped crystals in the presence of clay; face to face aggregates of gibbsite were previously described (Kavanagh et al., 1976, their Plate 1B). Nordstrandite in Guam exhibits crystals that may have been initiated by nucleation on clay surfaces (Hathaway and Schalanger, 1965) comparable to the observed radiating groups of crystals (Fig. 3d).

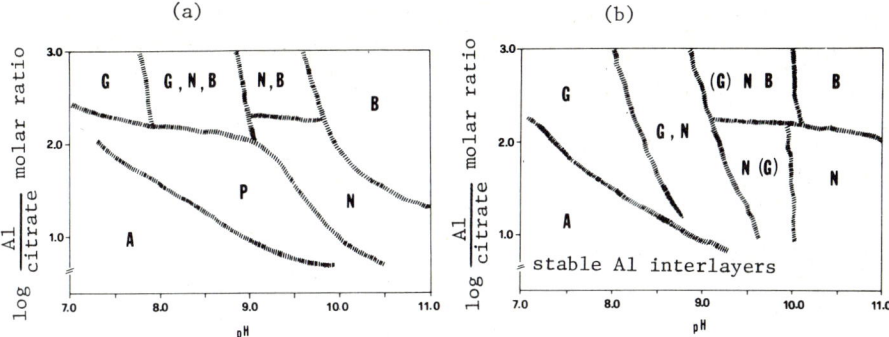

Fig. 4. Diagrams illustrating the influence of Al/citrate molar ratio and pH after 2 months aging at 25°C: (a) on formation in the absence of clay of amorphous aluminum hydroxide (A), pseudoboehmite (P), and Al(OH)$_3$ polymorphs, i.e. gibbsite (G), nordstrandite (N), and bayerite (B); and (b) on formation in the presence of Al(OH)$_x$-montmorillonite complexes of aluminum phase (as stable interlayers of montmorillonite) (A), and Al(OH)$_3$ polymorphs, i.e. gibbsite (G), nordstrandite (N), bayerite (B). (G) in (b) indicates low amounts.

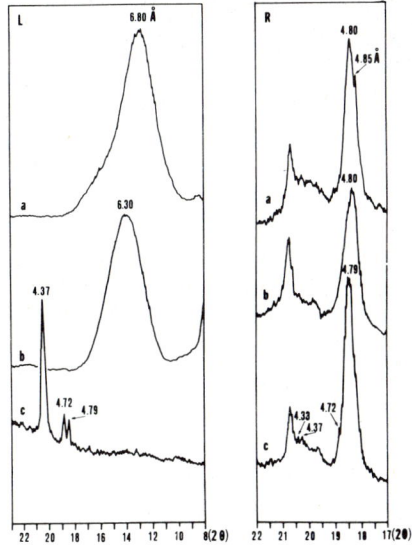

Fig. 5. L (left) X-ray diffractograms of Al(OH)$_3$ after 60 days: (a) at pH 8.5 and Al/citrate = 100, (b) at pH 9.0 and Al/citrate = 100, and (c) at pH 10.5 and Al/citrate = 80 (transition from pseudoboehmite to bayerite); R (right) Al(OH)$_x$-montmorillonite complexes after 60 days: (a) at pH 8.5 and Al/citrate = 100, (b) at pH 9.0 and Al/citrate = 100, and (c) at pH 10.5 and Al/citrate = 80.

CONCLUSIONS

(1) It appears clear from our study that nordstrandite crystallizes in the presence of citrate, in the absence or presence of clay, particularly at > pH 8.5 or 9, in keeping with the occurrence of nordstrandite in alkaline natural environments.

(2) From these experiments in citrate systems, it is possible to hypothesize that in soils the presence of clay and organic matter with its carboxylic and amine groups (analogous to citrate; or ethylendiamine, Hauschild, 1963) inhibits the crystallization of bayerite, by favoring the formation of gibbsite and/or nordstrandite. The almost complete absence of bayerite in natural environments and soils (Schoen and Roberson, 1970; Milton et al., 1975) is thus explained. Of course, other organic or inorganic anions such as sulfonate, sulfate, and carbonate (Goldberry and Loughnan, 1970) could have a similar influence.

ACKNOWLEDGEMENTS

This research was supported in part by the College of Agric. and Life Sci., Univ. of Wisconsin, Madison (U.S.A.) under project 1123; in part by the National Science Foundation under grant EAR76-19783-Jackson; and in part by NATO-Consiglio Nazionale delle Ricerche fellowship for the stay in Madison by A. Violante; through an International Consortium for Interinstitutional Cooperation in the Advancement of Learning (ICICAL). We appreciate the editorial suggestions by Dr. M. M. Mortland and Mr. C. H. Lim.

REFERENCES

Bosmans, H.J., 1970. Unit cell and crystal structures of nordstrandite, $Al(OH)_3$. Acta Crystallogr., B26: 649–652.

Barnhisel, R.I. and Rich, C.I., 1965. Gibbsite, bayerite and nordstrandite formation as affected by anions, pH, and mineral surfaces. Soil Sci. Soc. Am. Proc., 29: 531–534.

Davis, C.E. and Hill, V.G., 1974. Occurrence of nordstrandite and its possible significance in Jamaican bauxites. Yugoslavia Academy of Sciences, (Zagreb) Travaux ICSOBA, 11: 61–70.

Goldberry, R. and Loughnan, F.C., 1970. Dawsonite and nordstrandite in the Permian Berry Formation of the Sydney Basin, New South Wales. Am. Mineral., 55: 477–490.

Hathaway, J.C. and Schlanger, S.O., 1962. Nordstrandite from Guam. Nature, 196: 265–266.

Hathaway, J.C. and Schlanger, S.O., 1965. Nordstrandite ($Al_2O_3 \cdot 3H_2O$) from Guam. Am. Mineral., 50: 1029–1037.

Hauschild, U., 1963. Über Nordstrandite, $\gamma$-$Al(OH)_3$. Z. Anorg. und Allg. Chemie, 324: 15–30.

Kavanagh, B.V., Posner, A.M. and Quirk, J.P., 1976. The adsorption of polyvinyl alcohol on the gibbsite and goethite. J. Soil Sci., 27: 467-477.

Lippens, B.C., 1961. Structure and Texture of Aluminas. Proefschrift Delft, Netherlands, 177 pp.

Milton, C., Dwornik, E.J. and Finkelman, R.B., 1975. Nordstrandite, $Al(OH)_3$, from the Green River Formation in Rio Blanco County, Colorado. Am. Mineral., 60: 285-291.

Ng Kee Kwong, K.F. and Huang, P.M., 1975. Influence of citric acid on the crystallization of aluminum hydroxides. Clays Clay Minerals, 23: 164-165.

Ng Kee Kwong, K.F. and Huang, P.M., 1977. Influence of citric acid on the hydrolytic reactions of aluminum. Soil Sci. Soc. Am. J., 41: 692-697.

Papée, D., Tertian, R. and Biais, R., 1958. Recherches sur la constitution des gels et des hydrates cristallines d'alumine. Soc. Chim. France Bull., 81: 1301-1310.

Schwertmann, U., Fischer, W.R. and Papendorf, F.H., 1968. The influence of organic compounds on the formation of iron oxides. Trans. 9th Int. Cong. Soil Sci. (Adelaide), 1: 645-655.

Schoen, R. and Roberson, C.E., 1970. Structures of aluminum hydroxide and geochemical implications. Am. Mineral., 55: 43-77.

Van Nordstrand, R.A., Hettinger, W.P. and Keith, C.D., 1956. A new aluminum tri-hydrate. Nature, 177: 713-714.

Violante, P. and Violante, A., 1976. Morphological and chemical aspects of the surface relationships between bentonite and aluminum hydroxide. 2° Congresso Nazionale sulle Argille organized by Gruppo Italiano dell'AIPEA (in Italian) Bari, 1976.

Wall, J.R.D., Wolfenden, E.B., Beard, E.H. and Deans, T., 1962. Nordstrandite in soil from West Sarawak, Borneo. Nature, 196: 264-265.

NATURE OF HYDROLYTIC PRECIPITATION PRODUCTS OF ALUMINUM AS INFLUENCED BY LOW-MOLECULAR WEIGHT COMPLEXING ORGANIC ACIDS

K.F. NG KEE KWONG and P.M. HUANG

Department of Soil Science, University of Saskatchewan, Saskatoon (Canada)

ABSTRACT

The nature of the hydrolytic precipitation products of Al formed in the individual presence of 0, $10^{-6}$ and $10^{-4}$ M citric, malic, aspartic and p-hydroxybenzoic acids in systems at the initial Al concentration of either $1.10 \times 10^{-3}$ M or $1.10 \times 10^{-4}$ M and the OH/Al molar ratio of 2 and 3 was investigated.

X-ray diffraction analyses showed that the degree of structural distortion created within the precipitation products, as a result of the occupation of the coordination sites of Al by the organic ligands, increased with the molar ratio of organic acid to Al in the system and with the stability constant of the complexes formed between the organic ligands and Al. Thermogravimetric analyses concurred with the X-ray diffraction data. The inhibition of the formation of crystalline aluminum hydroxides by the organic acids was further illustrated by the electron micrographs of the hydrolytic precipitation products. Those products, which were non-crystalline to X-rays, precipitated in the presence of the organic acids, were fine shapeless colloids possessing rough fluffy and extensive surfaces as opposed to the well-defined crystals formed in their absence.

This study has therefore revealed the important role which low-molecular weight organic acids play in hindering the hydrolytic reactions of Al and in the subsequent formation of non-crystalline hydrous Al hydroxides. Attainment of the knowledge gained is of fundamental importance in understanding the genetic mechanisms of non-crystalline inorganic components in soils and sediments and would also pave the way for the study of the significance of these components in the transport and fate of nutrients and pollutants in terrestrial and aquatic ecosystems.

INTRODUCTION

The significance of the aqueous chemistry of Al in geochemical and soil processes and environmental protection has been established (Jackson, 1963; Rich, 1968; Luciuk and Huang, 1974; Huang, 1975; Hayden and Rubin, 1976; Hsu, 1977; Kwong and Huang, 1977, 1978). The existing literature shows that the formation of crystalline or non-crystalline Al hydroxides is influenced by the ionic environments in the system.

Organic acids are found in soil solutions and natural waters. Their importance in the weathering of primary minerals has been recognized and extensively investigated (Huang and Keller, 1970; Boyle et al., 1974; Razzaghe and Robert, 1975; Jorgensen, 1976). In soils and sediments, hydrolysis of Al is the rule rather than the exception (Thomas, 1977). Yet the role of the organic acids in affecting the nature of the hydrolytic reaction products of Al is not well understood. Kwong and Huang (1977) shows that citric acid at concentrations of $10^{-6}$ to $10^{-4}$ M hinders the formation of crystalline Al hydroxides and gives rise to fine colloidal precipitates.

The organic acids in soil solutions and natural waters are very diverse in chemical composition, structure, nature of functional groups, size and basicity. They may thus vary in their ability to affect the hydrolytic reactions of Al. The objectives of this investigation, therefore, were to ascertain the influence of low-molecular weight organic acids, varying in ability to dissolve Al from rocks and minerals, on the structural organization and surface morphology of the hydrolytic precipitation products of Al.

MATERIALS AND METHODS

Citric acid, L-malic acid, DL-aspartic acid and p-hydroxybenzoic acids which are commonly found in the natural environments were selected in this study. All chemicals used were of reagent grade quality.

Determination of stability constants of complexes formed between aluminum and organic acids

Stability constants of complexes formed between Al and citric, L-malic, DL-aspartic, and p-hydroxybenzoic acids were determined by the potentiometric method. Experimentally, 100 ml of 0.1M KCl solutions containing (i) 0.01M HCl, (ii) 0.01M HCl and an organic acid at concentrations varying between $10^{-4}$ M and $10^{-2}$ M, and (iii) 0.01M HCl, $10^{-3}$ M Al and an organic acid at the same concentration as in (ii) were titrated with continuous stirring against a 1M NaOH solution on a Metrohm Herisau E436 potentiograph at $20°C$. Purified nitrogen gas was passed through the solutions during the titration to minimize carbon dioxide absorption.

From the resulting titration curves, the average ligand number $\bar{n}$ and the free organic ligand concentrations $[L]$ were calculated (Irving and Rossotti, 1954) at the pH of 3.5 to avoid the hydrolysis of the complexes and the formation of hydroxy-Al polymers. The formation curves were then obtained by plotting $\bar{n}$ against the negative logarithm of $[L]$. The formation curves so obtained were used to determine the stability constants of the complexes (Rossotti and Rossotti, 1961).

## Formation of hydrolytic precipitation products of aluminum

For reaction systems containing no organic acids, 950 ml solutions of $1.16 \times 10^{-4}$M and $1.16 \times 10^{-3}$M $AlCl_3$ were titrated against 0.1M NaOH with continuous stirring at the rate of approximately 100 ml NaOH per hour to OH/Al molar ratios of 2 and 3. The Al concentrations in the suspensions were then adjusted to $1.10 \times 10^{-3}$M and $1.10 \times 10^{-4}$M by bringing the final volume of the solutions to 1000 ml with $H_2O$.

Reaction systems in the individual presence of citric, malic, aspartic and p-hydroxybenzoic acids were prepared in the same way except that the organic acid was added prior to titration against 0.1M NaOH. The organic acid concentrations in the suspensions after final volume adjustment were $1.0 \times 10^{-6}$M and $1.0 \times 10^{-4}$M, respectively.

After one and 40 day aging, the solid phase products were collected by ultrafiltration through a millipore MF filter of 0.025 μm pore size (Huang and Luciuk, 1974).

## Structural organization and surface morphology of hydrolytic precipitation products of aluminum

The hydrolytic precipitation products of Al were analyzed by X-ray diffraction (Luciuk and Huang, 1974) and observed under scanning electron microscope. The solid phase products, suspended in deionized water, were dried on aluminum stubs and given a gold coating to render their surfaces more conductive to the electron beam before they were viewed under the Cambridge scanning electron microscope Mark II.

Samples of hydrolytic precipitation products of Al formed in the absence and in the presence of organic acids were also examined by thermogravimetry according to the procedure outlined by Kwong and Huang (1977).

## RESULTS AND DISCUSSION

### X-ray diffraction

In systems at the initial Al concentration of $1.1 \times 10^{-3}$M, OH/Al molar ratio of 3.0 and citric acid concentration of $10^{-6}$M, the complexation of Al by citrate, which has the strongest affinity for Al as characterized by the stability constants (Table 1), was sufficient to cause the hydrolytic precipitation products formed to be non-crystalline to X-rays after one day aging (Fig. 1E). Malic, aspartic and p-hydroxybenzoic acids at the concentration of $10^{-6}$M, which have weaker affinity for Al than citric acid (Table 1), in the same reaction systems did not complex Al sufficiently for the subsequent structural distortion to delay the formation of bayerite after one day aging (Fig. 1).

When the concentration of citric, malic and aspartic acids was raised to $10^{-4}$M in systems at the initial Al concentration of $1.1 \times 10^{-3}$M, the degree of structural distortion caused by the complexation of Al by the organic ligands was sufficient

TABLE 1

Stability constants of the complexes formed between Al and citric, malic, aspartic and p-hydroxybenzoic acids

| Organic acid | Stability constants of the complexes | |
| --- | --- | --- |
| | log $K_1$ | log $K_2$ |
| p-hydroxybenzoic acid | 1.66 ± 0.13 | – |
| Aspartic acid | 2.60 ± 0.10 | – |
| Malic acid | 5.14 ± 0.03 | 8.52 ± 0.13 |
| Citric acid | 7.37 ± 0.24 | 13.90 ± 0.48 |

to maintain the hydrolytic precipitation products formed at the OH/Al molar ratio of 3.0 in a non-crystalline state to X-rays even after 40 day aging (Fig. 2, B, C, D). The importance of the affinity of the organic ligands for Al is emphasized by the fact that p-hydroxybenzoic acid, which formed the least stable complexes with Al (Table 1), did not complex Al extensively enough for the resulting structural distortion to cause the hydrolytic precipitation products of Al to be non-crystalline to X-rays (Fig. 2A).

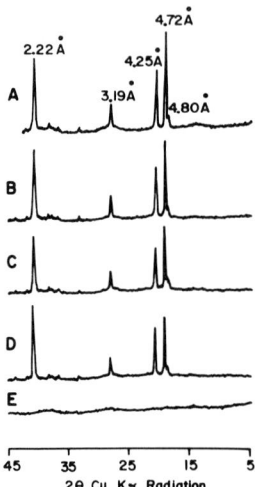

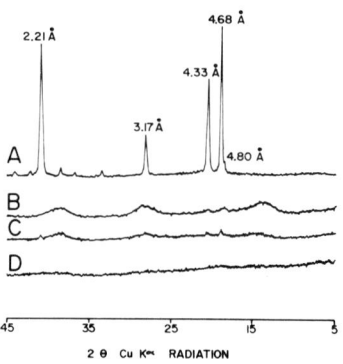

Fig. 1. X-ray diffraction patterns of hydrolytic reaction products of Al precipitated at the initial Al concentration of $1.1 \times 10^{-3}$M, OH/Al molar ratio of 3.0 and aged for one day in the presence of: (A) no organic acid, (B) $10^{-6}$M p-hydroxybenzoic acid, (C) $10^{-6}$M aspartic acid, (D) $10^{-6}$M malic acid, and (E) $10^{-6}$M citric acid.

Fig. 2. X-ray diffraction patterns of hydrolytic reaction products of Al precipitated at the initial Al concentration of $1.1 \times 10^{-3}$M, OH/Al molar ratio of 3.0 and aged for 40 days in the presence of $10^{-4}$M: (A) p-hydroxybenzoic acid, (B) aspartic acid, (C) malic acid, and (D) citric acid.

The reaction products formed at the initial Al concentration of $1.1 \times 10^{-3}$M in the presence of $10^{-6}$M malic, aspartic and p-hydroxybenzoic acids after 40 day aging were well crystallized gibbsite despite the fact that the OH/Al molar ratio was lowered to 2.0 (Fig. 3). On the other hand, $10^{-6}$M citric, malic and aspartic acids and $10^{-4}$M p-hydroxybenzoic acid were able to cause the solid phase products formed in systems at the initial Al concentration of $1.1 \times 10^{-4}$M and OH/Al molar ratio of 3.0 to be still poorly crystalline to X-rays after 40 day aging (Fig. 4). These findings indicate the significant influence of the molar ratio of organic acid to Al on the crystallization of Al hydroxides.

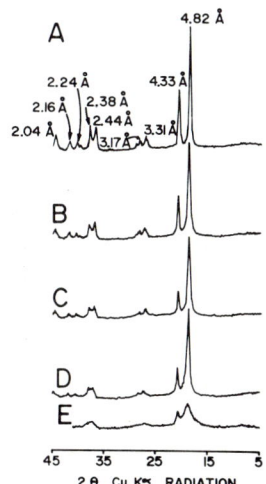
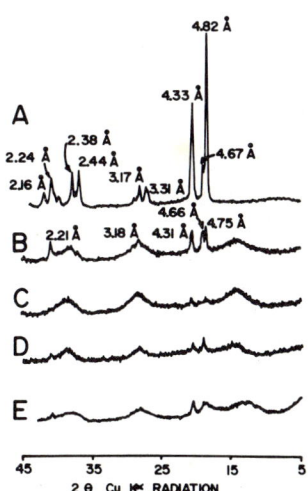

Fig. 3. X-ray diffraction patterns of hydrolytic reaction products of Al precipitated at the initial Al concentration of $1.1 \times 10^{-3}$M, OH/Al molar ratio of 2.0 and aged for 40 days in the presence of: (A) no organic acid, (B) $10^{-6}$M p-hydroxybenzoic acid, (C) $10^{-6}$M aspartic acid, (D) $10^{-6}$M malic acid, and (E) $10^{-6}$M citric acid.

Fig. 4. X-ray diffraction patterns of hydrolytic reaction products of Al precipitated at the initial Al concentration of $1.1 \times 10^{-4}$M, OH/Al molar ratio of 3.0 and aged for 40 days in the presence of: (A) no organic acid, (B) $10^{-4}$M p-hydroxybenzoic acid, (C) $10^{-6}$M aspartic acid, (D) $10^{-6}$M malic acid, and (E) $10^{-6}$M citric acid.

## Thermal analysis

The strong influence of the organic acids, even at the low concentration of $10^{-6}$M, on disturbing the structural organization of the hydrolytic precipitation products of Al is illustrated by the increase in quantity of the adsorbed water and the gradual loss of the structural water upon heating the precipitation products (Fig. 5, B,C,D,E). This is in contrast to the nature of the hydrolytic reaction products of Al in the absence of organic acids (Fig. 5A).

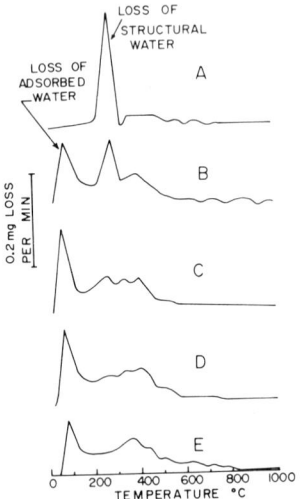

Fig. 5. DTG patterns of hydrolytic reaction products of Al precipitated at the initial Al concentration of $1.1 \times 10^{-4}$M, OH/Al molar ratio of 3 and aged for 40 days in: (A) absence of organic acids, (B) $10^{-4}$M p-hydroxybenzoic acid, (C) $10^{-6}$M aspartic acid, (D) $10^{-6}$M malic acid, and (E) $10^{-6}$M citric acid.

Electron optical observations

The inhibition or delay in the formation of crystalline Al hydroxides caused by the organic acids was further reflected by the scanning electron micrographs of the solid phase reaction products. The poorly crystalline hydrolytic reaction products of Al formed in the presence of $10^{-6}$M citric, malic and aspartic acids and $10^{-4}$M p-hydroxybenzoic acid in systems at the initial Al concentration of $1.1 \times 10^{-4}$M and OH/Al molar ratio of 3.0 and aged for 40 days (Fig. 4 and 5) were shown by the scanning electron micrographs to have extensive and damaged surfaces (Fig. 6) as opposed to the well-defined crystals formed in the absence of organic acids (Kwong and Huang, 1977).

GENERAL DISCUSSION AND CONCLUSIONS

The present study reveals that the structural distortion of the hydrolytic precipitation products of Al as influenced by the organic acids increases with the increase of the molar ratio of organic acid to Al and the stability constants of the complexes formed between Al and the organic acids (Fig. 1-5, Table 1). Both parameters lead to an increasing occupation of the coordination sites of Al by the organic ligands. The complexation reactions of Al ions by the organic ligands have

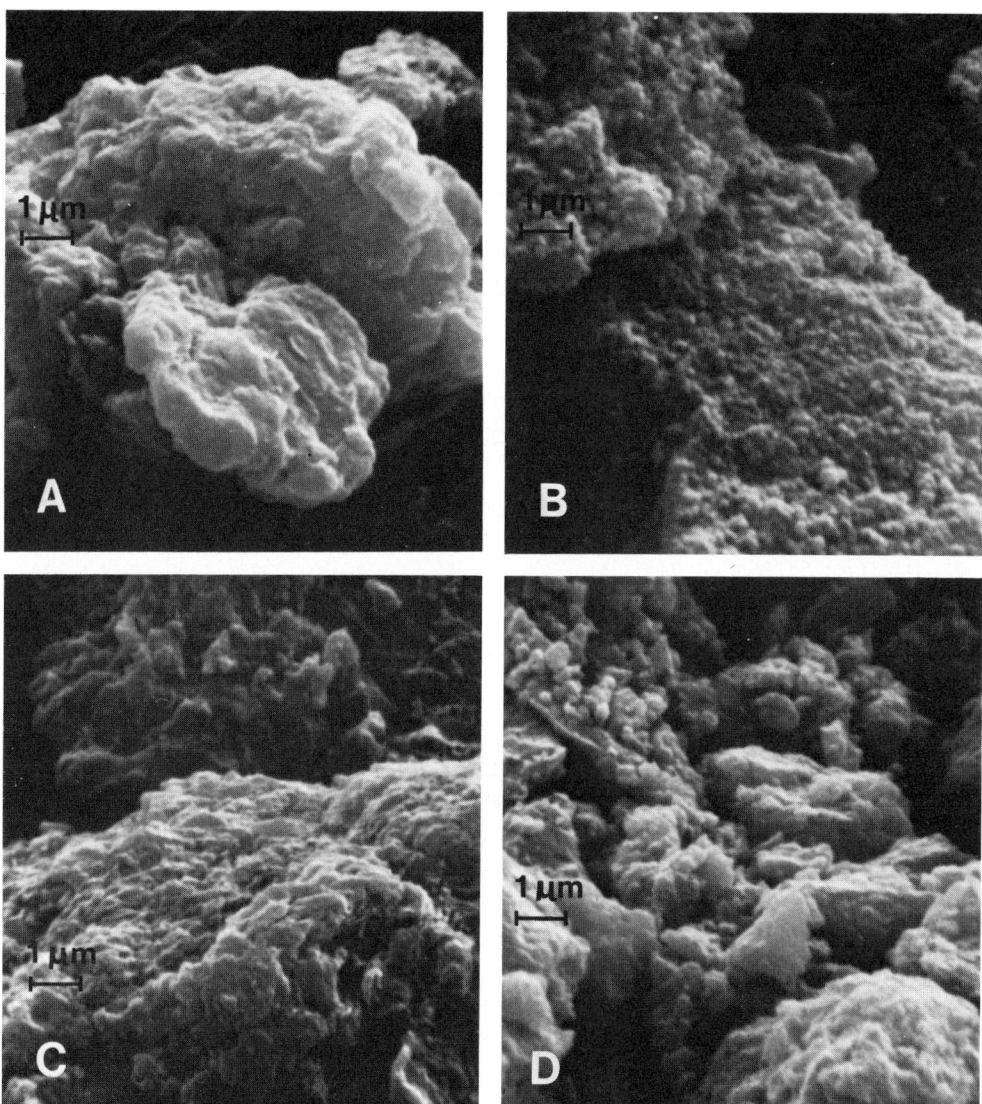

Fig. 6. Scanning electron micrographs showing the surface morphology of hydrolytic reaction products of Al precipitated at the initial Al concentration of $1.1 \times 10^{-4}$M, OH/Al molar ratio of 3.0 and aged for 40 days in the presence of: (A) $10^{-6}$M aspartic acid, (B) $10^{-4}$M p-hydroxybenzoic acid, (C) $10^{-6}$M malic acid, and (D) $10^{-6}$M citric acid.

two major consequences on the mechanisms of the transformations of Al. First of all, the hydrolysis of the positively charged edges of the hydroxy-Al polymers is hampered due to the occupation of the coordination sites of Al by the organic ligands instead of $H_2O$ molecules. Secondly, because of steric factors, the organic ligands occupying the coordination sites of Al distort the arrangement of the unit layers normally found in crystalline Al hydroxides (Fig. 7). If the structural distortion is sufficiently extensive, the hydrolytic precipitation products of Al are poorly crystalline to non-crystalline to X-rays. The structurally distorted precipitation products of Al possess rough and fluffy surfaces with numerous fissures (Fig. 6) and are of porous nature (the authors' unpublished data).

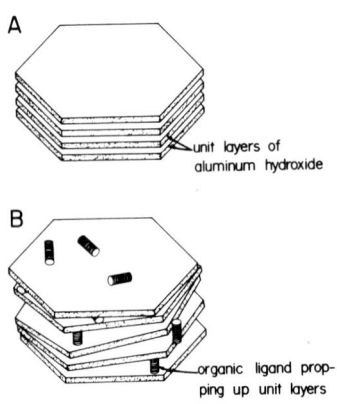

Fig. 7. Structural models showing the distortion of the arrangement of the unit layers of Al hydroxides by the organic ligands.

The findings obtained in this study clearly show the important role of low-molecular weight organic acids in affecting the nature of the hydrolytic reaction products of Al. These organic acids are commonly present in minute but measurable quantities in natural environments (Stevenson, 1967) and possess more functional groups for complexation than extensively polymerized humic materials (Razzaghe and Robert, 1975; Jackson, 1977). Furthermore, since the time required for the formation of soils and aquatic sediments can extend over a period of centuries, the cumulative effect in soils and sediments of even very small quantities of low-molecular weight biochemicals such as organic acids is considerable. The present study thus indicates that in addition to hydroxy-Al interlayering of expansible phyllosilicates (Jackson, 1963; Rich, 1968), the formation of crystalline Al hydroxides in soils and sediments can be hampered through the complexation reactions of hydroxy-Al with these organic acids, especially in environments where accumulation of organic acids tends to occur.

This study, moreover, provides experimental evidence to elucidate a mechanism by which the formation of non-crystalline hydrated aluminum hydroxides can be promoted in soils and sediments. The implications of this finding are evident in the formation of non-crystalline organo-mineral complexes in soils (Mitchell et al., 1964; Campbell et al., 1977). Since the levels and kinds of organic acids in soils are dependent upon the nature of the vegetation and the organic matter being added to soils (Kaurichev et al., 1963; Wang et al., 1967), the extent to which the formation of non-crystalline inorganic components is promoted in soils is expected to vary with land utilization practices. The fine colloidal products, which are subsequently formed in the phase transformations of Al, merit close attention in the study of the transport and fate of nutrients and pollutants in terrestrial and aquatic ecosystems.

ACKNOWLEDGEMENTS

This study was supported by National Research Council of Canada Grant A3248-, E1770- and G0042-Huang. Contribution No. R215, Saskatchewan Institute of Pedology.

REFERENCES

Boyle, J.R., Voight, G.K. and Sawhney, D.L., 1974. Chemical weathering of biotite by organic acids. Soil Sci., 117: 42-45.
Campbell, A.S., Young, A.W., Livingstone, L.G., Wilson, M.A. and Walker, T.W., 1977. Characterization of poorly-ordered aluminosilicates in a vitric andosol from New Zealand. Soil Sci., 123: 362-368.
Hayden, P.L. and Rubin, A.J., 1976. Systematic investigation of the hydrolysis and precipitation of aluminum (III). In: A.J. Rubin (Editor), Aqueous-Environmental Chemistry of Metals. Ann Arbor Science Publishers Inc., Ann Arbor, Michigan, pp. 317-381.
Hsu, P.H., 1977. Aluminum hydroxides and oxyhydroxides. In: J.B. Dixon and S.B. Weed (Editors), Minerals in Soil Environments. Soil Science Society America, Madison, Wisconsin, pp. 99-144.
Huang, P.M., 1975. Retention of arsenic by hydroxy-aluminum on surfaces of micaceous mineral colloids. Soil Sci. Soc. Amer. Proc., 39: 271-274.
Huang, P.M. and Luciuk, G.M., 1974. An investigation of techniques for precipitate collection in hydrolytic reactions of aluminum both in the presence and absence of monomeric silicic acid. Trans. 10th Int. Congr. Soil Sci., Moscow, II: 286-293.
Huang, W.H. and Keller, W.D., 1970. Dissolution of rock forming silicate minerals in organic acids: Simulated first stage weathering of fresh mineral surfaces. Amer. Mineral., 55: 2076-2094.
Irving, H.M. and Rossotti, H.S., 1954. The calculation of formation curves of metal complexes from pH titration curves in mixed solvents. J. Chem. Soc., 2904-2910.
Jackson, M.L., 1963. Aluminum bonding in soils: A unifying principle in soil science. Soil Sci. Soc. Amer. Proc., 27: 1-10.
Jackson, T.A., 1977. A relationship between crystallographic properties of illite and chemical properties of extractable organic matter in pre-Phanerozoic and Phanerozoic sediments. Clays Clay Miner., 25: 163-170.
Jorgensen, S.S., 1976. Dissolution kinetics of silicate minerals in aqueous

catechol solutions. J. Soil Sci., 27: 183-195.

Kaurichev, I.S., Ivanova, T.N. and Nozdrunova, E.M., 1963. Low molecular weight organic acid content of water soluble organic matter in soils. Soviet Soil Sci., 547-556.

Kwong, Ng Kee K.F. and Huang, P.M., 1977. Influence of citric acid on the hydrolytic reactions of aluminum. Soil Sci. Soc. Amer. J., 41: 692-697.

Kwong, Ng Kee K.F. and Huang, P.M., 1978. Sorption of phosphate by hydrolytic reaction products of aluminum. Nature 271: 336-338.

Luciuk, G.M. and Huang, P.M., 1974. Effect of monosilicic acid on hydrolytic reactions of aluminum. Soil Sci. Soc. Amer. Proc., 38: 235-244.

Mitchell, B.D., Farmer, V.C. and McHardy, W.J., 1964. Amorphous inorganic materials in soils. Adv. Agron., 16: 327-383.

Razzaghe-Karimi, M. and Robert, M., 1975. Alteration des micas et geochimie de d'aluminium: role de la configuration de la molecule organique sur l'aptitude a la complexation. C.R. Acad. Sci. Paris, 280D: 2645-2648.

Rich, C.I., 1968. Hydroxy-interlayers in expansible layer silicates. Clays Clay Miner., 16: 15-30.

Rossotti, F.J. and Rossotti, H., 1961. The Determination of Stability Constants. McGraw-Hill, New York.

Stevenson, F.J., 1967. Organic acids in soils. In: A.D. McLaren and G.H. Peterson (Editors), Soil Biochemistry. Marcel Dekker Inc., New York, pp. 119-146.

Thomas, G.W., 1977. Historical developments in soil chemistry: Ion exchange. Soil Sci. Soc. Amer. J., 41: 230-238.

Wang, T.S.C., Cheng, S.Y. and Tung, H., 1967. Dynamics of soil organic acids. Soil Sci., 104: 138-144.

STRUCTURAL FORMULAS OF ALLOPHANES

KOJI WADA
Faculty of Agriculture, Kyushu University 46, Fukuoka 812, Japan


ABSTRACT

Structural formulas for the possible two end-members of allophanes with the molar $SiO_2/Al_2O_3$ ratios of 1.0 and 2.0 (Table 1) were derived by modifying the kaolin structure. The major features of the allophane structure are the occurrences of defects both in the tetrahedral and octahedral sheets and of Al and the water molecule bonded to this Al in the tetrahedral sheet. The latter is considered to act as a Brønsted acid site and as a source of pH-dependent negative charge. These features are also broadly consistent with the data on the morphology, density, X-ray diffraction, infrared spectra, and dehydration-dehydroxylation of allophanes.


INTRODUCTION

Knowledge of the atomic arrangement in clay minerals has great value in predicting and understanding their behaviour. Attempts have also been made to obtain structural formulas or to develop structure models for allophane despite its non-crystalline nature.

A model of allophane, "permutite" core with Al in the 4-fold coordination plus hydroxy-Al cation coating, has been proposed by de Villiers and Jackson (1967), Cloos et al. (1969) and de Villiers (1971). Wada (1967) suggested a structure scheme common to allophane and imogolite with the chemical composition $2SiO_2 \cdot Al_2O_3 \cdot 3H_2O$ and $SiO_2 \cdot Al_2O_3 \cdot 2H_2O$, which consists of the silica tetrahedral and alumina octahedral chains. Iimura (1969) calculated an average structural formula for allophanes by equating the $Ba(OH)_2$ titre with the number of Si-OH groups and by assuming that Al is entirely in the 6-fold coordination:

$$(OH)^I_{0.86} O^I_{1.61} Si_{1.33} O^{II}_{1.23} Al_2 (OH)^{II}_{4.77}$$

where $(OH)^I$ and $(OH)^{II}$ represent OH groups bonded to Si and Al, and $O^I$ and $O^{II}$ represent oxygens in Si-O-Si and Si-O-Al bonds, respectively. Udagawa et al. (1969), on the basis of thermal and X-ray fluorescence measurements, indicated that allophane contains Al in both the 4- and 6-fold coordinations and that allophane has a kaolin-like layer structure rather than a chain structure. Brindley and Fancher (1969) considered various defect kaolin structures to yield compositions for allophanes, in which Al is in the 6-fold coordination and vacancie in the tetrahedral sites account for the low Si content. Modifications of the kaolin structures including Al in the tetrahedral sites have been proposed by Okada et al. (1975) on the basis of X-ray radial distribution, X-ray fluorescence and nuclear magnetic resonance analyses. However, as pointed out by Henmi and Wada (1976) and Brindley (1977), they did not consider the balance of the electric charges in the structures. The electron optical evidence, chemical composition, Al coordination and density data led Wada and Wada (1977) to a conclusion that allophane has a unique structural arrangement namely hollow, spherical 'structure units' and that the walls of the units consist of defect kaolin structures, as illustrated in Fig. 1.

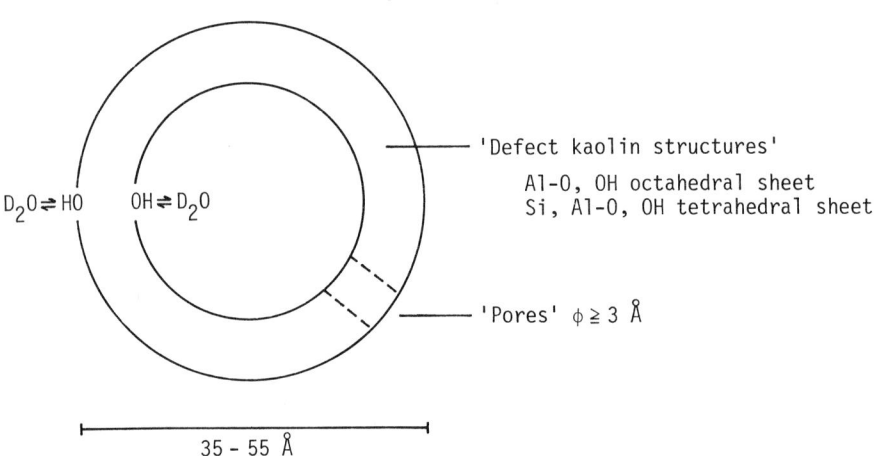

Fig. 1. A schematic presentation of spherical 'structure unit' of allophane.

The purpose of this paper is to consider the atomic arrangement in the defect kaolin structures of allophanes and to discuss the relationships between their structure and morphology, density, surface acidity and electric charge characteristics.

THE STRUCTURAL FORMULAS OF ALLOPHANES

Table 1 shows the structural formulas for the possible two end-members of allophanes with the molar $SiO_2/Al_2O_3$ ratios of 1.0 and 2.0 and those for imogolite and kaolinite for comparison. For allophanes, the kaolin structure is modified to yield both the molar $H_2O(+)/Al_2O_3$ ratios and to conform with the Al coordination data.

TABLE 1

Structural formulas for imogolite, allophane and kaolinite[1]

| | Imogolite | Allophane | | Allophane | | Kaolinite | |
|---|---|---|---|---|---|---|---|
| | 6OH (−6) | 4OH, 2$H_2O$ | (−4) | 2OH, $H_2O$ | (−2) | 6OH | (−6) |
| | 4Al (+12) | 3Al | (+9) | 1Al | (+3) | 4Al | (+12) |
| | 6O (−12) | 2O, 4OH | (−8) | 1O, 2OH, $H_2O$ | (−4) | 4O, 2OH | (−10) |
| | 2Si (+8) | 2Si, 1Al | (+11) | 2Si, 1Al | (+11) | 4Si | (+16) |
| | 2OH (−2) | 3O, 2OH, $H_2O$ | (−8) | 3O, 2OH, $H_2O$ | (−8) | 6O | (−12) |
| a[2] | 1.0 | 1.0 | | 2.0 | | 2.0 | |
| b | 2.0 | 2.5 | | 3.0 | | 2.0 | |

1) The numbers in parentheses indicate the number of electric charges.
2) a: $SiO_2/Al_2O_3$ molar ratio.  b: $H_2O(+)/Al_2O_3$ molar ratio.

Allophane usually coexists with imogolite, but the two can be distinguished easily by their morphological features. Thus, combining the chemical analysis and electron microscopy, Henmi and Wada (1976) showed that allophane has a $SiO_2/Al_2O_3$ ratio ranging at least from 1.0 to 2.0, whereas imogolite has a $SiO_2/Al_2O_3$ ratio close to 1.0. The values of the $H_2O(+)/Al_2O_3$ ratios for allophanes of established purities have not been available. However, Yoshinaga (1966) and Iimura (1969) gave the $H_2O(+)/Al_2O_3$ ratios ranging from 2.5 to 3.0 for the clays in which allophane predominates. Kitagawa (1974) gave much lower values 1.73 and 1.86 for two allophanes. These values are also lower than 2.00 for kaolinite and are not taken into account in the present study.

X-ray fluorescence spectroscopy showed that allophanic clays contain both $Al_{IV}$ and $Al_{VI}$ (Egawa, 1964; Udagawa et al., 1969; Okada et al., 1975), where $Al_{IV}$ and $Al_{VI}$ stand for Al in the 4- and 6-fold coordinations. Henmi and Wada (1976) indicated that the $Al_{IV}$ content of allophane increases with the increasing $SiO_2/Al_2O_3$ ratio and amounts to 50 % of the total Al in allophane with a $SiO_2/Al_2O_3$ ratio of close to 2.0, whereas imogolite contains only $Al_{VI}$. In the derivation of the structural formula, the $Al_{IV}$ contents of allophanes with the $SiO_2/Al_2O_3$ ratios of 1.0 and 2.0 are assumed to be 25 and 50 % of total Al, respectively. Some ambiguity is, however, involved in the determination of the $Al_{IV}$ and $Al_{VI}$ contents by X-ray fluorescence spectroscopy, and therefore, the structural formulas may be uncertain in this regard.

The formulas in Table 1 indicate possible, "average" compositions for the respective allophanes, and the formula unit does not indicate a regular repetition of a unit with this composition throughout the structure. Thus, the clustering of Al in both the octahedral and tetrahedral sheets is possible, and the former will result in the increase of the OH content and the decrease of the $H_2O$ content.

THE STRUCTURE - MORPHOLOGY RELATIONSHIPS

Kitagawa (1971) first indicated by high resolution electron microscopy that allophane consists of spherical, "hollow" unit particles with an average diameter of 55 Å. Henmi and Wada (1976) observed similar particles in sixteen allophanic clays, irrespective of the lithologic compositions, ages and origins of volcanic ashes and pumices from which the clays were separated. They estimated the external diameter of the particle to range from 35 to 50 Å and the thickness of the wall to be 10 Å or less. The thickness of the wall is broadly consistent with that of the defect kaolin structures.

The structural formulas indicate that more defect and/or more discontinuity of the atomic bonding occur in the octahedral sheet than in the tetrahedral sheet. This together with the presence of $Al_{IV}$ in the tetrahedral sheet would enhance the effect of the misfit between the octahedral and tetrahedral sheets in the kaolin structure (Brindley, 1961, pp.54-55), and cause the curling of the resulting layer due to the internal strains.

## THE STRUCTURE - DENSITY RELATIONSHIPS

Kitagawa (1971) determined the density of five allophanic clays by displacement with water, and reported the values ranging from 1.89 to 1.93 $g/cm^3$. Later, he corrected these data and reported the values 2.64 and 2.59 $g/cm^3$ (Kitagawa, 1976). Wada and Wada (1977) obtained the values 2.72 and 2.78 $g/cm^3$ for two allophanes with the $SiO_2/Al_2O_3$ ratios of 1.19 and 1.75. These values were slightly higher than 2.65 $g/cm^3$ for kaolinite and much higher than 2.44 $g/cm^3$ for volcanic glass. The higher density of allophane as compared with kaolinite and volcanic glass can be understood on the basis of the structural formula (Table 1). In allophane, water molecules have access to all vacancies which are large enough to accommodate them in either the tetrahedral or the octahedral sheets, whereas in kaolinite and volcanic glass there are such vacancies in the tetrahedral sheet and in the interior of the three dimensional silica framework, respectively, to which water molecules have no access. The ease and completeness of the OH-OD exchange in allophanes (Wada, 1966; Russell et al., 1969) supports the above structural concept.

Wada and Wada (1977) also reported that the density values of the allophanes by a float-sink test in $CH_3OH$-, $CH_3COCH_3$-, $C_6H_6$- and $CCl_4$-$C_2H_2Br_4$ mixtures are 1.84, 1.94~1.98, 2.35 and 2.35~2.39 $g/cm^3$, respectively. The lower densities measured by the float-sink test were interpreted in terms of a strong adsorption of water and of an exclusion of $C_2H_2Br_4$ molecules from the space where other organic liquid molecules were adsorbed. The very similar density values of allophanes suspended in $C_6H_6$ and $CCl_4$, which are markedly different in density, i.e., 0.879 vs 1.595 $g/cm^3$, suggest that there is no space which would retain $C_6H_6$ or $CCl_4$ in preference to $C_2H_2Br_4$. The infrared spectroscopy of the samples indicated the presence of water but not of $CH_3OH$ and $CH_3COCH_3$ with which the samples were washed prior to washing with $C_6H_6$ or $CCl_4$. The amount of water remaining in allophane suspended in the latter liquids was estimated to be 10.8 g per 100 g oven-dry clay, assuming that the solid part of allophane has the density of 2.75 g/cm$^3$ and the remaining water has the normal liquid density. This water was considered to be retained in the interior of the spherical particle. However, it is equally possible that the water coordinated to Al in the wall remains after repeated washings with the organic liquids. The structural formulas presented in Table 1 indicate that the amounts of the water coordinated to Al in the two allophanes are 13.0 and 19.6 g per 100 g oven-dry clay. The amount of the water, particularly the latter, is reduced to a considerable extent if the occurrence of the Al in the octahedral sheet is clustered. If the remaining water is not retained in the interior of the spherical particle, the density data indicate that the wall must admit the passage of not only water molecules but all the organic liquid molecules including larger $CCl_4$ and $C_2H_2Br_4$. This means that the defects in the kaolin structures should be clustered rather than dispersed. Further studies on the status of strongly retained water in allophane are necessary.

THE STRUCTURE - ELECTRIC CHARGE - ACIDITY RELATIONSHIPS

In the derived formula, part of Al occurs in the tetrahedral sheet (Table 1). In layer silicates such as montmorillonite and vermiculite, the presence of Al in the tetrahedral sheet creates negative charge. This excess of the negative charge is balanced by retention of exchangeable cations, and the amount of the retained cations is known to be independent of the pH and electrolyte concentration of an ambient solution. On the other hand, the retention of exchangeable cations by allophane has been known to increase with the increasing pH and electrolyte concentration (Wada and Harward, 1974; Wada, 1977). This has been taken into consideration in the derivation of the structural formula, in which the Al in the tetrahedral sheet is involved in three electron pair bonds to oxygens, as illustrated in Fig. 2:

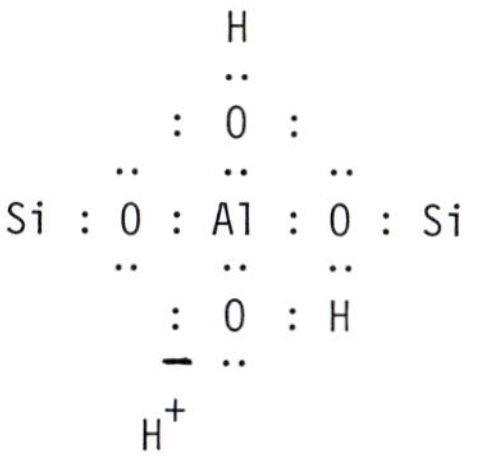

```
           H
           ··
         : O :
           ··
    ·· ·· ··
Si : O : Al : O : Si
    ·· ·· ··
         : O : H
         ─── ··
          H⁺
```

Fig. 2. A Brønsted acid site in allophane.

This Al will require an additional electron pair to complete its p-orbital, and therefore, has a strong tendency to coordinate molecules with lone-pair electrons such as water. The system is regarded as a Lewis acid and changes to a Brønsted acid by coordination with a water molecule. The water molecule then dissociates

$H^+$ depending on the extent of hydration under fairly dry conditions and on the pH and electrolyte concentration in the solution. In the imogolite formula, no such acid sites are present, and the dissociation of $H^+$ from the Si-OH groups probably accounts for the development of negative charge.

The critical data which might indicate the difference in the electric charge and acid characteristics between allophane and imogolite are scarce. Yoshida (1971 treated allophane and imogolite with 1 N $AlCl_3$ and found that all the exchange sites in these minerals were occupied by H and not by Al. This indicates that the strengths of all acid sites on allophane and imogolite are lower than that of $Al(H_2O)_6^{3+}$. The titration of the resulting H-clays with KOH up to pH 8.5 showed that the two minerals are different in their acid characteristics and in the amount of the dissociable $H^+$, i.e., 60.8 and 15.4 me/100 g oven-dry clay for allophane and imogolite, respectively.

Wada and Okamura (1977) found that the development of negative charge (CEC; me/100 g oven-dry clay) in soils in which imogolite and/or allophane predominate follows a general equation:

log CEC = $\underline{a}$ pH + $\underline{b}$ log C + $\underline{c}$

where C is the concentration of index cation (N) and $\underline{a}$, $\underline{b}$ and $\underline{c}$ are the coefficients, constant for the respective soils. The coefficients $\underline{a}$, $\underline{b}$ and $\underline{c}$ are 0.31~0.34, 0.25~0.27 and -0.72~-0.97 for two soils containing allophane and imogolite with the $SiO_2/Al_2O_3$ ratios of about 1.0, whereas the corresponding figures are 0.22, 0.17 and 0.22 for one soil containing allophane with the $SiO_2/Al_2O_3$ ratio of about 1.8.

The maximum development of negative charge arising from the $Al_{IV}-H_2O$ groups in the formula would amount to 242 and 362 me/100 g oven-dry clay for allophanes with the $SiO_2/Al_2O_3$ ratios of 1.0 and 2.0, respectively. The retention of Na from a hot 2 % $Na_2CO_3$ solution (pH 10.5) was reported to range from 145 to 212 me/100 g oven-dry clay for allophane with the $SiO_2/Al_2O_3$ ratio of 1.32 to 1.95 (Yoshinaga, 1966), and from 23 to 43 me/100 g oven-dry clay for imogolite with the $SiO_2/Al_2O_3$ ratio of 1.06 to 1.16 (Wada and Yoshinaga, 1969). Iimura (1969) determined the retention of Ba at 0.015 N $Ba(OH)_2$ and obtained the value ranging from 380 to 279 me/100 g oven-dry clay for allophane with the $SiO_2/Al_2O_3$ ratio of 0.89 to 1.43, though he considered that the Ba is retained by Si-OH groups.

An increase in acidity or proton donating properties of the clay surface with decreasing water content has been demonstrated by several investigators (Mortland, 1970). Henmi and Wada (1974) estimated the acid strengths of allophane and imogolite by observing the colouration of indicators adsorbed onto the samples in benzene, and compared the result with other clay minerals. H(Al)-saturated allophane behaved as a strong acid (pKa;-5.6~1.5) in a relatively dry environment (relative humidity; 10~55 %), but its acid strength was very much reduced either by increasing its water content or by saturating it with Na. Imogolite showed only a very weak acidity (pKa; 4.6~6.8) under medium dry to moist conditions (relative humidity; 30~100 %) and showed no response to Na saturation treatment. A marked enhancement of acidity occurred in both allophane (pKa; -5.6~-8.2) and imogolite (pKa; 1.5~-5.6) when they had been dried over $P_2O_5$ or heated to result in dehydroxylation. H(Al)-saturated montmorillonite, kaolinite and halloysite showed stronger acidities (pKa; -5.6~-8.2) in a relatively moist environment (relative humidity; >60 %), whereas gibbsite showed only a very weak acidity (pKa; 4.6~6.8) even when it had been dried over $P_2O_5$. The observed acid characteristics of allophane are in accordance with what predicted from the structural formula. The agreement remains qualitative, however, because an attempt to determine the number of acid sites in allophane by titration with n-butylamine has failed for an unaccounted ineffectiveness of n-butylamine as a neutralization reagent.

CONCLUSIONS

The structural formulas for allophanes are broadly consistent with the electron optical evidence and density data as described above. The modifications of the kaolin layer caused by the size and shape of the unit particles and by the occurrence of defects may account for the absence of both basal and two-dimensional X-ray reflections, the broadness and poor resolution of Si-O-(Al) absorption bands in the 1100~900 $cm^{-1}$ region of the infrared spectra, and the continuous loss of water with increasing temperature due to dehydration and dehydroxylation. The observed electric charge and acid characteristics of allophanes are also broadly consistent with what is predicted from the structural formulas. Determination of the contents of $Al_{IV}$, Si-OH and $Al_{VI}$-OH as well as the number of the acid sites with different acid strengths are important problems left to future studies. Recently, synthesis of non-crystalline, hydrous aluminosilicates which are very similar to allophane in their morphology, composition and properties has been found possible (Wada and Wada, 1978) by applying a heating procedure used for synthesis of imogolite (Farmer et al., 1977). These synthetic products are useful towards a better understanding of the nature and properties of allophane along the line described above.

REFERENCES

Brindley, G.W., 1961. Kaolin, Serpentine, and Kindred Minerals. In: G. Brown (Editor), The X-Ray Identification and Crystal Structures of Clay Minerals. Mineralogical Society, London, 51-131.

Brindley, G.W. and Fancher, D., 1969. Kaolinite defect structures; possible relation to allophanes. In: L. Heller (Editor), Proc. Int. Clay Conf. (Tokyo, Japan) 1969, 2: 29-34.

Cloos, P., Leonard, A.J., Moreau, J.P., Herbillon, A. and Fripiat, J.J., 1969. Structural organization in amorphous silicoaluminas. Clays Clay Miner., 17: 270-287.

de Villiers, J.M., 1971. The problem of quantitative determination of allophane in soil. Soil Sci., 112: 2-7.

de Villiers, J.M. and Jackson, M.L., 1967. Cation-exchange capacity variations with pH in soil clays. Soil Sci. Soc. Am. Proc., 31: 473-476.

Egawa, T., 1964. A study on coordination number of aluminum in allophane. Clay Sci., 2: 1-7.

Farmer, V.C., Fraser, A.R. and Tait, J.M., 1977. Synthesis of imogolite: A tubular aluminum silicate polymer. J. Chem. Soc. Chem. Comm., 13: 462-463.

Henmi, T. and Wada, K., 1974. Surface acidity of imogolite and allophane. Clay Miner., 10: 231-245.

Henmi, T. and Wada, K., 1976. Morphology and composition of allophane. Am. Mineral., 61: 379-390.

Iimura, K., 1969. The chemical bonding of atoms in allophane — The "structural formula" of allophane. In: L. Heller (Editor), Proc. Int. Clay Conf. (Tokyo, Japan) 1969, 1: 161-172.

Kitagawa, Y., 1971. The "unit particle" of allophane. Am. Mineral., 56: 465-475.

Kitagawa, Y., 1974. Dehydration of allophane and its structural formula. Am. Mineral., 59: 1094-1098.

Kitagawa, Y., 1976. Specific gravity of allophane and volcanic ash soils determined with a pycnometer. Soil Sci. Plant Nutr. (Tokyo), 22(2): 199-202.

Mortland, M.M., 1970. Clay-organic complexes and interactions. Adv. Agron., 22: 75-117.

Okada, K., Morikawa, S., Iwai, S., Ohira, Y. and Ossaka, J., 1975. A structure model of allophane. Clay Sci., 4: 291-303.

Russell, J.D., McHardy, W.J. and Fraser, A.R., 1969. Imogolite: a unique aluminosilicate. Clay Miner., 8: 87-99.

Udagawa, S., Nakada, T. and Nakahira, M., 1969. Molecular structure of allophane as revealed by its thermal transformation. In: L. Heller (Editor), Proc. Int. Clay Conf. (Tokyo, Japan) 1969, 1: 151-159.

Wada, K., 1966. Deuterium exchange of hydroxyl groups in allophane. Soil Sci. Plant Nutr. (Tokyo), 12(5): 176-182.

Wada, K., 1967. A structure scheme of soil allophane. Am. Mineral., 52: 690-708.

Wada, K., 1977. Allophane and Imogolite. In: J.B. Dixon (Editor), Minerals in Soil Environments. Soil Science Society of America, Inc., Madison, 603-638.

Wada, K. and Harward, M.E., 1974. Amorphous clay constituents of soils. Adv. Agron., 26: 211-260.

Wada, K. and Okamura, Y., 1977. Measurement of exchange capacities and hydrolysis as means of characterizing cation and anion retentions by soils. In: Proceedings of the International Seminar on Soil Environment and Fertility Management in Intensive Agriculture, Soc. Sci. Soil Manure Japan, Tokyo, pp. 811-815.

Wada, S. and Wad, K., 1977. Density and structure of allophane. Clay Miner., 12: 289-298.

Wada, S. and Wada, K., 1978. Formations of allophane and imogolite from solutions containing silica and aluminum ions. Abst. Pap. Soc. Sci. Soil Manure Japan, 24: 29.

Wada, K. and Yoshinaga, N., 1969. The structure of imogolite. Am. Mineral., 54: 50-71.

Yoshida, M., 1971. Acidic properties of kaolinite, allophane and imogolite. J. Sci. Soil Manure (Japan), 42: 329-332 (Japanese).

Yoshinaga, N., 1966. Chemical composition and some thermal data of eighteen allophanes from Ando soils and weathered pumices. Soil Sci. Plant Nutr. (Tokyo), 12(2): 47-54.

## SYNTHETIC IMOGOLITE, A TUBULAR HYDROXYALUMINIUM SILICATE

V.C. FARMER and A.R. FRASER

Department of Spectrochemistry, Macaulay Institute for Soil Research, Craigiebuckler, Aberdeen, AB9 2QJ SCOTLAND.

ABSTRACT

Infrared spectroscopy indicates the formation of a proto-imogolite by the interaction of hydroxyaluminium cations with orthosilicic acid in dilute solutions (Al:Si atomic ratio = 2) at pH values less than about 5. Heating such solutions to 96-100°C generates a dispersed synthetic imogolite with tube diameter 13-17% larger than natural imogolite, as indicated by electron microscopy, electron diffraction, and X-ray diffraction. Tube formation causes a marked fall in pH towards 3, indicating that the doubly coordinated hydroxyl groups of imogolite are less basic than the singly coordinated hydroxyl groups of its precursors, the hydroxyaluminium cations and the proto-imogolite.

The volume of gel formed on making imogolite solutions alkaline with ammonia was used as a guide to optimizing synthesis conditions. As anions inhibited tube formation ($Cl^- > NO_3^- > ClO_4^-$), the largest yields of synthetic imogolite in a single step were obtained in salt-free systems, by heating stock solutions prepared by hydrolysing aluminium s-butoxide and silicon tetraethoxide in perchloric acid. In such solutions, imogolite continued to form at Al concentrations up to at least 140 mM. The optimum temperature of synthesis is around 100°C.

PRELIMINARY EXPERIMENTS

The successful synthesis of imogolite, a widespread soil component, was achieved (Farmer et al., 1977a) through the integration of three pieces of evidence. Firstly, our interpretation of the electron-diffraction pattern of imogolite (Cradwick et al., 1972) indicated that the basic structure of an imogolite tube was a single gibbsite sheet bent round in the form of a cylinder, with orthosilicate groups attached to the inside of the cylinder, each group replacing three OH groups round an empty octahedral site (Fig. 1). Secondly, observations in Japan of the natural occurrence of imogolite in the form of gel films coating weathered pumice particles (Wada and Harward, 1974) convinced us that imogolite formed from solution, especially since only allophane spheres could sometimes be isolated from inside these weathered

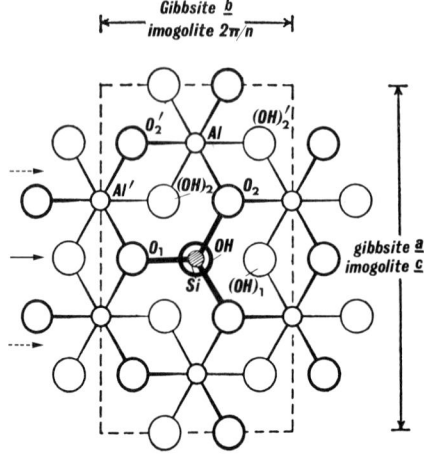 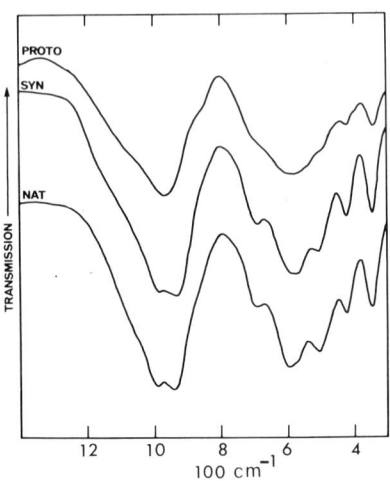

Fig. 1. Mode of attachment of a SiO$_3$OH group to the face of a gibbsite sheet, causing it to curl to form an imogolite tube. Reproduced from Cradwick et al., 1972.

Fig. 2. Infrared spectra of proto-imogolite (PROTO), synthetic imogolite (SYN) and natural imogolite (NAT).

pumice particles. The third piece of evidence was provided by the finding that imogolite has a very distinctive absorption band in its infrared spectrum (Fig. 2) at 348 cm$^{-1}$ (Farmer et al., 1977b). The importance of the infrared evidence is that it provided a tool with which the formation of incipient imogolite-like structures (proto-imogolite) could be recognized. The fact that kaolinite and halloysite also have strong absorption bands at 348 cm$^{-1}$ has proved no impediment: we have found no evidence for the formation of kaolin minerals or proto-kaolin structures under the conditions used.

Initial experiments, based only on the first and second lines of evidence, attempted to react orthosilicic acid with a soluble gibbsite precursor, the pre-gibbsite sol described by Gastuche and Herbillon (1962). This procedure gave extremely stable sols, from which only traces of boehmite separated over many months. Electron diffraction of the dispersed component, however, showed little evidence of structure. Further experimentation was suspended until the infrared characterizati tool became available, and from then progress was rapid.

THE CRYSTALLIZATION PROCESS

It immediately became clear that, when dilute solutions containing non-complexin salts of aluminium and orthosilicic acid with Al:Si atomic ratios of about 2:1 were adjusted to a pH of around 4.5, a proto-imogolite formed in the solution, and could be isolated by freeze-drying. The IR spectrum of the proto-imogolite was very similar to that of true imogolite (Fig. 2), although more diffuse. Heating such

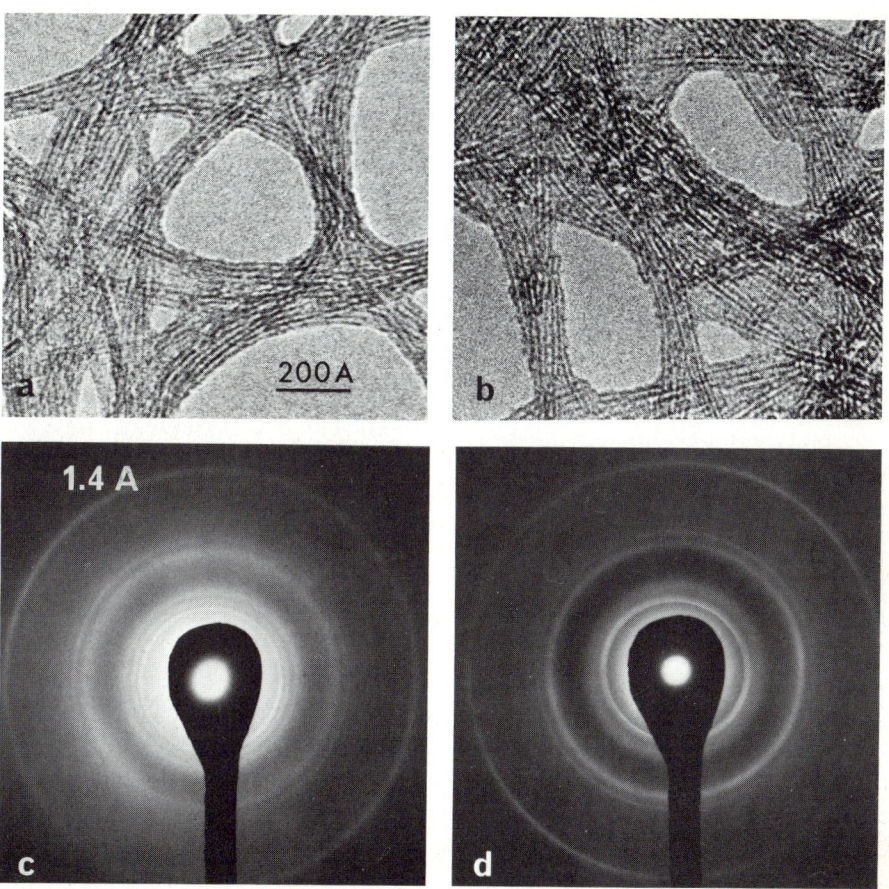

Fig. 3. Electron micrographs of (a) natural and (b) synthetic imogolite, kindly supplied by Prof. N. Yoshinaga, Ehime University, Japan. Electron diffraction of (c) natural and (d) synthetic imogolite, (microscopist, J.M. Tait).

solutions to 96-100°C yielded synthetic imogolite with a spectrum almost identical with that of natural imogolite (Fig. 2). At the same time, electron microscopy revealed the formation of imogolite-like tubes (Fig. 3 a,b), and electron diffraction showed the development of a diffraction pattern, again similar to, but not quite identical with, the natural material (Fig. 3 c,d).

The process of tube formation can be regarded as akin to crystallization, leading to a unidimensional crystal. Imogolite seeds (proto-imogolite) readily form by reaction between orthosilicic acid and hydroxyaluminium species generated by partial neutralization of the aluminium salts. This material probably consists initially of small rafts of imogolite-like structure, and gives only broad diffuse diffraction rings at 2.3 Å and 3.45 Å, which are nevertheless in the same position as pronounced features in imogolite diffraction patterns (Farmer et al., this conference). The progress of tube formation is indicated first by the appearance of a rather diffuse reflection around 1.4 Å which is associated with the repeat distance (8.4 Å) along the tube axis. With further development of the tube morphology, this reflection sharpens, and other orders appear at 2.1 Å and 4.2 Å. If, however, conditions are unfavourable for crystallization (tube formation) e.g. when excessive salt, silicic acid, or reagent concentrations are present, or the pH is too high, the imogolite seeds may condense in a more disordered form, and the diffraction pattern and infrared spectrum then remain those of proto-imogolite, or of very poorly ordered imogolite. Once such a disordered condensed structure has formed, recrystallization is difficult.

One marked difference between ordered and disordered condensation processes is that in the former the solution always drops in pH, often moving to near pH 3 from a starting pH of around 4.5 whereas in the latter little shift in pH occurs, and sometimes a slight increase is observed. It would seem, therefore, that imogolite formation liberates acid by a mechanism which eliminates ionized sites: i.e.

$$\text{Al}^+{\begin{matrix}OH_2\\OH_2\end{matrix}} + \text{H}_2O\begin{matrix}Si\\|\\O\\\end{matrix}\text{Al} \rightarrow \text{Al}\begin{matrix}Si\\|\\O\\OH\end{matrix}\text{Al} + \text{H}^+$$

In the final gibbsite-like sheet, the surface OH must have little affinity for protons. More disordered condensation processes presumably leave projecting aluminium octahedra with singly coordinated OH groups for which the equilibrium

$$\text{AlOH} + \text{H}^+ \rightleftharpoons \text{Al}^+ - \text{OH}_2$$

lies more to the right than for the doubly coordinated bridging OH groups of the gibbsite sheet.

## OPTIMIZATION OF SYNTHETIC CONDITIONS

The process of producing well-formed synthetic imogolite in substantial quantities is, like most crystallization processes, more art than science. To follow the progress of tube formation conveniently and rapidly, the gel-forming properties of the solution were assayed by measuring the volume of centrifuged gel precipitated by ammonia from 10 ml of a solution diluted to either 1 mM or 2.5 mM in Al. A drop of Indian ink was added to the solution before precipitation to make obvious the depth of centrifuged gel, which was otherwise often too transparent to see clearly. Centrifuge tubes of 13 mm internal diameter were used, and the gel was centrifuged at 2 500 rpm for 10 min. This bulky gel consists, presumably, of an open cross-linked framework of imogolite tubes. By this criterion, it was found that tube formation was inhibited by the presence of salts, or by solution pH values in excess of 5. The optimum temperature of synthesis was around 100°C, since the amount of

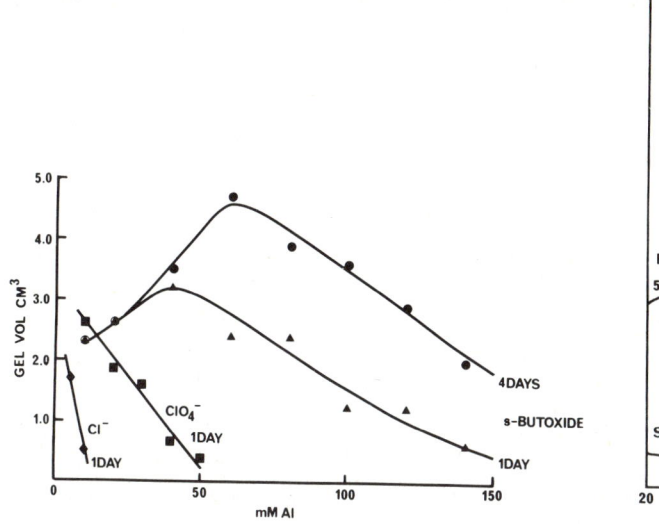 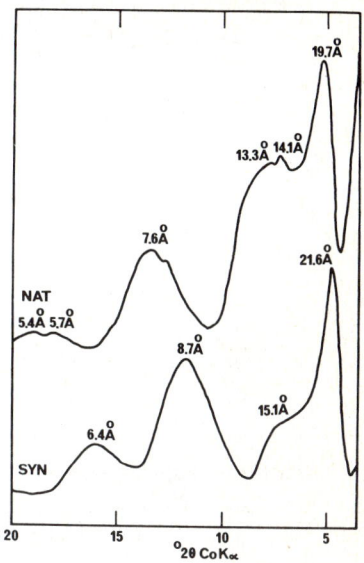

Fig. 4 (left). Formation of synthetic imogolite at 96°C in solutions containing Al:Si atomic ratios 2:1, indicated by the centrifuged gel volumes from 10 ml solution adjusted to 2.5 mM in Al. $Cl^-$ systems prepared from stock solutions containing 10 mM $AlCl_3$ and 5 mM $Si(OH)_4$, adjusted to pH 4.5; $ClO_4^-$ system prepared from 50 mM $Al(ClO_4)_3$, 25 mM $Si(OEt)_4$ (hydrolysed in situ) adjusted to pH 4.5; butoxide system prepared by premixing $Al(s-BuO)_3$ and $Si(EtO)_4$, then hydrolysing, with vigorous stirring in perchloric acid to give $Al:Si:HClO_4$ millimolar ratios of 150:75:75.

Fig. 5 (right). X-ray diffraction of natural and synthetic imogolite prepared by evaporating dispersions in dilute acetic acid on to glass microscope slides.

silica remaining in solution increased rapidly at higher temperatures, while the rate of formation dropped off rapidly below 90°C. Substantial yields of imogolite have been obtained at 60°C over three weeks, but only very low yields of gel at 40°C and below, even over some months.

Perchlorate, nitrate, and chloride were progressively more inhibiting, so that aluminium perchlorate was the most satisfactory inorganic salt, giving substantial gel yields at concentrations up to 30 mM compared with only 5 mM for chloride solutions (Fig. 4). Nitrate was only slightly superior to chloride systems. Salt-free systems, prepared by the hydrolysis of a mixture of aluminium s-butoxide with silicon tetraethoxide in perchloric acid were superior (Fig. 4), giving optimum yields of gel when diluted to 50 mM Al, and substantial yields even at 140 mM Al, although the development of gel yields became slower at higher concentrations. Other procedures for reducing salt content of the reacting solutions, for example by redispersing a freshly precipitated aluminosilicate gel in dilute $HClO_4$ ($Al:Si:HClO_4$ = 2:1:1) did not allow such high concentrations of reagents as did the use of alkoxides, possibly because of anions adsorbed by the gel. The conditions controlling gel yields, are, however, not yet fully understood. Often, it seems, weakly inhibiting conditions can favour high gel yields, possibly because the number of active seeds is then reduced. Thus, increasing reagent concentrations often initially retards gel formation but increases gel yields (compared after dilution to standard conditions), whereas still higher concentrations inhibit gel formation (compare 1- and 4-day curves for the butoxide system, Fig. 4).

COMPARISON OF SYNTHETIC AND NATURAL IMOGOLITE

Synthetic and natural imogolite are not identical. Electron microscopy, electron diffraction, and X-ray diffraction all indicate that the synthetic tubes have somewhat greater diameters (N. Yoshinaga, personal communication, 1977). Thus, in X-ray diffraction patterns of films dried at $100^{\circ}C$ (Fig. 5), natural samples exhibit a sharp peak at 19.7 Å, and a series of broad peaks at 13.3, 7.6 and 5.5 Å, whereas synthetic imogolite gives similar features at higher spacings, the increase ranging from 13-17% for different preparations. The sharp peak probably corresponds to the planes of a close-packed hexagonal array of tubes, and so indicates a centre-to-centre tube separation of 22.7 Å for natural and 26.2 Å for synthetic. The series of broad maxima correspond to the scattering of individual tubes, and Cradwick's calculations (Cradwick et al., 1972) indicate that the series of spacings for natural imogolite correlate well with those calculated for models of outer diameter (oxygen to oxygen) around 20 Å, consistent with inter-tube spacings of near 23 Å. This indicates a circumference of some 12 gibbsite unit cells in natural material and 14 unit cells in synthetic imogolite.

Like X-ray diffraction, electron diffraction patterns reflect the greater diameter of the synthetic material in the scattering maxima perpendicular to the tube axis. They also give evidence of a more regular packing of natural imogolite tubes, with adjacent tubes displaced relative to each other by nc/2, where c is the repeat distance (8.4 Å) along the tube axis and n is any integer. This is indicated by the resolution of the 02 layer line of natural imogolite into discrete reflections corresponding to 12, 32, and 42 planes (Cradwick et al., 1972) whereas synthetic imogolite gives only a continuous streak along this layer line, suggesting random displacements of the tubes relative to each other.

SIGNIFICANCE

The laboratory synthesis of imogolite carries a number of more general implications:
(1) Since imogolite has been found to form only in solution, its site of precipitation in nature can be no guide to its conditions of formation: i.e. the age, pH, and solution composition of the soil or pumice horizons in which it is found need not be those of its formation, which could have occurred in higher, younger horizons.
(2) Since hydroxyaluminium cations react so readily with orthosilicic acid, part of

the soluble aluminium in acid soils could well exist as hydroxyaluminium silicate complexes.

(3) The synthesis makes available a novel type of inorganic polymer, of regular structure, high surface area, defined porosity, and exceptional gel-forming properties. We look forward to the exploration of its potential uses.

REFERENCES

Cradwick, P.D.G., Farmer, V.C., Russell, J.D., Masson, C.R., Wada, K., and Yoshinaga, N., 1972. Imogolite, a hydrated aluminium silicate of tubular structure. Nature Phys. Sci. 240: 187-189.
Farmer, V.C., Fraser, A.R., and Tait, J.M., 1977a. Synthesis of imogolite. J. Chem. Soc. Chem. Comm., 462-463.
Farmer, V.C., Fraser, A.R., Russell, J.D., and Yoshinaga, N., 1977b. Recognition of imogolite structures in allophanic clays by infrared spectroscopy. Clay Miner., 12: 55-57.
Gastuche, M.-C., and Herbillon, A., 1962. Etude des gels d'alumine: cristallisation en milieu désionisé. Bull. Soc. Chim. Fr. 1404-1412.
Wada, K. and Harward, M.E., 1974. Amorphous clay constituents of soils. Adv. Agron. 26: 211-260.

APPLICATION OF MÖSSBAUER SPECTROSCOPY TO THE STUDY OF IRON OXIDES IN SOME RED AND YELLOW/BROWN SOIL SAMPLES FROM NEW ZEALAND

C.W. CHILDS,* B.A. GOODMAN
Macaulay Institute for Soil Research, Aberdeen, AB9 2QJ, Scotland
and G.J. CHURCHMAN
Soil Bureau, DSIR, Lower Hutt, New Zealand

ABSTRACT

Mössbauer spectroscopy has been used to characterise the secondary Fe oxides in whole samples of some red (Munsell hue in the range 10R-5YR) and yellow/brown (7.5YR-2.5Y) soil samples from New Zealand. Dithionite/citrate/bicarbonate treatment of all samples produced whitish-grey to grey residues indicating that the colouring material was present in the "free Fe oxide" fraction of the soils.

At room temperature the spectrum of each red sample has a magnetic hyperfine component attributable to hematite. This is absent from the spectra of all yellow/brown samples. At 77K further magnetic hyperfine components are evident in some red samples and in all yellow/brown samples, and are attributable to goethite of very small particle size (superparamagnetic) and/or akaganéite.

The Munsell designation can be used to relate colour to the concentration of free Fe oxides with reasonable accuracy ($\pm$ 25%) for the red samples, but not the yellow/brown samples. In addition, the increasing intensity of redness, judged subjectively, is closely related to increasing concentration of hematite in the red samples.

INTRODUCTION

Secondary Fe oxides and hydroxides are usually the dominant pigments in well-drained soils that contain little organic matter. Red coloration is traditionally associated with the presence of anhydrous oxides, especially hematite ($\alpha$-$Fe_2O_3$), and yellow/brown coloration with the presence of hydrous oxides, especially goethite ($\alpha$-FeOOH). These relationships had been inferred largely from the colours of pure synthetic minerals.

Schwertmann and Lentze (1966), in a study on a large number of natural Fe-rich samples, confirmed these colour/mineral relationships, but showed that, in some cases, components such as organic matter and Mn oxides could interfere. Segalen and Robin (1969), in a

*On leave from Soil Bureau, New Zealand

study of tropical soils, concluded that the colour of yellow soils seemed to be due to goethite, whereas "amorphous" Fe oxides were responsible for the colour in red soils. Several studies have been made on red and yellow soils from Australia: Isbell and Smith (1976) detected hematite in most red earths, but in none of the yellow earths they studied. Goethite was detected in some red earths and in some yellow earths. Davey et al. (1975) concluded that goethite, containing about 13-14 mol % AlOOH, was the dominant Fe oxide mineral in red and yellow podzolic soils and gave the yellow soils their characteristic colour. The red soils also contained finely divided hematite which masked the colour of the goethite. Taylor and Graley (1967) associated high goethite to hematite ratios with yellow/brown colours and low ratios with red to red/brown colours in a toposequence of basaltic soils. In these last three studies samples were boiled in 5M NaOH to concentrate the Fe oxides before identification. However there is evidence that such severe treatment can increase the crystallinity of poorly-ordered synthetic Fe oxides, and can cause lepidocrocite ($\gamma$-FeOOH) to transform to goethite (Kojima, 1963). The effect on natural Fe oxides and hydroxides in soils is unclear.

In an extensive investigation of coloration in Coastal Plain soils from North Carolina, Soileau and McCracken (1967) concluded that there was no consistent relationship between amounts of free Fe oxides and yellow-to-red Munsell hue of the B horizons nor was coloration due to the relative thickness of adsorbed iron coatings on the clay surfaces. Goethite was the only Fe oxide identifiable by X-ray diffraction but X-ray amorphous components probably accounted for a significant portion of the free Fe oxide. They suggested that the form of Fe oxide was the dominant factor in influencing hue.

We have studied a set of related red and yellow/brown soil samples from a range of sites. Mössbauer spectroscopy, rather than X-ray diffraction, has been used as the primary method for determining the nature of the secondary Fe oxides, and this has advantages in that (i) spectra are due to Fe nuclei only and non-Fe containing minerals do not interfere; (ii) spectra are obtained even from nuclei which are only involved in short-range order (amorphous to X-rays); (iii) chemical pretreatment of samples is not necessary.

SAMPLES AND METHODS

Eight pairs of soil samples were investigated. Each pair consists of a red sample and a yellow/brown sample taken close to each other from sites listed in Table 1. All are subsoils except the pair from site 8 which is in a very dry area with little to no vegetation. Site 6 has been described by Challis (1975). Samples were air-dried and ground to pass 2 mm.

Some characteristics of the samples are shown in Table 2. Red samples and yellow/brown samples are denoted "r" and "y", respectively, following the site numbers. Total Fe was determined colorimetrically (Scott, 1941). Dithionite Fe and dithionite Al, i.e. extractable by the dithionite/citrate/bicarbonate method of Mehra and Jackson (1960), were determined by emission flame spectrometry.

TABLE 1

Sampling sites

| Site no. | Location | Map ref.* | Parent material |
|---|---|---|---|
| 1 | Belmont | N160/463339 | greywacke/loess/colluvium |
| 2 | Belmont | N160/466339 | greywacke/loess/colluvium |
| 3 | Stokes Valley | N160/518358 | greywacke/loess |
| 4 | East of Papakura | N47/503330 | sandstone |
| 5 | Rama Rama | N47/483248 | andesitic ash |
| 6 | Upper Hutt | N161/625422 | greywacke/loess/rhyolitic ash |
| 7 | Pauatahanui | N160/475464 | loess |
| 8 | Butchers Dam | S143/147396 | schist |

*NZMS1 grid references

The predominant clay minerals were determined by X-ray diffraction of the clay-size (<2 μm) fraction of samples which had been extracted with dithionite to remove free Fe oxides. The procedure used is based, in part, on that of Avery and Bullock (1977), and will be described in a forthcoming paper by G.J. Churchman. Kaolinites and halloysites were distinguished according to the criteria of Churchman and Carr (1975). Kaolin phases having insufficient characteristics for this distinction are regarded as disordered kaolins. Red and corresponding yellow/brown samples are similar to each other, apart from colouring material, for sites 1, 3, 4, 5 and 7, but those from sites 2, 6 and 8 are considerably different (Table 2). However, all residues after deferration were whitish-grey to grey.

TABLE 2

Sample characteristics

| Sample no. | Munsell colour* | Total Fe(wt%) | Dithionite Fe(wt%) | Dithionite Al(wt%) | Predominant clay minerals† |
|---|---|---|---|---|---|
| 1r | 5YR 6/6, reddish yellow | 4.3 | 2.2 | 0.25 | Kn(dis.), IHM |
| 2r | 2.5YR 5/6, red | 4.4 | 2.5 | 0.20 | Kt, I |
| 3r | 2.5YR 6/6, light red | 2.5 | 1.4 | 0.22 | Kt, I |
| 4r | 10R 4/4, weak red | 5.5 | 2.1 | 0.40 | H, IHM |
| 5r | 10R 4/6, red | 7.7 | 4.3 | 0.30 | H, IHM, I |
| 6r | 10R 5/6, red | 5.3 | 3.6 | 0.25 | Kt, M |
| 7r | 5YR 6/6, reddish yellow | 3.3 | 1.8 | 0.20 | H, IHM, I |
| 8r | 10R 4/8, red | 8.8 | 7.1 | 0.18 | Kt, IHM |
| 1y | 10YR 7/6, yellow | 6.0 | 1.0 | 0.20 | Kn(dis.), I, IHM |
| 2y | 2.5Y 7/4, pale yellow | 5.3 | 1.3 | 0.20 | I, H, IHM |
| 3y | 10YR 7/6, yellow | 6.2 | 1.9 | 0.40 | Kt, I |
| 4y | 7.5YR 6/6, reddish yellow | 5.9 | 1.6 | 0.42 | H, IHM |
| 5y | 7.5YR 4/6, strong brown | 7.0 | 3.9 | 0.85 | I, IHM, H |
| 6y | 10YR 7/6, yellow | 3.3 | 2.1 | 0.35 | H, I, IHM |
| 7y | 10YR 7/6, yellow | 3.1 | 1.9 | 0.20 | I, H, IHM |
| 8y | 10YR 5/2, greyish brown | 2.8 | 0.7 | 0.12 | M, Kn (dis.) |

*On air-dry samples. Munsell Soil Colour Charts (1975).
†Other than Fe oxides. Abbreviations used: Kn(dis.) = kaolin (disordered); I = illite; IHM = interlayered hydrous micas; Kt = kaolinite; H = halloysite; M = mica.

Mössbauer spectra at room temperature (295 ± 3K) and 77K were obtained as described previously (Goodman and Berrow, 1976). Velocity scales were calibrated with respect to metallic Fe (Preston et al., 1962) and the thickness of each sample was limited to give not more than 10 mg Fe/cm$^2$.

RESULTS

Room temperature Mössbauer spectra of samples 1r, 1y, 5r, 5y, 8r and 8y (Fig.1) are similar to those obtained for all other pairs of samples. There is a clear distinction between red and yellow/brown samples. Each red sample has a spectrum consisting of a 6-line magnetic hyperfine component (labelled a) and a 2-line paramagnetic component from Fe (III) (b), whereas each yellow/brown sample has a spectrum showing the 2-line component (b'), similar to b, but no 6-line component. Further 2-line components (c'), due to Fe(II), are evident in some spectra.

Mössbauer spectra of some red samples following removal of free Fe oxides (Fig.2) show that the 6-line components are removed and the intensities of the central 2-line components are reduced by the treatment.

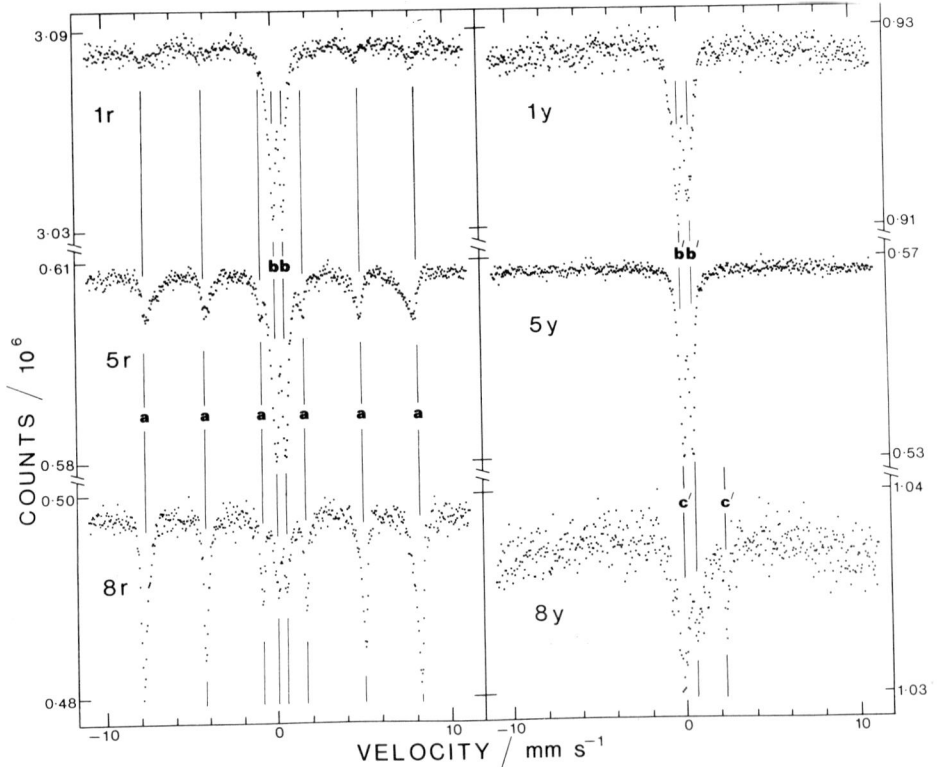

Fig.1. Mössbauer spectra of some samples at room temperature. Sample number is given beside each spectrum.

The parameters of the 6-line components (a, Fig.1) are consistent with those of hematite (Artman et al., 1968). They differ significantly from those due to antiferromagnetic goethite, and neither akaganéite ($\beta$-FeOOH) nor lepidocrocite have 6-line spectra at room temperature. Although maghemite ($\gamma$-Fe$_2$O$_3$) has similar room temperature Mössbauer parameters to those of hematite (Coey and Khalafalla, 1972), separations with a hand magnet and identification by X-ray diffraction showed that the only strongly magnetic material present was sand-sized grains of magnetite (Fe$_3$O$_4$). This occurred only in those samples where it was expected as a primary mineral and the amounts were too small to have been evident in the Mössbauer spectra or X-ray diffractograms of the whole samples.

The 2-line components (b, b') in Fig.1 may arise from (i) Fe (III) in aluminosilicate minerals, (ii) Fe (III) in the oxide-hydroxides akaganéite, lepidocrocite, superparamagnetic goethite, (iii) Fe (III) in a poorly-ordered pre-cursor to the oxide-hydroxides, labelled here "amorphous Fe(OH)$_3$" for brevity, (iv) Fe(III) in superparamagnetic hematite.

The 2-line component (c') in Fig. 1 is attributable to Fe (II) in aluminosilicate minerals.

Mössbauer spectra at 77K show 6-line spectral components for both red and yellow/brown samples (Fig.3). For some red samples two 6-line components, d and e, are evident, and e is similar to e' for the yellow/brown samples. Central 2-line components (f, f') are also evident in most of these spectra, and 2-line Fe (II) components (g') are evident in two of them.

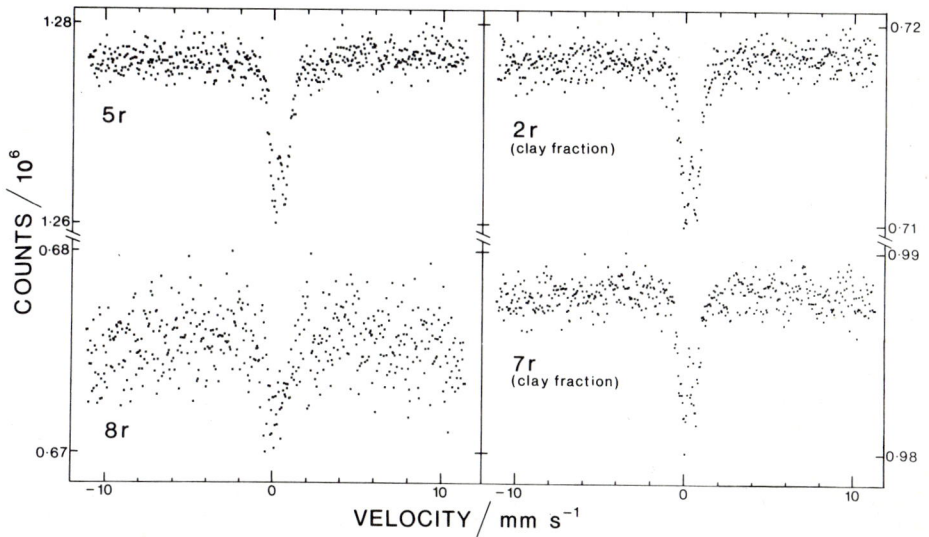

Fig. 2. Mössbauer spectra of some samples at room temperature following removal of free Fe oxides. Sample number is given beside each spectrum.

The component d in the red samples at 77K (Fig.3) has parameters attributable to hematite (van der Kraan, 1973) and corresponds to component a in Fig. 1. Components e and e' probably represent the same form of Fe and are attributable to akaganéite (Hanzel and Sevsek, 1976) and/or superparamagnetic goethite (Shinjo, 1966) but not lepidocrocite nor amorphous Fe(OH)$_3$ as these two forms are paramagnetic at 77

Recent work by Johnston and Logan (1978) and by ourselves (unpublished) has shown that the Mössbauer spectrum of synthetic akaganéite differs from that of superparamagnetic goethite at room temperature, and from that of antiferromagnetic goethite at 77K. However, the spectra of the soil samples we have studied are complex and akaganéite cannot be differentiated from goethite in Fig. 3.

Two yellow/brown samples were examined at room temperature on an expanded velocity scale (Fig.4). The spectrum of 5y resembles that of superparamagnetic goethite, particularly Al-substituted goethite, (B.A. Goodman and D.G. Lewis, unpublished), rather than akaganéite, whereas 1y resembles akaganéite. As discussed previously, however, other forms of Fe(III) may contribute to these central 2-line components and, on this evidence alone, these assignments must remain tentative.

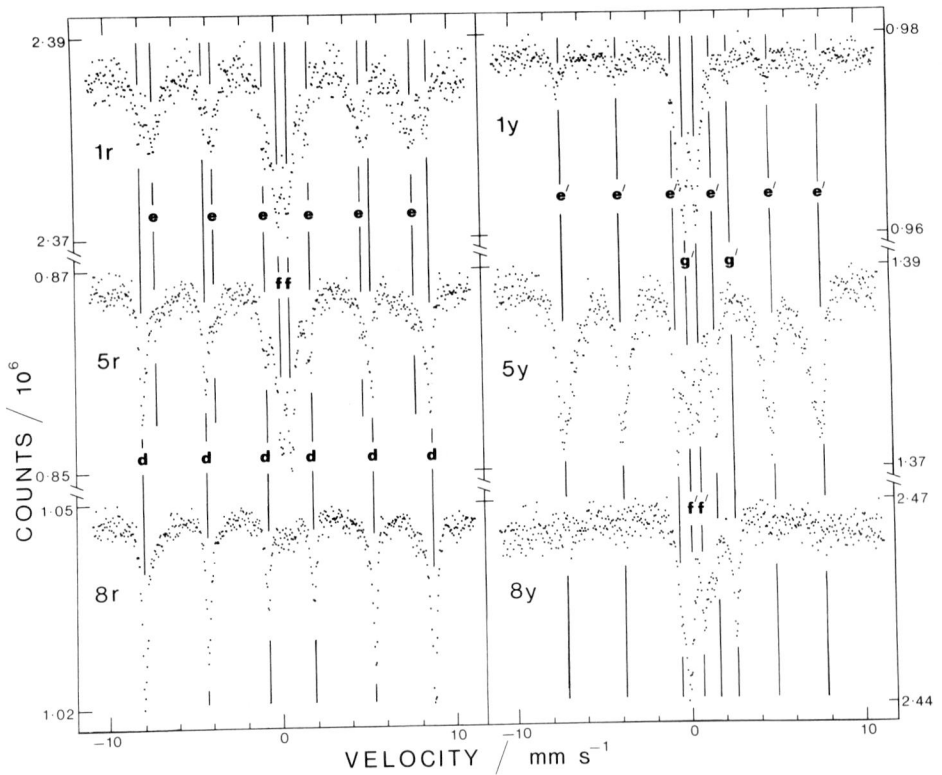

Fig. 3. Mössbauer spectra of some samples at 77K. Sample number is given beside each spectrum.

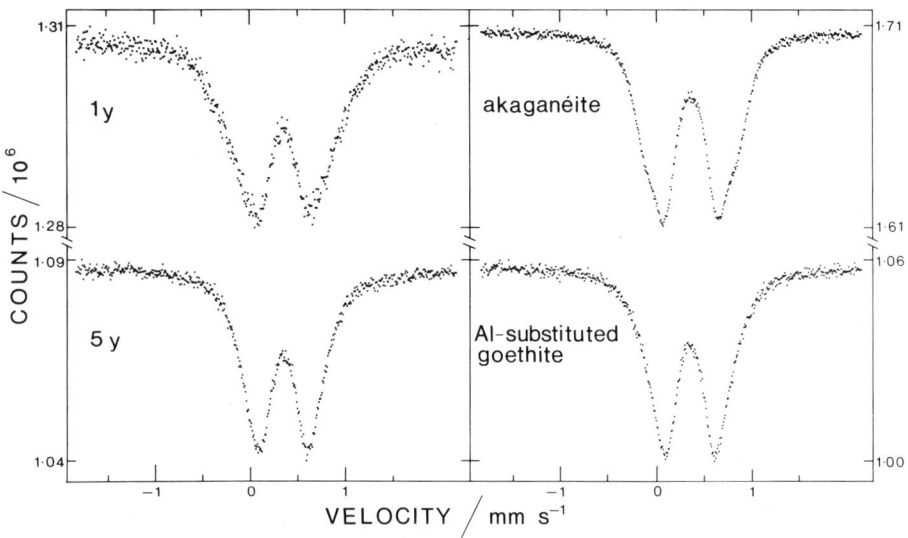

Fig. 4. Mössbauer spectra of two samples at room temperature on an expanded velocity scale, compared with spectra for synthetic oxide-hydroxides. The aluminous goethite contains one Al per nine Fe atoms.

The abundance of hematite in the red samples has been estimated from the areas of the 6-line components in the room temperature spectra. A number of Lorentzian doublets were fitted to each spectral envelope with the peaks of each doublet constrained to have equal areas and widths (Goodman and Berrow, 1976). In most cases the inner pair of magnetic peaks were swamped by the paramagnetic components, but with the assumption that the six magnetic peaks have areas in the ratio 3:2:1:1:2:3 (cf. Bancroft, 1973), the two outer pairs of magnetic peaks were used to calculate the areas attributable to hematite. The proportions of the total areas due to hematite, combined with the total Fe concentrations gave the concentrations of hematite shown later in Table 4.

DISCUSSION

The forms of Fe identified are summarised in Table 3. In addition to the forms shown, all of the samples probably contain Fe (III) in alumino-silicate minerals (see, for example, Fig.2), and they may contain some amorphous $Fe(OH)_3$. The total concentration of Fe(II) and Fe (III) in alumino-silicate minerals is indicated by the differences between total Fe and dithionite Fe in Table 2.

Although lepidocrocite is a possible contributor to the 2-line components b, b', f and f' (Figs. 1,3), it is not considered to be present because the conditions which normally lead to its formation (Schwertmann and Thalmann, 1976) do not occur in the soils sampled. Although superparamagnetic (at room temperature) hematite was listed as a possible contributor to the 2-line components, b and b' (Fig.1), the 77K spectra rule out its presence in the yellow/brown samples. Also, comparisons of room temperature and 77K spectra indicate that very little, if any, is present in the red samples.

Three other techniques were tried to help identify the Fe oxides present in the wh samples. X-ray diffraction, using CuKα and FeKα radiation, failed to reveal any rel evant peaks. Infra-red spectroscopy, apart from supporting the presence of goethite

TABLE 3
Forms of Fe identified

| Red samples | | Yellow/brown samples | |
|---|---|---|---|
| Sample no. | Forms* | Sample no. | Forms* |
| 1r | G/A, H | 1y | G/A, P (II) |
| 2r | H, G/A, P (II) (?) | 2y | G/A, P (II) |
| 3r | G/A, H, P (II) | 3y | G/A, P (II) (?) |
| 4r | H, G/A | 4y | G/A |
| 5r | H, G/A | 5y | G/A |
| 6r | H, G/A, P (II) | 6y | G/A, P (II) (?) |
| 7r | G/A, H, P (II) | 7y | G/A, P (II) |
| 8r | H | 8y | P (II), G/A |

*Abbreviations used: H=hematite; G=superparamagnetic goethite; A=akaganéite; P (II)=Fe (II) in alumino-silicate minerals; /=and, or; (?)=barely detectable. Forms are given in approximate order of decreasing abundance.

in 5y, was of no help, the concentrations of Fe oxides being too low and the absorpt by the aluminosilicate minerals too strong. Electron diffraction patterns were als of little help, apparently because of the very finely divided nature of the Fe oxide and their intimate association with other soil particles. However, hematite was identified in an electron diffraction pattern of 8r.

As the dithionite treatment produces whitish-grey to grey residues it is evident that the materials producing the red or yellow/brown colours are present in the free Fe oxide fractions of the soils. The essential difference, as far as colour is conc ed, appears to be the presence of hematite in the red samples and its absence in th yellow/brown samples. The pigment in the yellow/brown samples is superparamagnetic goethite (particle sizes less than about 10 nm (Shinjo, 1966)), and/or akaganéite, with, possibly, small amounts of "amorphous $Fe(OH)_3$" also present.

In some red samples (Table 3) the yellow/brown pigments are also present, but the tend to be masked by the hematite. Masking occurs even where the hematite-Fe/dithionite-Fe ratio is as low as 0.3 (Table 4), consistent with the findings of Dave et al. (1975) and Isbell and Smith (1976) that hematite and goethite may be present red soils, but hematite is always absent from yellow soils. It also confirms Soile and McCracken's (1967) observation that it is the form of the Fe oxides, rather than their total concentration, or relative thickness on particles, which is the dominant factor in influencing hue.

Hurst (1977), in an attempt to relate the colour of saprolite to the nature and total concentration of Fe oxides, has reduced each Munsell colour designation to a single number H*L/C. Here H* is the hue (H) expressed on a scale where H*=5 (H=5R), H*=10 (H=10R), H*=15 (H=5YR), H*=20 (H=10YR), etc., and L and C are the Munsell values for order and chroma respectively. When our samples are used to test Hurst's % free Fe oxides vs H*L/C vs H relationships (which are based on synthetic mixtures of minerals), the yellow/brown samples do not fit well and estimates of % free Fe oxides differ from the chemical values by a factor of up to 5. However, the red samples fit reasonably well, all estimates of % free Fe oxides agreeing within about $\pm$ 25%.

The H*L/C values for the red samples are shown in order (Table 4) together with a subjectively judged "increasing intensity of redness", which is the mean of assessments by six colleagues who had neither seen the samples previously nor were familiar with the project. A soil surveyor might be expected to make a similar judgement in the field. The six assessments were all similar to the order shown except for differences in the order of samples 1r, 3r and 7r, which have low concentrations of hematite. The order of the H*L/C values (decreasing with increasing redness) agrees very well with the subjectively judged "increasing intensity of redness". The relationship of both to the concentration of hematite is also good, the only "errors" being a lower concentration than expected in sample 3r, and a reversal of order for samples 4r and 2r.

TABLE 4

Relation between colour and hematite concentration in the red samples

| Sample no. | H*L/C [†] | Subjective [‡] rank | Hematite Fe (wt%) | Hematite Fe / Dithionite Fe |
|---|---|---|---|---|
| 1r | 15.0 (1) | (2) | 0.6 (2) | 0.3 |
| 2r | 10.4 (4) | (4) | 1.9 (5) | 0.8 |
| 3r | 12.5 (3) | (3) | 0.4 (1) | 0.3 |
| 4r | 10.0 (5) | (5) | 1.3 (4) | 0.6 |
| 5r | 6.7 (7) | (7) | 3.9 (7) | 0.9 |
| 6r | 8.3 (6) | (6) | 2.2 (6) | 0.6 |
| 7r | 15.0 (1) | (1) | 0.6 (2) | 0.3 |
| 8r | 5.0 (8) | (8) | 7.5 (8) | 1 |

(†) Hurst (1977), see text. Ranked orders given in parentheses.
(‡) "Increasing intensity of redness", see text.

CONCLUSIONS

1. The relationship between soil colours and Fe oxide mineralogy has usually been studied by X-ray diffraction, often following severe chemical pretreatment to concentrate the Fe oxides. Mössbauer spectroscopy offers a means of observing this relationship in untreated samples even when Fe concentrations are low and the oxide particles are very small.

2. Hematite is present in the red samples, but not in yellow/brown samples. The oxide-hydroxides, akaganéite and/or superparamagnetic goethite, are present in all yellow/brown samples and in all but one of the red samples. Some poorly-ordered pre-cursor to the oxide-hydroxides, labelled "amorphous $Fe(OH)_3$", cannot be excluded and may also be present.

3. The Munsell hues, together with the H*L/C values and the graphical relationships defined by Hurst (1977), can be used to estimate the % free Fe-oxide concentrations of the red samples to within ± 25%. The relationships are much less successful for the yellow/brown samples.

4. Increasing hematite concentrations, measured by Mössbauer spectroscopy, are closely related to the increasing intensity of redness in the samples.

ACKNOWLEDGEMENTS

We thank D.C. Bain, M.L. Berrow, J.D. Russell, J.M. Tait, and Miss Barbara Cosnett who assisted us in various ways.

REFERENCES

Artman, J.O., Muir, A.H. and Wiedersich, H., 1968. Determination of the nuclear quadrupole moment of Fe-57m from α-ferric oxide data. Phys. Rev., 173: 337-343.
Avery, B.W. and Bullock, P., 1977. Mineralogy of clayey soils in relation to soil classification. Soil Surv. Tech. Monogr. 10, Harpenden, 64 pp.
Bancroft, G.M., 1973. Mössbauer spectroscopy: An introduction for inorganic chemists and geochemists. McGraw-Hill, London, 252 pp.
Challis, G.A., 1975. Pyrite-hematite alteration as a source of colour in red beds and regolith. Nature, 255: 471-472.
Churchman, G.J. and Carr, R.M., 1975. The definition and nomenclature of halloysite. Clays Clay Miner., 23: 382-388
Coey, J.M.D. and Khalafalla, D., 1972. Superparamagnetic $\gamma$-$Fe_2O_3$. Phys. Status Solidi A, 11: 229-241.
Davey, B.G., Russell, J.D. and Wilson, M.J., 1975. Iron oxide and clay minerals and their relation to colours of red and yellow podzolic soils near Sydney, Australia. Geoderma, 14: 125-138.
Goodman, B.A. and Berrow, M.L., 1976. The characterization by Mössbauer spectroscopy of the secondary iron in pans formed in Scottish podzolic soils. J. Phys. (Paris) Colloq. C6, 37: 849-855.
Hanzel, D. and Sevsek, F., 1976. A study of thixotropic β-FeOOH by Mössbauer effect. J. Phys. (Paris), Colloq. C6, 37: 291-292.
Hurst, V.J., 1977. Visual estimation of iron in saprolite. Geol. Soc. Am. Bull., 88: 174-176.
Isbell, R.F. and Smith, G.M., 1976. Some properties of red, yellow, and grey massive earths in North Queensland. CSIRO Aust. Div. Soils Tech. Pap. No. 30, 1-52.
Johnston, J.H. and Logan, N.E., 1978. A precise Fe-57 Mössbauer spectroscopic study of ferric iron in the octahedral and channel sites of β-FeOOH (akaganéite). J. Chem. Soc. Dalton Trans., in press.
Kojima, M., 1963. Effects of concentrated NaOH treatment on the structure of synthetic iron minerals. Nippon Dojo-Hiryogaku Zasshi, 34: 331-334. (in Japanese)
Mehra, O.P. and Jackson, M.L., 1960. Iron oxide removal from soils and clays by a dithionite-citrate system buffered with sodium bicarbonate. Clays Clay Miner., 7: 317-327.
Preston, R.S., Hanna, S.S. and Heberle, J., 1962. Mössbauer effect in metallic iron. Phys. Rev., 128: 2207-2218.
Schwertmann, U. and Lentze, W., 1966. Soil colour and form of iron oxide. Z. Pflanzenernaehr. Dueng. Bodenkd., 115: 209-214.
Schwertmann, U. and Thalmann, H., 1976. The influence of [Fe(II)], [Si], and pH on the formation of lepidocrocite and ferrihydrite during oxidation of aqueous $FeCl_2$ solutions. Clay Miner., 11: 189-200.
Scott, R.O., 1941. The colorimetric estimation of iron with sodium salicylate. Analyst, 66: 142-148.
Segalen, P. and Robin, F., 1969. Colour of sesquioxide soils in the intertropical zone: yellow and red soils. Cah. ORSTOM Ser. Pedol., 7: 225-236.

Shinjo, T., 1966. Mössbauer effect in antiferromagnetic fine particles. J. Phys. Soc. Jpn., 21: 917-922.
Soileau, J.M. and McCracken, R.J., 1967. Free iron and coloration in certain well-drained Coastal Plain soils in relation to their other properties and classification. Soil Sci. Soc. Am. Proc., 31: 248-255.
Taylor, R.M. and Graley, A.M. 1967. The influence of ionic environment on the nature of iron oxides in soils. J. Soil Sci., 18: 341-348.
van der Kraan, A.M., 1973. Mössbauer effect studies of surface ions of ultra fine $\alpha$-$Fe_2O_3$ particles. Phys. Status Solidi A, 18: 215-226.

NATURAL AMORPHOUS MATERIALS, THEIR ORIGIN AND IDENTIFICATION PROCEDURES

J. RIMSAITE
Geological Survey of Canada, Ottawa, Canada

ABSTRACT

Comparative optical and x-ray diffraction studies of partly argillized and thermally altered rocks reveal the presence of amorphous phases that may easily be overlooked on x-ray diffraction patterns. The amorphous or poorly crystalline materials include either the major constituents or accessory minerals and discrete particles dispersed in a rock. The apparently amorphous materials form during the following processes:
1: hydration and argillization of primary igneous and metamorphic minerals;
2: precipitation of colloidal substances in fractures and voids in rocks;
3: chemical depletion of detrital minerals in jarosite-bearing oxisoil and clay;
4: transition of hydrous minerals to anhydrous phases; and
5: structural damage as a result of radiation.
Examples of amorphous and structurally disordered materials that cannot be identified by x-ray diffraction, pertaining to the above five processes are discussed. The apparently amorphous and poorly crystalline substances have been identified and studied by combining optical, differential thermal, thermogravimetric, x-ray diffraction and electron microprobe methods.

INTRODUCTION

Identification of amorphous and poorly crystalline materials depends largely on analytical methods. With the improvement of instrument sensitivity and the application of high resolution electron microscopy and diffraction methods coupled with electron microprobe analyses it is possible to determine the degree of structural ordering, homogeneity and chemical composition of materials that previously have been considered amorphous.

Among the important investigations in this field to date were studies of an amorphous transitional phase formed during process of kaolinization of andesitic plagioclase conducted by Delvigne and Martin (1970). Small amounts of newly-formed synthetic kaolinite, unnoticeable by x-rays have been confirmed using electron diffraction by De Kimpe et al.(1964).

Serpentine and talc-like minerals with 7Å and 10Å spacings have been identified among disordered or amorphous garnierites and illustrated in an electron micrograp at $1.5 \times 10^6$ magnification by Brindley and Phan Thi Hang (1972, Fig. 1).

Chukhrov et al. (1973) applied electron diffraction to study transitional forms of hydrous iron oxides from the lowest degrees of crystallinity (ferrierite) towar structurally ordered hematite in so-called 'brown amorphous hydroxide'.

Most of the apparently amorphous materials are heterogeneous in the degree of crystallinity and in chemical composition due to the presence of adsorbed and disseminated particles. In such heterogenous aggregates, poorly crystalline phase may easily be overlooked and masked by structurally ordered phases. Identification of fine-grained dispersed and/or adsorbed particles in clays and along fractures and grain boundaries in minerals is also difficult and requires special techniques Studies of apparently amorphous materials are difficult and scarce.

The purpose of this paper is to provide additional examples on the nature, genesis and methods of investigation of poorly crystalline and amorphous materials The following amorphous and poorly crystalline materials are discussed:

(1) alteration products of hydrated and argillized feldspars (Figs. 1a to 1d);
(2a) disintegrating accessory minerals in a phyllosilicate groundmass (Figs. 2a,2b
(2b) crusts, fracture fillings and adsorbed ions (Fig. 2c);
(3) baueritized or amorphous depleted biotite flakes in jarosite-bearing soil & cl
(4) transitional amorphous phases of partly-dehydrated materials (Fig. 3, JR-23);
(5) amorphous and poorly crystalline uraniferous hydrous oxides and phyllosilicate (Fig. 3, RKp-10).

PREPARATION OF SPECIMENS AND ANALYTICAL PROCEDURES

All specimens were examined in polished thin sections and in oil immersion powde mounts under the petrographic microscope to determine the content of optically different phases present in each sample. Coarse grained residual minerals were separated from fine grained alteration products by crushing and screening samples. The altered fragments were concentrated under the microscope by hand-picking, and t finest alteration material was separated from a -325 screen fraction (screen openir 45 microns) using centrifuge and sedimentation procedures. The coarse grained material was concentrated using heavy liquids and isodynamic separator.

All specimens were studied by x-ray diffraction(XRD) using single-crystal technic Weissenberg camera for coarse grained micaceous flakes and powder camera and x-ray diffractometer for fine grained fractions.

Differential thermal (DTA) and thermogravimetric (TG) analyses in air atmosphere using heating rates of $6°C$ and $10°C/min.$ were performed on selected concentrates(Fig Polished thin sections were studied in autoradiographs to locate radioactive mater and by electron microprobe analyses using quantitative spot analyses and qualitativ x-ray line scans for selected elements to locate high concentration spots (Table 1

TABLE 1

Electron microprobe analyses* of fresh feldspars and alteration products

| Analysis No. | Type of alteration (Fig. No.) | Mineral or aggregate | Weight per cent | | | | | | | | |
|---|---|---|---|---|---|---|---|---|---|---|---|
| | | | $SiO_2$ | $Al_2O_3$ | $TiO_2$ | $FeO$** | $MgO$ | $CaO$ | $Na_2O$ | $K_2O$ | $U_3O_8$ |
| 1 | Fresh(1a) | Plagioclase | 62.9 | 23.6 | 0.0 | 0.0 | 0.0 | 4.4 | 8.9 | 0.1 | 0.0 |
| 2 | Fresh(1a) | K-feldspar | 65.4 | 18.3 | 0.0 | 0.0 | 0.1 | 0.0 | 0.4 | 15.8 | 0.0 |
| 3 | I (1a) | Speckled Fsp*** | 75.4 | 12.9 | 0.0 | 0.1 | 0.3 | 0.0 | 0.3 | 11.0 | 0.0 |
| 4 | 1,2 (1b) | Red speckled Fsp | 38.8 | 18.6 | 0.0 | 25.8 | 1.5 | 0.4 | 0.5 | 0.5 | 0.0 |
| 5 | 1 (1c) | Kaolinite-like | 41.6 | 34.2 | 0.0 | 0.2 | 2.1 | 0.6 | 0.4 | 1.0 | 0.0 |
| 6 | 1 (1c) | Flaky in '5' | 29.3 | 15.2 | 0.1 | 6.0 | 14.0 | 0.3 | 0.4 | 0.6 | 0.0 |
| 7 | 1,2 (1c) | Fe-rich vugs | 22.2 | 22.0 | 0.2 | 44.6 | 1.9 | 0.6 | 0.0 | 0.3 | 2.0 |
| 8 | 1 (1d) | Halloysite-like | 42.7 | 36.2 | 0.0 | 0.0 | 0.5 | 0.0 | 0.4 | 0.2 | 0.0 |
| 9 | 1,2a,5(1d) | U-bands in '8' | 42.6 | 35.9 | 0.0 | 0.0 | 0.5 | 0.0 | 0.8 | 0.1 | 6.1 |
| 10 | 1 (1d) | Flaky PS*** in '8' | 41.3 | 23.5 | 0.0 | 0.4 | 10.4 | 0.4 | 0.2 | 1.7 | 0.0 |
| 11 | 2,2a,5(2a) | U-rich spots | 18.1 | 3.3 | 0.0 | 19.1 | 3.0 | 0.0 | 0.0 | 0.0 | 35.7 |
| 12 | 2 | Fe-rich specks | 21.0 | 13.0 | 0.2 | 53.0 | 2.0 | 0.2 | 0.0 | 0.2 | 0.0 |
| 13 | 2 (2b) | Ti-rich specks | 23.5 | 15.1 | 14.5 | 6.8 | 21.3 | 0.0 | 0.0 | 0.1 | 0.0 |
| 14**** | 3,4 (3) | Orange min.aggr /mineralization | 10.0 | 1.0 | 0.1 | 0.3 | 1.0 | 3.0 | 0.0 | 0.4 | 57.0 |
| 15**** | 3,4,5 | Amorphous min. aggr./min. | 0.5 | 0.0 | 0.5 | 0.0 | 0.0 | 2.4 | 0.0 | 0.0 | 70.2 |
| 16***** | 2 | Th-rich spots | 10.7 | 0.0 | 0.0 | 5.2 | 0.0 | 2.8 | 0.0 | 0.0 | 0.0 |
| 17 | 1,5 (3) | Groundmass PS | 39.1 | 16.1 | 0.1 | 18.5 | 12.4 | 0.7 | 0.0 | 1.4 | 0.0 |
| 18 | 5 (3) | U-PS in '17' | 44.6 | 21.3 | 0.0 | 5.6 | 10.8 | 0.7 | 0.2 | 3.8 | 5.8 |
| 19 | 3 (3) | Chloritized mica | 30.0 | 22.0 | 0.4 | 2.0 | 32.0 | 2.0 | 0.1 | 0.1 | 0.0 |
| 20 | 5 | U-bands in '19' | 26.0 | 13.0 | 0.4 | 4.0 | 20.0 | 0.2 | 0.0 | 1.5 | 35.0 |
| 21 | 2a | Ti-spots in '19' | 33.0 | 12.0 | 9.0 | 5.0 | 17.0 | 0.2 | 0.0 | 0.1 | 0.0 |
| 22 | 5 (3) | U-mixed-layer PS | 37.5 | 21.7 | 0.1 | 6.1 | 6.1 | 1.2 | 0.0 | 1.1 | 21.3 |
| 23***** | 5 (3) | U-mixed-layer PS | 27.4 | 3.0 | 0.6 | 24.7 | 1.7 | 3.9 | 0.0 | 0.3 | 23.5 |
| 24**** | 1,5 (3) | PS replace U min | 35.2 | 25.6 | 0.1 | 0.4 | 2.5 | 0.8 | 0.4 | 2.1 | 2.0 |
| 25 | 5 | U-hydrocarbon | 0.5 | 0.5 | 0.0 | 0.0 | 0.5 | 3.0 | 0.0 | 0.0 | 10.0 |

* Analysts: M. Bonardi, A.G. Plant, G.J. Pringle and G.R. Lachance (Geological Survey of Canada). The accuracy of this method is discussed by Rimsaite (1975).
** Total iron reported as FeO
*** Abbreviations used: Fsp = feldspar; PS = phyllosilicates; U- (Ti-) = U-(Ti-bearing; U-aggr. = uranoan aggregates; min. = mineral(s)
**** PbO in analyses '14', '15', '23 and '24' = 17.0, 0.2, 2.3 and 2.0 wt.%.
***** Other elements are: Ce= 0.6; Gd = 1.2; Nd = 1.0; P = 4.0;Pb = 0.6; Pr = 0.2; S = 2;Sm = 1.2;Th = 38;Y = 4.8 wt. %. Only maximum values are reported here.

Figs. 2a and 2b). The application of the most useful techniques will be discussed in the following examples of the apparently amorphous materials.

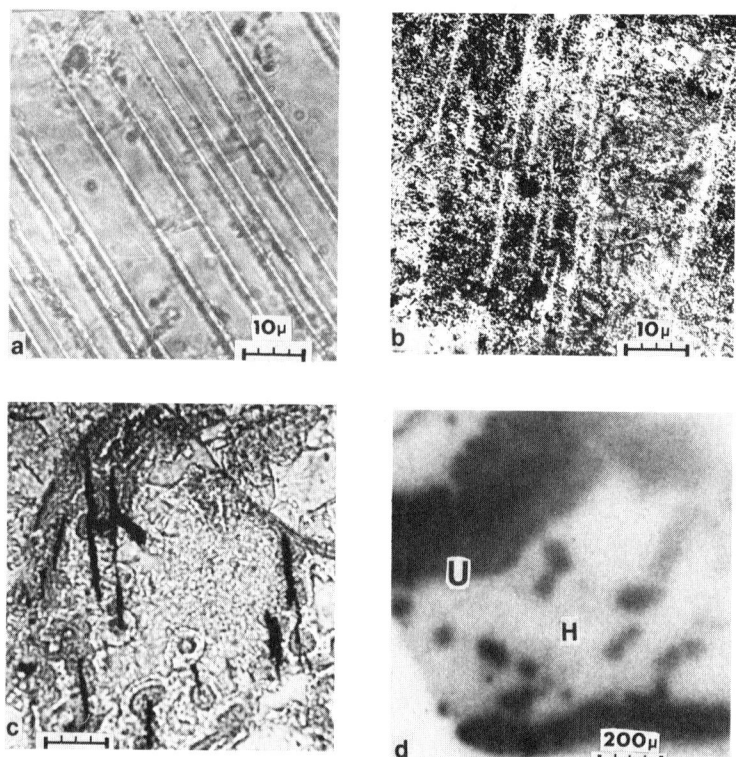

Fig. 1a. Plagioclase with thin lamellae of alkali feldspar. Initial alteration to speckled material can be seen along twinning lamellae (Table 1, analyses 1 to 3
Fig. 1b. Plagioclase clouded by white and red alteration specks with clear (white lamellae (Table 1, analysis 4).
Fig. 1c. White amorphous kaolinite-like and greenish colloform chlorite- and serpentine-like alteration of argillized rocks (Table 1, analyses 5 to 7).
Fig. 1d. An autoradiograph of white poorly crystalline halloysite(H) and radioacti (U) amorphous uraniferous halloysite filling fractures (Table 1, analyses 8 to 10)

EXAMPLES OF AMORPHOUS AND STRUCTURALLY DISORDERED MATERIALS AND THEIR OCCURRENCES
1: Alterations of feldspar to soft amorphous aggregates with red hematite stains

These alterations have been observed in brecciated Precambrian rocks in the Hidden Bay area, northern Saskatchewan (Wallis, 1971, p. 16). The author studied progressive alterations of uranium-mineralized occurrences in the same general area of Saskatchewan (Rimsaite, 1976, 1978).

At the initial stages of alteration, the plagioclase becomes clouded by fine-grained particles along the twinning planes, fractures, and thin antiperthitic lamellae of alkalifeldspar (Fig. 1a). At this stage of alteration, the fine-graine alteration products are dispersed and cannot be separated from unaltered host for positive identification by x-ray diffraction and by electron microprobe analyses. At the more advanced stages of alteration, the plagioclase host is clouded by an argillitic alteration with or without iron oxides, but thin lamellae of alkali feldspar remain clear (Fig. 1b). When the alteration film is very thin, it appears to be amorphous, and it is difficult to determine whether the fine-grained alteration products are too dispersed and masked by the relatively fresh host, or the newly-formed specks are truly amorphous to x-rays. With the progressive alteration, the fine-grained aggregates replace the feldspar and associated ferromagnesian minerals forming chemically and structurally heterogeneous residual clayey rocks. Depending on the stage of alteration and on the original composition of the host rock, residual argillites consist either of diverse crystalline and amorphous materials (Fig. 1c) or predominantly alumina-rich phyllosilicates, such as poorly-crystalline halloysite (Fig. 1d).

Chemical changes during the progressive alteration of feldspars and their host rocks are illustrated in Table 1, analyses 1 to 10. All the residual clays examined are heterogeneous and contain abundant residual quartz that is readily visible under the microscope and on XRD patterns. The other phyllosilicates include flaky chlorite, vermiculite, serpentine, talc, montmorillonite-like and mixed-layer varieties. The main chemical difference between fresh feldspars and their alteration products is in the concentrations of silica, alumina and alkalis. Concentrations of iron and magnesium vary depending on the effect of associated decomposed ferromagnesian minerals. All transformations from the anhydrous phases (feldpsar) to hydrous phyllosilicates are accompanied by marked losses of silica and alkalis, and relative increases of alumina in kaolinite-like aggregates and of magnesia in chlorite-like aggregates. Liberated silica produces spotty silicifications in altered rocks. Comparative optical, electron microprobe and XRD studies indicate that the argillized rocks consist of relatively well-crystallized (quartz, chlorite) and poorly crystallized phases (kaolinite, mixed-layer phyllosilicates, Fig. 3, speciments JR-44, RKp-10, JR-34).

The optical and electron microprobe studies of polished thin sections are very useful for the determination of residual minerals and apparently amorphous alteration products in argillized rocks.

2a: Altered residual accessory minerals in argillitic groundmass

In argillized rocks, accessory minerals alter along with the rock-forming minerals. Unless the altered rocks are entirely homogenized, residual spots containing high concentrations of Co, Cu, Fe, Mo, Th, Ti, U or rare earths remain after the original accessory minerals. These spots can be studied in X-ray scanning images and by a quantitative electron microprobe analyses (Figs. 2a and 2b), and analyses 11, 12, 13, 16 and 21 in Table 1). Some of the spots are smaller than the radius of the electron beam so that the groundmass contributes also to the spot analyses. When the spots are very small and dispersed in the phyllosilicate groundmass, x-ray identification of such spots is not possible. The specks may be either amorphous or their x-ray patterns are masked by the crystalline phases of the groundmass.

2b: Crusts and fracture fillings (Fig. 2c)

The argillized rocks are fractured and chemically diverse materials precipitate along the fractures and in the vugs from percolating aqueous solutions. These fracture fillings are commonly very fine-grained and poorly-crystallized or consist of very thin films that are firmly stuck to the host and cannot be separated for positive identification. The elements introduced by aqueous solutions are common allochthonous. These precipitates are apparently amorphous and the most useful method for their study is an electron microprobe analysis.

The radioactive substances in veins, fractures and spots can be identified in autoradiographs of polished sections (Fig. 1d, Table 1, analysis 9). The U-bearing bands and specks in the rock altered to halloysite are amorphous to x-rays. The U-bearing crusts are very thin, hardly visible at the $10^4$ magnification, and apparently coat and replace halloysite particles producing greenish to opaque coloration along the fractures and in radioactive spots. The electron microprobe analyses of the halloysite-like host and uraniferous fracture fillings are very similar, and the uranium apparently replaces the (adsorbed?) water (Table 1, analyses 8 and 9). The hydroxyl and water (by difference, 100-reported analyses) in halloysite (analysis 8) is 20 wt. per cent and in the amorphous 'uraniferous' halloysite (analysis 9) 15 wt. per cent.

The fracture filling illustrated in Fig. 2c is zoned. The uranophane (Uph) crystallizes along the periphery, whereas siliceous, apparently amorphous hydrous iron oxide (Fe) fills the central portion of the fracture.

Apparently amorphous crusts and impregnations form also in partly-altered mica (Table 1, analysis 20, and Rimsiate 1978, Fig. 58.2), and around radioactive minerals (Rimsaite and Lachance, 1966, p. 217). Red crusts coating fractures in rocks containing radioactive minerals are composed mainly of iron and radiogenic lead. In altered rocks, adjacent to mineral deposits, vugs and cavities are filled with phyllosilicates that are coated by Fe-rich specks associated with Cr, Mn, Ni U or other elements liberated from the adjacent mineral deposit (Table 1, analysis 7, Rimsaite, 1975, p. 250). Some vugs are filled with secondary uranium minerals that are mixed and partly replaced by montmorillonite-like minerals (Fig. 3, JR-3 and Table 1, analyses 14 and 24). Oxygenated hydrocarbons contain variable quantities and heterogeneous distribution of adsorbed ions, including uranium (Table 1, analysis 25, only maximum values are reported).

The disintegrating accessory minerals and the allochthonous ions precipitating from aqueous solutions along fractures and in vugs, can account for the presence of radioactive and rare earths elements and heavy metals in argillized rocks adjacent to mineral deposits.

3: Decomposition of detrital mica to amorphous flakes in oxisoils

The initial stage of decomposition of coarse-grained detrital chloritized mica

(>100 microns in diameter) involves the depletion of potassium and iron. Such bleached flakes consist mainly of silica, magnesia, alumina and hydroxyl (Table 1, analysis 19) and produce strong x-ray diffractometer patterns of chlorite (Fig. 3, JR-1). The amorphous flakes in oxisoil in the final stages of decomposition are heterogeneous and contain poorly crystalline areas within the apparently amorphous bleached flake (Rimsaite, 1975, p. 250). Changes of physical and chemical properties accompanying the transformation of fresh biotite to bleached amorphous flakes are as follows: (a) the refractive index decreases from 1.646 to 1.520; (b) specific gravity decreases from 3.10 to 2.20; and (c) chemical changes include apparent gains (in wt. %) of $SiO_2$ (88); $Fe_2O_3$ (275); CaO (100); $H_2O$ (488); and Cl (400); and losses (in wt. %) of $Al_2O_3$ (89); FeO (78); total iron as FeO (38); MgO (85); $K_2O$ (93); Rb (92); and F (84). It is important to point out that the relative incease in silica is accompanied by increases of ferric iron, that is similar to the association of ferric iron and silica in fracture fillings, and that the gain of water is similar in proportion to the gain of Cl, suggesting that the water of hydration contained chlorine. Losses of the potassium and rubidium are greater than 90 wt. per cent, whereas losses of the other constituents vary between 78 and 90 per cent. Some of the original constituents are probably retained in the remaining structurally partly-ordered areas of the amorphous flakes.

4: <u>Transitional amorphous phases in partly-dehydrated materials (Fig. 3)</u>

Amorphous partly-fused feldspar, phlogopite and pyroxene in extrusive rocks of Stromboli volcano and in eclogitic xenoliths in kimberlite have been studied by Rimsiate (1971, Fig. 1). Dehydration experiments of micas under laboratory conditions revealed the presence of amorphous phases during transition of phlogopite to leucite and olivine (above $1147°C$) and of muscovite to corundum and spinel (endothermal peak of dehydration at $820°C$ and exothermal peak of crystallization of $Al_2O_3$ at $1000°C$, Rimsaite, 1971, p. 693-694).

The fine-grained orange uranyl-bearing aggregates that consist of hydrous uranium and lead oxides with silicate impurities also pass through an amorphous transitional phase after heating at $550°C$ (Fig. 3, specimen JR-23 and Rimsaite, 1978, Fig. 58.6). These radioactive hydrous oxides lose water in two stages, below $280°C$ and at $723°C$, somewhat resembling DTA and TG curves of the associated montmorillonite (Fig. 3, speciments JR-34 and JR-29). The amorphous uranyl-bearing aggregate(Table 1, analysis 15) might possibly be a naturally dehydrated transitional phase of hydrous uranium oxide.

5: <u>Amorphous and poorly-crystalline uraniferous hydrous oxides and phyllosilicates</u>

Several principal U and Th ore minerals, such as brannerite, euxenite, uranothorite and many others commonly occur in nature as structurally disordered or 'metamict' substances as a result of radioactivity. The above minerals are coarse grained, euhedral and yield distinct XRD patterns after ignition in a vacuum. Robinson (1955) reported the occurrence of fine-grained radioactive substances that produced

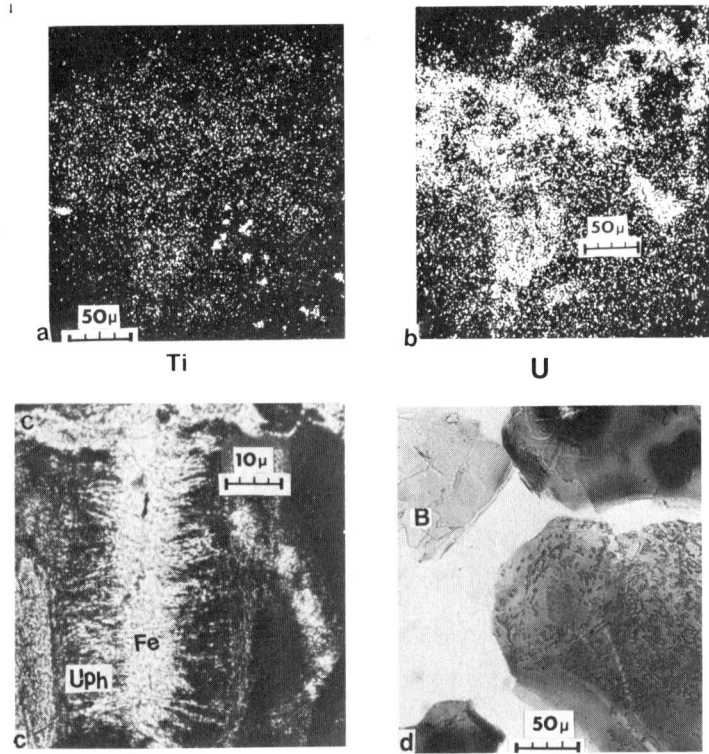

Fig. 2a. X-ray scanning image (Ti) showing small disintegrating Ti minerals (white spots) in a phyllosilicate groundmass (Table 1, analysis 13).
Fig. 2b. X-ray scanning image showing distribution of uranium in a phyllosilicate matrix (Table 1, analysis 11).
Fig. 2c. Fracture fillings with uranophane (Uph) along the walls and poorly crystalline siliceous goethite (Fe) in the centre.
Fig. 2d. Detrital flakes of biotite with prominent black pleochroic halos in the upper field, right; speckled partly-altered biotite with prominent basal fractures in the lower field, right; and a bleached biotite (B) altered to amorphous silica.

blackening (radioactive) spots on autoradiographs, but could not be positively identified by conventional methods.

This section deals only with the fine-grained poorly crystallized phyllosilicat and hydrous oxides that have been identified in polished thin sections using autoradiographs and electron microprobe analyses (Table 1, analyses 22, 23; Fig. 3 and Rimsaite, 1978, Fig. 58.6, specimen RKp-10). The untreated original material produces a very broad XRD reflection at 14Å and a faint trace at 10Å. These reflect become somewhat better defined in glycol-treated samples and the double 10Å-9Å pea becomes more prominent after heating at 550°C. These uraniferous phyllosilicates appear to be mixed-layer varieties consisting of mica, hydromica, chlorite and montmorillonite phases. The DTA and TG curves are very poorly defined: exothermal peaks indicate gradational dehydration at the temperatures below 200°C and at 540° and two endothermal reactions at 480°C and at 860°C indicate partial recrystalliz-

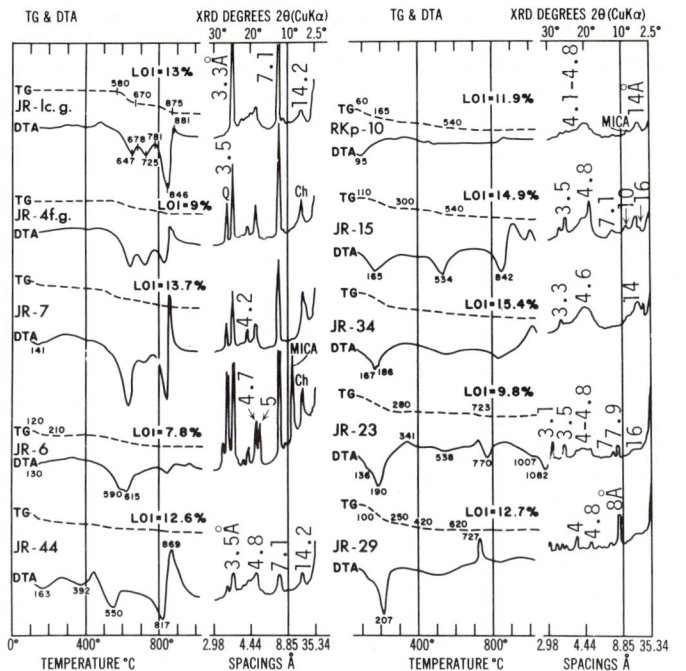

Fig. 3. DTA (at 10°C/min), TG (at 6°C/min) and XRD (Ni-filtered Cu-Kα radiation) analyses of crystalline chlorites, and chlorite (Ch), quartz (Q), mica mixtures, and glycol-treated poorly crystalline uraniferous mixed-layer phyllosilicates (RKp-10), associated montmorillonite (JR-34) and orange uranium-lead oxides JR-23 and JR-29 (Table 1, analyses 14, 22, 23, and 24). DTA and TG analyses by W.S. Bowman.

ation of anhydrous phases. The structural destruction of these U-bearing phyllosilicates is attributed at least in part to radioactivity. The radioactive, structurally disordered mixed-layer phyllosilicates ultimately alter to montmorillonitic clays (Table 1, analysis 24 and Fig. 3, JR-34).

The uranoan oxygenated hydrocarbon is an amorphous asphalt-like material associated with secondary carbonate in radioactive rocks. Electron microprobe analyses of the uranoan hydrocarbon indicated the erratic heterogenous distribution of Si, Al, Mg, Ca and U. Some of the uranium is present in form of thin pitchblende-like crusts and fracture fillings, but part of uranium might also be present in uranium hydrocarbon compounds. The maximum values for the impurities are given in Table 1, analysis 25.

SUMMARY AND CONCLUSIONS

Apparently amorphous and poorly crystalline materials have been studied applying optical, XRD, DTA TG and electron microprobe analyses. The examples described and the processes that account for the apparent amorphous state of these materials are as follows: (1) fine grained alteration products formed during progressive hydration and argillization of feldspar under weathering conditions; (2a) decomposition products of accessory minerals in argillized rocks and in phyllosilicate groundmass accompanied the decomposition of rock-forming minerals; (2b) fine-grained fracture fillings, crusts and adsorbed particles on phyllosilicates, oxides and hydrocarbons precipitated from allochthonous elements and soils in percolating aqueous solutions; (3) anion- and cation-depleted biotite flakes grading to amorphous hydrous silica forms in oxisoils under weathering conditions in oxidizing environment;

(4) transitional optically-amorphous phases of muscovite, phlogopite and hydrous uranyl-bearing aggregates formed at temperatures above dehydration and below crystallization of anhydrous phases; (5) amorphous and poorly crystalline uranium-bearing mixed-layer phyllosilicates and oxides having disordered structures as a result of radioactivity.

The most effective methods for identification of fine grained amorphous U-bearing materials are electron microprobe analyses in conjunction with DTA and TG analyses. Electron diffraction and infrared analyses are used to establish structural disorder

The uraniferous phyllosilicates, iron hydroxides and hydrocarbons that apparently contain adsorbed uranium and other elements, warrant more detailed studies applying the high resolution transmission electron microscopy.

ACKNOWLEDGEMENTS
The author wishes to thank W.S. Bowman of Canada Centre for Mineral and Energy Technology for DTA and TG analyses and M. Bonardi, G.R. Lachance, A.G. Plant and G.J. Pringle, Geological Survey of Canada for electron microprobe analyses.

REFERENCES
Brindley, G.W. and Pan Ihi Hang, 1972. The hydrous magnesium-nickel silicate minerals (so-called garnierites). Preprints Internat. Clay Conf. (Madrid) 1: 41-50.
Chukhrov, F.V., Zvyagin, B.B., Ermilova, L.F. and Gorshkov, A.I., 1973. New data on iron oxides in the weathering zone. Proc. Internat. Clay Conf. (Madrid), 333-3
De Kimpe, C.R., Gastuche, M.C. and Brindley, G.W., 1964. Low-temperature syntheses of kaolin minerals. Am. Mineral., 49: 1-16.
Delvigne, J. et Martin, H., 1970. Analyse à la microsonde electronique de l'alterat d'une plagioclase en kaolinite par l'intermédiaire d'un phase amorphe. Cah. O.R.S.T.O.M., Sér. Géol., 2: 259-295.
Rimsaite, J. and Lachance, G.R., 1966. Illustration of heterogeneity in phlogopite, feldspar, euxenite and associated minerals. Mineral. Soc. India, IMA Volume, Int. Mineral. Ass. Papers IV General Meeting 209-229.
Rimsaite, J., 1971; DTA, TG IR and isotopic analyses and properties of phlogpite, biotite, muscovite and lepidolite in temperature range of metamorphic reactions. Proc. Third ICTA 3: 683-695.
Rimsaite, J., 1975. Natural alteration of mica&reactions between released ions in mineral deposits. Clays and Clay Minerals 23: 247-255.
Rimsaite, J., 1976. Progressive alteration of pitchblende in an oxidation zone of uranium deposits. Abstracts. 25th Internat. Geol. Cong., 2: 594-595.
Rimsaite, J., 1978. Layer silicates and clays in the Rabbit Lake uranium deposit, Saskatchewan. Current Research, Geol. Surv. Can., Paper 78-1A: 303-315.
Robinson, S.C., 1955. Mineralogy of uranium deposits, Goldfields, Saskatchewan. Geol. Surv. Can., Bull. 31, 128pp.
Wallis, R.H., 1971. The geology of the Hidden Bay area, Saskatchewan. Sask.Dep. Min. Res., Rep. 137, 74pp.

OTHER PAPERS PRESENTED IN SECTION 6

LATTICE EXPANSION AND INTERCALATION OF LEWIS BASES IN LEPIDOCROCITE ($\gamma$-FeO.OH) and STRUCTURALLY RELATED COMPOUNDS

A. Weiss, J.H. Choy and H. Mittereder (Institut für Anorganische Chemie der Universität München, Meiserstrasse 1, 8000 München 2, Federal Republic of Germany.)

SOLID SOLUTION IN SYNTHETIC $Fe_2O_3$-$TiO_2$-$H_2O$ SYSTEMS PREPARED AT NEAR AMBIENT CONDITIONS

R.W. Fitzpatrick[1], J. le Roux[1] and U. Schwertmann[2] ([1] Department of Soil Science, University of Natal, South Africa. [2] Institut für Bodenkunde, Tech. Univ. München, Federal Republic of Germany.)

SURFACE REACTIVITIES OF HYDROLYTIC PRECIPITATION PRODUCTS OF ALUMINUM AS INFLUENCED BY LOW MOLECULAR WEIGHT COMPLEXING ORGANIC ACIDS

K.F. Ng Kee Kwong and P.M. Huang (Department of Soil Science, University of Saskatchewan, Saskatoon, Canada S7N 0W0.)

SYNTHESIS OF Al-GOETHITES AT NEUTRAL pH VALUES AND AMBIENT TEMPERATURES FROM THE Fe(II) SYSTEM

R.M. Taylor[1], U. Schwertmann[2] and D.G. Lewis[3] ([1] C.S.I.R.O. Division of Soils, Private Bag No. 2, Glen Osmond, South Australia. [2] Institut für Bodenkunde, T.U. München, 8050 Freising-Weihenstephan, Federal Republic of Germany. [3] Waite Agricultural Research Institute, Private Bag No. 1, Glen Osmond, South Australia.)

PREPARATION OF SYNTHETIC GOETHITES CONTAINING ALUMINIUM

D.G. Lewis[1] and U. Schwertmann[2] ([1] Waite Agricultural Research Institute, Glen Osmond 5064, South Australia. [2] Institut für Bodenkunde, T.U. München, 8050 Freising-Weihenstephen, Federal Republic of Germany.)

PROTON CONDUCTIVITY IN HYDROXIDES AS RELATED TO TRANSPORT PHENOMENA IN SOILS

F. Freund and H. Wengeler (Mineralogisches Institut der Universität, Zülpicher Str. 49, D-5000 Köln 1, Federal Republic of Germany.)

AMORPHOUS IRON AND ALUMINIUM OXIDES IN CLAYS FROM TROPICAL SOILS

M. Landu, L. Rodrique and A.J. Herbillon (Groupe de Physico-Chimie Minérale et de Catalyse, Université Catholique de Louvain, Place Croix du Sud 1, B-1348 Louvain-la-Neuve, Belgium.)

DEVIATIONS FROM THE IDEAL DISORDERED STRUCTURE IN MINERALS OF THE PYROAURITE GROUP

D.L. Bish (Dept. of Geological Sciences, Harvard University, Cambridge, MA, USA.)

ALLOPHANE DEPOSITED FROM A SPRING WATER

C.W. Childs, C.J. Downes, and N. Wells (Department of Scientific and Industrial Research, Wellington, New Zealand.)

SYNTHESIS AND MORPHOLOGY OF AKAGANEITE (β-FeOOH)

E. Paterson, J.M. Tait and D.R. Clark (The Macaulay Institute for Soil Research, Craigiebuckler, Aberdeen, Scotland.)

KINETICS AND MECHANISM OF THERMAL TRANSFORMATION OF GOETHITE TO HEMATITE

J. Cornejo (Centro de Edafología y Biología Aplicada del Cuarto, Apdo. 1052, Sevilla, Spain.)

OCCURRENCE OF A FIBROUS IRON MINERAL IN SOME SOILS FROM JAPAN AND SCOTLAND

M. Nakai and N. Yoshinaga (Faculty of Agriculture, Ehime University, Matsuyama, Japan.)

STRUCTURAL STUDY OF VENEZUELAN LATERITES BY INFRARED SPECTROSCOPY

E. Mendelovici and S. Yariv (Laboratorio de Física-Química de los Materiales, Instituto Venezolano de Investigaciones Cientificas, Apartado 1827, Caracas, Venezuela.)

THE STRUCTURE OF NATURAL AND SYNTHETIC ALUMINOSILICATE GELS

V.C. Farmer, A.R. Fraser and J.M. Tait (The Macaulay Institute for Soil Research, Craigiebuckler, Aberdeen, Scotland.)

# SECTION 7

# Kaolin Investigations

*Chairman's Introduction*

METHODS OF KAOLIN INVESTIGATION

W. D. KELLER
University of Missouri-Columbia, Columbia, Missouri, USA.

INTRODUCTION

I am honored to be asked to preside at this section and to make a few introductory comments on the subject designated for it, namely, "Methods of kaolin investigation". Much more should be said than the allotted time here, and space in publication, will permit. I have space to cite only too few names and publications in connection with the methods. Those cited will mainly recognize either priority in the use of the method, or an important or recent achievement, and/or one which provides a useful entry into the literature--those in addition to the well-known books by Grim (1962, 1968), Millot (1970), and Weaver and Pollard (1973). Please forgive my conscious but unavoidable omissions of important relevant papers.

Furthermore, I am forced by limitations of space to regretfully tabulate methods instead of discussing them. Methods of kaolin investigation may be organized in several ways, including the following one that I will use:
    Field Geology ($EFH_2$) and Pedology
    Laboratory (BBB)
        Physical
        Chemical
    Industrial and Engineering.

In case you are unfamiliar with the parenthesized acronyms, BBB stands for Brains and Black Boxes, referring to methods and equipment used in the laboratory, and for greatest appeal in grantsmanship. But I have never heard of a laboratory black box finding a kaolin dsposit (Bates, 1978)--and kaolin must be found before it can be further investigated.

To find kaolin, we also need $EFH_2$, referring to talents which my former student, Prof. "Al" E. J. Engel, a geologist member of the National Academy of Sciences, says honor geologists; it stands for Eyes, Feet, Hands, and Hammer. I venture that these were the abilities and method that a brainy, alert Chinese used, more than 300 years ago, to find and dig the white clay from Kao-ling (Anglicized to Kaolin) in Kiangsi Province.

FIELD METHODS OF INVESTIGATING AND FINDING KAOLIN

I cite for an example of successfully found kaolin, the relatively recent (1969) discovery of the large deposit, or multiple deposits, of kaolin on the Jari and Capim Rivers of Brazil. I am indebted to Volker Eisenlohr who told me that a German geologist (unnamed by him), while on geological exploration on a boat trip up the Jari River, discerned a scantily vegetated mountain top--an anomaly in the forested jungle. Thus he used his Eyes on the terrain (he was not playing poker, or reading the last issue of C and CMS). He stopped the boat ride, industriously took to his Feet to climb the mountain, and dug the kaolin with Hands and Hammer. The result: discovery of a deposit of kaolin, the reserves of which envious visiting geologists rate at over 200 million tons. A Brazilian anecdote (too long to repeat) about finding this deposit has as its punch line, "But you have to leave the office and laboratory and get out into the field to find a kaolin mineral deposit."

Geobotanical. The method of investigation used was geobotanical, applied in a negative sense here because it was the absence of plants. In the positive mode, geobotanical prospectors use:

- stands of trees vs. grass
- evergreens vs. deciduous trees
- remote sensing of varieties of vegetation
- silica accumulators (Lovering, 1959)
- alumina concentrators (Keller, 1949).

Other methods and techniques for field use include:

- geologic and geomorphic environment (as lacustrine terraces of Plio-Pleistocene age in the Amazon Basin), or sedimentary basins of ball clay
- field mapping of petrologic units (Keller, et al, 1954)
- aerial photos
- use of pollen and spores ("Georgia kaolin")
- geologic ages of rocks (Carboniferous, Cretaceous, Eocene, for example)
- sedimentologic facies (Keller, 1968)
- weathering crusts (as in central Europe)
- outcrop features (resistant bauxite vs. slaking clay)
- plate tectonics (Bardossy, 1977)
- earth resistivity (Cornwall; Missouri clay pits)
- seismic methods (short-range, low energy vibrations)
- gravity and magnetic methods (little used)
- auger drilling (hand and power)
- percussion drilling
- core drilling (air and water)
- A book on geologic prospecting by M. Kuzvart and M. Böhmer is (or has been) translated and published in English.

METHODS OF INVESTIGATING BULK OR LUMP KAOLIN FROM THE MINE

These methods emphasize properties of the kaolin in the natural-lump state, e.g., natural lump density, inter-crystallinity, cohesion etc., which are lost in the process of pulverizing the kaolin.

    bulk density (green and processed)
    size-reduction characteristics (crushing)
    slaking properties
    texture (rock)
    refractoriness (PCE; Seger cone, fusions)
    plasticity (qualitatively; Atterberg limits)
    volume change (in refractory service).

OPTICAL METHODS OF INVESTIGATION (ROSS AND KERR, 1930).
    identification
    solid impurities (exceedingly sensitive for carbonate minerals)
    textural relations to parent rock
    electrically oriented double-refraction
    dark-field microscopy.

X-RAY METHODS OF INVESTIGATING KAOLIN.

X-ray diffraction is the method that has been used in making the greatest strides toward an understanding of the crystallographic mineralogy of kaolin. It serves as the major method of identification of kaolin and interpretation of its crystal structure. In 1930, Ross and Kerr differentiated the kaolin minerals by X-ray (and optical) methods although their camera recorded d-spacings no wider than 5 Å. Amazing instrumentation has been developed since then. In pace with the hardware, far-reaching conceptual advances have been made, of which the pertinent literature has been assembled by Brindley (1951), updated by Brown (1961), and a latest volume is in preparation under auspices of the Mineralogical Society.

    identification
    polymorphs (Bailey, 1963)
    internal crystallinity (Hinckley, 1963; Range, et al, 1969)
    order-disorder (Brindley and Robinson, 1946; Murray and Lyons, 1956;
        Brindley, 1977)
    structural defects (Plancon and Tchoubar, 1977)
    mixtures of polymorphs of kaolin (Keller and Haenni, 1978)
    single crystal modes
    powder diffractometers and film
    low-angle and selected-angle range resolution
    estimation of particle size
    temperature and atmosphere-controlled specimen holders

multiple-specimen, automated, computerized, and print-out instruments
X-ray fluorescence (analytical units)

ELECTRON METHODS OF INVESTIGATING KAOLIN.

Apparatus employing electron beams can be exceedingly versatile and adaptable to kaolin. Late models permit a single crystal, or particle, to be (1) viewed by both transmission and scanning modes, and from that particular specimen (2) an electron diffraction pattern be taken, and (3) an elemental analysis made by energy or wave dispersive modes. Electron methods apparently are superseding X-ray.

    crystallographic determinations
    identification (as for halloysite, Chukhrov and Zvyagin, 1960)
    diffraction patterns
    estimation of particle size
    transmission electron microscopy (TEM), morphology
        direct specimen
        replicas
    high resolution with ultramicrotomy, 7 Å domains (Lee, et al, 1975)
    environmental cell for hydrated kaolin (Kohyama, et al, 1978)
    selected area diffraction, (SAD; SAED)
    scanning electron microscopy (SEM), morphology
        textural studies (Keller, 1977, 1978a, b)
        critical point drying (Jernigan and McAtee, 1975)
    electron micro-probe (analysis)
    energy-dispersive elemental analysis
    wave-dispersive elemental analysis
    TEM and SEM combination in one instrument
    Auger electrons (unused? method)
    electron excited, cathodoluminescence
        "Luminoscope" (Keller and Huang, CMS, Banff meeting, 1973).

DIFFERENTIAL THERMAL METHODS OF INVESTIGATING KAOLIN.

The literature on differential thermal analysis (DTA) is authoritatively and comprehensively discussed up to 1970 by Mackenzie (1970). DTA indirectly yields information to the kaolin-ceramist on important aspects of industrial firing schedules and other thermal changes not tabulated in detail below.

    identification of kaolin mineralogy
    determination of heats of reaction
    quality control in industry
    simple DTA analysis; derivative DTA and variations
    thermal gravimetric analyses, TGA, and varieties

elaborate instrumentation
   multiple sample
   automated, computerized.

INFRARED (IR) METHODS OF INVESTIGATING KAOLIN.

Current investigation of kaolin by IR is facilitated by 2 recent books which are authoritative and comprehensive in 2 topics of the field (Farmer, 1974; Van der Marel and Beutelspacher, 1976). The use of IR to study <u>powdered</u> minerals began with clay, and silica minerals (Keller and Pickett, 1950; API Reference Clay Minerals, 1950). IR supplements X-ray diffraction methods by yielding information on OH and other group configurations not discernible by X-ray.

   identification
   direct OH and group absorptions
   structure from intercalations (Ledoux and White, 1964, 1966)
      hydroxyl geometry
      hydrogen banding
      deuteration
      proton migration
      crystal defects and disorder.

CHEMICAL METHODS OF INVESTIGATING KAOLIN.

Ross and Kerr (1930) provided good elemental analyses of kaolin, and stated its ideal composition. Rarely are kaolins analyzed that have the ideal composition (Weaver and Pollard, 1973). As colloidal particles, kaolin chemistry is that of colloids (Van Olphen, 1977). Laboratory syntheses, genesis in nature, and thermodynamic stabilities and kinetics of kaolinization are important areas of chemical investigation.

   elemental analysis
   chemical treatment (Jackson, 1969)
   intercalation compounds (Wada, 1961; Miller and Keller, 1963; Weiss, et al, 1966; Range, et al, 1969; Barrios, et al, 1977; Jackson and Abdel-Kader, 1968)
   organic compounds with kaolin (Theng, 1975)
   laboratory synthesis
      high temperature-pressure (Roy and Osborn, 1954)
   low temperature (De Kimpe, 1969)
   organic compounds (Linares and Huertas, 1971)
   very dilute solutions (Harder, 1977)
   doping elements (Angel, 1977)
   laboratory weathering (Parham, 1969)
   free energy of formation (Kittrick, 1970, also in Dixon, 1977; Huang 1974)

chemical kinetics.

MISCELLANEOUS METHODS OF INVESTIGATING KAOLIN.
    Mössbauer spectroscopy (Weaver, et al, 196/; Jefferson, et al, 1975)
    electron spin resonance (ESR)
    nuclear magnetic resonance (NMR)
    proton magnetic resonance (PMR)
    mathematical calculation of lattice energies
    Raman spectroscopy
    imbibometry (Konta, 1961, 1963)
    staining
    texture and fabric related to genesis and classification (Keller, 1977, 1978a, b; with Hanson, 1975)
    observable kaolinization in nature (Keller, et al, 1966, 1971).

PEDOLOGIC METHODS OF INVESTIGATING KAOLIN.
    Representative literature by Dixon, 1977; Marshall, 1977.

INDUSTRIAL METHODS OF INVESTIGATING KAOLIN.
    Representative literature by Murray, 1977; AIME, 1975
    size range and distribution
    shape, area, discreteness, stacking of plates
    dispersion and sedimentation
    magnetic benefication
    flotation
    bleaching by oxidation and/or reduction
    whiteness, brightness, reflectivity, opacity
    absorption of pigment and poisonous compounds
    abrasiveness
    calcination
    ceramic properties listed under "bulk" category.

HIGHEST PRIORITY METHOD FOR INVESTIGATING KAOLIN.
  We must never forget that the highest priority method of investigating kaolin that also is indispensible, most versatile, and central to all other methods is the thought process of an imaginative investigator. It is a trite saying that advances in knowledge are made in the mind, and then demonstrated with materials.

FORECAST METHODS FOR INVESTIGATION.
  Since I have exhausted myself preparing this historical account of methods already employed, I would call upon you, the fertile, imaginative minds of the

listening and reading audience, to forecast the new methods of investigation. I shall try to guess only a few methods in forecast: I would (1) expect much more to be learned about structure with high resolution TEM; and (2) serious work with SEM to follow our start on textures. Electron diffraction (3) is taking over from X-ray. Look for discoveries of (4) natural organic reagents, including the (5) energies of bacteria to account for certain natural processes of kaolinization. Discovery should be made how, within apparent inorganic systems, kaolin is (6) dissolved and precipitated, possibly by complexing (?). Shifting emphasis from cations to (7) anions, what are the roles of $SO_4$ (Lesar, et al, 1946), and $PO_4$ (Loughnan and Ward, 1970), in substitution of $SiO_4$ in kaolin? The role of mineralizers, (8) volatile as well as (9) non-volatile, in promoting crystallization of (10) kaolin (raw), and (11) its high-temperature products (mullite, etc.) has scarcely been explored. Some weird mineralogy, apparently mineralizer-induced associations occur on a large scale in systems heated by combustion metamorphism (Keller, 1977, V, p. 361) or near igneous contacts (Brindley and Porter, 1977). Al-chlorite (12), and Li may be found in abundance in kaolins. You take it from here.

PROBLEMS FOR FURTHER INVESTIGATION OF KAOLIN.

In my opinion, the most glaring problems in kaolin mineralogy are those about the interrelationship of plates and elongates, i.e., the very well crystallized kaolinite (e.g., the Keokuk variety, Keller, et al, 1966), with the spectrum of decreasing-crystallinity "kaolinites" (Murray and Lyons, 1956), to the elongates, halloysite ($2H_2O$) and endellite ($4H_2O$), which may also be spherical. These problems encompass crystallinity, morphology, genesis, thermodynamic stability, and nomenclature of kaolin. Apparently kaolinite and dickite, in the other direction from kaolinite, also may be interlayered (Brindley and Porter, 1977). The solution of these problems will probably utilize most of the methods of investigation of kaolin cited in this summary, and those reported on in the papers to follow in this meeting.

ACKNOWLEDGMENTS.

I gratefully acknowledge aid on citations of publications from Professors Bailey Dixon, Huang, Jackson, Johns, Kittrick, Kuzvart, Mortland, Murray, Patterson, Weaver, White, and others. Continuing aid from the National Science Foundation Grant 76-188004 has been applied to presentation of this report.

REFERENCES

1 AIME, 1975. Industrial Minerals and Rocks; Patterson, S.H., and Murray, H., Clays. AIME, New York, 519-585.
2 Angel, B. R., Cutter, A. H., Richards, K. S., and Vincent, W. E. J., 1977. Synthetic kaolinites doped with $Fe^{2+}$ and $Fe^{3+}$ ions. Clays and Clay Mins., 25, 381-383.
3 API Reference Clay Minerals, Infrared Spectra, 1950. American Petroleum Institute Project 49, Preliminary Report No. 8, Columbia Univ., New York, 146 pp.
4 Bailey, S. W., 1963. Polymorphism of the kaolin minerals. Amer. Mineral., 48, 1196-1209.
5 Bardossy, G., 1977. World relationship of bauxite formation and tectonism. Abstract, Jamaica meeting of CMS and ICSOBA, p. 22.
6 Barrios, J., Plancon, A., Cruz, M. I., and Tchoubar, C., 1977. Qualitative and quantitative study of stacking faults in a hydrazine treated kaolinite--Relationship with infrared spectra. Clays and Clay Mins., 25, 422-429.
7 Bates, R. L., 1978. The geologic column. Geotimes, 23, 4, p. 46.
8 Brindley, G. W., 1951. X-ray identification and crystal structures of clay minerals. Mineralogical Society, London, 345 pp.
9 Brindley, G. W., 1977. Aspects of order-disorder in clay minerals--a review. Clay Science, 5, 103-112.
10 Brindley, G. W., and Porter, A. R. D., 1977. Occurrence of dickite in Jamaica--ordered/disordered forms. Abstracts, Joint meeting of CMS and ICSOBA, Jamaica, page 8.
11 Brindley, G. W., and Robinson, K., 1946. Randomness in the structures of kaolinitic clay minerals. Trans. Faraday Soc., 42B, 198-205.
12 Brown, G., 1961. X-ray identification and crystal structures of clay minerals. Mineralogical Society, London, 544 pp.
13 Chukhrov, F. V., and Zvyagin, B. B., 1966. Halloysite, a crystallo-chemically and mineralogically distinct species. Proceed. Int. Clay Conf., Jerusalem, vol. 1, 11-26.
14 De Kimpe, C. R., 1969. Crystallization of kaolinite at low temperature from an alumino-silicic gel. Clays and Clay Mins., 17, 37.
15 Dixon, J. B., 1977. Kaolinite and serpentine group minerals, 357-403, in Minerals In Soil Environments, J. B. Dixon, Editor, Soil Sci. Soc. Amer., Madison, WI, 948 pp.
16 Farmer, V. C., 1974. The infrared spectra of minerals. Mineralogical Society, London, 539 pp.
17 Grim, R. E., 1962. Applied Clay Mineralogy, McGraw-Hill, 422 pp.
18 Grim, R. E., 1968. Clay mineralogy. McGraw-Hill, New York, 596 pp.
19 Gritsaenko, G. S., and Samotoyin, N. D., 1966. The decoration method applied to the study of clay minerals. Proceed. Int. Clay Conf., Jerusalem, vol. 1, 391-400.
20 Hinckley, D. N., 1963. Variability in "crystallinity" values among the kaolin deposits of the coastal plain of Georgia and South Carolina. Clays and Clay Mins., 11th Nat'l. Conf., Pergamon Press, 229-235.
21 Huang, W. H., 1974. Stabilities of kaolinite and halloysite in relation to weathering of feldspars and nepheline in aqueous solution. Amer. Mineral., 59, 365-371.
22 Jackson, M. L., 1969. Soil chemical analysis--advanced course. Published by author, Madison, Wisc.
23 Jackson, M. L., and Abdel-Kader, F. H., 1978. Kaolinite intercalation procedure for all sizes and types, with X-ray diffraction spacing distinctive from other phyllosilicates. Clays and Clay Min., in press.
24 Jefferson, D. A., Tricker, M. J., and Winterbottom, A. P., 1975. Electron-microscopic and Mössbauer spectroscopic studies of iron-stained kaolinite minerals. Clays and Clay Mins., 23, 355-360.
25 Jernigan, D. L., and McAtee, J. L., Jr., 1975. Critical point drying of electron microscope samples of clay minerals. Clays and Clay Mins., 23, 161-162.
26 Keller, W. D., 1949. Higher alumina content of oak leaves and twigs growing over clay pits. Ec. Geol. 44, 451-454.

27. Keller, W. D., 1968. Flint clay and a flint-clay facies. Clays and Clay Mins., 16, 113-128.
28. Keller, W. D., 1977. Scan electron micrographs of kaolin collected from diverse environments of origin--IV and V, Georgia, Australia, Japan kaolins. Clays and Clay Mins., 25, 311-364.
29. Keller, W. D., 1978a. Classification of kaolins exemplified by their textures in scan electron micrographs. Clays and Clay Mins., 26, 1-20.
30. Keller, W. D., 1978b. Kaolinization of feldspar as displayed in scanning electron micrographs. Geology, 6, 184-188.
31. Keller, W. D., and Haenni, R. P., 1978. Effect of mixtures of kaolin minerals on their X-ray powder diffractograms. In press, Clay and Clay Mins.
32. Keller, W. D., and Hanson, R. F., 1975. Dissimilar fabrics by scan electron microscopy of sedimentary versus hydrothermal kaolins in Mexico. Clays and Clay Mins., 23, 201-204.
33. Keller, W. D., Hanson, R. F., Huang, W. H., and Cervantes, A., 1971. Sequential active alteration of rhyolitic volcanic rock to endellite and a precursor phase of it at a spring in Michoacan, Mexico. Clays and Clay Mins., 19, 121-127.
34. Keller, W. D., McGrain, P., Reesman, A. L., and Saum, N. M., 1966. Observations on the origin of endellite in Kentucky, and their extension to "indianite". Clays and Clay Mins., 13, 107-120.
35. Keller, W. D., and Pickett, E. E., 1950. The absorption of infrared spectra by clay minerals. Amer. Jour. Sci., v. 248, 2640273.
36. Keller, W. D., Pickett, E. E., and Reesman, A. L., 1966. Elevated dehydroxylation temperature of the Keokuk geode kaolinite--a possible reference mineral. Proceed. Int. Clay Conf., Jerusalem, vol. 1, 75-85.
37. Keller, W. D., Westcott, J. F., and Bledsoe, A. O., 1954. The origin of Missouri fire clays. Clays and Clay Mins., Proceed. 2nd Conf., Nat'l Acad. Aci.-Nat'l Res. Coun. Pbl., 327, 7-46.
38. Kittrick, J. A., 1970. Precipitation of kaolinite at $25°$ C and 1 atm. Clays and Clay Mins., 18, 261-268.
39. Kohyama, N., Fukushima, K., and Fukami, A., 1978. Observation of the hydrated form of tubular halloysite by an electron microscope equipped with an environmental cell. Clays and Clay Mins., 26, 25-40.
40. Konta, J., 1961. Imbibometry--a new method for investigation of clays. Amer. Mineral., 46, 289-303.
41. Konta, J., 1963. Identification of clay minerals and the study of argillaceous rocks by the imbibometric method. Clays and Clay Mins., 10th Nat'l. Conf., Pergamon Press, 42-58.
42. Ledoux, R. L., and White, J. L., 1964. Infrared study of selective deuteration of kaolinite and halloysite at room temperature. Science, 145, 47-49.
43. Ledoux, R. L., and White, J. L., 1966. Infrared studies of hydrogen bonding interaction between kaolinite surfaces and intercalated potassium acetate, hydrazine, formamide, and urea. Jour. Colloid and Interface Sci. 21, 127-152.
44. Lee, S. Y., Jackson, M. L., and Brown, J. L., 1975. Micaceous occlusions in kaolinite observed by ultramicrotomy and high resolution electron microscopy. Clays and Clay Mins., 23, 125-130.
45. Lesar, A. R., Krinbill, C., Keller, W. D., and Bradley, R. S., 1946. Effect of compounds of sulfur on reheat volume change of fire-clay and high-alumina refractories. Jour. Amer. Ceram. Soc. 29, 70-75.
46. Linares, J., and Huertas, F., 1971. Kaolinite: synthesis at low temperature. Science, 171, 896-897.
47. Lovering, T. S., 1959. Significance of accumulator plants in rock weathering. Bull. Geol. Soc. Amer., 70, 781-800.
48. Loughnan, F. C. and Ward, C. R., 1970. Gorceixite-goyazite in kaolinite rocks of the Sydney Basin. Jour. and Proceed. Roy. Soc. New South Wales, 103, 77-80.
49. Mackenzie, R. C., 1970. Differential thermal analyses. Academic Press, London, 775 pp.
50. Marshall, C. E., 1977. Physical chemistry and mineralogy of soils II. Wiley-Interscience, New York, 313 pp.

51 Miller, W. D., and Keller, W. D., 1963. Differentiation between endellite-halloysite and kaolinite by treatment with potassium acetate and ethylene glycol. Clays and Clay Min., 10th Nat'l. Conf., 244-253.
52 Millot, G., 1970. Geology of clays. Springer-Verlag., New York, 429 pp.
53 Murray, H., 1977. Kaolin--a versatile pigment, extender, and filler. Proceed. Canadian Pulp and Paper Assn., 63rd Ann. Meeting, A29-A34.
54 Murray, H., and Lyons, S. C., 1956. Correlation of paper coating quality with degree of crystal perfection of kaolinite. Proceed. 4th Clay Conf., NAS-NRC Pub. 456, 31-40.
55 Parham, W. E., 1969. Formation of halloysite from feldspar: Low temperature, artificial weathering versus natural weathering. Clays and Clay Mins., 17, 13-22.
56 Plancon, A., and Tchoubar, C., 1977. Determination of structural defects in phyllosilicates by X-ray powder diffraction-1. Principle of calculation of the diffraction phenomena,-II. Nature and proportion of defects in natural kaolinites, Clays and Clay Mins., 25, 430-450.
57 Range, K. J., Range, A., and Weiss, A., 1969. Fire-clay type kaolinite or fire-clay mineral? Experimental classification of kaolinite-halloysite minerals. Proceed. Int. Clay Conf., Tokyo, vol. 1, 3-14.
58 Ross, C. S., and Kerr, P. F., 1930. The kaolin minerals. U.S. Geol. Survey Prof. Paper 165E, 151-176.
59 Theng, B. K. G., 1974. The chemistry of clay-organic reactions. Halsted-Wiley-Hilger, New York and London, 343 pp.
60 Van der Marel, H. W., and Beutelspacher, H., 1976. Atlas of infrared spectroscopy of clay minerals and their admixtures. Elsevier, Amsterdam, 396 pp.
61 Van Olphen, H., 1977. An introduction to clay colloid chemistry. Interscience, John Wiley, New York, 318 pp.
62 Wada, K., 1961. Lattice expansion of kaolin minerals by treatment with potassium acetate. Amer. Mineral., 46, 78-91.
63 Weaver, C. E., and Pollard, L. D., 1973. The chemistry of clay minerals. Elsevier, Amsterdam, 213 pp.
64 Weaver, C. E., Wampler, J. M., and Pecuil, T. C., 1967. Mössbauer analysis of iron in clay minerals. Science, 156, 504-508.
65 Weiss, A., Thielepaper, W., and Orth, H., 1966. Neue kaolinit-einlagerungs verbindrengen. Proceed. Int. Clay Conf., Jerusalem, vol. 1, 277-293.

An inadvertent omission from the typescript is the investigative method of calculating the energies of interlayer bonds, hydroxyl orientation and stacking sequence, as illustrated by Giese, R. F., Clays and Clay Mins., 21, 145-150; 21, 102-104; Nriagu, J. O., Amer. Mineral., 60, 834-839.

AUSTRALIAN KAOLINS

A.J. Gaskin[1], P.J. Darragh[1] and F.C. Loughnan[2]

[1]Division of Mineralogy, CSIRO, Private Bag, P.O. Wembley, Australia 6014.
[2]Department of Applied Geology, University of New South Wales, P.O. Box 1, Kensington, Australia 2033.

ABSTRACT

The characteristics of the principal varieties of Australian kaolins are illustrated by descriptions of typical occurrences. Most of the residual deposits were formed during early Tertiary time when intense lateritic weathering of igneous and metamorphic components of the old peneplain surface generated deep pallid zones of kaolin, now exposed by erosion to an extent related to the climate and topography of the locality. Transport of these residual kaolins has not favoured the deposition of large sedimentary bodies of high grade clay. Mesozoic clay sequences in the inland regions contain vast amounts of halloysite. Residual kaolins of definitely hydrothermal origin are rare.

INTRODUCTION

At the close of the Mesozoic era the dominant feature of the Australian landscape was a vast peneplain sloping slightly in towards the central region. During early Tertiary time the warm and humid conditions which prevailed over most of the continent, even though it was then much further from the equator, caused deep and intense lateritization of the poorly drained land surface. Many of the kaolin deposits now evident represent the pallid zones of this lateritic profile.

By the end of the Miocene period, when lateritization had virtually ceased over the southern regions of the continent, the first of a series of episodes of differential uplift initiated dissection of the peneplain, particularly in the eastern and western coastal regions, exposing the residual kaolins to an extent depending on the degree of aridity to which the climate progressed. In very arid western and central areas much of the lateritic capping still remains, but in the more pluvial eastern areas the profile is now extensively dissected and kaolin from the pallid zone has been locally transported and deposited in flood plains. Climatic and topographic factors have not favoured the accumulation of transported kaolins in large well sorted sedimentary sequences and there are none at present known of a size and quality comparable with those of Devon or Georgia. The long standing flood and drought climatic cycle that affects most of the continent has generated large expanses of sedimentary clays contaminated with silt and has also accentuated the concentration of iron in the upper regolith. Deposits of industrially valuable kaolins of acceptable purity and plasticity are therefore very rare. Those of the Gulgong area, described below, are typical of the local sources which supply the market for medium grade clays.

Apart from the disadvantages mentioned in regard to the reworked kaolins of the dissected pallid zones, most of the clays from arid inland areas exhibit contamination with soluble salts, usually to a degree precluding application in industry. Alunite is a common component and is very deleterious because of its slow rate of dissolution in water. The Mesozoic clay sequences which formed the central part of the ancient peneplain have become highly alunitic in areas subjected to the influence of sulphated ground waters. An example is described from the Imbitcha district, where halloysite is the dominant clay mineral.

The scarcity of transported clay of high quality is in part offset by the relative abundance of residual kaolins of industrial or mineralogical interest. The general features of those formed by Tertiary lateritization will be evident from descriptions of deposits at Pittong, Gabbin and Egerton. Definite instances of hydrothermal kaolinization have been rarely established, but the Williamstown-Mt Crawford area provides examples.

There appear to be few consistent relationships between the mode of origin and the mineralogical characteristics of Australian kaolins. The morphology of transported clay flakes is usually poor whilst those of hydrothermal origin show good crystal outlines and those formed by lateritization show intermediate degrees of perfection. The crystallinity index tends to follow the same trend, but it can show wide variation within a single deposit for no evident reason, and it can be strongly influenced by the degree of deformation imposed on samples during preparation for X-ray examination. There is a suggestion of a correlation between this index and the ESR response of some kaolins, but it is difficult to obtain specimens sufficiently free from traces of iron minerals to obtain significant results. The viscosity of concentrated aqueous slips of Australian kaolins, a crucial factor in industrial exploitation, is generally unacceptably high for various reasons some of which are evident whilst others are still quite obscure.

## DESCRIPTIONS OF SELECTED DEPOSITS
### Gabbin

The Precambrian basement rocks of Western Australia were exposed, like those of the central and eastern regions of the continent, as a peneplain at the close of the Mesozoic and subjected to periods of lateritization during Tertiary time.

The Gabbin and Tallering Peak deposits are examples of the general kaolinization of granites, gneisses and metasediments that is evident down to depths of around 50m. Dissection of the old peneplain has not proceeded as far as in the eastern states and the deposits are usually obscured by sandplain, laterite and other cover.

At Gabbin, 280km northeast of Perth, the overburden includes layers of sand and laterite for the first 6m from the surface, then 1m of quartz grains, showing relict granitic texture, cemented by opaline silica. Kaolinized granite occurs from 7 to 40m grading down into fresh rock over a further distance of 1m with clear evidence of albite having decomposed more rapidly than microcline. The deposit near Tallering Peak, some 400km north of Perth is a similar example, with a thinner covering of sand and silcrete overlying at least 20m of kaolinized Archaean material, which in this case includes metamorphosed sediments, judging from the relict textures.

Kaolinite is the only significant mineral in either deposit, apart from coarse grained relict quartz. The $-2\mu m$ fraction at Gabbin, probably originating from the alteration of feldspars, contains moderately well shaped flakes, sometimes in books of notable thickness. Larger flakes, probably derived from mica, show irregular outlines. Tallering Peak kaolinite shows irregular morphology throughout the particle size range, a characteristic which correlates with the fact that it produces higher viscosity aqueous suspensions. Although the Gabbin material is somewhat variable throughout the deposit, much of it gives low viscosity results in the concentrated suspension range used for testing paper coating kaolins at high shear rates. In this respect it resembles kaolin from Pittong, the only other substantial deposit of paper coating clay known in Australia at present. Although the Gabbin area is semi-arid, the kaolin does not have a high soluble salt content and beneficiation by water levigation should be practicable. However, there is evidence of the presence of small amounts of slowly soluble aluminous and siliceous materials which could point to potential problems in the fields of dispersion and viscosity. Reserves are of the order of 10 million tonnes of recoverable kaolin.

Pittong

The extensive deposit of kaolinized granite at Pittong, 130km northwest of Melbourne, is a prime example of the deep and intense weathering that has affected the Lower Palaeozoic basement rocks of Central Victoria. Exposed as a peneplain at the close of the Cretaceous, the basement Silurian and Ordovician sediments and Devonian intrusives were lateritized in the early Tertiary and then subjected to swampy conditions followed by further lateritization. Leaching and oxidation can be seen to have extended to depths of 200m in areas where most of the old weathered profile has been preserved from the subsequent erosion that dissected the peneplain. At Pittong, and at Lal Lal, the kaolinized plutons have escaped erosion through the influence of local topographic factors whilst at other localities such as Bulla, Pleistocene basalt flows have provided protective cover.  The virtual absence of overburden at Pittong, coupled with the quality of the kaolin that can be extracted by levigation, the ready availability of water and the not excessive distance to a local market and shipping port have been factors promoting the exploitation of the deposit.  A treatment plant near the open-cast mining area is producing about 15,000 tonnes per annum of kaolin, including filler and coating grades.  Considerable variations in certain characteristics of the kaolin over lateral distances as small as 1 or 2m, particularly in properties such as slip viscosity, do not appear to have been caused by variations in the composition of the parent rocks but may reflect the influence of the original joint system on the kaolinization process.  In other parts of the area less well leached zones show that joint systems have functioned as exit channels for the removal of iron during the alteration of the rock. There are no indications of any significant layers of ferruginous or siliceous alteration products having been formed above the kaolinized zones, not unexpectedly in view of the long pluvial periods which followed the main episode of kaolinization and no appreciable amounts of salt or other evaporites occur in the kaolinized material.  The altered zone extends down to 40m in places but does not reflect the present topography.  Kaolinite is the only clay mineral of significance in the deposit.  The dominant $-2\mu m$ fraction consists of flakes of a moderate degree of morphological perfection, probably derived from the alteration of feldspars. A coarse kaolinite fraction, containing flakes up to $500\mu m$ across and exhibiting irregular outlines, occurs in minor amounts at Pittong and other analogous deposits of the region and is considered to represent the kaolinization of micas.  Reserves of several million tonnes exist in the Pittong area.

Egerton

Dykes throughout the basement complex of southeastern Australia were generally kaolinized, often to great depths when the region was exposed as a peneplain, by the intense weathering that began in the early Tertiary.  Quartz-bearing dykes associated with the suites of Lower Palaeozoic granitic intrusives were altered to gritty kaolin bodies, some of which have been exploited as fire-clays.  Suites of quartz-free dykes of Cretaceous to Palaeocene age, have been altered to gritfree kaolins of remarkable purity at some locations.

One of the best known of these is the kaolinized tinguaite dyke at Egerton near Ballarat.  It has been mined to depths of 150m over a distance of 1,000m along its N-S strike and for many years produced much of the fine kaolin used industrially in Victoria.  Production rates of up to 4,000 tonnes per annum were achieved although the width of the dyke is only 2 to 3m and appreciable sections of the deposit are affected, especially along joints, by hematitic staining which does not respond to bleaching treatment. Thinner dykes, otherwise analogous, occur throughout the area in the same strike direction, cutting the staurolite schist of the Ordovician basement. Soft, fairly plastic kaolin is only found near the water table in valley situations. The main deposit is at a higher level in the present dissected topography and is hard, compact material similar to many Australian kaolins that have existed for long periods in a dry environment.

There is no definite evidence of hydrothermal activity having been associated with the kaolinization, although the alteration has been essentially complete to a depth of at least 200m. Relict tinguaite occurs in small amounts at two locations and the alteration sequence can be traced over a distance of about 1m from the relatively fresh rock fragments. These consist of a fine grained feldspathic ground mass containing phenocrysts of aegerine, orthoclase and a mineral which had the typical outlines of nepheline but which is now halloysite. In a traverse away from the relict tinguaite position, aegerine first becomes corroded as slight alteration of the ground mass sets in, then disappears when the ground mass is completely kaolinized with only some feldspar phenocrysts remaining. Further out into the main body of kaolinite, alteration is so complete that none of the original rock texture is recognizable. Apart from traces of halloysite that occur along joint planes throughout the deposit, kaolinite is the sole clay mineral occurring in the $-10\mu m$ fraction as flakes showing quite good morphology even when derived from hard samples which can only be dispersed by grinding. The crystallinity index of 0.36 is lower than that of other Australian kaolinites produced by weathering and may indicate more intense conditions of hydration and leaching, even to the extent of suggesting that mild hydrothermal activity was involved in the genesis of the kaolin.

## Mt Crawford

Hydrothermal alteration of the Proterozoic basement appears to have been intense in the Mt Crawford-Williamstown district 35-40km northeast of Adelaide. Sillimanite schists, feldspathic dykes and portions of a pegmatite have been kaolinized to depths of at least 40m, probably during the last stages of the Lower Palaeozoic orogeny that metamorphosed the Adelaidean sediments of the area. Hydrothermal conditions for the leaching and hydration episode are suggested by the nature of the kaolinite and the unusual degree of alteration of most minerals in the adjacent formations. Zones of massive sillimanite have been largely converted to kaolinitic pseudomorphs, and in the pegmatite even the tourmaline has been kaolinized. Extremely fine-grained bright red rutile is common throughout the kaolinized sillimanite, and in an altered dyke finely recrystallized ilmenite occurs in thin veinlets. Masses of coarse hydromica, locally referred to as damourite, occupy a sheer zone adjacent to the sillimanite occurrence and appear to have replaced a kyanite-mica parent rock.

All this evidence indicates that hydrothermal rather than weathering processes have been locally responsible for this unusually intense degree of alteration. In many other localities within a radius of 300km the kaolinized igneous and sedimentary rocks of the Precambrian basement show no signs of alteration of comparable intensity, only the usual deep weathering effects associated with Tertiary lateritization.

Kaolinites from the hydrothermally altered zones at Williamstown and Mt Crawford are coarser and more even grained than most kaolinites produced by weathering and show more perfect crystal morphology. X-ray patterns however, indicate only a moderate degree of crystallinity. Material from the Williamstown deposit has been extensively used in the refractories industries especially when it contains a proportion of very fine residual sillimanite. The total kaolin reserves of the district are of the order of 1 million tonnes.

## Gulgong

Late Miocene differential uplift of the epi-Cretaceous peneplain led to the erosion of much of the lateritized and kaolinized basement, especially in eastern Australia. Stable conditions prevailed in the Pliocene and flood plain accumulations then filled the broad mature valleys until further uplift in the late Pliocene resulted in the erosion of much of this sedimentary material. Remnants that persist throughout the eastern region sometimes become industrially important as sources of white plastic clay.

The Gulgong-Home Rule-Stubbs district, some 300km northwest of Sydney provides typical examples. Here the Pliocene alluvials extend to depths of over 100m and contain lenses of kaolin up to 7m thick which have been a major source of semi-ball clay for many years. Kaolinite, the only significant clay component, has probably been transported from kaolinized zones of Palaeozoic granitic rocks which occur in the area. Although disordered and fine grained most of it being in the $-2\mu m$ fraction, the kaolinite is relatively uncontaminated by iron minerals, quartz and mica at some localities. In other areas beneficiation by air flotation is required to remove grit. Taken in conjunction with other deposits of similar material in the region, reserves of several million tonnes would be available.

## Imbitcha

During the episodes of Tertiary lateritization, the flat lying Mesozoic clay sequences which covered much of the central part of the old peneplain appear to have been subjected to less pluvial climatic conditions than the outer regions. Extensive development of thick silcrete cappings indicates a lesser degree of mobilization of silica than is normal in lateritization. In the region north and west of Oodnadatta dissection of the peneplain has exposed many sections through the silcrete and the underlying Lower Cretaceous clays which are remarkably white, low in iron, rich in alunite and of curious mineralogy. In the Imbitcha-Arkaringa district, 900km northwest of Adelaide, halloysite is the dominant mineral in the thick and extensive clay sequence which occurs below the silcrete and also contains montmorillonite, a mineral found at depth near the opal zones in sediments of comparable age in other districts. Attempts to utilize the halloysite in industry have been made because of its very high reflectivity, but the ubiquitous presence of alunite, in amounts up to 15%, preclude all normal applications. The alunite has been generated by the action of the sulphated ground waters which are characteristic of the eastern regions of the Great Artesian Basin and its presence is a general feature of many surficial clay deposits throughout the arid inland areas of South and Western Australia. It does not indicate hydrothermal activity. The quantity of white clay which occurs in the Imbitcha district is enormous. Beds up to 20m thick outcrop for many kilometres around the walls of valleys and mesas.

TABLE 1

Some characteristics of Australian kaolinites

| Location | Particle sizing Per cent less than | | | | G.E. Brightness | | Crystallinity index $0\bar{1}/060$ |
|---|---|---|---|---|---|---|---|
| | 10μm | 4μm | 2μm | 1μm | 457nm | 570nm | |
| Gabbin | 90 | 71 | 52 | 37 | 88.5 | 93.4 | 0.39 |
| Pittong | 100 | 97 | 83 | 69 | 90.7 | 94.3 | 0.45 |
| Egerton | 77 | 60 | 45 | 34 | 73.2 | 87.0 | 0.19 |
| Mt Crawford | 85 | 53 | 23 | 8 | 81.0 | 87.0 | 0.43 |
| Gulgong | 100 | 89 | 68 | 50 | 85.0 | 91.2 | 0.24 |
| Imbitcha | 100 | 88 | 76 | 62 | 90.5 | 92.8 | 0.0 |

ACKNOWLEDGMENT

The assistance provided by Mr. J.L. Perdrix in the determination of kaolin characteristics is gratefully acknowledged.

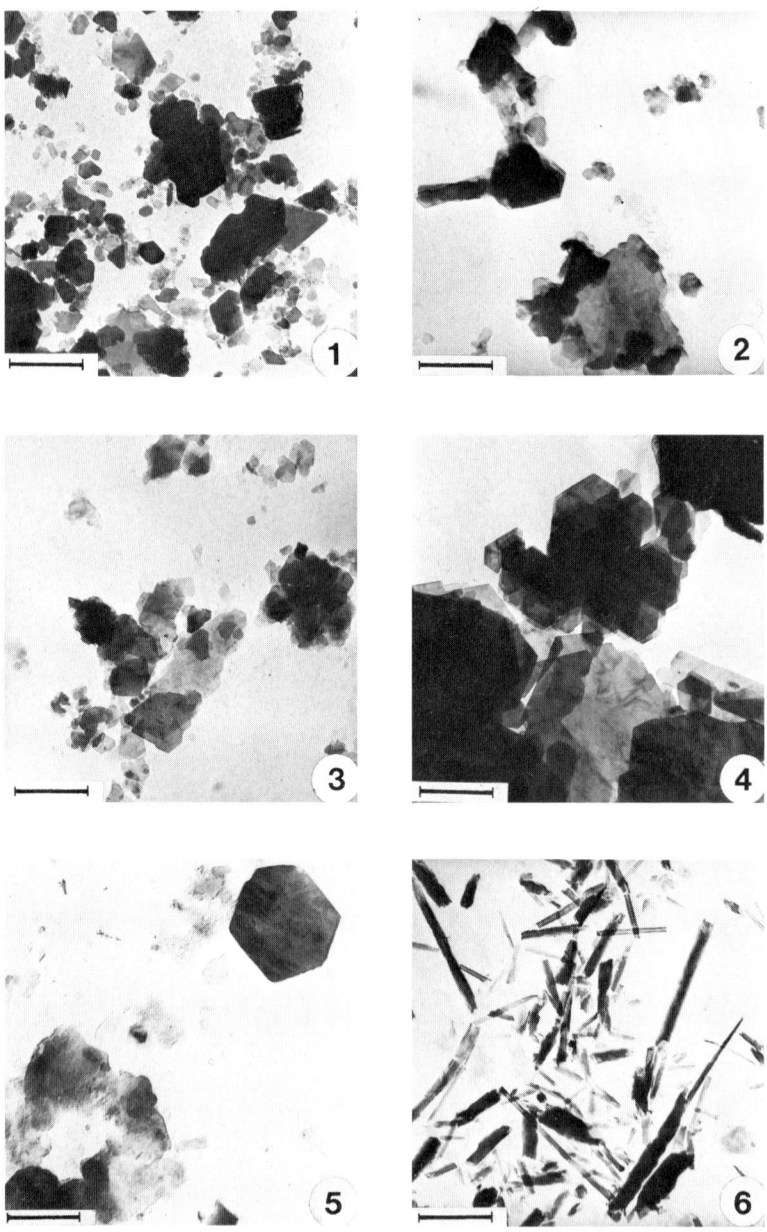

Figs 1 - 6.   Electromicrographs of:  1 Gabbin, 2 Pittong, 3 Egerton,
4 Mt Crawford, 5 Gulgong, 6 Imbitcha.   Bar represents 1 micron.

Fig. 7.

SOUTH AFRICAN KAOLINS

L.J. MURRAY and R.O. HECKROODT
Serina (Proprietary) Limited, P.O. Box 160, Fish Hoek, 7975, R.S.A.
University of Cape Town, Rondebosch, 7700, R.S.A.


ABSTRACT

Since the last paper on this subject was presented in 1968 by C.B. Coetzee at the 23rd International Geological Congress, there has been considerable development in the exploitation of local kaolins.

The Council for Scientific and Industrial Research has published a series of Technical Notes on 50 Ceramic Raw Materials, 14 of which are kaolins. Data is given on origin, chemical analysis, rheological properties, raw and fired characteristics, particle size distribution, reflectance and refractoriness of each kaolin.

An attempt has now been made to correlate all the available information and to give some idea as to the extent and reserves of the known kaolin deposits. To date the most intensive prospecting has been done in the Western Cape area and here a substantial reserve of high quality kaolin has been proved.


INTRODUCTION

Clay materials containing kaolinite as a major constituent are common in South Africa, but they are mostly heavily contaminated with iron compounds. The naturally white kaolins are nearly always residual or primary deposits and this review will consider mainly materials of this kind.

During the last decade there has been considerable development in the exploitation of the local kaolins, although a systematic investigation of the various occurrences is far from complete. The two general references are the Geological Survey Handbook: *Mineral Resources of the Republic of South Africa* and the National Building Research Institute Technical Notes: *Ceramic Raw Materials of Southern Africa*. Previously unpublished information is also included in this report.

In 1961 the whiteware ceramic industry accounted for nearly 60 per cent of kaolin consumption, with the insecticide industry being the second largest user, consuming just over 30 per cent. The remaining 10 per cent was taken mainly by the paper, rubber and paint industries. This distribution has not changed much since then, but

## ANALYTICAL DATA FOR SOUTH AFRICAN KAOLINS

| Material | A | K | C | DV | NN | BF | G | VZ | S1 | S2 | Zeb |
|---|---|---|---|---|---|---|---|---|---|---|---|
| **Chemical composition (%)** | | | | | | | | | | | |
| $SiO_2$ | 46,50 | 48,60 | 46,86 | 46,05 | 48,48 | 47,10 | 66,83 | 69,68 | 54,52 | 48,74 | 61,46 |
| $Al_2O_3$ | 39,00 | 35,20 | 36,46 | 38,35 | 36,91 | 38,30 | 21,52 | 19,16 | 28,81 | 33,97 | 25,96 |
| $Fe_2O_3$ | 0,54 | 1,22 | 0,54 | 1,15 | 0,49 | 0,34 | 0,45 | 0,62 | 1,05 | 1,54 | 0,34 |
| $TiO_2$ | 0,39 | 0,76 | 0,61 | 0,48 | tr. | 0,10 | 0,74 | 0,91 | 0,85 | 0,79 | 2,10 |
| $MgO$ | 0,14 | 0,43 | 0,10 | 0,26 | 1,18 | 0,21 | 0,22 | 0,42 | 0,37 | 0,31 | 0,01 |
| $CaO$ | 0,06 | 0,13 | 0,14 | 0,06 | 0,35 | 0,08 | 0,84 | 0,28 | 0,56 | 0,17 | 0,02 |
| $Na_2O$ | 0,05 | 0,11 | 0,50 | 0,12 | 0,06 | 0,07 | 0,80 | 0,53 | 0,58 | 0,23 | 0,08 |
| $K_2O$ | 0,76 | 1,84 | 0,85 | 0,90 | 1,08 | 0,60 | 2,75 | 2,65 | 0,70 | 0,69 | 0,32 |
| L.o.i. | 12,57 | 11,76 | 13,97 | 12,84 | 12,24 | 13,26 | 6,07 | 5,61 | 12,50 | 13,53 | 9,80 |
| **Rational analysis (%)** | | | | | | | | | | | |
| Clay substance | 92 | 73 | 79 | 88 | 87 | 84 | 22 | 20 | 60 | 77 | 62 |
| Quartz | 1 | 7 | 4 | 1 | 1 | 5 | 42 | 47 | 21 | 9 | 31 |
| Mica | 7 | 17 | 13 | 9 | 12 | 10 | 33 | 29 | 13 | 9 | 4 |
| **Particle size distribution** | | | | | | | | | | | |
| $-30\mu m$ | 100 | 99 | 96 | 94 | 93 | 100 | 97 | 89 | 95 | 93 | 94 |
| $-20$ | 100 | 97 | 94 | 91 | 91 | 100 | 95 | 80 | 80 | 91 | 82 |
| $-10$ | 98 | 89 | 87 | 81 | 87 | 97 | 83 | 57 | 54 | 71 | 66 |
| $-5$ | 88 | 75 | 76 | 65 | 72 | 87 | 67 | 38 | 35 | 52 | 50 |
| $-2$ | 61 | 51 | 58 | 42 | 49 | 66 | 47 | 22 | 21 | 33 | 35 |
| $-1$ | 44 | 37 | 43 | 25 | 34 | 52 | 32 | 14 | 14 | 24 | 22 |
| Whiteness (raw) | 85E | 85E | 88L | – | 86P | 88L | 79L | 72L | 82L | 83L | 75L |
| Fired colour (1300°C) | +90E | – | 82L | – | – | 96L | 87L | 80L | 87L | 84L | 82L |
| Refractoriness (°C) | – | – | 1780 | – | – | – | 1650 | 1620 | 1740 | 1760 | 1730 |
| Reference | JM | ROH | Ker13 | ROH | Ref 2 | JM | Ker11 | Ker12 | Ker 8 | Ker 9 | Ker35 |

the picture could well change in the near future, particularly regarding the supply of material suitable for the paper industry.

The more important analytical data of the kaolins are compared in table form. Because the information has been taken from a variety of sources, comparisons are not always straightforward, particularly regarding the whiteness of the materials, as different instruments or techniques were used. In the table the letters "L", "E" and "P" refer to Leukometer, Elrepho and Photovolt determinations, while "Reference: Ker 13", for example, refers to Technical Note Number 13 in reference 2.

The residual kaolin deposits of economic importance may be grouped in five areas, as shown on the map. These areas are:

1) Western Cape:    Cape Peninsula, Kuils River, Stellenbosch, Somerset West, Vredenburg.
2) Namaqualand:     Bitterfontein, Garies.
3) Eastern Cape:    Grahamstown, Mossel Bay, Riversdale, Albertinia.
4) Natal:           Inanda, Ndwedwe, Mkuze.
5) Transvaal:       Zebediela, Koster, Lawley.

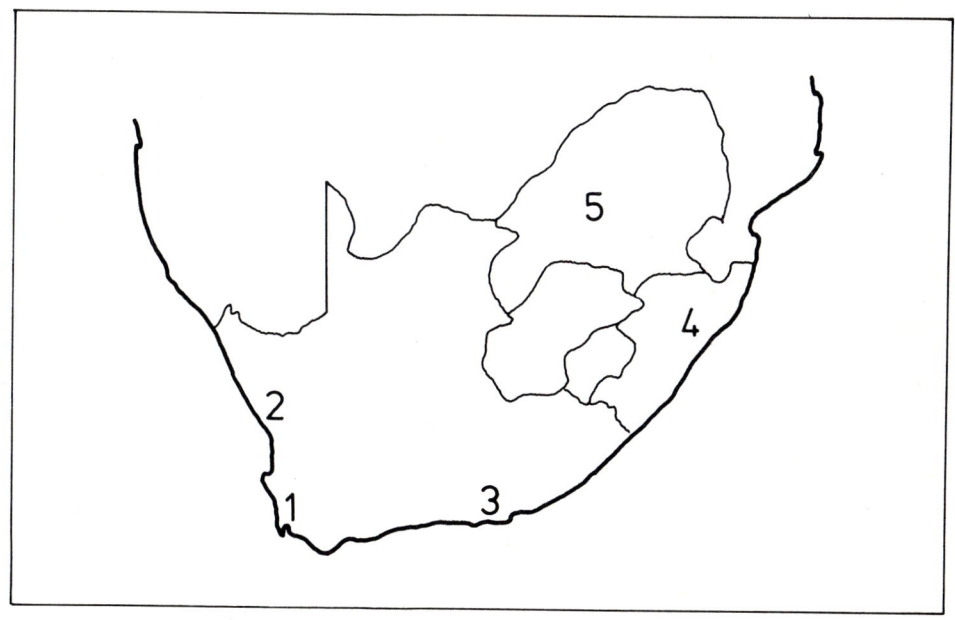

WESTERN CAPE

Numerous occurrences are known of residual kaolins derived from the weathering of the Cape granite. This Proterozoic granite is a coarsely porphyritic biotite type containing abundant large phenocrysts of orthoclase feldspar, with muscovite, pinite and tourmaline as accessory minerals. An interesting feature of these deposits is that they are practically indistinguishable from one another, and it is not possible to differentiate between materials that occur as far apart as 150 kilometres. The deposits are characterized by their high kaolin yields. The kaolinite content of the kaolinized granite may vary considerably within a deposit: from nearly 100 per cent kaolinized pegmatite veins to less than 30 per cent in quartz-rich areas, with an average of about 50 per cent.

The associated minerals are relatively coarse, and it is a feature of these materials that practically no quartz or mica are found in the minus 30 μm e.s.d. fractions. On the other hand, a montmorillonite-illite mixed-layer mineral occurs particularly in material near the surface of the deposit. The presence of the mixed layer mineral is reflected by the higher specific surface area of some of the beneficiated kaolins: up to 23 $m^2/g$, as compared with 15 $m^2/g$ for material in which the presence of the mixed-layer could not be detected by X-ray diffraction. The green and dry strength will also be appreciably higher. However, the rheological properties of the kaolin are not noticeably affected. Materials beneficiated by wet screening through a 45 μm sieve all show a critical concentration of between 60 and 65 per cent when using a sodium acrylate dispersant, or about 55 per cent when using sodium silicate.

In general there appears to be a definite relationship between weathering and fault zones. It is also of interest to note that the texture of the kaolin bodies almost identical to that of the fresh granite found close by, with the feldspar crystals having been kaolinized completely *in situ*. It is generally assumed that surface weathering played the major role in the genesis of these kaolins, but there seems to be increasing evidence that hydrothermal alteration also played an important part. The kaolinization - tourmalinization - greissenization trinity postulated by Bristow (1974), the low alkali content (Millot, 1970), signs of silver mineralization (B.J. Liebenberg, personal communication, 1963), depth of weathering (Treasure, 19 and the presence of mixed-layer expanding clay minerals in the upper zones (Kromer 1973) all tend to support the theory of a hydrothermal genesis.

The deposits in the *Cape Peninsula* are at present the most important sources of high-grade beneficiated kaolins. The Fish Hoek - Kommetjie - Noordhoek triangle, roughly 5000 hectares in extent, encloses at least four large bodies of *in situ* weathered Cape granite. These bodies are not surface related and the overburden thickness varies from nothing to more than 20 metres, while the depth of weathering may exceed 60 metres in places. In 5 of the 75 prospecting holes drilled to date this area, incompletely weathered granite was encountered at depths of 17 to 60 metres, while in others, signs of unaltered feldspar and biotite were found at depth of 30 to 50 metres (Treasure, 1977).

Typical analytical data on some of the beneficiated kaolins from this area are shown in the table. Material A is a recent production sample of minus 240 mesh (66 μm) washed kaolin, Material K a laboratory prepared minus 325 mesh (45 μm) kaolin while the data for Material C relate to a kaolin produced in 1967.

Although much of the kaolin is pure white in colour, there is unfortunately also some iron contamination. The chemical analysis of a sample of fresh Cape granite shows an iron content of 2.36% FeO, most of which is probably due to the biotite in the granite. Iron stained kaolin shows up as yellow and pink patches in a mottled pattern, often associated with quartz-rich veins. In some cases there are signs of upward migration of the iron, and pure white kaolin is found directly below an iron-rich overburden of ferricrete and sandstone, (Treasure, 1977). In general the iron content decreases with depth, but sometimes red kaolin is found beneath the white. Many iron-rich nodules are encountered, some filled with powdery ochre, others with marcasite crystals. However, by resorting to hand sorting in the quarry, it is possible to keep the iron content of the washed product below 0,5% $Fe_2O_3$ for ceramic and paper filler grade kaolin. The off-colour kaolin is suitable for other uses such as fibreglass, stoneware and insecticides where up to 1% $Fe_2O_3$ can be tolerated. The Leukometer reflectance of the washed kaolin can be as high as 90 per cent, and up to 95 per cent when fired to 1300°C. The actual yield of minus 240 mesh ceramic grade kaolin from one of the deposits, based on a 32 month production period, was 44,5 per cent. More than 50 per cent of this beneficiated material will have particles of less than 2 μm e.s.d. Materials suitable for the paper industry can be prepared from this kaolin, and beneficiation with a 40 mm hydrocyclone produced a material with a low abrasion value (1,63 mg/cm$^2$).

The importance of these deposits is underlined by the fact that the Department of Planning recently prohibited further township development in this area in order to preserve the kaolin resources for industrial use. At the present rate of production, the known resources should last for at least 30 years.

Near *Kuils River* several small occurrences of material similar to the Peninsula kaolins are known, except that this kaolin seems to be more plastic, with green strengths exceeding 2 MPa as compared with 1 MPa for the Peninsula kaolins. It is of interest to note that tin was once mined in this area.

In the *Stellenbosch* district practically all the good primary kaolin deposits occur on intensively cultivated viticulture farms and they are thus not available for exploitation. The size of the reserves has not been established, but they could conceivably be as large as those in the Peninsula area. Analytical data of a typical kaolin, beneficiated in the laboratory, are given in the table as Material DV.

The weathered Cape granite near *Somerset West* was at one time mined and beneficiated by the Marina Company. The reserves were estimated at about 150,000 tonnes, but there are other occurrences of the same material in this vicinity.

Good quality kaolin has been found in the *Vredenburg-Saldanha* area, but these occurrences have been prospected only haphazardly, and insufficient data make it impossible to estimate reserves.

## NAMAQUALAND

The area around *Bitterfontein* was thoroughly investigated in 1960 (Heystek, et al 1961). The kaolins are the product of surface weathering of feldspathic gneiss and granulite originating from the metamorphosis of Malmesbury sedimentary rocks. Kaolinization appears to have been controlled by post-Archaeac fault zones. The alteration of the feldspars directly to kaolinite, and not through an intermediate stage of sericite, was almost certainly caused by superficial weathering agencies, probably during a more humid era, when chemical weathering dominated mechanical weathering.

In all, 26 occurrences around *Bitterfontein* were examined, of which at least 7 are of economic interest. The prospecting results indicated a total of some 3 million tonnes of high grade white kaolin suitable for the paper and ceramic industries, nearly 8 million tonnes of medium quality and 9 million tonnes of poor quality kaolin. In general, the recovery of kaolin is between 40 and 60 per cent, the main contaminant being coarse quartz. The analytical data of a minus 10 micrometre fraction of one of these kaolins are given in the table as Material NN. It was possible to improve considerably the brightness of the poorer quality of kaolin by using 1 per cent of sodium hydrosulphite.

A survey carried out in 1975 of an occurrence in this area, which had not been previously investigated, revealed a reserve of a further 2 million tonnes of white kaolin. The data of kaolin from this deposit, prepared by screening and passing through a 40 mm hydrocyclone, are given in the table as Material BF.

Owing to the semi-arid climate of this region, the amount of ground-water is not only limited, but too saline for kaolin beneficiation by washing. However, the infra structure of this region is rapidly expanding, particularly through the Sishen-Saldanha development, and there is the prospect of a fresh-water supply becoming available.

The *Garies* deposits, some 60 kilometres north of Bitterfontein, have been discovered only recently, but they are already being exploited for the paper industry. The kaolin is the *in situ* weathered product of a granitic gneiss and it is associated with quartz veins and a silcrete topping. Prospecting pits and drill holes point to large tonnages. Some of the samples gave a yield of 70 per cent, but the beneficiated kaolin sometimes contained a good deal of fine silica. Because of the arid climate, the weathered material is trucked nearly 100 kilometres to the nearest railhead and then transported to the Transvaal, where it is beneficiated.

## EASTERN CAPE

Two kinds of kaolins are found in the Eastern Cape. At Mossel Bay, steeply dipping feldspathic dolerite dykes in granite have been kaolinized to a depth of about 30 metres. The dykes are 2 to 15 metres wide and are associated with silcrete cappings. The kaolins at Grahamstown are residual clays derived from shales and tillites of the Karroo and Cape Systems and they are usually unconformably covered by a silcrete capping.

The *Grahamstown* clays have been worked for many years, originally for pottery and later also for the paper industry as a filler grade. The total reserves appear to be very large, with some of the deposits up to 30 metres thick. Some of the clays contain appreciable amounts of organic material and they are quite plastic. The kaolins consist mainly of kaolinite, hydromuscovite and very fine-grained quartz; they are thus high in silica (see Material G). Attempts to up-grade these kaolins by separating the quartz out have not been successful, and they are thus merely milled to minus 300 mesh before use in the ceramic industry. Their rheological properties are very good and they are, in fact, more plastic than the beneficiated kaolins from the Cape Peninsula.

In the *Albertinia-Riversdale* area numerous deposits of residual clays are found, similar to those of the Grahamstown area, but with a much higher content of free quartz, (see Material VZ). The clays are derived from the weathering of shales of the Cape System and are associated with a silcrete capping. The reserves are very large, and the kaolins have been used for many years by the rubber, paint and other industries. The clays are not beneficiated but mining is selective and they are milled to minus 300 mesh. Their fired colour is unfortunately not very good.

The kaolins from *Mossel Bay* are also not beneficiated, but only milled to minus 300 mesh. The materials consist mainly of kaolinite, hydrous mica and quartz, with occasionally some pyrophyllite. The relative quantities of the various minerals vary considerably, as can be seen from the data for Materials S1 and S2. These kaolins fire to a cream colour.

NATAL

Residual kaolins derived from the weathering of the Archaean granite are found at *Inanda*, 48 kilometres north-west of Durban, and at *Ndwedwe*. The quality of these kaolins is high, and the reserves are probably large. These deposits are not exploited at present.

An interesting material, consisting of montmorillonite, kaolinite and cristobalite, and formed by the alteration of perlite and tuff, is found at *Mkuze*. This clay may possibly have ceramic applications because of its high dry strength, white fired colour and thermal expansion properties.

TRANSVAAL

Kaolinitic clays containing quartz and large amounts of hydromuscovite are mined at *Koster* for the ceramic industry. They are derived from the weathering of Pretoria shales. The thickness of the deposits is less than 2 metres and the reserves are limited. The *Zebediela* kaolin deposit, on the other hand, attains a thickness of 30 to 46 metres, and it is possibly a weathered tuff. The deposit is large but the reserves of white kaolin appear to be limited. The main consumer of this clay (Material Zeb) is the insecticide industry.

At *Lawley*, secondary clays of Karroo age are found in a large dolomite sinkhole or depression. The bottom layers are dark ball clays, but the top layers are light-coloured semi-plastic kaolins, containing some quartz as well as very small amounts of mica and montmorillonite; they are used particularly for tile manufacture.

SUMMARY

Deposits of primary kaolins, derived from the weathering of granitic and other feldspathic igneous rocks, as well as from sedimentary rocks, occur in five main area in South Africa. Many of these materials are very suitable for making high-quality whitewares, but only those formed by the weathering of granites can be beneficiated to a sufficiently low grit-content to be acceptable to the paper industry.

Unfortunately no reliable estimate has been made of the probable reserves of the various qualities of kaolin, but it does appear that there is more than sufficient material of the different grades to satisfy local demand.

REFERENCES

1. Bristow, C.M., 1974. The two trinities associated with kaolins of hydrothermal origin. International Geological Correlation Programme, Third Kaolin Symposium, Exeter, September, 1974.
2. Heystek, H., de Jager, D.H., de Waal, P. and Urli, G.L.P., 1961. The kaolin deposits of the area between Bitterfontein and Landplaas, Vanrhynsdorp District. Bulletin 36, R.S.A. Geological Survey, Pretoria.
3. Kromer, H. and Schüller, K.H., 1973. Eigenschaften von Kaolinen für die Keramie und als Füllstoff. Ber. dt. keram. Ges., 50: 15-19.
4. Millot, G., 1970. Geology of Clays. Springer-Verlag, New York/Heidelberg/Berlin
5. National Building Research Institute, Council for Scientific and Industrial Research, 1968-. Ceramic Raw Materials of Southern Africa (CSIR Technical Notes Pretoria.
6. Schmidt, E.R., 1976. Clays. In: Coetzee, C.B. (Editor), Mineral resources of t Republic of South Africa. Handbook 7, R.S.A. Geological Survey, Pretoria, pp. 275-288.
7. Treasure, P.A., 1977. Kaolin deposits of Noordhoek Valley, Cape Peninsula. Report Eg 6/238, R.S.A. Geological Survey, Pretoria.

THE CRYSTALLINITY INDEX OF KAOLINITE IN RELATION TO OTHER
PROPERTIES OF THE KAOLIN MASS OF KARLOVY VARY

J. KONTA
Department of Petrology, Charles University, Prague, Czechoslovakia

ABSTRACT

Kaolinite of thirteen petrographic kaolin types formed by the weathering of granite rocks in the Karlovy Vary area, Czechoslovakia, displays a degree of crystallinity in the washed material below 20 μm which rises with increasing bulk density of the raw kaolin. The bulk density increases with rising quartz content and/or with a lower degree of decomposition of potassium feldspar and biotite. In contrast to the kaolinite in clay pseudomorphs after potassium feldspar with a more or less abundant admixture of illite, the kaolinite in the pseudomorphs after acid plagioclase is substantially purer, always more readily dispersible, having a higher crystallinity index within the mining zone. The interrelations between the physical properties are shown in correlation diagrams.

INTRODUCTION

The Karlovy Vary kaolins in Czechoslovakia represent a very valuable source of raw materials for the manufacture of high quality porcelain. So far no relationships have been described between the crystallinity index of unground kaolinite in the individual petrographic kaolin types and other properties, such as the bulk density of raw kaolin, the dispersibility of clay mass (size fraction below 20 μm) and $K_2O$ content; neither have there been described any interrelations between non-traditionally determined physical properties. The aim of this paper is to explain the causes of variation in the crystallinity index of kaolinite in the mined petrographic kaolin types derived from various types of granite rocks, and to contribute to the theory of the genesis of raw kaolins. The relationships mentioned above and their interpretation have not yet been described in any other primary kaolins.

## RESULTS

### Experimental material

The kaolins in the Karlovy Vary area in West Bohemia display a well-preserved texture derived from the granite source rocks. Macroscopically, 13 petrographic kaolin types can be distinguished on the basis of their main textural features and mineral composition (Konta and Koscelník, 1968); the tint of the clay mass is an important additional characteristic. Thirty-four samples were taken from five opencast kaolin mines: Osmosa, Jimlíkov, Hájek, Podlesí and Otovice. In Fig. 1 the individual samples are plotted as experimental points expressing the interrelations between the lutite, arenite and rudite fractions obtained from raw kaolins under constant conditions. The coarse-grained kaolin types I and II, the medium-grained types III and IV and the fine-grained ones V, Va and VI contain a substantially higher amount of quartz than the very fine-grained types VII, VIII and IX, of which especially the last two are poor in quartz; the Tr type represents a kaolin transported for a short distance; the abbreviation Org. indicates a washed kaolin containing a combustible organic substance. The GR type is a slightly kaolinized and sericitized granite. Types I, III and Va contain, in addition, non-decomposed biotite and are characterized by higher values of bulk density and clay pseudomorphs after potassium feldspar, dispersible with difficulty. If more than one sample of the same petrographic type is taken from a deposit, it indicates that such a type occurs in several differently coloured varieties.

### Methods

All petrographic types of kaolin were studied in thin sections under a polarizing microscope.

From each size fraction below 63 μm the fraction not exceeding 20 μm was prepared by a three-times repeated sedimentation in distilled water at pH 9.5 (adjusted by $Na_2CO_3$ solution). This fine fraction roughly corresponds to the material obtained by the industrial washing of kaolin.

The crystallinity index of kaolinite in size fractions below 20 μm was determined by Hinckley's (1963) X-ray method using a Guini camera with $CuK_\alpha$ radiation under the usual conditions; a study was made of unground material (washed out only) and ground material (a batch of 0.06 g, ground in an agate mortar for 2 minutes under constant conditions). By evaluation of selected X-ray reflections of non-oriented preparations, the approximate contents of kaolinite, mica and quartz were determined.

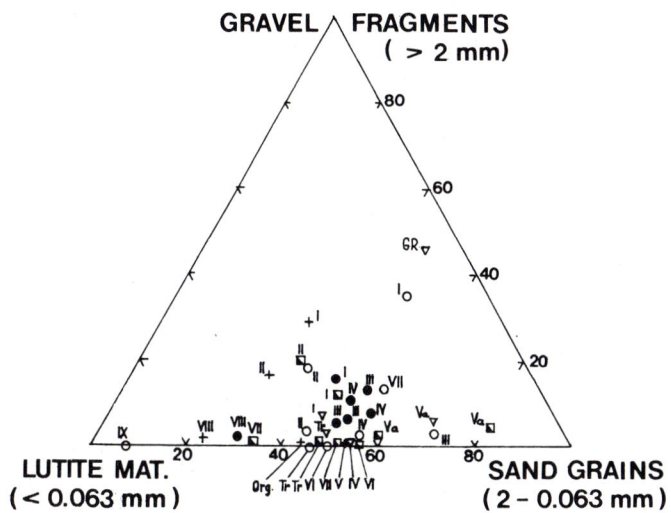

Fig. 1. Interrelations between size categories of the petrographic kaolin types designated by Roman numerals; Karlovy Vary area, Czechoslovakia. Deposits: ○ Osmosa, ▽ Jimlíkov, ◼ Hájek, ● Podlesí, + Otovice. (Sample VII Osmosa contains less desintegrable clay mass.)

The bulk density of raw kaolins was determined from 40-100 g rock fragments by weighing in air and in a liquid.

Imbibometric tests of all samples of size fractions below 20 µm in 0.5 g batches, compressed at 30 MPa under constant conditions into the form of small cylinders of base 100 $mm^2$ were obtained on an automatic imbibograph and yielded information on the dynamics of the absorption of distilled water at a temperature of 20 °C. For correlation, the water contents were ascertained as the number of milligrams sorbed over 15 minutes. The more kaolinite or quartz in the sample, and the greater the relative size of the crystals, the larger are the pores in the compressed aggregate, and the more rapid is the initial sorption of the liquid. The greater the admixture of illite and/or a mixed-layer illite-smectite, the slower is the initial sorption of water, and the content of water after 15 minutes is lower.

The amounts of $K_2O$ measured with a flame photometer in the size fractions below 20 µm provide an idea of the proportion of micaceous minerals, primary and secondary, as feldspar is not present in the kaolin mass studied, with the exception of a solitary GR sample from Jimlíkov.

Correlation diagrams

Fig. 2 represents the correlation between the bulk density of raw kaolin samples and the content of the size fraction below 20 μm separated by washing. The tendency to indirect dependence is marked; the relation is affected partly by the primary composition of the granite source rock, partly by the intensity of kaolinization. The higher the feldspar content and the lower the amount of quartz in the source rock (i.e. the two most important constituents in granite rocks before kaolinization), the lower was the resulting bulk density in the kaolinized rocks, as feldspars chemically altered to clay mass are more porous than quartz. Examples of kaolinized granite rocks of low bulk density are dike rocks originally rich in feldspars and poor in quartz (types VII, VIII, IX). The more intense

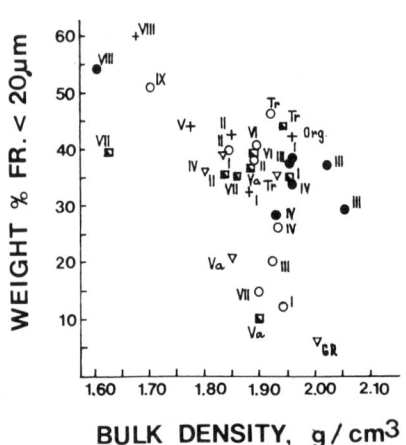

Fig. 2. Correlation between the bulk density of raw kaolin and the content of the size fraction below 20 μm separated from it; expressed in mass percent. (For key see Fig. 1.)

was the chemical decomposition, the greater was the removal of substances from feldspars, biotite and other less important minerals; the greatest porosity and thus the lowest bulk density is achieved when pseudomorphs of an almost pure kaolinite after feldspars are formed and when biotite also decomposes. Petrographic types II, IV and V of kaolins without biotite and with a high kaolin content are an example of this, while the kaolins with biotite with a relatively lower amount of kaolinite are exemplified by the petrographic types I, III, Va and GR.

From Fig. 2 it also follows that the content of the size fraction below 20 μm in this kaolin mass increases with decreasing bulk density. But the quality of kaolin rises with decreasing bulk density only in the area of the densest accumulation of experimental points, i.e. within a range of 1.78 - 2.05, where the proportion of the fraction below 20 μm exceeds 30 %. The samples whose experimental points fall within the area of bulk density values below 1.75 contain unfavourably high illite contents; the samples whose experimental points are at the opposite end of the field, where the fraction below 20 μm represents 20 % or less of the raw kaolin mass, are characterized by the least advanced decomposition of feldspars or contain very greatly increased amounts of quartz.

Fig. 3 shows that there exists a tendency to a direct interdependence between the bulk density of the raw kaolin and the crystallinity index of the kaolinite in the unground fraction below 20 μm. The crystallinity index of kaolinite and therefore its maturity in the washed mass of the under 20 μm fraction increases with the rising bulk density which in most Karlovy Vary kaolins in the mining zone increases partly with the increasing content of coarse-grained quartz, partly with the disintegrability of clay pseudomorphs after potassium feldspar becoming more difficult. Otherwise, it holds for the Karlovy Vary kaolins too that the crystallinity index of kaolinite decreases with depth in profile, and that with greater depth the proportion of 2:1 clay minerals and the bulk density also increases (Konta, 1969). In the sample designated GR we might theoretically expect a lower crystallinity index because of the low degree of decomposition of the original granite rock. The high value of the crystallinity index is due to the fact that the greater part of the kaolinite here formed from acid plagioclase which underwent complete kaolinization earlier, while the potassium feldspar remained fresh or was sericitized, as it follows from microscopic study of many thin sections of several profiles.

The highest degree of crystallinity of kaolinite exists in those samples, during whose washing the mass of white clay pseudomorphs after acid plagioclase becomes optimally washable and passes into the size fraction below 20 μm, and when illite-rich clay mass, arising from potassium feldspar, becomes more cohesive and is removed in coarser size fractions during washing (types GR, I, III and Va).

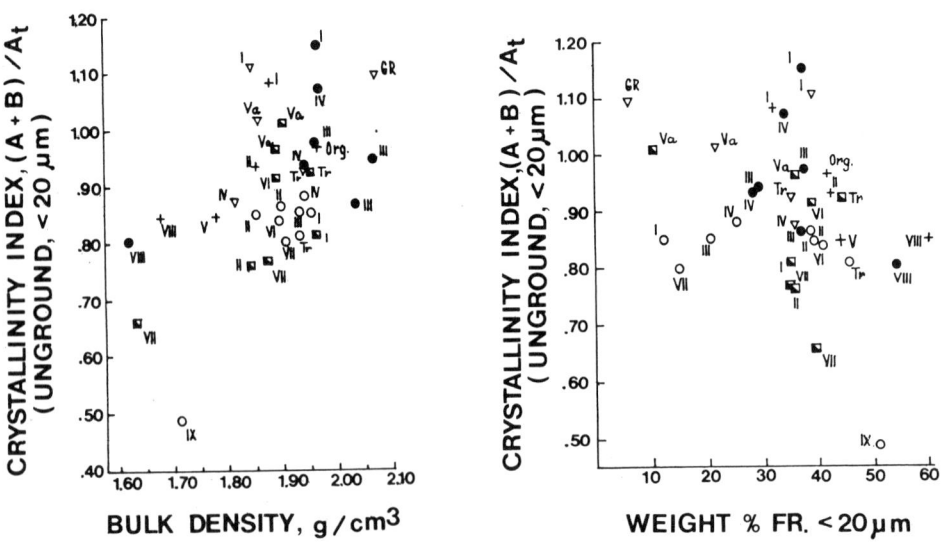

Fig. 3. (left) Correlation between the bulk density of raw kaolin and the crystallinity index of kaolinite in the unground size fraction below 20 μm. (For key see Fig. 1.)

Fig. 4. (right) Correlation between the crystallinity index of kaolinite in the unground size fraction below 20 μm and the content of the size fraction below 20 μm. (For key see Fig. 1.)

The above correlation, illustrated in Fig. 3, is supported by Fig. 4 which shows the tendency of the group of experimental points to indirect interdependence between the crystallinity index of the kaolinite in the unground size fraction below 20 μm and the content of size fraction below 20 μm washed out from raw kaolin. The greater the content of the size fraction below 20 μm washed out from raw kaolin, the more probable is a lower value of the kaolinite crystallinity index. In such cases not only do clay pseudomorphs of pure kaolinite after acid plagioclases easily disintegrate, but so also do clay pseudomorphs after potassium feldspar containing, besides illite, a less perfectly crystallized kaolinite.

Fig. 5 shows that the kaolinite in the unground size fractions below 20 μm always displays a crystallinity index higher in value than that of the ground fractions. Under constant conditions of 2-minute grinding, the layers in the kaolinite are shifted along the basal plane, in some samples to a greater extent (e.g. the types: Va Hájek, III Osmosa, GR and I Jimlíkov) and in others to a lesser extent (e.g. most samples from Podlesí and Osmosa and also the type VII Hájek). In about 75 % of the samples, the more marked is the lowering of the crystallinity index, the lower is the amount of quartz in the size fraction below 20 μm. An increasing admixture of quartz possibly protects the kaolinite from shearing forces, and so decreases the extent of layer displacements and delamination. The kaolinites in some samples from Podlesí and Osmosa are an exception; in these the kaolinite crystallinity index decreases relatively slightly during grinding (black and white circles in Fig. 5), even when the quartz content is low.

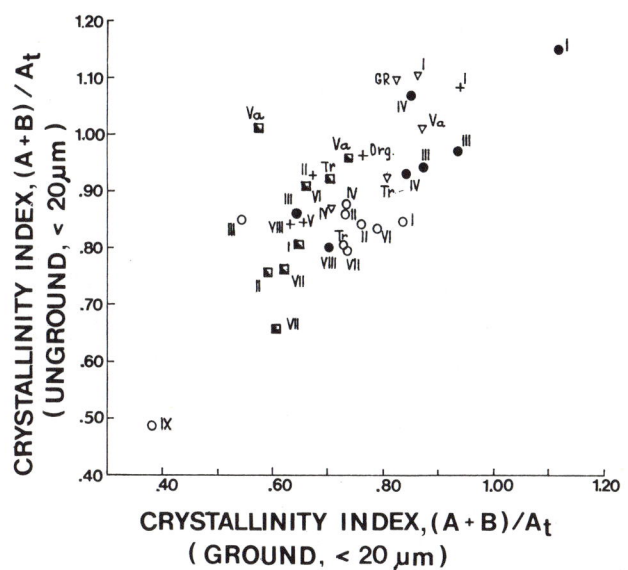

Fig. 5. Correlation between the crystallinity index of kaolinite in the unground and ground size fractions below 20 μm, separated from various petrographic types of kaolin. (For key see Fig. 1.)

Fig. 6 illustrates the correlation between the bulk density of raw kaolin and the $K_2O$ percentage in the size fraction below 20 μm. The overwhelming majority of the experimental points constitute a field in the lower right-hand part of the diagram with a weak tendency to direct interdependence. The samples with a low $K_2O$ content in this field of experimental points usually display the greatest concentrations of kaolinite. Four experimental points in the lower left-hand part of the diagram indicate a material of extremely low bulk density (with an extremely high output of the size fraction below 20 μm) and with various amounts of $K_2O$ and therefore also of mica. Four experimental points shifted upward in the diagram (Osmosa I, III, VII, Otovice I) suggest a material with relatively elevated $K_2O$ content bound particularly in mica (primary + illite). The correlation is complicated by the widely varying amounts of quartz in raw kaolin, which fact substantially influences the bulk density values, and also by the fact that $K_2O$ is bonded in physically and chemically different primary and secondary mica material.

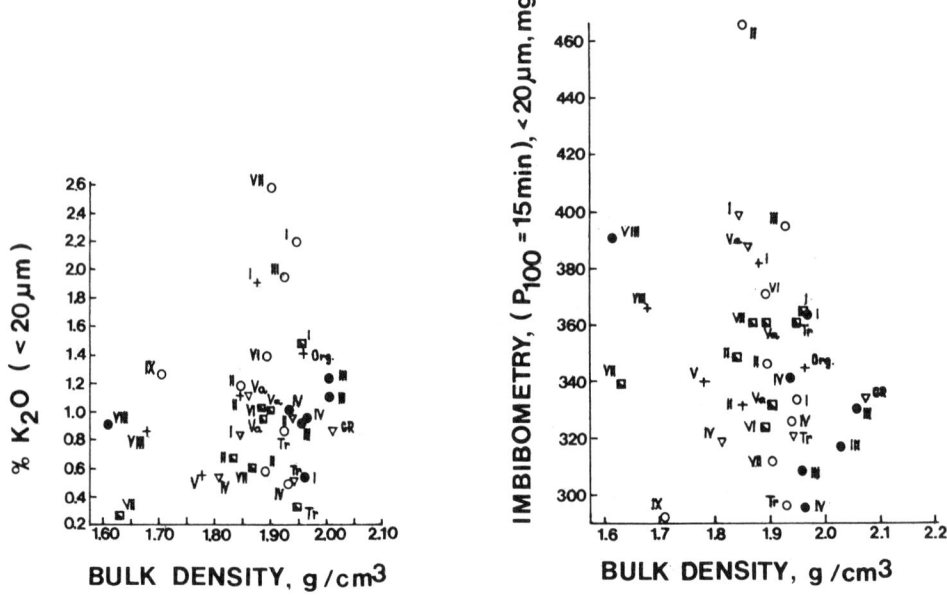

Fig. 6. (left) Correlation between the bulk density of raw kaolin and the percentage of $K_2O$ in the size fractions below 20 μm. (For key see Fig. 1.)

Fig. 7. (right) Correlation between the bulk density of raw kaolin and the imbibometric value representing the amount of water soaked during 15 minutes of the initial sorption. (For key see Fig. 1.)

Fig. 7 presents the correlation between the bulk density of raw kaolin and the imbibometric value, representing the amount of water sorbed during 15 minutes. Most experimental points are grouped into the field in the right-hand part of the diagram, showing some tendency to indirect interdependence. With increasing illite and the mixed-layer illite-smectite in the size fraction below 20 μm, the value of the sorbed water decreases (Konta, 1976), as inter-particulate swelling of the thin-leaf-shaped crystals of illite (whose specific surface area is many times larger than that of kaolinite) retards the initial penetration of water into the compressed aggregate. The extremely high value of the sorbed water (over 460 mg) in one petrographic variety of type II from Osmosa is due to the extremely large crystals of kaolinite and to a relatively very low percentage of illite. Four samples (VIII Podlesí, VIII Otovice, VII Hájek, IX Osmosa) whose imbibometric values are very low rank among the kaolins which arose from granite rocks with an extremely high content of feldspars and a very low amount of quartz; in a raw state they are rich in clay.

CONCLUSIONS

The methods applied and the correlation of the data obtained in this study of primary kaolins are experimentally simple and quick. They enable the recognition of the main physical properties of kaolins, on which depend such important ceramic properties as dry strength and flowability. The differences ascertained between the individual petrographic types of kaolin mined from five deposits in the Karlovy Vary area permit the control of the optimum mixing conditions of raw material in the production of various types of washed kaolin.

The crystallinity index of kaolinite may easily be influenced by short-term grinding which leads to an irregular lowering of its value. An elevated admixture of quartz appears to hinder the delamination of kaolinite and the lowering of the crystallinity index during dry grinding.

The crystallinity index of kaolinite in the separated size fraction below 20 µm, which in primary kaolins may be regarded as a measure of maturity, increases with rising bulk density in the upper mining zone of the Karlovy Vary kaolins where the bulk density increases partly with increasing quartz content, partly with lowering disintegrability of clay pseudomorphs after potassium feldspar and with an increasing admixture of undecomposed biotite. The mined petrographic types of kaolin with more quartz contain a clay mass richer in kaolinite and poorer in illite, possibly because acids in the permeating water were able to decompose more completely - at the same vertical level - the smaller volume of feldspars present. The kaolinite in pseudomorphs after acid plagioclase displays a crystallinity index higher than that in pseudomorphs after potassium feldspars. If the pseudomorphs after potassium feldspar are readily dispersed, then the mean crystallinity index of the kaolinite in the washed product of particle size below 20 µm is lower than when fragments of non-dispersed clay pseudomorphs with higher illite contents are removed into coarser fractions.

In the Karlovy Vary area it also holds that the crystallinity index of kaolinite decreases with depth (as has been affirmed in boreholes), associated with increasing bulk density of kaolin from granite rock of constant composition. A lowering of the kaolinite crystallinity index with depth was previously recognized in a kaolin profile formed by alteration of orthogneiss in the Kadaň area in Bohemia (Neužil, 1970).

REFERENCES

Hinckley, D.N., 1963. Variability in "crystallinity" values among the kaolin deposits of the Coastal Plain of Georgia and South Carolina. In: E. Ingerson (Editor), Clays and Clay Minerals, Proc. 11th Nat. Conf., Ottawa, Canada, pp. 229-235.

Konta, J., 1969. Comparison of the proofs of hydrothermal and supergene kaolinization in two areas of Europe. In: L. Heller (Editor), Proc. Internat. Clay Conf., Tokyo, Japan, Vol. 1: 281-290

Konta, J., 1976. Rapid industrial control of basic rheological properties of washed kaolins using imbibometry. Interceram, Nr. 4, pp. 249-251.

Konta, J. and Koscelník, Š., 1969. Petrographical types of kaolin in the Karlovy Vary granite massif. XXIIIth Internat. Geol. Congr., Vol. 14: 79-94.

Neužil, J., 1970. Petrology of kaolin profiles on crystalline schists in the environs of Kadaň (western Bohemia). In: J. Konta (Editor), Proc. Fifth Conf. Clay Min. and Petrology, Praha, Czechoslovakia, pp. 73-100.

RATE OF TRANSFORMATION OF HALLOYSITE TO METAHALLOYSITE UNDER HYDROTHERMAL
CONDITIONS

H. MINATO and M. AOKI
Institute of Earth Science and Astronomy, College of General Education,
University of Tokyo, Tokyo (Japan)

ABSTRACT

The rate of transformation of halloysite to metahalloysite under hydrothermal conditions has been compared with that under dry-air conditions and a possible mechanism for the transformation is discussed. The halloysite used occurred as a pale yellow mass in a vein 4-5 cm thick in an andesitic lava flow in Fukushima Prefecture. After pulverization the halloysite was treated with distilled water in closed vessels at 135°C, 150°C and 165°C for various time intervals and the amount of metahalloysite in the treated materials was measured from a working curve established by preliminary experiments using X-ray diffraction.

The rates of transformation of halloysite to metahalloysite ($v = -\Delta$[halloysite]$/\Delta t$) have been plotted against the residual amount of halloysite and at each temperature a curve showing two distinct slopes has been obtained. During the early part of the reaction three virtually straight parallel lines are obtained, v being almost proportional to the amount of halloysite. In the second part, after two hours reaction, no such proportionality is observed but the amount of metahalloysite gradually increases. X-ray diffraction shows that a sharp 10Å peak changes to a broad 7.9-7.6Å band and then to a sharp 7.5Å peak for the transformed material.

Weight losses were measured for two specimens, the massive halloysite from Fukushima Prefecture and clayey halloysite from Nagano Prefecture, under dry-air conditions after various times of treatment.

The mode of dehydration under hydrothermal conditions leaves large domains of halloysite after partial reaction. This result differs from previous work on dehydration of halloysite under dry-air conditions.

---

INTRODUCTION

Hydrothermal alteration as well as weathering processes is one of the important modes of formation of halloysite. Many locations of hydrothermal halloysite have been reported from Japan, Minato et al. (1969), Minato (1975, 1977), Mexico, Keller et al. (1971), Korea, Minato (1977) and etc.. The temperature stability range of halloysite in natural occurrence may be greatly influenced by the vapour pressure of water. Possible temperature limit for the formation of halloysite can be estimated by the transformation experiments in pure water at high temperatures.

EXPERIMENTAL METHODS

Massive halloysite from Akazurayama, Mafune, Nishigo-mura, Nishishirakawa-gun, Fukushima Prefecture in northern part of Japan was used for these experiments as well as clayey halloysite from Otaki, Otaki-mura, Kiso-gun, Nagano Prefecture, Central Japan (Minato et al., 1972). Both halloysite materials from Akazurayama and Otaki are free from metahalloysite and other minerals when they were analyzed by means of X-ray powder methods. Akazurayama halloysite occurs in veins of 4 or 5 cm thickness in an andesite lava flow and has a massive wax like appearance with a pale yellow colour. Otaki halloysite occurs in the lower part of a white tuffaceous bed composed of acidic rhyolitic pyroclastics and exhibits a clayey appearance with a pale yellow colour. Observations under the electron microscope show that the former material is mainly tubular with a small amount of spherical forms while the latter material is mainly spherical. Chemical analyses of the two materials are cited in Table 1.

TABLE 1

Chemical analysis of halloysites

|  | Akazurayama | | Otaki | |
| --- | --- | --- | --- | --- |
|  | Wt. % | Mol. prop. | Wt. % | Mol. prop. |
| $SiO_2$ | 37.97 | 0.6328 | 40.30 | 0.671 |
| $TiO_2$ | 0.13 | 0.0016 | non. | ----- |
| $Al_2O_3$ | 35.93 | 0.3523 | 33.23 | 0.326 |
| $Fe_2O_3$ | 0.65 | .0041 | 2.30 | 0.014 |
| MnO | tr. | ------ | tr. | ----- |
| MgO | 0.06 | 0.0015 | 0.27 | 0.007 |
| CaO | 0.09 | 0.0016 | tr. | ----- |
| $Na_2O$ | 0.09 | 0.0015 | 0.05 | 0.001 |
| $K_2O$ | tr. | ------ | 0.05 | 0.001 |
| $H_2O^+$ | 11.41 | 0.6339 | 13.59 | 0.754 |
| $H_2O^-$ | 13.34 | 0.7411 | 9.63 | 0.535 |
| Total | 99.67 |  | 99.42 |  |

(Analyst: Minato, 1970, Minato & Aoki, 1978)

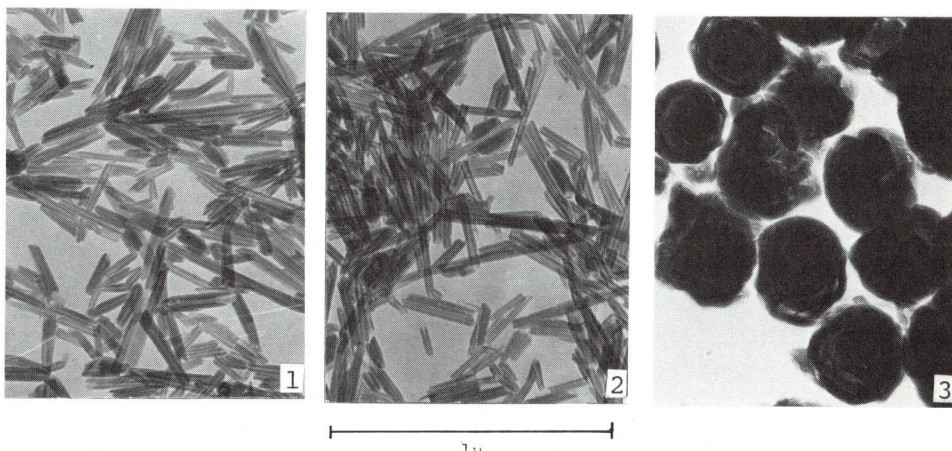

1: Halloysite from Akazurayama.
2: Treated halloysite of Akazurayama by hydrothermal method.
3: Halloysite from Otaki.

Fig. 1. Electron micrographs of halloysite.

Experiment 1

One gram of pulverized halloysite from Akazurayama was placed with distilled water in teflon vessels of about 25 cm$^3$ capacity enclosed in stainless steel bombs and maintained in an electric air bath at 100°C, 120°C, 135°C, 150°C and 165°C for various periods of time. The accuracy of the temperature control was believed to be less than $\pm$ 1°C. The vessels were quenched after the runs in ice-water mush. The treated materials were separated from water 15 minutes after the quenching. The wet materials were spread on a non-reflection holder made from a single quartz crystal, and analyzed by means of X-ray diffraction. Relative intensity was measured on the basal reflections of halloysite and that of metahalloysite. The amount of remaining halloysite and that of newly formed metahalloysite in the treated materials were determined using a working curve established by means of X-ray diffraction.

Preliminary experiments for the establishment of a working curve was carried out as follows: fully hydrated halloysite from Akazurayama and metahalloysite which was prepared from the same materials by heating at 150°C, was mixed in various ratio of amounts. Relative diffraction intensities of basal reflections of halloysite and metahalloysite were measured on these materials. The intensity measurement of basal reflection was made using the peak height in each basal reflections. The working curve is shown in Fig. 2 as the relation between the content of metahalloysite and I[metahalloysite]/(I[halloysite]+[metahalloysite]).

Experiment 2

About 10 milligrams of wet halloysite (pulverized Akazurayama material) was spread

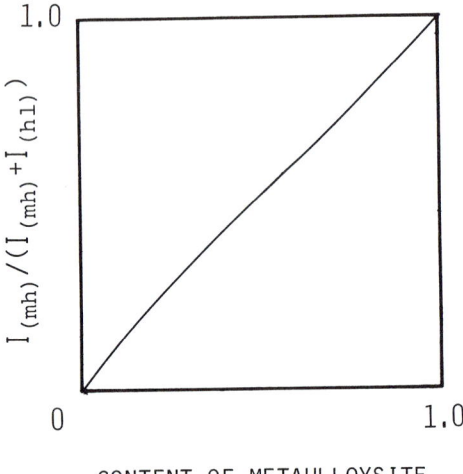

Fig. 2. Working curve for determination of the amount of metahalloysite.

on a non-reflection holder and analyzed by means of X-ray diffraction after various holding times in a desiccator with a desiccant of silica-gel. The amount of halloysite and that of metahalloysite were determined by the working curve mentioned above.

Experiment 3

The Otaki clayey halloysite was purified by sedimentation methods without crushing. Wet halloysite of the purified material was mounted on a non-reflection sample holder made of a single quartz crystal and analyzed by means of X-ray diffraction in the range from 15° to 5° 2θ, after being dried in room conditions at 60% relative humidity for 3 hours. The possible dehydration of the material during the course of X-ray analysis was checked in every experiment. The estimation of the degree of dehydration of the material was carried out by means of gravimetry after various holding times in a desiccator with a desiccant of silica-gel at 20°C. The weight loss and the change in the amount of metahalloysite determined by X-ray diffraction were plotted as function of holding time in the desiccator.

RESULTS AND DISCUSSION

The rate of dehydration under hydrothermal conditions was estimated by the rate of transformation of halloysite to metahalloysite, and the results are shown in Fig. 3 as the relationship between the reaction time and the amount of halloysite produced[+]. The hydrothermal treatments were carried out at 100°C, 120°C, 135°C, 150°C and 165°C, but obvious transformation was not observed in the materials treated at 100°C and 120°C. During the early stage of the reactions at 135°C, 150°C and

---

[+] Since there is no change in grainform on hydrothermal treatment (compare electron micrographs 1 and 2 in Fig. 1), the formation of kaolinite can be excluded.

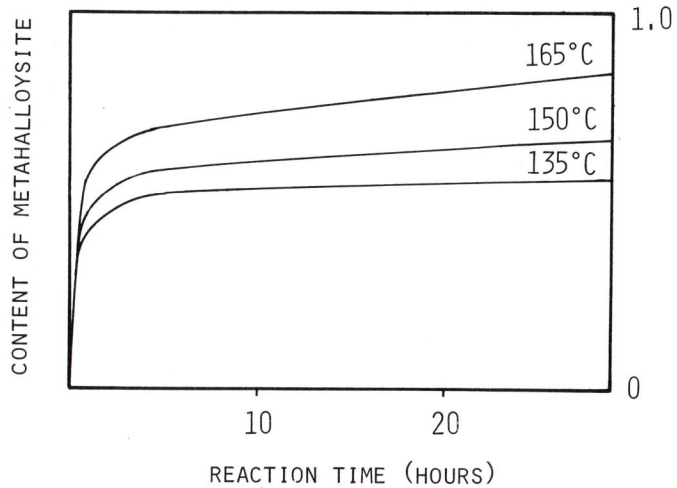

Fig. 3. Relationship between the reaction time and the content of metahalloysite in treatments under hydrothermal conditions.

165°C, rapid transformation by dehydration was observed initially and slow transformation at a later stage of the reactions. In general, the reaction rate is greater at high temperatures. In Fig. 4, the residual amount of halloysite was plotted against $-dv/dt$ where $v = -\Delta[\text{halloysite}]/\Delta t$. During the early stage of the reaction, the rate is directly proportional to the residual amount of halloysite, but not at later stages of the reaction. The reactions at 135°C, 150°C and 165°C show the same general tendencies.

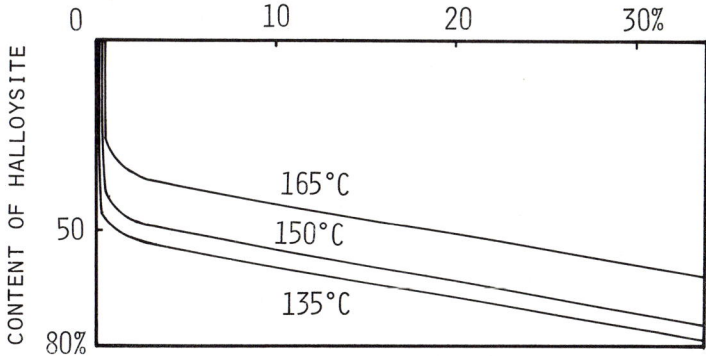

Fig. 4. Relationship between the residual amount of halloysite and the rate of decrease of halloysite transformation ($-dv/dt$) on treated materials under the hydrothermal conditions.

The mechanism of this dehydration reaction and the transformation to metahalloysite from halloysite can be explained mainly by the process of diffusion of water in a direction parallel to the sheet structure. A report by Reed et al. (1962) on experiments of the leaching process of $K^+$ from biotite and muscovite is instructive for the elucidation in process of the dehydration reaction of halloysite. In the release of $K^+$ from micas, the process of diffusion of $K^+$ from a flat circular disc to the outside of the mica through the particle periphery is a process of two dimensional interlayer diffusion with a moving boundary. The relation of degree of reaction as a function of the reaction time was shown to be:

$$Q/Q_o (1 - \ln Q/Q_o) = 1 - kt \qquad \ldots\ldots\ldots(1)$$

where $Q_o$: content of $K^+$ in original material at the time of start in the reaction,

$Q$: content of $K^+$ in the reaction material at t,

k: constant determined by crystallographic properties of the mineral, grain size, diffusion coefficient and concentration gradient.

In the transformation to metahalloysite from halloysite, the rate of diffusion of water in halloysite is influenced very much by shrinking in the interlayer space This being so, the diffusion process of water will differ slightly from the process of leaching out of $K^+$ from mica. When the degree of diffusion of water is plotted as a relation between $Q/Q_o(1 - \ln Q/Q_o)$ and t, there is some deviation from the equation (1), and a possible diffusion mechanism of water from halloysite is sugges below.

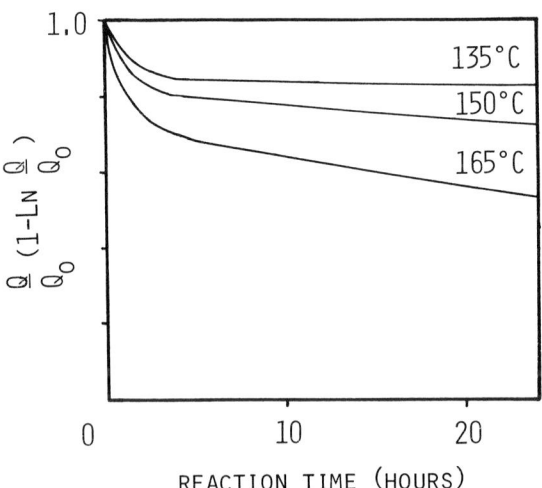

Fig. 5. Relationship between $Q/Q_o(1 - \ln Q/Q_o)$ and reaction time on the treatment of halloysite under the hydrothermal conditions.

In Fig. 5, the relationships with time are different straight lines at various reaction temperatures, in the later stage of the reaction, but in the early stage the plots are not proportional to the time. This indicates that the mechanism of diffusion of water approaches a steady state gradually, in the course of the dehydration.

The X-ray diffraction experiments revealed the mode of transformation as follows: metahalloysite is formed at the expense of halloysite without any decrease in sharpness of the basal reflection as shown in the right part in Fig. 6. It suggests that the dehydration process leaves sufficiently large domains to give sharp X-ray diffraction for halloysite. Thus following model is proposed for the dehydration process: the dehydration takes place in the direction of the halloysite tube through the interlayer space, followed by shrinking of the space. Thus this model accounts

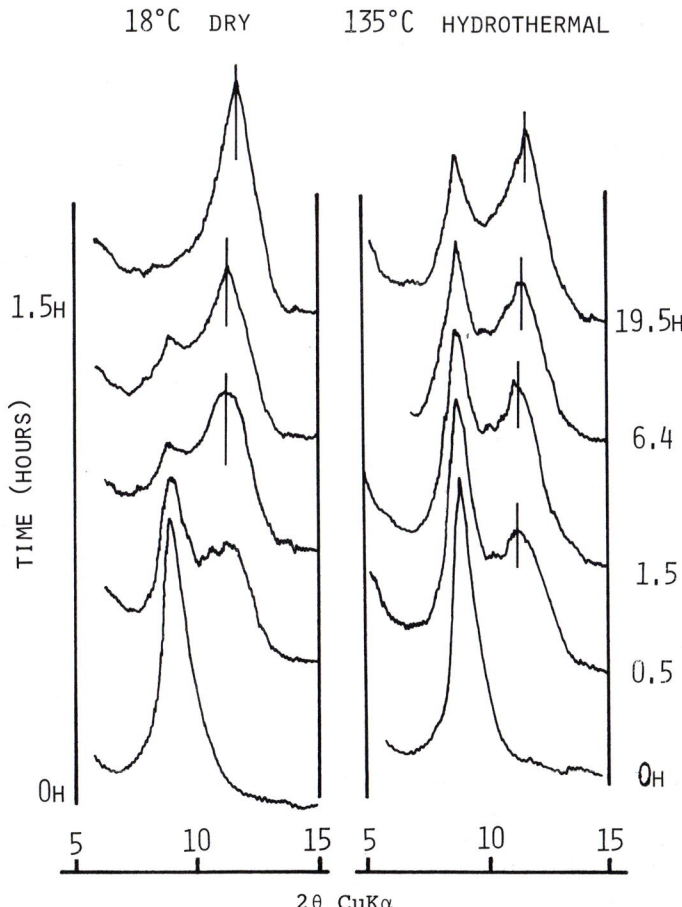

Fig. 6. X-ray diffraction patterns of halloysite from Akazurayama. Treated materials and untreated materials.

for the rapid diffusion of water in the early stage of the reaction, and the slow diffusion of water in the later stage of the reaction, because of the restrictions of interlayer space.

In a dry condition, the velocity of dehydration of halloysite and of formation of metahalloysite from halloysite was considerably greater than under hydrothermal conditions. The sharp X-ray diffraction peak of halloysite changes to a diffused peak with dehydration as shown in the left part in Fig. 6, and the change differed from the results of experiments under hydrothermal condition. It suggests that the halloysite domains were smaller than that in hydrothermal conditions.

The peak shift and the broadning of the basal reflection of the halloysite and metahalloysite which occur during the transformation can be explained as the result of the interstratification of these minerals, as have been discussed for minerals prepared under dry-air conditions by Brindley et al. (1948) and Churchman et al. (1972). The line broading and the shift of the halloysite peaks can barely be observed for the materials which were prepared under hydrothermal conditions. This indicates a large domain of halloysite remains together with the mixed layered portion.

The degree of dehydration of halloysite and the content of metahalloysite are plotted as a function of reaction time in Fig. 7. The loss of water from halloysite precedes transformation to metahalloysite. The distance between these two curves appears nearly constant. The content of metahalloysite is an approximate measure of the degree of the dehydration.

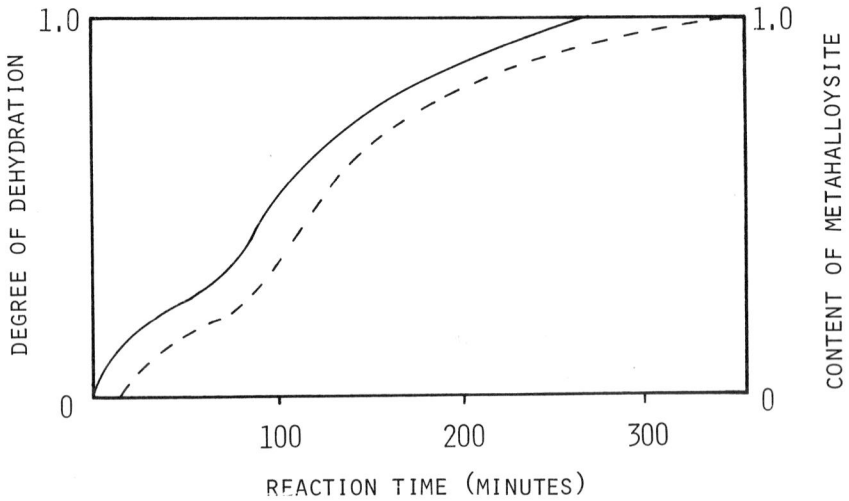

Fig. 7. Degree of dehydration of halloysite (solid line) and content of metahalloys (dashed line) as a function of reaction time.

CONCLUSION

The temperature limit of the stability of halloysite is estimated to be less than 135°C under the condition of 100% vapour pressure of water in the experiments performed under hydrothermal conditions. Transformation from halloysite to metahalloysite under hydrothermal conditions proceeds more heterogeneously than is the case under dry-air conditions.

REFERENCES

Brindley, G.W., and Goodyear, J., 1948. X-ray studies of halloysite and metahalloysite. Part II. The transition of halloysite to metahalloysite in relation to relative humidity. Mineral. Mag., 28: 407-422.

Churchman, G.J., Aldridge, L.P. and Carr, R.M., 1972. The relationship between the hydrated and dehydrated states of an halloysite. Clays and Clay Minerals, 20: 241-246.

Keller, W.D., Hanson, R.F., Huang, W.H. and Cervantes, A., 1971. Sequented active alteration of rhyolitic volcanic rock to endellite and a precursor phase of it at a spring in Michoacan, Mexico. Clay and Clay Minerals, 19: 121-127.

Minato, H. and Utada, M., 1969. Mode of occurrence and mineralogy of halloysite from Iki, Japan. Proceedings of the International Clay Conference Tokyo, 1969, 1: 393-402.

Minato, H. and Utada, M., 1972. A mode of occurrence of halloysite from Otaki, Nagano Prefecture, Central Japan. Kaolin Symposium, 1972 International Clay Conference, Madrid, 31-39.

Minato, H., 1975. Mineralogy and mode of occurrence of spherical halloysite from Japan. Contribution to Clay Mineralogy in honour of Professor Toshio Sudo, 1975. 73-81.

Minato, H., 1977. Kaolin deposits of Japan and Surrounded East Asia. Proceedings of the 8th International Kaolin Symposium and Meeting on Alunite. K-14: 1-13.

Reed, M.G. and Scott, A.D., 1962. Kinetics of potassium release from biotite and muscovite in sodium tetraphenylboron solutions. Soil Science Society Proceedings, 1962. 437-440.

# REACTIONS OF SALTS WITH KAOLINITE AT ELEVATED TEMPERATURES - PART 2

L. HELLER-KALLAI+ and M. FRENKEL*
+Department of Geology and *Casali Institute of Applied Chemistry,
The Hebrew University, Jerusalem

## ABSTRACT

Reactions of kaolinite and metakaolinite with salts of alkali metals were studied in the temperature range of 570-950°C, in air and in a nitrogen atmosphere, using thermal and chemical analyses and infra-red spectroscopy, supplemented by other methods. The mixtures react according to the equation

$$2SiO_2 \cdot Al_2O_3 \cdot xH_2O + 2yMA \longrightarrow 2SiO_2 \cdot Al_2O_3 \cdot yM_2O + (x-y)H_2O + 2yHA$$

The water required may be derived from clay hydroxyls or from the atmosphere. The rate is faster when the clay has not been entirely dehydroxylated, but even trace amounts of water in the surrounding atmosphere sustain the reaction. With Li, Na or K salts products resembling eucryptite, nepheline or kalsilite, respectively, tend to be formed. Molten $K_2CO_3$ reacts with dehydroxylated kaolinite even in the absence of detectable amounts of water. Prolonged heating with KF leads to anion exchange.

## INTRODUCTION

In Part 1 (Heller-Kallai, 1978) of this study mixtures of kaolinite with salts of alkali cations were heated for periods up to 60 minutes in the temperature range of dehydroxylation of the clay. It was shown that the reactions differed basically from those of smectites with salts. Smectites are deprotonated at relatively low temperatures and electrical neutrality is preserved by incorporation of alkali cations (Heller-Kallai, 1975a,b). The reactions with kaolinite commence with the dehydroxylation of the clay. With alkali halides it was assumed to follow the scheme

$$2SiO_2 \cdot Al_2O_3 \cdot 2H_2O + 2yMA \longrightarrow 2SiO_2 \cdot Al_2O_3 \cdot yM_2O + (2-y)H_2O + 2yHA \qquad (1)$$

The reaction proceeded more rapidly the greater the solubility of the salt and the smaller the radius of the cation.

Preliminary experiments indicated that the reaction continues after the principal dehydroxylation reaction of the clay, probably according to the equation

$$2SiO_2 \cdot Al_2O_3 \cdot xH_2O + 2yMA \longrightarrow 2SiO_2 \cdot Al_2O_3 \cdot yM_2O + (x-y)H_2O + 2yHA \qquad (2)$$

and with alkali carbonates

$$2SiO_2 \cdot Al_2O_3 \cdot xH_2O + yM_2CO_3 \longrightarrow 2SiO_2 \cdot Al_2O_3 \cdot yM_2O + xH_2O + yCO_2 \qquad (3)$$

The reaction with carbonates proceeded more rapidly than would have been expected on the basis of their solubility.

It is the purpose of the present study to establish whether equations (1)-(3) represent the course of the reactions, to determine the effect of the magnitude of x and to elucidate whether the water must be derived from the clay itself or whether it can be supplied from outside, after the clay has been partially or completely dehydroxylated.

Two series of experiments were performed. In the first, samples of kaolinite were heated with salts in open crucibles, as described in Part 1, but heating was continued for 22 hours. These experiments were initiated to test the effect of the initial hydroxyl content on long-term reactions, in the presence of water vapour, and to study the nature of the products. The second series, carried out in a thermal apparatus under controlled conditions, was designed to confirm that the reactions do, indeed, follow equations (1)-(3) and to find the minimum value of x for which they will still occur.

EXPERIMENTAL

Material

The kaolinite used was from Oneal Pit (supplied by Wards).

Three samples of heated kaolinite were prepared: M(A) : Metakaolinite A - kaolinite heated in air for one hour at $550°C$, M(B) : Metakaolinite B - kaolinite heated in air for one hour at $650°C$, $M^1(C)$ : kaolinite heated up to $950°C$ at $10°C$/min. in the thermal apparatus under a stream of $N_2$.

The salts used were of analar quality.

Procedure

Series 1: Aliquots of kaolinite or metakaolinite were heated with excess KBr, $K_2CO_3$ or KF in open crucibles for 22 hrs at $570°C$ (unless otherwise stated). A clay:salt ratio of 1:3 was found to be convenient. The products were washed with distilled water until, as applicable, no halide was detected in the washings or the infra-red spectra of the solids did not show $CO_3^{2-}$ absorptions. Aliquots of the washed samples were dissolved and analysed by atomic absorption spectroscopy. X-ray powder diffraction patterns and infra-red spectra of KBr disks of the products were recorded. The wave length of the Al $K\alpha$ emission was determined using a Phillips X-ray fluorescence unit with a PET crystal.

Other clay/salt mixtures (e.g. with LiCl or $NaHCO_3$) were also heated in open crucibles but were less thoroughly studied. Chemical analyses were confined to the alkali cations and to aluminium and, in addition, infra-red spectra were recorded.

Series 2: Mixtures of kaolinite or metakaolinite with KBr (1:3 and 1:3.5 respectively) were heated in a Mettler thermoanalytical apparatus a) under a stream of nitrogen, as supplied by Matheson, labelled "prepurified $N_2$", and b) with the same gas predried by passing through zeolite 3A cooled with liquid air, which will be designated "dry $N_2$". The flow rate was 88 ml/min. A uniform heating rate of 10°C/min. was employed and isotherms were run at selected temperatures. In some of the experiments the gases evolved were trapped in liquid air and subsequently extracted with 0.5N NaOH solution at room temperature. The tubing which connects the oven to the traps was also washed with 0.5N NaOH. The solutions obtained from the traps as well as the washings were analysed for potassium and bromine.

Mixtures of $M^1(C)$ with $K_2CO_3$ (1:3) were similarly heated under "dry $N_2$".

RESULTS AND DISCUSSION

Series 1. Mixtures heated in open crucibles

Heating kaolinite in the thermal apparatus showed that dehydroxylation was not complete until 950°C was reached. M(A) and M(B) differ in their hydroxyl content: on heating up to 950°C they liberated 9.5 and 1.2% of water respectively. No further weight loss was observed when samples were maintained at that temperature. $M^1(C)$ lost no water on further heating at 950°C.

Chemical analyses. Complete chemical analyses of kaolinite and of nine heated kaolinite- or metakaolinite-salt mixtures showed that the Al:Si ratio remained constant and that anion exchange occurred only with KF. The amount of K in the products obtained on heating with KBr and $K_2CO_3$ are shown in Table 1.

TABLE 1
Selected properties of reaction products

| Starting material | %K in product | molecular formula deduced | Infra-red absorptions | $cm^{-1}$* | |
|---|---|---|---|---|---|
| Ka+KBr | 14.94 | $2SiO_2.Al_2O_3.0.60\ K_2O$ | 1000 | 705 | |
| M(A)+KBr | 13.75 | $2SiO_2.Al_2O_3.0.56\ K_2O$ | 1000 | 710 | |
| M(B)+KBr | 13.22 | $2SiO_2.Al_2O_3.0.53\ K_2O$ | 1010 | 710 | |
| Ka + $K_2CO_3$ | 22.50 | $2SiO_2.Al_2O_3.0.91\ K_2O$ | 980 | 690 | |
| M(A)+$K_2CO_3$ | 22.90 | $2SiO_2.Al_2O_3.0.93\ K_2O$ | 980 | 690 | |
| M(B)+$K_2CO_3$ | 18.94 | $2SiO_2.Al_2O_3.0.77\ K_2O$ | 990 | 690 | |

Abbreviation: Ka - kaolinite
*The corresponding absorptions for M(A) and M(B) are 1060 and 795 $cm^{-1}$

Infra-red spectra (see below) indicate that the samples do not contain amorphous silica or alumina and that they are not mixtures of metakaolinite and a product containing alkali ions (Part 1). The molecular formulae (Table 1) were therefore deduced on the assumption that the products are composed of a single substance. With KF more K was incorporated (25-26.5%). It was previously shown (Part 1) that heating kaolinite with KF for 30 minutes leads to disintegration of the clay particles. This was attributed to the action of HF liberated in the course of the reaction. On heating for 22 hrs at 570°C products of the approximate composition $Si_2Al_2K_{2.7}O_{0.6}F_{15.5}$ were obtained.

X-ray diffraction and spectroscopic analyses. X-ray powder patterns of kaolinite and metakaolinite A heated with $K_2CO_3$ show a peak at 3.12Å, which corresponds to the strongest reflection of kalsilite. The metakaolinites used and the reaction products obtained with $K_2CO_3$, KBr and KF were amorphous. In Part 1 some X-ray patterns obtained from mixtures of kaolinite with potassium salts were identified as kaliophilite. These patterns are diffuse and cannot be interpreted unambiguously.

The infra-red spectra of metakaolinites A and B are identical and show that no residual kaolinite remained. The frequencies of the broad absorptions centred at 1060 and 795 $cm^{-1}$ (Fig. 1a) are reduced on heating with $K_2CO_3$, KBr or KF, (Fig. 1b,c), but no other absorptions appear in this region.

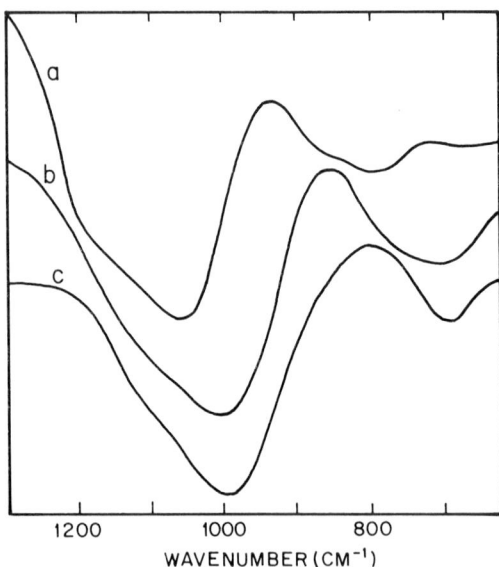

Fig. 1. Infra-red spectrum of KBr disk of a) M(A); b) M(A) + KBr, 570°C, 22 hrs; c) M(A) + $K_2CO_3$, 570°C, 22 hrs.

It was shown in Part 1 that there is a correlation between the amount of alkali ions incorporated into the clay and the position of the bands. This was also observed in the present study. Samples heated with $K_2CO_3$ absorb at lower frequencies than those heated with KBr, in agreement with their K content (Table 1). The spectrum of kaolinite heated with KBr for 6 days at 570°C resembles that of M(B) heated with $K_2CO_3$ for 22 hours at the same temperature. This indicates that the reactions with KBr and $K_2CO_3$ proceed at very different rates, but that the products formed are similar. The spectra resemble those of a fused glass of nepheline composition described by Sahama (1965) or of soda-alumino silicate glasses (Day and Rindone, 1962). Spectra of corresponding potassium alumino silicates were not available, but in view of the structural similarity between kalsilite and nepheline and the resemblance of their infra-red spectra, it seems reasonable to assume that the spectra of the poorly crystallised potassium containing phases resemble those of their sodium analogues.

Samples heated with KF gave rise to relatively sharp bands at 995-985 $cm^{-1}$ and at 695 $cm^{-1}$. An Si-O absorption at about 460 $cm^{-1}$, that appeared in all the other samples and in metakaolinite, was absent, probably due to the exchange of oxygen by fluorine.

The changes in infra-red spectra occurring when kaolinite is converted to metakaolinite and, through a spinel phase, to mullite, have been studied by several investigators. Particular significance was attached to the absorption at 800 $cm^{-1}$, which was taken to indicate the state of coordination of Al. It was inferred that Al in metakaolinite is present in 4 coordination only (Pampuch, 1966; Freund, 1967; Percival et al., 1974) or in 4 and 6 coordination (MacKenzie, 1969). Interpretation of the increase in frequency of the 800 $cm^{-1}$ band observed on heating metakaolinite to higher temperatures also differs. MacKenzie (1969) attributed it to conversion of octahedral to tetrahedral Al; Percival (1974) related it to the presence of "isolated" Al tetrahedra in mullite. The absorption appears at about 800 $cm^{-1}$ in the intermediate spinel phase (Freund,1967), which probably contains 4 and 6 coordinated Al.

On heating with salts of alkali metals the absorption at 800 $cm^{-1}$ is shifted towards lower frequencies and the spectra tend to resemble those of kaolinite, nepheline or $\beta$eucryptite (Kolesova, 1963) with K, Na or Li salts respectively. In all these structures Al is 4 coordinated. For nepheline, the band was shown to be independent of order-disorder effects, i.e. the degree of "isolation" of the tetrahedra (Sahama, 1965). It is also not significantly affected when oxygen is exchanged by fluorine. Comparison of the spectral changes occurring when metakaolinite is heated with and without salts may, with due consideration of the differences in chemical composition, throw some light on the nature of the various phases involved.

The wavelength of the Al K$\alpha$ emission of samples heated with KBr, $K_2CO_3$ or KF is identical, within the limits of experimental error, and $\Delta 2\theta$, the shift from the position of pure Al metal, is less than that found for metakaolinite

(0.090 and 0.081° $2\theta$ respectively). This suggests that the Al-O bond lengths are larger in the K containing compounds than in metakaolinite (Wardle and Brindley, 1971), a concept difficult to reconcile with tetrahedrally coordinated Al in a kalsilite-like arrangement. This requires further investigation.

Inference. Chemical, X-ray and infra-red analyses show that $K_2CO_3$ reacts more readily with both kaolinite and metakaolinite than KBr. Heating with KF leads to fluorination of the samples.

It was found that the reaction of KBr with kaolinite was faster than with M(A), which in turn reacted more rapidly than M(B). With $K_2CO_3$ the hydroxyls retained by M(A) sufficed to render its reaction similar to that of kaolinite. M(B) reacted more slowly with $K_2CO_3$, showing that the rate of reaction (3) is also affected by the presence of water.

Although doubts remain about the nature of the reaction products, it appears that they tend towards a kalsilite-like arrangement. Further study may not only provide information about these products but also about the structure of metakaolinite.

## Series 2. Thermal analysis

Clay-KBr mixtures. KBr was used for these experiments because the weight losses incurred in the course of the reactions are relatively large and can therefore be accurately measured and the gases evolved are conveniently analysed.

Some of the results obtained are summarised in Table 2. In experiments (1) and (2) the gases evolved were trapped and analysed. The amount of potassium in the traps was negligible. The washings from the tubing contained equivalent amounts of potassium and bromide.

On the assumption that the reactions with kaolinite proceed according to equation (1), the value of y was calculated from the amount of bromide found in the traps (column d). Using this value of y, the total weight losses expected due to evolution of $H_2O$ and HBr was derived, which, together with the KBr in the tubing, constitutes the calculated weight loss shown in column (e).

The difference between experiments (1) and (2) lies in the surrounding atmosphere - nitrogen was used as supplied in experiment (1) and "dry nitrogen" was used in experiment (2). The time for which the material was maintained at 950°C was different in the two experiments and the values of y (Table 2) are therefore not directly comparable. The thermogravimetric curves showed that the weight loss below 600°C was similar (229 and 205 mg/g clay respectively), but the weight loss between 600°C and 950°C differed significantly (227 and 108 mg/g clay respectively). This demonstrates that the reaction was independent of the surrounding atmosphere, as long as there were sufficient hydroxyl groups in the clay, but after most of the hydroxyls were lost, i.e. above 600°C, minor amounts of water vapour in the atmosphere, such as occur in commercially pure $N_2$, affected the rate of the reaction. It follows that some of the water in experiment 1 was derived from the atmosphere and the calculated weight loss should be increased accordingly. The correction is small but may explain the discrepancy between the calculated and observed weight loss in experiment 1 (Table 2).

TABLE 2
Thermal analysis of selected samples

| No. | (a) Sample | (b) Gas | (c) Thermal regime °C* | (d) y** calc. from Br$^-$ | (e) Weight loss (mg/g clay) calc*** | (f) Weight loss (mg/g clay) obs. | (g) Amt. of $H_2O$ (mg/g clay) required | (h) Amt. of $H_2O$ (mg/g clay) in clay |
|---|---|---|---|---|---|---|---|---|
| (1) | Ka+KBr 1:3 | $N_2$ | 950 1.67 hrs | 0.71 | 732 | 746 | 99 | 139 |
| (2) | Ka+KBr 1:3 | $N_2$(d)# | 950 0.1 hr | 0.31 | 311 | 314 | 43 | 139 |
| (3) | M(B)+KBr 1:3.5 | $N_2$(d)# | 710 | | | 134 | 15 | 12 |
| (4) | $M^1$(C)+KBr 1:3.5 | $N_2$(d)# | 710 | | | 9 | 1 | - |
| (5) | $M^1$(C)+KBr 1:3.5 | $N_2$ | 710 | | | 37 | 5 | - |

*Heating rate 10°C/min. up to temperature indicated, isotherm as stated;
**y - see equations (1) and (2); *** this includes KBr in tubing; # - "dry nitrogen".

In experiment (3) M(B) was heated with KBr to 710°C, i.e. below the melting point of KBr, under a stream of "dry nitrogen". Neither the traps nor the tubing contained any potassium, showing that at these temperatures no KBr was lost and that the entire weight loss may be attributed to loss of $H_2O$ and HBr. If the reaction proceeds according to equation (2) and is limited by the amount of water in M(B), i.e. x=y, the value of x (or y) corresponding to the observed weight loss may be calculated (column g). Comparison with column (h) shows that this is similar to the amount available from M(B). With $M^1$(C), which is anhydrous, the entire water must be derived from the atmosphere and the overall weight loss may be calculated from the equation

$$2SiO_2 \cdot Al_2O_3 + 2yKBr + (yH_2O) \longrightarrow 2SiO_2 \cdot Al_2O_3 \cdot yK_2O + 2yHBr \qquad (4)$$

i.e. it corresponds to 2yHBr-yO per formula unit of $M^1$(C). In experiment (4) (Table 2) the total weight loss up to 710°C was only 9 mg/g clay, indicating that the reaction with dry KBr under a stream of "dry $N_2$" was almost negligible. The small discrepancy between columns (g) and (h) for experiments (3) and (4) suggests that traces of moisture persisted either in the gas or adsorbed on the clay or in the apparatus. In experiment (5) the same mixture as in experiment (4) was heated in a stream of nitrogen, as supplied. The weight loss was much larger than with "dry nitrogen". The mixture was maintained at 710°C and the rate of the reaction was recorded by DTG. Bubbling the gas through water before it was introduced into the thermal apparatus caused no change in this rate. It appears that the very minor amounts of water present in the nitrogen gas sufficed to sustain the reaction.

Mixtures of $K_2CO_3$ with $M^1(C)$. Weight losses recorded on heating kaolinite or metakaolinite with carbonates are not as informative as those with KBr, because they may be partly due to loss of $CO_2$ from the carbonate. The value of y cannot, therefore, be deduced from the TG curves.

$M^1(C)$ heated with dry $K_2CO_3$ under a stream of "dry $N_2$" showed a low temperature weight loss, which was completed at 130°C and a second, commencing at 490°C, which continued up to 950°C, when the experiment was terminated. Another mixture was heated only up to 700°C. The sample heated up to 950°C had partially melted, the second persisted as a fine powder. Both were washed and infra-red spectra were recorded. The spectrum of the mixture heated up to 700°C was identical with that of metakaolinite, while the sample heated to 950°C gave rise to a spectrum resembling that of kalsilite.

Inference. The reaction between kaolinite and metakaolinite and KBr follows equations (1) and (2) respectively. When the water required for the reaction is derived from the clay, the rate is faster, but the reaction also occurs with very minor amounts of water vapour in the atmosphere. The presence of additional water vapour in the atmosphere does not accelerate the reaction appreciably. The rate becomes negligible when both the reagents and the atmosphere are dry.

The reaction between $M^1(C)$ and $K_2CO_3$ is inhibited by the absence of water below the melting point of the carbonate, but proceeds at higher temperatures.

CONCLUSION

When kaolinite or metakaolinite was heated with $K_2CO_3$, KBr or KF in air, potassium was incorporated without any change in the Si/Al ratio. With KF most of the oxygen was exchanged by fluorine. Samples heated with $K_2CO_3$ gave products resembling kalsilite; with KBr products intermediate between metakaolinite and the kalsilite-like material were obtained.

It was shown that reactions of kaolinite or metakaolinite with KBr, which may be regarded as a prototype of alkali chlorides and bromides, follow equations (1) and (2). They depend upon the presence of water, which may be derived from the clay or from the surrounding atmosphere. In a completely dry atmosphere the reaction is restricted by the amount of hydroxyl in the clay - when all the hydroxyls have been consumed, the reaction ceases. In the presence of trace amounts of water vapour, such as are contained in a stream of commercially "pure" nitrogen, the reaction occurs with a sample heated up to 950°C, which contains no hydroxyl groups, but the rate is faster when some structural hydroxyls are present. The amount of water vapour in the atmosphere, beyond a minimum value, does not seem to affect the rate of the reaction. The presence of water is also essential for reactions of kaolinite with $K_2CO_3$ below its melting point, but molten $K_2CO_3$ reacts with $M^1(C)$ even in the absence of detectable amounts of water, i.e. when x in equation (3) approaches 0. It follows that the mechanism of the reaction with molten $K_2CO_3$ differs from that with alkali halides.

Minor amounts of water are always present in natural systems. It may therefore be concluded that in nature or in industry, whenever kaolinite or metakaolinite is in contact with salts of alkali metals at elevated temperatures, reactions of type (1)-(3) occur. These may also be regarded as models for similar reactions with other reagents.

ACKNOWLEDGEMENTS

A grant from the Israel Commission for Basic Research, National Academy of Science, in partial support of this work, is gratefully acknowledged. The authors also wish to thank Prof. N. Lahav, of The Hebrew University, and Dr. Y. Nathan, Geological Survey of Israel, for critically reading the manuscript.

REFERENCES

Day, D.E. and Rindone, G.E., 1962. Properties of aluminosilicate glasses: Refractive index, density, molar refractivity and IR absorption spectra. J. Amer. Ceram. Soc., 45: 489-496.
Freund, F., 1967. Infrarotspektren von Kaolinit, Metakaolinit und Al-Si-Spinell. Ber. Dtsch. Keram. Ges., 44: 392-397.
Heller-Kallai, L., 1975a. Montmorillonite-alkali halide interaction - a possible mechanism for illitization. Clays and Clay Minerals, 23: 462-467.
Heller-Kallai, L., 1975b. Interaction of montmorillonite with alkali halide. Proc. Int. Clay Conf. Mexico, 361-372.
Heller-Kallai, L., 1978. Reactions of salts with kaolinite at elevated temperatures, Part 1. Clay Minerals, in press.
Kolesova, V.A., 1963. Synthetic and natural eucryptite. IZV. Akad. Nauk. SSSR. Khim. Nauk., 187-190.
MacKenzie, K.J.D., 1969. An infra-red frequency shift method for the determination of the high-temperature phases of aluminosilicate minerals. J. Appl. Chem., 19: 65-67.
Pampuch, R., 1966. Infra-red study of thermal transformations of kaolinite and the structure of metakaolin. Polska Akademia Nauk, Prace Mineralogiczne, 6: 53-70.
Percival, H.J., Duncan, J.F. and Foster, P.K., 1974. Interpretation of the kaolinite-mullite reaction sequence from infra-red absorption spectra. J. Amer. Ceram. Soc., 57: 57-61.
Sahama, Th.G., 1965. Infra-red absorption of nepheline. Comptes Rendue de la Société Géologique de Finlande, 37: 107-117.
Wardle, R. and Brindley, G.H., 1971. The dependence of wavelength of AlK$\alpha$ radiation from alumino-silicates on the Al-O distance. Amer. Mineral., 56: 2123-2128.

CRITICAL ASSESSMENT OF THE JOINT USE OF VARIOUS PHYSICO-CHEMICAL TECHNIQUES IN THE STUDY OF THE THERMAL TRANSFORMATION OF KAOLIN.

B. DELMON, A.J. HERBILLON[*], A.J. LEONARD, M. BULENS[**]
Groupe de Physico-Chimie Minérale et de Catalyse,
Université Catholique de Louvain, Louvain-la-Neuve (Belgium).

ABSTRACT

The communication is based on our recent work on the transformations of kaolinite and shows how the simultaneous use of several techniques and methods can contribute to the elucidation of problems related to the transformations of kaolin at medium and high temperatures. Two main problems will be examined, namely the identification of the various crystalline and noncrystalline phases which form and the identification of the factors which influence reactivity and reaction paths.

INTRODUCTION

Over the last ten years, several members of the Groupe de Physico-Chimie Minérale et de Catalyse have been involved in a reevaluation of the mechanisms of phase transformations occurring when metakaolinite is fired above 900°C. The present paper aims at recalling, in a synoptic form, the major results and the main conclusions of this work. It also includes an assessment of the methodology followed with respect to its possible application to other clay minerals.

Generally speaking, two different types of problems are encountered in such studies. The first and preliminary step deals with an accurate and, when possible, quantitative identification of the phases produced during the firing. If, as is the case when metakaolinite is fired, some transient phases are ill-crystallized, suitable methods of identification should enable their detection. The second type of problem concerns the reaction paths which lead from the transient phases to the final decomposition products and to the parameters determining the reactivity. As we shall see, ill-ordered phases formed by the breakdown of the

---

[*] Section de Physico-Chimie Minérale du Musée Royal de l'Afrique Centrale
[**] Aspirant, Fonds National de la Recherche Scientifique

metakaolinite network exhibit such a high reactivity that the subsequent sequence of transformations can be controlled almost at will. There is no unique thermal reaction pathway beyond the metakaolin stage.

NATURE OF THE PHASES FORMED DURING THE 950°C TRANSFORMATION OF KAOLINITE.

Literature dealing with the identification of the various phases occurring when metakaolinite is fired at about 950°C is rather confusing. The exothermic peak which accompanies the transformation was attributed to quite different phenomena : recrystallization of $\gamma$-alumina, nucleation of mullite, recrystallization of a mixed Si-Al spinel, or recrystallization of silica into high-quartz (Comeforo et al., 1948; Colegrave and Rigby, 1952; Richardson and Wilde, 1952; Roy et al., 1955; Lundin, 1958; Brindley and Nakahira, 1959; Nicholson and Fulrath, 1970 and Chaudhuri, 1977).

In particular, one problem was to decide whether a well-defined reconstructive path was followed by the reaction. The hypothesis of a progressive reconstructive transformation, leading to the growth of mullite, was indeed fascinating and satisfactory in principle. It confirmed ideas of mineralogists on stepwise reaction and ideas of solid state chemists on topochemical reactions or, more generally, on reactions involving only gradual structural changes, all ideas which, even now, are quite valuable. If this hypothesis had been substantiated, it would have given rise to generalizations in many clay systems. Actually, our conclusion about the transformation was quite different (although possibly it was not valid for all kaolinites, especially those forming extremely large crystals (Brindley and Nakahira, 1959)); rather, it emphasized the extremely high reactivity of the metakaolinite lattice and, hence, an inherent lack of selectivity in its transformation.

Using the crystallinity index (C.I.) of Hinckley (1963) as an indicator of crystallinity, we investigated mainly representative kaolinites of medium (Kolloid, Karlovy Vary, C.I. = 0.56) (Lemaître et al., 1976a and b and Bulens and Delmon, 1977b) or high crystallinity (Zettlitz, C.I. = 0.85) (Bulens and Delmon, 1977b and Léonard, 1977).

The properties of the expected phases were not very different: in addition, it was suspected that a large amount of amorphous material, the characterization of which is intrinsically difficult, might be formed. The identification of amorphous phases is possible by R.E.D. (radial electron distribution) density, and, sometimes, I.R. measurements (if a suitable bond gives a characteristic band). Absolute measurements of the physical density have been essential in our work, and were strongly confirmed by R.E.D..

Density measurements

Three different reaction paths had to be examined :
Path A
$$Al_2O_3 \cdot 2SiO_2 \rightarrow \frac{1}{2} SiO_2 \text{(amorph.)} + \frac{6}{32} Si_8Al_{10.67}O_{32} \text{ (mixed spinel)}$$
Path B
$$Al_2O_3 \cdot 2SiO_2 \rightarrow (1-3y)\, \gamma\text{-}Al_2O_3 + y(3Al_2O_3 \cdot 2SiO_2)\text{(mull.)} + x\, SiO_2 \text{ (cristob.)}$$
$$+ (2-2y-x)\, SiO_2 \text{ (amorph.)}$$
Path C
$$Al_2O_3 \cdot 2SiO_2 \rightarrow x\gamma\text{-}Al_2O_3 + y(3Al_2O_3 \cdot 2SiO_2) + [(1-x-3y)Al_2O_3 + (2-2y)SiO_2]\text{(amorph.)}$$

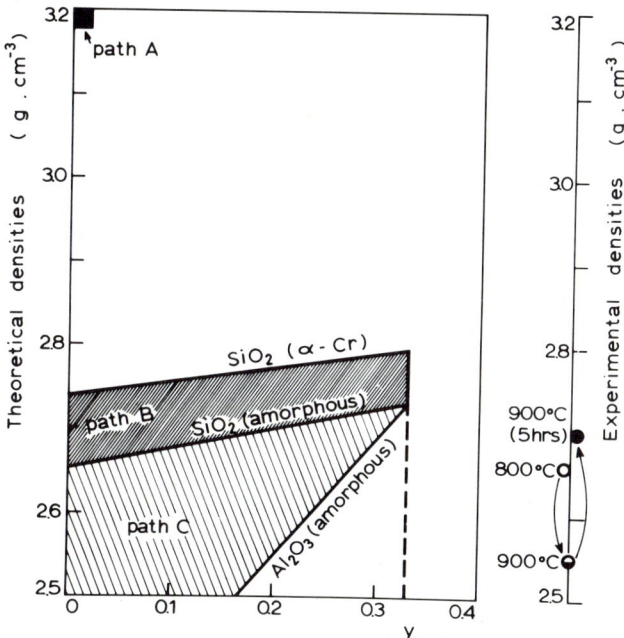

Fig.1. Comparison of experimental density of metakaolinite with theoretical densities calculated assuming different reaction paths. The theoretical density corresponding to the hypothesis of Brindley and Nakahira is the point indicated "path A".
Paths B and C give theoretical densities represented by the shaded areas. In both cases, the lower limit of each area corresponds to the zero value of the respective coefficient x appearing in the corresponding stoechiometric equation. The representative points move towards the upper limit, when the corresponding x increases. This means that the upper left corner of each area corresponds both to complete crystallization of $SiO_2$ to α-cristobalite (path B) and $Al_2O_3$ to $\gamma\text{-}Al_2O_3$ (path C) respectively and to complete segregation of $Al_2O_3$ and $SiO_2$ (y=0).
Experimental densities correspond to the following firing cycles:
○  heating to 800°C (5°C/min) and immediate cooling (30°C/min);
◐  heating to 900°C (5°C/min) and immediate cooling (30°C/min);
●  heating to 900°C (5°C/min), 5 hrs at 900°C, and cooling (30°C/min).

Path A gives a product of density 3.20 (Lemaître et al., 1976b) (or even higher, if $SiO_2$ is crystallized). This density is obtained on the basis of the parameters of the mixed spinel indicated by Brindley and Nakahira (a = 7.886 Å) and its composition (mol.weight : 1024). Actual values for metakaolinite fired at 800°C (short time), 900°C (short time) or 900°C (5hrs) are 2.67, 2.56 and 2.70, respectively. These results are completely incompatible with the hypothesis that a mixed Si-Al spinel phase forms.

The density of the products in reactions following paths B and C depends on the numerical values of x and y. Figure 1 (Lemaître, 1976) indicates the possible range of variation of density for various mechanisms. If quartz is formed, instead of cristobalite, in a reaction similar to path B, densities beyond the upper limit of the B domain would be obtained. It is clear that the only acceptable hypothesis is C, which implies that the reaction leads to the extensive formation

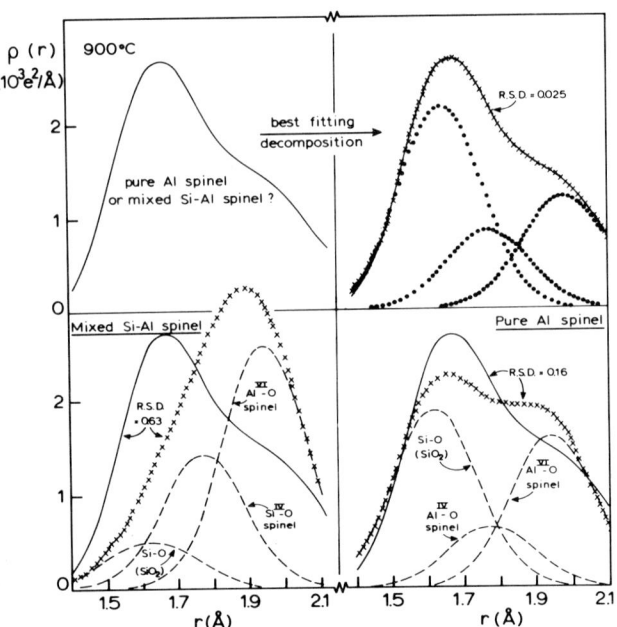

Fig.2. Comparison of the experimental radial electron distribution (R.E.D.) curve (upper left) and the components fitted to it (upper right). Zettlitz metakaolinite (Czechoslovakia, C.I. = 0.85) heated at 900°C for 24 hrs. Curves corresponding to the mixed Si-Al spinel hypothesis (lower left) and to the hypothesis of a segregation of $SiO_2$ and Al spinel (lower right). Relative Standard Deviation (R.S.D.) = $\Sigma (D_{obs.} - D_{calc.})^2 / \Sigma D_{obs.}^2$. For the decomposition, the only constraints imposed to the computer were the approximate position of the maximum of the components and their half band width (0.273 Å) related to the intrinsic breadth of the convolution product of the electron distribution of the concerned atoms.

of amorphous material, together with lesser amounts of $\gamma$-$Al_2O_3$ and mullite. Our results indicate that the proportion of mullite in the crystalline phases does not exceed y ≃ 0.2, but its presence cannot be excluded. The important point is that <u>silica cannot be extensively crystallized</u>, if the overall average density is to be kept within the range of the observed values. This extensive formation of amorphous material is certainly the most conspicuous result. It should be noted that the density <u>falls below that of metakaolinite</u> (2.56 in comparison with 2.67 for metakaolinite), which suggests that a segregation of silica and alumina in an amorphous state may be the very first step of the reaction sequence following the metakaolin stage.

To confirm the validity of this conclusion, it was crucial to evaluate the accuracy of the density measurements. Complementary measurements, including pore-size distribution (Lemaître et al., 1976b), showed that, even assuming a large inaccuracy (100 % error) on the evaluation of the first layer of adsorbed molecules, the above conclusions remained unaffected.

R.E.D. measurements

Radial Electron Distribution measurements confirm that the formation of a mixed Si-Al spinel must be ruled out (Léonard, 1977). Further evidence is provided by figure 2. It is not possible to reconstruct the experimental R.E.D. profile in the region 1.4 - 2.1 Å, when assuming the presence of a mixed Si-Al spinel. Conversely, the figure shows that R.E.D. measurements very satisfactorily support the hypothesis of an extensive segregation of $SiO_2$ and $Al_2O_3$ (without indicating the degree of crystallinity of these phases).

Discussion

The two important conclusions of our work summarized above are :
(1) the absence of any mixed Si-Al spinel and
(2) the occurrence of extensive segregation of silica and alumina. Although the formation of some primary mullite cannot be completely excluded, segregation is the major process. This, of course, implies that later on in the reaction sequence mullite must be formed by <u>recombination</u> of silica and alumina. Actually, it was observed (Lemaître et al., 1976a) that segregated silica disappeared upon further heating at 900°C.

It was also observed that the crystallinity of the kaolinite influences the extent of the segregation process. Kaolinites with low crystallinity suffer more intense segregation (Bulens and Delmon, 1977b). In our explanation, this merely means that the metakaolinite framework inherited from poorly ordered kaolinites has a still higher free energy content and can therefore segregate more easily.

REACTION PATHS IN THE 950°C TRANSFORMATION OF METAKAOLINITE

The above results suggest that the transformation of metakaolinite is intrinsically nonselective, but we had no complete proof of this. A real proof would be that the reaction can be directed at will in two or several directions. It was thus logical to try mineralizers, in order to see whether they would promote certain reaction paths at the expense of others. On the other hand, the only hope to analyze successfully the mechanism of each individual reaction (e.g. direct reaction to mullite, or reaction to $\gamma$-$Al_2O_3$ + $SiO_2$) was to isolate it, because there is no practical possibility of studying a mixed phenomenon. We speculated that mineralizers could help us <u>isolate</u> individual reaction paths or, at least, promote one with respect to another. A third reason is also of a methodological nature. When several phases are present in a mixture, one might find clues as to the nature of these phases if it is possible to obtain mixtures of widely different phase compositions. It was expected that this information could be provided by using samples prepared at <u>identical temperatures</u>, but containing different proportions of phases because they had been mineralized.

Induced nucleation

If a reaction is intrinsically nonselective, it should be very sensitive to artificial nucleation, namely to the presence of particles of the product of one reaction path, which usually act as nuclei for the growth of that phase. Addition of mullite was shown to increase the formation of mullite (Bulens et al., 1974). This was a first proof. More subtle experiments, involving comparative firing cycles, intended to promote natural nucleation of mullite or $\gamma$-alumina, support this picture (Bulens and Delmon, 1977a) : $\gamma$-$Al_2O_3$ additions promote $\gamma$-$Al_2O_3$ formation just as mullite promotes further mullite formation.

Systematic studies using reactions oriented by mineralizers.

A logical continuation of the above ideas was to add small amounts of substances known to promote the formation of certain phases. Magnesium oxide added to an aluminium containing reactant was an excellent candidate for nucleating spinels, (first $MgAl_2O_4$ and subsequently, $\gamma$-$Al_2O_3$). The idea turned out to give the expected result. Generalizing, we systematically investigated many mineralizers. Calcium oxide was found to promote mullite formation. We thus arrived at the following reaction scheme (Lemaître et al., 1976a and b and Bulens and Delmon 1977a).

$$\text{metakaolinite} \begin{cases} \xrightarrow{+ \text{ CaO}} \text{mullite} + SiO_2 \\ \text{pure} \\ \xrightarrow{+ \text{ MgO}} \gamma\text{-}Al_2O_3 + SiO_2 \end{cases}$$

Although they are not the only mineralizers to impart selectivity to the reaction, CaO and MgO can be regarded as typical.

In the case of the transformation of metakaolinite, the detection of supposed effects on selectivity cannot rest solely on the detection of phases which are readily characterized by X-ray techniques. These phases develop only at high temperature or upon prolonged heating. It is necessary to detect the changes in phase composition at early stages of the transformation.

R.E.D. measurements (Lemaître et al., 1976b) give clear indications that MgO promotes a spinel phase even in poorly crystallized samples. The intensity of the effect exceeds that which could be brought about by $MgAl_2O_4$, suggesting that $MgAl_2O_4$ nucleates $\gamma$-$Al_2O_3$.

The displacement of the Al $K_\alpha$ line, which varies with the coordination of Al atoms, also gives information about the phases formed, even when they are poorly crystallized. The information, as in the case of R.E.D., is averaged, and weights all Al atoms.

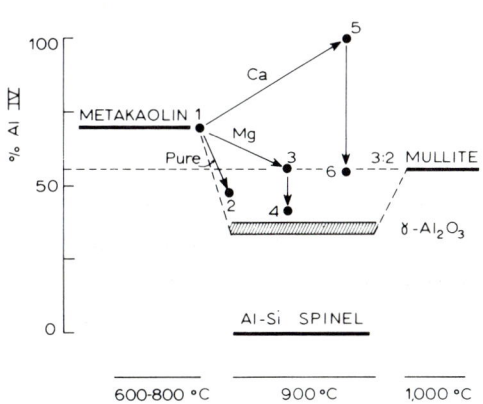

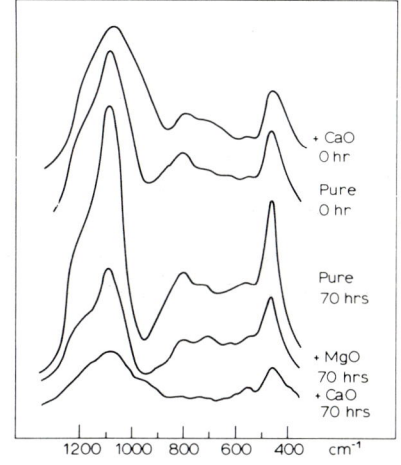

Fig.3. Variation upon firing of $Al^{IV}$ content, determined by the shift of the Al $K_\alpha$ line.
1, pure kaolinite, 800°C, 24 hrs.
3, 5, respectively, mineralized with MgO or CaO, 900°C, 0 hr.
2, 4, 6, respectively, pure, mineralized with MgO or CaO, 900°C, 124 hrs.

Fig.4. I.R. spectra of pure or mineralized metakaolinite fired for different times at 900°C.

Figure 3 (Bulens et al., 1978) symbolizes the variation of the coordination of Al, as a function of firing temperature and time, for metakaolin or MgO and CaO mineralized preparations. Clearly, CaO prevents the formation of octahedral Al, whereas MgO favors it. With pure and MgO mineralized samples, the product contains more Al in octahedral coordination than can be present in mullite. This series of measurements incidentally confirms again that no mixed Si-Al spinel forms.

Discussion

The above results indicate how powerful nucleation is for directing the reaction. This is not surprising; the metakaolinite structure is extremely unstable around 950°C. Any small effect, (even the small free energy contribution of an extraneous nucleus), can trigger the reaction in a given direction. Following this idea, one could speculate that the main energy parameter is the free energy stored in the metakaolinite lattice, and that the formation of products makes little difference to that energy. This conclusion seems substantiated by the modest variation of the area of the 950°C D.T.A. peak with the nature of the mineralizers (and, hence, the prefered reaction path) (Lemaître et al., 1976a). Along these lines, one could expect the crystallinity of the original kaolinite to influence strongly its D.T.A. behaviour. Such an effect is actually observed (Bulens and Delmon, 1977b), but the results are not consistent.

The reaction path leading to alumina formation is relatively clear, in the sense that it is promoted when alumina formation is made easy. The other reaction path is also clear as far as the effect of extraneous mullite is concerned. But the role of CaO is less obvious. Figure 3 suggests that it prevents the formation of octahedral Al. I.R. measurements suggest the formation of a silicate : bands appear in the 900 - 1,000 $cm^{-1}$ region in CaO mineralized samples (Figure 4) (Bulens, 1976), which are not visible in pure or MgO mineralized sample (where only the broad 1,000 - 1,150 $cm^{-1}$ band, corresponding to three-dimensional silica is observed). In particular, CaO is known to promote the formation of gehlenite ($2CaO.Al_2O_3.SiO_2$) (De Keyser, 1954; Biehl, 1962; Segnit and Gelb, 1971 and Lemaître, 1976).

GENERAL DISCUSSION

It is not surprising that studies dealing with clay thermal transformations have first focussed on the crystalline products produced and therefore that, when one tried to establish reaction pathways, the specific structures of these transient crystalline products have provided the main basis of the argument (Brett et al., 1970). When kaolinite is fired, however, the metakaolinite phase, which persists over a broad range of temperatures, finally decomposes into poorly ordered phases. The existence and the high reactivity of these phases appear to be the two steppingstones from which the further reaction pathways stem. Recombination of the segregated phases will take place along ways which are going to be very dependent on specific conditions. In other words, the high reactivity of the transient segregated phases (which is in line with their "amorphous" nature) allows the system to be "manipulated". From a methodological point of view, this characteristic in turn implies that the different steps of the reaction can be isolated and therefore better analyzed. Other than in kaolin, few attempts have been made to evaluate the extent to which amorphous, reactive, and possibly segregated phases are also produced during the thermal transformations of other clay minerals. Freund (1974) reports that a "meta"-chrysotile phase occurs during the thermal transformation of 1:1 Mg-bearing clay minerals. Similarly Brindley and Hayami (1965) show that amorphous and segregated silica is an essential constituent in the reaction pathway leading from serpentine to enstatite. In Ni-bearing serpentines (Pham Thi Hang and Brindley, 1973) the disappearance of its characteristic X-R.peaks and the crystallization of high temperature phases are separated by about 200°C. The low specific gravity measurements, reported by Grim (1968) for several clay minerals, also suggest this point. I.R. measurements also indicate the presence of large amounts of amorphous phases. This would imply that the commonly reported crystalline phases in these systems would only represent a small portion of the reaction products. If such disordered phases are formed, the experiments reported above suggest that they could be detected by appropriate methods including those outlined in this paper.

From a more practical point of view, the consequence of that is that the reaction can only be directed at the very early stages of the crystalline or pseudo-crystalline framework breakdown. The very efficient role of mineralizers, especially when they are well dispersed, can be explained by the fact that they are able to direct the reaction, during the whole succession of transformation.

REFERENCES

Biehl, N., 1962. Examination and determination of new mineral formation during firing of low temperature hydrates. Silkat. Tech., 13:433-438.
Brett, N.H., McKenzie, K.J.D. and Sharp, J.H., 1970. The thermal decomposition of hydrous layer silicates and their related hydroxides. Quart. Rev., XXIV(2):185-207.
Brindley, G.W. and Hayami, R., 1965. Mechanism of formation of forsterite and enstatite from serpentine. Min. Mag., 35:189-195.
Brindley, G.W. and Nakahira, M., 1959. The kaolinite-mullite reaction series, I, II, III. J. Amer. Ceram. Soc., 42:311-324.
Bulens, M.,1976. Study of the reactivity factors controlling thermal transformations in metakaolinite. Thesis, Université Catholique de Louvain, Louvain-la-Neuve.
Bulens, M. and Delmon, B., 1977a. Kinetic control of the formation of the high temperature phases in the kaolinite-mullite reaction sequence. Bull. Soc. Chim. Belg., 86:413-418.
Bulens, M. and Delmon, B., 1977b. The exothermic reaction of metakaolinite in the presence of mineralizers. Influence of crystallinity. Clays and Clay Min., 25:271-277.
Bulens, M., Lemaître,J. and Delmon, B.,1974. Mise en évidence du rôle de la germination dans la formation de mullite à partir de kaolinite. C.R. Acad. Sci., Paris, Sér. C, 279:275-277.
Bulens, M., Léonard, A.J. and Delmon, B., 1978. Spectroscopic investigations of the kaolinite-mullite reaction sequence. J. Amer. Ceram. Soc., 61:81-84.

Chaudhuri, S.P., 1977. Induced mullitization of kaolinite : a review. Trans. J. Brit. Ceram. Soc., 76:113-119.

Colegrave, E.B. and Rigby, G.R., 1952. Decomposition of kaolinite by heat. Trans. Brit. Ceram. Soc., 51:355-367.

Comeforo, J.E., Fischer, R.B. and Bradley, W.F., 1948. Mullitization of kaolinite. J. Amer. Ceram. Soc., 31:254-259.

De Keyser, W.L., 1954. Réactions à l'état solide dans le système ternaire $SiO_2$, CaO, $Al_2O_3$. Bull. Soc. Chim. Belg.,63:40-58.

Freund, F., 1974. Ceramics and thermal transformations of minerals. In : V.C. Farmer (Editor), The infrared spectra of minerals. Mineralogical Society, London, Chap. 20, p.478.

Grim, R.E., 1968. Clay Mineralogy. Mc Graw-Hill, $2^{nd}$ Ed., Tab. 9-1, p.308.

Hinckley, D.N., 1963. Variability in "crystallinity" values among the kaolin deposits of the coastal plain of Georgia and South Carolina. In : E. Ingerson (Editor), Proc. $11^{th}$ Nat. Conf. on Clays and Clay Min.. Clays and Clay Min., Pergamon Press, Oxford, pp. 229-235.

Lemaître, J., 1976. Durcissement des céramiques à base d'argile. Aspects chimiques, mécaniques et texturaux. Thesis, Université Catholique de Louvain, Louvain-la-Neuve.

Lemaître, J., Bulens, M. and Delmon, B., 1976a. Influence of mineralizers on the 950°C exothermic reaction of metakaolinite. In : S.W. Bailey (Editor), Proc. Int. Clay Conf., Appl. Publ., Wilmette, Ill., pp.539-544.

Lemaître, J.,Léonard, A.J. and Delmon, B., 1976b. The sequence of phases in the 900-1,050°C transformation of metakaolinite. In : S.W. Bailey (Editor), Proc. Int. Clay Conf., Appl. Publ., Wilmette, Ill., pp. 545-552.

Léonard, A.J., 1977. Structural analysis of the transition phases in the kaolinite-mullite thermal sequence. J. Amer. Ceram. Soc., 60:37-43.

Lundin, S.T., 1958. Formation of mullite. Geol. Föreningens Förhand., 80:458-480.

Nicholson, P.S. and Fulrath, R.M., 1970. Differential thermal calorimetric properties of kaolinite. J. Amer. Ceram. Soc., 53:237-240.

Pham Thi Hang and Brindley, G.W., 1973. The nature of garnierites. III. Thermal transformations. Clays and Clay Min., 21:51-57.

Richardson, H.M. and Wilde, F.G., 1952. X-ray study of the crystalline phases that occur in fired clays. Trans. Brit. Ceram. Soc., 51:387-400.

Roy, R., Roy, D.M. and Francis, E.E., 1955. New data on thermal decomposition of kaolinite and halloysite. J. Amer. Ceram. Soc., 38:198-205.

Segnit, E.R. and Gelb, T., 1971. Kaolinite-$CaCO_3$, kaolinite-MgO and kaolinite-ZnO reactions below 1,000°C. J. Aust. Ceram. Soc., 7:1-6.

OTHER PAPERS PRESENTED IN SECTION 7

PAPER-COATING KAOLINITES OF AUSTRALIA

P.J. Darragh and A.J. Gaskin (Division of Mineralogy, CSIRO, Private Bag, P.O., Wembley, Western Australia 6014.)

THE PROPERTIES OF THE CAPE KAOLINS OF SOUTH AFRICA

R.O. Heckroodt (Department of Metallurgy and Materials Science, University of Cape Town, Rondebosch, South Africa.)

EFFECT OF REPEATED POTASSIUM ACETATE INTERCALATION ON A KAOLIN CLAY: ROLLED FORMS OF KAOLINITE

Pérsio de Souza Santos[1,2], Helena de Souza Santos[3], Adriana M.V. Avezedo[2] and Gilberto M. Figueiredo[2] ([1] Dpto. Engenharia Química, Escola Politécnica, Universidade de São Paulo, Brazil. [2] Instituto de Pesquisas Tecnológicas, São Paulo, Brazil. [3] Instituto de Física, Universidade de São Paulo, Brazil.)

SOME FACTORS INFLUENCING THE ISOELECTRIC POINT OF THE EDGE SURFACE OF KAOLINITE

H.M.M. Diz and B. Rand (Department of Ceramics, Glasses and Polymers, University of Sheffield, England.)

STUDY OF THE 00$\ell$ REFLECTIONS OF THE PARTIALLY DISORDERED KAOLINITES ON AN ABSOLUTE SCALE

J. Ben Brahim[1], C.H. Pons[2], C. Tchoubar[2] ([1] Université de Tunis, ENS, Tunisie; C.R.S.O.C.I., CNRS, Orleans, France. [2] Laboratoire de Cristallographie, Université d'Orleans, C.R.S.O.C.I., CNRS, Orleans, France.)

THE EFFECT OF IRON ON THE PHYSICOCHEMICAL, CRYSTALLOGRAPHIC AND MORPHOLOGICAL PROPERTIES OF KAOLINS

O. Lietard[1], J.M. Cases[2] and J. Yvon[2] ([1] Union Minérale, Chemin de Halage, 60460 Precy, France. [2] E.N.S.G., B.P. 452, 54000 Nancy, France.)

TONSTEINS, FLINT CLAYS AND THE KAOLINITE CLAYROCK FACIES

F.C. Loughnan (School of Applied Geology, University of New South Wales, Box 1 P.O., Kensington, Australia 2033.)

KAOLINITIC BENTONITE PARTINGS (TONSTEIN) IN COALS OF THE WESTERN UNITED STATES

Bruce F. Bohor (U.S. Geological Survey, Box 25046, MS 972, Denver, CO 80225, USA.)

THE KAOLIN AND ALUNITE MINERALIZATIONS OF LATIUM, ITALY

G. Lombardi[1] and P. Mattias[2] ([1] Istituto di Mineralogia e Petrografia, Università degli Studi di Roma, Città Universitaria, Rome, Italy. [2] Istituto di Geologia Applicata della Facoltà di Ingegneria, Università degli Studi di Roma, Via Eudossiana, Rome, Italy.)

GEOCHEMICAL STUDY OF TRACE ELEMENTS IN KAOLIN DEPOSITS OF DIFFERENT GENETIC TYPE

A.V. Dangić (Faculty of Mining and Geology, Djušina 7, Belgrade, Jugoslavia.)

ROLE OF THE NATURE AND MODE OF INTRODUCTION OF MINERALIZERS ON SOME PRACTICAL CHARACTERISTICS OF FIRED CLAY BODIES

J. Lemaitre, R. Dervaux, B. Delmon (Groupe de Physico-Chimie Minérale et de Catalyse, Université Catholique de Louvain, Place Croix du Sud, 1, 1348 Louvain-la-Neuve, Belgium.)

DIFFERENTIATION OF HYDROTHERMAL KAOLINITE FROM WEATHERED KAOLINITE BY IR SPECTRA IN THE OH REGION

E. Kato and S. Kanaoka (Government Industrial Research Institute, Nagoya, Hirate-cho, Kita-ku, Nagoya, Japan.)

MÖSSBAUER STUDIES OF A FERROUS IRON-DOPED SYNTHETIC KAOLIN

A.H. Cuttler (Department of Mathematical Sciences, Plymouth Polytechnic, Drake Circus, Plymouth PL4 8AA, Devon, England.)

MINERALOGY OF SEDIMENTARY KAOLIN DEPOSITS AROUND NAGOYA WITH SPECIAL REFERENCE TO THE GENESIS OF KAOLINITE AND HALLOYSITE

Keinosuke Nagasawa[1] and Hiroshi Shimizu[2] ([1] Geoscience Institute, Shizuoka University, Oya, Shizuoka, Japan. [2] Department of Earth Sciences, Kobe University, Nada, Kobe, Japan.)

A STUDY OF CHROMIUM-BEARING KAOLINITE FROM TESLIC, YUGOSLAVIA

Zoran Maksimovic[1], J.L. White[2], and M. Logar[1] ([1] Faculty of Mining and Geology, University of Belgrade, Belgrade, Yugoslavia. [2] Department of Agronomy, Purdue University, West Lafayette, Indiana, USA.)

WEATHERING OF LABRADORITE TO KAOLIN MINERALS IN AN ANORTHOSITE IN SOUTHERN BRAZIL

M.L.L. Formoso and D. Pintaude (Department of Geology, Federal University of Rio Grande du Sul, Porto Alegre, Brazil.)

EFFECT OF THERMAL PRETREATMENT ON THE FORMATION OF INTERCALATION COMPLEXES OF SOME KAOLIN MINERALS

J. Pascual Cosp, A. Justo Erbez, J. Poyato Ferrera and J.L. Pérez Rodriguez (Centro de Edafología y Biología Aplicada del Cuarto, C.S.I.C., Apartado 1052, Sevilla, Spain.)

LOOKING FOR CLAY-BASED PRIMARY LIFE

A.G. Cairns-Smith (Department of Chemistry, University of Glasgow, Glasgow, G12 8QQ, Scotland.)

AUTHOR INDEX

Abbott, B.M. 334
Abed, A.M. 333
Alcover, J.F. 95
Alexander, E. 333
Almon, W.R. 236, 333, 334
Anderson, W.L. 75
Andrews, R.H. 447
Annabi-Bergaya, F. 235
Aoki, M. 619
Appelo, C.A.J. 95
Aragón de la Cruz, F. 96
Arakawa, Y. 424
Ashby, D.A. 311
Avezedo, A.M.V. 649

Bain, D.C. 65
Bain, J.A. 437, 487
Basham, I.R. 334, 423
Belviso, R. 334
Ben Brahim, J. 96, 649
Berrier, J. 236
Besson, G. 45, 95
Beyme, B. 177
Bish, D.L. 577
Bjørlykke, K. 281
Bland, D.J. 301
Bohor, B.F. 649
Bond, R.D. 423
Breeuwsma, A. 333
Brindley, G.W. 423
Brunner, P. 235
Brusewitz, A.M. 333
Bulens, M. 639
Burchill, S. 236

Cairns-Smith, A.G. 650
de la Calle, C. 37, 95

Canesson, P. 217
Carbajal, H. 236
Carvalho, A. 424
Casal, B. 236
Cases, J.M. 235, 649
Cebula, D.J. 111
Cecconi, S. 235
Chaussidon, J. 141
Cherubini, C. 334
Childs, C.W. 555, 577
Chilingar, G.V. 96
Choy, J.H. 577
Chukhrov, F.V. 55
Churchman, G.J. 555
Clark, D.R. 578
Cole, W.F. 424
Copin, E. 95
Cornejo, J. 578
Cotecchia, V. 334
Cremers, A. 207, 237
Cruz(-Cumplido), M.I. 217, 235
Cuttler, A.H. 650

Dangić, A.V. 650
Daniele, E. 235
Darragh, P.J. 591, 649
Dayal, R. 423
Delmon, B. 639, 650
Dervaux, R. 650
Dietrich, D. 334
Dillard, J.G. 153, 237
Diz, H.M.M. 649
Downes, C.J. 577
le Dred, R. 236
Dristas, J.A. 96
Drits, V.A. 55

Eberl, D. 375, 423
Ermilova, L.P. 55
Estéoule, J. 291
Estéoule-Choux, J. 291

Fan, P.F. 369
Farmer, V.C. 96, 547, 578
Federico, A. 334
Feldkamp, J.R. 187
Fernandez, M. 27
Fernandez Alvarez, T. 236
Fernandez-Caldas, E. 424
Figueiredo, G.M. 649
Fitz, S. 235
Fitzpatrick, R.W. 577
Formoso, M.L.L. 650
Foscolos, A.E. 261
Fraser, A.R. 547, 578
Frenkel, M. 629
Freund, F. 577
Frey, E. 131
Fripiat, J.J. 235
de la Fuente, C. 487

García Ramos, G. 334, 487
Gaskin, A.J. 591, 649
Gatineau, L. 95, 235
Gaultier, J.P. 45, 167
Gibson, J.A. 447
Giese, R.F., Jr. 95
Glaeser, R. 37
Goilo, E.A. 55
Gonçalves, N.M.M. 424
González-García, F. 334, 487
Goodman, B.A. 65, 555
Gorshkov, A.I. 55
Gunal, A. 333
Güzel, N. 423

Hall, P.L. 121
Harder, H. 423
Hayes, M.H.B. 121, 236

Heckroodt, R.O. 601, 649
Hein, J.R. 333
Heller-Kallai, L. 95, 629
Herbillon, A.J. 423, 577, 639
Hermosín Gaviño, M.C. 227
Highley, D.E. 301, 437
Holm, C. 235
Homshaw, L.G. 141
Honda, S. 424
Horváth, I. 96
Huang, P.M. 527, 577
Huertas, F. 334
Huff, W.D. 333

Inoue, A. 235

Jackson, M.L. 517
Jacobs, P.S. 487
Jones, B.F. 333
Jørgensen, P. 281
Justo Erbez, A. 650

Kahr, G. 487
Kalt, A. 509
Kanaoka, S. 650
Kanaris-Sotiriou, R. 334
Karlsson, W. 281
Kato, E. 650
Kayama, I. 423, 424
Keller, W.D. 581
Konta, J. 609
Koppelman, M.H. 153, 237
Kraus, I. 333
Kristmannsdóttir, H. 359
Kromer, H. 333
Kwong, K.F.N.G. 527, 577

Lagaly, G. 131, 235
Lancucki, C.J. 424
Landu, M. 577
Latouche, C. 271
Lehmann, R. 237

Léonard, A.  639
Lemaitre, J.  650
Letellier, M.  235
Lewis, D.G.  577
Lietard, O.  649
Linares, J.  334
Lippmann, F.  424
Logar, M.  650
Lombardi, G.  649
Louail, J.  291
Loughnan, F.C.  591, 649
Lüddeke, H.  237

McCreery, R.A.  423
McHardy, W.J.  424
Mackenzie, R.C.  1, 424
Maes, A.  207, 236, 237
Maksimović, Z.  650
Mamy, J.  45, 167
Martinez, S.  487
Mashhady, A.S.  424
Matsui, N.  423
Mattias, P.  649
Melfi, A.J.  424
Mendelovici, E.  578
Mesa López-Colmenar, J.  487
Mestdagh, M.M.  423
Meunier, A.  405
Meyer, H.  237
Minato, H.  235, 619
Mingelgrin, U.  236
Mittereder, H.  577
Morgan, D.J.  301, 334, 423
Mortland, M.M.  237
Müller-Vonmoos, M.  487
Murray, L.J.  601
Muter, R.B.  96

Nagasawa, K.  650
Nakai, M.  578
Nishiyama, T.  85

Okuma, M.  424
Oinuma, K.  85
Otsuka, R.  96

de Pablo Galán, L.  475, 487
Pascual Cosp, J.  650
Paterson, E.  501, 577
Pearson, M.J.  311, 333
Pédro, G.  235
Peigneur, P.  207
Pekenc, E.  236
Pelgrims, F.  237
Pérati, B.  509
Pérez-Rodríguez, J.L.  227, 334, 487, 650
Pézérat, H.  37, 95
Pinnavaia, T.J.  237
Pintaude, D.  650
Pirlet, H.  323
Plançon, A.  45
Poinsignon, C.  235
Poncelet, G.  487
Pons, C.H.  95, 96, 649
Powell, T.G.  261
Poyato Ferrera, J.  650
Pozzuoli, A.  334
del Prete, M.  334

Rand, B.  236, 649
Rasquin, E.  237
Rausell-Colom, J.A.  27
Rautureau, M.  95
Reda, M.  424
von Reichenbach, H. Graf  177
Rieke, H.H.  96
Rioche, D.  465
Rimsaite, J.  567
Ristori, G.G.  235
Roaldset, E.  334
Robert, M.  236, 375
Rodrique, L.  577
Rodriguez-Hernandez, C.M.  424

Rodriguez-Pascual, C. 424
Rønningsland, T.M. 334
Rosenqvist, I.Th. 334
Ross, C.R. 333
Ross, D.K. 121
Roth, C.B. 236
le Roux, J. 577
Rozenson, I. 95
Rudnitskaya, E.S. 55
Ruiz Amil, A. 96
Ruiz-Hitzky, E. 235, 236

Saehr, D. 236
Saltzman, S. 236
Sánchez Camazano, M. 237
Sánchez Martín, M.J. 237
Sanz, J. 95, 235
Sarkisyan, S.G. 333
Sato, M. 85
Sayin, M. 177
Scafe, D. 487
Scandone, P. 334
Schoonheydt, R.A. 237
Schwertmann, U. 491, 577
Scott, A.D. 17
Serna, C.J. 197
Serratosa, J.M. 27, 99
Seto, H. 235
Shaw, I.M. 447
Shimizu, H. 650
Siffert, B. 337, 465
Soggetti, F. 334
de Souza Santos, H. 96, 649
de Souza Santos, P. 427, 649
Spears, D.A. 334
Środoń, J. 251
Stacey, F.R. 487
Stefanovits, P. 349
Stepkowska, E.T. 457
Stucki, J.W. 75
Stul, M. 223, 235

Sudo, T. 96, 241
Suquet, H. 95,
Suzuki, Y. 424

Tait, J.M. 577
Tan, K.H. 423
Tauler, E. 487
Taylor, R.M. 423, 577
Tazaki, K. 415
Tchoubar, C. 45, 95, 96, 649
Tchoubar, D. 95, 96
Tejedor-Tejedor, I. 501
Tessier, D. 235, 236
Thomas, R.K. 111
Thorez, J. 323
Traynor, M.F. 237
Tsutsumi, S. 96
Tuck, J.J. 121

Uytterhoeven, J.B. 235, 236, 237, 487

Valero-Saez, A. 334
Varju, E.M. 349
Van Damme, H. 217
Van Leemput, L. 235, 236
VanScoyoc, G.E. 197
Velde, B. 95, 333, 395, 405
Velghe, F. 237
Veneau, G. 375
Veniale, F. 334
Violante, A. 517
Vollset, J. 281

Wada, K. 537
Weir, A.H. 333, 395
Weiss, A. 235, 237, 577
Wells, N. 577
Wengeler, H. 577
Wey, R. 236, 509
White, J.L. 187, 650
White, J.W. 111
Whitney, G. 423

Wilson, M.H. 423, 424

Yada, K. 96
Yamanaka, S. 423
Yariv, S. 578
Yoshinaga, N. 578
Youssef, A.F. 17
Yucel, A. 95, 96
Yvon, J. 649

Zvyagin, B.B. 55
Zkhus, I.D. 333

SUBJECT INDEX

This index is compiled from key words drawn from titles and abstracts, and refers to the first page of the paper concerned. Underlining indicates papers listed by title only, abstracts of which are available: see Preface.

Accessory minerals  491
Acidity, surface  187, 235
Adsorption (see also Water)
  applications  427
  of alcohols  197, 235
  of alkanes  236
  of alkines  237
  of ammonia  197
  of bases  187
  of crown ethers  236
  of heavy metals  207, 237, 427
  of isonitriles  237
  of organic acids  333
  of pesticides  187, 227, 236, 237
  of polyelectrolytes  465
  of polyethylene glycol  236
  of polyamine complexes  207, 237
  of porphyrins  217
  on palygorskite and sepiolite  99, 197
Akaganeite  578
Alcohol-smectite complexes  235
Alkanes on clays  236
Alkines on clays  237
Allophane  415, 491, 537, 577, 578
Allunite  649
Alumina from kaolin  487
Aluminium hydroxides  342, 491, 517, 527, 577, 578
Aluminosilicate synthesis  337, 549, 578
Amine complexes
  in interstratified smectite  96
  in phyllodisilicate  509
  in smectite  207, 237

Ammonium, quaternary, in smectite  236
Amorphous minerals  415, 491, 527, 537, 567, 578, 639
Analysis  333, 447
Applications  427, 437, 601
Arabia  333, 424
Archeology  2
Atlantic Ocean  271

Basalt alteration  359, 369, 424
Bayerite  517
Beidellite  131, 395
Belgium  323
Beneficiation  427, 437
Bentonite: see K-bentonite and Montmorillonite
Biotites
  iron oxidation in  17, 177
  K-depletion  17
  octahedral ordering  27
  OH vibrations  27
  weathering  405, 424
Bipyridyl complexes  237
Brazil  424, 650
Bricks  2
Britain  301, 311, 334, 423, 424
Buserite  501

Cadmium sorption  207
Calorimetry  141
Canada  261, 487
Carboniferous  251, 323, 334
Catalysts  427, 487
Cement  427

Cenamonian 291
Cenozoic 271
Ceramics 1, 423, 438, 487, 601, 650
Chlordimeform sorption 227
Chlorite
  thermal decomposition 65
  formation 251, 359
  montmorillonite interstratified 96
  Mössbauer 65
  vermiculite interstratified 85
  weathered 85
Chrysotile 427
Classification 5
Coal 251, 311
Coagulation 131, 487, 649
Cobalt sorption 237
Colloidal properties 45, 95, 111, 131, 187, 235, 236, 437, 465, 487, 581, 649
Compression effects 457
Copper sorption 207, 237
Cretaceous 291, 301, 475
Crown ethers sorption 236
Crystallinity index 609, 639
Czechoslovakia 609

Definitions 2
Dehydration 65, 85, 95, 96, 261
Density measurements 609, 639
Diagenesis 241, 251, 261, 311, 333, 334
Double layer, electrical 465, 649
Drilling muds 236, 465

Electron diffraction 8, 55, 547
Electron microprobe 567
Electron microscopy
  chrysotile 55
  clay-water structure 235, 236
  ferripyrophyllite 55
  halloysite 415
  imogolite 547

Electrophoresis 465
England: see Britain
Epidote 359
EPR 9
ESCA 9, 153, 217, 237
Estuary sediments 333
Evaluation of clays 427, 437

Felspar weathering 323, 415, 609, 650
Ferripyrophyllite 55
France 291, 405

Garnierite catalyst 487
Genesis 5, 301, 333, 337, 423, 424, 609, 649, 650
Geochemistry 333, 334, 650
Geology 241-334, 359, 405, 415, 423, 475, 591, 601, 649
Germany 55, 333
Gibbsite 491, 517
Goethite 491, 555, 577, 578

Haematite 491, 555, 578
Halloysite
  metahalloysite transformation 619
  morphology 415
  occurrence 591
  water in 235
Hawaii 369
Hectorite bipyridyl complexes 237
History 4
Hungary 349
Hydrobiotite 405
Hydrothermal alteration 359, 369, 375, 395, 423, 591, 619
Hydroxyl
  orientation 95
  vibrations 27, 95, 650

Iceland 359

Illite
  formation  251, 261, 349, 395
  pastes  235
  smectite interstratified  241, 251, 323, 334, 395
  vermiculite interstratified  323
Imogolite  491, 547
Intercalation
  in kaolinite  236, 649, 650
  in lepidocrocite and related structures  577
  in phyllosilicic acids  235, 509
Intersalation of bipyridyl complexes  237
Interstratified clays  85, 96, 131, 241, 251, 261, 323, 334, 375
Ion exchange
  and phase separation  167, 235
  bipyridyl complexes  237
  buserite  501
  heavy metals  207
  chlordimeform  227
  crown ether complexes  236
  polyamine complexes  207, 237
  vermiculite  236
Iron oxidation
  in biotites  17, 177
  in chlorites  65
Iron oxides and hydroxides  491, 555, 577, 578, 649
IR spectra  8
  celadonite and glauconite  96
  far, of cation motions  236
  imogolite  547
  OH in biotites and phlogopites  27
  OH in kaolinite  650
  OH in smectites  95
  palygorskite and sepiolite  197
  quartz analysis  447
  Si-O (apical) stretch  95
  smectite  95
  water on smectite  235

Italy  334, 649

Jurassic  333, 475
Japan  85, 415, 424, 619, 650

Kaolinite  581-650
  alumina from  487
  applications  427, 601, 609
  Australian  591, 649
  chromium containing  650
  crystallinity  609, 649
  formation  301, 405, 423, 457
  geology  475, 591, 601, 649, 650
  genesis  649, 650
  intercalation complexes  236, 649, 650
  IR spectrum  650
  iron content  649, 650
  methods of investigation  581, 639
  Mexican  475
  Mössbauer study  650
  pastes  235
  processing  427
  reaction with salts  629
  smectite interstratified  423
  stacking faults  649, 650
  surface charge  465, 649
  suspensions  465
  thermal transformations  629, 639
  trace elements in  650
Kilns  2
K-bentonites  323, 333, 334

Lake, saline  333
Layer silicate
  Pb-rich  96
  phyllodisilicate  509
Layer stacking
  kaolinite  649, 650
  montmorillonite  45, 95 167
  vermiculite  95
Lepidocrocite  491, 577

Lepidomelane oxidation 17

Magadiite, water in 235
Manganese oxides and hydroxides 491, 501
Marine deposits 271, 281, 333, 334
Mercury adsorption 207
Mesozoic 475
Mexico 475
Mica
  electrostatic calculations 95
  expanded 17, 177
  hydroxyl orientation 95
  IR spectra 27, 95
  oxidation 17, 177
  uses 427
Minor element content 333
Miocine 281
Mixed layer minerals 85, 96, 131, 241,
  251, 261, 323, 334, 359, 375, 395
Montmorillonite (see also Smectite)
  aggregate porosity 141, 457
  alkine and isonitrile complexes 237
  amine complexes 96, 207, 237
  chlordimeform complex 227
  crown ether complexes 236
  coagulation 131
  genesis 424
  heavy metal trapping 207
  hydroxide complexes 423
  imidan on 237
  interstratified 96
  ion exchange 207
  K-Ca segregation 167, 235
  K-fixation 45, 167, 235
  layer stacking 45, 95, 167
  mixed layer charge 131
  Mexican 475
  parathion on 236
  pastes and suspensions 111, 141, 235,
    236, 457
  polyethylene glycol sorption 236

  porphyrin complexes 217
  quaternary ammonium 236
  surface acidity 187, 235
  surface area 235
  uses 427
  water complex 111, 121
Mössbauer spectra 9, 65, 555, 650

Neutron scattering 111, 121
New Zealand 555, 577
Nickel sorption 207, 237
NMR 9, 235
Nomenclature 4, 6
Nontronite reduction 75
Nordstrandite 491, 517
North Atlantic 271
North Sea 281

Oil formation 241, 261, 333
Oil well 281
Ordivician 333
Origin of life 650
Oxidation
  of biotite 17, 177
  of chlorite 65
  of muscovite 177

Pacific Ocean 333
Paleozoic 475
Palygorskite
  applications 432
  Mexican 475
  occurrence 487
  surface properties 99, 197
Parathion on clays 236
Pastes and suspensions 95, 111, 235, 2
  437, 487
Permian 334
Pesticide-clay complexes 187, 227, 236
  237

Phlogopite
  OH vibrations  27
  -talc intermediate  423, 424
Poland  251
Pore structure of clays  141, 236, 457
Porphyrin complexes  217
Potassium fixation  45, 167, 235, 349
Prehnite  359
Pyroaurite  577
Pyrometry  487
Pyrophyllite
  application  427
  ferric analogue  55
  muscovite intermediates  395
  smectite interstratified  375

Quartz analysis  447

Radial electron distribution  639
Reduction of nontronite  75
Refractory clays  487
Rheology  236, 437, 457, 465, 487, 601

Saponite
  iron rich  359
  b-dimension  95
Scandinavia  333, 334
Scanning calorimetry  141
Scotland: see Britain
Sea water  423
Sedimentology  241-334, 591
Selective extraction  7
Sepiolite
  applications  432
  domain shapes  95
  surface properties  99, 197
Serpentine  334
Silicic acid  235, 423, 509
Silurian  334
Silver sorption  177, 237

Smectite (see also Beidellite, Hectorite,
    Montmorillonite, Saponite)
  alcohol complexes  235
  beidellite-montmorillonite mixed
    layers  131
  crown ether complexes  236
  dehydroxylation  95
  IR spectra  95
  kaolinite interstratified  423
  pyrophyllite interstratified  375
  stacking  95
  water on  235
Soil  349, 385, 423, 424, 555
South Africa  601, 649
Spain  334, 424, 487
Spectra: see IR, Mössbauer, Visible, etc.
Stability  385, 424
Stacking faults  45, 650
Stoneware clays  487
Strength  457, 487
Surface acidity  187, 235
Surface area  235, 236
Surface charge  465, 649
Surface properties  10, 99-237
  aluminium hydroxides  577
  buserite  501
  lepidocrocite  577
  phyllodisilicates  509
Surite  96
Suspensions and pastes  95, 111, 235, 236,
    649
Sweden  333
Swelling properties
  montmorillonite  65
  vermiculite  37
Synthesis
  akaganeite  578
  glauconite  423
  goethite  577
  illite  395

imogolite 549
nordstrandite 517
pyrophyllite-smectite 375
review 337
smectite-kaolinite 423
talc-phlogopite 423

Talc
  applications 427
  genesis 424
  -phlogopite intermediates 423
Techniques 7, 437, 581, 639
Tertiary 281, 333, 475, 591
Thermal analysis 8, 14, 457, 567
Thermal transformations
  chlorite 65
  chrysotile 96
  goethite 578
  halloysite 619
  kaolinite 629, 639
  metal hydroxides 577
  smectite 95, 96
Titanium oxides 491, 577
Tonsteins 323, 334, 649
Trace elements 650
Tremolite 334
Triassic 475
Turkey 423

USA 333, 423, 649
USSR 55

Vermiculite
  applications 427
  chlorite-interstratified 85
  crown ether complex 232
  from chlorite 65
  interlamellar water 95, 121
  interlayer cation ordering 95, 236
  layer stacking 95
  swelling behaviour 37, 95

Mexican 475
Visible spectrum nontronite 75

Wales: see Britain
Water
  in buserite 501
  in clay aggregates 141, 457
  in vermiculite 95, 121
  on halloysite 235
  on magadiite 235
  on montmorillonite 95, 111, 121
  on palygorskite and sepiolite 197
  in smectite 235
  sorption and retention 141, 457
Weathering 385, 405, 423, 424, 650

X-ray diffraction 8
  disordered layer minerals 96
  montmorillonite-chlorite complexes
  montmorillonite-mica complexes 96
  montmorillonite suspension 95
  sepiolite domains 95
X-ray irradiation effects 236
X-ray photoelectron spectroscopy 9, 1
  217, 237

Yugoslavia 650

Zinc sorption 207
Zeolite genesis 259, 424